33rd International DAAAM Symposium on Intelligent Manufacturing and Automation 2022

Vienna, Austria
27-28 October 2022

Editor:

B. Katalinic

ISBN: 978-1-7138-6557-5

Printed from e-media with permission by:

Curran Associates, Inc.
57 Morehouse Lane
Red Hook, NY 12571

Some format issues inherent in the e-media version may also appear in this print version.

Printed with permission by Curran Associates, Inc. (2023)

For permission requests, please contact DAAAM International Vienna
at the address below.

DAAAM International Vienna
Bahnallee 10
A-2120 Wolkersdorf
Austria, EU

Phone: +436646424398

http://www.daaam.info

Additional copies of this publication are available from:

Curran Associates, Inc.
57 Morehouse Lane
Red Hook, NY 12571 USA
Phone: 845-758-0400
Fax: 845-758-2633
Email: curran@proceedings.com
Web: www.proceedings.com

ISSN 1726-9679

DANUBE ADRIA ASSOCIATION FOR AUTOMATION & MANUFACTURING

DAAAM INTERNATIONAL VIENNA

www.daaam.info

PROCEEDINGS

OF THE 33RD INTERNATIONAL DAAAM SYMPOSIUM
"INTELLIGENT MANUFACTURING & AUTOMATION"
27-28TH OCTOBER 2022
HOSTED FROM VIENNA UNIVERSITY OF TECHNOLOGY, VIENNA, AUSTRIA, EU

ORGANIZED BY:
DAAAM INTERNATIONAL VIENNA & VIENNA UNIVERSITY OF TECHNOLOGY

IN COOPERATION WITH:
UNIVERSITY OF MOSTAR, ÖIAV 1848, INTERNATIONAL ACADEMY OF ENGINEERING AND UNIVERSITY OF APPLIED SCIENCES – TECHNIKUM WIEN

UNDER THE AUSPICES OF THE DANUBE RECTORS' CONFERENCE &
RECTORS' AND PRESIDENTS' HONOR COMMITTEE OF DAAAM INTERNATIONAL

EDITOR: B.[RANKO] KATALINIC

IMPRESSUM 2022

Proceedings of the 33rd International DAAAM Symposium "Intelligent Manufacturing & Automation"
27-28th October 2022, hosted from Vienna University of Technology, Vienna, Austria, EU, Volume 33, No.1,
Abbreviated key title: ***Ann. DAAAM 2022***

ISSN 1726-9679

Ed. B. Katalinic,

Published by DAAAM International, Vienna, Austria, EU, 2022
Layout & Design: B[ranko] Katalinic
Technical Editors: I[lija] Zec / L[uka] Zec

Publisher:

DAAAM International Vienna
TU Wien, Getreidemarkt 9, 1060 Wien, Austria
E-mails: president@daaam.info / branko.katalinic@tuwien.ac.at
DAAAM homepage: http://www.daaam.info

CONTENTS

AUTHOR INDEX

C

D

E

F

G

Z

ORGANIZING COMMITTEE FOR 2022

RECTORS' AND PRESIDENTS' HONOR COMMITTEE

International Scientific Committee

Alander, Jarmo Tapani (Fin)
Arai, Tamio (Jpn)
Ares, Enrique (Spa)
Arndt, Guenter (Aus)
Attard, Mario M. (Aus)
Bachynski, Morrel (Can)
Belajev, Gennady J. (Blr)
Bilberg, Arne (Den)
Bliedtner, Jens (Deu)
Bohez, Erik L. J. (Tha)
Bright, Glen (Rsa)
Browne, J. (Irl)
Celar, Stipe (Hrv)
Coelho, Reginaldo T. (Bra)
Covic, Dragan (Bih)
Cruz, Fernando M. M. (Por)
Dar-El, E. (Isr)
De Filippi, A. (Ita)
Devahastin, Sakamon (Tha)
Dincmen, M. (Tur)
Diniz, Anselmo Eduardo (Bra)
Dohda, Kuniaki (Jpn)
Dong, Jianhua (Chn)
Dornfeld, David (Usa)
Dove, Rick (Usa)
Dudas, Illes (Hun)
Egorov, Sergey B. (Rus)
Endter, R. (Deu)
Eversheim, W. (Deu)
Galantucci, Luigi Maria (Ita)
Giacaglia, G. E. (Bra)
Gibson, I. (Aus)
Girot, Frank (Fra)
Goldfarb, Venyamin Y. (Rus)
Grabenweger, Johann (Bra)
Gusev, Boris V. (Rus)
Habchi, Georges (Fra)
Han, Qing Yao (Chn)
Hock, Seo Swee (Sng)
Hofmann, Peter (Deu)
Hryu, Jee-Hwan (Kor)
Ibishi, Ismet Maliq (Kwt)
Iltschev, Velko (Bgr)
Irfan, Mohammad A. (Pak)
Itoh, Kiyoshi (Jap)
Ivan, Nicolae-Valentin (Rom)
Kals, H. J. J. (Ndl)
Katalinic, Branko (Aut)
Kimura, Fumihiko (Jpn)
Kinnander, Anders (Swe)
Knosala, Ryczard (Pol)
Koch, Jan (Pol)
Kondratenko, Yuriy (Rus)
Kuosmanen, Petri (Fin)
Kyttner, Rein (Est)
Le, Cung (Vtn)
Lin, Chyi-Yeu Jerry (Twn)
Llorente, J. I. (Esp)
Lombardi, F. (Ita)
Low, Kin Huat (Mal)
Lu Chang, Tao (Chn)
Lu, S. S. (Twn)
Luong, Lee Hong Son (Aus)
Marcos, Mariano (Esp)
Meghdari, Ali (Iri)
Mertins, Kai (Deu)
Michelini, R. C. (Ita)
Mikhaylov, Yuri O. (Rus)
Munteanu Radu (Rom)
Nanasi, Jozef (Svk)
Neumann, H. J. (Deu)
Nicoletto, G. (Ita)
Novakovic, Branko (Hrv)
Oczos, Kazimierz E. (Pol)
Otto, Tauno (Est)
Park, Hong-Seok (Kor)
Pegden, Denis (Usa)
Peterka, Jozef (Svk)
Porter, Brian (Hkg)
Pritschow, Guenter (Deu)
Pryanichnikov, Valentin (Rus)
Przybylski, W. (Pol)
Radermacher, F. J. (Deu)
Radhakrishnan, V. (Ind)
Ramayah T. (Mla)
Rungruang, Channarong (Tha)
Schmid, Steven (Usa)
Segui Llinares, V. Jesus (Esp)
Sladek, Jerzy (Pol)
Stazhkov, Sergey (Rus)
Stefanescu, D. Michael (Usa)
Tabucanon, Mario T. (Tha)
Takakuwa, Soemon (Jap)
Takala, J. (Fin)
Tikka, H. (Fin)
Topalova, Irina (Bgr)
Torims, Toms (Lva)
Torres-Leza, F. (Esp)
Torvinen, S. (Fin)
Tsai, B. S. Y. (Twn)
Udiljak, Toma (Hrv)
Uglesic, Ante (Cro)
Uhlmann, Eckart (Deu)
Uyar, Erol (Tur)
Van Brussel, Hendrik (Bel)
Van Oudheusden, Dirk L. (Bel)
Varga, Gyula (Hun)
Vavrin, Petr (Cze)
Venkatesh, V. C. (Sgp)
Vetturi, David (Ita)
Vitecek, Antonin (Cze)
Waldir, P. O. (Bra)
Wertheim, Rafi (Isr)
Wessely, Emil (Svk)
Westkaemper, Engelbert (Deu)
Wiendahl, Hans-Peter (Deu)
Williams, D. J. (Gbr)
Winston, Patrick H. (Usa)
Zalzala, Ali M. S. (Gbr)
Zhao, Daozhi (Chn)

International Program Committee

Acko, Bojan
Ahuja, Harish
Albers, Albert
Alekseyev, Wladimir A.
Amberg, Michael
Banyai, Agota
Banyai, Tamas
Baritz, Mihaela Ioana
Barreiro, Joaquin
Begic, Derzija
Bijelonja, Izet
Bleicher, Friedrich
Levin, Alexander
Li, Guihua
Lica, Dumitru
Lipa, Zdenko
Liu, Huaming
Lo Valvo, Ernesto
Loghin, Carmen Maria
Lopez De L., Luis Norberto
Low, Kin Huat
Luca Motoc, Dana I.
Ludovico, Antonio D.
Lujic, Roberto
Perniu, Dana
Perutka, Karel
Peters, Jacques
Petkovic, Darko
Petre, Cristian
Petre, Marian
Petrescu, Valentin
Pinca-Bretotean, Camelia
Pintilie, Gheorghe
Pirtea, Gabriel Marilen
Pivonka, Petr
Pjero, Elenica

BLEY, HELMUT
BOTEZATU, CORNELIA PAULINA
BOTEZATU, MIHAI
BRAJLIH, TOMAZ
BREZAK, DANKO
BREZOCNIK, MIRAN
BUCHMEISTER, BORUT
BUCURESTEANU, ANCA
CAMPBELL, R. IAN
CEKIC, AHMET
CELAR, STIPE
CHAN, CHEE SENG
CHAUS, ALEXANDER
CHRAMCOV, BRONISLAV
COHODAR, MAIDA
COLEMAN, J. J.
COSIC, ILIJA PETAR
COSMA, SMARANDA
COTET, COSTEL EMIL
CUBONOVA, NADEZDA
CURTU, IOAN
DAMIC, VJEKOSLAV
DENKENA, BEREND
DEVAHASTIN, SAKAMON
DOLGA, LIA
DOLGA, VALER
DOLINAY, JAN
DOLINAY, VILIAM
DOSTAL, PETR
DOSTALEK, PETR
DRAGCEVIC, ZVONKO
DRAGICEVIC, MARIJA
DRSTVENSEK, IGOR
DUHOVNIK, JOZE
EJUB DZAFEROVIC,
FIRST ROGALE, SNJEZANA
FUJII, NORIO
GHEORGHIU, HORIA
GODEC, DAMIR
GOMEZ, ANTONIO
GORDIN, DOINA MARGARETA
HLAVA, JAROSLAV
HOFREITER, MILAN
HRUBINA, KAMIL
HRUSKA, FRANTISEK
HRUSKOVA, ERIKA
IBISHI, ISMET M.
JANACOVA, DAGMAR
JANSSENS, THIERRY
JARVIS, CHRISTINE W.
JAVOROVA, ANGELA
JERBIC, BOJAN
JURAGA, IVAN
JURUM-KIPKE, JASNA
KARAULOVA, TATYANA
KATALINIC, BRANKO
KAWALEC, ANDRZEJ
KHANNA, ASHUTOSH
KHOMCHENKO, VASILY
KIRK, CHRIS. S.
KISS, IMRE
KITTL, BURKHARD
KLIMENKO, S. A.
KNEGO, NIKOLA
KNEZEVIC, PETAR
KOHLERT, HELMUT
LUKIC, IVANA
LUPEA, IULIAN
LUPU, CIPRIAN
LUPULESCU, NOURAS BARBU
MACHEDON PISU, TEODOR
MACKU, LUBOMIR
MACUROVA, ANNA
MAGHETI, IOAN
MAJETIC, DUBRAVKO
MALISA, VIKTORIO
MALOPOLSKI, WALDEMAR
MAN, MARIANA
MANDIC, LIDIJA
MANDRU, DAN
MANESCU, TIBERIU STEFAN
MARASCU KLEIN, VLADIMIR
MARCOS, MARIANO
MARCU, MARIUS
MAREK, KUBALCIK
MARIAN TOLNAY
MARIN, DUMITRU
MARIN, MIHNEA
MARINESCU, NICULAE ION
MARN, FERDINAND
MARONEK, MILAN
MARSLAND, DAVID
MARTA, CONSTANTIN
MARTINEZ, SUSANA
MASTACAN, LUCIAN
MATUSOVA, MIRIAM
MATUSU, RADEK
MEHTA, NARINDER KUMAR
MEIER, HORST
MENZ, PETER
MICHAL, BLIZNAK
MICIETA, BRANISLAV
MICLOSI, VIOREL
MOHORA, CRISTINA
MOK, H.
MORARU, SORIN AUREL
MORIWAKI, T.
MORTOIU, DOINA
MOTOMANCEA, ADRIAN
MUNTEAN, MIHAELA
MUNTEAN, SEBASTIAN
MUNTEANU, ANCA
MUSCA, GAVRIL
MUSTAFIZUL, KARIM
MYSHLYAVTSEV, ALEXANDR
NAGI, MIHAI
NAGLA, K. S.
NAKAMOTO, NOBUYUKI
NEUMANN, PETR
NICA COTET, GABRIELA BEATRICE
NIKOLIC, GOJKO
NKUMBWA, RODGER
NOH, SANG DO
NOVAC, OVIDIU
NOVOSELOVIC, DANIEL
NURUL AMIN A.K.M.
OANCEA, GHEORGHE I.
OKAMOTO, J.
OLARU, ADRIAN, DUMITRU
OLARU, ONISIFOR
OLEKSIK, VALENTIN
OLTEANU, CIPRIAN
PLESA, ANGELA
PODURAEV, JURY
POKORNY, PAVEL
POMAZAN, VALENTINA
POPA, CICERONE LAURENTIU
POPPEOVA, VIERA
PRODAN, DAN
PROKOP, ROMAN
PRYANICHNIKOV, VALENTIN
PYTKO, STANISLAW
QUINN, ROGER D.
RAMAJ, VEHBI XH.
RAVAS, RUDOLF
RAY, T. S.
RIKALOVIC, ALKSANDAR
RIZESCU, SABIN VASILE
ROGALE, DUBRAVKO
ROHDE, ULRICH L.
ROMERO, FERNANDO
ROS, OLIMPIA
ROSADO, V.O.G.
RUDAS, IMRE J.
RUGESCU, DRAGOS RADU DAN
RUJA, IOAN
RUSU ANGHEL, NICOLAE
RUSU, LIGIA
RUSU-CASANDRA, AURELIA
RUTEMI, GANI BRAHIM
RUZIC, DRAGO
SAKULSKI, DUSAN
SALIHU, AVDI HAJDIN
SAMEK, DAVID
SCHREIBER, PETER
SELVAKUMAR, ARUL LAWRENCE
SEONG, POONG HYUN
SERCER, MLADEN
SHIBASAKA, TOSHIROH
SHIH, MING CHANG
SIEMIATKOWSKI, M.
SIHN, WILFRIED
SINAY, JURAJ
SISEK, BORIS
SITUM, ZELJKO
SKOCZYNSKI, WACLAW
SKOLUD, BOZENA
SLUGA, ALOJZ
SMIRNOV, ALEXANDER
STANKOVSKI, STEVAN
STAROVEŠKI, TOMISLAV
STEFANIC, NEDELJKO
STOPPER, MARKUS
STRGAR KURECIC, MAJA
SUBRAMANIAN, ARUNKUMAR
SUGAR, PETER
SYROVA, LIVIA
SYSALA, TOMAS
TADAMURA, KATSUMI
TAKALA, JOSU
TAMASILA, MATEI
TAN, KOK KIONG
THIERHEIMER, WALTER WILHELM
TORIMS, TOMS
ULRICH, GUENTHER
VAARIO, JARI
VASEK, VLADIMIR
VELISEK, KAROL

REVIEWING COMMITTEE

33RD DAAAM INTERNATIONAL SYMPOSIUM ON INTELLIGENT MANUFACTURING AND AUTOMATION

Preface

The 33rd DAAAM International Symposium on Intelligent Manufacturing and Automation was organised as virtual online conference hosted by the Vienna University of Technology, Vienna, Austria, between the 28th and 29th October 2022, during the DAAAM International Week. The Symposium was organized by DAAAM International Vienna and Vienna University of Technology in cooperation with University of Mostar, ÖIAV 1848, International Academy of Engineering and University of Applied Sciences – Technikum Wien and Under the Auspices of the Danube Rectors' Conference & Rectors' and Presidents' Honor Committee of DAAAM International for 2022. The Symposium was hosted from Vienna, Austria. This year's symposium aimed at continuing the success of the previous years, focusing on the five-fold traditional objectives of the symposium: the presentation of the most recent high-quality results, support of development of young scientists and researchers, organization of international (summer) doctoral school, inauguration of new members of Central European Branch of International Academy of Engineering and the provision of the necessary setting for stimulating discussions, brainstorming and networking among European and international researchers coming both from the academia government agencies and industry

DAAAM International Symposium has been established as the leading Central European conference devoted to the cutting-edge advancements in the field of modern production and automation. Symposium provides an international forum for engineers, scientists and industrial managers to present and discuss the current status and impact of advanced manufacturing and automation. The international series of Symposium was initiated by the first meeting in Vienna (1990) and continued in Štrbske Pleso (1991), Budapest (1992), Brno (1993), Maribor (1994), Cracow (1995), Vienna (1996), Dubrovnik (1997), Cluj - Napoca (1998), Vienna (1999), Opatija (2000), Jena (2001), Vienna (2002), Sarajevo (2003), Vienna (2004), Opatija (2005), Vienna (2006), Zadar (2007), Trnava (2008), Vienna (2009), Zadar (2010), Vienna (2011), Zadar (2012), Zadar (2013), Vienna (2014), Zadar (2015), Mostar (2016), Zadar (2017), Zadar (2018), Zadar (2019), Mostar (2020), Vienna (2021), Vienna (2022).

Regular organization of the annual Symposium is one of the central activities of the DAAAM International association. The association was founded on 5. November 1990 in Vienna during a meeting organised by Professor B. Katalinic in order to celebrate the 175 Years of Vienna University of Technology. The main idea of the association was to establish a network for a new kind of international scientific and academic cooperation in the fields of intelligent manufacturing and automation. Our primary focus is on: all aspects of production of technical products made out of parts and corresponding automation and knowledge, trans-disciplinary application of methods and all kinds of education.

DAAAM International activities, during the last thirty-three years, have been supported by scientists and experts from over fifty countries all over the world. In the course of the last decade, DAAAM International has focused its activities on the regional and global long-term co-operation between (technical) universities. The result of this initiative is the founding of the University Network of DAAAM International Vienna and DAAAM International Network for Advanced Technologies which was founded on 2002-10-25 in Vienna.

The main activities of DAAAM are:

- Organisation of annual DAAAM International Symposium.
- To maintain and improve the University Network of DAAAM International Vienna as a strategic alliance and permanent international open platform for long-term co-operation and partnership in all fields of technology, research and education.
- Organisation of long-term projects, initially known as Inter-University Doctoral Studies.
- Organisation of long-term projects, initially known as the Virtual University.
- Publishing of international journals, books, and proceedings.
- Promotion of the exchange of people, scientific, and technical information.
- Promotion of international co-operation among scientists and engineers.
- Promotion of R & D, education, and industrial application.
- To co-operate with other national and international organisations in matters of common interest.
- Organisation of International Doctoral School as a platform for the support of the next generation of engineers at the highest level.

This proceeding contains 77 refereed papers from 226 authors presented at the 33rd DAAAM 2022. In DAAAM proceedings, since the first symposium in 1990, we published papers from 9545 authors.

All the papers included in these proceedings are reviewed by experts in double blind peer review way, and finally revised by the authors for the improvement according to referee's reports and advices.

The papers discuss many aspects of modern manufacturing and automation such as: Algorithms, Artificial Intelligence, CAX, Computer Integration, Control, Cutting Tools, Design, FEM, Invited Lectures, Knowledge, Management, Manufacturing System, Mechatronics, Methodology, Methods, Modelling, Optimization, Robotics, Simulation, Technical Solutions, Technology and Trends.

Appreciation is forwarded to reviewers who contributed to improve the quality of papers reviewing the papers, providing constructive critical comments, improvements, and corrections to the authors. We also recognize the support of reviewers, who were in charge of acceptance/rejection process. We wish to thank the Scientific, Program and Organizing Committees, Session Chairs, student helpers and administrative assistants, for contributing to the success of this Conference. They did an excellent job.

Finally, we wish to thank all the authors who submitted papers, making this conference possible and the authors of accepted papers for making appropriate modifications and corrections with care and patience and for preparing the final manuscript. It is the quality of their presentations and their passion to communicate with other participants that really makes this conference series a success.

Special thanks to FESTO company for support of our activities and support of young researchers and scientists in the form of Festo Prize for the participants of the 11th DAAAM International Doctoral School, which was held parallel to the symposium and was positively integrated.

We would like to thank all members of the International Academy of Engineering – Central European Branch, Rector of the Vienna University of Technology Professor Dr Sabine Seidler, and Rector of the University of Mostar Professor Dr. Sc. Zoran Tomić and his team, management and all those nice colleagues for their excellent support in the organization of 33rd DAAAM International Symposium and the 11th DAAAM International Doctoral School.

Vienna, 2022-12-05
Editor
Chair of the 33rd DAAAM 2022

Branko Katalinic

Professor rtd Dr.sc. Dr.mult.h.c.Prof.h.c. Branko Katalinic
President of DAAAM International
TU Wien, IFLT-IMS, Getreidemarkt 9
1060 Wien, Austria, Europe
URL: http://www.daaam.info
E-mail: president@daaam.info
branko.katalinic@tuwien.ac.at

DOI: 10.2507/33rd.daaam.proceedings.001

APPLICATION OF LINEAR PROGRAMMING TO INCREASE THE EFFICIENCY OF CONSTRUCTIONAL VENEER PRODUCTION

Aleksandra Kostic, Izet Horman, Valentina Timotic & Melisa Kustura

This Publication has to be referred as: Kostic, A[leksandra]; Horman, I[zet]; Timotic, V[alentina] & Kustura, M[elisa] (2022). Application of Linear Programming to Increase the Efficiency of Constructional Veneer Production, Proceedings of the 33rd DAAAM International Symposium, pp.0004-0011, B. Katalinic (Ed.), Published by DAAAM International, ISBN 978-3-902734-36-5, ISSN 1726-9679, Vienna, Austria
DOI: 10.2507/33rd.daaam.proceedings.001

Abstract

The paper presents the application of linear programming to increase production efficiency in a furniture factory based on constructional veneer technology. Veneer production classifies as novel wood processing technology and it is mainly an automated process. Insufficient quantities of delivered thermal energy for veneer drying result in discontinuity in the production process. Inadequate veneer humidity causes production problems and exploitation of veneer-based products. The problem of using human resources in the newly created situation was also considered. Algorithms for solving problems are given in the paper.

Keywords: linear programming; integer programming; wood; algorithm; veneer drying.

1. Introduction

Wood and wood-based raw materials are becoming increasingly popular materials in product design. Therefore, high performance with high dimensional stability and durability are expected from these materials. Veneers are of particular importance because they have high elasticity even if they are obtained from a hard type of wood [12].

Veneer production historically belongs to newer technologies for wood processing. According to the method of production and purpose, there are three types of thin sheets of wood or veneers. These are peeled, sliced and constructional. Sliced veneer is mainly used for surface finishing of wood-based panels. Structural or peeled veneer is used for the production of structural elements such as various panels based on veneer, seats or chair backs, armrests and the like.

In this paper, the focus is on the analysis of the technological process of production of structural veneer. The advantage of using products based on this veneer compared to solid wood is in improving strength and especially elastic properties. The problems with elements made of solid wood are: in smaller or larger radii of curvature in wooden elements and relatively small percentages of wood utilization. Also, the problem with massive wood bending is seen in smaller diameters and the loss of elastic properties with dynamic loading. In the production of structural veneers, as well as other types of veneers, the drying phase plays a very important role. The consequences of insufficiently dried veneer appear in the later stages of processing and exploitation of the product.

That is why many authors have dealt with the problem of veneer drying. In the paper [10], a prediction of the optimal drying temperatures of veneers for the production of pressed veneers is given. Some authors analyzed the influence of veneer drying temperature on certain physical properties and formaldehyde emissions in birch plywood [3]. In [1], Baldwin investigated the process of veneer drying and preparation for the veneer pressing phase. An interesting experimental investigation and calculation of the influence of ambient air parameters during drying on temperature, moisture content and resulting deformations and stresses in wood with the aim of determining the optimal drying speed is given in [9]. In the paper [5], the method of finite volumes was used to obtain non-stationary fields of temperature, displacement and stress in wood, which are the result of changes in the parameters of the surrounding medium, i.e. the regime of heat treatment of wood.

In the production of veneers, it is important to provide a sufficient amount of thermal energy in order to ensure the application of appropriate drying regimes. The discontinuity of production is caused by insufficient heat energy, which prevents the application of appropriate modes. Therefore, there is an imperative to make a certain investment in technological equipment, i.e. means of production. In order for the investment to be financially justified, an expert analysis of technological processes and appropriate mathematical decision-making tools are needed.

The research is focused on the application of certain scientific knowledge in the field of production organization and management. The challenge is to transfer knowledge when managing certain processes [11]. The paper will use integer programming as a subtype of linear programming. Integer programming, even though it is a mathematical method, is used today as a tool for business decision-making where it is necessary to make the right decision based on the given criteria, taking into account the availability and limited resources in the factory. Thus, optimization can be understood both as a way of setting an engineering problem and as a specialized mathematical decision support tool. By introducing linear programming as a mathematical tool for solving certain production problems, it has a professional contribution and scientific contribution to the profession. The results of the research will help in making an objective assessment of the investment.

The paper will also focus on human resources. The issue of human resources has particularly come to the fore in the COVID19 pandemic. A mathematical model for the redistribution of executors to other operations in order to maintain the existing production capacity in changed conditions as well as to overcome production discontinuity based on the redistribution of human resources was discussed in [7].

The paper is organized as follows. In section 2, the physical model of the problem considered in the paper is given. The basics of linear programming are presented in section 3. The mathematical model for the specific factory is presented in section 4. The analysis of the results is presented in section 5. In section 6, the results of the implementation research are presented and recommendations for further research are given.

2. Physical model

The physical model includes the following stages of veneer production:

- heat treatment of wood;
- peeling logs;
- cutting the veneer to a certain width and removing errors in construction and wood processing;
- veneer drying;
- disposal of dry veneer.

Figure 1. shows the principle scheme of the technological process of structural veneer manufacturing. Focus of this paper is on the veneer drying.

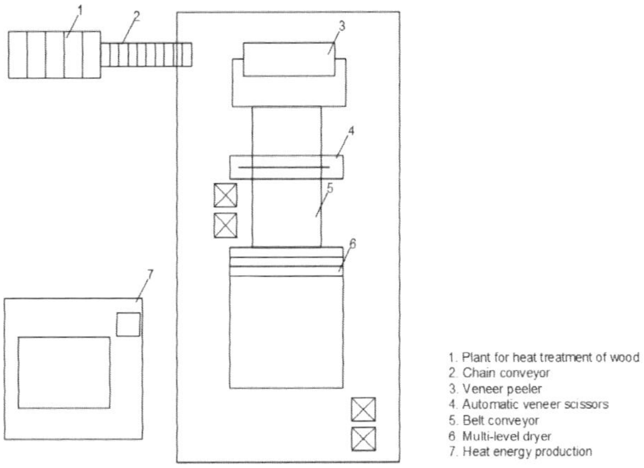

Fig. 1. The principle scheme of the technological process

3. Linear programming

Optimization plays a very important role in the production process. It is actually a new approach to engineering synthesis, i.e. process or product development based on optimality. Optimization procedures are applied in a whole range of linear and non-linear problems, of which we will mention only the most important ones: the problem of optimal investment decisions, the transport problem, the problem of resource allocation, the problem of optimizing mechanical structures, the problem of optimizing heat flows, the problem of optimal management of systems and others. Process optimization in the engineering sense is not a simple procedure and the optimization scheme is shown in Figure 2.

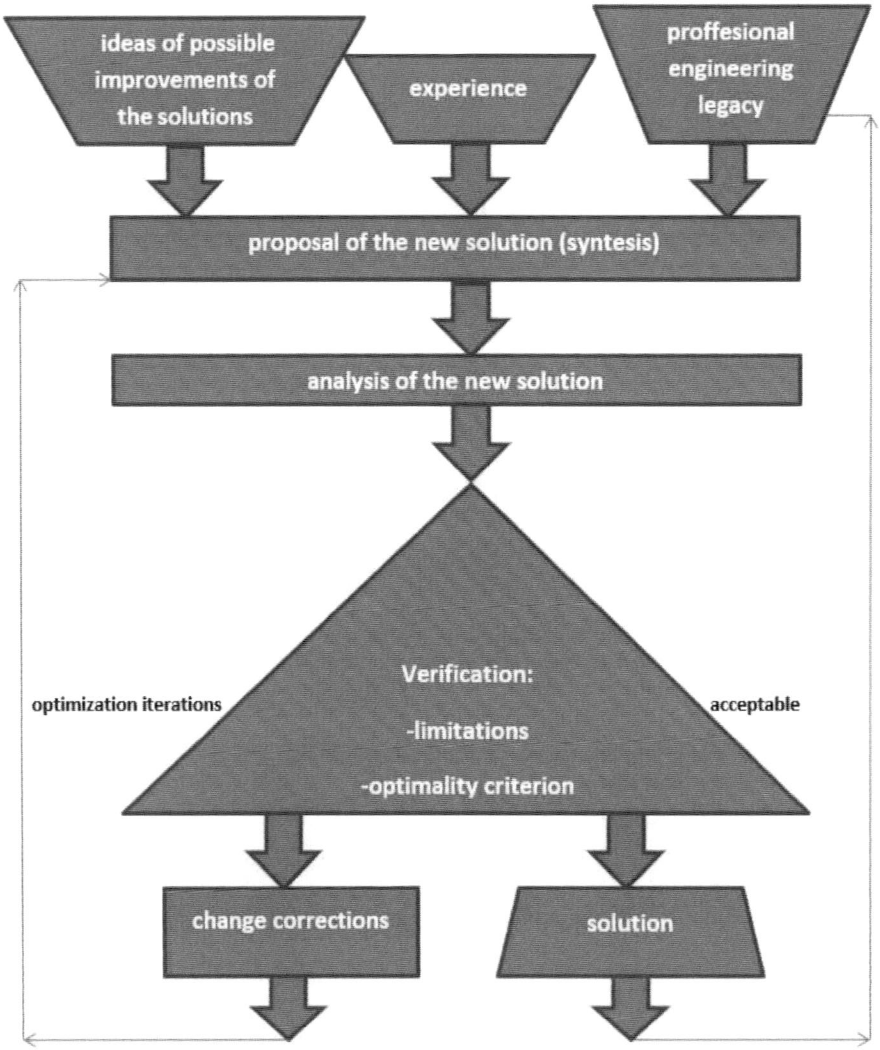

Fig. 2. The Optimization Scheme

In order to choose the most suitable solution and scientifically verify that solution, it is necessary to pose the problem mathematically. Optimization or mathematical programming is a branch of mathematics that studies the maximization and minimization of real functions of a real variable. Linear programming is a very important field in optimization. We can find its beginnings with Fourier, after whom one of the methods for solving linear programming problems got its name. Namely, it is about Fourier-Motzkin elimination. In 1939, Leonid Kantorovich developed the linear programming technique with the aim of reducing military costs and increasing enemy losses in World War II. Until 1947. linear programming methods were not available to the general public. In 1947. George B. Dantzig published the simplex method, which is one of the most important methods in linear programming. After the Second World War, linear optimization methods are used in everyday practice, because many practical problems in operations research can be expressed as linear programming problems.

Linear programming is a mathematical method for minimizing or maximizing a linear objective function with constraints in the form of linear inequalities, and with non-negativity constraints for the variable. We can write the general form of the minimum problem as follows:

Minimize the objective function

$$Min\ z = \sum_{j=1}^{n} c_j x_j \tag{1}$$

with constraints

$$\sum_{j=1}^{n} a_{ij} x_j \leq b_{i,} \quad i = 1,2, \ldots, m \tag{2}$$

$$\sum_{j=1}^{n} d_{ij} x_j \geq e_{i,} \quad i = 1,2, \ldots, l \tag{3}$$

$$\sum_{j=1}^{n} f_{ij} x_j = g_{i,} \quad i = 1,2, \ldots, p \tag{4}$$

and the negativity condition

$$x_j \geq 0,\ j = 1,2, \ldots, n,$$

where
c_j- the coefficient of the objective function j -th variable;
x_j- structural variables;
b_i, d_i, g_i- the quantity of the i -th constraint, i.e. the coefficients on the right side of the equations, i.e. the inequalities;
a_{ij}, d_{ij}, f_{ij} – the amount of the i -th restriction required for the unit of the j -th variable.

In order to write down the problem in matrix form, we introduce the following notations:

$$\boldsymbol{c} = \begin{bmatrix} c_1 \\ \vdots \\ c_n \end{bmatrix}, \quad \boldsymbol{x} = \begin{bmatrix} x_1 \\ \vdots \\ x_n \end{bmatrix}, \quad \boldsymbol{A} = \begin{bmatrix} a_{11} & \cdots & a_{1n} \\ \vdots & \ddots & \vdots \\ a_{m1} & \cdots & a_{mn} \end{bmatrix}, \quad \boldsymbol{D} = \begin{bmatrix} d_{11} & \cdots & d_{1n} \\ \vdots & \ddots & \vdots \\ d_{l1} & \cdots & d_{ln} \end{bmatrix}, \quad \boldsymbol{F} = \begin{bmatrix} f_{11} & \cdots & f_{1n} \\ \vdots & \ddots & \vdots \\ f_{p1} & \cdots & f_{pn} \end{bmatrix},$$

$$\boldsymbol{b} = \begin{bmatrix} b_1 \\ \vdots \\ b_m \end{bmatrix}, \quad \boldsymbol{e} = \begin{bmatrix} e_1 \\ \vdots \\ e_l \end{bmatrix}, \quad \boldsymbol{g} = \begin{bmatrix} g_1 \\ \vdots \\ g_p \end{bmatrix}.$$

where $\boldsymbol{c}$ objective function coefficient vector, $\boldsymbol{x}$ vector of unknowns and structural variables, $\boldsymbol{A}, \boldsymbol{D}, \boldsymbol{F}$ matrices of structural coefficients, $\boldsymbol{b}, \boldsymbol{e}, \boldsymbol{g}$ are vectors of free members.

We can write the linear programming problem in matrix form as follows:

$$Min\ \boldsymbol{z} = \boldsymbol{c}^T \boldsymbol{x} \tag{5}$$

$$\boldsymbol{Ax} \leq \boldsymbol{b} \tag{6}$$

$$\boldsymbol{Dx} \geq \boldsymbol{e} \tag{7}$$

$$\boldsymbol{Fx} = \boldsymbol{g} \tag{8}$$

$$\boldsymbol{x} \geq \boldsymbol{0} \tag{9}$$

Vector $\boldsymbol{x}$, which satisfies the constraints (6), (7) and (8) represent a solution to the problem. A problem solution that additionally satisfies condition (9) is called a possible problem solution. A possible solution to the problem that satisfies condition (5) is called an optimal solution to the problem.

Two methods are used to solve linear programming problems:

- the graphical method, which is the simplest, but which is limited to two-dimensional problems;
- simplex method, which is used when we have more than two variables.

The graphic method consists in the graphic representation of the lines that determine the conditions in the plane $x_1 0 x_2$. The conditions determine some polygon whose vertices are possible solutions. By checking the value of the objective function at the vertices of the polygon, we obtain optimal solutions from a set of possible solutions.

The simplex method is a finite, iterative and general method for solving linear programming problems. In the case of the simplex method, one starts from some basically possible solution and based on the criteria of optimality, one examines whether that solution is optimal. If the solution is optimal, the procedure is over. In case we did not get the optimal solution, the simplex method finds a new basic possible solution whose solution corresponds to the neighboring extreme point and at that point the objective function receives a better value. If the minimum problem is being solved, then the objective function takes on a smaller value. We repeat the process until we get the optimal solution. When solving a problem, non-basic variables are introduced. Non-basic variables are a set of variables whose value is set to zero. Basic variables are a set of variables whose value is calculated, and as a rule, they are different from zero. When solving problems using the simplex method, special tables are introduced, which make it easier to find the optimal solution. More information on linear programming can be found in [2], [4], [6] and [8].

A special subtype of linear programming is integer linear programming which offers a solution in integer form. Three types of problems that require the application of integer linear programming are: the fixed cost problem, the optimal investment decision problem, and the knapsack problem. Every problem that needs to be solved in such a way is solved using the standard simplex method. After obtaining an optimal solution that is not integer, it is necessary to use methods of integer forms or methods of branching and fencing to obtain a final, integer solution that is treated as optimal. However, there are simpler examples from practice where, due to the integer number of variables, it is easy to write appropriate algorithms that quickly provide an optimal solution to the problem. We will present two such algorithms in the next section.

4. Mathematical model for a specific factory

From the physical model shown in the technological scheme in Figure 1, it can be seen that in the production of structural veneer, a multi-level dryer and a plant for the production of thermal energy play a significant role. In the specific case, the multi-level dryer consists of 6 sections, and the plant for the production of thermal energy consists of a 4 MW boiler. In the production process, production discontinuity occurs, because for the production of raw veneer from one shift, it is necessary for the dryer to work in three shifts. There are two ways to solve this problem, namely the installation of additional dryer sections and the purchase of an additional boiler for the production of thermal energy. The number of possibly supplied dryer sections is limited by the money for investment and the size of the place where the acquired sections would be located without disrupting the production process. A maximum of 3 sections can be purchased for a specific case. On the other hand, there are occasional stoppages in production related to boiler malfunctions, the losses are large and in the order of 50,000 euros. It means that the possible purchase of a 1MW or 2MW backup boiler should be considered. Reduction of production costs is an important data for making an investment decision. New investments will affect both possible reductions in the number of shifts and costs per shift. Since the investments are planned for the calendar year, it is more likely that a failure will occur during the year than during the month, the costs of labor production were observed on an annual basis. Therefore, the objective function to be minimized has the following form

$$z = z_I + z_L + z_{SH} \tag{10}$$

where

$$z_I = 3000x_1 + 15000x_2 \tag{11}$$

$$z_L = 12500(2 - x_2) + 25000 \tag{12}$$

$$z_{SH} = 36000(3 - 0.\dot{3}\,x_1 - 0.5x_2) \tag{13}$$

z_I, z_L i z_{SH} are direct investment function, function of losses due to boiler failure and dryer costs due to workers shift changes, respectively.x_1 is the number of acquired dryer sections and x_2 power of the new boiler. The coefficients in formula (11) along with x_1 and x_2 correspond to the price of a section or a 1MW boiler on the market.

Formula (10) can be written in another form

$$z = 3000x_1 + 15000x_2 + 12500(2 - x_2) + 36000(3 - 0.\dot{3}\,x_1 - 0.5x_2) + 25000 \tag{14}$$

After sorting we obtain the objective function z in form of

$$z = 158000 - 9000x_1 - 15500x_2 \tag{15}$$

The linear programming problem for the specified veneer factory has the form

$$\min z = 158000 - 9000x_1 - 15500x_2 \tag{16}$$

$$x_1 + x_2 \leq 5 \tag{17}$$

$$0 \leq x_1 \leq 3 \tag{18}$$

$$0 \leq x_2 \leq 2 \tag{19}$$

Conditions (17), (18) and (19) are determined by the size and place for installing the purchased new sections of the dryer and the funds that the factory has decided to invest in solving production optimization. In matrix form, the linear programming problem looks like:

$$\min z = 158000 - 9000x_1 - 15500x_2 \tag{20}$$

$$\begin{bmatrix} 1 & 1 \\ 1 & 0 \\ 0 & 1 \end{bmatrix} \begin{bmatrix} x_1 \\ x_2 \end{bmatrix} \leq \begin{bmatrix} 5 \\ 3 \\ 2 \end{bmatrix} \tag{21}$$

$$\begin{bmatrix} x_1 \\ x_2 \end{bmatrix} \geq \begin{bmatrix} 0 \\ 0 \end{bmatrix} \tag{22}$$

Since this is a linear programming problem with two variables, it is suitable for solving by the graphical method. The graphical method is shown in Figure 3.

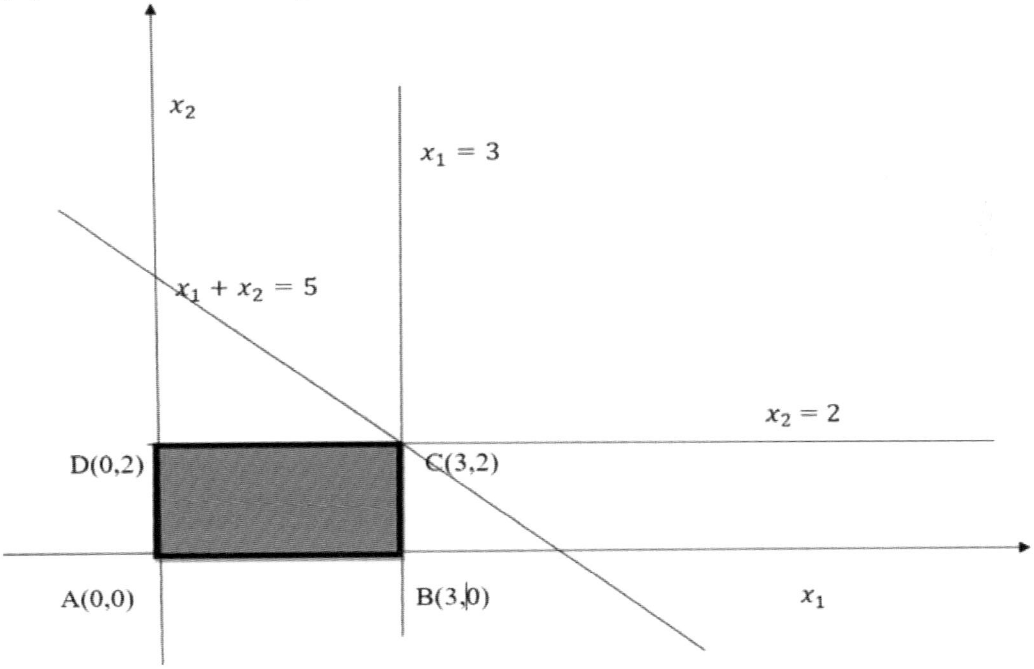

Fig. 3. The graphical method

So, it is necessary to check the value of the function z at the points *A(0,0)*, *B(3,0)*, *C(3,2)* and *D(0,2)*, respectively, and choose the smallest value among them. Since *z(0,0)=158000*, *z(3,0)=131000*, *z(3,2)=100000* and *z(0,2)=127000*. Comparing the values of the function at the vertices of the quadrilateral, we conclude that the optimal solution is obtained at the point *C(3,2)* for which the objective function has the value *z=100000*. This means that $x_1 = 3$ and $x_2 = 2$, that is, it is necessary to acquire three dryer sections and a 2MW boiler. We should note that in this case the dryer will work only in one shift.

Let us now state the corresponding algorithm

Algorithm 1

```
m=158000
for i =0:3
    for j=0:2
        if i+j>0
            z=158000-9000*i-15500*j;
            if z<m
                m=z;
                x1 = i;
                x2 = j;
            end
        end
    end
end
x1,, x2, m
```

In this example, we have a dryer that is not fully automated. If we want to check that the acquisition of equipment for automatic filling veneer, the price of which is 3000 euros, is economically worthwhile, we introduce a new variable x_3 which can take the integer values 0 and 1. On this occasion, we do not want to invest more funds for the acquisition of means of production than in the previous case. Now the new integer linear problem will take form of

$$minz = 158000 - 3480x_1 - 8300x_2 - 43200x_3 \quad (23)$$

$$\begin{bmatrix} 1 & 1 & 1 \\ 1 & 0 & 1 \\ 1 & 0 & 0 \\ 0 & 1 & 0 \\ 0 & 0 & 1 \end{bmatrix} \begin{bmatrix} x_1 \\ x_2 \\ x_3 \end{bmatrix} \leq \begin{bmatrix} 5 \\ 3 \\ 3 \\ 2 \\ 1 \end{bmatrix} \quad (24)$$

$$\begin{bmatrix} x_1 \\ x_2 \\ x_3 \end{bmatrix} \geq \begin{bmatrix} 0 \\ 0 \\ 0 \end{bmatrix} \quad (25)$$

Appropriate algorithm is

Algorithm 2

```
m=158000
for i =0:3
    for j=0:2
        for k=0:1
            if i+j>0
                if i+k<4
                    z=158000-3480*i-8300*j-43200*k;
                    if z<m
                        m=z;
                        x1 = i;
                        x2 = j;
                        x3 = k;
                    end
                end
            end
        end
    end
end
x1,, x2, x3, m
```

Here we have obtained an optimal solution $x_{1,} = 2, x_2 = 2, x_3 = 1$ a $minz = 91240$.

5. Result analysis

In this section, the optimization results from the previous section are analyzed. Due to the discontinuity in the production of structural veneer, a decision was made to purchase dryer and boiler sections of 1MW or 2MW. At the beginning of the operation, for the operation of the dryer, which works in three shifts and receives heat energy through a 4MW boiler, at least 158000 euros were needed for the costs of possible loss and the costs of working in the dryer's shifts. The problem was solved by using integer linear programming. By applying Algorithm 1, it was found that the optimal solution is the purchase of three dryer sections and a 2MW boiler. Then the costs of possible losses and labor costs in the dryers are reduced to 100000 euros. So for an investment in production assets of 24000 euros, we get an improvement of 26000 euros in the first year, and the improvement starting from the second year is as much as 50000 euros. It should be pointed out that the work of the dryer was reduced to one shift, while it used to work in three. So the entire production can work in one shift. The labor of workers in one shift is also very important for their psychophysical health, and this directly reduces the costs of sick leave. On the other hand, workers who worked in multiple shifts can now be moved to other work tasks with proper scheduling.

In the second part, we dealt with the justification of purchasing equipment for automatic filling veneer, the price of which is 3000 euros. By applying integer programming with the same investment for production assets of 24000 euros, the optimal solution is obtained in the form of the purchase of 2 dryer sections, a 2MW boiler and equipment for automatic filling veneer, the price of which is 3000 euros. Then the costs of possible losses and work in shifts amount to 91240 euros. It means that by applying mathematical programming, we have found high-quality optimal solutions.

6. Conclusion

Veneer drying is the most important but also the most demanding phase in the production of structural veneers. Production discontinuity usually occurs due to an inadequate amount of heat energy in the dryer. This is solved in two ways: by acquiring additional sections of the dryer and by using a boiler of a certain power. In order to estimate the required number of dryer sections and the power of the boiler to be purchased, we used integer linear programming. Through linear programming, we proved that the number of shifts in the dryer can also be reduced.

If the dryer is not fully automated, which includes the supply of veneers to the drying sections; it has been proven here that investing in automation significantly reduces costs in the long term. The use of linear programming to improve production efficiency was considered in the specific example of a factory for the production of the structural veneers. Linear programming gave us two suitable algorithms for providing help in the making of investment decisions. Further research goes in the direction of applying multi-criteria linear programming and methods of fuzzy multi-criteria linear programming for the optimization of technological procedures and the production of structural veneer.

7. References

[1] Baldwin, R.F. (1995). Veneer Drying and Preparation. In: Plywood and Veneer-Based Products, Miller Freeman Books, San Francisco

[2] Barkovic, D. (2001). Operational Research, Josip Juraj Strossmayer University of Osijek - Faculty of Economics, Osijek

[3] Bekhta, P.; Sedliacik, J. & Bekhta, N. (2020). Effect of Veneer-Drying Temperature on Selected Properties and Formaldehyde Emission of Birch Plywood, Polymers, Vol. 12, No. 3, 593

[4] Dantzig, G. (1963). Linear Programming and Extension, Princeton University Press, Princeton, N. J.

[5] Horman, I. & Martinovic, D. (2001). Numerical Simulation of the Process of Thermal Preparation of Wood for Veneer Production, Proceding of the 2. International meeting Trends in the Development of the Wood Industry System - Privatization and Development, pp. 39–47, Bihać, 26–27 October.

[6] Gass, S.I. (1958). Linear Programming: Methods of Applications, Mc grow-Hill, New York

[7] Kostic, A.; Maric B.; Kustura, M. & Timotic, V. (2021). Mathematical Model for Human Resource Planning in the Production Process, Proceedings of the 32nd DAAAM International Symposium, ISSN 1726-9679, ISBN 978-3-902734-33-4, Katalinic, B. (Ed.), pp. 4-9, Published by DAAAM International, Vienna, Austria

[8] Lukac, Z.; Neralic, L. (2012). Operational Research, University of Zagreb, Zagreb: Element d.o.o.

[9] Martinovic, D. & Horman, I. (2001). Determining the Appropriate Wood Drying Regime, Mechanical engineering, Vol. 43, (1-3), pp. 5–16

[10] Ozsahin, S. & Aydin I. (2014). Prediction of the Optimum Veneer Drying Temperature for Good Bonding in Plywood Manufacturing by Means of Artificial Neural Network, Wood Science and Technology, Vol. 48, pp. 59-70

[11] Pistun, Y.; Fedoryshyn, R.; Zagraj, V.; Nykolyn, H. & Kokoshko, R. (2019). Experimental Study and Mathematical Modelling of Nonlinear Control Plant, Proceedings of the 30th DAAAM International Symposium, ISSN 1726-9679, ISBN 978-3-902734-22-8, Katalinic, B. (Ed.), pp. 967-975, Published by DAAAM International, Vienna, Austria

[12] Slabohm, M.; Mai, C. & Militz, H. (2022). Bonding Acetylated Veneer for Engineered Wood Products-A Review, Materials, Vol. 15, No. 10, 3665

DOI: 10.2507/33rd.daaam.proceedings.002

SITE SELECTION FOR INDUSTRIAL WOOD PROCESSING

Aleksandra Kostic, Izet Horman, Melisa Kustura & Valentina Timotic

This Publication has to be referred as: Kostic, A[leksandra]; Horman, I[zet]; Kustura, M[elisa] & Timotic, V[alentina] (2022). Site Selection for Industrial Wood Processing, Proceedings of the 33rd DAAAM International Symposium, pp.0009-0016, B. Katalinic (Ed.), Published by DAAAM International, ISBN 978-3-902734-36-5, ISSN 1726-9679, Vienna, Austria
DOI: 10.2507/33rd.daaam.proceedings.002

Abstract

The wood resource in Bosnia and Herzegovina, due to its nature and structure, and significant natural regeneration, represents a key resource for the development of the wood processing industry. Different types of wood are represented, of which the most important from the industrial aspect are softwood and hardwood types of technical wood. Populations of these species are spread depending on geographical and climatic conditions. For the proper selection of the location of the factory for industrial wood processing, it is important to take into account several parameters that confirm or do not confirm a certain location in advance. Choosing a location for any investment, including investment in facilities and equipment for wood processing, is a complex process that includes, in addition to technical and physical, economic, social, ecological and other conditions. The paper will use the fuzzy logic methodology and MATLAB as a tool for evaluating location selection. It is necessary to create a model that provides the most desirable location for investment. The parameters that limit the optimal choice of location are the length of raw material transportation, human resources, road infrastructure and energy infrastructure.

Keywords: Fuzzy logic; Factory for industrial wood processing; Benefits and costs; Site selection; Mamdani fuzzy inference systems.

1. Introduction

Wood is a natural material and belongs to renewable resources, and is therefore interesting from the aspect of good physical and mechanical properties and natural decomposition. These properties give it an advantage over other synthetic materials, especially when it comes to environmental protection. Bosnia and Herzegovina is one of the richest European countries with this resource. Traditionally, wood was processed and furniture was produced in the area of the Western Balkans. This is one of the reasons why this paper considers the possibility of investing in this sector. However, it is not enough to have a raw material resource in order to have a successful production of the final product. It is necessary to take into account some other key parameters for the selection of locations for future production. In the first place is the availability of professional workforce. Then, no less important is the road and electric power infrastructure.

Certain regions in Bosnia and Herzegovina were traditionally oriented towards various production, such as mining, agriculture, metal industry, textile industry, food industry, etc. Previously, the start of production was based on assessments of the narrow or wider local community. These assessments were generally confirmed to be correct, although in some cases some subjective influence could be objected to. Wood processing, either primary or final, in the earlier period was mainly located where the highest concentration of raw materials was. The prevailing opinion was that the most important condition for successful business is the proximity of raw resources. Other influential factors were not neglected either, but there was no objective assessment of the justification of the choice of location. In order to achieve a formalization of the subjective assessment that would tend towards an objective assessment, it was necessary to choose a certain mathematical apparatus that analyzes weight influences.

Problems of planning, management and optimization in the production process from the aspect of available human resources are one of the most important factors in the production process. In the paper [6], the case of reducing the executor in production while maintaining the achieved level of production was considered. This is especially evident in emergency situations such as pandemics. The paper [7] presents a new approach of simulated modeling for the assessment of ecological and economic aspects for sustainable business. The problem is defined and then a model is presented that includes the economic and ecological aspects of the production system. The authors in [3] present criteria for choosing a location based on a literature review for a case study conducted on a Swedish company that is present on the global market. Article [8] is focused on the selection of criteria for location selection, based on 111 companies. Priority criteria for selecting a location for a specific production were taken into account. A quantitative model for location selection for some high-tech manufacturing companies can be found in [5]. In the past, site selection was usually qualitative in nature and highly subjective. It is an attempt to introduce a more objective quantitative approach. The most important location selection parameters for high-tech manufacturing companies have been identified, ranked and weighted based on research.

The paper is organized as follows: In Section 2, the basics of fuzzy logic are given. Fuzzy logic methodology for decision making and construction fuzzy rules are given in Section 3. Defuzzification and results are given in Section 4, conclusion and recommendation for further research are in Section 5.

2. Fuzzy Logic

Fuzzy logic is a branch of mathematics that is a generalization of Boolean logic, and has the greatest application in engineering disciplines. For example, the application of fuzzy logic in energy, more precisely for the selection of a location for the use of wind energy as well as renewable energy sources, is given in works [1], [2].

Fuzzy logic is based on Fuzzy-Sets, which were defined as a new term in 1965 [9] with the aim of presenting and modeling uncertainties in linguistics in a mathematically formalized way. Fuzzy-Set is a generalization of the classic set, because with classic sets the membership of the set is extremely distinct, that is, the element belongs or does not belong to the set, and with fuzzy-sets, the element's membership in the fuzzy-set can be characterized by a number from the interval [0,1]. This means that the fuzzy-set membership function maps each element of the universal set to the mentioned segment of real numbers. More information about this can be found in [4].

The first step in applied fuzzy logic is fuzzification. By applying the appropriate membership function, the phaser converts the exact value into the degree of membership. The most frequently used forms of membership function are: triangular, trapezoidal and Gaussian curve. Due to the uncomplexity of the calculation and sufficient accuracy, the use of the triangular rule is preferred. Mamdani and Sugeno are the most frequently used phase inference techniques. Mamdani fuzzy inference process will be used in this paper. The Mamdani phase system is a simple method that can apply the "if-then" rule to control the system. Defuzzification is the reverse process of fuzzification, where the combination of all fuzzy outputs results in a precisely determined numerical value of the output quantity.

The concepts of fuzzy model and probabilistic model are often mistaken one for another. The essential difference between these two terms is as follows. Fuzzy membership function represents the similarity of objects in the context of imprecise definition of features, while similarity provides information about the frequency of repetition. So, fuzzy logic formalizes the subjectivity of human thinking, feelings, language, and probability provides objective statistics in natural and technical sciences

3. Application of Mamdani Fuzzy Inference System

Fuzzy logic is a form of artificial intelligence. We used Fuzzy Inference System-FIS in MATLAB as a complete control system for decision making. We used Mamdani's method because it is the method most similar to human thinking. For decision-making, we had 12 input variables, which we divided into two groups: benefit and cost. We also had two output variables, so that the first six input variables were tied to the first output variable, and the other six to the second output variable. We also had 30 rules for the decision, the first 15 for the first output variable and the second 15 for the second output variable. We used the *trimf* membership function because it covers all cases well and is easy to compute. The linguistic variables used in the fuzzy methodology are: L- Low, N- Normal, H- High. The division of variables into benefits and costs is shown in Table 1. We used two locations to test our model. Location 1 is a rural location and Location 2 is an urban location. Fuzzy sets are given in Table 4.

Benefits	Costs
B1=Raw material availability	C1=Location cost
B2=Proffesional staff	C2=Construction land cost
B3=Road infrastructure	C3=Environmental concerns
B4=Electrical infrastructure	C4=Distance of administrative headquarters
B5=Telecomunication connectivity	C5=Place of residence of the employee
B6=Educational institutions	C6=Nearby of the protected zone and cultural and historical monuments

Table 1. Classification of input variables

	B1	B2	B3	B4	B5	B6	Result
Location1	0.8	0.3	0.4	0.7	0.6	0.2	0.4485
Location2	0.2	0.9	0.8	0.9	1	0.9	0.6272

Table 2. Overall fuzzy weights for the selected sites based on benefits

	C1	C2	C3	C4	C5	C6	Result
Location1	0.2	0.7	0.8	0.7	0.6	0.6	0.5981
Location2	0.9	0.2	0.6	0.2	0.3	0.9	0.6799

Table 3. Overall fuzzy weights for the selected sites based on costs

Parametres	**Symbol**	**Variable type**	**Linguistic variables**			**Range**
			1	**2**	**3**	
Raw material availability	B1	INPUT	L	N	H	0 - 1
Proffesional staff	B2	INPUT	L	N	H	0 - 1
Road infrastructure	B3	INPUT	L	N	H	0 - 1
Electrical infrastructure	B4	INPUT	L	N	H	0 - 1
Telecomunication connectivity	B5	INPUT	L	N	H	0 - 1
Educational institutions	B6	INPUT	L	N	H	0 - 1
Location cost	C1	INPUT	L	N	H	0 - 1
Construction land cost	C2	INPUT	L	N	H	0 - 1
Environmental concerns	C3	INPUT	L	N	H	0 - 1
Distance of administrative headquarters	C4	INPUT	L	N	H	0 - 1
Place of residence of the employee	C5	INPUT	L	N	H	0 - 1
Nearby of the protected zone and cultural and historical monuments	C6	INPUT	L	N	H	0 - 1
	B	**OUTPUT**	**L**	**N**	**H**	**0 - 1**
	C	**OUTPUT**	**L**	**N**	**H**	**0 - 1**

Table 4. Fuzzy sets

4. Results

In this section, we approach the defuzzification process, because we can directly make a decision from it. In Fig. 1. shows the dependence of costs on location costs and environmental protection costs. It was noticed that environmental protection costs affect the total costs much more than the price of the location, as can be seen from Fig. 1.

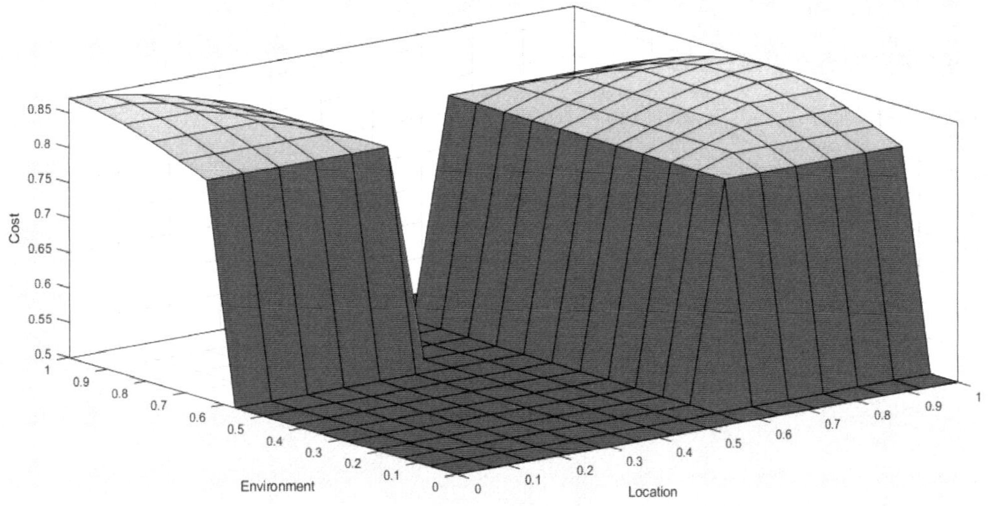

Fig. 1. Fuzzy interface

Our task was to make a decision by applying Fuzzy logic on choosing a location for the construction of a wood factory. We have seen that the defuzzification result for the rural location, Location 1, from the point of view of benefits is $b_1 = 0.4485$, and from the point of view of costs it is $c_1 = 0.5981$, which indicates that when choosing a rural location, the benefits are still far less than the costs. When choosing the urban location, Location 2, the benefit value is $b_2 = 0.6272$, and the cost value is $c_2 = 0.6799$. We can see that the benefit value of the urban location, Location 2 is still less than the cost value, but it is closer to the cost value. In order to make an impartial decision, we will compare the quotients $\frac{b_1}{c_1}$ and $\frac{b_2}{c_2}$.

Since $\frac{b_1}{c_1} = 0.74987 < \frac{b_2}{c_2} = 0.922249$. The decision to build a wood processing factory is in a urban location, which reflects practical experience.

5. Conclusion

In our paper, we used the MATLAB program and its FIS part to select a location for the construction of a wood processing plant. When forming the fuzzy set, we had 12 input parameters, which we divided into 6 parameters representing benefits and 6 parameters representing costs. We used a total of 30 rules for decision making. In order to decide the optimal location for the wood processing factory, we used relevant parameters. The question of determining the site for the wood industry is very important for Bosnia and Herzegovina, because due to the wealth of wood resources, a good part of the gross income is based on the wood industry. To verify our results, we used two sites, one rural and one urban, and we proved that the urban site is much more suitable, because the focus in the modern world is on the human factor and its education. It was also shown that when choosing a location, investment in environmental protection is more important than the price of the land itself or the proximity of wood resources. In further research, it is necessary to collect more specific locations, where it is possible to build a factory for wood processing, and to apply our proposed model to it. It should also be considered whether some other parameters enter the decision-making process and possibly expand the Fuzzy set.

6. References

[1] Aljicevic, Z.; Kostic, A.; Dautbasic, N. & Karli, G. (2016). Model of fuzzy logic for selection infrastructural investment project of wind farm locations, Proceedings of the 27th DAAAM International Symposium, ISSN 1726-9679, ISBN 978-3-902734-08-2, Katalinic, B. (Ed.), pp. 743-748, Published by DAAAM International, Vienna, Austria

[2] Aljicevic, Z.; Kostic, A. & Dautbasic, N. (2017). Selecting location for infrastructural investment project in renewable sources of energy using matlab and fuzzy logic, Proceedings of 13th International Conference DEMI 2017, Banja Luka, ISBN: 978-999938-39-72-9, Tansic, Z. (Ed.), pp. 255-261

[3] Bruch, J.; Wiktorsson, M. & Bellgran, M. (2014). On the production location decision: A case study on process and criteria, International Journal of Manufacturing Research, Vol. 9, No. 1, pp. 74-91

[4] Cintula, P.; Fermuller, C.G. & Noguera, C. (2021). Fuzzy logic, In: Stanford Encyclopedia of Philosophy, Zalta, E. N. (Ed.), page numbers (1-10), The Metaphysics Research Lab, ISSN: 1095-5054, Stanford

[5] Davies, B. & Lazniarz, M. (1985). A site selection model for high technology manufacturing firms in the United States, Scientometrics, Vol. 8, pp.103–116

[6] Kostic, A.; Maric, B.; Kustura, M. & Timotic, V. (2021). Mathematical model for human resource planning in the production process, Proceedings of the 32nd DAAAM International Symposium, ISSN 1726-9679, ISBN 978-3-902734-33-4, Katalinic, B. (Ed.), pp. 4-9, Published by DAAAM International, Vienna, Austria

[7] Ojstersek, R.; Palcic, I. & Buchmeister, B. (2020). Simulation modelling of environmental and economic aspects in manufacturing systems, Proceedings of the 31st DAAAM International Symposium, ISSN 1726-9679, ISBN 978-3-902734-29-7, Katalinic, B., (Ed.), pp. 436-443, Published by DAAAM International, Vienna, Austria

[8] Spalanzani, A.; Ageron, B. & Iskander, Z. (2016). Manufacturing operations location decision: What are the main criteria, Supply Chain Forum: An International Journal, Vol. 17, No. 4, pp. 205-217

[9] Zadeh, L.A. (1965). Fuzzy sets, Information and Control, Vol. 8, No. 3, pp. 338-353., ISSN 0019-9958, DOI: 10.1016/S0019-9958(65)90241-X

DOI: 10.2507/33rd.daaam.proceedings.003

ICT IN EDUCATION DEVELOPMENT

Budić Hrvoje, Đuro Horvat & Ružić Ivan

This Publication has to be referred as: Budic, H[rvoje]; Horvat, D[uro] & Ruzic, I[van] (2022). ICT in Education Development, Proceedings of the 33rd DAAAM International Symposium, pp.0017-0023, B. Katalinic (Ed.), Published by DAAAM International, ISBN 978-3-902734-36-5, ISSN 1726-9679, Vienna, Austria
DOI: 10.2507/33rd.daaam.proceedings.003

Abstract

The introductory part of the paper emphasizes the importance of using ICT in the educational process and emphasizes its fundamental developmental determinants. The second part of the paper emphasizes the concept of digital literacy, which is superior to the concepts of information and IT literacy. Also here, the differences between distance learning and online learning are highlighted, as well as the characteristics of both models. Regarding the increasing number of users of mobile devices, data on the use of mobile and PDA devices as well as their elementary advantages in learning are quantitatively presented. The practical part of the work is based on the assessment of the distinctiveness of the use of ICT, and with regard to the type of scientific educational institution, it was tested with relevant statistical tests, and the basic hypothesis was confirmed. The conclusion presents an overview of the entire work, recommendations and assumptions for the future.

Keywords: ICT; Digital literacy; Distance learning; E - learning; Mobile learning (m - learning)

1. Introduction

Today society is heavily influenced by the development and implementation of modern information and communication technologies (ICT). Last fifty years it was changing from industrial to informational, which creates a completely new society, society - based knowledge. In a such society, the economy and the quality of individual life depend on knowledge and on the finding, processing and application of information. The result of that is a continuous growth of human knowledge which creates the need for acquiring increasing volume and quality of knowledge of every individual and institution in a very short time [1]. Learning through the use of technology cannot be identified with classical, frontal, face - to - face learning in the classroom. The ultimate goal is the same, but the form of acquiring knowledge in the learning process is completely different. A different approach to learning implies the necessity of thinking about the best possible use of technology in the learning process to achieve the best possible results. Beforehand, it is necessary to know the possibility of the used media for knowledge transfer, know its limitations, and know the answer to the question of why this particular media is the best for improving learning outcomes. Taking into account the definition of educational technology, we can say that it is a complex process that includes people, procedures, ideas, tools and organization for analyzing problems, selecting tools and evaluating different viewpoints of human learning. Following the above, educational design and technology include the analysis of learning and teaching, design, development, use, evaluation and organization of teaching and other processes and materials for improving learning and action in different environments, especially in educational institutions and workplaces.

The study of education as an object of research, the development of its scope and methods beyond the concrete limits and goals of science with the latest scientific and technological achievements and the increased demands of education is in the program of the perception of today's education. Educational technology actually represents an act of systematized transformation of scientific knowledge into its application. Many authors who have dealt with the impact of technology on learning so far emphasize that it is an important element for effective learning. First of all, the use of technology, in addition to other indisputable factors that it possesses and implies, encourages the development of critical thinking of the fundamental participants of the educational process. Computers successfully provide effective learning related to nature and learning techniques. Effective technology enabled learning involves principles that may require new learning environments [2]. The concept of information and communication technology, which is applied in education, grew out of the previous concept of information technology and new technologies and represents a huge area of rapid changes and rapid growth. In this sense, ICT also contributes highlighting new concepts such as information and communication literacy and digital literacy. When talking about ICT in education, the acronym ITLET (Information Technology for Learning, Education and Training) is used, which represents a key element in all forms and models of distance learning [3].

Today, learning through the application of information and communication technology is a generally recognized and accepted way of knowledge transfer, both at the national and international level. This is evidenced by numerous documents (strategies) such as Europe 2020 and the Education, Science and Technology Strategy of the Republic of Croatia [4]. The Europe 2020 strategy, in its analysis of key issues states that higher education institutions are agents of change, i.e., catalysts of systemic changes in improving and raising the quality level of new educational models [5]. Under the European Digital Agenda, the need to take advantage of the transformational advantage of information and communication technology is stated, all to improve teaching processes. Also, it is suggested to encourage greater diversity in the ways of studying, continuing education for people who have stopped education, and taking advantage of information and communication technology for more effective learning and new research methods [6]. Due to the increasing level of digital literacy and the implementation of ICT in institutions of higher education and science authors considered it appropriate to investigate the assessment of the volume of its use. As the level of digital literacy is generally increasing, especially among the younger population living in the digital age, and taking into account the level of their education, as well as the characteristics of the teaching staff, authors wanted to assess how similar it is, or not, i.e. different. For this purpose, a hypothesis was set up where relevant statistical tests are used to determine the above.

2. New educational model based on ICT

Technology opens a new perspective for education in the transfer of knowledge. The correct use of technology can give exceptional results, can enrich teaching process, and make education accessible even in the most inaccessible parts of the world, recognized in the term of distance learning. Computer and information literacy is necessary in order to use ICT in the correct way. Although they have similar names, information and IT literacy are fundamentally different. Informatics, or computer literacy, represents the ability to use computers and associated computer programs, while the term information literacy refers to the ability to find and use information, manage and analyze it in order to turn it into knowledge. The use of information and communication technology also implies the acquisition of digital literacy, which is related precisely to the use of textual material in the form of hypertext, or multimedia on which educational materials in the process of e - learning and in mobile applications (m - learning) are largely based. Digital literacy skills include the ability to make judgments about Internet sources, search, content management, and online communication. However, the process of searching for information and evaluating the necessary information belong to the field of information literacy, while the use of multimedia materials and communication belong to computer literacy. Searching for information and making judgments are competencies of information literacy, while managing multimedia content and communicating via the network belong to computer literacy skills.

2.1. Distance learning

Distance learning as a term is primarily related to the situation when the participants in the teaching process are not physically located in the same locations, so their mutual interaction is carried out via computers, i.e., ICT. However, in addition to spatial dislocation there is also temporal dislocation, and depending on according to the mode of interaction, distance learning is categorized into synchronous and asynchronous [7]. Although synchronous and asynchronous learning are fundamentally different, and synchronous communication provides direct real - time communication, asynchronous also has certain advantages:

1. Accessing educational content from anywhere at any time,
2. Allowing enough time to consider and solve the problem. In this way, it is possible to have much better quality discussions than in classical classes. A typical representative of asynchronous communication are distance learning systems (Learning Management Systems - LMS like Loomen, or Moodle),
3. Possibility of anonymity of the user,
4. There is no time limit (eg time zones),
5. Does not require investments in expensive hardware and fast communication networks [8].

2.2. Internet learning (e - learning)

Instead of the previous practice where teachers traveled to other faculties, ideas are exchanged using the Internet, and the teacher is located at the home university, and the educational content is transmitted to the students via the Internet. The Internet is first of all:

- easily accessible - it is not technically limited to a narrow circle of IT specialists, it is not exclusive in the sense of restricting access to information, it is financially available to a wide range of users;
- interactive - does not filter communication through political or economic mediators, but quite the opposite enables direct communication between individuals, individuals and groups, and the groups themselves;
- diverse - in relation to easy accessibility, individuals and groups of the most diverse viewpoints can express their opinions, ideas and attitudes. The Internet is connected to contemporary culture, but it is not its substitute and it enables the creation of new cultural forms by combining contemporary culture and communication technology;
- experimental - it is open to new ideas, uses and processes, and is determined by its cultural, social and political benefit, not just a commercial basis [9].
- The system of adoption of educational content via the Internet must enable the following:
- communication between participants of the educational process via computer;
- presentation and adoption of educational content in a certain order;
- the organization of learning in connection with the generation of various educational data about students and educational content;
- knowledge tests with various tests and self - check quizzes;
- storing educational content on the web server (server), and connecting them and creating tests and discussions about the above [10].

E - learning implies multiple forms, and some of them use technology with a description as shown in Figure 1.

<table>
<tr><th colspan="2">Shape</th><th>Description</th><th>Technology</th></tr>
<tr><td colspan="2">Classic teaching (face - to - face)</td><td>Teaching in classroom</td><td>Not used (except preparing the lesson before teaching process</td></tr>
<tr><td rowspan="3">E - learning</td><td>Teaching without the help of ICT</td><td>Technology to improve classical teaching in classroom</td><td>Presentations, multimedia, CD - ROM, Web sites, interactive posters, online quizes, e - mail, forum, blog, wiki, e - portfolio, etc., web 2. 0 tools, webinars…</td></tr>
<tr><td>Mixed education</td><td>Classical teaching with the help of ICT</td><td rowspan="2">Educational contents are delivered through the Website, or using the Learning Management System, video conference...</td></tr>
<tr><td>Online education</td><td>Learning and teaching exclusively with the help of ICT; no face - to - face classes</td></tr>
</table>

Table 1. Technology and e - learning [11]

Internet applications used by educational institutions in the Republic of Croatia have different purposes and use different technological platforms. As an example of the use of Internet applications for educational purposes in the Republic of Croatia, the following main categories can be cited:

- Administration and administrator applications,
- Control and monitoring applications,
- Applications for viewing content,
- Applications for research purposes (data collection),
- Applications for submitting content,
- Websites, web content management systems
- Databases - digital repositories,
- Applications for distance learning
- Cloud applications
- Internal information services,
- Other special purpose applications [12].

2.3. Significance of mobile applications for learning processes

Mobile learning can be simply described as learning using portable devices anytime and anywhere. The basic platform for the development of mobile learning is distance learning in conjunction with the use of mobile technology, i.e. portable devices. The use of mobile technology basically means the use of mobile, or "smartphones", personal digital assistants (PDA), and laptops. Today, there are various internet services through which the user can download the desired mobile application, and install and use it for various purposes, including educational ones. Some of the leading services are Google Play adapted for Android operating systems, App Store adapted for iOS operating system, and Microsoft Store adapted for Windows Mobile operating system. Also, the main feature of today's websites is the so - called Responsiveness that enables a display adapted to the screens of portable devices. According to data obtained from the Quora website, the number of educational content applications in the App Store is approximately 176,000 out of a total of 2.2 million, which represents a share of 8.5%. The same source states that the number of applications intended for user education on Google Play is approximately 500,000 out of a total of 2.8 million, which represents a share of 17.8%. According to the content, educational - type applications highlight differences in the concepts of learning approaches so some have already predefined material, and some are intended to repeat previously mastered material.

It should be noted that the application has the ability to adapt to the size of educational material, time dynamics, and adapt to one's own preferences [13]. According to data obtained from the website We Are The Social, the number of unique mobile phone users is 5.15 billion with a penetration of 67%, while the number of Internet users is 4.47 billion with a penetration of 58%. [14]. A kind of technological revolution in its first appearance in 2007 mostly included changes in the way of using mobile phones, which was marked by the appearance of the first mobile phones with iOS and Android operating systems [15]. The use of mobile applications in teaching enables the collection and processing of data even during current teaching lessons, emphasizing the technological component as a useful supplementary tool in the hybrid model of teaching. Ozdamli believes that it is more useful to improve and enrich standard methods of learning and teaching with additional activities aimed at the use of mobile phones, the Internet and mobile applications outside the classroom. [16]. The development of mobile technologies marked an evolutionary progress in application of technology, both in everyday life and in the business environment.

Mobile platforms offer greater flexibility, and give the possibility of using services on the go, which opens up new possibilities. New mobile platforms and operating systems such as Android and iOS are recording a significant growth in mobile application downloads. The market shares of popular operating systems for mobile devices are shown in table 1. The market reach of the most popular applications according to categories at the world level are shown in table 2. It should be noted that 85% of the holding share belongs to the Android mobile platform, while the most popular category is communication applications with the market with a reach of 99.39%. Categories of applications for the Android platform in the field of education have a market reach of 30% at the global level. Market shares of mobile applications according to operating systems are shown in table 1.

Year	2017	2018	2019
Android	85.1%	85.1%	85.0%
iOS	14.7%	14.9%	15.0%
Others	0.2%	0.0%	0.0%
Total	100%	100%	100%

Table 2. Global indicators of the market share of mobile applications by operating system [17]

Year 2019th										
Category	Tools	Communication	Society	Fun	Photography	Books	Video	Shopping	Education	Sport
Reach %	99.81	99.39	95.02	83.85	75.77	70.74	96.63	35.79	29.28	30.44

Table 3. Market reach of some of the most popular Android app categories as of September 2019 [18]

3. Research sample and methodology

The primary research was conducted at 3 public polytechnics, 2 private higher education institutions and at 2 public universities, (3 faculties within public universities), and includes two target populations - teaching staff and students of the 2nd and 3rd year of undergraduate studies at the mentioned scientific - teaching institutions of higher education. A scale from 1 = "extremely low" to 5 = "extremely high" volume was used to assess the volume of use of information and communication technologies in the teaching process and extracurricular communication in scientific and teaching institutions of higher education. The program package IBM SPSS Statistics for Windows, version 23.0 was used to process the research results. The research was conducted using an online questionnaire with a purposive convenience sample. The results are presented at the level of the target population, i.e. the responses of 70 teachers and 474 students who participated in the research.

The empirical research is based on the following **hypothesis:**
The difference in the level of use of modern communication devices and Internet applications in the academic community for the needs of the teaching process and extracurricular communication is visible depending on the type of scientific and educational institution (public polytechnic studies, high private business schools, university studies), although it is about the same educational level. The total sample of teaching staff included 42.9% of teachers at public polytechnic studies, 34.3% at public universities and 22.9% at private universities. The total sample of students included 63.7% of public higher education students, 24.5% of public university students and 11.8% of private higher education students.

Note: Students could choose more than one answer to the questions related to the undergraduate studies they are studying and therefore the sum of the percentages within a single question may be greater than 100%.

				Total		Very high volume	High volume	Neither high nor low volume	Low volume	Extremely low volume
				Count	Column N %	Row N %	Row N %	Row N %	Row N %	Row N %
Target group	Teaching staff	**Total**	.	**70**	100,0%	14,3%	28,6%	34,3%	22,9%	0,0%
		Type of scientific - teaching institutions	Public higher education institution	30	42,9%	0,0%	26,7%	40,0%	33,3%	0,0%
			Private higher education institution	16	22,9%	25,0%	25,0%	25,0%	25,0%	0,0%
			Public university	24	34,3%	25,0%	33,3%	33,3%	8,3%	0,0%
	Students	**Total**		**472**	100,0%	11,0%	43,2%	36,4%	7,6%	1,7%
		Type of scientific - teaching institutions	Public higher education institution	302	64,0%	9,3%	39,1%	43,7%	7,3%	0,7%
			Private higher education institution	56	11,9%	32,1%	53,6%	10,7%	3,6%	0,0%
			Public university	114	24,2%	5,3%	49,1%	29,8%	10,5%	5,3%

Table 4. Assessment of the volume of use ICT in the teaching process and extracurricular communication of higher education institutions according to target groups and type of scientific and teaching institutions [19]

4. Research results

The normality of the distribution of the variable assessing the volume of ICT use was tested with the Kolmogornov - Smirnov test and the results showed that the distribution of the results of the degree of agreement with the observed statement in both target groups significantly deviated from the normal distribution ($p < .05$). Because of the above, non - parametric data processing methods were used in the analysis.

	M	IQR	Skew	Skew Se	Kurt	Kurt Se	Statistic	df	K - S p
Teaching staff **(Total)**	3,00	1,00	0,176	0,287	-0,980	0,566	0,207	70	p <.05
Students **(Total)**	4,00	1,00	-0,382	0,112	0, 246	0,224	0,247	472	p <.05

Table 5. Descriptive parameters of the variable for assessing the volume of ICT use in higher education institutions and testing the normality of the distribution according to the target groups [20]

Legend: M - median, IQR - interquartile dispersion, Skew - skewness of distribution, Kurt - flattening of distribution (kurtosis), SE - standard error, Statistic - result of the performed test, df - degrees of freedom, K - S p - level of significance Kolmogorov - Smirnov test

The difference in the average assessment of the volume of ICT use in the teaching process and extracurricular communication in relation to the target groups and types of scientific and educational institutions for each individual target group was tested with the Kruskal - Wallis test. No statistically significant differences were found in the average perceived volume of ICT use in relation to the target group.

Statistically significant differences were found in the average perceived volume of ICT use according to the type of institution and in the group of teachers ($X2 = 9{,}039$, $df = 2$, $p < .05$) and in the group of students ($X2 = 33{,}719$, $df = 2$, $p < .05$) .

The difference between the average perceived volume of use of individual pairs of groups of teachers and students according to the type of institution was tested using the Mann - Whitney U test. The results indicate that teachers at public universities estimate the volume of ICT use significantly higher than teachers at public universities ($U = 194.00$, $p < .05$). Furthermore, students at private universities estimate the volume of ICT use to be significantly higher than students at public universities ($U = 1752.00$, $p < .05$) and public polytechnics ($U = 4712.00$, $p < .05$).

The obtained results confirm the hypothesis according to which the difference in the level of use of modern communication devices and internet applications in the academic community for the needs of the teaching process and extracurricular communication is visible depending on the type of scientific and educational institution (public polytechnic studies, high private business schools, university studies), although it is about the same educational level. The difference in the average volume of use was determined both in the group of teachers and in the group of students according to the type of scientific and educational institution, and the assumed direction of the difference was confirmed for the group of teachers, but rejected for the group of students. A more effective use of modern communication devices and internet applications was assumed among teachers and students of university studies, which is true for the group of teachers. However, in the group of students, the use is significantly higher among those who attend private universities.

5. Conclusion

The introductory part of the paper presented the basic premises of ICT in education, and distinguished the concepts of information and IT literacy. In this sense, the concept of digital literacy is created. Furthermore, the educational model from distance learning, e - learning to the use of mobile and PDA devices is presented. Here is a thorough presentation of the aforementioned and the basic characteristics of certain models, as well as their advantages and disadvantages. The practical part of the paper contains the testing of the set hypothesis with appropriate statistical tests. In this sense, the hypothesis was confirmed because the authors assumed that the assessment of the ICT used in the given sample was different. The authors assumed that the level of communication and digital literacy is higher in university studies than in public polytechnic studies and private business schools. In support of this is the fact that students at university studies have a higher previous level of knowledge and information and communication literacy, although this is not entirely necessary, so they use ICT more efficiently, just like the teaching staff at the mentioned institutions, which have stricter criteria for selection in scientific teaching professions and higher quality permanent education, so it is assumed that they should have the specified levels of knowledge at a higher level in order to be able to implement and use the specified applications and devices. Speaking of this, the hypothesis was confirmed because a different level of ICT use was proven, although not entirely according to the author's presumptions in the direction of use, but a statistically significant difference was proven.

Thinking about the future, it is impossible not to think that ICT in education will not have its role, in fact, it could play a key role, but it can not completely replace the teacher in the classroom. The connection of various tools, dynamism and progressiveness will be indisputable. Also, the use of virtual worlds and simulations in learning is assumed. Likewise, ICT in education will also help marginalized groups, for example by developing experimental interfaces for brain - computer communication without the intervention of limbs, which represents a fantastic advantage for people with impaired motor skills.

6. References

[1] Budic, H.; Marinac, A. & Filipovic, F. (2015). Technology and media in education, Proceedings of the ENTRENOVA - ENTerprise Research InNOVAtion Conference, Vol 1., No. 1., 10. - 11.09.2015., Kotor, Montenegro, ISSN 2706 - 4735, Bacovic, M.; Milkovic, M. & Pejic Bach, M.; Pekovic, S. (Ed.), pp. 130 - 137, Udruga za promicanje inovacija i istraživanja u ekonomiji "IRENET", Zagreb, Croatia.

[2] Akbiyik, C. & Simsek, N. (2009). Accelerated learning in classroom and computer environments, Eurasian Journal of Educational Research, Vol. 37., (October, 2009.), pp. 32 - 52, ISSN 1302-597X.

[3] Breslauer, N. (2011). Obrazovanje uz pomoć informacijsko - komunikacijskih tehnologije. Education with the help of information communication technology. Zbornik radova Međimurskog veleučilišta u Čakovcu, Vol. 2, No. 2., (October, 2011), pp. 27 - 31, ISSN 1847 - 702X, eISSN 1849 - 1138.

[4] https://vlada.gov.hr/europa-2020/19454 (2020). Europa 2020, Accessed on: 2022 - 20 - 09.

[5] Host, A.; Antonic Jardas, J.; & Skender Pavlic, H.; Racki, Z.; Sokolic, D. (2018). Uvođenje novih tehnologija u nastavni proces, Introducing new technologies into the teaching process, Ekonomski fakultet Sveucilista u Rijeci, ISBN 978 - 953 - 7813 - 39 - 0., Rijeka, Croatia [6]

[6] https://euroalert.net/publication/272/digital-agenda-for-europe-rebooting-europe-s-economy (2015). European Commission, Directorate - General for Communication, Digital Agenda for Europe. Rebooting Europe's economy, Accessed on: 2022 - 20 - 09 .

[7] King, Frederick B.; Young, Michael F.; & Drivere - Richmond, K.; Schrader, P. G. (2001). Defining distance learning and distance education, AACE journal, Vol. 9., No. 1., pp. 1 - 14, ISSN 1065 - 6901.

[8] Kung - Ming, T. & Khoon - Seng, S. (2005). Asynchronous vs. synchronous interaction, In: Encyclopedia of Distance Learning, pp 104 – 113, IGI Global, ISBN 1 - 59140 - 557 - 2.

[9] Budic, H. & Mario, H. (2014). Primjena suvremene obrazovne tehnologije u nastavi, Application of modern educational technology in teaching, Proceedings of the 4th International Conference "Vallis Aurea", Focus on Regional & Innovation Development, Pozega, Croatia, 18. - 20.09.2014. Katalinic, B. (Ed.), pp. 65 - 73, Polytechnic in Pozega, Croatia & DAAAM International Vienna, Austria, ISBNN 978 - 3 - 902734 - 02 - 0.

[10] Ruzic, D. & Budic, H. (2012). Utjecaj informacijsko komunikacijske tehnologije na poboljšanja nastavnog procesa, The impact of ICT on the improvement of the educational process, Proceedings of the 3rd International Conference "Vallis Aurea", Focus on Regional Development, Pozega, Croatia, 05.10.2012. Katalinic, B. (Ed.), pp. 895 - 903, Polytechnic in Pozega, Croatia & DAAAM International Vienna, Austria, ISBN 978 - 3 - 901509 - 78 - 0.

[11] https://www.carnet.hr/wp-content/uploads/2019/09/Digitalni_nastavni_materijali_-_prezentacija.pdf (2019). Digitalni nastavni materijali, Digital teaching materials, Accesed on: 2022 - 20 - 09.

[12] https://www.aaiedu.hr/statistika-i-stanje-sustava/web-aplikacije (2019). Statistika i stanje sustava, Statistics and system status, Accesed on: 2022 - 20 - 09.

[13] https://www.quora.com/How-many-educational-apps-are-on-the-App-Store (2019). How many educational apps are on the App Store?, Accessed on: 2022 - 20 - 09.

[14] https://wearesocial.com/blog/2019/10/the-global-state-of-digital-in-october-2019 (2019), Accessed on: 2022 - 20 - 09.

[15] https://www.cnet.com/reviews/t-mobile-g1-review/ (2017). T - Mobile G1 review: CNET reviews the first-ever Android phone, Accessed on: 2022 - 20 - 09.

[16] Ozdamli, F. & Cavus, N. (2011). Basic elements and characteristics of mobile learning, Procedia – Social and Behavioral Sciences, Vol. 28., (December, 2011)., pp 937 - 942, ISSN 1877 - 0428.

[17] https://www.cnet.com/reviews/t-mobile-g1-review/ (2019). Smartphone Market Share, Accessed on: 2022 - 20 - 09.

[18] https://www.statista.com/statistics/200855/favourite-smartphone-app-categories-by-share-of-smartphone-users/ (2019). Market reach of the most popular Android app categories worldwide as of September 2019, Accessed on: 2022 - 20 - 09.

[19] Budic, H. (2021). Marketing aspects of using modern internet application in academic community, Dissertation thesis, p. 163., https://repozitorij.efos.hr/islandora/object/efos:4211, Accessed on: 2022 - 20 - 09.

[20] Budic, H. (2021). Marketing aspects of using modern internet application in academic community, Dissertation thesis, p. 160., https://repozitorij.efos.hr/islandora/object/efos:4211, Accessed on: 2022 - 20 - 09.

DOI: 10.2507/33rd.daaam.proceedings.004

DESIGN AND DEVELOPMENT OF A FLEXIBLE, CABLE-CONTROLLED AND HYPERREDUNDANT ROBOT WITH STRENGTH-ADJUSTABLE BALL JOINTS

Dominik Diek, Kemajl Stuja & Mohamed Aburaia

This Publication has to be referred as: Diek, D[ominik]; Stuja, K[emajl] & Aburaia, M[ohamed] (2022). Design and Development of a Flexible, Cable-Controlled and Hyperredundant Robot with Strength-Adjustable Ball Joints, Proceedings of the 33rd DAAAM International Symposium, pp.0024-0028, B. Katalinic (Ed.), Published by DAAAM International, ISBN 978-3-902734-36-5, ISSN 1726-9679, Vienna, Austria
DOI: 10.2507/33rd.daaam.proceedings.004

Abstract

The universal industrial robot is the 6-axis articulated robot, which consists of six serially connected joints and has six degrees of freedom. Hyper-redundant robots have additional, redundant degrees of freedom thanks to additional serially connected link axes, resulting in increased flexibility that allows them to better manoeuvre curved paths. Hyper-redundant robots either feature precise control of each joint axis using only one actuator per joint or can be manipulated in at least two spatial directions in each joint axis. However, these single systems do not exhibit these two properties in combination. The goal of this work is to develop a concept in the form of a prototype that combines these features. By combining them, an optimization in terms of size and flexibility can be achieved. In this sense, a blocking (holding) system has been developed which is based on strength adjustable ball joints. The 3D-printed hyper-redundant robot is controlled by cables. To validate this system concept, after development of the kinematics and control of the robot, tests were carried out regarding the payload of a joint and the entire robot arm, with subsequent absolute accuracy testing of the robot arm. The results of this work in terms of the concept are presented in tabular form and discussed in detail. An outlook for future development is also given.

Keywords: hyper-redundant robot; strength adjustable ball joint; holding system; absolute accuracy; 3D printing.

1. Introduction

Industrial robots cannot be used for certain applications with special requirements. An example of this is inspections of rooms or rescue operations after a natural disaster (earthquake), when we are dealing with a destroyed, disorderly environment. Here the snake-like robots with several degrees of freedom (several axes) are used to move objects through the obstacles (curved paths). Against it, the industrial robot is the 6-axis articulated robot and consists of six serially connected link axes. Due to their arrangement, these six articulated axes allow the end effector a maximum degree of freedom of six. Three degrees of freedom are required for obtaining the spatial position and three are required for orientation [1]. This means that the objects in the work area of the industrial robot can be manipulated in cartesian space.

On the other hand, the number of joint axes thus limits the maximum degree of freedom of the industrial robot. In contrast, hyperredundant robots have redundant degrees of freedom, resulting in an increase in flexibility that gives them the ability to manoeuvre curved paths. The resulting kinematic redundancy enables robots of this type to achieve a pose in multiple robot configurations [2]. Applications for robots of this class can be found, for example, in surgery, cleaning and inspection operations, and as manipulators during rescue operations [3].

To expand these application areas, an optimization regarding size and flexibility is to be carried out in the course of this work. The joint axes of the hyperredundant robot should be individually controllable in two degrees of freedom, using only one actuator per joint. The design process is based on the V-model according to VDI 2206 [4] and methodical approach is related to works [5] and [6]. After the conceptual design of a holding system that meets the requirements of the mechatronic system, a 3D-printed prototype of the robot is created based on this, so that the elaborated concept, which enables the individual control of joint axes, can be tested, and validated too.

2. Problem Statement – Materials and Methods

Hyperredundant robot systems either have precise control of each joint axis or can be manipulated in at least two spatial directions in each joint axis. However, these individual systems do not exhibit these two characteristics when only one actuator per joint axis is used. The aim is to investigate how these features can be combined so that flexible control of the kinematic structure can be realized with maximum reduction of active components or actuators, both in the robot base and directly on the manipulator. The reduction of installed actuators should extend the range of application, entail financial savings, and positively influence flexibility and size.

2.1. Design of the holding system and the hyperredundant robot

The designed hyperredundant robot (see Figure 1 left) is cable-controlled and consists of a kinematic chain of four strength-adjustable ball joints connected in series (see Figure 1 right). The robot base is the static part of the system that houses the electronics, pneumatics, and motors. There are three motors whose flanges are wrapped cables that run through holes in all joints of the robot (see the figure 3 für better understanding). The 3 holes on joints are designed symmetrically at 120 degrees. The dynamic part of the system is the manipulator. The working area of the robot is the area outlined with blue colour. At the end of the kinematic chain is tool centre point (TCP). Sectional view of the joint is shown at the right side. When pressure is applied, a pneumatic short-stroke cylinder presses a stop buffer made of elastomer against the joint head. The resulting friction causes the joint to lock. By blocking all joints except the one to be manipulated, an individual control of a joint can be achieved by the hyperredundant robot. When tightening or loosening the ropes attached to the TCP, only the deblocked joint is manipulated due to the freedom of movement.

2.2 Design of the hyperredundant robot

The cables or steel ropes are wound or unwound in the base of the robot by three stepper motors. Figure 2 shows the chain of relationships for determining the forward kinematics. The motor steps change the cable lengths. This change influences the respective change of the joint angles of the joint to be controlled, which determines the pose of the end effector.

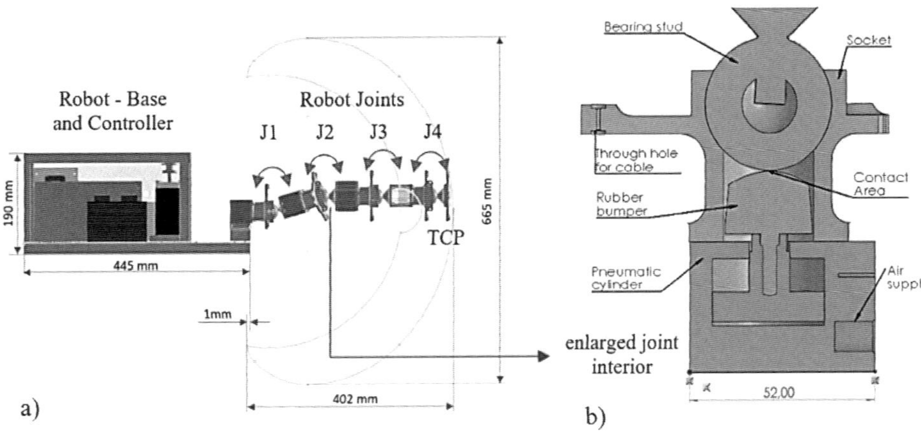

Fig. 1. Hyperredundant robot work area (a) and sectional view of the joint with pneumatic cylinder (b) to illustrate the principle of holding

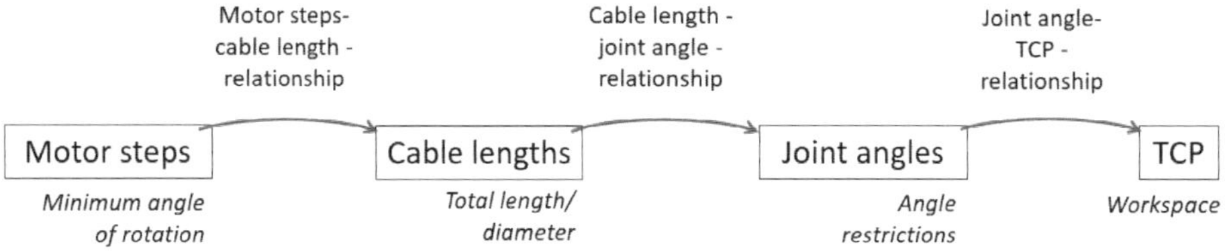

Fig. 2. Relations and limitations for the realization of forward kinematics

The prototype will be subjected to tests on load-bearing capacity and absolute accuracy. For its execution, the control is implemented based on the models of the calculations from the forward kinematics using DH table [7]. These joint angles are calculated similarly to the concept of Xu [8] based on the rope lengths between the joints, and then transferred to the corresponding number of motor steps. By constructing the through hole for the cable at the height of the rotary axis of the bearing stud the construction space could get minimized.

3. Design of experiments, results and discussion

To perform the tests, the prototype (cf. Fig. 3) of the robot is needed, as well as a digital and a more sensitive analog force gauge. In addition, length measuring instruments (here: rulers) are required for measuring the TCPs of different configurations. The payload of the robot depends on the applied air pressure and is 1.5 kg at 10 bar. With a dead weight of the robot arm of 1.3 kg, a high power-to-weight ratio can thus be defined.

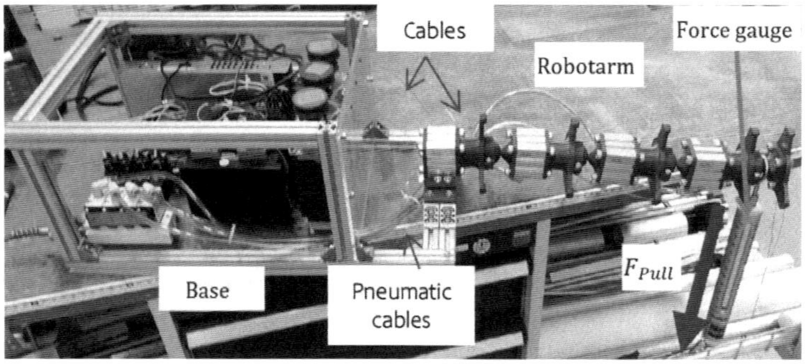

Fig. 3. Prototype and test setup for the investigation of the maximum payload in the stretching position

For the validation of the developed strength-adjustable ball joint, Figure 4 shows the experimental setup to determine the maximum load on it and Table 1 shows the values of the resulting friction coefficients.

As a result of the tests, the mean values of the ratios (coefficient of static friction) of applied force (measured tensile force) to piston force are relatively constant at 0.54 (softer elastomer: 45 Shore A) and 0.47 (harder elastomer: 60 Shore A). For rotations about the longitudinal axis of the socket, these ratios decrease by 0.21 and 0.34, respectively, due to distortions in the material under load. Accordingly, the softer elastomer achieved higher coefficients of friction in the series of measurements and is therefore more suitable for this robot application. During the tensile test at 8.0 bar and the stop buffer of the 45 Shore A material, the material failed at the 9.2 mm thick taper, resulting in separation of the ball. Thus, 339.2 N can be considered as the maximum load (without torsion) of the ball joint.

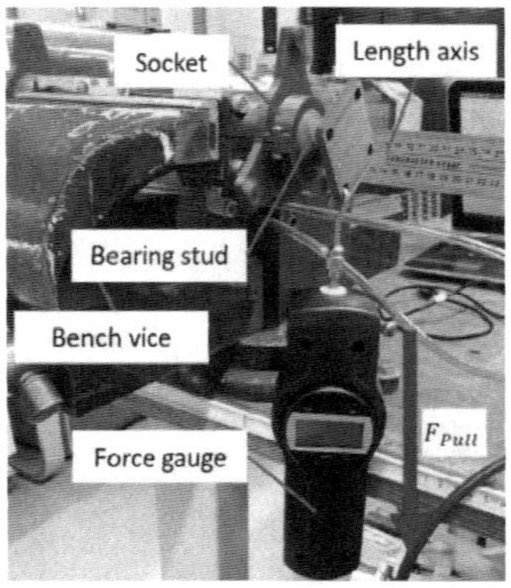

Fig. 4. Test setup for the investigation of the maximum payload of the ball joint

Air pressure [bar]	Piston force [N]	Applied force [N]		Torque [Nm]		Coefficient of static friction	
		45 Shore A	60 Shore A	45 Shore A	60 Shore A	45 Shore A	60 Shore A
0.0	0,0	0,0	0,0	0,00	0,00	0,00	0,00
0.5	40,2	2,0	0,0	0,07	0,00	0,05	0,00
1.0	80,4	30,7	27,6	1,01	0,91	0,38	0,34
1.5	120,6	77,7	74,1	2,56	2,45	0,64	0,61
2.0	160,8	96,7	90,0	3,19	2,97	0,60	0,56
2.5	201,1	112,5	106,3	3,71	3,51	0,56	0,53
3.0	241,3	137,6	126,4	4,54	4,17	0,57	0,52
3.5	281,5	148,6	142,4	4,90	4,70	0,53	0,51
4.0	321,7	164,7	155,2	5,44	5,12	0,51	0,48
4.5	361,9	191,7	172,4	6,33	5,69	0,53	0,48
5.0	402,1	230,6	190,0	7,61	6,27	0,57	0,47
5.5	442,3	240,5	201,7	7,94	6,66	0,54	0,46
6.0	482,5	248,7	229,8	8,21	7,58	0,52	0,48
6.5	522,8	293,3	247,4	9,68	8,16	0,56	0,47
7.0	563,0	295,4	237,1	9,75	7,82	0,52	0,42
7.5	603,2	329,6	240,7	10,88	7,94	0,55	0,40
8.0	643,4	339,2	241,6	11,19	7,97	0,53	0,38

Table 1. Maximum load for a strength-adjustable ball joint for rotations orthogonal to the longitudinal axis

To understand the kinematics and motion of the hyperredundant robot, the manipulator was placed in different configurations. Figure 5 shows 10 different configurations of the robot. Depending on how the 3 motors move (which of 3 cables is pulled or released) and by selectively blocking axes, different configurations can be approached. For example, Figures 1 and 4 show the movement of the first axis. Figures 2, 3 and 5 show the movement of the second axis. Figures 8 and 9 show the movement of all axes.

To control the movement of robot is necessary to synchronize the movement of motors and with switching on/off the pneumatic. The inverse kinematics was not done in this work; therefore, no statement can be made here about the positioning accuracy, trajectory accuracy and other important characteristics for the qualification of robots. This task then remains to be done for the future.

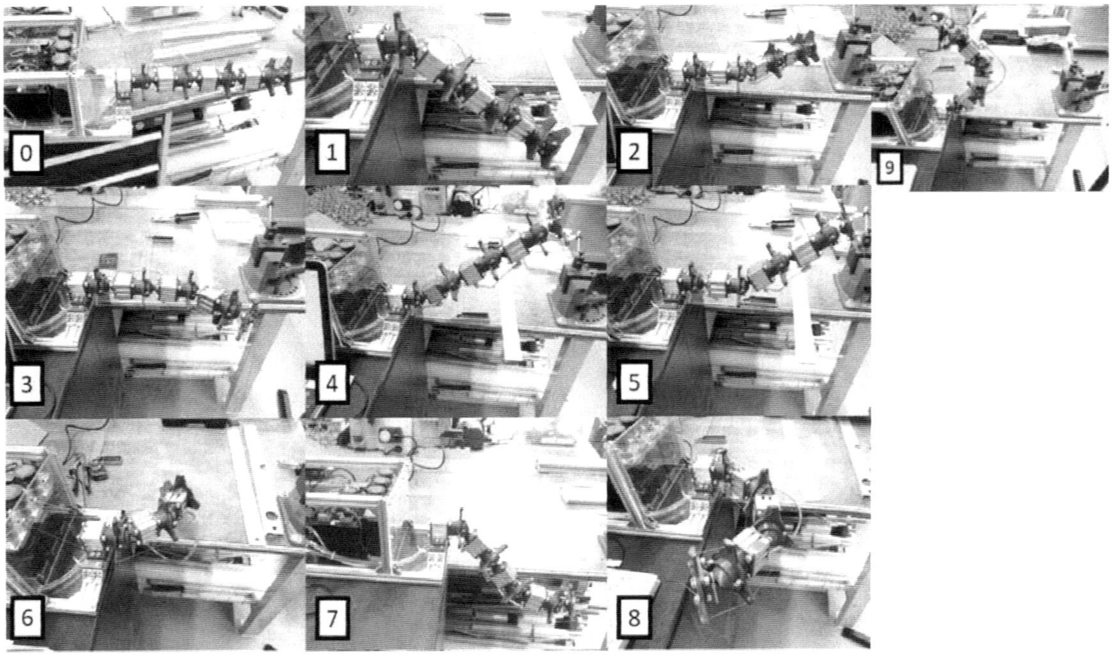

Fig. 5. Moving robot to different configurations.

4. Summary, conclusion and outlook

The design of a hyperredundant robot is a very extensive engineering task. To find an optimal design and a creative solution for this kind of robot kinematics, both a methodical approach and different software tools are required. In this work, a well-known methodological approach of VDI-2206 for the design of mechatronic systems was used. This methodological approach allows to use the creativity of the user to find different solutions for subsystems of the robot. From the different partial solutions, the best solution was selected for the design of the robot links/joints according to the requirements.

The main part of hyperredundant robot is the "holding system", which enables robot to remain to the achieved position- axes configuration. The holding system accurately provides the blocking of all joints except the one to be actuated, was solved with the aid of strength-adjustable ball joints. A link consists of a pneumatic cylinder, a stop buffer, the 3D-printed joint socket and bearing stud. The base of the joint socket is a hollow cylinder, which allows the stop buffer bolted to the pneumatic cylinder to press against the joint head via this cavity. Based on this, a prototype was built to allow validations to be completed. The kinematic model was created, and the calculations based on it for the forward kinematics and the determination of the rope length as well as the corresponding motor steps were implemented for the control system.

To improve the design and functionality of strength-adjustable ball joints a real test stand stations were built. On test stand robots were driven with different movement parameters to validate the presumptions of friction model. These parameters and results are presented tabular in this work. This experiment has shown that the robot can lift its own weight and act flexibly with the aid of one built-in actuator per joint. The fully functional, flexible, cable-controlled and hyperredundant robot with strength-adjustable ball joints confirms the theoretical concept and fulfils the requirements that were defined based on the objective.

However, during experiments with real robots, it was found that specific components have potential for further optimization. Optimization approaches exist in the surface processing of the 3D-printed parts. the principal solution is based on the friction of components; therefore, the surface condition of contacted surfaces is very important. Clean Surface reduces the pressure needed to block (hold) the robot joints, prolong the life of components and are more energy efficient.

Further research should focus on realistic use cases with mounted grippers or cameras on the end effector. Further, the robot should be tested outdoors to see if the robot structure is negatively affected by environmental parameters such as temperature, humidity, etc. The inverse kinematics and the dynamics of flexible robot axes will be worked out to qualify the robot for different applications.

5. Acknowledgments

Our special thanks go to the University of Applied Sciences Vienna -FH Technikum Wien- for providing the equipment and the laboratory as well as for sponsoring this work. Furthermore, we would like to thank the colleagues of the engineering departments for their support.

6. References

[1] Bender, D. G. B. (2020). Dubbel Taschenbuch für den Maschinenbau 2: Anwendungen, Springer Berlin Heidelberg, 978-3-662-59712-5, Berlin.

[2] Rost, S. (2016). Entwurf, Analyse und Regelung einer kinematisch redundanten Roboterstruktur mit hydraulischen Antrieben variabler Nachgiebigkeit, Ph.D. Dissertation, Department of Electrical Engineering, Information Technology, Physics, Technische Universität Carolo-Wilhelmina zu Braunschweig, Braunschweig, Germany.

[3] Tang, L.; Wang, J.; Zheng, Y.; Gu, G.; Zhu, L. & Zhu, X. (2017). Design of a cable-driven hyper-redundant robot with experimental validation, International Journal of Advanced Robotic Systems, Vol. 14, No. 5, pp. 1-12, DOI: 10.1177/1729881417734458.

[4] Ponn, J & Lindemann, U. (2011). Konzeptentwicklung und Gestaltung technischer Produkte, Springer Berlin Heidelberg, 978-3-642-20579-8, Berlin.

[5] Stuja, K.; Katalinic, B.; Aburaia, M. & Pajaziti, A. (2021). Design of a Robot Application with Regard to Energy Efficiency, Proceedings of the 32nd DAAAM International Symposium, pp.0033-0039, B. Katalinic (Ed.), Published by DAAAM International, ISBN 978-3-902734-33-4, ISSN 1726-9679, Vienna, Austria.

[6] Stuja K., Bruqi, M., Markl E. & Aburaia, M. (2016). Lightweight 4 -Axis Scara Robot for Education and Research, Proceedings of the 27th DAAAM International Symposium, pp. 0102-0108, B. Katalinic (Ed.), Published by DAAAM International, ISBN 978-3-902734-08-2, ISSN 1726-9679, Vienna, Austria.

[7] Denavit, J. & Hartenberg, R. S. (1955). A Kinematic Notation for Lower-Pair Mechanisms Based on Matrices, Journal of Applied Mechanics, Vol. 22, No. 2, pp. 215–221, DOI: 10.1115/1.4011045.

[8] Xu, W.; Liu, T. & Li, Y. (2018). Kinematics, Dynamics, and Control of a Cable-Driven Hyper-Redundant Manipulator, IEEE/ASME Trans. Mechatron., Vol. 23, No. 4, pp. 1693–1704, DOI: 10.1109/TMECH.2018.2842141.

DOI: 10.2507/33rd.daaam.proceedings.005

DEVELOPMENT OF THE INTELLIGENT ROBOTIC SYSTEMS

Pryanichnikov V.E., Ksenzenko A.Ya. & Plotnikov A.V.

This Publication has to be referred as: Pryanichnikov, V[alentin]; Ksenzenko, A[lexander] & Plotnikov, A[leksey] (2022). Development of the Intelligent Robotic Systems, Proceedings of the 33rd DAAAM International Symposium, pp.0029-0033, B. Katalinic (Ed.), Published by DAAAM International, ISBN 978-3-902734-36-5, ISSN 1726-9679, Vienna, Austria

DOI: 10.2507/33rd.daaam.proceedings.005

Abstract

The article presents the results of the creation of robotic complexes (RTC) and the development of the technology of their prototyping, replication. These works are carried out in Keldysh Institute of Applied Mathematics (Russian Academy of Sciences), Volgograd State Technological University and Moscow State Technological University "STANKIN" within the framework of the project "Intelligent Robotronics". Several models of service mobile tracked robots weighing up to 20 kg of the Amur series have been developed, and an underwater walking robot weighting 200 kg. In parallel, a cycle of theoretical and experimental research on the creation of means and methods for the functioning of information-measuring and control interaction of a coalition of underwater floating drones with an uninhabited walking robot moving along the bottom as a base station.

Keywords: Service mobile tracked robots; underwater walking supervisory controlled robot; patrol robot.

1. Introduction

This article is based on the results of the project "Intelligent robotronics", and dealing with two main problems in robotics – developing the integrating software and fast prototyping. When creating mobile service robots, there is a natural problem of using software of various authors for intellectualization, processing of sensory flows and control. Since the configuration of hardware and software and robot designs are also very diversed, the central task becomes the development of integration software and testing tools for individual packages and devices being combined. The implementation of this project required to make a training system for specialists, using the real tasks of developing service and mobile robotic complexes also used for teaching various specialities: IT technologies, mechatronics, sensors and robotics, industrial automation.

The developed samples of robotic complexes (RTC), within the framework of the Intelligent Robotics project can be delivered to customers fairly quickly, taking into account the modernization necessary for them and with additional sensor equipment. The main problems solved were the creation of technology for the rapid replication of RTC, as well as the creation of reliable, quickly deployable software for the implementation of intelligent supervisory and autonomous control [1], [2], [3]. That is, on our opinion, an interesting experience in the mainstream in robotics.

2. Amur series robots

Mobile autonomous service robot Amur has been developed on the basis of a small transport platform. Amur robot include supervisory control, using radio channels in real time and with autonomous execution of a number of functions. This model is designed for automated transportation of various items and materials in medical and other premises with a dangerous environment for humans, and can also provide inspection and patrolling of such premises.

The intelligent control system ensures safety motion. Control system includes automatic obstacle avoidance under the supervisory control by a remote operator. Operators can control robots in comfortable conditions, without the need to be presented in intensive care wards, infectious sanitary units and isolation wards, in chemically infected areas and other places of restricted access.

The robot design allows medical devices and instruments to be placed on the transport platform using a loader-manipulator with two grips. Amur robot is able to provide technological assistance to both doctors and patients in automatic mode. An autonomous navigation system, using lidar, ultrasonic sensors and rangefinders ensures confident mo of the robot. Navigation system can operate with the presence of unknown obstacles and moving objects, for example, people. A special version of the platform has been developed, which is capable avoid stairways. There is also a version of robot with partial sealing. It is possible to combine a group of robots into multi-agent hierarchical and heterogeneous complexes. The developed technology significantly reduces the cost of equipment for facilities with a harmful environment for humans. The Sensorika International Laboratory in Moscow has deployed a framework for the manufacturing of small series of such RTCs.

The main characteristics of the Amur-307 robots:

- total mass of the base platform, less than 25 kg;
- maximum total weight of transported items is 15 kg;
- dimensions of the platform, excluding the manipulator, are no more, than 650x600x350mm;
- dimensions of the transported items on the lifting table, are no more, than 300x300x600 mm;
- power consumption of the base platform, is no more, than 75 watts;
- average operating time without recharging the batteries is 2 hours;
- distance between the links of the double parallel manipulator is at least 350 mm;
- speed range is from 0.1 to 2.5 km/h;
- turning radius up from 0 m;
- surmountable slopes / ramps — up to 30 degrees;
- a complex of rangefinder sensors, a wide-angle TV camera;
- information interface Ethernet (1 Gb/s) or wireless Wi-Fi, RC (up to 1 km).

a)

b)

Fig. 1. Amur-307 mobile platform (a) and loader manipulators (b)

3. Underwater robot

An underwater cargo walking robot with high cross-country capability has been developed and manufactured. The scope of application of this transport platform with a walking mover covers the performance of work on the movement of significant loads in extreme conditions on the unprepared bottom surface of the offshore coastal zone, in the surf zone, in the presence of significant currents and surf waves, moving through swampy terrain, as well as with the possibility of going ashore. Studies have shown, that the choice of a walking type of actuator, in comparison with wheeled and tracked, provided the best performance, when overcoming obstacles on an unprepared bottom surface, as well as the ability to create a significant traction force exceeding its own weight by almost 1.5 - 2 times (unlike wheeled or tracked actuators, where this parameter is less than 0.55).

The complex consists of a walking robot and an operator's control station located on the shore or on a servicing watercraft. The robot and the control station are connected by a cable. The power supply of the robot is carried out via the cable as well, with the transmission of video images and control commands The robot is equipped with the necessary devices (television cameras, ultrasonic sensors, spotlights, manipulators, measuring instruments for various purposes, etc.). The power supply is organized from a powerful batteries included in the control console or from a gasoline generator.

The main characteristics of the underwater RTC with a walking propulsion:

- the total mass of the base platform, not more, than 250 kg;
- the maximum total weight of transported items is up to 200 kg;
- the dimensions of the platform, excluding the manipulator, are not more, than 2000x1000x800 mm;
- the dimensions of the transported items are not more, than 700x700x1000 mm;
- the distance between the links of the double parallel manipulator is up to 600 mm;
- average power consumption of the base platform, no more, than 2000 watts;
- diving depth - up to 90m, can work on shore;
- speed ranges from 0.1 to 6 km/h;
- turning radius from 0 m, surmountable slopes — up to 30-35 degrees;
- the average working time is not limited;
- a complex of rangefinder sensors, a wide-angle TV camera;
- Ethernet information interface (0.1 Gb/s).

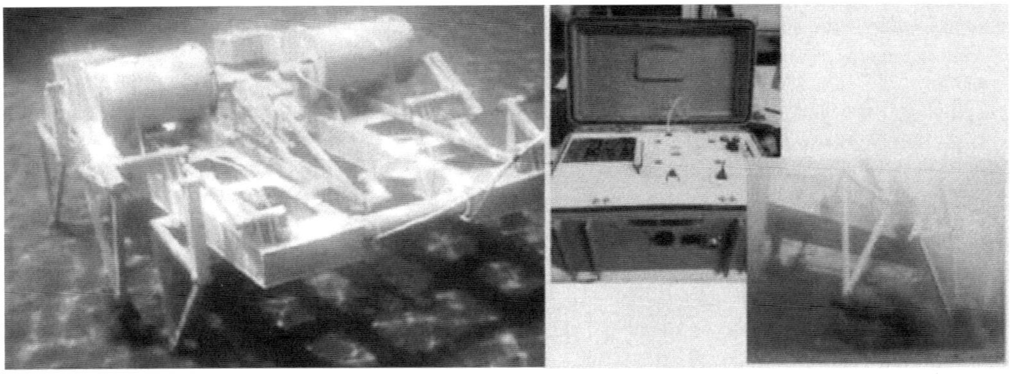

a) b)

Fig 2. A six-legged underwater vehicle based on a tunable Chebyshev mechanism (a) and its microprocessor control system (b)

4. Comparison with similar robots

Prototypes of walking vehicles and comparison with other other companies. The analysis of the exhibits of the International Military-Technical Forum "Army-2022" [4], materials of the conferences "Extreme Robotics", other sources of scientific and technical information showed an exceptionally high level of development and a versatile approach to solving scientific and technical problems, when creating a variety of developed RTCs. At the same time, a small number of competitors for our developments were found, when solving similar tasks, in particular, when comparing both overall and cost indicators, as well as some important operational characteristics, including load capacity and the ability to overcome obstacles, when moving along an unprepared bottom. The advantages of using a walking propulsion system in underwater systems of a similar purpose, for example, are beyond any competition. For comparison, we will give examples of several of the most interesting developments, but actually related to near classes of RTCs. We present this information from a popular reviews, to show areas of applications.

1. *Amphibian Kalan-M robotic complex*. Fully autonomous (up to 70 km range, up to 2 hours operational time), crawler chassis, satellite controlled, max depth – 5 meters, payload – up to 500 kg. The main disadvantage is low terrain passability and traction characteristics compared to the proposed walking propulsion, an interference-dependent control system, high complexity and cost. Saint Petersburg, Russia [5].
2. *Remote-controlled uninhabited light underwater vehicle "Marlin-350"*. Cable control, two video cameras, propulsion – six propellers, max depth – up to 530 meters, manipulator, robot weight - 60 kg, surface unit weight 85 kg, manual winch 90 kg. The main disadvantage is significantly lower traction characteristics compared to the proposed walking propulsion, high cost. Developer – JSC "TETHYS PRO", Moscow, Russia [6].

3. *Remote-controlled uninhabited underwater vehicle RING PACKET-3000.* Composition: control post (on board or on shore), fixed bottom base station, mobile robot with four propellers. All elements of the complex are connected to each other by cables. The weight of the mobile robot is 5000 kg, the depth of immersion is up to 3000 meters. The main disadvantage is significantly higher complexity and cost, high weight (incomparable applications). The developer is JSC "Central Design Bureau of Marine Equipment "RUBIN", St. Petersburg, Russia [7].
4. *Remote controlled by cable, walking six-legged robot Crabster CR200.* Possible applications are studying the sea bottom near the coast, taking soil samples. Six legs with 30 mechanical joints. Independence from undercurrents. Dimensions 2.42 m X 2.42 m X 2m, Weight 600 kg, plus a 1000 kg support device, immersion depth – up to 200 meters, cable power from the service vessel. Two video cameras, mechanical capture. Control is operated by four operators. Disadvantages – complexity of construction and maintenance. Developed in the South Korean Institute of Technology and Oceanology KIOST [8].
5. Remote-controlled walking robot SILVER-2. Remote control with the help of an operator on an accompanying vessel via a radio channel with a floating buoy. The dimensions of the robot are 60 x 30 cm, weight 9 kg. The diving depth is up to 200 meters. Operator control from the accompanying vessel via radio channel from a floating buoy, from it via cable with the robot. Shuffles along the bottom on six legs by stepping or jumping, overcomes obstacles with a height of up to 10 cm. Disadvantages – insecurity of the radio control channel, complexity, weak traction. Development of the School of Advanced Studies named after St. Anna, Pisa, Italy [9]

In the papers [10-15] we have published our results in the field of software integration and robotic design.

5. Conclusion

It should be noted, that the main problems of creating mobile robotics are now concentrated around the development of methods for automatic recognition of types of equipment, on specific materials using systems with artificial intelligence, as well as the development of industrial technologies, workplace automation and the improvement of production systems. The complex of research and development carried out by us provides a solution to these problems within the framework of two classes of robotic complexes created by us.

6. References

[1] A.Ya. Ksenzenko, Ju.S. Marzanov, E.A. Prysev, V.E. Pryanichnikov,V.V. Chernyshev Prototyping of a Contactless Data Exchange and Energy Supply For A Group Of Underwater Robots-Satellites With Walking On The Bottom The Base Station. Extreme Robotics. // Abstracts of the International Scientific and Technological Conference. – Saint-Petersburg: «Gangut», 2017. – p.268-269, ISBN 978-5-85875-522-7 / [Ксензенко А.Я., Марзанов Ю.С., Прысев Е.А., Пряничников В.Е., Чернышев В.В. Прототипирование бесконтактного обмена данными и энергообеспечения группировки подводных роботов-сателлитов с шагающей по дну базовой станцией. Сборник тезисов Международной научно-технической конференции «Экстремальная робототехника», Санкт-Петербург, 2 – 3 ноября 2017 г., с. 268-269 ISBN 978-5-85875-522-7].

[2] V. Arykantsev, A.A. Aryskin, O.O. Belyaev, A.Ya. Ksenzenko, E.A. Prysev, V.E. Pryanichnikov, V.V. Chernyshev, S.R. Eprikov. Supervisory Control of the Underwater Legged Vehicle. Extreme Robotics and Conversion Tendencies. // Abstracts of the International Scientific and Technological Conference. – Saint-Petersburg: Company «Politekhnika-Print», 2018. p. 203 / [Арыканцев В.В., Арыскин А.А., Беляев О.О., Ксензенко А.Я., Прысев Е.А., Пряничников В.Е., Чернышев В.В., Эприков С.Р. Супервизорное управление подводным шагающим аппаратом. Сборник тезисов международной научно-технической конференции «Экстремальная робототехника», Санкт-Петербург, июнь 2018 г., с. 203].

[3] Pryanichnikov V.E., Chernyshev V.V., Arykantsev V.V., Aryskin A.A., Ksenzenko A.Ya., Petrakov M.S., Travushkin A.S., Eprikov S.R. Supervisory Control of Underwater Walking Platform with Autonomous Satellite-Robots. Intelligent Adaptive Robots (Special issue of the journal Information Measuring and Control Systems), vol. 16, No. 1-2, 2019 ISSN, 2070-0814, pp. 49-56, DOI 10.18127/j2070-0814-201812-05 / [Пряничников В.Е., Чернышев В.В., Арыканцев В.В., Арыскин А.А, Ксензенко А.Я., Петраков М.С., Травушкин А.С., Эприков С.Р. Супервизорное управление шагающей подводной базой с автономными роботами-спутниками. Интеллектуальные адаптивные роботы (специальный выпуск журнала Информационно-измерительные и управляющие системы), т. 16, № 1-2, 2019 ISSN, 2070-0814, с. 49-56, DOI 10.18127/j2070-0814-201812-05].

[4] www.rusarmyexpo.ru/business_program/41019/43167, (2022) Modern methods of automatic recognition of types of equipment on specific materials using artificial intelligence systems. Accessed on 11.09.2022

[5] work.vk.com/wall-61195360_39644, (2022). Amphibious vehicles. Accessed on 01.10.2022.

[6] www.tetis-pro.ru, (2022). Remote-controlled underwater robots, Accessed on 05.10.2022.

[7] dzen.ru/media/mashnews/ckb-mt-rubin-vpervye-prodemonstriroval-novye-modeli-podvodnyh-teleupravliaemyh-apparatov-62fbb19d67dd375fe2bd28d7, (2022). Remote-controlled uninhabited underwater vehicle of the working class RING PACK-3000. Accessed on 10.09.2022.

[8] edition.cnn.com/2014/04/01/tech/innovation/giant-six-legged-robot-crab,(popular review). Remote-controlled Walking Six-legged Robot Crabster CR200. Accessed on 08.10.2022.

[9] techxplore.com/news/2020-05-silver2-aquatic-robot-seabed.html, (2022). Remote-controlled walking robot SILVER-2. Accessed on 11.10.2022.

[10] Plotnikov, A[leksey] & Pryanichnikov, V[alentin] (2021). Kinetostatic Analysis of the Service Mobile Robot with Manipulator, Proceedings of the 32nd DAAAM International Symposium, pp.0446- 0453, B. Katalinic (Ed.), Published by DAAAM International, ISBN 978-3-902734-33-4, ISSN 1726-9679, Vienna, Austria DOI: 10.2507/32nd.daaam.proceedings.065

[11] Pryanichnikov, V[alentin]; Kolesov, Y[uriy]; Tarasov, R[adomir] & Plotnikov, A[leksey] (2021). Integration Software for a Distributed Group of Mobile Service Robots, Proceedings of the 32nd DAAAM International Symposium, pp.0311-0315, B. Katalinic (Ed.), Published by DAAAM International, ISBN 978-3-902734-33-4, ISSN 1726-9679, Vienna, Austria DOI: 10.2507/32nd.daaam.proceedings.045

[12] Pryanichnikov V.E., Chernyshev V.V., Arykantsev V.V., Aryskin A.A., Eprikov S.R., KsenzenkoA.Ya., Petrakov M.S.(2018). Enhancing the Functionality of the Groups of Autonomous Underwater Robots // Proceedings of the 29th DAAAM International Symposium, 2018. — DAAAM International, Vienna, Austria, 2018. — P.1319-1325. DOI:10.2507/29th.daaam.proceedings.190.

[13] Pryanichnikov V. E., Aryskin A. A., Ksenzenko A. Ya., Petrakov M. S., Ignatiev V. A., Khelemendik R. V. Construction of a service Autonomous mobile robot based on control using logical analysis of the realizability of operations // Scientific service on the Internet: proceedings XX All-Russian scientific conference (September 17-22, 2018, Novorossiysk). Moscow: KIAM, 2018. DOI:10.20948/abrau-2018.

[14] Pryanichnikov V.E., Aryskin A.A., Eprikov S.R., Kirsanov K.B., Khelemendik R.V., KsenzenkoA.Ya., Prysev E.A., Travushkin A. S. (2017). Technology of Multi-Agent Control for Industrial Automation with Logical Processing of Contradictions, Proceedings of the 28th DAAAM International Symposium, pp.1202-1207, B. Katalinic (Ed.), Published by DAAAM International, ISBN 978-3-902734-11-2, ISSN 1726-9679, Vienna, Austria. DOI:10.2507/28th.daaam.proceedings.167.

[15] Pryanichnikov V.E., Aryskin A.A., Eprikov S.R., Kirsanov K.B., Khelemendik R.V., KsenzenkoA.Ya., Prysev E.A., Travushkin A. S. (2017). Technology of Multi-Agent Control for Industrial Automation with Logical Processing of Contradictions // Proceedings of the 28th DAAAM International Symposium, 2017. — DAAAM International, Vienna, Austria, 2017. — P.1202-1207. DOI:10.2507/28th.daaam.proceedings.167.

DOI: 10.2507/33rd.daaam.proceedings.006

ANALYSIS OF PHYSICAL CAUSES OF MACHINED SURFACES IN SELECTED MACHINING METHODS

Jozef Peterka*, Jozef Martinovic, Thomas de Silva Wallyson & Vitalii Kolesnyk

This Publication has to be referred as: Peterka, J[ozef]; Martinovic, J[ozef]; Wallyson, T[homas] d[e Silva] & Kolesnyk, V[italii] (2022). Analysis of Physical Causes of Machined Surfaces in Selected Machining Methods, Proceedings of the 33rd DAAAM International Symposium, pp.0034-0041, B. Katalinic (Ed.), Published by DAAAM International, ISBN 978-3-902734-36-5, ISSN 1726-9679, Vienna, Austria
DOI: 10.2507/33rd.daaam.proceedings.006

Abstract

In this paper, we will try to theoretically calculate, and experimentally determine the roughness of the machined surface in terms of the physical causes. The surface roughness has been experimentally obtained for two turning methods, which are challenging both in terms of setting the cutting conditions and in terms of the economics of the experiments performed. These are the deep turning of hardened steel internal holes and the internal turning of internally shaped surfaces. The article will point out the theoretical calculation of machined surfaces' roughness by the cutting theory and the experimental results obtained.

Keywords: Machining; Surface roughness; Cutting processes; The deep turning of internal holes; Broaching of internal Shaped surfaces.

1. Introduction

The roughness of the machined surface is one of the important parameters for the evaluation of the quality of machined surfaces It is therefore important to know not only the methods for the evaluation of surface roughness [1] but especially to know what are the causes of the roughness of machined surfaces. From the theory of cutting we know the following causes of the roughness of machined surfaces: copying of the shape of the cutting part of the tool into the workpiece, copying of the roughness of the cutting edge into the workpiece, the existence of vibration of the cutting tool or the technological tool-workpiece junction, existence of accretion on the machined surface, accidental causes. These physical causes act simultaneously and thus each of them contributes a certain amount to the resulting roughness of the machined surface. The problem is which of these causes is dominant and contributes most to the overall magnitude of the roughness of the machined surface.

In our article, we will discuss for selected technological methods which of the above physical causes is most involved in the formation of the roughness of the machined surface. In scientific journals, there are countless articles describing results obtained by experimental research and calculations for example for machining technologies [2] but also for additive technologies [3] and most recently also by prediction and simulation [4].

All articles are focused on the evaluation of the roughness of machined (already additive) surfaces under different changes in technological and process conditions, cutting environment (air, emulsion, oil, ...), machined or additive material, etc. In particular, research is focused on different technological methods. In [5], the authors predicted the roughness of the machined surface in circumferential milling technology. The influence of the cutting edge radius on the resulting roughness in milling was described by the authors in [6]. In [7], the authors analyzed the roughness of the machined surface in HSS tapping technology using different tap coatings. The paper [8] is to evaluate the influence of selected cutting parameters (cutting speed and feed rate) on the surface topography of drilled WPC material and in the [9] study, surface characteristics of the samples of experimentally manufactured wood plastic composites (WPC) were determined. The turning process was used to produce surfaces by removing material from a rotating workpiece.

In our article, we will focus not on the experimental results in terms of the achieved value of the roughness of the machined surface but on the theoretical analysis and experimental proof, which cause (of the mentioned physical causes) has the greatest influence on the given technological method. For this strategy of evaluation of the roughness of the machined surface, those technological methods were selected for which we can clearly say which physical causes cannot contribute to the roughness of the machined surface. For these reasons, we selected the following technological methods: broaching of internally shaped grooves and machining of hardened steel.

1.1 Broaching of internally shaped grooves

The broaching machines perform a straight-line movement. The tools used are broaching tools - so-called broaching mandrels. The shape, design, and material of the broaching tools guarantee high precision and quality of the machining of parts using this metal chip technology. The technology of broaching the internally shaped surfaces of workpieces consists in producing an internal profile that is machined on parts with a continuous circular hole. In practice, this hole 'd' is usually machined by turning or drilling as a previous manufacturing operation. The workpiece is placed, without clamping, in a positioning jig attached to the machine, where it is centered by a broaching tool during the straight-line movement of the machine. The cutting speed in the broaching process is the speed at which the tool passes through the workpiece. By moving the broaching tool in a straight line through the workpiece, the desired shape is produced. This is the negative contour of the tooth shape of the broaching tool on the workpiece. The workpiece, after the production operation of broaching is completed, may have, in terms of further requirements, either the final design of the internal form surface or this form surface is machined in further production operations. The technology of stretching internal molded surfaces enables high productivity to be achieved in the production process. Procurement costs for broaching tools are higher. For these reasons, it is suitable for use in series and high-volume production. A depiction of the principle of the internal contour surface stretching technology and an example of an internal contour surface is shown in Fig. 1. Inner forming surface - is a flange-shaped hub. The inner hole was first turned.

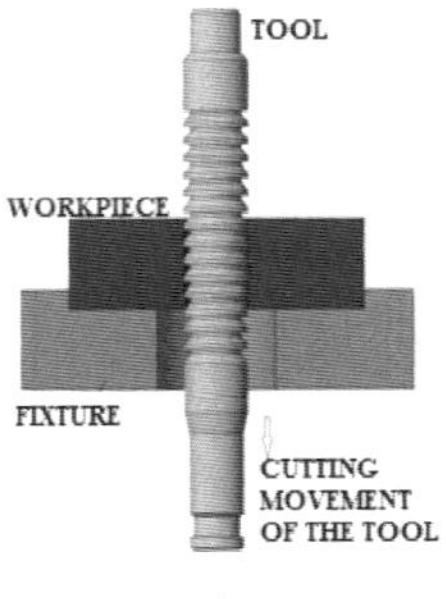

a)

b)

Fig. 1. Broaching of the inner-shaped grooves
a) Illustration of the principle b) Example of an internal shaped face on a girth

Experimental conditions

The experiment was carried out on a 7B66 broaching machine, under series production conditions in an emulsion liquid environment. The workpiece material was C45 (or 1.0503 according to DIN standard). The chemical composition of the material in [%] is as follows: C 0.43-0.5, Mn 0.5-0.8, P max 0.045, Si max 0.4, Ni max 0.4, S max 0.045, Cr max 0.4, Mo max 0.1. The tensile strength Rm is defined for the workpiece blank at 750-900 [N/mm^2].

Comparison of the roughness of the calibration teeth of the broaching mandrel and the workpiece. The relative position of the tool relative to the workpiece was marked on the broaching mandrel and the workpiece. After performing the broaching operation of the internal form surface, the cutting edges of the broaching mandrel Fig. 2 (a) and the internal form surfaces of the workpiece after broaching Fig. 2 (b) were subsequently numbered. The roughness of all the teeth of the last row of teeth was measured on the broaching mandrel, Fig. 4. Roughness measurements were performed in the middle part of the teeth of the broaching mandrel and also on the workpiece in the middle part.

a) b)

Fig. 2. Numbering of teeth and grooves
a) numbering of the last row of teeth, b) numbering of internal form surfaces after broaching on the workpiece

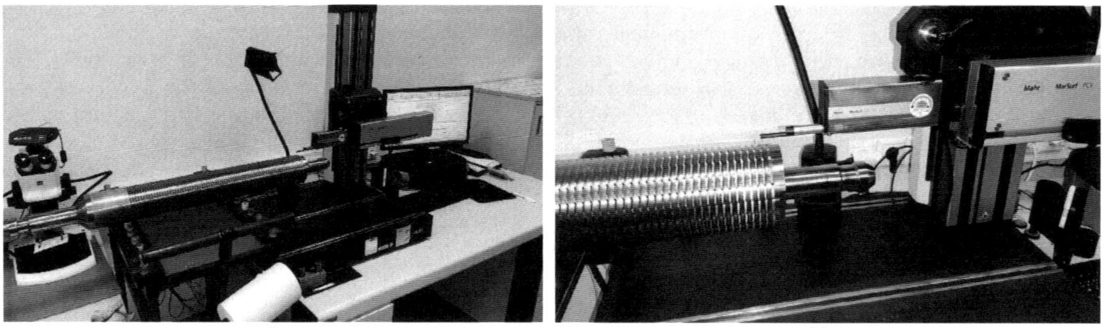

a) b)

Fig. 3. Preparation and measurement of the teeth of the broaching mandrel
a) preparation of the measurement of the roughness of the teeth of the broaching mandrel
b) measurement of the roughness of the last row of teeth

Type of broaching machine, manufacturer	7B66, Minsk Maschine Company
Cutting speed	2.5 [m.min^{-1}]
Length of broaching mandrel	1100 [mm]
Workpiece materiál	C45
Number of shaped workpiece surfaces	20
Emulsion liquid	Blaser 4000 Strong
Emulsion fluid concentration	15 [%]
Roughness measuring device	Mahr Marsurf XCR20

Table 1. The experimental conditions of broaching

Results

The measurements of the roughness of the internally shaped surfaces of the workpiece were carried out in the central part of the grooves. A detail of the contact tip of the roughness measuring device is shown in Fig. 4. The results of the roughness measurements of the mandrel teeth and the workpiece are shown in Table 2.

a) b)

Fig. 4. Measuring the roughness of the internal form surface of the workpiece
a) measurement of the workpiece in the central part of the groove, b) detail of the measurement

Broaching mandrel		Workpiece			
Tooth serial number	**Roughness Rz**	**Groove serial number**	**Roughness Rz**	**Difference = Column 2 - Column 5**	**Max a Min**
[--]	[μm]	[--]	[μm]	[μm]	[μm]
Column 1	Column 2	Column 3	Column 4	Column 5	Column 6
1.	4.34	1.	5.35	-1.01	1.01
2.	3.18	2.	3.19	-0.01	0.01
3.	5.05	3.	5.66	-0.61	
4.	4.57	4.	5.21	-0.64	
5.	6.32	5.	6.42	-0.10	
6.	5.47	6.	5.90	-0.43	
7.	6.75	7.	6.84	-0.09	
8.	4.55	8.	5.32	-0.77	
9.	6.86	9.	7.10	-0.24	
10.	4.01	10.	4.05	-0.04	
11.	5.88	11.	5.98	-0.10	
12.	5.75	12.	5.97	-0.05	
13.	5.97	13.	5.28	0.69	
14.	8.08	14.	8.44	-0.36	
15.	7.13	15.	7.47	-0.34	
16.	6.81	16.	7.18	-0.37	
17.	5.74	17.	6.06	-0.32	
18.	5.71	18.	5.78	-0.07	
19.	5.21	19.	5.44	-0.23	
20.	6.12	20.	6.25	-0.13	

Table 2. Resulting roughness parameter Rz

Discussion

From the table Table 1. we can see the following:

- the roughness of the broaching mandrel, i.e. its teeth (except for one tooth) was less than the roughness of the manufactured grooves,
- the maximum roughness deviation between the roughness of the tooth of the broaching mandrel and the roughness of the manufactured groove was 1.01 μm and the minimum was 0.01 μm,
- no increase was observed on the machined surfaces,
- no traces of tool oscillation were observed on the machined surfaces (the design of the machine and the tool practically do not allow oscillation to occur).

We can say that in this case of machining both physical causes of the roughness of the machined surface, namely chatter and surge, are eliminated. Further, since this technological method is implemented with only one movement, namely the straight-line movement of the broaching mandrel, there is also no first (usually the greatest) physical cause of roughness of the machined surfaces, namely the copying of the shape of the cutting part of the tool into the workpiece. In this case, we are left with the only physical cause of roughness and that is the copying of the cutting-edge roughness itself into the workpiece. Experiments have shown that, given the above, and the results of the roughness measurements in Table 2, we can sufficiently prove that in this method of machining a new surface is formed only by copying the roughness of the cutting edges of the cutting part of the tool itself.

This statement is significant in that, for cases where high-quality machined surfaces are required after broaching in terms of roughness, we as technologists have in our hands a tool for improving the roughness of machined surfaces. In this case, to achieve the best roughness on the workpiece, we ensure the best roughness on the cutting wedge surfaces of the broaching mandrel teeth.

1.2 Turning a deep internal hole in hardened steel

The second technological method to investigate the physical causes of the roughness of the machined surface was the internal turning of hardened steel. The reader will ask why this particular technology was chosen. This technology was chosen because it is a technology where the cutting tool operates in the so-called non-incremental cutting zone, i.e. where no growth is formed, and for two reasons. The first reason is that sufficiently high cutting speeds are used (use of CBN cutting materials) at which, from a theoretical point of view, no build-up is formed. The second reason is the thermal condition of the workpiece - it is hardened steel with a high proportion of martensitic structure, which, as we know, again does not form a surge when cutting.

These assumptions allow us to say that the following physical causes will have a decisive influence on the resulting roughness of the machined surface: copying the shape of the cutting part of the tool, copying the roughness of the cutting edge, and the existence of chatter. Since the technological method of turning internal cylindrical surfaces implies the use of a cutting tool with certain greater unloading, its rigidity is the weakest point of the technological system. This gives the assumption that it is the physical cause - the existence of tool oscillation - that will have a decisive influence on the formation of the roughness of the machined surface.

Experimental conditions

The main concern of this experiment consisted of the effect of the L/D ratio on the roughness of the workpiece while using different internal turning tools [10]. All the internal turning experiments of this project were performed using a DMG CTX alpha 500 CNC turning center with 20/27 kW (100/40 % ED) of power in the main motor and a maximum spindle rotation of 5000 rpm. The selected tools for the experiments were 16 mm in diameter: one with high hardenability ANSI 4140 steel – code A16R SCLCR 09-R, one with carbide – code E16R SCLCR 09 R and the last one was an antivibration tool (Silent tool) code 570-SCLCL-16-06 for the support and code 570-3C 16 156 for the anti vibrational bar – It was supplied by Sandvik Coromant [11]. As for the tool insert, an adequate insert for finishing operations on smooth surfaces of hardened steels was chosen. It was composed of CBN (50 % wt.) and a ceramic phase of TiCN and Al2O3; it is the ISO code is CCGW09T308S01020F 7015 (class ISO H10) for Steel and Carbide tools while for the Silent tool the code is CCGW 060208 S01030F 7015 [11].

The advantage of the chosen tool inserts, when compared to others with a greater CBN content, is their chemical stability about iron. Besides, its toughness is enough to preserve its cutting edge, even though it is reduced when compared to other inserts with a greater CBN content. The DIN EN 1.2842 (ISO 90MnCrV8) steel used in the fabrication of the test specimens is a widely employed material in the metal mechanical industry. It presents high hardenability, bad weldability, and reasonable machinability, as well as good resistance to torsion and fatigue –it's hardened to hardness 58 HRC. The cutting conditions and the machine setup (tool overhang) were tested in maximum flank wear (VBmax) of 0.2 mm according to ISO 3685 [12]. The DOI 10.2478/rput-2022-0003 25 tests were carried out without a cooling system and constant cutting parameters, with a cutting speed (v_c) of 360 m/min, feed rate (f) of 0.14 mm/revolution, and depth of cutting (ap) 0.1 mm. The conditions for clamping the boring bar in the turret were determined considering the distance from the tool tip to the beginning of the tool holder. The condition of clamping the tool was according to the following distribution in the L/D ratio parameter (ratio between the length "L" and the diameter "D" of the tool): 3 to 8 with intervals of 0.25. The roughness of the cylindrical surface of the workpiece was measured 5 times and verified at three equidistant points (120°) at a controlled laboratory temperature of 20 °C.

Therefore, to quantify the finish obtained during the turning process, a profilometer SURFCOM 1900SD2 model, Accretech brand, was used, with resolution on the X axis and Z axis equal to 0.04 µm, measuring force of 0.75 mN and expanded uncertainty of 0.01 µm (k=2), following the DIN 4768 (1990-2005) [13] standards. In addition, to treat the roughness profile, a Gauss filter was used, and the sampling length (cut-off) used was equal to 0.8 mm, following the ISO 4288:1996 [14] parameters. The arithmetic means deviation of the profile (Ra) and the total height of the profile (Rz) were chosen.

Results

In the case of internal turning, the tool overhang is, together with the cutting parameters, fundamental to obtaining a high finish surface on the workpiece. Figure 5 shows the behavior of the roughness represented by parameters Ra and Rz, as a function of the L/D ratio for three different tool types (steel, carbide, and silent tool (antivibration tool).

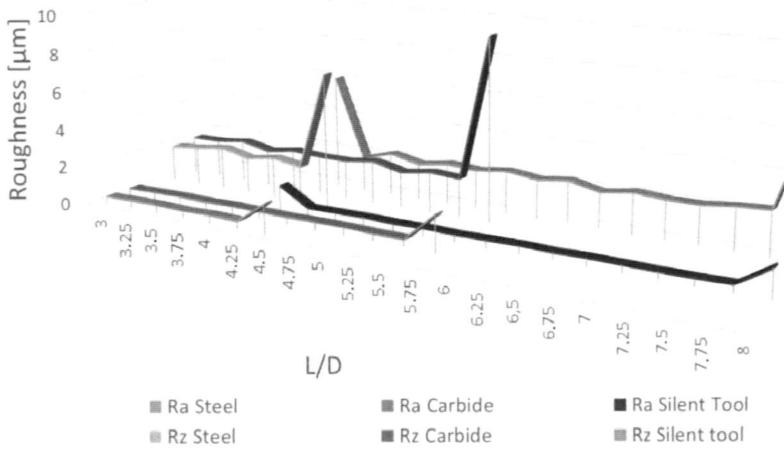

Fig. 5. Effect of the L/D ratio over the roughness in the workpiece with different types of boring bars with a standard deviation of Ra ±0.02 µm and Rz ±0.06 µm [13]

As a general result of this experiment, Table 3 summarizes the findings describing the overhang limits of each tool during internal turning operation for constant cutting parameters. To define the critical L/D ratio of the boring bar, the dynamic stability of internal turning was experimentally observed and classified as stable: arithmetic average roughness (Ra) was lower than 0.8 µm and no chatter marks in the workpiece surface or unstable: arithmetic average roughness (Ra) was higher than 0.8 µm and chatter marks were present in the workpiece surface due to the self-excited vibration, as illustrated by Figure 6.

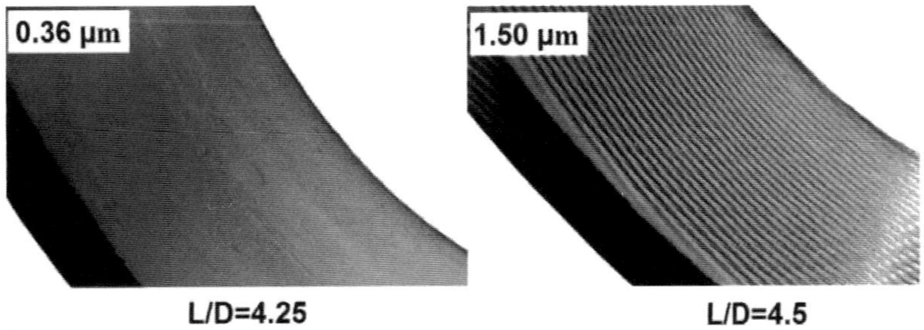

Fig. 6 The surface textures in boring with the same cutting conditions, steel boring bar, and different L/D ratios

Tool	Overhang (L/D)	Overhang [mm]	Stability	Ra [µm]	Rz [µm]
Steel	4.25	68	Stable	0.36	1.86
	≥ 4.50	≥ 72	Unstable	1.50	7.00
Carbide	5.50	88	Stable	0.35	2.00
	≥ 5.75	≥ 92	Unstable	1.83	9.44
Antivibration (Silent Tool)	$L/D_{min} = 4.50$	72	Stable	0.36	1.73
	$L/D_{max} = 7.75$	124		0.41	2.29
	$L/D_{min} \leq 4.25$	≤ 68	Unstable	1.30	6.01
	$L/D_{min} \geq 8.00$	≥ 128		1.48	7.00

Table 3. Resulting values of the roughness parameter of the machined surface Ra, Rz

Discussion

It can be seen in Figure 2 that, for measured parameters, the workpiece roughness is independent of the tool overhang up to a certain value, from which the roughness increases abruptly.

It was also observed that in the range of overhangs in the stable limits (Steel L/D = <4.25, Carbide L/D = 5.5, and Silent tool 4.5≥L/D ≥7.75), the roughness parameter Ra and Rz are very close and below 0.8 μm, while for unstable limits Steel L/D ≥ 4.5, Carbide L/D ≥ 5.75, and Silent tool 4.25≤ L/D≥8) is where the influence of tool vibration over the roughness. Since the feed rate, depth of cut, and the tip radius of the tool, which form the theoretical roughness of the part (geometric contribution of the feed rate and tip radius to the roughness), were constant during the tests, the vibration was observed. The main factor that affects the stability of the machining process is the static stiffness of the toolbar which was tested in a variety of overhangs. Therefore, it is possible to say that the stability of the machine-tool-workpiece system has a huge impact on the surface finish and not, for this time the cutting conditions.

One recommendation to keep the roughness values to a minimum is to adjust the tool clamp to the smallest overhang (L/D = 3) possible for the operation. As pointed out by Thomas et al. [15], among other efficient techniques to increase the stability of the system and reduce vibration in internal turning operation with deep holes.

2. Conclusion

This study is based on which of the physical causes contributes decisively to the roughness of the machined surface for selected technological methods of machining. Two technological methods were selected: broaching the internal groove and turning a deep internal hole in hardened material. In conclusion:

- We have confirmed the theoretical assumptions with experimental results,
- For the technology of internal groove broaching, the physical cause of the roughness of the machined surface is decisive: the copying of the roughness of the cutting edge of the broaching mandrel tooth,
- For the technology of turning an internal hole in hardened material, the decisive physical cause of the roughness of the machined surface is the existence of chatter.
- These statements are a very important guideline for technological practice. They give us practical guidance on how to achieve the prescribed roughness of machined surfaces in these technologies and how to improve (deteriorate) or control the achieved roughness of machined surfaces.
- Conclude as follows:
- the problem addressed was the confirmation/non-confirmation of the theoretical physical causes of the roughness of machined surfaces with a defined cutting edge. For this purpose, we selected two very specific machining methods - broaching of internal grooves and turning of internal holes in hardened steel.
- The results of the research confirmed the existence of two physical causes of the roughness of machined surfaces, namely: the copying of the roughness of the cutting edge itself into the machined surface and also the effect of chatter (if any).

In the future, we want to extend our research to other very specific machining methods, in the first step will be the technology of copy milling with different lengths of clamping part unloading.

3. Acknowledgments

The project is supported by the national research project - VEGA 1/0019/20 "Accurate calculations, modeling, and simulation of new surfaces based on physical causes of machined surfaces and surfaces of additive technologies in the conditions of the machine and robotic machining".

4. References

[1] Kubatova, D. (2020). Problems of Defining and Subsequently Measuring Surface Roughness in Aperiodic Surfaces, Proceedings of the 31st DAAAM International Symposium, pp.0236-0243, B. Katalinic (Ed.), Published by DAAAM International, ISBN 978-3-902734-29-7, ISSN 1726-9679, Vienna, Austria, DOI: 10.2507/31st.daaam.proceedings.033

[2] Peterka, J. (2004). A new approach to calculating the arithmetical mean deviation of a profile during copy milling, Strojniski Vestnik-Journal of Mechanical Engineering, Vol. 50, Issue 12, pp. 594-597

[3] Milde, J., Peterka, J., Jurina, F., Dobrovszky, P., Hrbal, J., Pitek, P. & Martinovic, J. (2021). Influence of Selected Photopolymers on the Resulting Accuracy and Surface Roughness of the Component in Digital Light Processing Technology, Proceedings of the 32nd DAAAM International Symposium, pp.0235-0245, B. Katalinic (Ed.), Published by DAAAM International, ISBN 978-3-902734-33-4, ISSN 1726-9679, Vienna, Austria, DOI: 10.2507/32nd.daaam.proceedings.034

[4] Nexhat Qehaja, Ilir Doci, Mirlind Bruci, Fitore Abdullahu, Kaltrine Jakupi & Fatlume Zhujani. (2016). Mathematical Modelling of Surface Roughness Through Machining Parameters and Machining Time During the Dry Milling Process, Proceedings of the 27th DAAAM International Symposium, pp.0187-0194, B. Katalinic (Ed.),

Published by DAAAM International, ISBN 978-3-902734-08-2, ISSN 1726-9679, Vienna, Austria, DOI: 10.2507/27th.daaam.proceedings.028

[5] Piska, M. & Kolomy, S. (2021). On the effective substitution of turning by peripheral milling, Proceedings of the 32nd DAAAM International Symposium, pp.0016-0021, B. Katalinic (Ed.), Published by DAAAM International, ISBN 978-3-902734-33-4, ISSN 1726-9679, Vienna, Austria, DOI: 10.2507/32nd.daaam.proceedings.003

[6] Hronek, O. & Zetek, M. (2017) The influences of cutting edge radius on surface roughness when milling nickel alloy. Proceedings of the International DAAAM Symposium. pp. 1037-1043. Katalinic, (Ed.). ISSN 1726-9679, ISBN 978-3-902734-08-2. Vienna, Austria. DOI:10.2507/28th.daaam.proceedings.144

[7] Piska, M. & Sliwkova, P. (2015). Surface parameters, tribological tests and cutting performance of coated HSS taps, Proceedings of the 25th DAAAM International Symposium, vol. 100, pp.125-124, Published by Elsevier Ltd. This is an open-access article under the CC BY-NC-ND license (http://creativecommons.org/licenses/by-nc-nd/4.0/). Peer-review under responsibility of DAAAM International Vienna, doi: 10.1016/j.proeng.2015.01.350

[8] Hutyrova, Z., Zajac, J., Michalik, P., Mital, D., Duplak, J. & Gajdos, S. (2015). Study of Surface Roughness of Machined Polymer Composite Material, International Journal of Polymer Science, vol. 2015, Art. Num. 303517, pp. 1-6, DOI: 10.1155/2015/303517

[9] Hutyrova, Z., Harnicarova, M., Zajac, J., Valicek, J. & Mihok, J. (2014). Experimental Study of Surface Roughness of Wood Plastic Composites after Turning, Material Science and Engineering Technology II, Book Series: Advanced Materials Research, Vol. 856, pp. 108+, DOI: 10.4028/www.scientific.net/AMR.856.108

[10] Peterka, J.; Walyson, T. de S.; Straka, R.; Vopát, T.; Hrbál, J. & Kritikos, M. (2022). Analysis of form error and roughness of hardened steel workpieces internally turned with different tools in long overhangs. Research Papers, Vol. 30, No. 50, pp. 21-29, ISSN 1338-0532 (online), DOI 10.2478/rput-2022-0003

[11] Sandvik, 2017. Silent tool for turning: overcome vibrations in internal turning. 4 Feb. 2021 <https://tibp.blob.core.windows.net/coromant/a9d341b5-16cc-4f7c-8e60-2f927ea651fa.pdf?sv=2014-02-14&sr=b&sig=gd6yrZv9dnCJrT4OJpY7FuBrp2rINAFabaFTmn70nBo%3D&st=2021-02-04T15%3A08%3A41Z&se=2031-02-02T15%3A13%3A41Z&sp=r&rsct=application%2Fpdf&rscd=inline%3B%20filename%3DC2920-40.pdf>.

[12] ISO 3685:1993Tool-life testing with single-point turning tools.

[13] DIN 4768, 1990-2005. Determination of values of surface roughness parameters Ra. Rz, Rmax using electrical contact (stylus) instruments Concepts and measuring conditions.

[14] ISO 4288:1996 Geometrical Product Specifications (GPS) - Surface texture: Profile method - Rules and procedures for the assessment of surface texture - revised by ISO 21920-3:2021

[15] Wallyson, T. de S.; Fulop Z. & Szilágyi, A. 2021. Passive Damping Techniques for Vibration Suppression in Boring Operation with Long Overhangs. In: Vehicle and Automotive Engineering 3, Károly Jármai & Katalin Voith (Ed.), pp. 256 – 264, Springer, ISSN 2195-4356 ISSN 2195-4364 (electronic), Lecture Notes in Mechanical Engineering ISBN 978-981-15-9528-8 ISBN 978-981-15-9529-5 (eBook), https://doi.org/10.1007/978-981-15-9529-5.

DOI: 10.2507/33rd.daaam.proceedings.007

CLASSIFICATION OF TWO-DIMENSIONAL MECHANICAL PARTS USING A CONVOLUTIONAL NEURAL NETWORK

Irina Topalova & Yuliana Staeva

This Publication has to be referred as: Topalova, I[rina] & Staeva, Y[uliana] (2022). Classification of Two-dimensional Mechanical Parts Using a Convolutional Neural Network, Proceedings of the 33rd DAAAM International Symposium, pp.0042-0049, B. Katalinic (Ed.), Published by DAAAM International, ISBN 978-3-902734-36-5, ISSN 1726-9679, Vienna, Austria
DOI: 10.2507/33rd.daaam.proceedings.007

Abstract

Image search, object recognition and classification are emerging as key components in modern automated and autonomous production systems that integrate artificial intelligence. The accuracy in recognizing and classifying these parts, regardless of their geometric transformations, determines to a high degree the accuracy of their manipulation and positioning by the flexible assembly. Based on a comparison of some modern methods for classification of machine parts, the choice of a method with in-depth training of a convolutional neural network is justified. In the presented article a model for classification of machine parts is proposed, which is based on deep training of a convolutional neural network. A model was presented and tested, with a variety of training strategies for the purpose of increased efficiency. The proposed model was based on application of Batch normalization, Gaussian Noise, Weight regularization, Image normalization and Early stopping. High classification accuracy has been achieved for a large training and testing sample. The experiment was conducted for four classes of machine parts having different spatial position and orientation, as well different shapes of the objects belonging to the same class. The parts are grouped in an appropriate training, validation and testing sample. The stability and efficiency of the model under variations of the hyper-parameters of the model have been proven, supported by experimental results.

Keywords: Mechanical parts; Classification; Convolutional neural networks; Deep learning.

1. Introduction

The application of computer vision in the automated production processes in the field of additive manufacturing, predictive maintenance, packaging inspection, quality control in product and components assembly, is emerging as a necessary factor in achieving innovative solutions. Image search, object recognition and classification are emerging as key components in modern automated and autonomous production systems that integrate artificial intelligence, aiming to overcome the challenges of implementing innovations in the modern European economy [1]. Implementing machine learning and artificial intelligence is a key stone of contemporary innovations in production automation as discussed in [2]. One of the main factors determining the efficiency of operations in the automated assembly of products is the high-precision recognition of mechanical parts, located in a spatial disorder on a production working area.

The accuracy in recognizing and classifying these parts, regardless of their geometric variations, determines to a high degree the accuracy of their manipulation and positioning by the flexible assembly [3]. In order to achieve high accuracy of recognition and classification of such objects, as well as in order to create an adaptive and easily reprogrammable environment, the methods of deep machine learning are applied.

In this study, we apply deep learning of a Convolutional Neural Network (CNN), using a chosen training strategy, in order to achieve high efficiency, applying the simplest model. The study was conducted for a large training, validation, and test sample of various mechanical parts. The investigated mechanical parts belonging to the same class, differ in shape, size, profile and spatial orientation. Different cases were studied, representing different hyper-parameter setups for the CNN. The analysis of the results and recommendations for future development and implementation of the model have been made.

2. State of the Art

There are various methods for classifying mechanical parts, integrated in real-time working hardware systems. Most of them use the "template matching" method, which "searches" for matching parts in the image with a predefined pattern/template [4]. The main disadvantage of this method is the need to pre-define the patterns and the delay due to the calculation of the correlation between the template and the corresponding area of the scanned image. The methods using neural networks are preferred, due to their adaptive abilities, the possibilities for deep training, and retraining in case of change in the setting of its parameters or in case of need for training with new mechanical parts. The question is what type of neural network is appropriate to use and optimise in the classification of mechanical parts? Submitting the entire image to the input neurons of a Multi-Layer Perceptron (MLP) network requires the design of a neural network with a large number of computational resources. It is necessary to pre-process the image in order to reduce its size, which in turn in many cases leads to loss of information. In this case, it is more appropriate to derive certain geometric or spectral features of the image, in order to constitute a feature vector to be used for training the MLP network [5]. It would be more appropriate to apply CNN, because the image size reduction is a basic idea that is implemented in the functional algorithm itself. Also, when training CNN, there is no need to group in advance different instances of mechanical parts, that belong to the same class (for example, class "bolt", class "nut", etc.) but differ in shape, size, profile, orientation, etc.

There are innovative solutions for the implementation of CNN neural networks for the classification of mechanical parts. In the implementation given in [6] in the pre-processing step, a quantum layer is added before the main CNN, which processes the data using quantum particle concepts. But this solution requires additional computing resources and proper data preparation before being submitted to a conventional CNN. The authors of [7] say CNNs are challenging to learn efficiently if the given dimension or amount of data or model become too large. We show in our research that training with a huge number of data/objects samples could give very good accuracy results together with a stable model work. The authors of [8] claim that CNNs trained on different popular datasets fail to detect objects when they see them under different lighting conditions and from different view angles. In this study, we show how by optimizing the parameters of the CNN, a very good result can be achieved even for such objects.

In our study, we aim to achieve maximum accuracy of recognition and classification of different classes of mechanical parts, without pre-processing of images and with conventional CNN. Various hyper-parameter setups have been applied, aiming to achieve the most simplified model while obtaining maximum efficiency. The research refers to training and testing with mechanical parts having different size, form, spatial orientation/different view angles, for exemplars belonging to the same class. The model is developed with coding in Python, using a cloud based platform.

3. General description, limitations and author's approach

With this research we aim to solve a classification problem in the field of computer vision and object recognition systems, applied to machine parts with different 3D spatial orientation. Our proposed methodology includes the usage of Deep Learning Convolutional Neural Networks, aiming to reduce the necessity of image preprocessing. We introduce a variety of training strategies for the purpose of increased efficiency during the training process in terms of higher classification accuracy on the testing set of images, but achieving the most simplified model. The large training, validation, and test sample of various mechanical parts, belonging to the same class, but different in shape, size, profile and spatial orientation, is a good prerequisite for the generalization of the method and the model. The following research limitations were applied: no variant of CNN with a maximum or minimum number of convolutional filters was investigated, the optimal number was determined experimentally; objects with changed contrast, brightness or different illumination of the working area are not included in the training and testing samples, considering that in the general case these parameters are kept constant.

3.1 Studied objects

In order to implement a solution to the mechanical parts recognition problem, we are using an image dataset, which consists of 7404 images of mechanical parts. The dataset consists of four image classes, each representing samples of mechanical parts. The classes are: Bolt, Locating pin, Nut, and Washer. There is an equal amount of samples for each image class in the dataset. All samples were divided into training, validation, and testing sets, following the ratio:

Training samples - 70 % of total samples; Validation samples - 20 % of total samples; Testing samples - 10 % of total samples. Some sample images are provided in Table 1.

Bolt

Locating pin

Nut

Washer

Table 1. Samples of the studied mechanical parts

3.2 The proposed Deep learning CNN model

The algorithm of choice for the solution of the image classification problem is a Deep Learning CNN. There are a total of four cases studied in this experiment, whereby each case represents a different hyper-parameter setup for the CNN. As the image samples are of size 224 x 224 pixels, we have decided to include only one convolutional layer, followed by one pooling layer before proceeding to flatten the convolved images. All four experimental cases are based on a simple topology due to the small image sizes. An increased number of convolution and pooling layers would result in greater dimensionality reduction of the images and loss of feature information. An overview of the primary network topology is shown in Fig. 1.

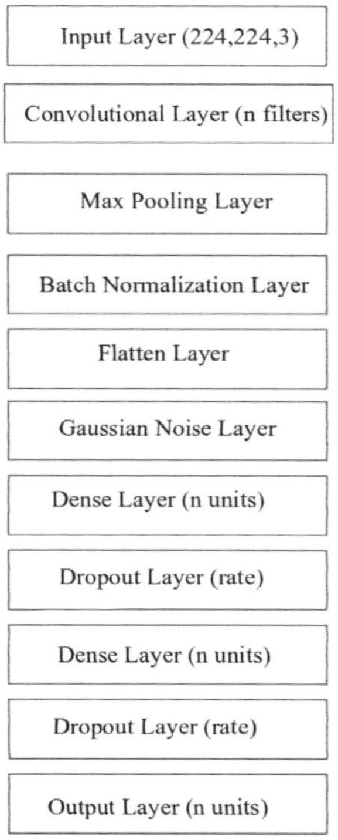

Fig. 1. Primary CNN topology

In order to increase the classification accuracy of all four CNNs, we have used the following training strategies:
Batch normalization: Consists of normalizing the input to each layer of the network, such that the inputs conform to a Gaussian distribution with a mean of zero and a standard deviation of one. This results in accelerated training for deep CNNs [9]. In this experiment the constructed CNN is not complicated in terms of depth, however, the usage of the batch normalization layer significantly decreases the number of training epochs required.
Gaussian Noise: this training technique is used to add random Gaussian noise to the image features with the aim to promote robust feature extraction. Gaussian noise is normally distributed and it has a mean of zero and a standard deviation of one [9].
Dropout: this training strategy resembles the bootstrap-aggregation training technique, whereby a number of simpler classifiers is used to solve a complicated problem, by dividing it into several less complicated problems. In the context of Neural Networks, a dropout layer is placed between two fully connected dense layers and it forces them to purposely ignore a random percentage of the input weights. This percentage is referred to as the dropout rate. The usage of a dropout rate results in preventing the neural network dense layers from trying to correct mistakes in previous layers, which in turn prevents the model from overfitting to the training data [9].
Weight regularisation: this training technique penalizes large weights by calculating either the sum of the values of the absolute weights (L1 norm) or calculating the sum of the squared weight values (L2 norm). The objective is to reduce the complexity of the produced decision boundary of the neural network and to keep the model as simple as possible to prevent overfitting. For this experiment, we have decided to use the L2 norm for weight regularisation [9].
Image normalization: this training technique reduces the computational complexity of training a Deep Learning CNN classifier, by rescaling the input data. In this case, the input data is in the form of images with pixel values in the interval [0, 255]. During the rescaling process, each pixel value is divided by 255. The output image has standardized values between [0,1] [9].
Early stopping: this training strategy includes monitoring the CNNs performance based on a chosen performance metric on each training iteration and interrupting the training process in case of successive poor performance [9]. For this experiment, the performance metric of choice is the validation accuracy.
The algorithm is implemented in Python in a cloud environment. The main parameters of the model are coded as follows.

```
cnn = Sequential([
        InputLayer(input_shape=(224,224,3)),
        Conv2D(filters =15 , kernel_size = (3,3), padding = 'valid', activation = 'relu', kernel_regularizer =
l2(0.0005),strides=(2,2)),
        MaxPool2D(pool_size=(3,3)),
        BatchNormalization(),
        Flatten(),
        GaussianNoise(0.001),
        Dense(units = 20, activation='tanh', kernel_regularizer = l2(0.0005)),
        Dropout(rate = 0.1),
        Dense(units = 10, activation = 'relu', kernel_regularizer = l2(0.0005)),
        Dropout(rate = 0.1),
        Dense(units = 4, activation = 'softmax')
early_stop = EarlyStopping(monitor='val_accuracy',verbose = 1, patience = 30)
model_checkpoint = ModelCheckpoint('best_mech_cnn_2.h5', monitor = 'val_accuracy', verbose = 1, save_best_only =
True).
```

3.3 Experimental setup

The experimental setup described in this article consists of the construction of four CNNs, each with a different hyper-parameter setup. The hyper tunning of each CNN is given in Table 2:

Hyper-parameter setup	**Case 1**	**Case 2**	**Case 3**	**Case 4**
Convolutional filters	15	20	15	15
Kernel regularizer	0.0005	0.0005	0.0005	0.0005
Gaussian noise	0.001	0.001	0.001	0.001
Dropout rate	0.2	0.2	0.2	0.5
Learning rate	0.00007	0.00007	0.00007	0.00007
Strides	(2,2)	(2,2)	(1,1)	(1,1)
Total parameters	411454	548514	1643554	1643554

Table 2. Hyper-parameter setup for all experimental cases

4. Experimental results

In this section we present the experimental results. The results contain a graphical representation of the learning process of the CNN, with respect to accuracy and loss functions, as well as a table summary of highest to lowest classification accuracy.

4.1 Hyper-parameter influence

This section is focused on evaluating the graphical results from the neural network training process. The results from each case study are shown in Fig 2.

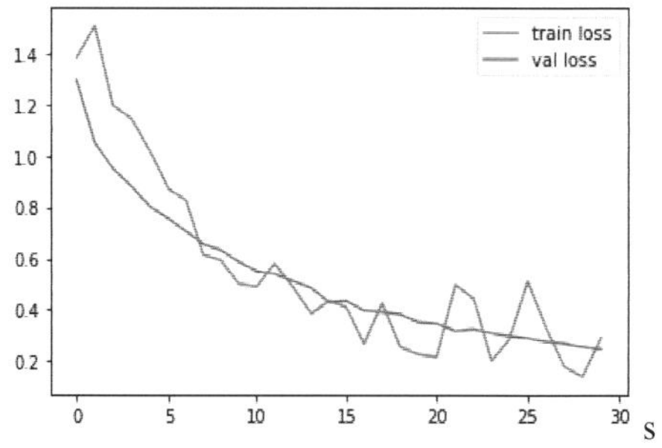

Fig. 2/a Loss function in case 1

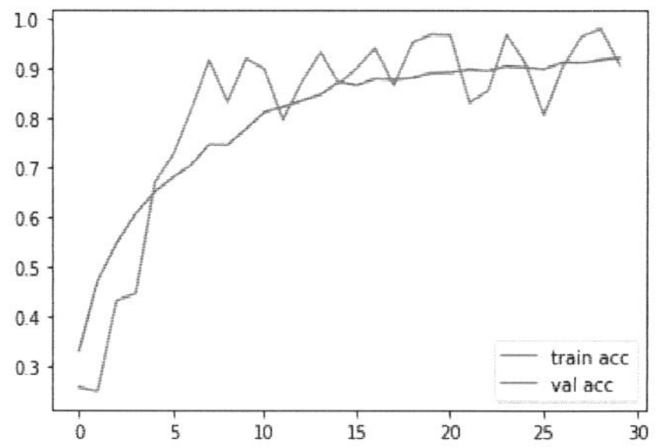

Fig. 2/b Accuracy in case 1

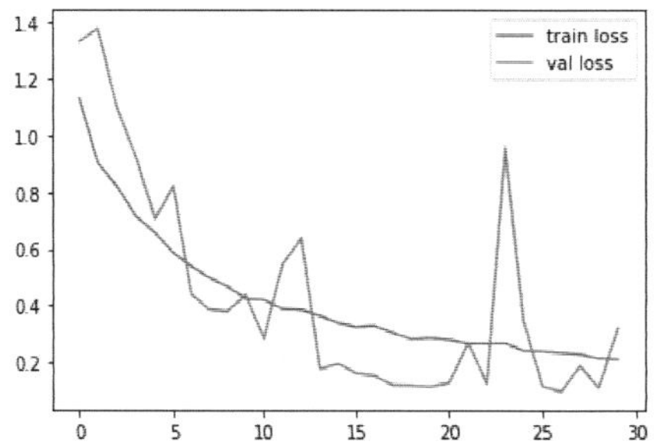

Fig. 2/c Loss function in case 2

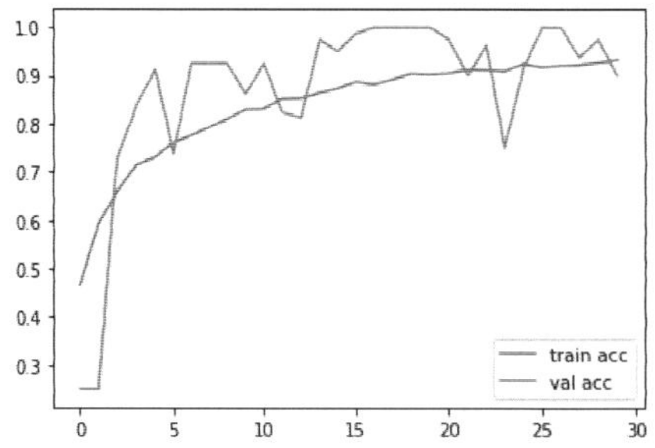

Fig. 2/d Accuracy in case 2

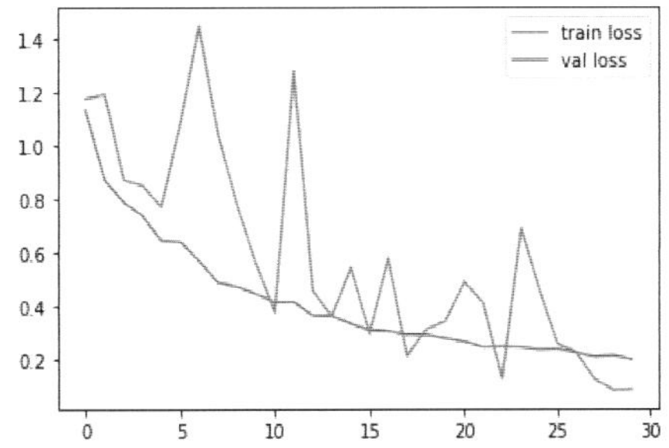

Fig. 2/e Loss function in case 3

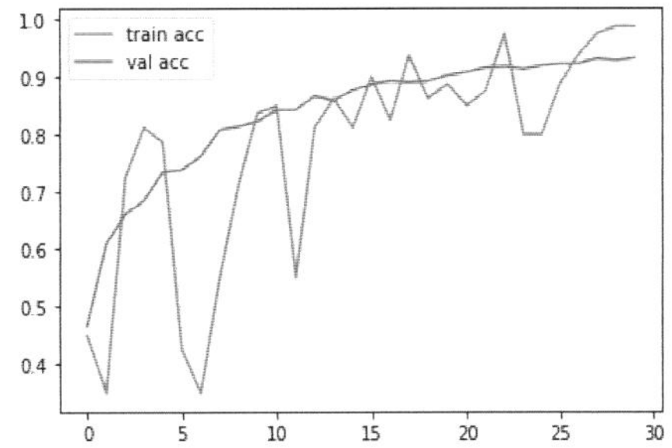

Fig. 2/f Accuracy in case 3

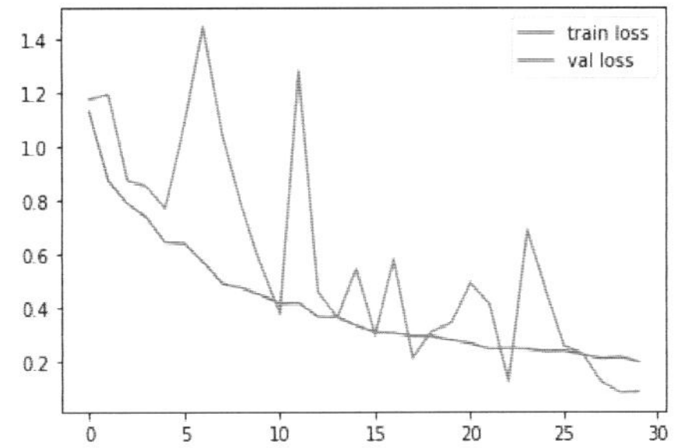

Fig. 2/g Loss function in case 4

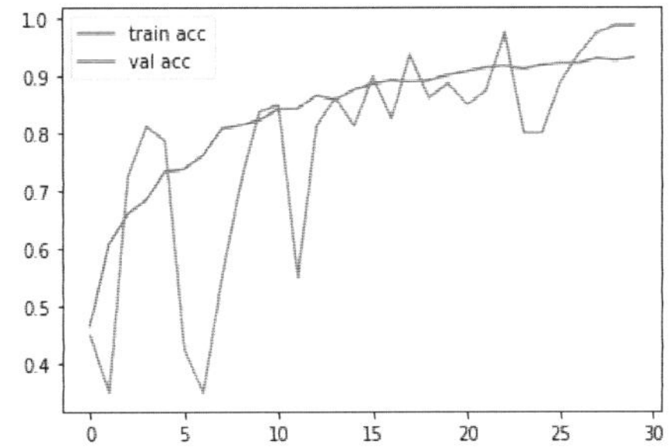

Fig. 2/h Accuracy in case 4

The first observation we have on the experimental results is that the stability in the training process decreases as the number of model parameters grows. In case 1 and case 2, both loss and accuracy functions of the validation set are similarly distributed to the loss and accuracy functions of the training set. This tendency decreases in cases 3 and 4, whereby the number of model parameters is increased due to the reduced strides of the convolutional filters. As a result, the validation loss and accuracy spike throughout the training process, making it unstable and less certain of a good outcome for both training and validation data.

Further observation in cases 1 and 2, shows that the increased number of filters in the convolutional layer has led to the model overfitting on the validation data in case 2. We assume that this is due to the fact, that the given image data is less complicated, with a clear contrast between the object to be recognized and the image background. Objects are visible, have clear edges and multiple angles of each object are included in the dataset. The simplicity of the images allows for effective feature extraction with fewer filters in the convolutional layer. Furthermore, this results in simpler network topology and therefore, fewer model parameters. The model has a reduced chance of overfitting the training data in case 1.

Performance comparison of the models in cases 3 and 4 reveals, that by increasing the dropout rate in case 4, the number of training epochs required to learn the training dataset is increased. Both models have the same amount of model parameters and a number of training epochs, however, in case 4 the accuracy on the training data only did reach 93 %, meaning that the model has not learned completely the training data.

4.2 Accuracy evaluation

In this section, we have provided a summary of validation accuracy for all four experimental cases. The result summary is shown in Table 3.

Case number	Accuracy
Case1	98.228 %
Case2	100 % (Overfitting)
Case3	98.750 %
Case4	93.750 %

Table 3. Result summary on validation accuracy for all 4 experimental cases

In order to study the model performance further, we have created a confusion matrix on each experimental case using the test dataset. The classification accuracy on each case is measured with the formula (1):

$$\text{Accuracy} = Number\ of\ correctly\ classified\ samples/\ Number\ of\ classified\ samples \quad (1)$$

The classification accuracy on the test dataset in all four cases is provided in Table 4.

Case 1	Case 2
Accuracy: 97.31 %	Accuracy: 98.29 %
[[355 25 1 0]	[[370 11 0 0]
[1 376 2 2]	[7 370 1 3]
[0 7 374 0]	[0 2 379 0]
[2 1 0 378]]	[0 0 2 379]]
Case 3	**Case 4**
Accuracy: 99.47 %	Accuracy: 97.05 %
[[380 1 0 0]	[[381 0 0 0]
[4 374 2 1]	[15 355 4 7]
[0 0 381 0]	[6 1 365 9]
[0 0 0 381]]	[3 0 0 378]]

Table 4. Confusion matrix and classification accuracy on test data in four experimental cases

5. Conclusion

The analysis of the results, obtained when testing the proposed model, shows that the application of Batch normalization, Gaussian Noise, Weight regularization, Image normalization and Early stopping, allows simultaneously significantly reducing the number of training epochs required, promoting robust feature extraction and preventing the model from overfitting to the training data, at the same time producing a model as simple as possible to prevent overfitting. The achieved stable high accuracy of classification for the four cases of given hyper-parameters, shows that the proposed CNN model achieves the highest accuracy in case 3, although at the expense of increased number of parameters due to the reduced "stride" value. The achieved stability of the model and the high accuracy achieved for the variation of the parameters in the four presented cases show its efficiency in recognizing machine parts and can be recommended for classification of other two-dimensional objects.

It was proven, that training with a huge number of data/objects samples could give very good accuracy results together with a stable model work, when appropriate CNN's hyper-parameters are adjusted. The represented experimental results show, that by optimizing the parameters of the CNN, a very good result can be achieved even for images of mechanical parts, having low resolution and different form, size, view angle and spatial orientation within the same class. As a future development of the method, we envisage conducting a larger number of tests with additional different classes of mechanical parts, as this would show further generalization of the model. Also, other possibilities for optimizing the parameters of the CNN may be sought. The same object samples could be tested for recognition and classification with adding a quantum layer before the main CNN and processing the data using quantum particle concepts, for benchmarking purposes.

6. References

[1] Pavlov, P. (2015). Innovation management, Book, University Publishing House of VFU (Varna Free University) "Chernorizets Hrabar", Varna, ISBN 978-954-715-645-6; Павлов, П. (2015). Иновационен мениджмънт, Учебник, Университетско издателство на ВСУ „Черноризец Храбър, Варна 2015, ISBN 978-954-715-645-6.

[2] Tekic, Z; Cosic, I & Katalinic, B. (2019). Manufacturing and the Rise of Artificial Intelligence: Innovation Challenges, Proceedings of the 30th DAAAM International Symposium, pp.0192-0196, B. Katalinic (Ed.), Published by DAAAM International, ISBN 978-3-902734-22-8, ISSN 1726-9679, Vienna, Austria, DOI: 10.2507/30th.daaam.proceedings.025

[3] Nerakae, Ph., Uangpairoj, P. & Chamniprasart, K. (2016) Using Machine Vision for Flexible Automatic Assembly System, Conference on Knowledge Based and Intelligent Information and Engineering Systems, KES2016, pp. 428 – 435, 5-7 September 2016, York, United Kingdom, https://doi.org/10.1016/j.procs.2016.08.090.

[4] Topalova, I. (2016). Automated Visual Inspection System in Praline Industry, 27th International DAAAM SYMPOSIUM "Intelligent Manufacturing & Automation: Theory, Practice & Education", pp. 175-179, ISBN: 978-3-901509-70-4, ISSN: 17269679, Mostar, Bosna, October.

[5] Topalova, I. & Tatarska, V. (2018). Automated Classification of Similar Two-dimensional Objects with Neural Network Application, Proceedings of the 29th DAAAM International Symposium, B. Katalinic (Ed.), Published by DAAAM International, pp. 0225-0231, ISBN 978-3-902734, ISSN 1726-9679, Vienna, Austria, 24th-27th October, DOI: 10.2507/29th.daaam.proceedings.

[6] Khullar V., Tiwari R.G. & Agarwal A.K. (2022). Quantum Layer-Inspired Deep Learning for Mechanical Parts Classification. In: Agarwal B., Rahman A., Patnaik S., Poonia R.C., Proceedings of International Conference on Intelligent Cyber-Physical Systems. Algorithms for Intelligent Systems. Springer, Singapore, ISSN 978-981-16-7135-7, ISBN 978-981-16-7136-4.

[7] Seunghyeok Oh, Jaeho Choi & Joongheon Kim A. (2020). Tutorial on Quantum Convolutional Neural Networks (QCNN), 2020 International Conference on Information and Communication Technology Convergence (ICTC), 21-23 October 2020, Jeju, Korea (South), INSPEC Accession Number: 20244515, DOI: 10.1109/ICTC49870.2020.9289439, Publisher: IEEE.

[8] Manav Mandal. (2021). Introduction to Convolutional Neural Networks (CNN), Advanced Computer Vision Deep Learning Image Analysis, Published On May 1, 2021 and Last modified on july 23rd, 2021 https://www.analyticsvidhya.com/blog/2021/05/convolutional-neural-networks-cnn/#h2_6.

[9] Goodfellow I., Bengio J. & Courville A. (2016). Deep Learning, Published by MIT Press, ISBN: 0262035618.

DOI: 10.2507/33rd.daaam.proceedings.008

ANALYSIS OF PARAMETERS INFLUENCING THE INDOOR AIR QUALITY

Halima Hadziahmetovic, Rejhana Blazevic, Mirela Alispahic,
Raza Sunulahpasic & Sanda Midzic Kurtagic

This Publication has to be referred as: Hadziahmetovic, H[alima]; Blazevic, R[ejhana]; Alispahic, M[irela]; Sunulahpasic, R[aza] & Midzic Kurtagic, S[anda] (2022). Analysis of Parameters Influencing the Indoor Air Quality, Proceedings of the 33rd DAAAM International Symposium, pp.0050-0057, B. Katalinic (Ed.), Published by DAAAM International, ISBN 978-3-902734-36-5, ISSN 1726-9679, Vienna, Austria
DOI: 10.2507/33rd.daaam.proceedings.008

Abstract

In modern society, when people spend most of their time in indoor environment, it is very important to ensure the best possible indoor climate in order to achieve general satisfaction with the users of the space. The paper provides an overview and description of the key parameters required for achieving a quality indoor climate. Also, the measurement of indoor air quality (IAQ) parameters was performed in one of the classrooms in the building of the Faculty of Mechanical Engineering in Sarajevo in the period from May 7 2019 till May 24 2019. The results obtained by measurement are compared with the values prescribed according to EN 15251, EN 16798, EN 13779 and ASHRAE 62.1 standards in order to determine the level of air quality in the room.

Keywords: indoor air quality; european standards; thermal comfort; ASHRAE standard

1. Introduction

The concept of quality indoor air is quite broad and there are many parameters that affect the achievement of thermal comfort. However, indoor air quality (IAQ) is most often determined using three key parameters: carbon dioxide (CO_2) concentration, indoor air temperature and relative humidity. In modern society, people spend 90% of their time indoors, which is why it is very important to achieve the appropriate microclimate and air quality in those spaces. [1]. A lot of research that has been conducted has shown that all aspects of comfort affect the overall satisfaction of occupants, including the subjective feeling of comfort, health and on-the-job performance. The research has also shown that the concentration of carbon dioxide depends on the level of activity and the number of people present inside. The analyses have revealed that a lower level of activity causes the concentration of CO_2 to decrease and scale back to night-time outdoor concentration levels [2]. The CO_2 level rises in poorly ventilated and oxygen-deficient spaces, therefore people staying or working in closed spaces with poor air quality commonly suffer from headaches, drowsiness, a drop in energy and concentration levels [3], [4], [5]. The comfort associated with spending time indoors is also closely related to energy consumption because unfavourable micro-climate conditions adversely affect people's mental and physical condition and in many cases act as a motivating factor for occupants to take steps to increase the level of comfort, which in turn can lead to higher energy consumption levels.

Authors in their work [6] have used a dynamic building simulation tool to compared the energy use of two classrooms designed to meet various indoor environmental quality (IEQ) standards. Simulations were performed using the criteria for heating and cooling setpoints, ventilation rates, humidity levels and illuminances for each category in the EN, ISO, ASHRAE and ISHRAE standards in order to identify how the criteria from each standard affect energy use and IEQ. In their research, they came to the conclusion that it is less energy expensive to improve IEQ parameters by moving from setpoints of the Category III to Category II than from Category II to Category I. The group of authors in their research [7] gave a critical review of standards for indoor thermal environment and air quality. Increasing the building's energy efficiency very often leads to poor comfort conditions in buildings. Therefore, future standards and regulations for defining energy performance should be focused on indoor comfort and air quality in the space, in addition to reducing energy consumption in the building. If the standards are focused only on the reduction of energy consumption in the building, it can lead to problems for the users and their poor productivity related to the disturbed indoor comfort. The most common problems occur due to the restrictive requirements of the standard for the minimum number of air changes in the room.

A number of international standards provide for recommended parameters for indoor air quality. This paper will consider BAS EN ISO 7730:2010, EN 15251:2007 which was modified with the standard EN 16789-1:2019 and ASHRAE 62.1 standards, and the results obtained by measurement will be matched against the values recommended in those standards.

2. Assessment of thermal comfort using BAS EN ISO 7730:2010

The BAS EN ISO 7730:2010 standard serves to analytically determine and interpret thermal comfort using the calculation of PMV (*PMV - Predicted Mean Vote*) and PPD indices (*PPD - Predicted Percentage of Dissatisfied*) and local thermal criteria. The standard determines the PMV and PPD indices depending on the level of activity and clothing. The PMV index is derived from the human heat balance equation and is expressed as (1) [8]:

$$PMV = (0{,}028 + 0{,}303 \cdot e^{-0{,}036 \cdot M})[(M - W) - H - E_c - C_{res} - E_{res}] \quad (1)$$

where: M – metabolic rate, W – effective mechanical power, H – sensitive heat exchange (dry heat loses), E_c – heat exchange by evaporation on the skin (wet heat losses), C_{res} – heat exchange by convection in breathing, E_{res} - evaporative heat exchange in breathing.

Expression (1) shows that the PMV index can be calculated for different values of the metabolic rate, clothing insulation, air temperature, mean radiant temperature, air velocity and humidity [8], [9]. The PMV index has a seven-point thermal sensation scale, from -3 (too cold) to +3 (too hot), with 0 translating as thermal neutrality. Even when the PMV index is 0 dissatisfaction with thermal comfort will still be expressed by some people, because even if all the occupants are dressed in a similar fashion and engage in similar activity levels the comfort sensation varies from one person to another. Once the PMV index is known, the PPD index that predicts the percentage of thermally dissatisfied occupants, i.e. feeling too cold or too hot, can be determined. The PPD index is determined by means of the mathematical expression (2) as a function of the PMV index [8], [10]:

$$PPD = 100 - 95 \cdot e^{-(0{,}03353 \cdot PMV^4 + 0{,}2179 \cdot PMV^2)} \quad (2)$$

Expression (2) shows that in case of thermal neutrality the minimal percentage of dissatisfied respondents is 5%, whereas for the maximum 10% of dissatisfied respondents, which is acceptable under the ISO 7730 standard, PMV can range from -0.5 to +0.5, which is regarded as an acceptable thermal environment. The predicted number of dissatisfied respondents is also illustrated in the diagram shown in Fig. 1 [3], [8], [9], [10].

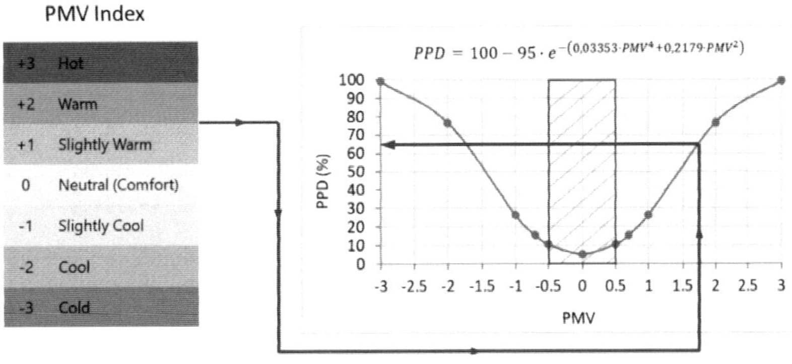

Fig. 1. Predicted percentage of dissatisfied as a function of predicted mean vote [3],[8],[9],[10]

PMV and PPD indices can be directly used as a criterion for the assessment of indoor thermal comfort. This is because they take into account the impact of key parameters determining the indoor thermal condition as well as those that are descriptive of a person (their physical activity and clothing) [8], [10].

In educational buildings where occupants (students and teachers) were an important source of air contamination, it was recommended to calculate the percentage of occupants dissatisfied with IAQ, as a function of pollutant concentrations. For the main representative pollutant, which is CO_2, the PD = f(cCO_2) function is expressed by equation [11]:

$$PD_{CO_2} = 407 \cdot e^{\left(-15{,}05 \cdot C_{CO_2}{}^{-0{,}25}\right)} \tag{3}$$

3. Recommended values of indoor air quality parameters according to EN 15251, EN 16789, EN 13779 and ASHRAE 62.1 standards

The European standard for indoor environment EN 15251:2007 sets permitted carbon dioxide concentration levels for indoor air compared to outdoor carbon dioxide concentrations for different categories of buildings, as shown in Table 1 [12].

Category	CO_2 concentration above outdoor concentration (ppm)	Fresh air per person (l/s/person)	Relative humidity	Air flow for pollution emission in buildings (l/s/m²)		
				Very low polluted buildings	Low polluted buildings	Polluted buildings
I	350	10	30 – 50	0.5	1.00	2.00
II	**500**	**7**	**25 – 60**	**0.35**	**0.70**	**1.40**
III	800	<4	20 – 70	0.2	0.40	0.80
IV	>800	-	-	-	-	-

Table 1. Permitted carbon dioxide concentrations according to EN 15251 [12]

According to the EN 15251 standard, the recommended carbon dioxide concentration in educational institutions is assigned Category II, which means that CO_2 concentrations should not exceed the upper limit of 500 ppm relative to outdoor CO_2 concentrations. For non-residential buildings, ISO 17772, EN 16798 and EN 15251 have similar requirements for ventilation rates, but EN 15251 does not specify any values for Category IV. Moreover, recommended indoor air temperature in educational institutions according to EN 15251 (Category II) is 20°C in the winter period and 26°C in summer [12].

Standard EN 16789-1 published in 2019, is update of the EN 15251 from 2007, covering indoor environmental criteria for the design of buildings, room conditioning systems, and lighting systems. The major change from EN 15251 is that the standard is split into a normative part, EN 16798-1, and a technical report part EN 16798-2 TR. The standard establishes three, or sometimes four categories of thermal comfort. Category I is recommended for users with less thermal adaptation (e.g. the elderly, sick, or children), while categories II and III are recommended for new and existing buildings, respectively. Category IV is with poor quality air and unacceptable regarding health. The listed CO_2 values can also be used for demand controlled ventilation according to standard EN 16789, as shown in Table 2 [13].

Category	CO_2 concentration above outdoor concentration (ppm)	Fresh air per person (l/s/person)	Relative humidity	Air flow for pollution emission in buildings (l/s/m²)		
				Very low polluted buildings	Low polluted buildings	Polluted buildings
I	550	10	30 – 50	0.5	1.00	2.00
II	**800**	**7**	**25 – 60**	**0.35**	**0.70**	**1.40**
III	1350	4	20 – 70	0.3	0.40	0.80
IV	>1350	<4	<20, >70	0.25	0.3	0.6

Table 2. Default design CO_2 concentrations above outdoor concentration assuming a standard CO_2 emission of 20 L/(h per person) according to EN 16789 [13]

The acceptable level for the Category I according to EN 16798 is 550 ppm, which is 200 ppm higher than in EN 15251 standard, and Category II is the same as Category III in above mentioned standard. According to EN 16789 standard, II category is recommended for educational buildings with corresponding value of 800 ppm CO_2 above outdoors for energy calculation.

The European standard EN 13779 classifies indoor air quality into four categories (Table 3) ranging from IDA 1 to IDA 4 to indicate high and low indoor air quality respectively. Also, according to the EU, the typical CO_2 indoor air concentration ranges from 600 to 1000 ppm relative to outdoor CO_2 concentrations.

Category	Quality of indoor air	CO_2 concentration above outdoor concentration (ppm)		Fresh air rate (l/s/person)
		Typical range	Default value (ppm)	
IDA 1	High	≤400	350	>15
IDA 2	Medium	400-600	500	10-15
IDA 3	Moderate	600-1000	800	6-10
IDA 4	Low	>1000	1200	<6

Table 3. Indoor air quality classification according to EN 13779 [14]

According to ASHRAE 62.1, the permitted upper limit concentration of CO_2 in the office space is 700 ppm relative to outdoor CO_2 concentrations [15]. Since outdoor CO_2 concentrations typically range from 300 ppm to 500 ppm (as the typical outdoor concentration of CO2 were taken 400 ppm value.), the recommended CO_2 concentration for business premises should be in range from 1000 ppm to 1200 ppm. Also, according to the ASHRAE 62.1 standard, the relative air humidity in educational buildings should be between 30% and 60%, and the indoor air temperature between 20 and 24°C in the winter period, and between 23 and 26°C in the summer period [15]. EN 16798 and EN 15251 require relative humidity to be maintained within 30-50% for Category I, 25-60% for Category II, and 20-70% for Category III. No criteria for the relative humidity for Category IV are stated meaning that any values beyond limits of the Category III are unacceptable [12], [13].

4. Description of the measuring instrument and analysis of indoor air quality parameters

The goal of this paper is to analyse of CO_2, indoor air temperature and relative humidity level and their variability during analysed period. The instrument used for the measurement of indoor air parameters is the OPUS 20-TCO data logger, manufactured by Lufft, Fig. 2. It consists of three internal sensors for recording air temperature (°C), relative humidity (%) and carbon dioxide concentration (ppm). The measuring range for carbon dioxide concentration is 0-5000 ppm, with ±50 ppm accuracy, for air temperature it is from -20 to +50°C with 0.3°C accuracy and for relative humidity the measuring range is 10-95%, with 0.2% accuracy. The measuring range is set to do measurements every 10 seconds, with values being recorded every 30 minutes. The sensor is placed inside a room, 1 meter above the floor and at a 1-meter distance from the user, as illustrated in Fig. 3.

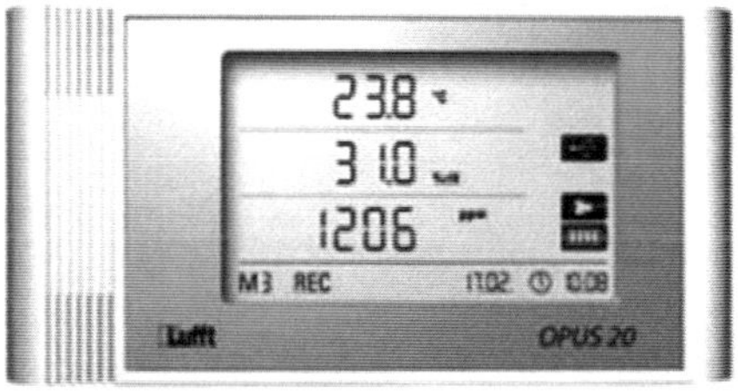

Fig. 2. Instrument for measuring air quality

Fig. 3. Positioning of the instrument

Measuring indoor air quality parameters was performed in one of the 2nd floor classrooms at the Faculty of Mechanical Engineering in Sarajevo. The classroom has 148.6 m^2 and the volume of 498.8 m^3. The glazing faces south and east. The classroom is equipped with sunscreen vertical blinds that can be moved according to need. There is no mechanical ventilation in the classroom, and the classrom is naturally ventilated. Carbon dioxide concentration, relative humidity and air temperature measurements were performed outside the heating season, from 7 May 2019 to 24 May 2019. Fig. 4 shows CO_2 values recorded in one of the 2nd floor classrooms in the analysed period.

The results have shown that during each working day the CO_2 concentration rises from the start of the lectures and reaches its maximum values at the end of the day, that is, when students leave the classroom. At the end of the working day the indoor CO_2 concentration begins to decrease until its value becomes approximately equal to the value of the external CO_2 concentration, which usually ranges from 300 ppm to 500 ppm. Fig. 4 shows that for some parts of the day CO_2 concentrations do not comply with the prescribed ASHRAE and EN 15251/EN 16789 standards. Higher CO_2 concentrations were particularly pronounced from 13 May 2009 to 17 May 2019. On those days, according to class schedule, teaching was performed from 9 a.m. to 3 p.m. or even 5 p.m.

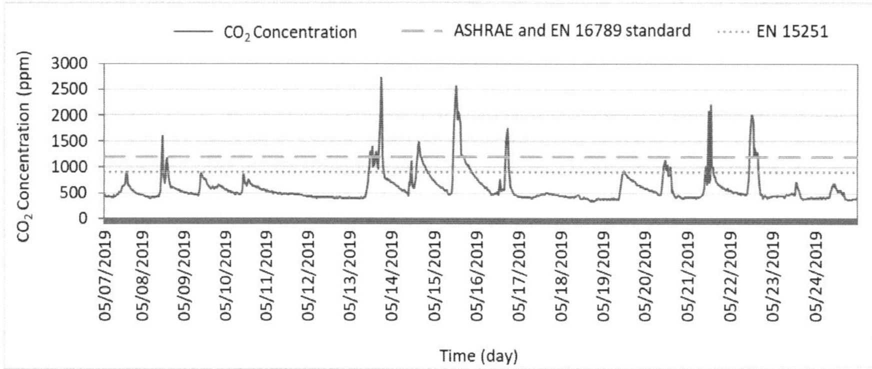

Fig 4. Change of carbon dioxide concentration in one of the classrooms at the Faculty of Mechanical Engineering in Sarajevo in the analysed period

Also, Fig. 4 indicates that the highest CO_2 concentration was recorded on 13 May 2019 at 6:00 p.m. (2729 ppm) and on 15 May 2019 at 12:30 p.m. (2560 ppm), both cases being 56% above the average values recommended by standards. The measurement results obtained for 13 May 2019 and 15 May 2019 show that the quality of air belongs to IDA 4 Category (concentration over 1000 ppm) according to EN 13779 standard. This category is used in reference to buildings with poor or low indoor air quality. When the values obtained by measurement are compared with the values set in the ASHRAE and EN 16789 standard, the mean daily concentration of CO_2 for 13 May 2019 and 15 May 2019 falls within the 1000 ppm - 2000 ppm range, the real-time CO_2 concentration at different parts of the day even exceeding 2000 ppm, which means that air quality is inadequate, causing people to experience pronounced lack of focus and sleepiness. It is common knowledge that CO_2 concentration levels and the amount of fresh air in indoor environments depend on the number of the people present, their activity level and the way the analysed space is ventilated. When the measurements were carried out in analysed classroom the number of students attending the lectures exceeded 50, which, in addition to inadequate classroom ventilation, represents a leading cause behind higher CO_2 concentration levels. Table 4 shows the overall measured values of IAQ parameters in one of the classrooms during it full occupancy for the analyzed period.

Date	CO_2 max (ppm)	CO_2 min (ppm)	CO_2 average (ppm)	Indoor air temperature (°C)	Relative humidity (%)
May-07	911	428	605	23.3	32.2
May-08	1604	434	857	24.6	29.3
May-09	1857	586	733	23	36.8
May-13	2729	414	1153	22.9	49.4
May-14	1483	449	914	21.7	44.3
May-15	2560	502	1614	23.5	42.5
May-16	763	455	535	22.5	41.7
May-20	1127	414	738	21.9	40.9
May-21	2207	457	945	21.9	42.9
May-22	2014	504	1258	23.2	47
May-23	703	450	529	23.5	43.8
Analysed period	**2729**	**414**	**898**	**22.9**	**41.0**

Table 4. Measured IAQ values in one of the classrooms at the Faculty of Mechanical Engineering in Sarajevo in the analysed period

During most period of time when students were in the classroom the CO_2 concentration was slightly higher than the recommended one by standards. The average temperature during occupied period was 22,9°C, with a minimum of 20,6°C and a maximum 25,7°C. Average humidity during classes was around 41%, with a minimum value of 26,3% and a maximum of 55,9%, which is the highest value measured on May 13, at the same time when the highest CO_2 concentration of 2729 ppm was measured. The percentage of occupants dissatisfied with indoor air quality, as a function of average CO2 concentrations according to expression (3) would be 24,8%.

Also, based on the obtained measurement results, a correlation analysis of the CO_2 concentration with relative humidity and indoor air temperature have been performed. Since the measurements were carried out outside the heating season the indoor CO_2 concentrations are strongly influenced by the outdoor CO_2 due the fact that windows are opened more frequently than in heating period. Consequently, the correlation was performed only for two days during the analyzed period when the classroom was fully occupied and outdoor air temperature was low in order to reduce the possibility of opening the window for a longer period of time because in case of naturally ventilated educational buildings many indoor air parameters depend significantly on outdoor air conditions. As it can be seen from Figures 5 and 6, high correlation was observed for these days, i.e. temperature and relative humidity values follow the change in CO_2 concentration. The similar investigation had been done by group of authors in [5].

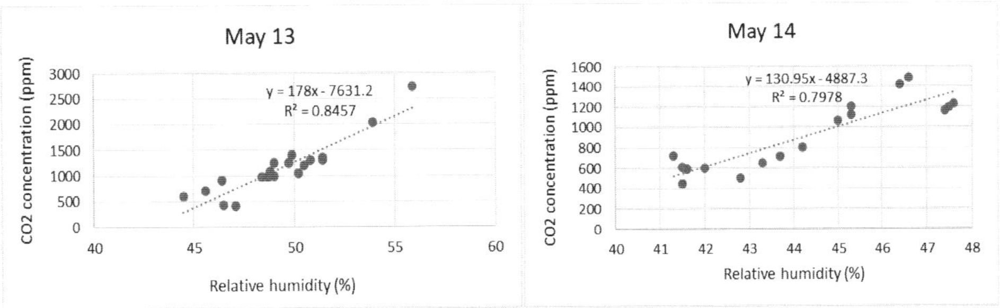

Fig. 5. Correlation coefficient between relative humidity and CO_2 during two occupied days

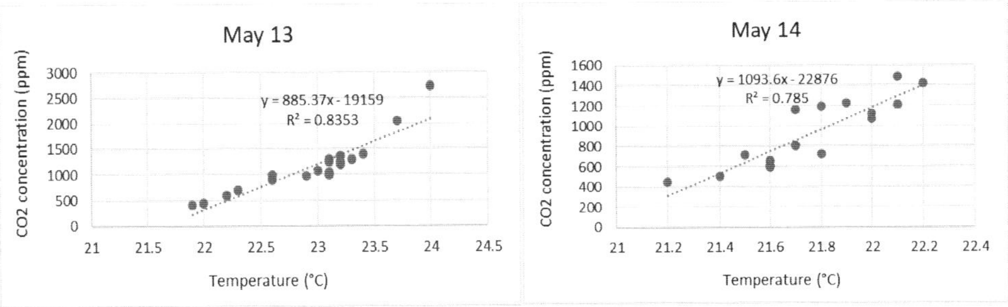

Fig. 6. Correlation coefficient between indoor air temperature and CO_2 during two occupied days

The level of relative humidity of indoor air also represents one of the key parameters that bears upon the quality of indoor air. Moisture in the building envelope often causes many problems, including low indoor air quality. Similarly, people endure high indoor temperatures more easily when humidity is low, because sweating regulates the body's temperature and in that way helps the body cool down. If indoor air is saturated with water vapour it cannot hold the extra vapour from sweating, which can lead to heat stroke. Likewise, low air humidity levels are not good as they can cause discomfort, respiratory problems and increase the risk of infections. Fig. 7 shows the change of temperature and relative humidity in one of the classrooms in the analysed period.

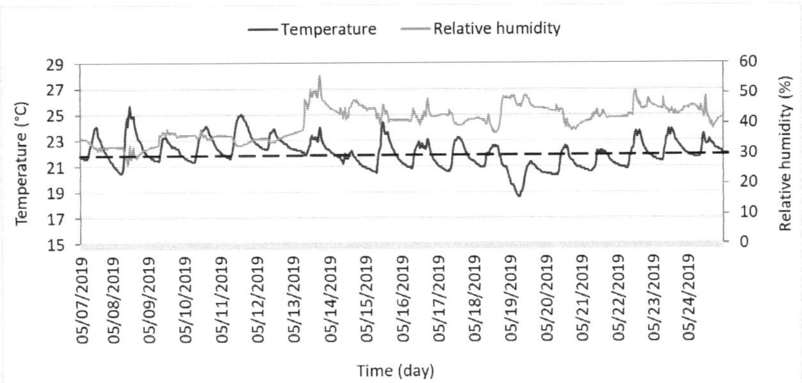

Fig. 7. Change of temperature and relative humidity in one of the classrooms at the Faculty of Mechanical Engineering in Sarajevo in the analysed period

Fig. 7 indicates that humidity concentration levels reached their peak of 55.9% on 13 May 2019 at 6:00 p.m. However, the relative humidity never registered levels above the Category II upper limit of 60% nor below lower limit of 25%, as defined by EN 16798-1:2019. After matching the relative humidity levels recorded in analysed classroom against the recommended values, relative humidity can be said to fall within the limit of recommended values. The lowest temperature levels were recorded over the weekend, namely on Sunday, 19 May 2016, at 6:00 a.m., when the temperature was 18.6°C, while the maximum temperature (25.7°C) was recorded on Wednesday, 8 May 2019. During working days, the indoor air temperature ranged from 20.5°C in the early morning hours to 25.7°C in the afternoon. Since the measurements were performed in May, outside the heating season, the results were compared using the summer season values. According to EN 16798-1:2019, the recommended indoor air temperature in educational facilities in summer is 26°C, while ASHRAE 62.1 recommends that indoor air temperature values be in the range between 23°C and 26°C. Bearing in mind the recommended values it can be concluded that the indoor air temperature in the analysed period remained within the permitted limits.

5. Conclusion

This paper cites the parameters necessary for achieving a quality climate in indoor environments, describes the impact these parameters have on human health, and gives their limit values as recommended under EN 15251, EN 16798, EN 13779 and ASHRAE 62.1 standards. The values recorded in one of the classrooms at the Faculty of Mechanical Engineering in Sarajevo indicate the presence of a high concentration of carbon dioxide, which at different parts of the day exceeds by 56% the levels recommended under the standards governing permissible concentrations in educational institutions. Inadequate ventilation is the main cause of high CO_2 concentration levels in educational institutions. Furthermore, the measurements performed lead to the conclusion that air temperature rises and air humidity decreases on working days and when people stay indoors. The results of this research indicate the need to make a plan and find solutions to control CO_2 concentration levels, as well as the need for additional ventilation to ensure a sufficient number of air exchanges in the form of mechanical ventilation as a complement to natural ventilation. Therefore, installing CO_2 sensors, implementing night cooling and ventilation arrangements in buildings in the summer period and using mixed mode or hybrid ventilation are some of the possible solutions for raising the level of indoor air quality.

In this paper, it was analyzed indoor air quality, however indoor environmental quality (IEQ) is characterized by four environmental categories such as thermal comfort, indoor air quality, lighting, and acoustics whereby equirements for each category are often split up in different standards. Since it is very important to consider all IEQ factors next research steps would be providing and implementing an IAQ model which can be used to determine the value of the overall indoor environmental quality index (IEQ) including these components: thermal comfort, indoor air quality, acoustic comfort and lighting quality. Also, it would be interest to evaluate the influence of different ventilation systems, air rates and schedules on the thermal comfort and indoor air quality of a refurbished educational building compared with unrefurbished one through measurement of IAQ parameters.

6. References

[1] World Health Organisation (WHO). (2000). Air Quality Guidelines for Europe, Second Edition, Regional Publications, European Series No. 91, Copenhagen.

[2] Hadziahmetovic, H.; Zecevic, N. & Neimarlija, A. (2021). Monitoring of indoor air quality in classroom, Proceedings of the 32nd DAAAM International Symposium, pp.0040-0047, B. Katalinic (Ed.), Published by DAAAM International, ISBN 978-3-902734-33-4, ISSN 1726-9679, Vienna, Austria, DOI: 10.2507/32nd.daaam.proceedings.006.

[3] Wolf, I. (2009). Thermal comfort models. Mechanical Engineering: Journal of Theory and Practice in Mechanical Engineering, 51(5), 507-517. Available from: https://hrcak.srce.hr/file/77596, Accessed: 2022-04-15.

[4] Bajc, T. S.; Todorović, M. N., & Papadopoulos, A. M. (2016). Indoor environmental quality in non-residential buildings-experimental investigation. Thermal Science, 20(5), pp. 1521-1529.

[5] Lazović, I.; Stevanović, Ž. M.; Jovašević-Stojanović, M.; Živković, M. M., & Banjac, M. J. (2016). Impact of CO2 concentration on indoor air quality and correlation with relative humidity and indoor air temperature in school buildings in Serbia. Thermal Science, 20(1), pp. S297-S307.

[6] Khovalyg, D.; Kazanci, O.; Gundlach, I.; Bahnfleth, W.: Toftum, J.; & Olesen, B. W. (2020). Impact of indoor environmental quality standards on the simulated energy use of classrooms. In Proceedings of the 11th Windsor Conference, pp. 848-860.

[7] Khovalyg, D.; Kazanci, O. B.; Halvorsen, H.; Gundlach, I.; Bahnfleth, W. P.; Toftum, J.; & Olesen, B. W. (2020). Critical review of standards for indoor thermal environment and air quality. Energy and Buildings, 213, 109819. DOI: https://doi.org/10.1016/j.enbuild.2020.109819

[8] BAS EN ISO 7730-2010: Ergonomics of the thermal environment– Analytical determination and interpretation of the thermal comfort using calculation if the PMV and PPD indices and local thermal comfort criteria.

[9] Kulić, E. (1989). Principi projektovanja sistema grijanja, Mašinski fakultet Univerziteta u Sarajevu, Sarajevo.
[10] ASHRAE Standard 55 – Thermal Environment Conditions for Occupancy
[11] Piasecki, M.; & Kostyrko, K. B. (2019). Combined model for IAQ assessment: Part 1—Morphology of the model and selection of substantial air quality impact sub-models. Applied Sciences, 9(18), 3918. DOI:10.3390/app9183918
[12] EN 15251:2007. (2007). Indoor environmental input parameters for design and assessment of energy performance of buildings addressing indoor air quality, thermal environment, lighting and acoustics
[13] EN 16798-1:2019. (2019). Energy performance of buildings- Ventilation for buildings- Part 1: Indoor environmental input parameters for design and addressing indoor air quality, thermal environment, lighting and acoustics- Module M1-6.
[14] EN 13779 Standard: Ventilation for non-residential buildings — Performance requirements for ventilation and room-conditioning systems
[15] ASHRAE Standard 62.1-2010. (2010). Ventilation for Acceptable Indoor Air Quality

DOI: 10.2507/33rd.daaam.proceedings.009

ASSESSMENT AND ANALYSIS OF WIND ENERGY POTENTIAL IN THE MOUNTAIN AREA AROUND SARAJEVO

Rejhana Blazevic, Halima Hadziahmetovic,
Sanda Midzic Kurtagic & Mirela Alispahic

This Publication has to be referred as: Blazevic, R[ejhana]; Hadziahmetovic, H[alima]; Midzic Kurtagic, S[anda] & Alispahic, M[irela] (2022). Assessment and Analysis of Wind Energy Potential in the Mountain Area Around Sarajevo, Proceedings of the 33rd DAAAM International Symposium, pp.0058-0067, B. Katalinic (Ed.), Published by DAAAM International, ISBN 978-3-902734-36-5, ISSN 1726-9679, Vienna, Austria
DOI: 10.2507/33rd.daaam.proceedings.009

Abstract

The main goal of this paper is to evaluate and discuss the energy potential of wind in the mountain area around Sarajevo. The analysis of wind energy potential was performed based on measured wind data for Bjelasnica in a period from 2000 to 2017. Measured data for wind velocity were obtained from the Federal Hydrometeorological Institute of Bosnia and Herzegovina. The results of this investigation showed that the analysed place falls under Class 7 according to the international system of wind classification, as the mean long term wind velocity recorded in the analysed area was 11 m/s and the corresponding mean long term power density was estimated about 2013 W/m^2 at 10 m height. Taking into account the power coefficient of the wind turbine as well as the utilization factor of the turbine and generator, it is obtained that the mean long term electrical power density is about 624 W/m^2. For a better insight into the characteristics of the occurrence of high-wind velocity, the calculation of expected maximum wind velocity for a return period of 50 years was carried out. It was shown that for the return period of 50 years, the reference extreme wind velocity is 64.6 m/s.

Keywords: Weibull distribution; renewable energy sources; wind potential; wind power density; wind velocity

1. Introduction

Need to promote and research the possibility of using renewable energy sources has been imposed for a long time as a consequence of the energy crisis, which is reflected in the increasing prices of fossil fuels and the increasing need for additional amounts of energy. One of the strategic goals of the European Union (EU) energy policy is to encourage the use of renewable energy sources with the aim of reducing the negative impact on the environment as well as reducing dependence on energy imports. Renewable energy sources, especially wind energy and solar energy, are playing an increasingly important role in the energy industry considering efforts to reduce the negative impact on nature. According to data from the International Energy Agency (IEA), since 1990 global renewable energy sources have grown at an average annual rate of 2.2% [1]. Only in 2020 renewable energy use increased 3% and the main reason for that is growth in electricity generation from renewable sources. Renewable electricity generation in 2020 has been expanded almost 7% where solar PV and wind represent two-thirds of renewables growth [2].

The International Renewable Energy Agency (IRENA) has established a huge potential for renewable energy sources in the area of Southeast Europe. As part of the report [3], the great technical potential of renewable energy sources in the region, which amounts to 740 GW, was confirmed. It is estimated that 50% of the world's land surface has wind velocities greater than 6 m/s at a height of 10 m, while when it comes to the area of Bosnia and Herzegovina, 10% of the surface has wind velocities greater than 5 m/s at the same height. Fig. 1 and Fig. 2 show the wind atlas for the area of Bosnia and Herzegovina with the average annual wind velocity and average annual wind power density at a height of 100 m.

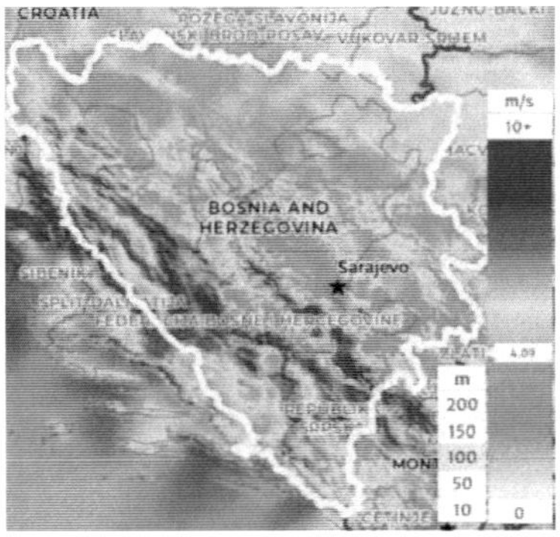

Fig. 1. Mean wind velocity in Bosnia and Herzegovina at 100 m height [4]

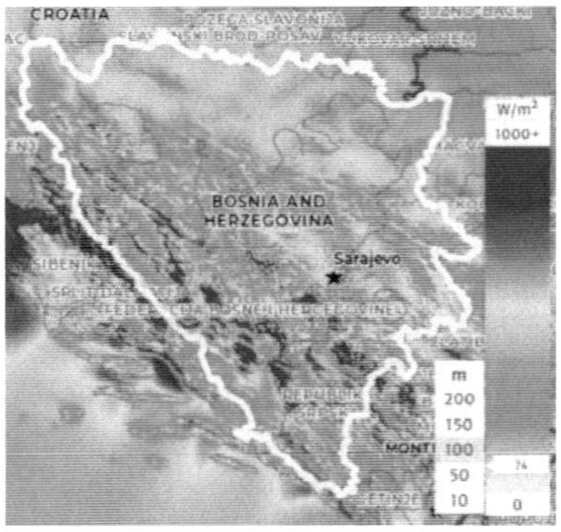

Fig. 2. Mean power density in Bosnia and Herzegovina at 100 m height [4]

From Fig. 1 and Fig. 2 it can be seen that the highest wind potential is located in the southwestern part of the country. Almost 50% of the territory of Bosnia and Herzegovina has a wind velocity of over 6 m/s. For 10% of the area with the highest wind potential, the mean wind power density is estimated to be 797 W/m^2, and the mean wind velocity is estimated to be 7,77 m/s at a height of 100 m, which represent good parameters regarding to qualitative assessment of the quality of wind resources.

Many articles have been published whose research topic was assessment and statistical analysis of wind velocity data. The most commonly recommended and used probability density functions in the analysis of wind velocities records are Weibull and Rayleigh distributions. For extreme wind analysis the Gumbel and generalized extreme value (GEV) distributions are also widely used. Yan et al. in their investigation [5] were analysed extreme wind velocities based on 40 years of the wind fields from the ERA5 dataset and parametric wind data in the northern South China Sea. They have been calculated extreme wind velocities by different distributions at three sites. The conclusion of their research was that the data have the greatest influence on the calculated result, followed by the distribution function and parameter estimation method. Furthermore, Parajuli [6] used a wider range of the wind data to estimate the wind power density based on Weibull and Rayleigh models of Jumla, Nepal. Similarly, the wind characteristics and wind energy potential on Hatiya Island in Bangladesh were also investigated. The wind characteristics of Antakya area in Hatay province were investigated and statistically analysed by Mert and Karakus [7]. For mountainous areas, which represent complex terrain areas, relevant investigation has also been carried out in the recent years. Kim et al. presented a methodology for evaluating suitable sites for wind farm development by identifying suitable ridges in mountainous terrain using morphometric analysis [8]. Medugu et al. (2020) analysed and discussed the wind resource potential available in the mountainous region at Mandara [9].

The most important feature of the wind potential at the analysed location is the mean annual wind velocity at a certain height above the ground. However, determining the mean annual wind velocity is only the first step in determining the wind potential [10]. After that, it is necessary to consider wind velocity distribution during the year. For example, the annual average may contain a large number of hours with a wind velocity above 30 m/s or below 3.5 m/s, which is actually not suitable for exploitation. In this paper, data obtained from the Federal Hydrometeorological Institute of Bosnia and Herzegovina related to the wind velocity in the area of Bjelasnica, for the period from 2000 to 2017, were analysed. The obtained data will be used to calculate the mean and maximum monthly and annual wind velocity values, as well as the mean wind power density for the mentioned period. Also, an assessment of the extreme wind velocity at the Bjelasnica location for different return periods will be made.

2. Materials and method

2.1. Meteorological station Bjelasnica

Bjelasnica is a mountain in the central part of Bosnia and Herzegovina, which belongs to the Dinaric mountain system, and is located southwest of the city of Sarajevo. Bjelasnica is covered with snow from November to May, and sometimes also in the summer months. Since the Mediterranean and continental climates meet in Bjelasnica, it is very interesting for meteorological observations. Meteorological station Bjelasnica is located at 43°43" latitude and 18°16" longitude, Fig. 3.

Fig. 3. Meteorological station Bjelasnica [11]

The meteorological station Bjelasnica started operating in 1895 and observes and measures temperature, precipitation, wind velocity, cloudiness, air pressure, relative humidity and insolation. Measurements at the station are made in term observations at 07:00, 14:00 and 21:00 during the day. Given that the measuring station is placed at an altitude of 2067 m above sea level, during the late autumn, winter period and early spring, the measuring poles and mounted equipment are exposed to low temperatures, snow blows and ice deposits. The average annual temperature in Bjelasnica is 0.7°C due to extremely low winter temperatures.

2.2. Wind potential analysis

The variation of wind velocity in a certain area is usually described by the Weibull and Rayleigh distributions. Weibull probability density function and cumulative density function as a function of wind velocity v are given as follows:

$$f(v) = \frac{k}{c}\left(\frac{v}{c}\right)^{k-1} \cdot e^{-\left(\frac{v}{c}\right)^k} \tag{1}$$

$$F(v) = \int_0^v f(v)dv = 1 - e^{-\left(\frac{v}{c}\right)^k} \tag{2}$$

Where k is the dimensionless shape factor, and c is the Weibull scale factor with units of velocity (m/s). Energy Pattern Factor Method was used to determine the parameters k and c. To find these parameters, the first thing to find is the energy pattern factor from the expression (3) [6], [7], [12], [13]:

$$E_{pf} = \frac{(v^3)_m}{(v_m)^3} = \frac{\Gamma\left(1+\frac{3}{k}\right)}{\Gamma\left(1+\frac{1}{k}\right)^3} \tag{3}$$

where $(v^3)_m$ represents the mean of the cube of wind velocity and $(v_m)^3$ represents the cube of the mean wind velocity. After calculating energy pattern factor, k and c can be determined by the following expressions [6], [7], [12], [13]:

$$k = 1 + \frac{3.69}{{E_{pf}}^2} \tag{4}$$

$$c = \frac{v_m}{\Gamma\left(1+\frac{1}{k}\right)} \tag{5}$$

Rayleigh's distribution is a simplified case of the Weibull distribution in which the shape factor is taken as $k = 2$. The general equation relating wind power density is [6], [7], [12], [13]:

$$\bar{P} = \frac{1}{2}\rho_m (v^3)_m \tag{6}$$

where $\bar{P}$ represents the mean wind power density, ρ is the air density values (a value of 1,225 kg/m^3 is used in this work), and v is the wind velocity.

When it comes to electricity production, only a certain proportion of the kinetic energy of the wind can be converted into electric. Electric power generated can be expressed as:

$$\bar{P}_{el} = \eta_{el}\eta_m C_p \bar{P} \tag{7}$$

where $\bar{P}_{el}$ represent the amount of electric power generated, η_{el} is the electric conversion efficiency of the wind turbine, η_m is the mechanical efficiency, and C_p is the power coefficient [12].

2.3. Assessment of extreme wind velocities

In order to make the correct selection and foundation of wind turbines, it is very important to consider as precisely as possible the extreme winds that can occur in the location where the construction of the wind power plant is planned [13].

The Gumbel distribution is often used to describe the statistics of extreme wind velocity values. Gumbel's function of probability density distribution of extreme wind velocities $F(v)$ is given by the following expression [13]:

$$F(v_R) = exp\left\{-exp\left[-\frac{v_R - b}{a}\right]\right\} \tag{8}$$

Where v_R is extreme wind velocity that occurs on average once every R year (usually said with a return period of R years), b is the location parameter, a is the scaling parameter.

The previous expression can be written in the following mathematical form:

$$v_R = a \cdot \{-ln[-\ln(F(v_R)]\} + b \tag{9}$$

Parameters a and b are determined by linear regression, i.e. by the method of least squares based on multi-year data on extreme wind velocities.

The relationship between the return period R of the occurrence of extreme wind velocities v_R and the corresponding probability is given by the following expression:

$$R = \frac{1}{1 - F(v_R)} \tag{10}$$

Combining relations (9) and (10), an explicit expression can be obtained for the calculation of the reference wind velocity v_R which occurs on average once every R year:

$$v_R = -a \cdot ln\left[-ln\left(1 - \frac{1}{R}\right)\right] + b. \tag{11}$$

The reference wind velocity v_R can be calculated according to the expression (11) for any return period. Of a particular interest for the selection of wind turbines is the return period of 50 years (assumed minimum lifetime of the structure). A return period of 50 years is prescribed as a standard in European norms that define the wind load on building structures [14].

The basic task of the statistical analysis of extreme wind velocities is to calculate the parameters a and b based on the available set of measurement data, and then, using expression (11), to calculate the requested reference extreme wind velocity. There are several methods used to estimate the coefficient of the Gumbel statistical function, one of the most common methods is the annual maximum method [13]. This method requires wind velocity measured data for a period of at least ten years. From the data set for each year, which is available for 365 days, the time interval in which the maximum wind velocity occurred in that year is selected. That data is declared as extreme velocity in that year. Such a procedure is carried out for all years in the available set of measurement data covering N years, and based on this, a corresponding table with extreme values of wind velocities is formed [13]. After creating a table with extreme values of wind velocities, it is necessary to sort these data by placing them in descending order and each type is assigned the corresponding rank m=1,2,..,N. For each type, the corresponding value of the cumulative probability distribution function can be estimated according to the following relation [13], [15], [16]:

$$F(v_R) = \frac{m}{N + 1}. \tag{12}$$

In practice, it has been shown that a lower dispersity of points, i.e. a better fit, can be achieved if instead of relation (12) relation (13) is used, which was proposed by Gringorten (1963) [17], [18]:

$$F(v_R) = \frac{m - 0{,}44}{N + 1 - 0{,}88} = \frac{m - 0{,}44}{N + 0{,}12}. \tag{13}$$

3. Results and discussion

One of the most important parameters in determining the energy power obtained from the wind is wind velocity. Fig. 4 and 5 show the mean monthly wind velocity values as well as the extreme wind velocity values that occurred at least one day during the considered period.

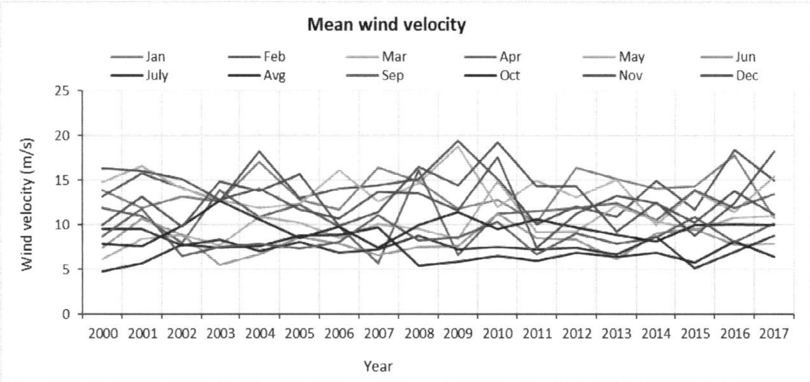

Fig. 4. Mean monthly wind velocities in the area of Bjelasnica for the period from 2000 to 2017

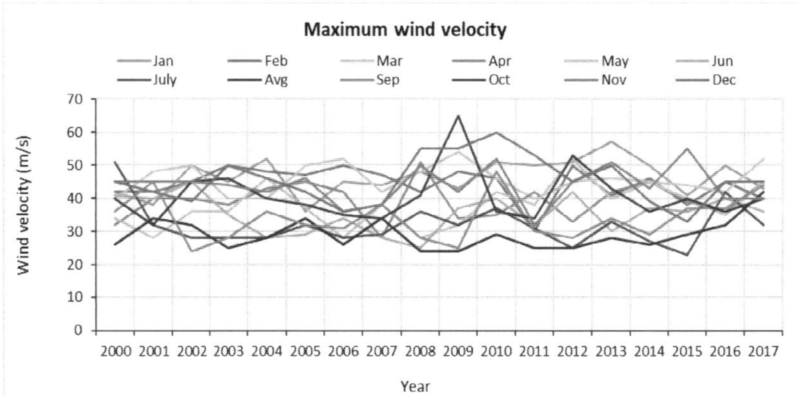

Fig. 5. Maximum monthly wind velocities in the area of Bjelasnica for the period from 2000 to 2017

From Fig. 4, it can be concluded that the values of the mean wind velocity range from 4,8 m/s to 19,4 m/s, which is a sufficient condition for the continuous operation of the wind turbine and the production of the maximum amount of electricity. On the other hand, when it comes to the maximum wind velocity, it ranges from 23 m/s to 65 m/s, which occurred in October 2009 and only for one day. When extreme wind velocities occur, i.e. velocities that exceed the permitted threshold, which usually ranges from 25 m/s to 30 m/s, it is necessary to ensure that the operation of the wind turbine is stopped in order to prevent damage to some of its parts. Maximum wind velocities that exceed the value of 30 m/s occur every year, however, when looking at the number of days during the year when these extremes occur, it is an average of 60 days a year. Which means that for the remaining 305 days of the year, the maximum wind velocities do not exceed 30 m/s.

When assessing the long-term potential of wind in a certain area, it is not enough to include only seasonal variations in wind velocity, because it is necessary to take into account that every year is not equally windy. For this reason, it is very important to take into account the variability of wind energy potential at the analysed location from year to year. Fig. 6 and 7 show the mean annual wind velocities as well as the deviation of the mean annual wind velocities from the long term average wind velocity in the Bjelasnica area.

From Fig. 6 and 7, it can be concluded that the measurements carried out during one year cannot be a reliable representative for the long-term period for which the wind power plant is designed. For example, for the analysed location, if the conclusion about the wind potential was made on the basis of the measurements carried out in 2011, the estimate of the wind energy potential would be lower than the long term average, because the average annual wind velocity in that year was about 12% lower than the long term average wind velocity. If 2010 were to be considered as the reference year when assessing the wind potential, since the average annual wind velocity in that year was about 12% higher than the long term average, the wind potential would be overestimated.

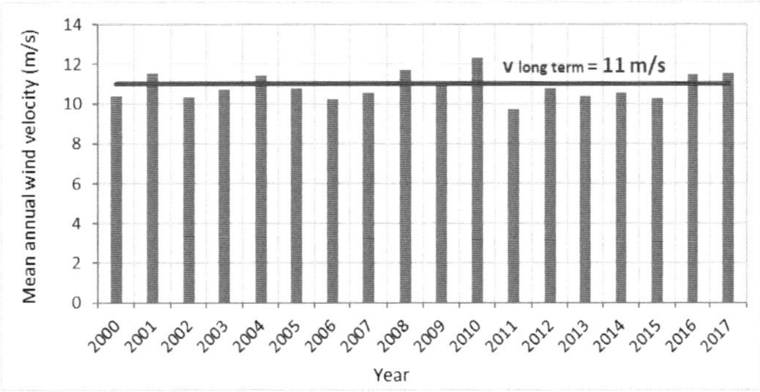

Fig. 6. Mean annual wind velocities in the area of Bjelasnica for the period from 2000 to 2017 with long term wind velocity indicated

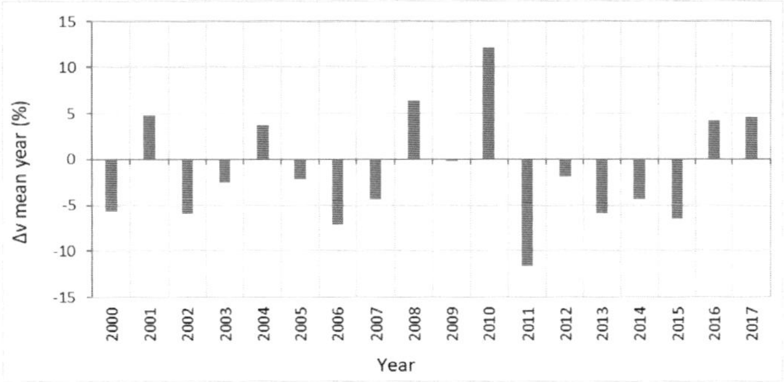

Fig. 7. Deviation of mean annual wind velocities from the long term average wind velocity in the area of Bjelasnica

Different types of wind dominate in Bosnia and Herzegovina, depending on the climate area. The shape of the wind rose (frequency, direction and mean wind velocity) depends on the orography of the terrain, so it is different for each station. The wind frequency rose and the mean wind velocity rose for the weather station Bjelasnica for the period 2000 to 2017 are shown in Fig. 8. From Fig. 8 it can be seen that the prevailing wind directions are north (N) in 31,2% of cases with a calculated wind velocity of 12,1 m/s, south (S) with a share of 16,5% and south-southwest (SSW) with a share of 16,8%.

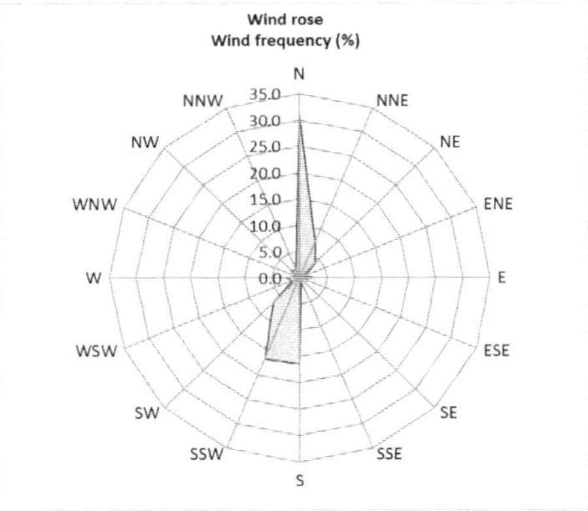

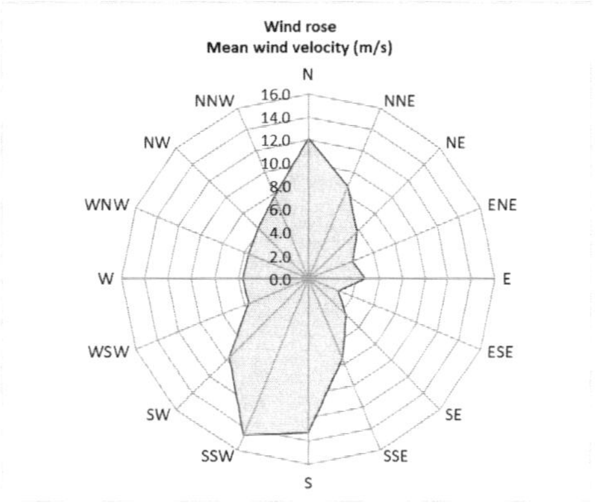

Fig. 8. Wind rose for meteorological station from 2000 to 2017 in the area of Bjelasnica

The long term average wind velocity is the basic indicator of the wind energy potential at a location, however, it is not sufficient to estimate the possible electricity production of a wind turbine. It is very important to know the wind velocity distribution, i.e. how long the wind blows at a certain intensity. This information is provided by the histogram of wind velocity duration. Fig. 9 shows a normalized histogram of wind velocities for the Bjelasnica area, obtained on the basis of eighteen-year measurements of daily wind velocities.

At the analysed location, according to the Fig.9, the most frequent wind velocity is around 4 m/s, i.e. 7,06% of the time (about 464 days during the analysed period of 18 years, i.e. an average of 26 days per year), and velocity of 6 m/s, i.e. 6,9% of the time (about 455 days during the analysed period of 18 years, i.e. an average of 25 days per year). The total duration of silence ($v \leq 0{,}5\ m/s$) is 0,58%, that is, about 38 days in the analysed period of 18 years, which is an average of 2 days per year. One of the ways of statistical representation of the wind potential is the arranged diagram of the cumulative duration of wind velocities. Fig.9 shows, also, the normalized diagram of the cumulative duration of wind velocities obtained on the basis of eighteen-year measurements of wind velocities in the analysed area at a height of 10 m above the ground. The cumulative probability distribution of wind velocities shows the percentage of time or probability that the wind velocity is less than or equal to v.

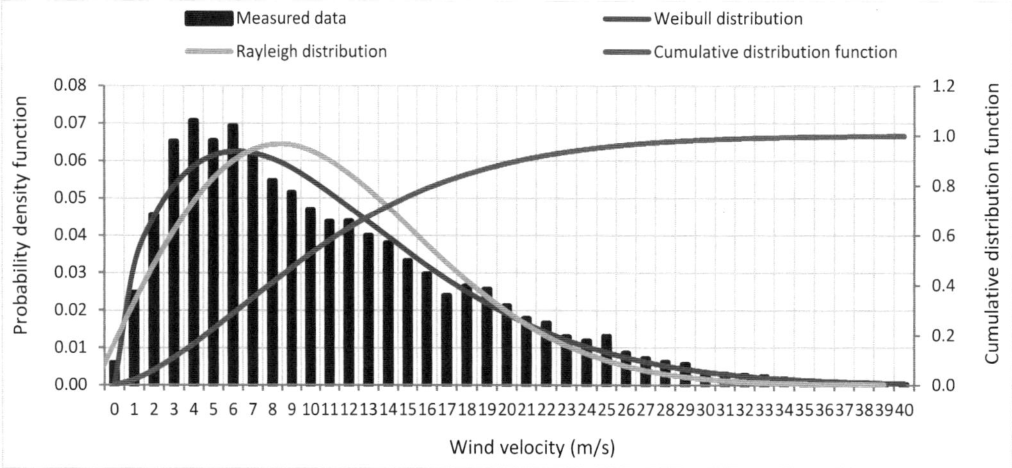

Fig. 9. Probability density distribution and cumulative distribution function of the of annual wind velocities

From Fig. 9 it can be seen that approximately 90% of the recorded wind velocities are below 21 m/s, and the frequency of occurrence of wind velocities greater than 25 m/s, i.e. the cut-out wind velocity is about 5%. The probability of occurrence of velocities below 3,5 m/s is about 10%.

For assessment the maximum theoretical wind power density, it is necessary to take into account the power coefficient of the wind turbine, and maximum utilization rate of turbine and generator which is about 0,65 and 0,8, respectively, to determine the maximum electrical wind power density. Taking into account all the above coefficients, it is possible to use about 31% of the total wind energy for electricity production. Fig.10 show the mean wind power obtained from the wind flow, the maximum theoretical power that can be obtained taking into account the upper limit for the power coefficient and electric power depending on the wind velocity for an area of 1 m^2.

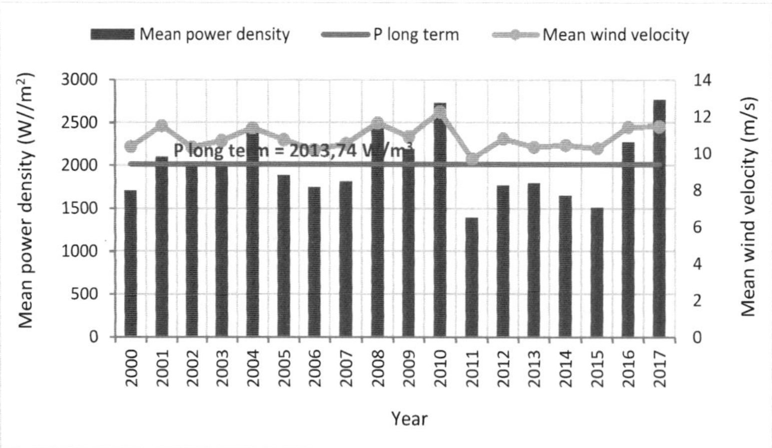

Fig. 10. Yearly variation of the mean power densities and mean wind velocity

Based on the calculated maximum theoretical wind power density in the Bjelasnica area for the period from 2000 to 2017, it is obtained that the long term average theoretical wind power density is 1194 W/m^2, while the long term average electrical wind power density is 624 W/m^2.

Considering that wind turbine safety design and wind turbine selection are primarily based on the annual average wind velocity, turbulence intensity and extreme wind condition it is very important to look more precisely at the extreme winds, that can occur at the location, where the construction of the wind power plant is planned. Table 6 shows the number of days with strong, stormy and strong stormy wind at the analysed location.

Year	Date of appearance	Maximum annual wind velocities (m/s)	Direction of maximum wind	Number of days with strong wind (10,8-17,1 m/s))	Number of days with stormy wind (>=17,2 m/s)	Number of days with strong stormy wind (>=25 m/s)
2000	13.07.2000.	51	N	265	189	100
2001	04.03.2001.	48	S	309	220	127
2002	06.01.2002.	50	N	280	184	105
2003	04.02.2003.	50	SSW	285	186	114
2004	17.01.2004.	52	SSW	285	193	118
2005	11.03.2005.	50	SSW	284	183	117
2006	05.03.2006.	52	SSW	292	170	80
2007	13.02.2007.	47	N	278	173	104
2008	01.12.2008.	55	SSW	295	210	122
2009	13.10.2009.	65	S	293	188	114
2010	06.12.2010.	60	SSW	305	229	152
2011	16.12.2011.	53	SSW	283	170	84
2012	27.10.2012.	53	S	325	200	107
2013	25.09.2013.	57	S	297	194	130
2014	26.01.2014.	50	NE	313	204	119
2015	20.11.2015.	55	S	308	196	114
2016	11.01.2016.	50	SSW, S	305	211	138
2017	05.03.2017.	52	SSW	296	192	144

Table 1. Number of days with strong, stormy and strong stormy wind at the analysed location

Based on the available measurement data for N=18 years, data and calculated values of statistical parameters of extreme wind velocities were sorted and presented in Table 1.

Year	Extreme wind velocity v_R (m/s)	Rank, m	$F(v_R) = \frac{m - 0,44}{N + 0,12}$	$-ln(-ln(F(v_R)))$
2000	47	1	0,031	-1,246
2007	48	2	0,086	-0,897
2001	50	3	0,141	-0,671
2006	50	4	0,196	-0,487
2003	50	5	0,252	-0,322
2002	50	6	0,307	-0,167
2009	50	7	0,362	-0,016
2008	51	8	0,417	0,135
2005	52	9	0,472	0,288
2004	52	10	0,528	0,447
2010	52	11	0,583	0,616
2011	53	12	0,638	0,800
2012	53	13	0,693	1,004
2013	55	14	0,748	1,238
2014	55	15	0,804	1,520
2015	57	16	0,859	1,882
2016	60	17	0,914	2,408
2017	65	18	0,969	3,461

Table 2. Statistical data for the calculation of Gumbel parameters of extreme wind velocities at the location of Bjelasnica using the annual maximum method

Based on the data from Table 2, more precisely the data from the second and last column, a linear regression diagram can be drawn, Fig. 11, and determine the parameters of the linear regression using the method of least squares, which is mathematically defined by expression (9).

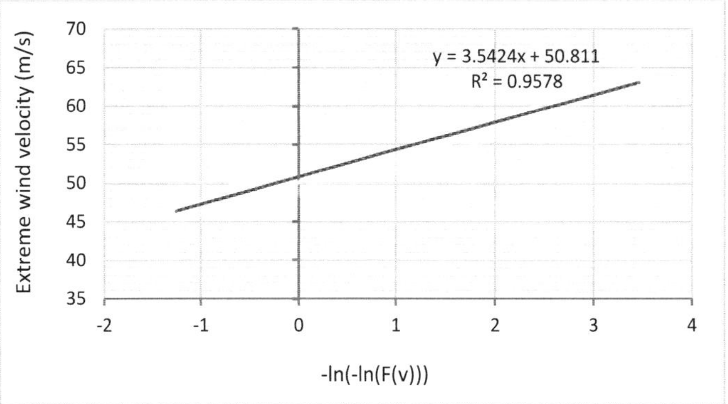

Fig. 11. Estimation of the parameters of the Gumbel distribution of extreme wind velocities using the annual maximum method for the analysed location

The analytical expression for the linear regression from Fig. 11 is:

$$v_R = -3{,}5424 \ln\left[-\ln\left(1-\frac{1}{R}\right)\right] + 50{,}811. \quad (14)$$

Using relation (14), the extreme value of the wind velocity for any return period can be calculated. Fig. 12 graphically shows the value of extreme wind velocity for the return period from 1 to 100 years.

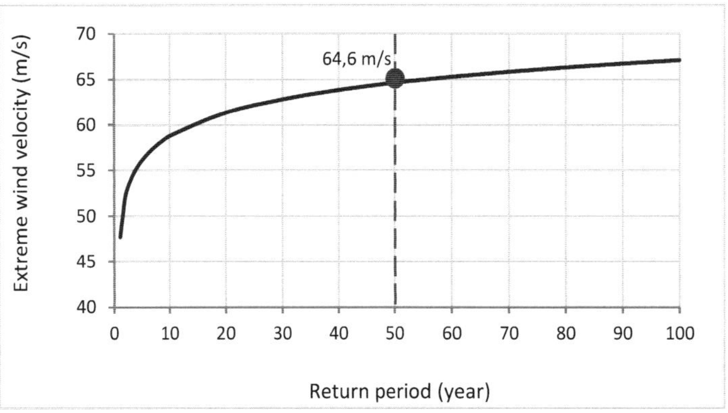

Fig. 12. Estimated extreme wind velocities at the Bjelasnica location for different return periods

For the return period of 50 years, the reference extreme velocity is 64,6 m/s, i.e., it can be expected that once in 50 years the wind velocity exceeds the value of 64,6 m/s, while the maximum average daily wind velocity that occurs on average once a year is around 47 m/s.

4. Conclusion

In this paper, it was analysed wind energy potential in the mountain area around Sarajevo. For that purpose, wind power density and wind velocity distribution parameters were investigated using wind velocity data obtained from the Federal Hydrometeorological Institute of Bosnia and Herzegovina. The measured data used to analyse the wind energy potential in this paper refer to an eighteen-year measurement period. The seasonal wind velocity profile was drawn and mean long term wind velocity was found to be 11 m/s at 10 m height. The mean wind power depending on the wind velocity for an area of 1 m^2 obtained from the wind flow was estimated to be 2013 w/m^2, the maximum theoretical power that can be obtained taking into account the upper limit for the power coefficient and electrical power were estimated to be 1194 w/m^2 and 624 w/m^2, respectively.

The results show that the available wind energy potential to generate electricity in Bjelasnica is high. However, due to the small number of observations it was not possible obtain comprehensive data for wind velocity. In this sense, it can be concluded that the available measurement data provide the possibility of roughly locating and estimating the potential of wind energy at the analyzed location, so they can serve for indicative purposes. Therefore, the assessment of possible electricity production can be considered as an indicative result that points to the existence of wind potential in the analyzed area, and not a reliable data on how much of this potential can be used. The next step of this work would be achieving more accurate results for wind power potential, and giving more insight of the power output reduction of wind farm under extreme wind condition. In order to achieve better and more accurate results it is necessary to have measured data in shorter time intervals, minimum 10-minute wind velocity values.

5. References

[1] https://euagenda.eu/publications/key-renewables-trends, (2016). International Energy Agency, Key Renewable Trends: Excerpt from Renewables Information, Accessed on: 2022-09-11

[2] https://www.iea.org/reports/global-energy-review-2021?utm_source=newsletter utm_medium=email&utm_campaign=newsletter_axiosgenerate&stream=top, (2021). International Energy Agency, Global Energy Review, Accessed on: 2022-09-15

[3] Ścigan, M.; Gonul, G.; Türk, A.; Frieden, D.; Prislan, B. & Gubina, A. (2017). Cost-Competitive Renewable Power Generation: Potential across South East Europe, IRENA.

[4] https://globalwindatlas.info/area/Bosnia%20&%20Herzegovina, Accessed on: 2022-09-20

[5] Yan, Z.; Pang, L. & Dong, S. (2020). Analysis of extreme wind velocity estimates in the northern South China Sea. Journal of Applied Meteorology and Climatology, 59(10), pp. 1625-1635, DOI: https://doi.org/10.1175/JAMC-D-20-0046.1

[6] Parajuli, A. (2016). A statistical analysis of wind velocity and power density based on Weibull and Rayleigh models of Jumla, Nepal. Energy and Power Engineering, 8(7), pp. 271-282, DOI: 10.4236/epe.2016.87026

[7] Mert, I. & Karakuş, C. (2015). A statistical analysis of wind velocity data using Burr, generalized gamma, and Weibull distributions in Antakya. Turkish Journal of Electrical Engineering & Computer Sciences, 23, pp. 1571-1586. DOI: https://doi.org/10.3906/elk-1402-66

[8] Kim, H. G.; Kang, Y. H. & Kim, J. Y. (2017). Evaluation of wind resource potential in mountainous region considering morphometric terrain characteristics. Wind Engineering, 41(2), pp. 114-123, DOI: https://doi.org/10.1177/0309524X166894

[9] Medugu, D. W.; Umar, A. S. & Waida, J. (2020). Evaluation of wind resource potential in mountainous region: A case study of Mandara mountains. International Journal of Sciences, 9(04), pp. 33-39, DOI: 10.18483/ijSci.2313

[10] Hadziahmetovic, H.; Dzaferovic, E.; Ahmovic, I.; Blazevic, R. (2019). Analysis of wind velocity data in the area of Bjelasnica in period from 2000-2010, Proceedings of the 30th DAAAM International Symposium, pp.0223-0231, B. Katalinic (Ed.), Published by DAAAM International, ISBN 978-3-902734-22-8, ISSN 1726-9679, Vienna, Austria, DOI: 10.2507/30th.daaam.proceedings.029

[11] Federal Hydrometeorological Institute of Bosnia and Herzegovina

[12] Shi, J. & Erdem, E. (2017). Estimation of wind energy potential and prediction of wind power - Chapter 3. Wind Energy Engineering: A Handbook for Onshore and Offshore Wind Turbines, pp. 25-49, Academic Press, DOI: https://doi.org/10.1016/B978-0-12-809451-8.00003-5

[13] Đurišić, Ž. (2019). Vjetroelektrane, Akademska misao, ISBN 978-86-7466-768-2, Beograd, Srbija.

[14] Holmes, J. D. (2007). Wind loading of structures. CRC press, London.

[15] Palutikof, J. P.; Brabson, B. B.; Lister, D. H. & Adcock, S. T. (1999). A review of methods to calculate extreme wind velocitys. Meteorological applications, 6(2), pp. 119-132, DOI: https://doi.org/10.1017/S1350482799001103

[16] Guo, S. L. (1990). A discussion of unbiased plotting positions for the general extreme value distribution. Journal of Hydrology, 121(1-4), pp. 33–44, DOI: https://doi.org/10.1016/0022-1694(90)90223-K

[17] Gringorten, I. I. (1963). A plotting rule for extreme value probability paper. Journal of Geophysical Research, 68(3), pp. 813-814, DOI: https://doi.org/10.1029/JZ068i003p00813

[18] Hannah, P.; Rainbird, P.; Palutikof, J. P. & Shein, K. (1996). Prediction of extreme wind velocities at wind energy sites. ETSU Report W/11/00427/REP, Energy Technology Support Unit, Harwell.

DOI: 10.2507/33rd.daaam.proceedings.010

PERCEPTION OF INDOOR AND OUTDOOR NOISE AT URBAN SETTLEMENTS

Mirela Alispahic, Halima Hadziahmetovic, Sanda Midzic Kurtagic, Rejhana Blazevic, Azrudin Husika & Kerim Druskic

This Publication has to be referred as: Alispahic, M[irela]; Hadziahmetovic, H[alima]; Midzic Kurtagic, S[anda]; Blazevic, R[ejhana]; Husika, A[zrudin] & Druskic, K[erim] (2022). Perception of Indoor and Outdoor Noise at Urban Settlements, Proceedings of the 33rd DAAAM International Symposium, pp.0068-0072, B. Katalinic (Ed.), Published by DAAAM International, ISBN 978-3-902734-36-5, ISSN 1726-9679, Vienna, Austria
DOI: 10.2507/33rd.daaam.proceedings.010

Abstract

This paper aims to assess the perception of residents about the level of noise caused by indoor and outdoor activities. The survey was also conducted to assess the level and type of negative emotions caused by noise. The survey included different groups of residents, as well as different types of buildings in which they live. 201 respondents were included in the survey, and the survey focused on outdoor sources of noise such as traffic, children's playgrounds, sports fields and cafes and restaurants, and indoor types of noise such as loud music, renovation of apartments, family quarrels and children's games. 72 respondents stated that they were most irritated by noise from construction work in neighbouring apartments, 51 of them cited arguments and noise from neighbouring apartments and 29 respondents mentioned loud music. However, the respondents express empathy towards children, because the noise caused by children's play does not bother them.

Keywords: indoor noise; outdoor noise; noise perception; acoustic comfort

1. Introduction

Noise pollution has been generally described as noise coming from road or railway traffic, industrial activities and public construction work as the most common sources of pollution, [1, 2]. However, larger cities are increasingly faced with the problem of noise arising from recreational activities located in city centres. More and more frequent and massive forms of socializing outdoors, in cafes, restaurants or parks, on walks, at music or sports events, especially in summer and at night. In addition to this problem, residents of buildings are also exposed to sources of noise resulting from various activities of their neighbours, especially in multi-story buildings. The level of indoor noise is recognized as a significant factor that affects the acoustic comfort, health and psychological state of building occupants, but the problem has not been sufficiently investigated.

Several recent studies dealt with this issue and the methodology of recreational noise research. Ljunggren *et al.* [3] investigated a possible correlation between sound insulation and subjective noise perception of residents. The research methodology was based on i) field measurements (noise caused by typing on a typewriter, noise manifested by the use of a rubber ball, according to ISO 16283-2), ii) characterization of buildings (in terms of materials and construction period) and a iii) questionnaire aimed at identifying the subjective perception of noise from neighbour's activities (noise caused by daily activities in the apartment: footsteps, moving furniture, etc.). Ottoz et al. [4] designed a questionnaire to study a socio-economic and health (self-reported health consequences of sleep) perspective of the subjective perception of outdoor noise by residents caused by recreational activities in Milan and Turin accompanied with a field measurement of a noise level. More recent works also treated the possible impact of forced stay in residential units during the COVID-19 quarantine on the subjective perception of noise and the connection with the acoustic comfort of the residents, [5], [6].

In Bosnia and Herzegovina, as well as in the Sarajevo Canton, the level of noise in buildings is regulated and refers to outdoor stationary and non-stationary noise sources. However, the issue of noise caused by sources inside residential buildings has not been the subject of research so far. Therefore, this paper aims to assess the perception of residents about the level of noise caused by outdoor sources of noise as traffic, children's playgrounds, sports fields and cafes and restaurants, and indoor types of noise as loud music, renovation of apartments, family quarrels, footsteps and children's games.

2. Questionnaire survey

The questionnaire was made of an introductory section including the information sheet and two more sections focusing on: basic demographic information (age, sex, educational level, basic employment information), respondents' home characteristics (individual family houses, old apartment buildings, renewed or newly built ones, number of apartments on the floor), sound sources information relevant for indoor soundscapes (traffic noise and other noise from outside) and personal description of different noise perception.

Information about the housing context that was collected ware: the ownership status, the house typology (single-family house, apartment block with two or more apartments on a floor), other people at home (children, spouse, parents, roommates), the number of people living at home. Also, the respondents were asked to identify the nearness of some recreational facilities in the area, such as children's playgrounds, stadiums and similar facilities, cafes and restaurants. They were requested to indicate whether they have to close the windows because of the noise from the street in the summer, as well as to state what they feel about it (irritation, annoyance, indifference or empathy).

In the following, the participants were asked to describe which of their neighbours' activities disturb them the most (home renovation, loud music, arguments or loud speech, footsteps from neighbours, children's noises or moving furniture).

2.1. Quantitative analysis of collected data

201 respondents participated in the online survey, 128 women and 73 men. The predominance was the working-age population: between 30 and 45 years (105 or 52.2 %), between 45 and 60 years (43 or 21.4 %) and only eight respondents were over 60 years old. The level of education of the respondents is above the statistical average in Bosnia and Herzegovina as follows: master's degree or doctorate for 37 respondents (18.4 %), university degree for 75 respondents (37.3 %), secondary school for 82 respondents (42.3 %) and the rest with a lower level of education (7 respondents in total). As for the respondents' age and educational distribution, this can be explained by implementing an online questionnaire because this methodology is closer to a younger and more educated population. Also, it can be confirmed based on the respondents' employment status: out of a total of 201 respondents, 145 were employed (71.6 %).

The majority of respondents own an apartment where they live (87.1 %), of which 43 % or 86 of respondents live in individual family houses, and the rest live in residential buildings. Also, almost 59 % of the respondents stated they lived in the same home for more than ten years, and 57 % of those said they lived there with more than two members, including children. 36.3 % of respondents have been living with someone else without children, and 6.7 % of those have been living alone.

The environment of respondents who lived in individual family houses has been generally described as quiet: 57 respondents from this group stated that their home is located further from the airport, main road, tram or railway, but also from children's playgrounds, restaurants and recreation centres, while only 26 of them pointed out that is located near the road. Similarly, 63 respondents replied that they were not exposed to any source of noise from human activities in neighbouring apartments or courtyards. It was noticed that there were no respondents in this group who reflected on their feelings about the noise that irritated them the most.

For the respondents in apartment buildings, the reflection ware quite opposing. Only 26 respondents from this group stated that their homes are located further from the airport, main road, tram or railway, and 22 said they are far from children's playgrounds, restaurants or sports centres. Also, 72 respondents stated they were exposed to renovation sounds from neighbouring apartments, followed by arguments (51 respondents) and loud music (29 respondents). As for the home typology, 75 % of the respondents stated they lived in multi-storey buildings with more than two apartments on the same floor, and almost 70 % said they lived in old apartment buildings.

So, the following discussion will be focused on this group of respondents.

2.2. Qualitative data analysis of observed group

The sound environment while being at home was reported to be dominated by sound generated by the renovation of neighbouring apartments, followed by outdoor sounds, by pets from neighbouring apartments and by sound generated by other human beings present at home (loud music and TV played, footsteps or moving furniture). Also, many respondents reflected on their perceptions of different sources of noise: "It irritates me I can't sleep and rest"; "I can't sleep, I have to turn up the television"; "The ceiling tolls from the footsteps in the bedroom"; "First I get nervous, and after all that drilling, a headache comes without warning"; etc. It was noticed that the words "sleep", "rest", and "bedroom" were frequently repeated in the comments left by the respondents. The previous answers could indicate the nature and consequences of the noise perceived at home, generated by other people's activities.

The majority of the respondents who mentioned complaints caused by noise from activities in other apartments lived in buildings with more than two apartments on the same floor, almost 96%. There was no significant difference between old buildings and newly.

Further, 60 or 52 % respondents replied that they have to close the windows in the summer, and 90 % were irritated with this, as detailed in Figure 1.

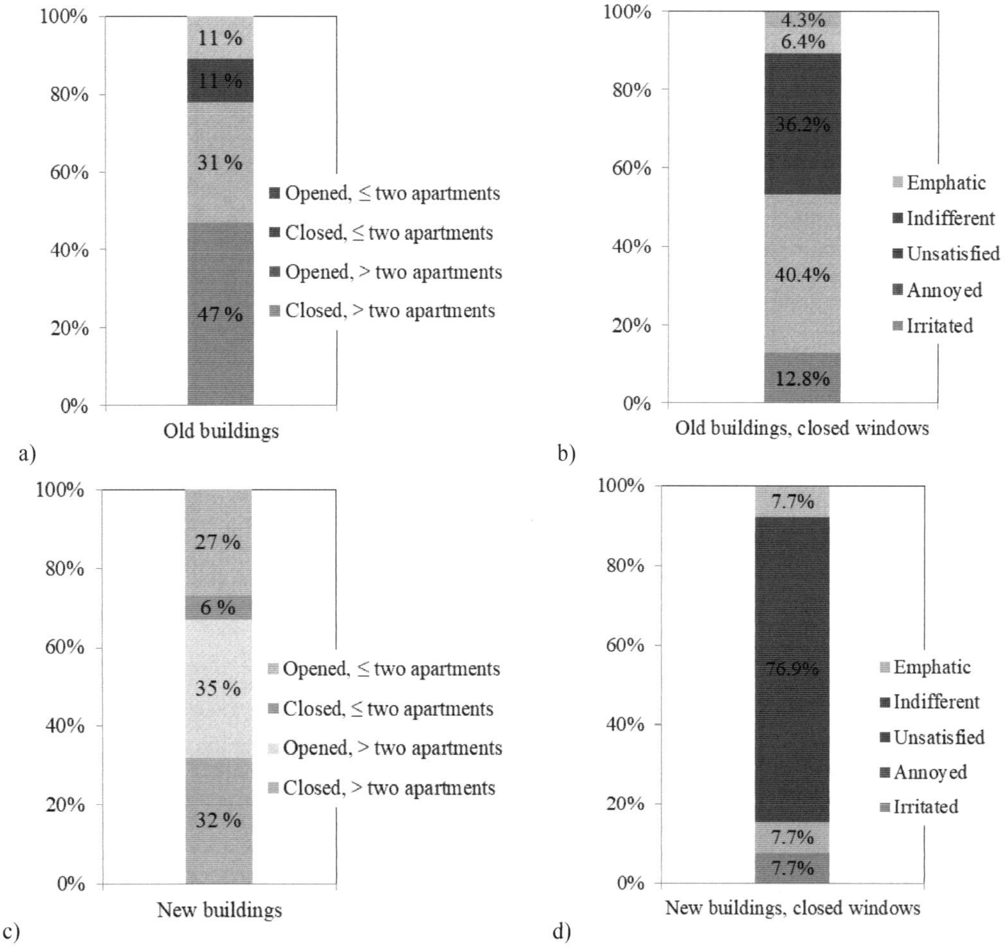

Fig. 1. Percentage values of responses to: a) closed windows in summer for the old buildings concerning the number of apartment units on the same floor; b) annoyance levels due to closed windows-old buildings; c) closed windows in summer for the new buildings concerning the number of apartment units on the same floor; d) annoyance levels due to closed windows-new buildings.

Furthermore, reported annoyance levels were compared across the socio-demographic variables (gender, age, educational level) and the building's characteristics. Reported indoor and outdoor noise annoyance levels differ significantly by respondents who lived in the building with more than two apartment units on the same floor, period of construction, educational level, and age group (test Anova, p-value > 0.05).

The results showed that the reported annoyance level is related to the level of education of the respondents, and it did not differ significantly according to gender, although females reported higher annoyance and anger ratings. Also, younger respondents showed higher empathy, especially those in the 18 to 30 age group. Respondents with a higher level of education showed more empathy towards loud music from the neighbours and noise from nearby cafes.

At the same time, they showed the lover empathy towards the noise generated by the apartment renovation, followed by the noise made by pets. However, they reported the level of annoyance as moderate (those irritancy ratings were associated with more than two apartments on the building's floor). Contrary to this, respondents with a lower level of education reported higher irritancy ratings. The respondents showed empathy towards noise associated with children, both from outdoor and indoor sources; there was no significant difference across the socio-demographic variables, although younger respondents showed slightly higher empathy.

Similarly, a survey in the UK showed that the most annoying neighbouring noise sources were voices, pets, and radio/television, whereas neighbours' footsteps and banging on walls or floors were the least annoying [7], [8]. In contrast, a study in Korea [9] found that the most noise irritation was associated with other people's footsteps, children running, moving furniture, and dropping small items, almost 78 %. At the same time, 3.7 % of respondents said that were irritated by renovation works in the neighbouring apartment (plumbing, hammering).

The outdoor and indoor noise exposure levels ranging from one to five in the preceding Figure 2 have been presented as exposure level 1: neutral/no exposure; level 2: slight; level 3: moderate; level 4: considerable; level 5: extreme based on respondents' reflection.

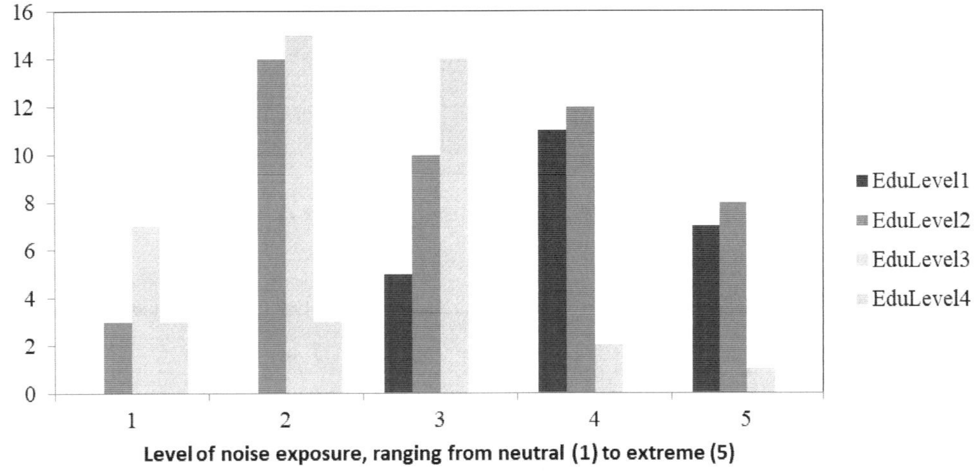

Fig. 2. The reported outdoor and indoor noise exposure levels ranging from one to five: EduLevel1 - master's degree or doctorate, EduLevel2- university degree, EduLevel3-secondary school, EduLevel4-lovest level

3. Conclusion

As already mentioned, the problem of exposure to indoor and outdoor noise sources is quite complex (it depends on a large number of parameters, both objective and subjective), so it is difficult to investigate it. As for recreational noise sources, it is related to specific urban areas, and exposure to it can be variable, depending on the period of the year when people's outdoor activities are more pronounced. Therefore, it is necessary to analyse the problem using a different methodology like online surveys and field research of the selected location and residential building. Current research is divided into two parts: the first part of the research concerning residents' perception is discussed in this paper, and the second part concerning field research is planned in the following.

The first part of the research was related to the online survey of residents living in apartment buildings and family houses. The purpose of the questionnaire was to determine whether there is a certain level of annoyance with the quality of life in apartment buildings, and it refers to outdoor (both usual and recreational) and indoor noise sources (possible sources of frustration arising from continuous exposure to noise generated by neighbours' everyday activities). The results of the survey showed that the residents are most disturbed by the noise generated by the renovation of the neighbouring apartments, the by sound generated by pets, and the least by loud music and noise made by children. However, many respondents indicated that noise disturbs their rest and sleep, which affects their health and quality of life. As for outdoor noise sources, the problem wares the proximity of the road and public places such as the markets. Children's playgrounds, cafes and restaurants were not indicated as sources of frustration.

The limitation of this type of research is reflected in the online questionnaire, which included younger groups of respondents with a higher level of education. The survey should be conducted in the field to include respondents of different socio-economical statuses, educational levels and ages. In the same way, the survey should include those respondents from locations in the city of Sarajevo, who are more exposed to sources of outdoor noise, originating from the activities of people.

In the following, it is planned to select a specific residential building and conduct a field analysis of the building's micro-location (it is necessary to specify where the building is located to identify possible sources of noise inside the building or outside the building: proximity to roads, railways, recreation facilities (stadiums, cafes, restaurants, children's playgrounds, markets and similar), analysis building characteristics (old building with or without renovation, newly built one, condition of the building envelope, number of floors in the building, number of apartments on the floor, and if possible acoustic characteristics of the space). Field measurements of indoor noise levels are also planned. Based on the information concerning both subjective and objective noise obtained in the analysis, it would be possible to estimate the actual situation and make appropriate recommendations accordingly.

4. References

[1] Badida, M.; Herczner, P.; Konkoly, J. & Lumnitzer, E. (2011). Annals of DAAAM for 2011 & Proceedings of the 22nd International DAAAM Symposium, Vol. 22, 1, B. Katalinic (Ed.), Published by DAAAM International, Vienna, Austria, DOI: 10.2507/daaam.scibook.2020.03

[2] Navrud S. The State of the Art on Economic Valuation of Noise," Report prepared for the European Commission, DG Environment, 1-38, available at Navrud S. 2010. "Economic Valuation of Transportation Noise in Europe," 34 Rivista Italiana Di Acustica 3; 2002. p. 15–25.

[3] Ljunggren, F.; Simmons, C. & Öqvist, R.. (2017). Correlation between sound insulation and occupants´ perception – Proposal of alternative single number rating of impact sound, part II. Applied Acoustics, Vol. 123, pp. 143-151, https://doi.org/10.1016/j.apacoust.2017.03.014

[4] Ottoz, E.; Rizzi, L. & Nastasi, F. (2018). Recreational noise: Impact and costs for annoyed residents in Milan and Turin. Applied Acoustics, Vol. 133, pp. 173–181, doi.org/10.1016/j.apacoust.2017.12.021

[5] Torresin, S.; Albatici, R.; Aletta, F.; Babich, F.; Oberman, T.; Stawinoga, A.E. & Kang, J. (2021). Indoor soundscapes at home during the COVID-19 lockdown in London – Part I: Associations between the perception of the acoustic environment, occupants activity and well-being. Applied Acoustics, Vol. 183, 108305, doi.org/10.1016/j.apacoust.2021.108305

[6] Torresin, S., Albatici, R., Aletta, F., Babich, F., Oberman, T., Stawinoga, A.E., Kang, J., 2022. Indoor soundscapes at home during the COVID-19 lockdown in London – Part II: A structural equation model for comfort, content, and well-being. Applied Acoustics, Vol. 185, 108379, doi.org/10.1016/j.apacoust.2021.108379

[7] Defra. Survey of noise attitudes (SoNA) 2013, Report Ref: 47067932.NN1501.R1/02. AECOM Infrastructure & Environment UK Limited; 2015

[8] Park, S. H., Lee, P. J., 2019. Reaction to floor impact noise in multi-storey residential buildings: The effects of acoustic and non-acoustic factors. Applied Acoustics, Vol. 150, p. 268–278, doi.org/10.1016/j.apacoust. 2019. 02. 021

[9] Park, S. H., Lee, P. J., Lee, B. K., 2017. Levels and sources of neighbor noise in heavyweight residential buildings in Korea. Applied Acoustics, Vol. 120, p. 148–157, doi.org/10.1016/j.apacoust.2017.01.012

DOI: 10.2507/33rd.daaam.proceedings.011

WORKFLOW AUTOMATION PATTERNS IN NEXT GENERATION MICROSERVICES ENVIRONMENTS

Bernhard Angerer, Markus Stopper & Branko Katalinic

This Publication has to be referred as: Angerer, B[ernhard]; Stopper, M[arkus] & Katalinic, B[ranko] (2022). Workflow Automation Patterns in Next Generation Microservices Environments, Proceedings of the 33rd DAAAM International Symposium, pp.0073-0076, B. Katalinic (Ed.), Published by DAAAM International, ISBN 978-3-902734-36-5, ISSN 1726-9679, Vienna, Austria
DOI: 10.2507/33rd.daaam.proceedings.011

Abstract

This paper explores the requirements, design and implementation challenges of workflow automation in the light of modern microservices based software architectures. It thereby points to specific design patterns and approaches which function as a guidance or rule book to successfully overcome the challenge of mapping workflows to multiple microservices with its different sizes and granularities in terms of providing functionality and the ability of recombination. In this mapping the top down view looks at end-to-end business processes in a holistic way and meets the bottom up view in which domains need to be defined under technical aspects like transactional security and the construction of interfaces to facilitate the communication and coordination between the microservices.

Keywords: Microservices; Workflow Automation; Business Process Modelling; Bounded Context; Domain Driven Design

1. Introduction

It is now state of the art to design and architect systems as an ensemble of services which run independently in virtual machines. These services then are glued together through a communication mechanism – typically through a message-bus which is provided by a separate server or through http/Rest communication endpoints which reside within the service. The objectives are better scalability, better modularity and maintainability as well as enhanced robustness of the overall system [1]. Concurrently the world of business process modelling (BPM) aims to deploy BPM engines which allow for a runtime execution of workflows thereby giving great flexibility in the world of workflow automation. Both worlds are now meeting each other for the first time and are supposed to converge and interact. A design approach is needed.

2. Description of the Problem

Up to this point the world of microservices is under intense debate and there is no general design approach existing which offers a reliable and incremental path of how to bring BPM engines and microservices together. In theory the creation of software through the combination and interaction of services creates a situation of flexibility because components had been identified which then can be mapped to the various "planes" (different abstraction levels) within the BPM flow charts. Since the business process modelling notation is standardized and is offering explicit graphical notations for specifying business processes it is perfectly complementary to the world of microservices.

Moreover, it could greatly utilize and help in the process of designing and operating microservices because it would bridge the world of process experts and software architects. In reality these two worlds speak very different languages on a human level as well as on a technical level.

The analysis, design and implementation of microservices are heavily determined by technical constraints. Transactional security and persistence in workflows or state-machines which encompass several services hence in-process and out-of-process communication are instantly facing a variety of new challenges which do not exist in a single process or monolithic system. A classic situation is a flight booking scenario where subsequent business transactions (the booking of a hotel or rental car) need to be cancelled as well if the main transaction (the flight) gets cancelled. In these situations so-called "compensate actions" need to be implemented where "long-lived" transactions which spawn several services need to be rolled back [3]. Another challenge in the design of microservices but especially when linking microservices and BPM is functional decomposition. Software architect usually decompose the system by functions [4]. This works well for the implementation of technical components like an email or a printer service. In the world of business transactions and workflows this can be described as an anti-pattern. The different nature of (long living) business transaction will inevitably collide with technological need of a transaction creating a reliable dialog with a specific resource (e.g. a database or a device).

3. Suggested Solution

A generic design pattern must be established which enables the mapping and linking of BPM workflows with specific services provided by one or several microservices. Moreover, in order to achieve agility and flexibility this technique must inherently support an incremental process where refactoring and redesign are first class citizens. The following characteristics need to be met:

- Orchestration needs to be at the core of the microservices design
- "Manager" services as docking facilities into the BPM world
- Coarse granularity of the services
- Volatility based decomposition of the services

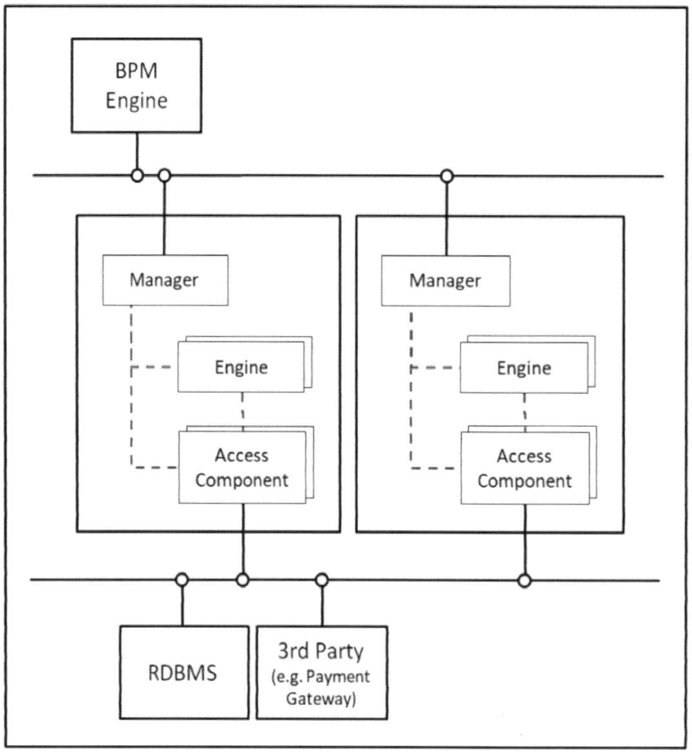

Fig. 1. The principle of the workflow automation pattern in microservices architectures

Orchestration aims to achieve automated configuration, management and coordination of computer systems, applications and services. In the proposed method orchestration needs to be implemented on a macro- (run-time orchestration through the BPM engine) as well as on a micro- (design-time orchestration due to implementation details of the microservice) level.

Figure 1 shows the principle of this approach through the decomposition of a service into managers, engines and access components. Access components simply abstract a specific resource like a database or a 3rd party component. Engines contain re-useable business logic. Managers implement the needed functionality through orchestrating the functionality provided by one or many engines and access components. These three types of components constituting a microservice reside within the same process and transactional context. The manager components are responsible of exposing the functionality of the microservice to the outer world through a network interface (e.g. a Rest API) hence they are the only "speaking partner" for the BPM engine. This overall approach leads to a coarse granularity of microservices and therefore greatly reduces complexity and works in favour of the system's communication topology [5].

The proposed design pattern also entails the decomposition method which functions as a guideline on which functionality should be wrapped into a manager (and therefore be contacted by the BPM engine) in the first place. Functional decomposition will collide with the business process context of the BPM workflows. The solution is the identification of these parts of the system which are most likely to undergo repeated change. The so-called "volatility based decomposition" identifies those domains in the system and tackles them with the manager-, engines-, access-component patterns as described above. It clusters all the areas of high volatility within the manager components. This way new functionality can be introduced more easily through design-time orchestration as well as run-time orchestration with the help of BPM.

4. Further Challenges

With microservices architectures the industry stepped into the world of distributed systems and thereby facing a whole new set of challenges when designing, implementing and operating enterprise systems. The decoupling of the semantics of distributed computing from those of the problem domain has many aspects and facets. In the context of the design patterns described in this paper the following points stand out in terms of further research which needs to be conducted:

- The boundary of BPMN's execution semantics
- Visibility into the manager, engine and access components interplay

With the version 2.0 the Business Process Model Notation (BPMN) introduced "execution semantics" to further describe the inner workings of BPM elements [6][7]. Alongside came the capability to extend BPMN itself as well as extended capabilities of dealing with events. Further thoughts in this context are outlining a so-called Service Contract Implementation (SCI) Layer which consists of BPM planes for stateful and stateless integration processes [8]. This clearly overlaps with the domain of the manager components and the objective of volatility based decomposition. A clear distinction is needed which defines the boundaries of both worlds.

On the contrary there are no standards which could bring visibility and traceability into the design and operation of the manager- engine- and access-components. Currently the methods of object oriented programming with its SOLID principles are the only methods existing in terms of meta-descriptive structures. On this aspect additional research is needed to facilitate the creation of correct domain models with its interplay to the BPM layer to achieve process automation in a flexible and sustainable way.

5. Conclusion

The interconnection of BPMN engines with microservice environments is postulating the next logical step in process automation. In particular the human machine interface provided by BPMN when developing and operating microservices is brought to the next level and enables new kinds of capabilities which are untapped so far. The key to this is a holistic course of action which revisits every layer in the technology stack. A pure top down approach like historically implied by the BPM community is not sufficient.

This paper presented design patterns which form a practical approach for the BPM as well as the middleware and microservices community to fruitfully work together. Those patterns are needed to bridge the gap between both worlds and future BPM engines as well as microservices environments may actively support those patterns. In return this will help both communities to fashion opinionated approaches in their design methodologies.

6. References

[1] Jamshidi P.; Pahl C.; Mendonça N.; Lewis J. & Tilkov S. (2018). Microservices: The Journey So Far and Challenges Ahead, IEEE Software, https://ieeexplore.ieee.org/abstract/document/8354433, DOI: 10.1109/MS.2018.2141039

[2] Rademacher F.; Sorgalla J. & Sachweh S. (2018). Challenges of Domain-Driven Microservice Design: A Model-Driven Perspective, IEEE Software, https://ieeexplore.ieee.org/document/8354426, DOI: 10.1109/MS.2018.2141028

[3] Stefanko M.; Chaloupka O. & Rossi B. (2019). The Saga Pattern in a Reactive Microservices Environment, Scitepress, https://www.scitepress.org/Papers/2019/79187/79187.pdf, DOI: 10.5220/0007918704830490

[4] Tyszberowicz S.; Heinrich R.; Liu B. & Liu Z. (2018). Identifying Microservices Using Functional Decomposition, https://www.researchgate.net/publication/326901590_Identifying_Microservices_Using_Functional_Decomposition

[5] Beschastnikh I.; Liu P.; Xing A.; Wang P.; Brun Y. & Ernst M.; (2020). Visualizing Distributed System Executions, https://dl.acm.org/doi/abs/10.1145/3375633, https://doi.org/10.1145/3375633

[6] Lemrabet Y.; Clin D.; Bigand M. & Bourey P.; (2010). From BPMN 2.0 to the Setting-Up on an ESB – Application to an Interoperability Problem, Springer, Berlin, Heidelberg, https://doi.org/10.1007/978-3-642-15961-9_85

[7] Fdhila W.; Gall M.; Rinderle-Ma S.; Mangler J. & Indiono C. (2016). Classification and Formalization of Instance-Spanning Constraints in Process-Driven Applications, Lecture Notes in Computer Science(), vol 9850. Springer, Cham. https://doi.org/10.1007/978-3-319-45348-4_20

[8] Schäffer E.; Stiehl V.; Schwab P.; Mayr A.; Lierhammer J. & Franke J. (2021). Process-Driven Approach within the Engineering Domain by Combining Business Process Model and Notation (BPMN) with Process Engines, Procedia CIRP, https://www.sciencedirect.com/science/article/pii/S2212827121001037, https://doi.org/10.1016/j.procir.2021.01.076.

DOI: 10.2507/33rd.daaam.proceedings.012

IDENTIFICATION OF THE COW'S NIPPLES USING A 3D CAMERA WITH TOF TECHNOLOGY

Denis Shilin, Pavel Ganin, Dmitry Shestov, Andrey Novikov, Nikolay Grebenshchikov, Dmitry Pavkin, Semyon Ruzin & Sergey Yurochka

This Publication has to be referred as: Shilin, D[enis]; Ganin, P[avel]; Shestov, D[mitry]; Novikov, A[ndrey]; Grebenshchikov, N[ikolay]; Pavkin, D[mitry]; Ruzin, S[emyon] & Yurochka, S[ergey] (2022). Identification of the Cow's Nipples Using a 3D Camera with ToF Technology, Proceedings of the 33rd DAAAM International Symposium, pp.0077-0084, B. Katalinic (Ed.), Published by DAAAM International, ISBN 978-3-902734-36-5, ISSN 1726-9679, Vienna, Austria
DOI: 10.2507/33rd.daaam.proceedings.012

Abstract

The paper considers the use of a 3D camera with ToF technology to identify biological images in the form of a cow's udder. An algorithm has been developed that is autonomously able to recognize the desired point clouds related to individual nipples. The algorithm has been tested and can be used as a basis for further search for positioning points of the working bodies of the milking manipulator.

Keywords: Data analysis; Milking machine; 3D ToF Camera; Udder of a cow; Point cloud.

1. Introduction

A three-dimensional time-of-flight camera or 3D ToF is a type of non-scanning lidar (a device for detecting, identifying and determining range using light), which uses powerful optical pulses of several nanosecond durations to capture depth information (usually, at short distances) within the region of interest. Time-of-Flight cameras are most often used as follows: navigation in a confined space, gesture recognition, scanning objects, tracking objects, measuring volumes, monitoring the target area, fast accurate indication of the distance to the target, augmented reality, evaluation of the size and shape of objects [1], [2], [3], [4], [5], [6], [7], [8], [9]. So, the first smartphone with a ToF camera was the iPhone X, released in the fall of 2017. It is a key element of the FaceID system that scans the user's face.

In 2015, with the help of a drone and a camera with ToF technology, a three-dimensional image of the statue of Christ the Savior in Brazil was successfully created. Prior to this, an exact model of this statue did not exist. ToF technology is used not only on drones and telephones, but also in various areas, such as: logistics, surveillance and security, robotics, medicine, games, photos, film production, auto industry, archeology, environmental projects, and farm automation [10]. Sensors such as lidars (used in Lely milking robots), lidar flashes, 360-degree 3D lidars and TOF cameras (used in Lely, GEA, DeLaval milking robots) are built on TOF technology.

The principles of TOF depth measurement are divided into pulse modulation, CW modulation, and pseudo noise modulation. TOF are used when it is necessary to analyze the volume, shape of an object, to determine the position of an object in three-dimensional space relative to the world coordinate system. Use as a head module to control a self-propelled or self-propelled object. Choi [11] developed a navigation system for a combine harvester using pulse modulation technology. The system has been tested under static and dynamic conditions with 0.02 m and 0.07 m rms errors. Ifm [12] has developed an intelligent sensor with continuous wave modulation, where the emitter is located in a separate unit from the receiver. The sensor is specially designed for outdoor use, which does not react to interference such as sunlight, materials with various reflective effects (glare resistant). The system captures contour lines, which makes it possible to recognize up to 20 different objects at a distance of up to 35 m. In studies [13], a system was developed based on TOF technology (modulation of continuous waves), which was protected from wind and sunlight. The authors found the superiority of TOF over stereo imaging sensors, as well as low TOF resolution and small viewing angle. Studies [14] used a combination of a two-dimensional camera with TOF (continuous wave modulation) to assess apple yield. The authors were able to recognize up to 88% of apples, noting that a high percentage was provided by images from multiple angles. The remaining 12% were not identified due to the fact that they were covered with branches and leaves. Lidar systems are more often used to solve geodetic problems. Large corporations use lidars to develop self-driving cars. In studies [15] Akhloufi M.A it was found that the use of the lidar system for scanning the teat of a cow's udder is not 100% effective. In studies [16], images of maize plants were obtained using lidar. It was found that for the correct obtaining of the results, the angle of installation of the lens of the three-dimensional sensor in relation to the object under study is important. In studies [17], the authors used a three-dimensional lidar -360 ° as a navigation sensor. Factors such as reaction of cows to the presence of a robot, remote control and tracking were positively assessed. In the framework of this article, the question of using a 3D camera to detect a biological image in the form of nipples of an udder of a cow is considered. Automatic identification is necessary for automated milking of an animal.

2. Automated Milking Plants

In existing automatic milking systems of various manufacturers, including Lely, Gea Farm, DeLaval, there are drawbacks associated with alternately putting on milking cups with a long time interval, which contribute to reducing the release of oxytocin due to delays in the stimulating effect on the nipple receptors, which leads to to incomplete elimination of milk from the alveoli. Moreover, the advantage of such systems is the absence of need to search and analyze all biological objects at the same time. The relevance of this work is the development of an identification algorithm for all the nipples of a cow with the aim of further analysis and the formation of positioning points for the working bodies of the milking robot, the kinematics of which allow positioning at all desired points at the same time.

Typically, such installations consist of multi-link manipulators driven in motion by a pneumatic or electric motor. In the work of Dorokhov A., et al. the question is just being considered of choosing the kinematic model of such manipulators and the location of the drive equipment to ensure the required reachable areas of the working body [18]. The solution of the kinematics problems is considered by Ganin, P.E., et al. [19]. The authors developed an algorithm for solving kinematics problems, which provides for scaling when changing the design parameters and the number of manipulator links. The search for a solution to kinematic problems is based on a neural network approach in combination with numerical refinement algorithms.

An important part of the automatic milking system for cows is the automatic nipple recognition system for the cow. This task is not effectively solved with the help of technical vision, since in this case it is necessary to know for sure the necessary threshold value for the preparation of the original image in order to detect the desired object, so Figure 1 shows the result of the preparation of the original image.

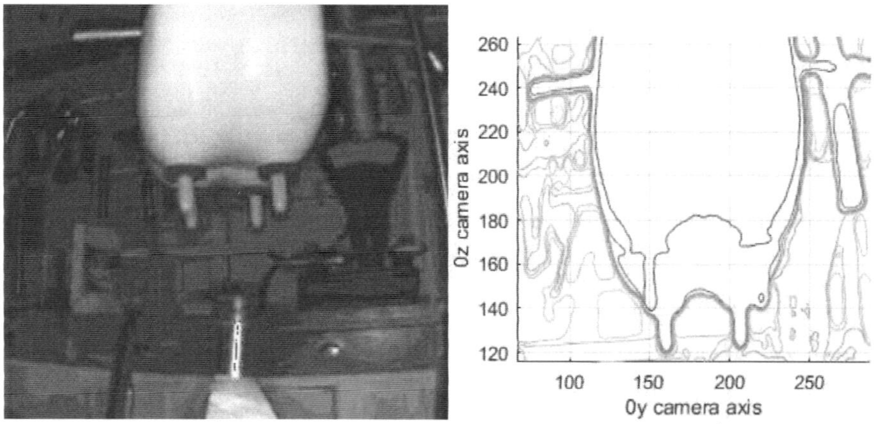

a b

Fig. 1. The result of the preparation of the original image: a) the original image, b) the binarized image with various threshold values

As you can see, when a threshold value changes in a certain range, they lead to the formation of many contours; accordingly, not knowing the exact threshold value for a particular case makes it impossible to accurately determine the coordinates of the desired object [20]. Nevertheless, there are algorithms where the aforementioned problem is solved by introducing intuitive parameters that provide a compromise between undersegmentation and excessive segmentation [21], [22] or segmentation of objects using the evaluation method core density, after which the points are divided into objects [23]. Nevertheless, when controlling robotic systems, it is recommended to use 3D ToF cameras as feedback [10].

3. Description of the experimental laboratory stand

The testing of the developed algorithms was carried out at the laboratory bench shown in Figure 2 in the 0zx plane. A plastic udder model was used as the object under study, and a 3D ToF camera was used as a camera. There are a huge number of cameras with ToF technology, their detailed comparison is presented by Deckers, N. and Reulke, R. [24]. In this work, we used IFM O3D303 capable of capturing an image in the format [264,352] in real time with a frequency of 25 Hz. The device has a working range from 0.3 m. to 8 m. And provides distance information in real time. A detailed discussion of the camera can be found in Bogaerts et al. [25]. Various techniques and techniques for processing signals from a ToF camera are disclosed in Kolb et al. [26]. Recognizing the objects in the image alone is not enough for the full functioning of the system, it is also necessary to know the location of the objects found relative to a certain reference frame. This is another reason to use an industrial camera with the ability to visually assess the distance - a 3D camera with ToF technology.

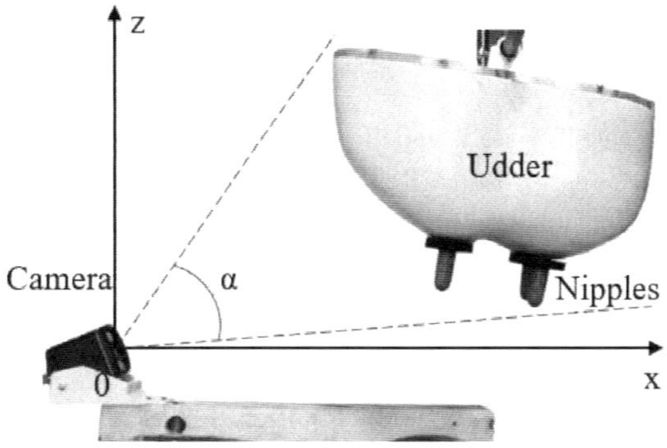

Fig. 2. The appearance of the laboratory bench in the 0zx plane

Knowing the aperture angles α (see Figure 2) and β along the axes 0x and 0y of the camera allow to build a point cloud in the space 0xyz. In order to automatically recognize biological objects in the form of nipples of the udder of a cow, it is necessary, first, to define a cloud of points of the udder itself.

4. Defining an udder point cloud

In the existing and developing milking machine robot, a 3D camera, which serves as feedback when controlling the positioning of the working bodies, is installed near the object, as shown in Figure 3. This arrangement ensures the location of the udder points in the center of the entire cloud of points, and at the closest possible distance to the camera.

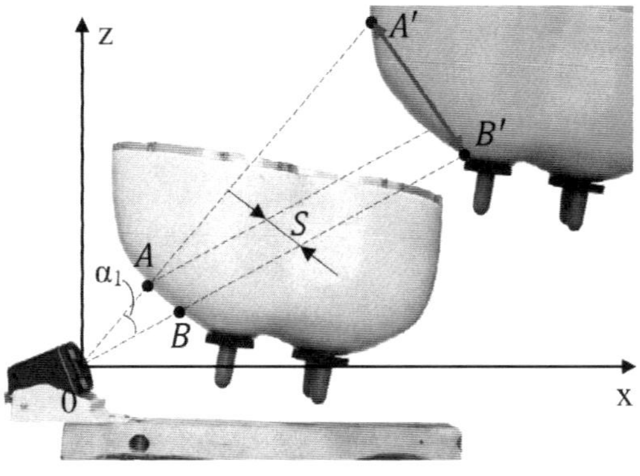

Fig. 3. The dependence of the distance between points on the location of the object on the plane 0zx

In Figure 3, points A and B are adjacent points, the distance **S** between them increases with increasing distance between the camera and the tracking object. According to the foregoing, an experimental range of values [0, **S**] was selected, which eliminated all points, the distance between which exceeded the permissible value of **S**. The range set by the operator [0, **S**] guaranteed no screening of the udder points. The distance between point A and B is calculated using the following equation:

$$\mathbf{S} = \sqrt{(\mathbf{x_A} - \mathbf{x_B})^2 + (\mathbf{y_A} - \mathbf{y_B})^2 + (\mathbf{z_A} - \mathbf{z_B})^2}, \quad (1)$$

where $(\mathbf{x_A}, \mathbf{y_A}, \mathbf{z_A})$ – coordinates of point A, m; $(\mathbf{x_B}, \mathbf{y_B}, \mathbf{z_B})$ – coordinates of point B, m.

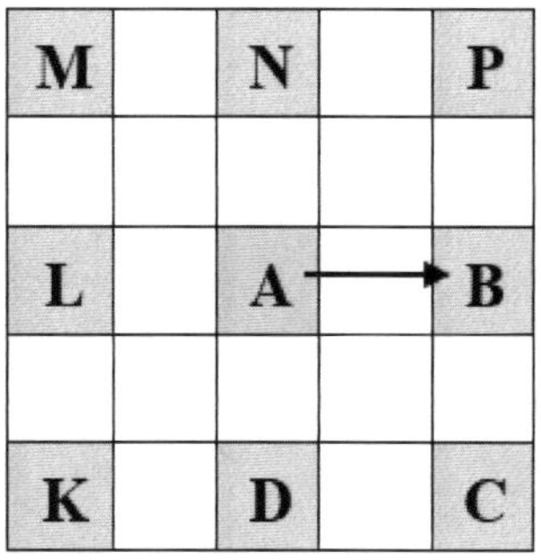

Fig. 4. The location of all adjacent points with point A

If the value of **S** is within the admissible value that is set by the operator, the entire series of points from A to B will enter the object and the verification of the point's belonging to the cloud is terminated, in this case, the cloud number that was indicated earlier at point B is assigned, or a new number is assigned (see figure 4). If the value of S is not in the acceptable range [0, **S**], then a similar verification of the distance to the point B, C, D, K, L, N, M, P follows.

Figure 5 shows an example of the work of the approach for searching point clouds. Checking points is carried out diagonally, provided that earlier the check was not carried out and not a single cloud of points was formed. For the first point, the admissible value of S was calculated to point B (see Figure 3), the second to the point D, the third to the point P, the fourth to the point L, and the fifth - the admissible value **S** was not found in all available options. Figure 5 shows that as a result of testing the algorithm, three-point clouds were found, the largest of which is marked with a red frame.

Figure 6 shows all the points that fit the specified requirement, where the purple color indicates the points that do not exceed the allowable value of S on the diagonals, blue - on the horizontal and vertical. Since the 3D camera is installed at the smallest distance to the udder, the largest point cloud (cluster of points) will be a point cloud consisting of cow udder points. Figure 7 shows in black the largest point cloud that was detected in the experiment.

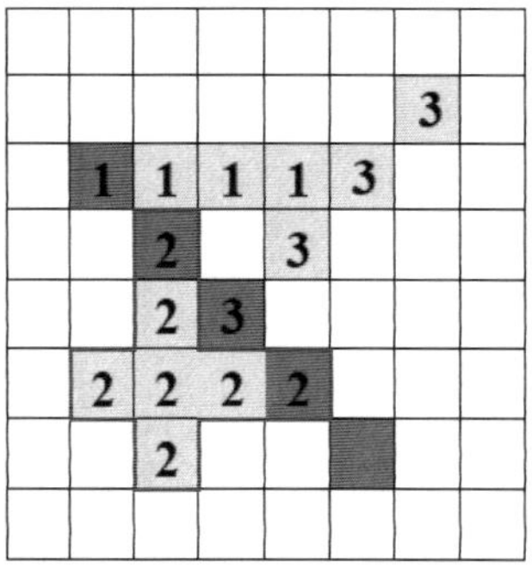

Fig. 5. An example of a point cloud search algorithm

For further analysis of the obtained cloud of the udder points of the cow, it is necessary to clarify with all the points that were not included according to the verification of permissible distances, but according to biological features probably related to this cloud of udder points. To do this, it is necessary to include in the cloud of udder points of the cow all the points from the original cloud of points that fall into the range $[\mathbf{x_{min}}, \mathbf{x_{max}}]$, $[\mathbf{y_{min}}, \mathbf{y_{max}}]$ and $[\mathbf{z_{min}}, \mathbf{z_{max}}]$, where $\mathbf{x_{min}}$, $\mathbf{y_{min}}$, $\mathbf{z_{min}}$, are the minimum values of the coordinates of the points along the x, y, and z axes, respectively, $\mathbf{x_{max}}$, $\mathbf{y_{max}}$, $\mathbf{z_{max}}$, are the maximum values of the coordinates of the points $\mathbf{x}$, $\boldsymbol{y}$, and $\mathbf{z}$ axes, respectively. Figure 7 shows in red a refined cloud of cow udder points.

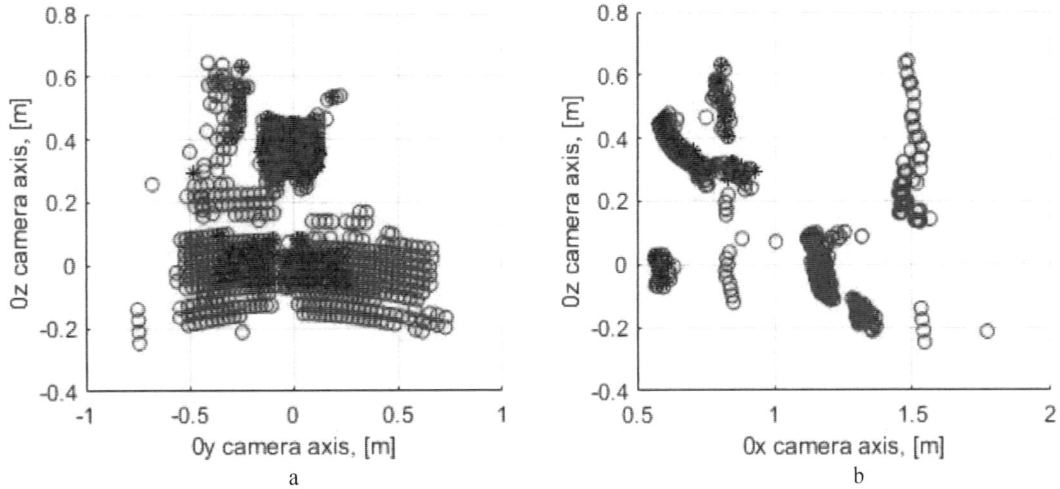

Fig. 6. The cloud of points, the distance between which does not exceed S: a) a front view, b) a side view

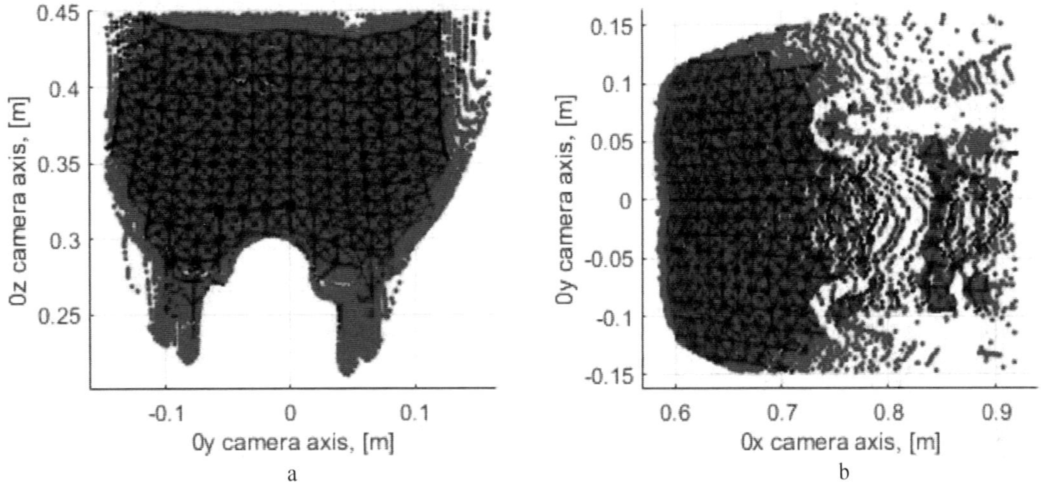

Fig. 7. The cloud of udder points: a) front view, b) top view

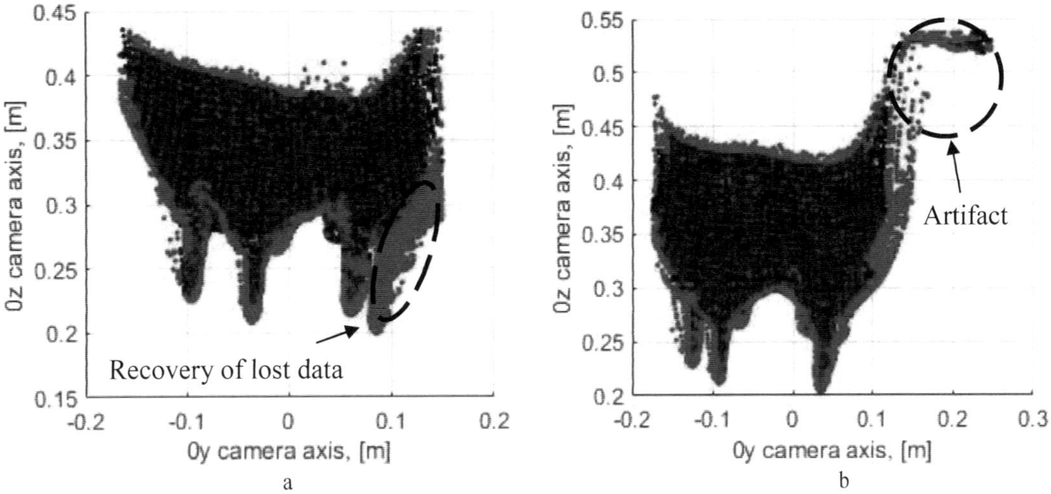

Fig. 8. Positive and negative impact of refinement of the point cloud: a) restoration of lost data, b) strengthening of artifacts

The aforementioned refinement has an important merit - for example, Figure 8a shows that after identifying the largest point cloud, information about the 4 nipples did not fit into the cloud, while the rectification of this cloud made it possible to fill in the lost data. In this case, there may be cases with a negative effect (see Figure 8b), when the point cloud contains artifacts.

5. Defining cow nipple point clouds

Knowing the biological features of the structure of the cattle udder and the generalized location of the nipples on the udder, we can conclude that of all the received points of the udder cloud of the cow, those points that are located below the udder, or with the smallest coordinate along the 0z axis, are of most interest. To eliminate unnecessary points, their total number was divided into 10 identical sections along the 0z axis to calculate them and determine the section with the largest content (see Figure 9). In graphic display (see Figure 9a), it is seen that the area in the center of the image is the most massive in terms of the number of points in it. Figure 9b shows the dependence of the number of points in a section on the number of sections where you can determine the exact section with the maximum number of points. Also, this section can be determined by taking the derivative of the function that describes this dependence, and equating it to zero:

$$\begin{aligned} &\mathbf{x} = -81{,}6 \cdot y^2 + 858{,}5 \cdot y - 505{,}2, \\ &\mathbf{x}' = -163{,}2 \cdot y + 858{,}5, \\ &\mathbf{x}' = 0, \\ &163{,}2 \cdot \mathbf{y} = 858{,}5, \\ &\mathbf{y} = 5{,}26 \rightarrow \mathbf{x} = -81{,}6 \cdot 5{,}26^{\,2} + 858{,}5 \cdot 5{,}26 - 505{,}2 = 1752 \text{ points.} \end{aligned} \tag{2}$$

In order to free up operational space, all data about points located on top is deleted, since they do not contain any information about the nipples.

By the identification of biological images is meant the detection of nipples and the formation of individual point clouds. In order to identify the desired objects, as in determining the cloud of the udder of a cow, the distances between the points were calculated by the equation (1). Next, a point cloud was determined, which includes the largest number of points and calculated its average location coordinates using the following equations:

$$\mathbf{x_{mid}} = \left[\sum_{i=1}^{n} \mathbf{x}\right] \Big/ \mathbf{n}, \mathbf{y_{mid}} = \left[\sum_{i=1}^{n} \mathbf{y}\right] \Big/ \mathbf{n}, \mathbf{z_{mid}} = \left[\sum_{i=1}^{n} \mathbf{z}\right] \Big/ \mathbf{n}, \tag{3}$$

where **n** – the number of points in the cloud, pcs; $(\mathbf{x_{mid}}, \mathbf{y_{mid}}, \mathbf{z_{mid}})$ – coordinates of the center of this point cloud, m.

When determining the subsequent biological images, it is also necessary to compare the found coordinates of the center of the current cloud so that they do not enter the zone described by the radius R relative to the center of another cloud of points (see Figure 10):

$$(\mathbf{x} - \mathbf{x'_{mid}})^2 + (\mathbf{y} - \mathbf{y'_{mid}})^2 + (\mathbf{z} - \mathbf{z'_{mid}})^2 > \mathbf{R}^2, \tag{4}$$

where (x, y, z) - coordinates of the center of the current point cloud, m; R is the radius described zone relative to the center of another cloud of points, m; $(\mathbf{x'_{mid}}, \mathbf{y'_{mid}}, \mathbf{z'_{mid}})$ - coordinates of the center of the previous point cloud, m.

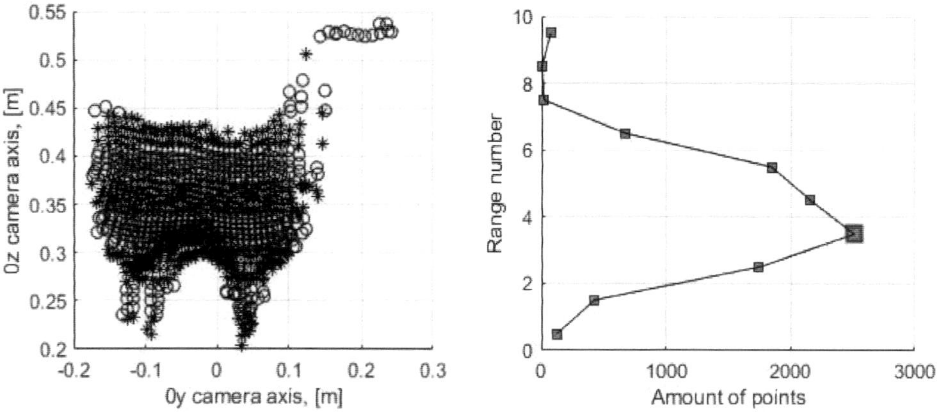

a b

Fig. 9. Search for an area with a maximum number of points: a) dividing a point cloud into equal sections, b) a graphical representation of the dependence of the considered point cloud segment on the number of points in it

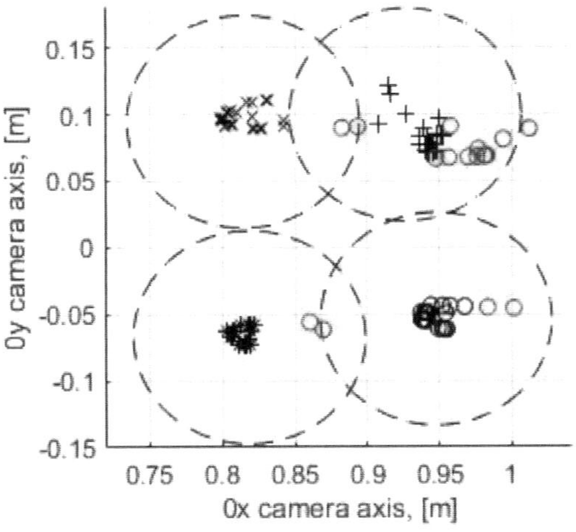

Fig. 10. Zones described around the centers of point clouds found

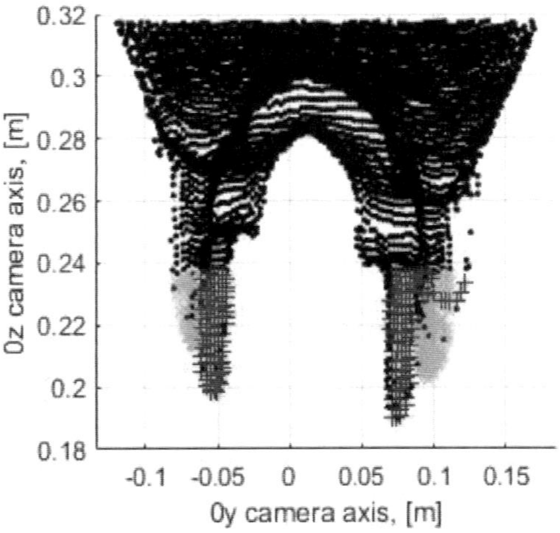

Fig. 11. Fixation of point clouds, characterizing each separately the nipple of the animal

Figure 10 shows the red points that are not included in any of the found point clouds. In Figure 11, on the 0yz plane, biological images are shown in red and green dots that satisfy the requirements of (4), and are the desired point clouds that characterize each individually nipple of the animal. As a further continuation of ongoing research, the authors propose the use of a methodology for finding the dimensions, orientation and position parameters of a parabolic ellipsoid, which was proposed by Min Dai, et al. [27]. It is assumed that this approach will determine the parameters of the elliptical paraboloid that describes the cow's nipple.

6. Conclusions

Based on the work done, the following conclusions can be drawn: This approach for identifying given biological images was tested and in 97.5% of all cases, clouds of nipple points of the cow were formed. Analyzing the work of the developed identification algorithm, it was noticed that the algorithm worked out too long, reaching up to 6 seconds. This is due to the fact that at the entrance to the developed algorithm a matrix with data of more than 90 thousand points arrives, which entails an increase in the processing time of the input information. To reduce operating space costs, it is recommended to reduce the number of points for analysis in those places where their number does not affect the content of the resulting point clouds. Although the experimental results demonstrated the reliability of the presented approach, further testing is required to determine the ability of the system to cope with various scenarios existing in natural conditions.

7. References

[1] Weingarten J., Gruener G., and Siegwart R. (2004). A state-of-the-art 3D sensor for robot navigation, in Proc. IEEE/RSJ Int. Conf. Intell. RobotsSyst., vol. 3, Sendei, Sep. 2004, pp. 2155–2160. DOI: 10.1109/IROS.2004.1389728.

[2] Kowalski, M., Piszczek, M., Szustakowski, M., Ciurapinski, W. (2012). Single-pixel camera - simulation tool and the first results of analysis of image quality. Proceedings of the 23rd International DAAAM Symposium, Volume 23, No.1, pp. 309–312.

[3] Hussmann S. and Liepert T. (2007). Robot vision system based on a 3D-ToFcamera, in Proc. 24th IEEE Instrum. and Meas. Tech. Conf., vol. 1-5, Warsaw, May 2007, pp. 1405–1409. DOI: 10.1109/IMTC.2007.379356.

[4] Dellen B., Aleny'a G., Foix S., and Torras C. (2009). 3D object reconstruction from Swissranger sensors data using a spring-mass model, in Proc. 4thInt. Conf. Comput. Vision Theory and Applications, vol. 2, Lisbon, Feb.2009, pp. 368–372.

[5] Foix S., Aleny'a G., Andrade-Cetto J., and Torras C. (2010). Object modeling using a ToF camera under an uncertainty reduction approach, in Proc. IEEE Int. Conf. Robot. Automat., Anchorage, May 2010, pp. 1306–1312. DOI: 10.1109/ROBOT.2010.5509197.

[6] Liu X. and Fujimura K. (2004). Hand gesture recognition using depth data, in Proc. 6th IEEE Int. Conf. Automatic Face Gesture Recog., Seoul, May 2004, pp. 529–534. DOI: 10.1109/AFGR.2004.1301587.

[7] Gokturk S. B. and Tomasi C. (2004). 3D head tracking based on recognitionand interpolation using a time-of-flight depth sensor, in Proc. 18th IEEE Conf. Comput. Vision Pattern Recog., vol. 2, Washington DC, June-July 2004, pp. 211–217. DOI: 10.1109/CVPR.2004.1315166.

[8] Nanda K., Fujimura H. (2004). Visual tracking using depth data, in Proc. IEEE CVPR Workshops, vol. 3, Washington, D. C., June 2004, pp. 37–37. DOI: 10.1109/CVPR.2004.202.

[9] Paredes, J.A., Álvarez, F.J., Aguilera, T., Aranda, F.J. (2020). Precise drone location and tracking by adaptive matched filtering from a top-view, Expert Systems with Applications, Vol. 141, 1 March 2020. DOI: 10.1016/j.eswa.2019.112989.

[10] Bourhane Kadmiry, Chee Kit Wong, P.P.K. Lim. (2014). Vision-based approach for the localisation and manipulation of ground-based crop, International Journal of Computer Applications in Technology, January 2014, DOI: 10.1504/IJCAT.2014.063909.

[11] Choi, J.; Yin, X.; Yang, L.; Noguchi, N. (2014). Development of a laser scanner-based navigation system for a combine harvester. Eng. Agric. Environ. Food 2014, 7, 7–13

[12] IFM Electronic 3D Smart Sensor – Your Assistant on Mobile Machines. Available online: http://www.ifm.com (accessed on 22 January 2016).

[13] Nakarmi, A.; Tang, L. (2012). Automatic inter-plant spacing sensing at early growth stages using a 3D vision sensor. Comput. Electron. Agric. 2012, 82, 23–31.

[14] Gongal, A.; Amatya, S.; Karkee, M. (2014). Identification of repetitive apples for improved crop-load estimation with dual-side imaging. In Proceedings of the ASABE and CSBE/SCGAB Annual International Meeting, Montreal, QC, Canada, 13–16 July 2014; Volume 7004, pp. 1–9.

[15] Akhloufi M. A. (2014). 3D vision system for intelligent milking robot automation // The conference Intelligent Robots and Computers Vision XXXI: Algorithms and Techniques. 2014. №. 9025.

[16] Garrido, M.; Paraforos, D.S.; Reiser, D.; Arellano, M.V.; Griepentrog, H.W.; Valero, C. (2015). 3D Maize Plant Reconstruction Based on Georeferenced Overlapping LiDAR Point Clouds. Remote Sens. 2015, 7, 17077–17096.

[17] Underwood, J.; Calleija, M.; Nieto, J.; Sukkarieh, S. (2013). A robot amongst the herd: Remote detection and tracking of cows. In Proceedings of the 4th Australian and New Zealand spatially enabled livestock management symposium; Ingram, L., Cronin, G., Sutton, L., Eds.; The University of Sydney: Camden, Australia, 2013; p. 52.

[18] Dorokhov A., Kirsanov V., Pavkin D., Shilin D., Shestov D., Ruzin S. (2019). Calculation of the Manipulator's Kinematic Model and Mounting Points of the Drive Equipment. in Proc. 2nd International Conference on Intelligent Computing and Optimization, vol. 1072, Koh Samui, Thailand, Oct. 2019, pp. 339–348. DOI: 10.1007/978-3-030-33585-4_34.

[19] Ganin, P.E., Shilin, D.V., Shestov, D.A. (2019). Building Unified Control System for Multi-Link Industrial Manipulators, Proceedings - 2019 International Russian Automation Conference,September 2019, DOI: 10.1109/RUSAUTOCON.2019.8867705.

[20] Sappa A., Restrepo-Specht A. and Devy M. (2001). Range Image Registration by using an Edge–based Representation. In: Proceedings of the 9th International Symposium on Intelligent Robotic Systems, pp. 167-176. DOI: 10.1007/11848035_98.

[21] Rabbania T., van den Heuvelb F. A., Vosselmanc G. (2006). Segmentation of point clouds using smoothness constraint, Int. Arch. Photogramm. Remote Sens. Spat. Inform. Sci. 2006.

[22] Felsberg, Forssén, & Scharr. (2006). Channel Smoothing: Efficient Robust Smoothing of Low-Level Signal Features. IEEE Transactions on Pattern Analysis and Machine Intelligence, Vol. 28 , Issue: 2 , Feb. 2006, pp. 209-222, DOI: 10.1109/TPAMI.2006.29.

[23] Oytun Akman, Neslihan Bayramoglu, A. Aydın Alatan, Pieter Jonker. (2010). Utilization of spatial information for point cloud segmentation, Conference: 3DTV-Conference: The True Vision - Capture, Transmission and Display of 3D Video (3DTV-CON), 2010. DOI: 10.1109/3DTV.2010.5506339.

[24] Deckers, N., Reulke, R. (2019). Comparison of Three-Dimensional Scanning Devices. International Conference Image and Vision Computing New ZealandVolume, December 2019. DOI: 10.1109/IVCNZ48456.2019.8961017.

[25] Bogaerts Boris, Penne Rudi, Sels Seppe, Ribbens Bart, Vanlanduit Steve. (2016).A simple evaluation procedure for range camera measurement quality, Conference: ACIVS, October 2016, Vol. LNCS 10016. DOI: 10.1007/978-3-319-48680-2_26.

[26] Kolb, A., Barth, E., Koch, R. and Larsen, R. (2009). Time-offlight sensors in computer graphics, in Proc. Eurographics State of the Art Reports, pp.119–134.

[27] Min Dai, Timothy S. Newman, Chunguang Cao. (2008). Least-squares-based fitting of paraboloids. Pattern Recognition 40, pp. 504-515. DeLaval. (2008). DOI:10.1016/j.patcog.2006.01.016.

DOI: 10.2507/33rd.daaam.proceedings.013

REPLACEMENT OF THE REFRIGERANT IN A VAPOR-COMPRESSION COOLING SYSTEM; CASE STUDY OF THE DATA CENTER

Džana Kadrić, Sanda Midžić-Kurtagić, Edis Kurtović. Aida Šabić & Nijaz Delalić

This publication has to be referred as: Kadric, D[zana]; Midzic Kurtagic, S[anda]; Kurtovic, E[dis]; Sabic, A[ida] & Delalic, N[ijaz] (2022). Replacement of Refrigerant in Vapour-Compression Cooling System: Case Study of Data Center, Proceedings of the 33rd DAAAM International Symposium, pp.0085-0091, B. Katalinic (Ed.), Published by DAAAM International, ISBN 978-3-902734-36-5, ISSN 1726-9679, Vienna, Austria
DOI: 10.2507/33rd.daaam.proceedings.013

Abstract

Many refrigerants traditionally used in vapor-compression cooling systems are designated for their negative impact on the environment with a high Global Warming Potential and an Ozon Depletion Potential. Replacement of refrigerants faces many environmental and technical issues, including reducing the negative impact on the environment, improving the parameters of the system, and addressing the minimum requirements for the alteration of components. The case study presented in this article is focused on the analysis of the performance parameters of the vapor-compression cooling system of the information technology company Data Center and explores the benefits of replacing the current refrigerant. The installed power of the computer equipment in the server room is 48 kW and represents space with significant cooling loads throughout the year. The calculation of the theoretical cooling load of the Data Center is performed and the results show good alignment with current system parameters. Using the data collected from the current system with R410A as the refrigerant parameters, the system was assessed. To reduce the impact of the system on the environment, the R454B refrigerant is accessed as an alternative to R410A and new system parameters are calculated. The analysis showed that refrigerant replacement is technically feasible and will result in improvement of system parameters, reduction of energy consumption, and reduction of the negative environmental impact.

Keywords: Vapor-compression cooling system; Refrigerants; Global Warming Potential; Data Center.

1. Introduction

The refrigerants represent the working substance that circulates through components of cooling systems and heat pumps. The refrigerant absorbs heat from the cooled space or medium during evaporation at low temperature and pressure, and is released to the atmosphere at high temperature and pressure [1]. The negative effect of cooling systems on the environment is present through a double effect; leakage into the atmosphere and indirect emissions of greenhouse gases as a result of system electricity consumption [2]. The effect of refrigerants on the environment is marked via the Global Warming Potential (GWP) and the Ozon Depletion Potential (ODP), where higher values of GWP and OPD represent more harmful refrigerants. Bosnia and Herzegovina signed the provisions of the Montreal Agreement, which refer to the reduction in the use and elimination of such refrigerants.

It primarily refers to chlorofluorocarbons (CFC) and hydrochlorofluorocarbons (HCFC). As an alternative, the fourth generation of synthetic cooling fluids is generated, with the main goal of reducing its GWP [3]. As a contribution to the topic, the analysis of a Data Center vapor-compression cooling system with R410A refrigerant is presented. Taking into account its high GWP, refrigerant replacement is analysed and new system parameters are evaluated, showing that system parameters can be improved and the negative effect of the cooling system on the environment can be reduced.

2. Refrigerant replacement

When selecting a refrigerant, one should primarily pay attention to its environmental impact parameters, system purpose, and type of installation. The use of the refrigerant should be optimized in the best possible way. Table 1. lists the sectors of application and the most common cooling fluids used within individual sectors, as the well as most commonly used alternative refrigerants.

Refrigerant	Chemical composition	Substitution for	GWP	Application	Notation
R134a		R12, R22	1300	Household appliances	Suitable for substitutions
R404a	143a/125/134a 52/44/4	R502, R22	3260	Mobile refrigerators for frozen goods	Pseudo azeotropic
R417a	600/134a/125 3,5/50/46,5	R22	2138	Water coolers, refrigerated cabinets	Temperature sliding
R410A	32/125 50/50	R22	2088	Split systems	High pressure
R23		R13	11700	Cascade cooling devices	High GWP

Table 1. Sectors of application of refrigerants [4]

Replacement of R22 with R410A was the focus of studies in the previous period [5], but better alternatives for R410 are investigated [5]. R410A is a chemically stable non-toxic mixture of two refrigerants (50 / 50%) with a temperature slide of 0.1 ° C at atmospheric pressure. The use of R410A is most common in vapor-compression split systems. R410A does not damage the ozone layer with zero ODP; however, it has rather high GWP, therefore, it is not a best long-term solution as a replacement for R22. According to [6] the best alternatives to R410A are the R466A and R452B with a GWP lower by 22-33%. RS-53 (R470A) is a new non-flammable replacement for R410A with a low GWP of less than half that of R410A. RS-53 (R470A) has a thermodynamic performance similar to R410A with the corresponding system efficiency and cooling load. R452B results in a system cooling load of R410A, with operating pressures approximately the same, and the replacement of existing system components is minimized. The R452B and the R447B are tested, and results have shown that these two refrigerants can offer almost the same cooling characteristics as the R410A under conditions where the outside temperature is less than 35 ° C [7]. Some of the mixtures that have been proven to be good replacement fluids for R410A are R454B, R452B, R447A, R459A and HPR2A. In this article, R454B is analysed as an alternative refrigerant for the Data Center for cooling system, instead of R410A.

3. Data Center cooling system/case study

The characteristics of the system, the calculation of the required cooling load, and the proposed measure to replace the cooling fluid are presented.

3.1. Data Center

The Data Center contains many computer servers and supporting equipment. In addition to computer servers, which are the basis of every Data Center, some other important components are installed, namely: internet equipment, telecommunication equipment, cooling systems, UPS systems along with a fire protection system and an access control system, which makes it very modern and safe. The layout of the Data Center rooms and the cooling system is shown in Figure 1.

Architectural data include values of surface area and construction of external walls, windows, roof, ceiling toward unheated space, and interior walls towards non air-conditioned adjacent spaces. These data are used to calculate the cooling load of the Data Center. The outer wall consists of solid brick 26 cm, lime and gypsum plaster 2 cm, cement screed 10 cm, and thermal insulation installed with polystyrene plates 10 cm. Floor construction is concrete slabs made of stone aggregates with a total thickness of 21 cm, polystyrene slabs with a thickness of 1 cm for thermal insulation, and a surface layer of homogeneous PVC material with a thickness of 2 cm. The ceiling consists of 35 cm of stone aggregate concrete slabs, 2 cm of gypsum and lime gypsum plaster, 2 cm thick polystyrene slabs, and 1.5 cm thick bituminous plaster. The windows are made of PVC material with "IZO" glass, partitioned with gypsum boards from the rest of the room to minimize the penetration of light and better insulation.

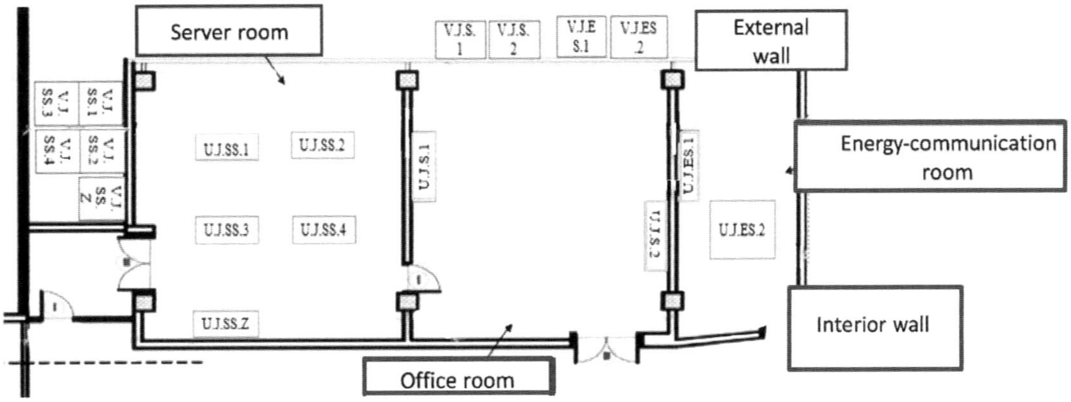

Fig. 1. Layout of the rooms and cooling system of the Data Center

Floor surface area, the volume of the room and designed room temperatures for the Data Center are shown in Table 2.

Room	Parameter	Value
Server room	Floor surface area	58,14 m^2
	Room volume	148,26 m^3
	Designed temperature	22 °C
Office room	Floor surface area	56,43 m^2
	Room volume	143,90 m^3
	Designed temperature	20 °C
Energy-communication room	Floor surface area	28,19 m^2
	Room volume	71,88 m^3
	Designed temperature	20 °C

Table 2. Architectural data of the Data Center

3.2. Analysis of the current cooling system

In Table 3 the parameters of the vapor-compression cooling system installed in the Data Center are shown.

No.	Location	Refrigerant	Marked in Figure 1
1.	Server room	R410A	U.J.SS.1
2.	Server room	R410A	U.J.SS.2
3.	Server room	R410A	U.J.SS.3
4.	Server room	R410A	U.J.SS.4
5.	Server room	R32	U.J.SS.Z
6.	Energy- communication room	R410A	U.J.ES.1
7.	Energy -communication room	R410A	U.J.ES.2
8.	Office room	R410A	U.J.S.1
9.	Office room	R410A	U.J.S.2

Table 3. Cooling equipment of the plant according to the premises

A total of five cooling units are located in the server room, four of which are currently active, and one serves as a backup unit. All five units are under-ceilinged, and each of the indoor units has its outdoor unit (marked as VJ in Fig. 1.). The active outdoor cooling units are Super Digital Inverters, connected to ceiling-type indoor units, and they use the R410A refrigerant as the working medium. The spare outdoor unit uses R32 refrigerant R32 as the working medium. In the office room, there are 2 internal cooling units that are used when needed. In the energy-communication room, 2 cooling units are installed, one of which is a wall unit, and the other is a ceiling unit (cassette).

3.3. Calculation of the theoretical cooling load

The theoretical cooling load is calculated using the IntegraCAD computer program [8], that is in line with the standard [9]. Input data are climatic data for location of Data Center (Sarajevo), architectural data (orientation, surface area and construction of external walls, windows, roof, ceiling towards unheated space and interior walls towards non-air-conditioned adjacent spaces) and designed temperatures. Climatic data is integrated in the computer program. The maximum calculated cooling load for cooled spaces is shown in Table 4.

Room	Theoretical cooling load Q_C [W]
Server room	43.825
Meeting room	3.446
Energy-communication room	22.812

Table 4. Calculated cooling load

From Table 4.3, it can be seen that the total cooling effect according to the current loads for the server room is 43,8 kW, and for the complete Data Center it is 70,08 kW.

3.4. Analysis of the selected cooling unit system

The detailed analysis of the performance characteristics of one vapor-compression cooling unit with R410A refrigerant is presented. The installed units have the configuration of a cooling system as shown in Figure 2, with an evaporator, two compressors, a condenser, two expansion valves, a flash chamber (steam separator) and a direct contact heat exchanger. The evaporator unit is installed inside the cooled space, and other components are installed outside (noted in Figure 1. as VJ). Due to the mode of operation, there is no additional overheating of the saturated steam and subcooling of the condensate, as can be noticed in Figure 2, showing the logp-h diagram of the cooling process.

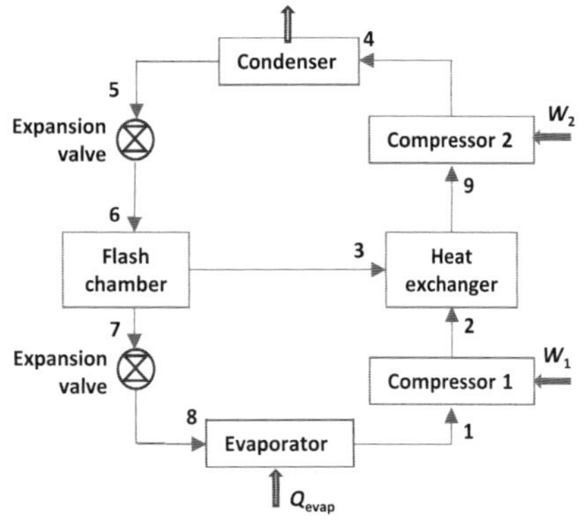

Fig. 2. Schematic representation of the cooling system

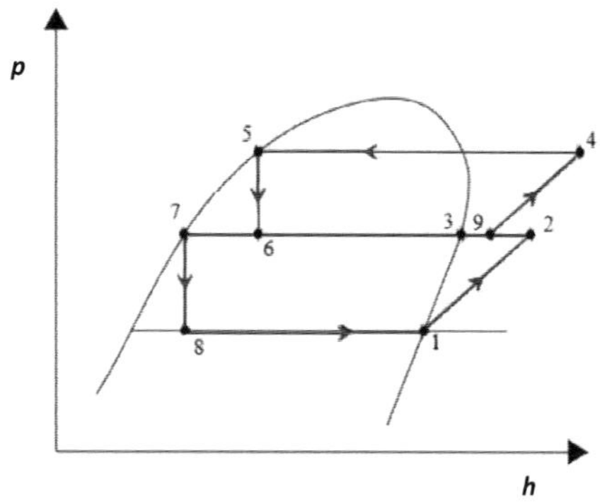

Fig. 3. Log *p-h* diagram of the cooling process

Using the logph diagram of R410A, the process parameters with associated temperatures, pressures, and enthalpies are determined for each state, as shown in Table 5.

System state	Temperature, [K]	Pressure, [bar]	Enthalpy, [kJ/kg]
1	285	11,42	424
2_i	305	16,97	434,7
2	307	16,97	436
3	299	16,97	426,4
4_i	329	25,21	446,9
4	330	25,21	448,8
5	315	25,21	268,9
6	299	16,97	268,9
7	299	16,97	241,2
8	285	11,42	241,2
9	307	16,97	434,6

Table 5. State points of the cooling process with R410A

The actual compression that takes place in the compressor is not isentropic but polytropic, which results in an additional increase in temperature of the vapor at the compressor outlet. As an example, in order to determine the parameters of actual state 2, opposed to ideal state (noted as 2i in Table 5.) [10], compressor efficiency of 90 % is used. Using the following expression, with enthalpies at 1 and 2i and compressor efficiency as known values, the actual enthalpy at state point 2 can be calculated:

$$\eta = \frac{h_{2i} - h_1}{h_2 - h_1} \tag{1}$$

The mass flow through the evaporator is noted as m_{evap} and represents the mass flow circulating on the low-pressure side. The mass flow that passes through the condenser and circulates on the high-pressure side is noted by m_{cond}. Vapor flow rate from the flash chamber is noted as m_3, and following mass flow balances apply:

$$m_{\text{evap}} = m_7 = m_8 = m_1 = m_2 \tag{2}$$

$$m_{\text{cond}} = m_9 = m_4 = m_5 = m_6 \tag{3}$$

The mass rate balance for direct contact heat exchanger encounters the mixing of the vapor mass flow from evaporator and compressor 1 (m_{evap}) and the mass flow of vapor from the flash chamber (m_3) as follows:

$$m_{\text{evap}} + m_3 = m_{\text{cond}} \tag{4}$$

The mass fraction of water vapor for state 6 is x_6=0,15, so $m_3 = 0{,}15 \cdot m_{\text{cond}}$. Therefore following equation can be derived from the previous equations:

$$m_{\text{evap}} = (1 - 0{,}15) \cdot m_{\text{cond}} \tag{5}$$

To determine parameters for state 9, along with previous mass balance equations, the energy balance for direct contact heat exchanger is set as:

$$m_{\text{cond}} h_9 = m_{\text{evap}} h_2 + m_3 h_3 \tag{6}$$

where:
m is mass flow rate [kg/s]
h is enthalpy [kJ/kg].

The specific cooling load (specific cooling load of the evaporator) is the following:

$$q_{\text{evap}} = \frac{m_{\text{evap}} \cdot (h_1 - h_8)}{m_{\text{cond}}} = (1 - x_6) \cdot (h_1 - h_8) = 155{,}4 \text{ [kJ/kg]} \tag{7}$$

The specific compressor power is calculated as:

$$w_{\text{tot}} = w_1 + w_2 = (1 - x_6)(h_2 - h_1) + (h_4 - h_9) = 24{,}4 \text{ [kJ/kg]} \tag{8}$$

Specific heat transferred to the environment through the condenser is calculated as:

$$q_{\text{cond}} = h_4 - h_5 = q_{\text{evap}} + w_{\text{tot}} = 179{,}9 \text{ [kJ/kg]} \tag{9}$$

For a theoretical cooling load per one cooling unit of 10,96 kW, particular mass flow rates can be calculated, and the total compressor power is 1,72 kW. Using the calculated data, the Energy Efficiency Ratio (EER) is calculated as:

$$\text{EER} = \frac{q_{\text{i}}}{w_u} = 6{,}36 \tag{10}$$

4. Refrigerant replacement

The refrigerant currently used is R410A. It is a mixture of refrigerants R125/R32 (50%/50%) with a negligible slip temperature of 0,1 °C at atmospheric pressure. The R454B refrigerant was selected as a refrigerant suitable for replacement of R410A. R454B has a 78 % lower GWP compared to R410A (Table 6). The most important parameters of both refrigerants are presented in Table 6.

Parameter	R410A	R454B
Composition (%)	R32/R125 (50/50 %)	R32/R1234yf (67/7/26 %)
Molecular mass (kg/kmol)	72,6	62,6
Critical temperature (°C)	70,17	78,1
Critical pressure (bar)	47,7	52,67
Boiling point at atmospheric pressure (°C)	-51,4	-50
Slip temperature	0,1	1,5
ODP	0	0
GWP	**2088**	**467**
ASHRAE Classification Group	A1	A2L

Table 6. Refrigerant parameters R410A and R454B [11]

To compare the system parameters using the two refrigerants, it is necessary to calculate the system parameters with the same input data and methodology as presented in the previous chapter and using the log*p*-*h* diagram of refrigerant R454B. The temperatures of the cooled space, the ambient temperatures and the required cooling load are adopted as input data. R454B is characterized by a sliding temperature of 1,5 K. Therefore, the evaporation and condensing temperatures are adopted as a mean temperature value. The input data are: evaporation temperature: T_{evap} = 285 K, condensation temperature: T_{cond} = 315 K, evaporation pressure: p_{evap} = 10,8 bar, condensation pressure: p_{cond} = 23,64 bar and compressor efficiency η = 0,9. All relevant system parameters for both refrigerants are shown in Table 7.

Refrigerant	GWP	$\frac{p_{\text{cond}}}{p_{\text{evap}}}$	q_{evap} [kJ/kg]	w_{tot} [kJ/kg]	q_{cond} [kJ/kg]	t_4 [°C]
R410A	2088	1,486	155,4	24,4	179,9	56,85
R454B	467	1,477	192,1	23,2	215,4	57,85

Table 7. Comparison of cooling fluids

The pressure ratio for the R454B, compared to the R410A system, changed by a very small percentage (0,6 %). The temperature in the second compressor did not increase significantly. Specific cooling capacity increased, and as a result, mass flow rate decreased, which is beneficial to the environment [12], [13]. The total specific work of the compressor for the R454B refrigerant system decreased by 4,78%. The cooling loads for both refrigerants are the same, but the R454B system has a smaller compressor power, as shown in Table 8. Taking into account that four cooling units in the Data Center server room are installed, with additional four units in other spaces, all with substantial annual working hours, replacement of the refrigerant will ensure the decrease of the electrical energy consumption by 23,5 %. EER for the R454B, for same cooling load is higher than for R410A, indicating better performance.

Refrigerant	Cooling load [kW]	Compressor power [kW]	EER
R410A	10,96	1,91	6,63
R454B	10,96	1,46	8,27

Table 8. The cooling load on the evaporator, compressor power, and EER for both refrigerants

From the analysis presented, it can be concluded that refrigerant R454B offers better working parameters than R410A, with a substantial reduction in GWP factor and a smaller negative influence on the environment without any system modification.

5. Conclusions

Many refrigerants used in vapor-compression cooling systems induce negative impact on the environment due to their high GWP and an ODP. Data from the case study are used to present the environmental benefits and improvement of system parameters resulting from refrigerant replacement. For the case study, an analysis of the cooling system of Data Center is presented with significantly high cooling loads imposed from installed electronic devices. Computer equipment with a power of 48 kW is installed in the Server Room of the Data Center, which is in operation 24 hours a day throughout the year.

A thermodynamic calculation and analysis of the cooling system of the Data Center were presented for the currently used refrigerant R410A. Although it has been used for a long time as a replacement for R22, R410A, with a GWP factor of 2.088, induces a significant negative impact on the environment. Therefore, replacement of R410A with a refrigerant with significantly lower GWP, R545B, is analysed.

System parameters are compared for R410A and the R545B and results have shown that R545B offers better system parameters, lower compressor power, lower refrigerant mass flow rates, and higher EER. The analysis showed that the replacement of the refrigerant is technically feasible, and it will result in improved system parameters, reduction of energy consumption, and reduction of negative environmental impact of the system.

Further research will include the analysis of system parameters for other refrigerants to find the optimal solution. Also, installation of central cooling system should be explored, with local and central regulation. This will be investigated as an option to ensure the further reduction of system energy consumption.

6. References

[1] B. Y. J. M. Calm and G. C. Hourahan. (2001). "Refrigerant data summary," Eng. Syst., vol. 18. No. 11, pp. 74-88

[2] J. M. Calm. (2002). "Emissions and environmental impacts from air-conditioning and refrigeration systems," Int. J. Refrig., vol. 25, no. 3, pp. 293–305, doi: 10.1016/S0140-7007(01)00067-6

[3] Risto Ciconkov. (2018). "Refrigerants: There is still no vision for sustainable solutions," Int. J. Refrig., vol. 86, pp. 441–448

[4] A. Sočev. (2009). "Diploma thesis; Novelty in the field of cooling technique" University of Zagreb, Croatia

[5] D. E. Mitu and F. S. Memet. (2011). "Considerations on adopting R 404A instead of R22 on reefers," Ann. DAAAM Proc. Int. DAAAM Symp., vol. 22, no. 1, pp. 661–662, doi: 10.2507/22nd.daaam.proceedings.324

[6] A. G. Devecioğlu. (2017). "Seasonal performance assessment of refrigerants with low GWP as substitutes for R410A in heat pump air conditioning devices," Appl. Therm. Eng., vol. 125, pp. 401–411

[7] Sethi and S. F. Y. Motta. (2016). "Low GWP Refrigerants for Air-conditioning and Chiller Applications," Int. Compress. Eng. Refrig. Air Cond. High Perform. Build. Conf., vol. 2, no. 2013, pp. 1–8

[8] Impuls ing d.o.o., "IntegraCAD." https://integracad.com/

[9] VDI-Standard. (2015). 'VDI 2078 - Calculation of thermal loads and room temperatures (design cooling load and annual simulation),' B. Verlag

[10] C. L. Townsend and J. M. Varnum. (1986). Fundamentals of engineering thermodynamics. John Wiley & Sons, Inc.

[11] B. Shen and M. R. Ally. (2021). "Comparative performance of low global warming potential (Gwp) refrigerants as replacement for r-410a in a regular 2-speed heat pump for sustainable cooling," Sustain., vol. 13, no. 15, doi: 10.3390/su13158199

[12] C. Guzzardi, M. Azzolin, and S. Lazzarato. (2021). "Effect Of the Refrigerant Charge On The System Performance And Mass Distribution In Air-To-Water Heat Pumps and mass distribution in air-to-water systems,"

[13] M. Rusko, I. Pavlova, J. Vanova, and R. Kralikova. (2017). "Connectivity of quality management systems and environmental management systems," Ann. DAAAM Proc. Int. DAAAM Symp., pp. 1213–1220, doi: 10.2507/28th.daaam.proceedings.169

DOI: 10.2507/33rd.daaam.proceedings.014

EFFECT OF RADIAL AND TANGENTIAL CUT ON THE STRENGTH WIDTH-JOINT BONDED SOLID WOOD

Obucina Murco, Hajdarevic Seid, Ibrisevic Alen & Smajic Selver

This Publication has to be referred as: Obucina, M[urco]; Hajdarevic, S[eid]; Ibrisevic, A[len] & Smajic, S[elver] (2022). Effect of Radial and Tangential Cut on the Strength Width-Joint Bonded Solid Wood, Proceedings of the 33rd DAAAM International Symposium, pp.0092-0097, B. Katalinic (Ed.), Published by DAAAM International, ISBN 978-3-902734-36-5, ISSN 1726-9679, Vienna, Austria
DOI: 10.2507/33rd.daaam.proceedings.014

Abstract

The anatomical structure of wood and cutting direction is an important parameter in determining the strength of the bonded samples. The main aim of this paper is to research the effect of radial and tangential grain cuts on bonding strength width-joint bonded samples. The combinations of the bonded surfaces were tangential-tangential, tangential-radial, and radial-radial. Beech-wood, steamed beech-wood, and thermally modified beech-wood samples were planed referring to a radial and tangential surface structure of wood. After the machining process samples were bonded with PVA-c class D3 glue and researched. The planing machine was Weinig Unimat 500 and the planing tool had four blades at a feed speed of 15 m·min-1. The cutting depth of the machining grip was 1.00 mm and the rake angle of the tool blade was bio $\gamma = 15°$, at 6000°/min. The diameter of the cutting blade was $\Phi = 125$ mm. The results obtained in this research show that the lowest bonding strength had the combination of two tangential surfaces thermally modified beech wood and the highest was at the combination of tangential beech wood.

Keywords: anatomical directions; width-joint; fracture; tensile strength.

1. Introduction

The strength of bonded wood joints depends not only on the type and quality of the adhesive and technological process but also on the condition of the wood material and especially on its surface [1]. Özçifçi and Yapici (2008) determined that there was a greater adhesion strength for beech and Scotch pine woods bonded with PVA-c adhesive along the tangential direction than the radial one [2]. Burdurlu et al. (2006) obtained similar results for Calabrian pine wood bonded with PVA-c and polyurethane (PUR) adhesives and recommended performing the bonding process on tangential surfaces with higher pressures [3]. The quality of the machined surfaces of wood depends on the anatomical structure. Feed speed has an important influence on the roughness of the machined surface [4], [5], [6]. Obućina et al., (2010) concluded that better results of the machining performance have been obtained with the decreasing rotation speed [7]. Also, hardwoods' surface structure (ring-porous oak wood and diffuse to semi-ring-porous beech) might affect the bonding quality. Adhesive penetration into the wood was reviewed in detail by Kamke and Lee (2007) [8], who identified several important factors. These are the fluid properties of the resin, anatomical characteristics and permeability of wood, and processing conditions.

The main goal when gluing wood is to achieve a combination of high strength and long-lasting strength, which in principle should not be less than the strength of the wood. The strength of the bonded joint depends on the interaction between wood and adhesives (adhesion). The stability of the joints during indoor and mainly outdoor exposures is important. The most essential parameters influencing the overall bonding quality of wood products include the main factors such as wood's species, density, chemical and anatomical structure, physical and strength characteristics, machining parameters determining the surface roughness, grain orientation, moisture content, adhesive's chemical structure, weight solid and viscosity, surface tension, mechanism of hardening, and the bonding technology's pressure, time, and temperature. Aicher et al. (2018) found that the adhesion strength of bonded wood is not always most prominently connected with its density [9]. PVA-c adhesives are commonly used in the wood industry for general assembly applications, film overlay and high-pressure lamination, edge gluing, wood veneer, and edge bonding [10]. They are safe, non-toxic, non-combustible, easily cleanable, without pollution, cured at room temperature, colourless, transparent, and tough after curing and give a high adhesion strength to bonded wood elements.

Obućina et al., (2014) found that the quantity of deposited adhesive and the deformation of wood under the influence of external pressure significantly affect the temperature at which bonding is performed, which is directly or indirectly reflected in the necessary amount of glue and internal stress in the joints [11]. In the case of glued laminated beams for building construction should be applied (170 ÷ 220) g/m^2 glue. Sernek et al. (1999) found better penetration of the water-based urea-formaldehyde (UF) adhesive into beech wood in the tangential direction at pressure application, while no significant difference between penetrations in the tangential and radial directions occurred when pressure was not applied [12]. Mihulja et al. (2006) show that the testing method, the form of the test piece, and how force is applied have the biggest effects on the shared strength [13]. This paper will determine and investigate the influence of different cutting directions on the bonding strength of width-joint elements of the solid wood panel (Figure 1).

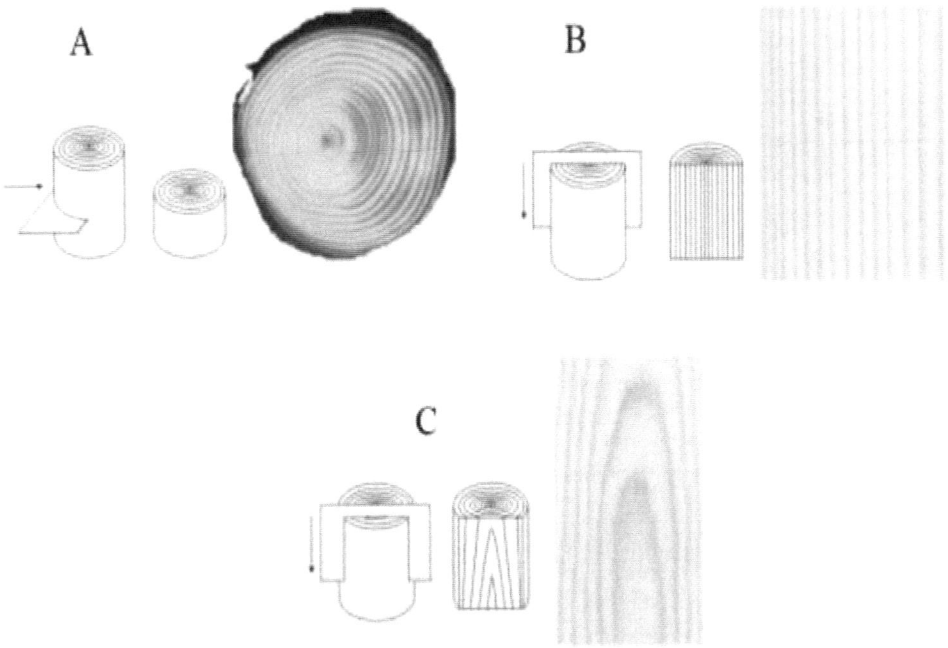

Fig. 1. Samples of anatomical directions (A-cross, B-radial, and C-tangential).

2. Material and experiment

Wood samples for research were prepared from beech wood (*Fagus sylvatica* L.) in three different ways of hydrothermal treatment, beech wood, steamed beech wood, and thermally modified beech wood. Dimensions of the samples before planing were 300×60×20 mm (Figure 2). The average moisture content of samples was 9,5 %, they were conditioned before planing at 21 °C and relative humidity of 60 ± 5%. The difference in moisture content between elements was not greater than 3%. It was prepared in a total of 180 beech wood samples which were divided into groups of 10 samples for each kind of treatment (Figure 2). The samples were bonded, after the machining process, on different cutting sides with a fluent one-component PVA-c Kleibit 314 bond without additives of solvent and admixture of formaldehyde (Figure 3). The adhesive application was 150 g/m2 and the press pressure was 0.7 MPa. Testing of bending strength of samples, as a function of different cutting directions, is done by creating six different groups of samples for testing.

Fig. 2. Samples prepared for cutting to the required dimensions

The size of the samples was 80x40x20mm and the divided groups were:

- Radial-radial beech wood samples for dry testing, 10 samples (RRBW)
- Radial-radial steamed beech wood samples for dry testing, 10 samples (RRSBW)
- Radial-radial for dry testing, 10 samples (RRTBW)
- Radial-tangential beech wood samples for dry testing, 10 samples (RTBW)
- Radial-tangential steamed beech wood samples for dry testing, 10 samples (RTSBW)
- Radial-tangential thermally modified beech-wood samples for dry testing, 10 samples (RTTBW)
- Tangential-tangential beech wood samples for dry testing, 10 samples (TTBW)
- Tangential-tangential steamed beech wood samples for dry testing, 10 samples (TTSBW)
- Tangential-tangential thermally modified beech-wood samples for dry testing, 10 samples (TTTBW)
- Radial-radial beech wood samples for soaked testing, 10 samples (RRBWS)
- Radial-radial steamed beech wood samples for soaked testing, 10 samples (RRSBWS)
- Radial-radial for soaked testing, 10 samples (RRTBWS)
- Radial-tangential beech wood samples for soaked testing, 10 samples (RTBWS)
- Radial-tangential steamed beech wood samples for soaked testing, 10 samples (RTSBWS)
- Radial-tangential thermally modified beech-wood samples for soaked testing, 10 samples (RTTBWS)
- Tangential-tangential beech wood samples for soaked testing, 10 samples (TTBWS)
- Tangential-tangential steamed beech wood samples for soaked testing, 10 samples (TTSBWS)
- Tangential-tangential thermally modified beech-wood samples for soaked testing, 10 samples (TTTBWS)

Fig. 3. Samples of thermally modified beech-wood bonded on different cutting surfaces

Testing of samples was performed in the climatic conditions in accordance with BAS EN 204-2016 temperature (21 ± 2) ° C and relative humidity (60 ± 5) % [14]. During the testing process of bending, strength was recorded visual assessment of the destruction of samples. The first set of samples was tested as dry, after curing for seven days in constant atmospheric conditions (Figure 4). The second set of samples was tested after curing for seven days and then soaked in water for 96 hours before testing (Figure 5). Testing was performed on a Zwick universal testing machine (Figure 6), equipped with a 5 kN load cell. The samples were tested to failure at a cross-head speed of 10 mm/min [15].

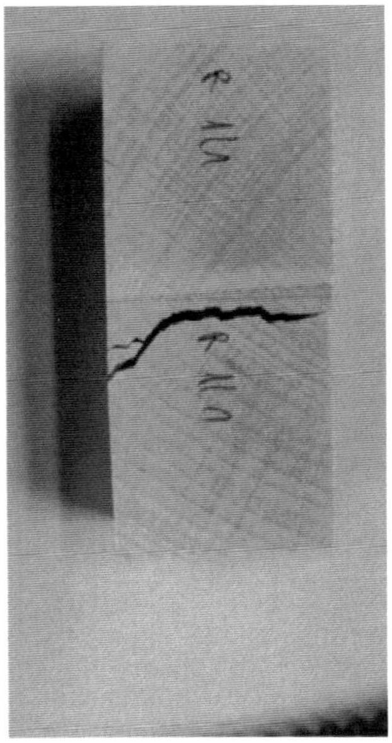

Fig. 4. Sample for dry testing

Fig. 5. Sample for soaked testing

Fig. 6. Testing machine with adapter

Values obtained for bending strength were performed and statistical analysis of data using the student's t-test by testing the equality of means two basic a set of as double-sided distribution (t-Test: Two-Sample Assuming Equal Variances).

3. Results and discussion

In the process of testing bending strength, as the function of the different cutting directions of width-joint bonded samples experiences some deformation and destruction. Destruction occurred in different forms, such as the destruction of wood, the destruction of bonded joints, or a combination of previous forms of destruction. Table 1 shows the mean values of measured bending strength on samples of beech wood. The values of standard deviation are also given in the same table.

Cutting direction	Wood species	Number of samples	σ (MPa)							
			Dry testing				Soaked testing			
			σ mean	σ min	σ max	Std.Dev.	σ mean	σ min	σ max	Std.Dev.
R-R	BW	10	8,912	7,358	10,546	0,820	0,292	0,0785	6,377	0,158
T-T	BW	10	13,006	6,180	12,508	16,225	0,099	0,0196	0,216	0,080
R-T	BW	10	7,922	5,346	9,859	1,515	0,138	0,0589	0,343	0,092
R-R	SBW	10	7,559	9,467	3,483	1,918	0,105	0,0185	0,144	0,073
T-T	SBW	10	8,851	10,693	6,867	1,590	0,248	0,0196	0,500	0,205
R-T	SBW	10	7,436	9,862	4,813	1,589	0,241	0,0392	0,549	0,159
R-R	TBW	10	0,771	1,052	0,428	0,225	0,060	0,0125	0,198	0,063
T-T	TBW	10	0,830	1,177	0,478	0,256	0,230	0,0294	0,461	0,151
R-T	TBW	10	0,602	0,915	0,375	0,214	0,033	0,0162	0,047	0,012

Table 1. Mean values of measured bending strength on samples of beech wood (R-R, radial-radial; T-T, tangential-tangential; R-T, radial-tangential combination. BW-beech wood sample, SBW-steamed beech wood, TBW-thermally beech wood samples)

Figure 7 shows a diagram with summary results of measured bending strengths of BW-beech wood samples, SBW-steamed beech wood samples and TBW-thermally beech wood samples, treated in two different ways and tested in dry and soaked conditions.

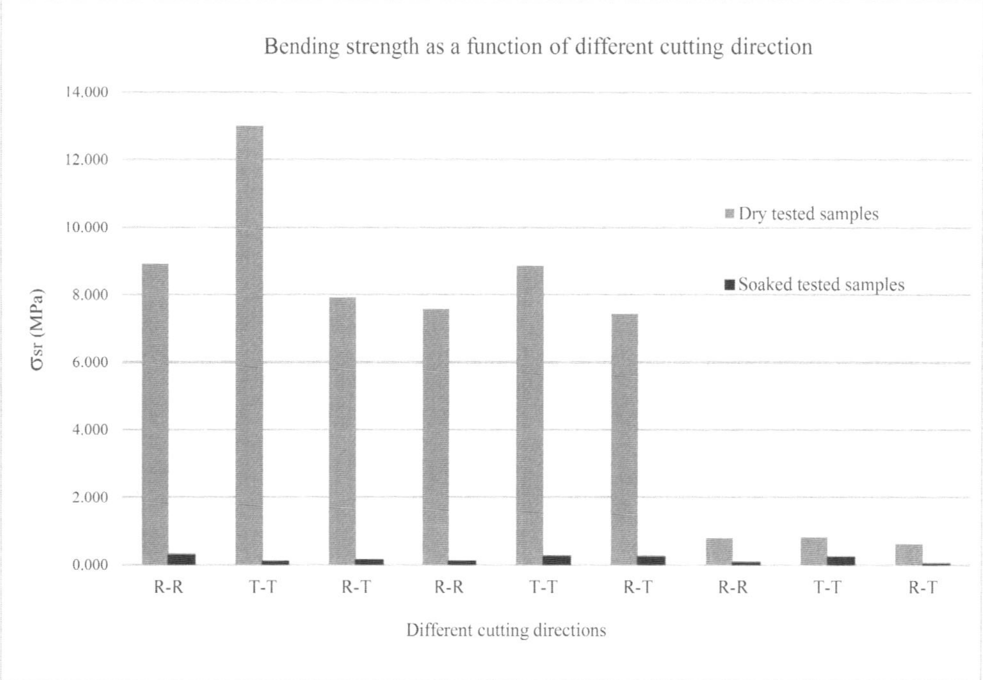

Fig. 7. Summary results of measured bending strengths of beech wood samples (R-R, radial-radial; T-T, tangential-tangential; R-T, radial-tangential combination)

Based on the presented results and diagrams (Figure 7) it is shown that significant differences between the bending strengths of the same anatomical directions, but different tests, are noticeable. The smallest difference is in thermally-treated beech wood. This phenomenon can be explained through the analysis of thermally-treated wood as a material and its characteristics. Considering that the thermally treated wood samples had a significantly lower moisture content, the wetting of the bonding surface resulted in better penetration of the adhesive into the pores of the wood and quality anchoring.

For a better understanding, a student's T-test was performed. Higher values of the t-number indicate that there is a large difference between the two sets of samples. T-test for two groups, the first group was beech, tangential direction, and steamed beech tangential direction (dry test). The second group was beech tangential direction and steamed beech tangential direction (wet test). The T value for strength in the first group was 0,175, and in the second group was 0,469. Other tested groups were beech radial direction with steamed beech radial direction (dry test) with a T value of 0,463 and beech radial direction and steamed beech radial direction (wet test) with T results of 0,645. There is a difference in strength for different types of tests.

These results show that during dry testing there are no big differences in strength between steamed beech and common beech with the same anatomical directions, while in wet testing there are significantly greater differences in strength. During the experiment and based on the results of the dry test, we can notice that the fractures in the tangential direction of steamed beech are mostly due to the adhesive. At the common beech with tangential direction, breaks are mostly over the wood. At the common beech and steamed beech with radial direction, most of the breaks are over the wood. On the soaked tests, all breaks were due to adhesive.

4. Conclusion

The main scope of this paper was to research the effect of radial and tangential grain cuts on bonding strength width-joint bonded samples. Based on this study, it can be concluded that bending strength in relation to the different cuts during bonding is the highest in common beech wood samples bonded in tangential directions. The lowest values of bending strength had samples of thermally-treated beech wood bonded in radial-tangential combination. Also, it can be concluded that there are significant differences between the bending strengths of the same anatomical directions. The smallest difference is in thermally-treated beech wood. Based on the presented results, further research will be continued, involving new parameters of machining, and a new type of adhesive with an increased using the number of samples.

5. References

[1] Živković, V. Muhulja, G. Gabarek, G. (2019). Influence of Natural Surface Ageing on Bonding Quality of Thermally Modified Oak and Beech Wood; Drvna Industrija, 70 (3) 273-278.
[2] Özçifçi, A., Yapici, F. (2008). Effects of machining method and grain orientation on the bonding strength of some wood species. J. Mater. Process. Technol., 202, 353–358
[3] Burdurlu, E. Kilic, Y. Elibol, G.C. Kilic, M. (2006). The shear strength of Calabrian pine (Pinus brutia Ten.) bonded with polyurethane and polyvinyl acetate adhesives. J. Appl. Polym. Sci. 99, 3050–3061
[4] Carrano, A.L.J.B.; Taylor, R.; Lemaster, R. (2002). Parametric characterization of peripheral sanding. Forest Products Journal 52 (9): pp. 44-50.
[5] De Moura, L.F.; Hernandez, R.E. (2006). Effects of abrasive mineral, grit size and feed speed on the quality of sanded surfaces of sugar maple wood. Wood Science and Technology 40 (6): pp. 517-530.
[6] Smajić, S. Jovanović, J. Beljo Lučić, R. (2020). Effect of different machining on the surface roughness of Oak and Beech Wood, Chapter 18 in DAAAM International Scientific Book, pp.217-226 Mostar, Bosna i Hercegovina.
[7] Obućina, M. Škaljić, N. Beljo Lučić, R. Smajić, S. (2010). Effect of Rotation Speed and Wood Species on Roughness of Machined Surface, Zadar, Croatia,
[8] Kamke, A. F.; Lee, J. N. (2007). Adhesive penetration in wood – a review. Wood and Fiber Science, 39: 205-220.
[9] Aicher, S. Ahmad, Z. Hirsch, M. (2018). Bond line shear strength and wood failure of European and tropical hardwood glulam. Eur. J. Wood Prod, 76, 1205–1222.
[10] Ülker, O. (2016). Wood adhesives and bonding theory. In Adhesives–Application and Properties, 1st ed.; Rudawska, A., Ed.; Intech Open: London, UK, pp. 271–288
[11] Obućina, M. Gondžić, E. Smajić, S. (2014). The Influence of Amount of Layer on the Bending Strength by Longitudinal Finger-Jointing Wood Elements, Elsevier, doi: 10.1016/j.proeng.2014.03.096.
[12] Sernek, M. Resnik, J. Kamke, F.A. (1999) Penetration of liquid urea-formaldehyde adhesive into beech wood. Wood Fiber Sci. 31, 41–48.
[13] Mihulja, G.; Bogner, A.; Župčić, I. (2006) Quality testing of PVAC adhesives by standard methods, Proceedings, 17th International Scientific Conference: European Union - challenges and perspectives for the wood-processing industry, Zagreb, Croatia, 13th October 2006, pp.59-69 ref.28
[14] EN 204: Timber structures – Structural timber and glued laminated timber – Determination of some physical and mechanical properties.
[15] Hajdarević, S.; Obućina, M. (2017). Mehanička svojstva drveta, Mašinski fakultet Sarajevo, BiH.

DOI: 10.2507/33rd.daaam.proceedings.015

THE IMPACT OF THE UTILITY-SCALE PHOTOVOLTAIC POWER PLANT ON THE TRANSMISSION GRID IN TERMS OF VOLTAGE AND REACTIVE POWER

Vladimir Budimir & Petar Marić

This Publication has to be referred as: Maric, P[etar] & Budimir, V[ladimir] (2022). The Impact of the Utility Scale Photovoltaic Power Plant on the Transmission Grid in Terms of Voltage and Reactive Power, Proceedings of the 33rd DAAAM International Symposium, pp.0098-0104, B. Katalinic (Ed.), Published by DAAAM International, ISBN 978-3-902734-36-5, ISSN 1726-9679, Vienna, Austria
DOI: 10.2507/33rd.daaam.proceedings.015

Abstract

In the time of high demand for electricity and the transition from the era of dominance of conventional coal-fired power plants to the era of power plants with large installed power whose resources are renewable energy sources, the connection of such power plants directly to the transmission network is becoming increasingly common. The connection of such energy sources to the high voltage grid represents a certain challenge in regulating the power system. Along with the production of useful active power, there is an increase in the network voltage in the connection hub due to the production of reactive power, which must be maintained within strictly prescribed limits. Some of the ways of regulating reactive power and voltage in the transmission network are transformers with transmission ratio regulation, reactive power compensation devices, ballasts, condenser batteries, synchronous generators, and devices based on power electronics. Power electronics-based devices such as inverters with controlling the amount of reactive power in the node are used in photovoltaic power plants.

Keywords: photovoltaic power plant; voltage regulation; power electronics; inverter.

1. Introduction

It is estimated that in the world by 2050 solar photovoltaic (photovoltaic – PV) power plants will produce about 25% of electricity [2]. Most will be connected to the high-voltage transmission grid directly as large power plants. The voltage in the transmission system is most often regulated by synchronous generators in power plants, which usually receive information about voltage conditions through the transmission system operator. The case study transmission network is underloaded, which means that the power system must deal with frequent network overvoltage using different methods of voltage regulation and reactive energy, which is a challenge when connecting large power plants to the transmission network. In the past, various power generation plants were considered exceedingly minor compared to electricity production in conventional power plants, and they were usually referred to as induction generators (wind power plants) or line inverters (photovoltaic power plants) that cannot regulate voltage and reactive power.

However, the increasing rate of penetration of non-traditional renewable energy, especially from wind and sun, has led to the need to produce electricity from renewable sources to significantly contribute to the regulation of voltage and reactive power of the power system. [1]

This paper deals with the example of connecting a solar photovoltaic power plant with an installed capacity of 8 MW to a 110 kV transmission network. Modern photovoltaic inverters have a significant dynamic ability to regulate reactive energy, which can be further improved with other regulation equipment to meet the requirements for fitting a new power plant into the power system.

2. Case study PV power plant integration to power system

The power plant case study in this paper refers to the actual photovoltaic power plant group PVPP 1-8. The location of the photovoltaic power plants is in Mediterranean climate conditions with an average annual temperature of 16° at 417 m above sea level. Surfaces of building particles amount to approx. twenty thousand square meters. The useful surface of one photovoltaic power plant is 13,000 m2, while the rest of the area will be free. Photovoltaic power plants will be built on dirt terrain with a slope of 6.6° out of 2480 solar modules placed on an aluminium fixed substructure at an angle of 30°.

To receive the electricity produced in the new photovoltaic power plant, a 110/20 kV transmission ratio substation will be built near the 110 kV transmission line as the simplest way of connecting the power plant to the transmission network. In addition to the photovoltaic panels, eight inverters with an installed AC power of 150 kW will be used in PVPP 1-8. The output power of the inverters will need to be limited to 124,875 kW so that their total output power corresponds to the installed power of the power plant (output of 8 x 999 kW).

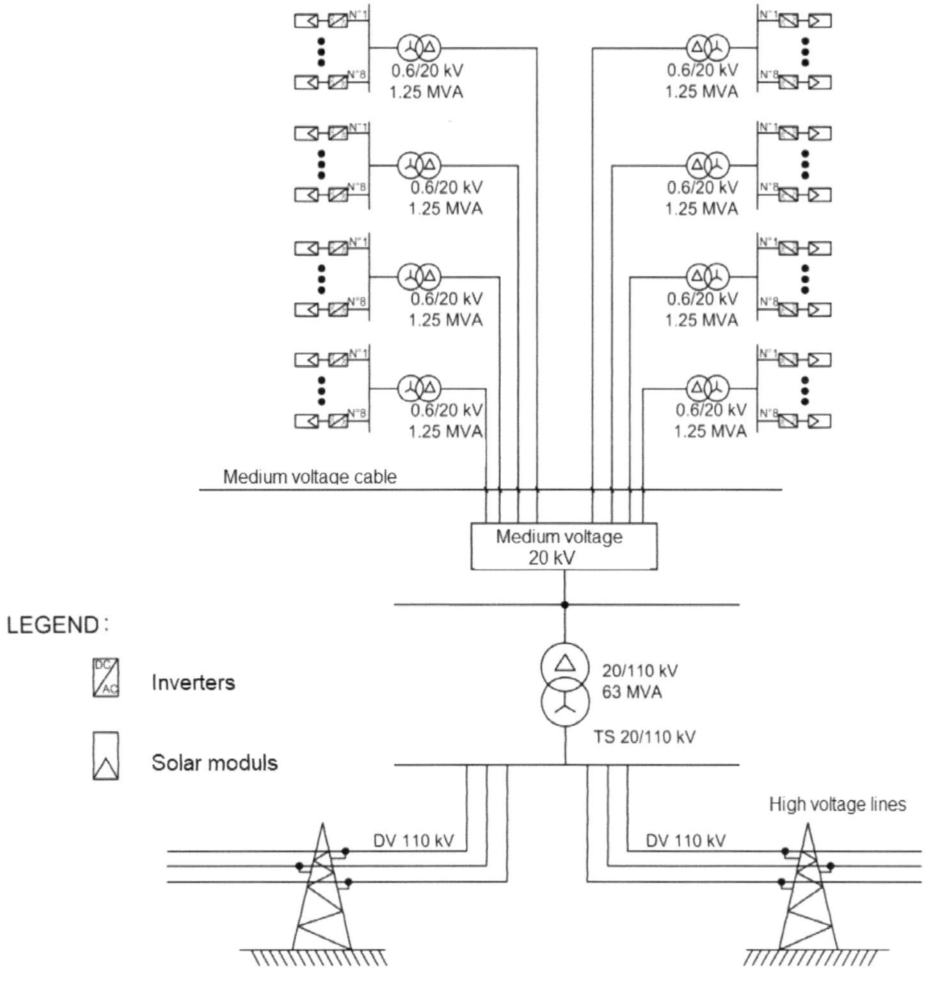

Fig. 1. Block diagram of the internal cable 20 kV grid

2.1. Reactive capability and interconnecting requirements for PV power plants

The difference between inverters and synchronous machines commonly used for reactive power compensation is that they are not power-limited but are limited by internal voltage, temperature, and electric current limitations. PV inverters were originally designed for installation in the distribution system, where applicable interconnection standards do not allow voltage regulation. Inverters for that application are designed to operate at a unity power factor and are sold with a kilowatt unit designation [kW], as opposed to kilovolt-amperes [kVA].

With the increased use of PV inverters on the transmission grid, the industry is moving towards the ability to provide voltage regulation capabilities through reactive power. Some PV inverters can absorb or inject reactive energy, as needed, provided the current and voltage are not exceeded. Considering that the cost of the converter is related to the rated current, providing reactive power at full capacity means that the converter power must be higher for the same plant power. Inverters could produce or absorb reactive power in power states lower than P1 (P2).

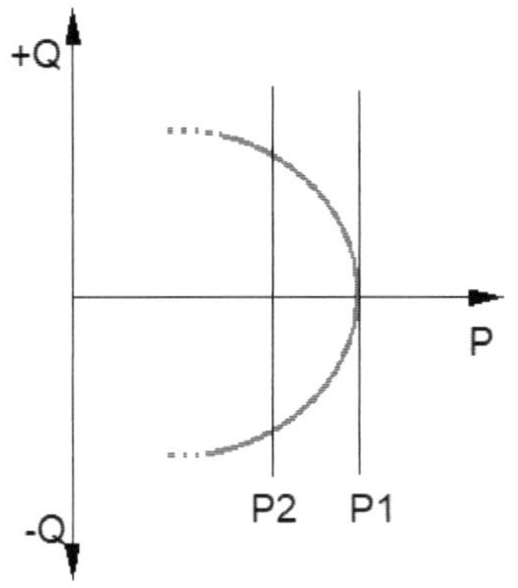

Fig. 2. Inverter reactive power capability (orange) based on electric current limitation

PV inverters are usually disconnected from the grid at night, in which case the inverter-based reactive power voltage regulation capability is not available. The characteristics of the power system cannot be significantly influenced if connects power plants as in this case. To produce reactive power, it is possible to act by controlling the inverter and the transmission ratio of the connected one's transformer. For this reason, in calculations and simulations, a constant value of the short-circuit voltage is most often used and the influence of the change in the transfer ratio of the connection transformers on the different operating modes of the inverter is analysed. Figure 3. defines the conditions regarding power and voltage at the point of connection of the power plant to the transmission network. In this way, it is not necessary to separately analyse individual states of the system in analysed regimes, but only the configuration of the photovoltaic power plant.

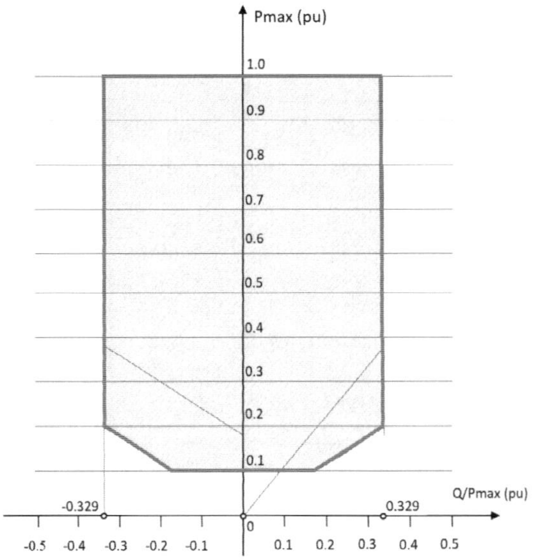

Fig. 3. Inverter reactive power capability (orange) based on electric current limitation

2.2. Inverters

Solar inverters convert the direct voltage of the solar modules into the alternating voltage of the regulated amount and frequency, synchronized with the voltage of the distribution network. The characteristics of the solar grid inverter *SMA Sunny Highpower* 150-20 (150 kW) are given in Figure 3. The minimum requirement for parallel operation is that the inverter protection works and isolates the PV system from the grid if there is a deviation from the voltage (overvoltage or undervoltage) or frequency (over and under), and deviations from factory settings (programmed for each inverter). SMA Sunny Highpower 150-20 inverters meet the requirements for protective functions and inverter adjustment limits according to EU standards and network rules of the distribution system operator. The inverters are a renowned product, and they combine all the circuits necessary for safe and reliable operation on the low-voltage grid. Eight inverters types like SMA Sunny Highpower 150-20 (150 kVA) will be used.

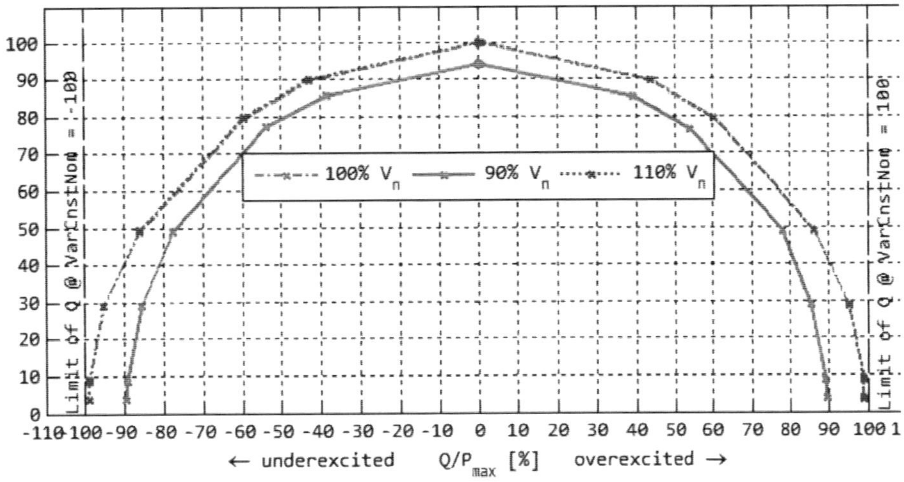

Fig. 4. Operating map SHP 150k-20 SMA (PQ characteristics)

The operating map shows the complex relationship between active power and voltage, and the reactive capability of the inverter. Given that the mentioned characteristics are modelled and slightly approximated to the database provided by the inverter manufacturer, the following tables are presented delivered measured values based on which the characteristics are given and modelled.

P_{DC}/P_{max}	U [V]	P/Sn [%]	Q_u/S_n [%]	P/S_n [%]	Q_o/S_n [%]	VarCnstNom
5%	599.68	3.57	-98.96	3.54	98.77	-99.9/99.9
10%	599.61	8.63	-98.90	8.6	98.78	-99.5/99.5
30%	599.59	28.84	-95.17	28.66	95.29	-95.4/95.4
50%	599.59	49.06	-86.06	48.81	86.63	-86.6/86.6
80%	599.56	79.64	-59.58	79.27	60.39	-60.0/60.0
90%	599.57	89.73	-42.74	89.58	43.88	-43.6/43.6
100%	599.57	99.83	0.14	99.75	0.87	0.0/0.0
5%	539.73	3.55	-89.84	3.59	89.65	-89.9/89.9
10%	539.67	8.63	-89.39	8.64	89.31	-89.5/89.5
30%	539.64	28.83	-85.57	28.82	85.71	-85.8/85.8
50%	539.63	48.98	-77.38	48.91	77.93	-77.9/77.9
80%	539.61	77.17	-53.60	76.57	54.36	-54.0/54.0
90%	539.63	85.67	-38.39	85.24	39.46	-39.2/39.2
100%	539.65	93.98	0.13	93.93	0.71	0.0/0.0
5%	659.70	3.82	-98.99	3.85	98.73	-99.9/99.9
10%	659.65	8.89	-98.92	8.89	98.74	-99.5/99.5
30%	659.63	29.11	-95.18	28.92	95.26	-95.4/95.4
50%	659.64	49.32	-86.05	49.06	86.61	-86.6/86.6
80%	659.60	79.89	-59.61	79.38	60.40	-60.0/60.0
90%	659.61	89.98	-42.80	89.64	43.88	-43.6/43.6
100%	659.62	99.87	0.16	99.80	0.81	0.0/0.0

Table 1. Reactive production limits of the inverter for different values of voltage

3. Case study model

The first step in modelling the connection of the power plant to the grid is the creation of a credible model of the local transmission network. The Neplan software tool was used for the model. [4] By entering the length of transmission lines, associated impedances, existing production, and a load of the local network, a model of the 110 kV ring to which the modelled 8 MW power plant will be connected was made.

A model consists of eight substations connected by 110 kV transmission lines. A feeding element has been added, as well as loads that simulate loads in real-time and production from a nearby hydroelectric power plant (120 MW). The case study power plant model used in this paper is being connected to similar grid model that the authors have already worked on and concerns a local 110 kV ring network. [3] The model calculation requires data on the observed transmission lines such as conductor type, transmission line length, line constants, voltage level, etc. [3][5][6].

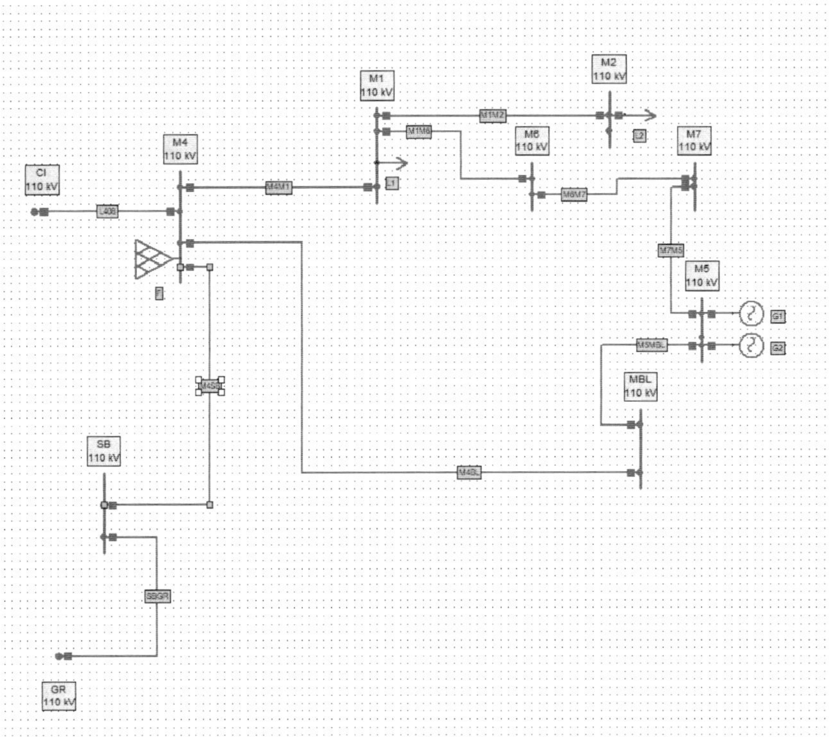

Fig. 5. 110 kV grid model

After all the parameters have been adjusted and all the alerts requested by the program have been eliminated, the calculation was performed. What was important to read from this iteration were the voltage conditions in the nodes. This data is needed so that after the connection of the power plant with the possibility of reactive power regulation via inverter as a utility, the new voltage conditions could be compared with those before the connection of the photovoltaic power plant and certain conclusions could be drawn about this type of regulation.

Busbar/Node	Voltage Level [kV]	The voltage before connecting PVPP 1-8 [kV]
CI	110	112.45
GR	110	112.40
M1	110	113.50
M2	110	113.06
M4	110	113.28
M5	110	113.14
M6	110	113.13
M7	110	113.05
MBL	110	113.22
SB	110	112.57

Table 2. Voltages per node before connecting PVPP 1-8

After extracting the necessary data of the grid model in the maximum mode of operation, the network model was further created in such a way that a new 110 kV substation was integrated into the existing 110 kV transmission line, through which the produced electricity from the photovoltaic power plant PVPP 1-8 will be injected directly into the 110 kV grid.

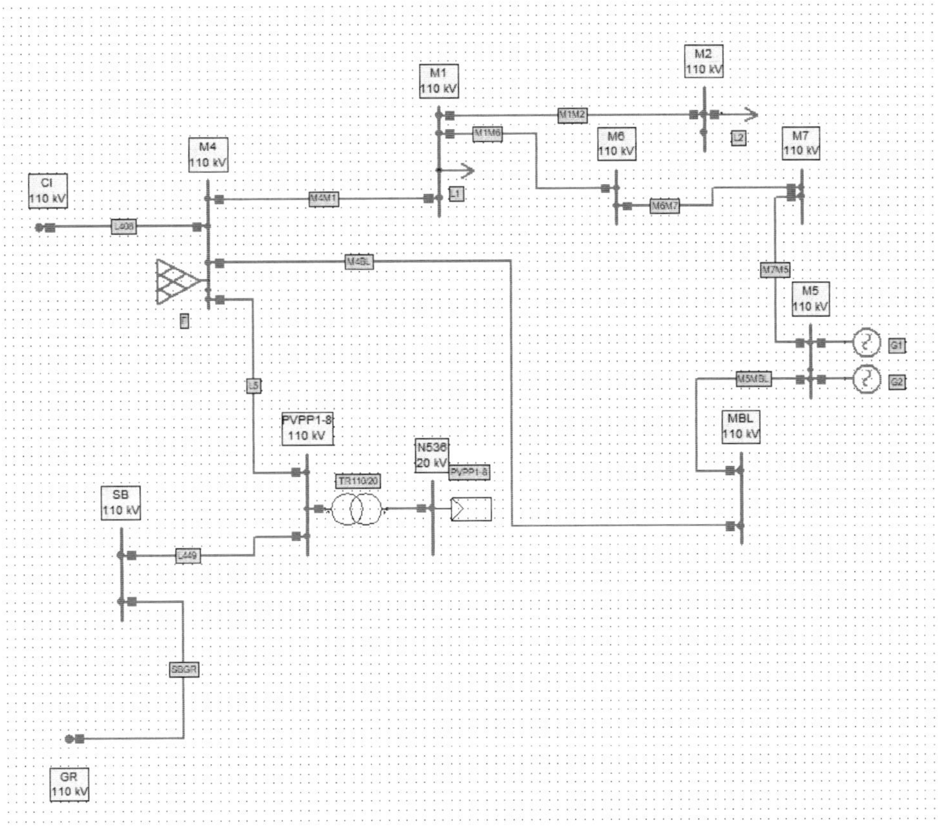

Fig. 6. PVPP 1-8 integrated into local 110 kV power grid

Busbar/Node	Voltage Level [kV]	Voltage after connecting PVPP 1-8 [kV]
CI	110	112.43
GR	110	112.47
M1	110	113.48
M2	110	113.04
M4	110	113.26
M5	110	113.11
M6	110	113.12
M7	110	113.03
MBL	110	113.19
SB	110	112.55
PVPP 1-8	110	113.30

Table 3. Voltages per node after connecting PVPP 1-8

The voltages in all nodes after connecting PVPP 1-8 to the transmission 110 kV level are above the nominal value but are within the permitted deviations. After the connection of the case study PV power plant into the power system, there is an insignificant reduction in voltage at most substations (Table 4., except in GR) in the local EES compared to the state before connection. The voltage on the 110 kV buses of the new substation PVPP 1-8 is 113.3 kV. The connection of the case study PVPP into the local power system's normal regime has a negligible impact on the transmission network because most of the power is placed in node M4, and because of that there is an insignificant change in the voltage conditions.

Busbar/Node	The voltage before connecting PVPP 1-8 V_B [kV]	Voltage after connecting PVPP 1-8 V_A [kV]	$\Delta V_A V_B$
CI	112.45	112.43	-0.02
GR	112.40	112.47	0.07
M1	113.50	113.48	-0.02
M2	113.06	113.04	-0.02
M4	113.28	113.26	-0.02
M5	113.14	113.11	-0.03
M6	113.13	113.12	-0.01
M7	113.05	113.03	-0.02
MBL	113.22	113.19	-0.03
SB	112.57	112.55	-0.02

Table 4. Voltage difference before and after the connection of PVPP 1-8

4. Conclusion

By analysing the model created as a substitute local 110 kV ringnetwork in the Neplan tool according to the data used and the structure of the local transmission 110 kV system in the vicinity of the future photovoltaic power plants, it can be concluded that the transmission network of the local electricity power system provides reliable and adequate power supply to the connected consumers. The analysis of the junction voltage shows that the voltage changes in the nearby 110 kV junctions are minor.

Voltages in most 110 kV nodes in the local power system are slightly reduced but remain within the prescribed limits. Based on the performed analysis of the fulfilment of (n-1) safety criteria, it can be concluded that the connection of PVPP 1-8 to the transmission network does not violate the safety requirements according to this criterion. Of course, the analysis of this paper refers to voltage regulation via built-in inverters and does not consider other types of voltage and reactive power regulation.

5. Acknowledgments

The authors thank engineering colleagues from Elektroprijenos BiH and Independent System Operator in Bosnia and Herzegovina (NOS BiH) for providing the data needed to write this paper.

6. References

[1] https://www.businessinsider.com, (2009). By 2050 25% Of The World's Electricity Will Be Solar, Accessed on: 2022-08-13

[2] Ellis, A., Nelson, R., Von Engeln, E., Walling, R., McDowell, J., Casey, L., Seymour, E., Peter, W., Barker, C. and Kirby, B. (2012.). Reactive Power Interconnection Requirements for PV and Wind Plants – Recommendations to NERC. Sandia Report, sand2012-1098, (February 2012) page numbers (1-40)

[3] Budimir, V., & Marić, P. (2020). Electric Load Modelling of the Power Transmission Lines and Comparison with SCADA Measurements. 31st DAAAM International Symposium on Intelligent Manufacturing and Automation

[4] NEPLAN – Power System Analysis, NEPLAN AG, Zurich, Switzerland

[5] Jozsa, L.: Parametri nadzemnih vodova, University of Josip Juraj Strossmayera in Osijek, Faculty of Electrical Engineering Osijek, Osijek, 2006

[6] Saadat, H.: Power System Analysis. WCB McGraw-Hill, New York, 1999.

DOI: 10.2507/33rd.daaam.proceedings.016

DETERMINING THE INFLUENCE OF MODEL PARAMETERS ON THE CHOOSING OF AN OPTIMAL SIZE RANGE OF PNEUMATICALLY ACTUATED LINEAR MODULES FOR SPRAYER ROBOTS

Ivo Malakov, Velizar Zaharinov, Stilian Nikolov,
Reneta Dimitrova & Grigor Stambolov

This Publication has to be referred as: Malakov, I[vo]; Zaharinov, V[elizar]; Nikolov, S[tilian]; Dimitrova, R[eneta] & Stambolov, G[rigor] (2022). Determining the Influence of Model Parameters on the Choosing of an Optimal Size Range of Pneumatically Actuated Linear Modules for Sprayer Robots, Proceedings of the 33rd DAAAM International Symposium, pp.0105-0114, B. Katalinic (Ed.), Published by DAAAM International, ISBN 978-3-902734-36-5, ISSN 1726-9679, Vienna, Austria
DOI: 10.2507/33rd.daaam.proceedings.016

Abstract

The paper presents results from a sensitivity analysis on the solution of the problem for size range optimization of pneumatically actuated linear modules. To that end, the optimization problem is solved for different values of chosen parameters included in the mathematical model of the problem, while keeping the values of the other parameters constant. An assessment is made regarding the influence of the studied parameters on the optimal solution. Sensitivity analysis (also called post-optimal analysis) is of a substantial practical significance, and is one of the important stages of an approach for designing optimal size ranges.

Keywords: sensitivity analysis; post-optimal analysis; size ranges; optimization; sprayer robots.

1. Introduction

The choice of an optimal size range, predefines in great deal, the good economic results for the manufacturer, as well as for the users, of mass-produced technical products from different industry branches [1], [2], [3], [4], [5], [6]. It is related to defining the elements of such a size range, that with minimum costs and/or maximum efficiency (profit) in the area of manufacturing and operation, completely satisfies given product demand regarding quantity [1], [2], [7], [8], [9]. Solving this problem is related to numerous difficulties [1], [7].

One of the main difficulties comes from the necessity for forecasting demand for products with different values of their main parameters, and building a functional dependency between a chosen optimality criterion/criteria and the influencing factors. Solving these problems is related to collecting and processing of a vast quantity of heterogeneous technical and economic information, referring to the connections of the designed size range with its environment throughout the different stages of the size range's lifecycle, and requires substantial expenditure of time, labour, and highly qualified specialists from different fields – marketing, design, manufacturing, operation, etc.

These activities are carried out in the earliest design stages, which stages are characterized by uncertainty and incomplete information regarding some of the influencing factors, and uncertainty increases with increase in product complexity [2], [10], [11], [12], [13]. Moreover, the level of uncertainty can rise additionally, because defining costs for materials, energy, salaries, etc. is made in conditions of dynamically changing domestic and international economic situation. As a consequence of this, the created model may not reflect in full the real conditions, and the optimal solution to the problem may have an approximate value [2]. On the other hand, for companies it is particularly important to study different possible situations related to the change in demand of products, production volume, costs, manufacturing conditions, operation, etc.

All of this determines the necessity for studying the dependency between the optimal solution and the change in the problem's parameters (coefficients and elements of the objective function, demand, production volume, feasible application areas, etc.) [14], [15]. To that end, the optimization problem is solved while one or several parameters are changed in certain limits, and the rest of the parameters remain constant. The analysis of the obtained results will allow for [16], [17], [18]:

- studying the influence of the individual parameters of the mathematical model on the solution of the optimization problem, and determining the most important ones, for which the most accurate information must be obtained. This will allow for less resources to be spent when determining the values of the rest of the model's parameters;
- obtaining information regarding the behaviour of the problem's solution in different situations, related to change in some model's elements – demand, production volume, etc.;
- analysing the workings of the developed mathematical model, and estimating if it is properly built;
- modifying (calibrating) the mathematical model for the purpose of adequately reflecting the real conditions, etc.

Subject of examination is the optimal size range of pneumatically actuated linear modules for sprayer robots (Fig. 1), determined after solving the problem by means of a proposed approach [2], [7]. The aim of the paper is to show the results from a sensitivity study of the solution to the problem for choosing an optimal size range of pneumatically actuated linear modules for sprayer robots (Fig. 1).

Fig. 1. Sprayer robot using a pneumatically actuated linear module for vertical motion [19]

2. Mathematical model

The mathematical model of the problem for choosing an optimal size range has the form:

For a given demand set $\bar{N} = \{\bar{N}^1, \bar{N}^2, \dots, \bar{N}^l, \dots, \bar{N}^{\bar{L}}\}$, $l = 1 \div \bar{L}$, of products $\bar{Z} = \{\bar{x}_1, \bar{x}_2, \dots, \bar{x}_l, \dots, \bar{x}_{\bar{L}}\}$, $l = 1 \div \bar{L}$, and their allowable possibilities for satisfying the different demand types, i.e. the elements of the applicability matrix $\Phi_{\bar{L}\times\bar{L}} = \|\varphi_m^p\|_{\bar{L}\times\bar{L}}$, find L^*, $Z^* = \{x_{l_1}^*, x_{l_2}^*, \dots, x_{l_j}^*, \dots, x_{l_{L^*}}^*\}$, $Z^* \subseteq \bar{Z}$, $j \in \{1,2,\dots,\bar{L}\}$, $N^* = \{N^{*l_1}, N^{*l_2}, \dots, N^{*l_j}, \dots, N^{*l_{L^*}}\}$, $L^* \leq \bar{L}$, for which a chosen effectiveness (optimality) criterion has a minimum value:

$$minR(L, \bar{x}_1, \dots, \bar{x}_l, \dots, \bar{x}_{\bar{L}}, \bar{N}^1, \dots, \bar{N}^l, \dots, \bar{N}^L, \Phi_{\bar{L}\times\bar{L}}) = \sum_{j=1}^{L} G\left\{x_{l_j}, N^{l_j}(\varphi_m^p, N^p)\right\} \quad (1)$$

for the following conditions:

$$\textstyle\sum_{j=1}^{L} N^{l_j} = \sum_{j=1}^{L^*} N^{*l_j} = \sum_{l=1}^{\bar{L}} \bar{N}^l = N_0, \quad (2)$$

$$x_{l_L} = x_{l_{L^*}}^* = \bar{x}_{\bar{L}}, \quad (3)$$

$$x_{l_j} \in \bar{Z} = \{\bar{x}_1, \bar{x}_2, \dots, \bar{x}_l, \dots, \bar{x}_m, \dots, \bar{x}_{\bar{L}}\}, \forall j = 1 \div \bar{L}, \quad (4)$$

where R is total costs for all elements in the size range; $\bar{L}$ - the number of elements of the initial size range $\bar{Z} = \{\bar{x}_1, \bar{x}_2, \dots, \bar{x}_l, \dots, \bar{x}_m, \dots, \bar{x}_{\bar{L}}\}$, determined after demand research, $dim\bar{Z} = \bar{L}$; L - the number of elements in the currently analysed size range $Z = \{x_{l_1}, x_{l_2}, \dots, x_{l_j}, \dots, x_{l_L}\}$, $j \in \{1,2, \dots, \bar{L}\}$. There exists a simple representation between the elements of the current and initial size range, whereby to each element x_{l_j} corresponds an element $\bar{x}_m$. The elements of the current size ranges are a combination of L, $L = 1 \div \bar{L}$, elements of $\bar{L}$ number of possible elements of the initial size range, taking into account their allowable application ranges; L^* - the number of elements in the optimal size range $Z^* = \{z^{*l_1}, z^{*l_2}, \dots, z^{*l_j}, \dots, z^{*l_{L^*}}\}$, $j = 1 \div L^*$, $L^* \le \bar{L}$; $\Phi_{\bar{L}\times\bar{L}} = \|\varphi_m^p\|_{\bar{L}\times\bar{L}}$ - applicability matrix, which elements give information about the possible allowable application ranges for the products in the initial size range $\bar{Z} = \{\bar{x}_1, \bar{x}_2, \dots, \bar{x}_{\bar{L}}\}$ for satisfying the separate demand types $\bar{N} = \{\bar{N}^1, \bar{N}^2, \dots, \bar{N}^{\bar{L}}\}$, where $\varphi_m^p = 1$, if the product of m-th type, $m = 1 \div \bar{L}$, can satisfy the demand $\bar{N}^p$ of product $\bar{x}_p$, $p = 1 \div \bar{L}$, and $\varphi_m^p = 0$ otherwise; $N^{l_j}(\varphi_m^p, N^p)$ - the demand of product x_{l_j} element of the currently analysed size range $Z = \{x_{l_1}, x_{l_2}, \dots, x_{l_j}, \dots, x_{l_L}\}$, whose index l_j corresponds to the index m of product $\bar{x}_m \in \bar{Z}$, element of the initial size range, where $x_{l_j} = \bar{x}_m$, and $N^{l_j}(\varphi_m^p, N^p) = \sum_{p=l_{j-1}}^{l_j} \varphi_{l_j}^p N^p$; N_0 - total production quantity of all elements in the size range; N^{l_j} - the quantity of l_j-th product size, $N^{l_j} \in N = \{N^{l_1}, N^{l_2}, \dots, N^{l_L}\}$, where N is the demand set for the products of the current size range.

Condition (2) means, that all analysed size ranges, including the optimal, must satisfy all demand in terms of quantity, condition (3) means, that every size range, including the optimal, must include the element from the initial size range with the maximum value of its main parameter $\bar{x}_{\bar{L}}$, and condition (4), that the elements of all size ranges are chosen from the set of elements of the initial size range, determined after demand research.

The demand for each clamp size included in the initial size range $\bar{Z} = \{\bar{x}_1, \bar{x}_2, \dots, \bar{x}_l, \dots, \bar{x}_{\bar{L}}\}$, is determined after market research, i.e., the elements of the set $\bar{N} = \{\bar{N}^1, \bar{N}^2, \dots, \bar{N}^l, \dots, \bar{N}^{\bar{L}}\}$, $l = 1 \div \bar{L}$, where $\bar{N}^l$ is the quantity of the required products of size $\bar{x}_l$. The obtained results are shown in Fig. 2. The product's main parameter values are input along the abscissa – the stroke length, and along the ordinate - the corresponding quantity.

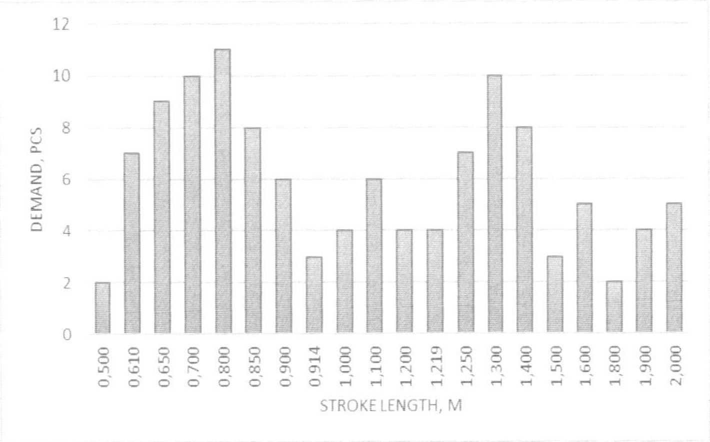

Fig. 2. Product demand from market research

3. Solving the problem – choosing an optimal size range

For choosing an optimal size range it is necessary to determine the functional relationship between the components of the optimality criterion R, and the influencing factors. The following relationship is obtained:

$$\mathrm{R} = \sum_{j=1}^{L}\left[\left(10271{,}7 + 925{,}7x_{l_j} + 8{,}651\right)\left(\frac{4}{N^j}\right)^{0{,}32} + 9304{,}32x_{l_j}^2 + 77{,}47\sqrt{30x_{l_j}} + 154{,}945x_{l_j} + \right.$$
$$\left. +66{,}67\right]N^j + \left(4323{,}9 + 308{,}57x_{l_j} + 2{,}88\right)\left(\frac{4}{N^j}\right)^{0{,}32} \quad (5)$$

It is determined on the basis of production analysis, and production and operational costs data for pneumatically actuated linear modules. The relationship for determining total costs is obtained through the application of methods for statistical data analysis and regression analysis. With the aid of developed tools (mathematical model of the optimization problem, cost models for production and operational costs, algorithm and application software) the problem (1) – (4) is solved.

The results are shown in Tab. 1 [20], where R^* are the minimum (optimal) total costs, EUR; L^* - - the number of elements in the optimal size range; $R_{\bar{L}}^{\bar{L}}$ - - the total costs of the size range including all possible sizes, i.e., of the initial size range, EUR.

Indicators	R^*	L^*	$R_{\bar{L}}^{\bar{L}}$	$\frac{R_{\bar{L}}^{\bar{L}}-R^*}{R_{\bar{L}}^{\bar{L}}}$
Optimal size range	2 531 731,07	9	2 796 363,88	9,46%

Table 1. Results from solving the problem

The data for the chosen optimal size range are shown in Tab. 2. The size range includes $L^* = 9$ sizes.

N	1	2	3	4	5	6	7	8	9
Size $x_{l_j}^* \in X^*$, $j = 1 \div 9$	$x_{l_1}^*$	$x_{l_2}^*$	$x_{l_3}^*$	$x_{l_4}^*$	$x_{l_5}^*$	$x_{l_6}^*$	$x_{l_7}^*$	$x_{l_8}^*$	$x_{l_9}^*$
Size $\bar{x}_m \in \bar{X}$, $l \in \{1 \div \bar{L}\}$	$\bar{x}_1 = 0{,}65$	$\bar{x}_6 = 0{,}80$	$\bar{x}_9 = 1{,}00$	$\bar{x}_{12} = 1{,}25$	$\bar{x}_{14} = 1{,}20$	$\bar{x}_{15} = 1{,}40$	$\bar{x}_{17} = 1{,}60$	$\bar{x}_{19} = 1{,}90$	$\bar{x}_{20} = 2{,}00$
Application range	50 - 80	180 - 360	360 - 560	560 - 650	750 - 840	1000	1050 - 1400	1600 - 2200	2200 - 2500

Table 2. Optimal size range

The production quantities of each element of the optimal size range (1), and the elements of the initial size range (2) are shown in Fig. 3. The total costs curve, determined in the process of solving the problem, is shown in Fig. 4.

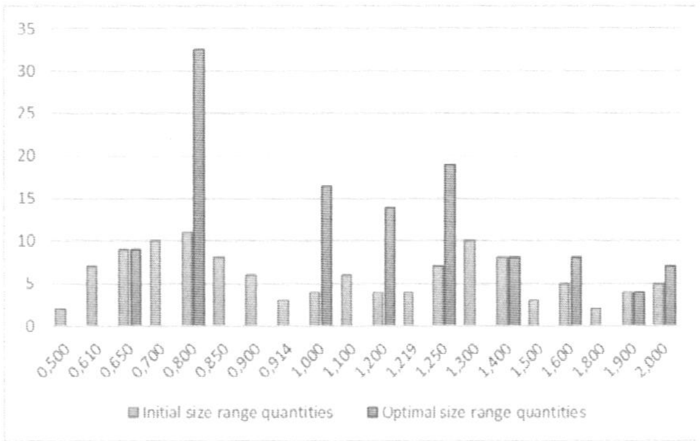

Fig. 3. Production quantity of the elements of the optimal size range (red), and the initial size range (blue)

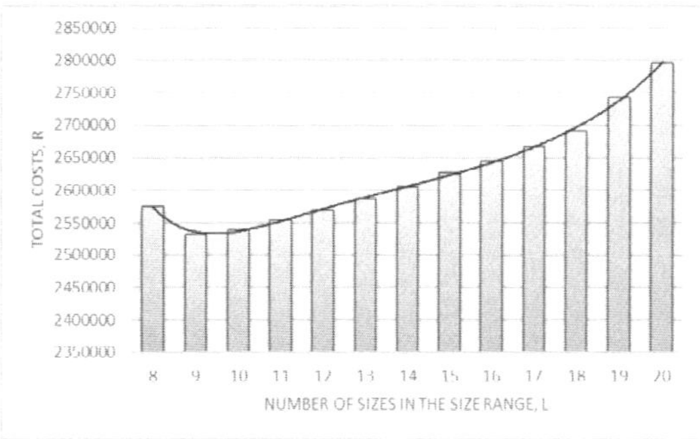

Fig. 4. Change in the total costs

The found optimal size range of the product "Pipe clamp" is characterized by:

- reduction in the number of sizes from 20 to 9, i.e., with 55%;
- reduction in the total costs in comparison to the size range including all possible sizes with 281 815,82 EUR, i.e. with 9%;

4. Sensitivity analysis of the optimal solution

A sensitivity analysis of the optimal solution is carried out in relation to the change in the values of the mathematical model's parameters. The main goal is to determine the influence of the separate components and coefficients on the optimization problem's solution, and to determine the most important ones, for which the most accurate information must be obtained.

To that end, numerical experiments are carried out, which consist of solving the problem (1) – (4) for different values of the main coefficients and components, included in the mathematical model.

The sensitivity analysis includes the following experiments:

A. Changing the demand function set $\bar{N} = \{\bar{N}^1, \bar{N}^2, \dots, \bar{N}^l, \dots, \bar{N}^{\bar{L}}\}$ while keeping the total production quantity N_0;

B. Changing the quantity of production N_0;

C. Changing the application ranges of the separate elements of the initial size range, given by means of the applicability matrix $\Phi_{\bar{L}\times\bar{L}} = \|\varphi_m^p\|_{\bar{L}\times\bar{L}}$;

D. Changing the coefficient, which takes into account the production scale factor;

E. Changing the coefficient, which takes into account the production learning rate.

4.1. Changing the demand function

Problem (1) – (4) is solved for different demand functions as follows: experiment A0 – demand shown in Fig. 2 (initial problem); A1 – demand shown in Fig. 5, A2 – demand shown in Fig. 6, and A3 - demand shown on Fig. 7. For these experiments the overall production quantity is not changing, i.e., the sum of the demands for the separate sizes for each experiment has the same value. The solutions found are shown in Tab. 3.

The aim is to study different market states, where there is a higher demand for larger sizes (A1), average in magnitude sizes (A2), and small sizes (A3). From Fig. 2 can be seen, that the demand function for the initial problem has two peaks – towards the small and average sizes.

In Fig. 8 the total costs are graphically represented in relation to the number of sizes for different demand functions. It is obvious, that the curves depicting the total costs function for different demand distribution, significantly change in value. The shape of the curves changes near the optimum, and afterwards it is similar. It can be seen, that when changing demand, the number of sizes included in the optimal solution can change, and from Tab. 3 it can be determined, that the composition of the size range also changes. Therefore, deviations of the demand distribution from the actual one, leads to significantly different problem solutions. Thus, for obtaining reliable result from the optimization, it is necessary to carefully and in detail collect data regarding customer orders, i.e., market research, and to correctly define the demand function.

Experiment	**A0**	**A1**	**A2**	**A3**
L^*	9	10	9	9
R^*	2 531 731,07	3 343 715,18	2 719 746,87	2 309 722,77
$R_{\bar{L}}^{\bar{L}}$	2 796 363,88	3 549 897,56	2 959 954,02	2 546 364,56
$x_{i_j}^* \in X^*$	0,65; 0,80; 1,00; 1,25; 1,20; 1,40; 1,60; 1,90; 2,00	0,65; 0,80; 1,00; 1,25; 1,20; 1,50; 1,60; 1,80; 1,90; 2,00	0,65; 0,80; 1,00; 1,25; 1,20; 1,50; 1,60; 1,90; 2,00	0,65; 0,80; 0,914; 1,25; 1,20; 1,50; 1,60; 1,90; 2,00

Table 3. Results from solving the problem with different demand functions

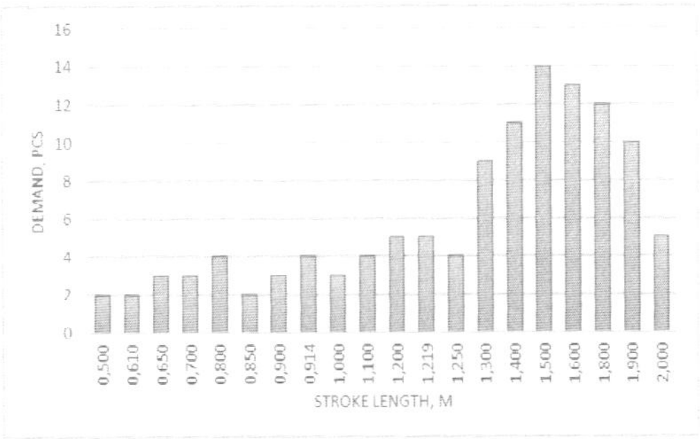

Fig. 5. Demand function for A1

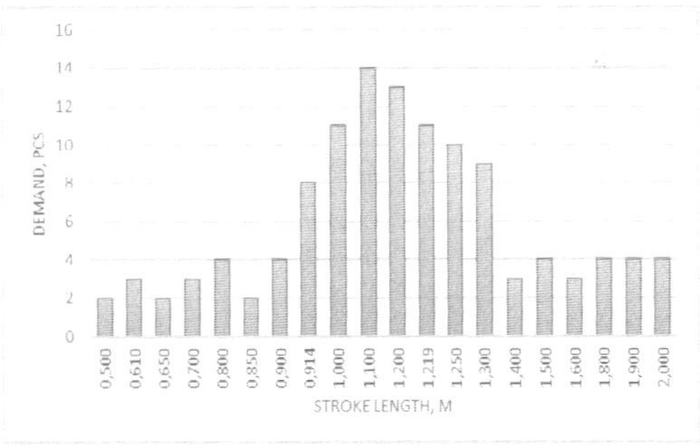

Fig. 6. Demand function for A2

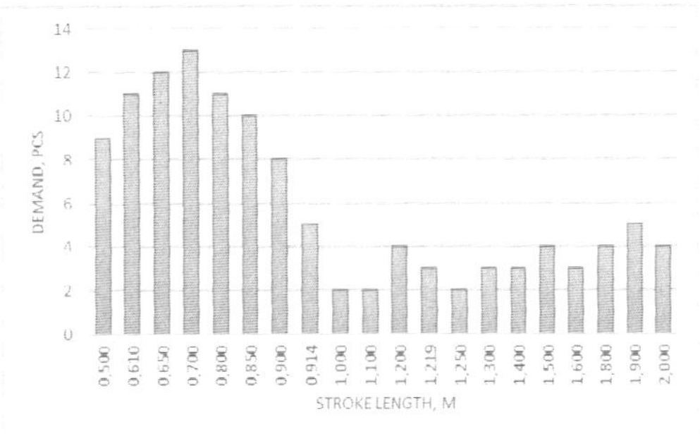

Fig. 7. Demand function for A3

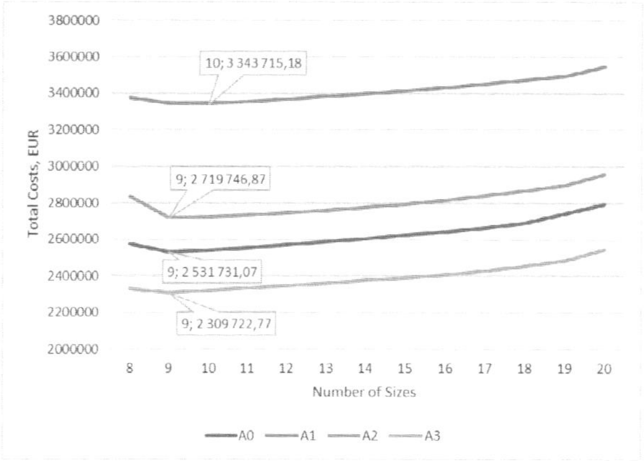

Fig. 8. Change in total costs for different demand functions

4.2. Changing the production quantity

Problem (1) – (4) is solved for different production quantities as follows: B0 – production quantity of the initial problem; B1 – doubling the production quantity; B2 – tripling the production quantity; B3 - quadrupling the production quantity; B4 – halving the production quantity. For these experiments the demand distribution does not change. The found solutions are shown in Tab. 4.

Experiment	**B0**	**B1**	**B2**	**B3**	**B4**
L^*	9	10	10	10	9
R^*	2 531 731,07	4 676 012,12	6 730 565,95	8 924 015,64	1 489 262,16
$R_{\bar{L}}^{\bar{L}}$	2 796 363,88	5 030 333,79	7 165 510,28	9 434 768,02	1 699 298,41
$x_{l_j}^* \in X^*$	0,65; 0,80; 1,00; 1,25; 1,20; 1,40; 1,60; 1,90; 2,00	0,65; 0,80; 0,914; 1,10; 1,25; 1,20; 1,40; 1,60; 1,90; 2,00	0,65; 0,80; 0,914; 1,10; 1,25; 1,20; 1,40; 1,60; 1,90; 2,00	0,65; 0,80; 0,914; 1,10; 1,25; 1,20; 1,40; 1,60; 1,90; 2,00	0,65; 0,80; 1,00; 1,25; 1,20; 1,40; 1,60; 1,90; 2,00

Table 4. Results from solving the problem with different production quantities

The following results are obtained for the total costs R^* (Fig. 9):

- Doubling, tripling, and quadrupling the production quantity leads to increase in the total costs accordingly with 85%, 166%, and 253% with respect to the total costs for the optimal size range of the initial problem;
- Halving the production quantity leads to lowering of the total costs with 41% with respect to the total costs for the optimal size range of the initial problem.

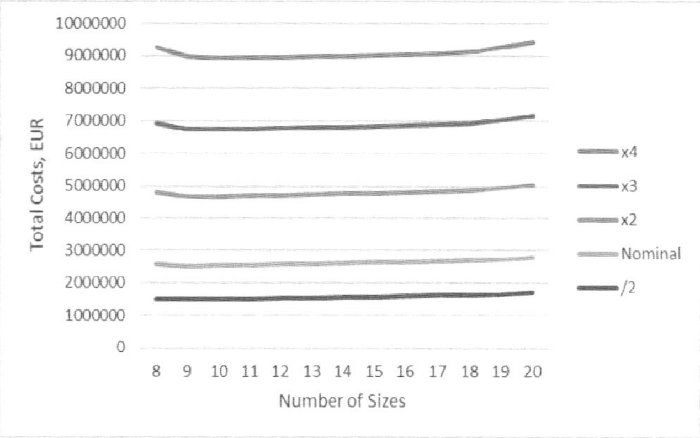

Fig. 9. Change in total costs for different production quantities

Additionally, when increasing the production quantity, the number of sizes in the optimal size range increases with one, and its composition changes. Lowering the production quantity does not change the composition of the size range in comparison to the optimal one.

4.3. Changing the application ranges of the separate elements of the initial size range

Problem (1) – (4) is solved for different application ranges of the separate elements of the initial size range. The change is made in two directions: adding and removing application fields. The changes made are shown in Fig. 10 for each of the experiments C0-C6, and are colour coded in the figure. The experiment C0 is the initial problem. For experiments C1-C3 fields are added starting with the initial problem and gradually adding one field for every following experiment until the smallest range covered with one size becomes four fields. For experiments C4-C6 fields are removed starting with the initial problem and gradually removing one field for each following experiment, until the largest range covered with one size becomes three fields.

The experiments' results are shown in Tab. 5.

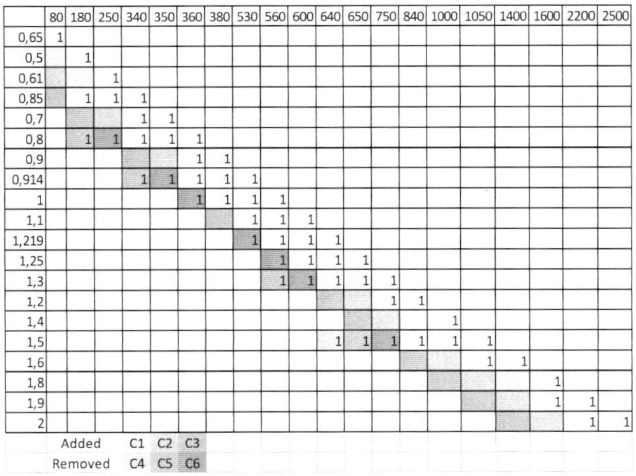

Fig. 10. Changing application ranges

Exp.	C0	C1	C2	C3	C4	C5	C6
L^*	9	9	8	8	9	10	11
R^*	2 531 731,07	2 511 562,97	2 485 160,05	2 452 522,33	2 531 731,07	2 549 321,07	2 571 879,25
$x^*_{i_j} \in X^*$	0,65; 0,80; 1,00; 1,25; 1,20; 1,40; 1,60; 1,90; 2,00	0,50; 0,80; 1,00; 1,25; 1,20; 1,40; 1,60; 1,90; 2,00	0,50; 0,70; 0,914; 1,219; 1,20; 1,40; 1,60; 2,00	0,50; 0,70; 0,914; 1,10; 1,20; 1,40; 1,60; 2,00	0,65; 0,80; 1,00; 1,25; 1,20; 1,40; 1,60; 1,90; 2,00	0,65; 0,50; 0,80; 1,00; 1,25; 1,20; 1,40; 1,60; 1,90; 2,00	0,65; 0,50; 0,61; 0,80; 1,00; 1,25; 1,20; 1,40; 1,60; 1,90; 2,00

Table 5. Results from solving the problem with different application ranges

Exp.	C0	C1	C2	C3	C4	C5	C6
C0	0	1	5	5	0	1	2
C1	1	0	3	4	1	1	2
C2	5	3	0	1	5	5	6
C3	5	4	1	0	5	5	6
C4	0	1	5	5	0	1	2
C5	1	1	5	5	1	0	1
C6	2	2	6	6	2	1	0

Table 6. Differences between the elements of the optimal size ranges, found by changing the application ranges, number of different elements

The results show, that when increasing the application ranges, the value of the total costs for the separate experiments decreases compared to the initial problem – for C1 with 1%, for C2 with 2%, and for C3 with 3%. When decreasing the application ranges the total costs increase compared to the initial problem's solution – for C4 the solution coincides with C0 and there is no change, for C5 the increase is 1%, and for C6 it is 2%. The total costs for the size range including all elements do not change. These increases and decreases are relatively small (less than 5%) and their significance can be argued.

On the other hand, the type and number of elements included in the optimal solution, significantly changes with changing the application ranges. This can be seen from Tab. 6, in which are presented the differences between the elements of the found optimal size ranges for the separate experiments.

4.4. Changing the coefficient, which takes into account the production scale factor

The coefficient taking into account the production scale factor in the initial problem (5) is equal to 4 (the value in the numerator of the expression $\frac{4}{Nj}$). The optimal solution's sensitivity is studied related to the change in this coefficient, expressing the production scale factor. The experiment is carried out for the following cases: D0 – coefficient value 4 (initial problem); D1 – coefficient value 20; D2 – coefficient value 40. The solutions found are shown in Tab. 7.

Experiment	D0	D1	D2
L^*	9	9	8
R^*	2 531 731,07	3 143 959,47	3 514 311,91
$R_L^{\bar{L}}$	2 796 363,88	3 630 495,55	4 145 115,41
$x_{l_j}^* \in X^*$	0,65; 0,80; 1,00; 1,25; 1,20; 1,40; 1,60; 1,90; 2,00	0,65; 0,80; 1,00; 1,25; 1,20; 1,40; 1,60; 1,80; 2,00	0,65; 0,80; 0,914; 1,30; 1,50; 1,60; 1,80; 2,00

Table 7. Results from solving the problem with different coefficients taking into account the production scale factor

Solving the optimization problem when changing the coefficient taking into account the production scale factor, shows the following results (Fig. 11): when changing the value of the coefficient from 4 to 20, the total costs for the optimal size range increase with 24%, and when changing the coefficient to 40, they increase with 39%. The optimal number of sizes changes from 9 when the value of the coefficient is 4 to 8, for the experiment with coefficient of 40.

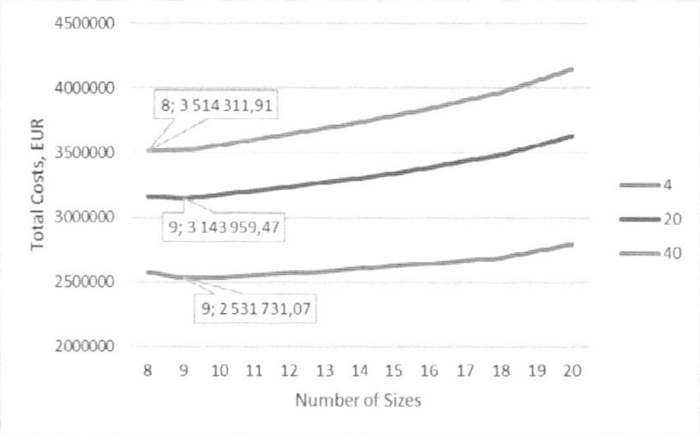

Fig. 11. Change in the total costs for change in the coefficient taking into account the production scale factor

The type of the elements constituting the optimal solution changes the most between experiments D1 and D2. Between experiments D0 and D1 the change is minimal (in one element). Increasing the scaling factor leads to increase in total costs and corresponding decrease in the number of sizes included in the optimal size range, shifting the optimal solution to the left (Fig 11).

4.5. Changing the coefficient, which takes into account the production learning rate

The coefficient, which takes into account the production learning rate has a value of 0,32 for the initial problem (the power in the expression $\left(\frac{4}{Nj}\right)^{0,32}$). The sensitivity of the optimal solution is studied relative to changes in the coefficient for the following cases: E0 – coefficient value of 0,22; E1 - coefficient value of 0,27; E2 - coefficient value of 0,32 (initial problem); E3 - coefficient value of 0,37; E4 - coefficient value of 0,42. The solutions found are shown in Tab. 8.

Solving the optimization problem with different values of the production learning rate coefficient, shows the following results (Fig. 12): when changing the coefficient's value from 0,22 to 0,27 the total costs decrease with 2,38%. When changing the coefficient to 0,32 total costs decrease with 2,25%. When changing the coefficient to 0,37 the costs decrease with 2,13%, and for a value of 0,42 the decrease is 2,01%. The optimal number of sizes remains the same for these changes. The elements included in the optimal solutions for the experiments do not change. Thus, when changing the production learning rate, the production costs decrease.

Experiment	E0	E1	E2	E3	E4
L^*	9	9	9	9	9
R^*	2 652 968,41	2 589 931,75	2 531 731,07	2 477 943,50	2 428 186,13
$R_{\bar{L}}^{\bar{L}}$	2 849 280,97	2 822 208,09	2 796 363,88	2 771 706,05	2 748 194,29
$x_{i_j}^* \in X^*$	0,65; 0,80; 1,00; 1,25; 1,20; 1,40; 1,60; 1,90; 2,00	0,65; 0,80; 1,00; 1,25; 1,20; 1,40; 1,60; 1,90; 2,00	0,65; 0,80; 1,00; 1,25; 1,20; 1,40; 1,60; 1,90; 2,00	0,65; 0,80; 1,00; 1,25; 1,20; 1,40; 1,60; 1,90; 2,00	0,65; 0,80; 1,00; 1,25; 1,20; 1,40; 1,60; 1,90; 2,00

Table 8. Results from solving the problem with different production learning rate coefficients

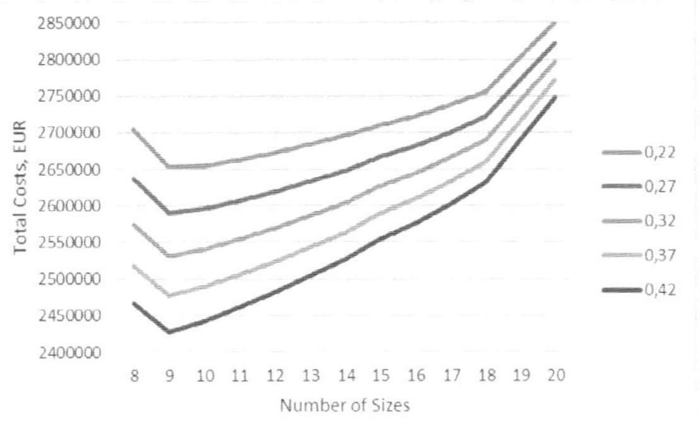

Fig. 12. Change in the total costs for change in the production learning rate coefficient

5. Conclusion

The following important results are presented in the current paper:

- A mathematical model is proposed of the problem for choosing an optimal size range for a technical product with constraints on the applicability of the elements in the size range.
- The optimal size range of pneumatically actuated linear modules is found for predefined application ranges for the elements of the size range.
- A study is made of the influence of the mathematical model's parameters on the optimal solution, and the following is established:
- the mathematical model adequately reflects the conflicting interests of manufacturer and users. Through it a size range is found, that satisfies all demand with minimum costs. The solution is a compromise between the users' requirement for bigger size range density that satisfies in full their demand, and the manufacturer's requirements for lower density size ranges aiming for bigger batches and lowering the costs for the manufactured products;
- the demand function must be determined with the greatest precision, in order to obtain reliable results;
- increasing the production quantity changes the number and type of sizes in the optimal size range, and decreasing the production quantity leads to no changes. The total costs change as such: when production quantity increases, they increase, and when production quantity decreases – costs also decrease;
- the application range significantly influences the optimal solution;
- increasing the production scale factor leads to increase in total costs, and a decrease in the number of sizes constituting the optimal size range;
- increasing the production learning rate decreases total costs, but do not change the composition and number of elements of the optimal solution.
- Through the conducted experiments the sensitivity of the elements included in the optimal solution can be studied and analysed regarding changes in the different components of the mathematical model. Even in cases with the same number of elements, different combinations are possible, having close values of their main parameter and with completely or partially overlapping application ranges. This confirms the necessity for post-optimal analysis for studying the optimum and possible close to the optimum solutions. The decision maker can have a greater preference for these solutions, due to the presence of secondary considerations and factors, which are not included, or are difficult to be taken into account, in the mathematical model.

A further development is conducting sensitivity analysis on size range optimization problems with multiple parameters. The inclusion of multiple main parameters raises significantly the mathematical model's complexity, and consequently the complexity of the experiments, that have to be conducted, both algorithmically and for solving.

6. Acknowledgments

This work was supported by the European Regional Development Fund within the Operational Programme "Science and Education for Smart Growth 2014 - 2020" under the Project CoE "National center of mechatronics and clean technologies" BG05M2OP001-1.001-0008

7. References

[1] Chivarov N., Stoev P., Yovkov S., Stoimenov N., Chivarov S., (2021) Specialized Gripper-Dispenser for Dosing Laboratory Mill With Grinding Bodies, 20th IFAC Conference on Technology, Culture and International Stability, TECIS 2021, Volume 52, Issue 25, September 14-17, IFAC-PapersOnLine, Publisher: IFAC Secretariat, ISSN:2405-8963, pp. 539-542

[2] Malakov, I. (2015). Optimization of parametric ranges of technical products, In: Complex automation of the discrete production, Malakov, I. & Nikolov, S., (Eds.), pp. 25-48, TU-Sofia, ISBN 978-619-167-153-3, Sofia

[3] Dichev, D., Zhelezarov, I., Karadzhov, T., Madzharov, N., Diakov, D. (2019) Method for measuring motion parameters of moving objects. Vide. Tehnologija. Resursi - Environment, Technology, Resources, 3, pp. 27-31

[4] Lotz, J. (2018). Beherrschung von Unsicherheit in der Baureihenentwicklung, Ph.D. Dissertation, Mechanical Engineering, TU-Darmstadt, Darmstadt, Germany

[5] Pahl, G., Beitz, W. & Feldhuzen, J. (2007). Konstruktionslehre. Methoden und anwendung, Springer-Verlag, ISBN 354-0-340-60-2, Berlin/Heidelberg

[6] Simpson, T. (2004). Product platform design and customization: status and promise, Artificial Intelligence for Engineering Design, Analysis and Manufacturing, Vol. 18, No. 1, (February 2004), pp. 3-20, DOI: 10.1017/S0890060404040028

[7] Malakov, I., Zaharinov, V. & Tzenov, V. (2015). Size ranges optimization, Procedia Engineering, Vol. 100, (January 2015), pp. 791-800, DOI: 10.1016/j.proeng.2015.01.433

[8] Malakov, I. & Boyadjiev, I. (2001). Choice of optimal range of parameters of technical products, In: Proceedings of the 12th International DAAAM Symposium, Katalinic, B., (Ed.), pp. 285-286, Published by DAAAM International, ISBN 3-901509-19-4, Vienna

[9] Mueller, D. (2011). A cost calculation model for the optimal design of size ranges, Journal of Engineering Design, Vol. 22, No. 7, (July 2011), pp. 467-485, DOI: 10.1080/09544820903449218

[10] Lotz, J., & Kloberdanz, H. (2014). Scaling under dynamic uncertainty using laws of growth, In: Proceedings of the International Symposium on Robust Design - ISoRD14, Howard, T. & Eifler, T., (Eds.), pp. 17-27, DOI: 10.4122/dtu:2083, Copenhagen

[11] Lotz, J.; Freund, T., Würtenberger, J. & Kloberdanz, H. (2016). Uncertainty in Size Range Development - an Analysis of Potential for a new Scaling Approach, International Design Conference - DESIGN 2016, , pp. 341-350, Dubrovnik

[12] Todorov, G.; Kamberov, K.; Ivanov, T. (2021) Parametric optimisation of resistance temperature detector design using validated virtual prototyping approach, Case Studies in Thermal Engineering, DOI: 10.1016/j.csite.2021.101302

[13] Todorov, G., Sofronov, Y., Dimova, K. (2022). Comparison Analysis Between Different Technologies for Manufacturing Patient-Specific Implants. In: Rackov, M., Mitrović, R., Čavić, M. (eds) Machine and Industrial Design in Mechanical Engineering. KOD 2021. Mechanisms and Machine Science, vol 109. Springer, Cham. https://doi.org/10.1007/978-3-030-88465-9_27

[14] Borgonovo, E. (2017). Sensitivity analysis, Springer, ISBN 978-3-319-52257-9

[15] Saltelli, A., Chan, K. & Scott, E. (2009). Sensitivity analysis, John Wiley & Sons, ISBN 978-0-470-74382-9, New York

[16] Rai, R. & Allada, V. (2003). Modular product family design: agent-based pareto-optimization and quality loss function-based post-optimal analysis, International journal of production research, Vol. 41, No. 17, (November 2003), pp. 4075-4098, DOI: 10.1080/0020754031000149248

[17] Razavi, S. & Gupta, H. (2016). A new framework for comprehensive, robust, and efficient global sensitivity analysis: Theory, Water Resource Research, Vol. 52, No. 1, (January 2016), pp. 423-439, DOI: 10.1002/2015WR017559

[18] Pianosi, F., Beven, K., Freer, J., Hall, J., Rougier, J., Stephenson, D. & Wagener, T. (2016). Sensitivity analysis of environmental models: a systematic review with practical workflow, Environmental Modelling & Software, Vol. 79, (May 2016), pp. 214-232, DOI: 10.1016/j.envsoft.2016.02.008

[19] BORUNTE Product Catalogue (2020), Vertical Sprayer, Guangdong BORUNTE Intelligent Equipment Co., Ltd.

[20] Zaharinov, V., Malakov, I., Nikolov, S., Dimitrova, R. & Stambolov, G. (2020). Choosing an Optimal Size Range of Pneumatically Actuated Linear Modules for Sprayer Robots, Proceedings of the 31st DAAAM International Symposium, pp.0159-0167, B. Katalinic (Ed.), Published by DAAAM International, ISBN 978-3-902734-29-7, ISSN 1726-9679, Vienna, Austria

DOI: 10.2507/33rd.daaam.proceedings.017

CLASSIFICATION OF AUTOMATIC DOORS

Velizar Zaharinov, Hasan Hasansabri & Ivo Malakov

This Publication has to be referred as: Zaharinov, V[elizar]; Hasansabri, H[asan] & Malakov, I[vo] (2022). Classification of Automatic Doors, Proceedings of the 33rd DAAAM International Symposium, pp.0115-0122, B. Katalinic (Ed.), Published by DAAAM International, ISBN 978-3-902734-36-5, ISSN 1726-9679, Vienna, Austria
DOI: 10.2507/33rd.daaam.proceedings.017

Abstract

The present paper proposes a classification of automatic doors. It includes research on automatic doors, that are used in different application areas. The classification is composed of six classification criteria, each having further branching, forming a tree-like hierarchy. In the definition of the classification criteria, an aim at maximizing orthogonality is sought. The criteria are explained in detail, examples of existing solutions are given and classified according to the criteria. The proposed classification is with an open structure, so it can be easily extended. The application area of the classification is in engineering design, aiding in the early stages of design, when market research, requirements list creation and concept generation are carried out.

Keywords: automatic doors; classification; hierarchy; automation; engineering design.

1. Introduction

The automatic doors are part of the architectural systems, which have major application in building. They are found at malls, shops, institutions, offices, hotels, restaurants, gas stations, etc., in practice, there is almost no public building that does not boast a door of this kind. Besides in public buildings, automatic doors are used in private homes, usually when there are big open areas, and when using heavy leaves with relatively big overall dimensions.

According to the current European law, the automatic doors are considered "machines", which means, that every new door of this kind must comply with the European directive on machinery [1]. In fact, in the harmonised standards [2], [3], [4] and the related [5] standard, the term "automatic door" is not used, but a distinction between manual and power operated doors is made. In [6] the following types of doors are indicated: revolving, balanced, swing, folding, and sliding.

In the standard [7] of the American national standards institute (ANSI), the term "power operated door" is defined, and in brackets, after this term, is added "automatic door". The following definition is given: "The combination of door, operator and controls constituting the system. Also called an Automatic Door." [8]. In this standard, four types of automatic doors are indicated: swing, sliding, balanced, and folding. In contrast to the mentioned European standards, where the term "power operated" door is encountered, and the American standard, in which power operated and automatic door coincide as concepts, in the current paper the term "automatic door" is adopted, i.e., a distinction is made between "power operated door" and "automatic door".

In this way a certain clarity is achieved, and is taken into account the fact that the term "power operated" [8], [9], suggests the need of an operator during the whole operating cycle, and the term "automatic" suggests partial or full level of autonomy during one or more operating cycles [10]. Thus, in general, the distinction between three types of doors can be made: manually operated, power operated, and automatic. In the specialized literature there is relatively small number of publications dealing with classification of automated doors. In [11] are indicated three design types of sliding doors: bi-parting, single slide, and telescopic. In [12] is proposed a classification of revolving doors according to the criterion "drive", separating this type of doors into three groups: manually operated, manually operated with auxiliary drive, and automatic.

At the moment of the paper's writing, to the authors' knowledge, there is no publication in the specialized literature that proposes a classification of automatic doors. In addition, the classification of sliding doors proposed in [11] is based only on one classification criterion, as the focus of the publication is completely different: simulation of a control system for sliding doors. The creation of a classification of automatic doors, will facilitate their design process and will aid the systematic study of the separate functions, systems, and components used in automatic doors building. The aim of the present paper is to propose such a classification.

2. Classification of automatic doors

The proposed classification of automatic doors is shown in Fig. 11. It is composed of six classification criteria (Fig. 1): energy used for the drive, type of motion for the leaf, leaf material, application, activation method, and leaf count.

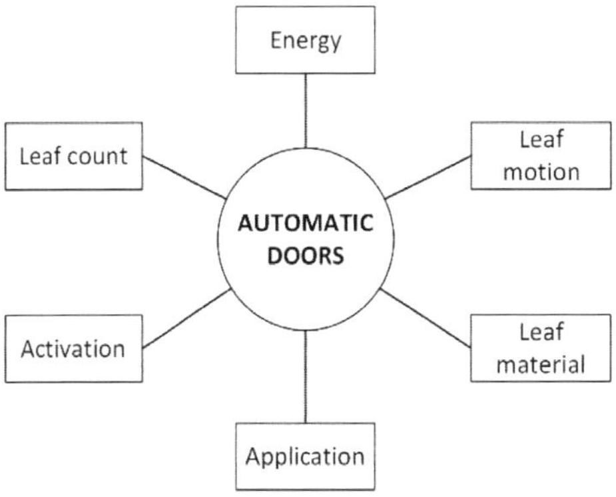

Fig. 1. Automatic doors classification – classification criteria

2.1. Energy

To eliminate the need for manual drive, and ensure autonomy, it is required to use motors for driving the mechanical system of automatic doors. The application of three types of motors is common: electrical [13], pneumatic [14], and hydraulic [15] (Fig. 2).

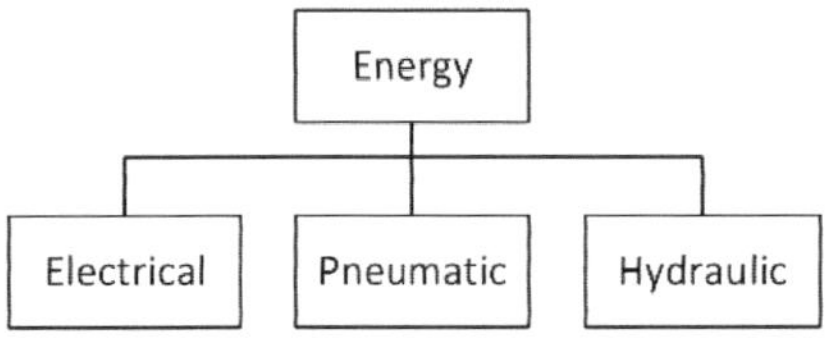

Fig. 2. Automatic doors classification – energy classification criterion

The most common drive is the electric drive. This is due to a number of advantages of those drives: accessible power source – buildings are electrified; precision control for position and speed [16], [17], [18]; the control devices and sensors also use electrical energy, therefore, there is no need for mixing power sources; high efficiency, which is important when considerning environmental impact [19], [20]; low noise levels.

Pneumatic and hydraulic drives, also called fluid drives, are used in special cases. Fluid drives are appropriate where high forces are required, i.e., very heavy leaves or high opening forces. In addition, fluidic drives are more compact than the electrical drives, for the same force/torque requirements, which in some situations can be a decisive decision-making factor, i.e., small spaces and integration into existing structures. When linear motion of the leaves is required, as per number of structural variants of automatic sliding doors, fluidic drives offer economic solution without the need of mechanism for transforming motion. There are also situations in which using electrical energy is hazardous, i.e., applications in explosive or flammable environments.

Use of fluid drives in automatic doors, can be justified also, because of its low cost (especially pneumatic drives) compared to electric drives. Some significant disadvantages of the fluid drives, hinder its widespread use with automatic doors, such as: leaks having negative environmental impact; higher energy cost; complicated precision control of position and velocity; higher noise levels; have to buy energy source and build infrastructure, as fluid systems are common in a small number of buildings – mainly industrial.

2.2. Leaf motion

A variety of motions used for moving the door leaves can be observed with automatic doors. Those motions can be classified in three major groups: rotation, translation, and planar (Fig. 3).

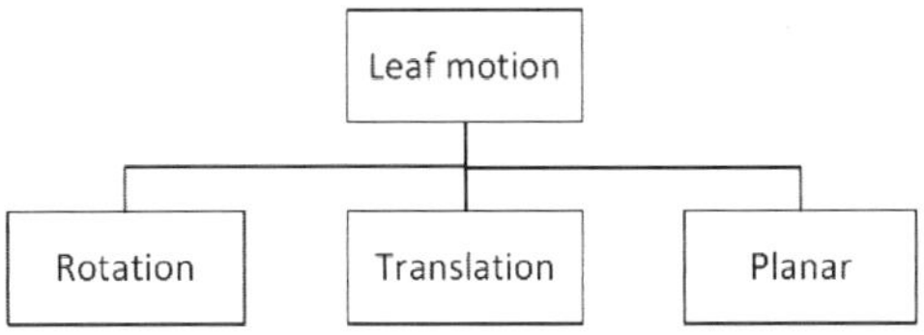

Fig. 3. Automatic doors classification – leaf motion classification criterion

In the rotation group there are three types of automatic doors: swing [21], revolving [22], and pivot [21] (Fig. 4).

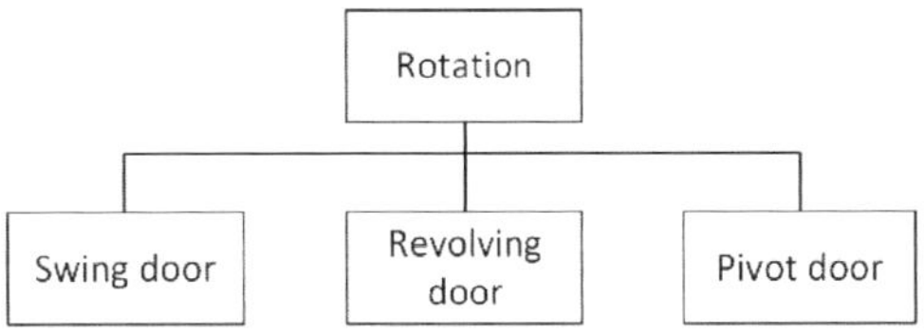

Fig. 4. Leaf motion rotation – door types

The swing automatic doors are similar to the traditional manual doors, but with added automatic drive and control. The motion is rotation around an axis situated on the left or right side of the leaf, provided with a corresponding number of hinges. A common automatic drive for this type of doors is by a motor and linkage mechanism, often integrated into one assembly unit, offered as a module [13]. Characteristic feature is that the drive unit is visible.

The automatic revolving doors are offered in a great variety of structural variants [6]. This type of automatic doors realizes the rotational motion of the leaves through a central axis. In contrast to the swing doors, here the drive module usually is integrated, and not visible. The application of these doors is in public buildings with large pedestrian traffic. The pivot doors are an alternative to swing doors in the cases where the leaf is large and heavy. In those cases, the use of hinges is not eligible due to the large forces and the risk of leaf displacement. In this type of doors, the rotation is provided by an axis, which is placed close to the right or left side of the leaf, and in contrast to the revolving door, it is not central.

Translational motion is realized in the sliding automatic doors. The latter are not only very common in public buildings, but also in vehicles and safety applications, i.e., safety systems in metro stations. Some common types are: sliding panoramic systems [24], lift and slide [25], and folding doors [26] (Fig. 5). The sliding panoramic systems are becoming very popular. They are not only utilized in public buildings, but also in private residences. These systems provide maximum concealment of the aluminum structure for achieving maximum transparency and good heat insulating and noise suppressing characteristics.

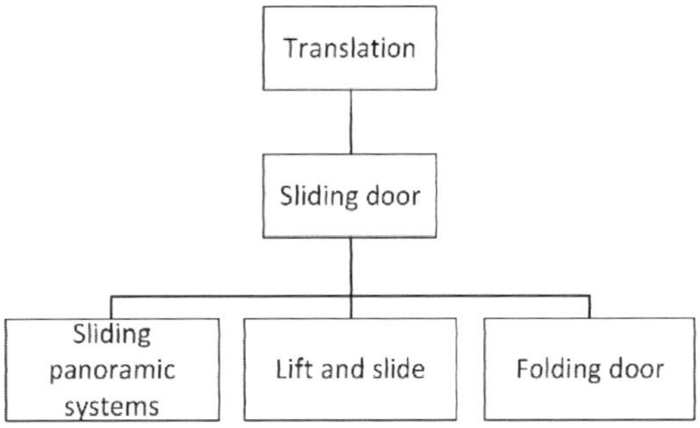

Fig. 5. Leaf motion translation – door types

Lift and slide are a combination of two translational motions of the leaf in the same plane. This type of doors provides better insulation (thermal and noise) characteristics with regard to the normal methods for insulating sliding doors. Usually, the drive is concealed and not visible. With folding doors, the leaves can be compactly packed (folded) during the translational motion. To enable folding, the leaf is divided into separate, flexibly linked parts, which rotate and orient parallel to each other when the leaf is slid.

Under planar motion it is understood movement of the leaf, which is composed of several simple (rotations and/or translations) motions, executed in one plane. The following door types are considered belonging to this group: roller doors [27], sectional doors [28], up-and-over doors [29], and volkswagen [30] (Fig. 6). Automatic roller doors combine translational motion (sliding) and rotation (rolling) in a longitudinal to the leaf plane. Usually, the translation is made to lift or lower the leaf, which necessitates that the rotation (the rolling) is executed over the door opening. Here, similar to folding doors, the leaf is composed of smaller elements, which can flexibly move relative to one another, which makes possible to roll them. The activation of this door type is usually from a distance or using a control panel. Used in applications for garage doors, storage rooms, and also as part of security systems, i.e., in shops.

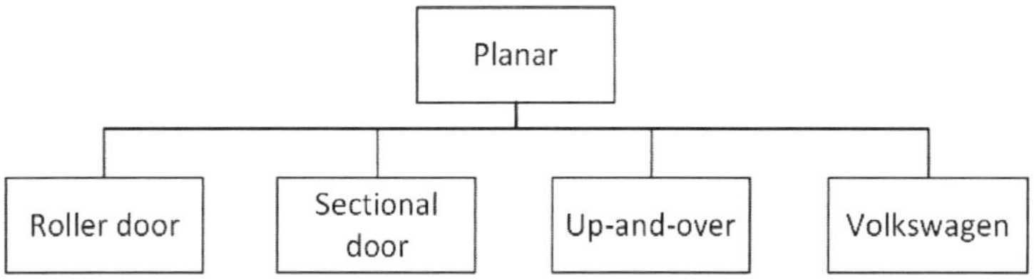

Fig. 6. Leaf motion planar – door types

Characteristic for the leaf's motion type, in sectional doors, is the complex planar movement, combining translation and rotation. By this movement the leaf is lifted and rotated, so that from vertical position it assumes horizontal position. The leaf of a sectional door is similar to the leaf of a roller door in that it is built from separate sections, which can move relative to each other. The difference is that the sections are bigger, and because of this the leaf cannot be rolled. Thus, instead of rotating for opening the leaf, it is lifted above the door opening, and assumes horizontal position.

Up-and-over doors have the same planar movement like sectional doors, but they are not sectioned – the whole leaf moves as one, from vertical to horizontal position. Automatic volkswagen doors use a combination of two translations in a transverse to the leaf plane. The first translation is in "front-back" direction, and the second – "left-right". Thus, the first translation moves the leaf away from or in direction of the door opening, and the second translation moves the leaf to the side of the door opening so it is possible to go through the door. This type of doors is used in automobiles, hence the name. In buildings, only manually operated volkswagen doors are used.

2.3. Leaf material

Automatic doors can be distinguished also by the material, from which the door leaves are made. According to this classification criterion they are made from wood [31], polymer (PVC) [32], steel [33], aluminum [22], glass [34], and hybrid (aluminum-wood) [35]. The automatic doors with wooden leaves have good thermal and noise insulation parameters. They can be renovated if damaged. Main disadvantage is the difficult maintenance, and high price.

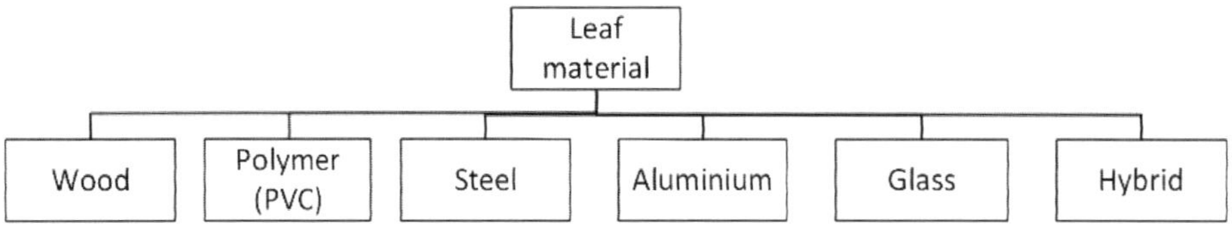

Fig. 7. Automatic doors classification – leaf material classification criterion

Polymer (PVC) automatic doors have their place in architectural systems. They are cheaper than wooden automatic doors and have good noise and thermal insulation. Their maintenance is easier in comparison to the wooden automatic doors. Can be recycled. Polymer doors are not suitable where big sized leaves are required.

Aluminum automatic doors are one of the most commonly used in practice. Aluminum is good heat conductor, and because of this, for exterior automatic doors, insulated (warm) profiles are used, in which the inner and outer parts of the profile are divided by polyamide or another polymer. The aim is to disrupt the thermal flow and to provide better thermal insulation. Uninsulated (cold) aluminum profiles are suitable for inner premises. These doors are also recyclable.

Steel doors provide greater anti-theft protection than other door materials. Steel doors are also fire resistant, and do not allow fire to spread in other rooms. Steel profiles can be also insulated and uninsulated. Steel doors allow for large leaves, and therefore large doors to be produced. These doors are recyclable. One disadvantage is the not so good thermal insulating properties.

Glass automatic doors are used for maximizing transparency. They can be with one or two sliding leaves. Mainly used in inner premises of buildings. Hybrid automatic doors (aluminum-wood) combine wooden door from the interior side, and aluminum material on the exterior side of the building, for better weather resistance. Limitation of such doors is the size in which they can be produced.

2.4. Application

As mentioned, the application of automatic doors is not limited to only buildings. According to application automatic doors can be divided into: entrance doors, garage doors, access portal, vehicle doors, and safety doors (Fig. 8). As entrance doors sliding and revolving doors are more frequently used, and less frequently automatic swing doors. Activation of the automatic entrance doors is usually contactless – by camera or proximity sensor. Most common are the electrically driven automatic entrance doors.

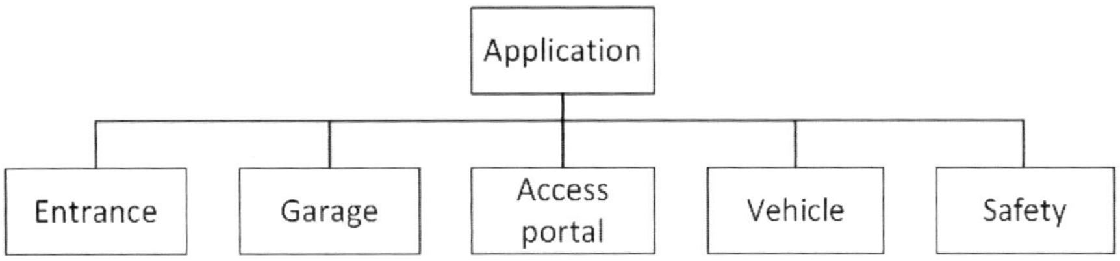

Fig. 8. Automatic doors classification – application classification criterion

Mostly sectional and roller doors are used as garage doors. Frequently used are also sliding doors, usually as a fence, obstructing the access to the area of a yard intended for use as a garage. Usually, activation is remote, and the drive is electric. Access portals are used for limiting access to certain rooms or whole buildings. In such applications the automatic door is usually integrated with a system for access control and/or security system. Access systems for public facilities, with sensor control and automatic elements, are offered in a variety of designs. They provide convenient, contactless entry to a variety of public places such as swimming pools, SPA centers, entertainment areas, museums, and public transportation.

Automatic doors are used also in vehicles. Often, this type of doors is sliding, and in a number of applications they are pneumatically actuated. Activation is typically through contact, but in automatic (robotic) vehicles it can be contactless. Automatic doors can be used also as safety devices. An example for such application is safety doors at metro stations and in automatic railway transport (robotic trains). The openings in the safety fences of these applications are opened and closed by automatic doors.

2.5. Activation

One of the main factors for autonomy of automatic doors is the method of detecting when to open and close the door by the control system. According to the activation method, automatic doors are contact or contactless activated (Fig. 9). Contact type activation can be done by sensitive mat [36] and by control panel [37], [38]. Contactless activation methods include remote activation [37], camera activation [39], [40] or presence sensor [39].

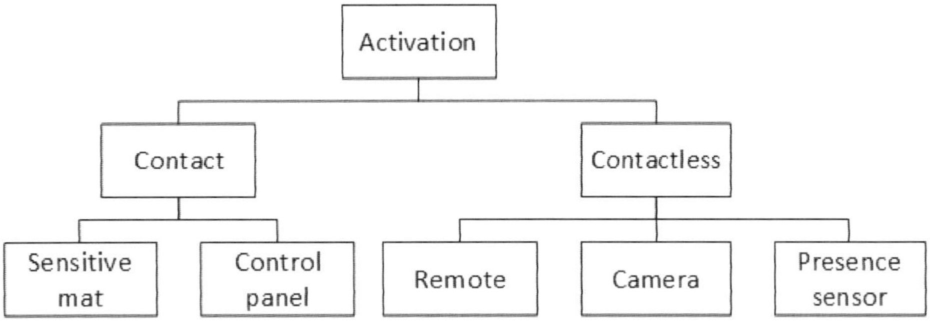

Fig. 9. Automatic doors classification – activation classification criterion

Activation through a sensitive mat is an activation method used also in safety systems. There are three operation principles of this type of device: electrical, pneumatic, and optical. For electric type, stepping onto the mat leads to contact between conductive elements, changing the electric characteristics of the circuit in which they are part of. These characteristics are monitored from the control system. For the pneumatic type, inside the mat there are flexible tubes, arranged in a particular layout. The tubes conduct air with low pressure, which when stepped on, produce pulses in the airflow of the system. For the optical type, stepping on the mat leads to the displacement of optical fibers that are inside the mat, which leads to a change in the received light from a photodetector.

Control panel activation for automatic doors can be done in the case where the door control system allows for autonomous operation after activation. Otherwise, the door is not automatic, but power operated, and would require user operation for the duration of opening and closing of the door. Variety of panels are utilized, which usually provide manual mode of operation for the automatic door. The integration of additional functions is possible, such as requirement for identification, and other procedures for authorized access, which places the application of such activation type in security systems.

Contactless activation is more common method of activation for automatic doors. Remote activation assumes the presence of activating device called a remote control. This device can be made as a conventional remote control – mobile control panel, which is wired or wirelessly connected to the door control system. Current systems for remote activation allow activation from mobile devices such as smartphones.

Activation through presence sensor allows door activation when a moving object is detected in the sensor sensing range. It is necessary to adjust the system before operation in order to define the sensing area and the size of the sensed objects. Infrared, microwave, and other presence sensors are used. Camera activation presents the possibility for recognition of characteristic features of the person approaching the door. This type of activation can be used for authorized access. In addition to this supplementary function, an advantage over using a presence sensor is the ability to include more decision-making parameters for the door activation (whether it opens or not in front of the approaching person or object). With a camera, objects that are not moving can also be detected.

2.6. Leaf count

Great variety, not only for automatic doors, but for doors in general, introduces the leaf count. There are automatic door designs with one [41], two [41], and more than two [26] leaves (Fig. 10). Most common are designs with one or two leaves. Automatic doors with more than two leaves typically are used for big openings, and for panoramic views, i.e., for terraces and balconies.

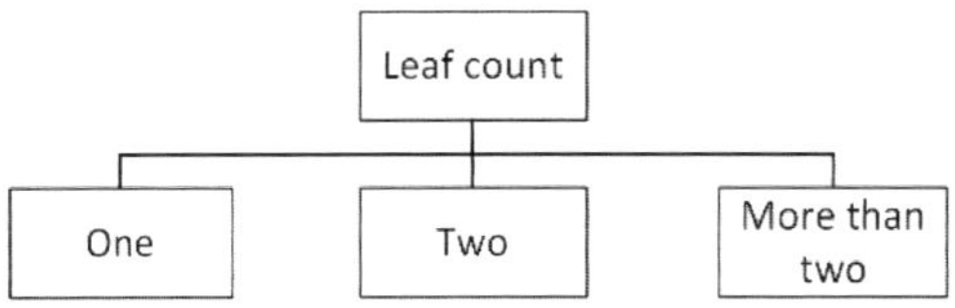

Fig. 10. Automatic doors classification – leaf count classification criterion

3. Conclusion

A classification of automatic doors is proposed including six classification criteria. The classification has a treelike structure (Fig. 11), with each of the classification criteria branching into a number of levels. The last level of each branch indicates particular automatic door design.

This classification structure is designed to be open, allowing for adding more criteria to it, and classifying new types of doors, easy and fast.

Each classification criterion and branch are explained in detail, and examples of existing designs are given with corresponding citations based on market research. The proposed classification is intended as a tool for aiding the design process of automatic doors in the following ways:

- Standards – a number of standards are cited, having relation to the design, building, and safety of automatic doors;
- Design – quick reference into the existing designs, and finding examples of such from the cited manufacturers;
- Drive – reference into the methods of driving automatic doors;
- Application areas – situations in which automatic doors find common application and are appropriate solution;
- Methods of activation – quick reference into the sensors and activation methods commonly used with automatic doors.

Accepting the proposed classification as a design aid in the design process of automatic doors, will have a positive impact on the design effort, speeding it up by providing plenty of information regarding posible structures, drives, door types, etc. The classification is also useful when trying to design innovative automatic doors, by showing what is already available, and thus aiding the idea generation process.

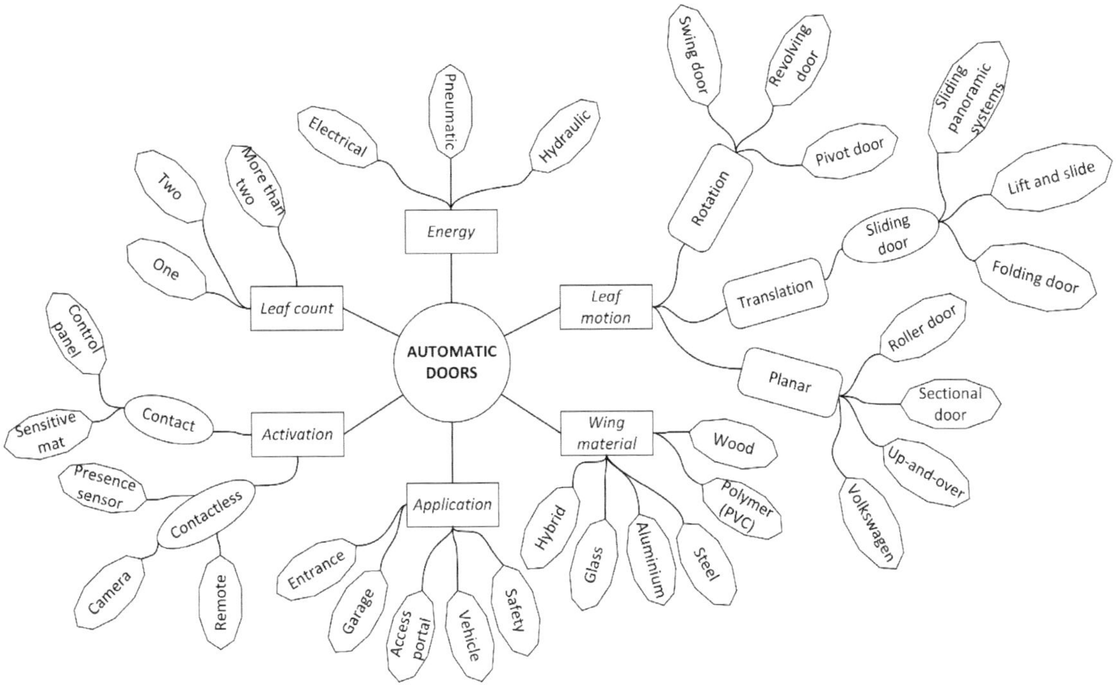

Fig. 11. Automatic doors classification

Some of the negatives of using classification schemes in the design process, are the tendency of going in the direction of the already known, and somewhat lowering the potential for new ideas. Nevertheless, the benefit of shortening the time for the initial idea generation, often is a more significant factor in practice.

The proposed classification is the first part of a research on sliding automatic doors, on defining the main systems and functional structures of this type of doors, their place amongst other types of automated doors, and common structures that they share. Future work is the use of the classification in the design of a base structure for a new sliding door of the type of the panoramic sliding systems, and a corresponding size range.

4. Acknowledgments

This work has been accomplished with financial support by the Grant No BG05M2OP001-1.002-0011 "MIRACle (Mechatronics, Innovation, Robotics, Automation, Clean technologies)", financed by the Science and Education for Smart Growth Operational Program (2014-2020) and co-financed by the European Union through the European structural and Investment funds.

5. References

[1] Directive 2006/42/EC (2006) Machinery (MD). European Parliament, Council of the European Union.
[2] EN 13241 (2003). Industrial, commercial, garage doors and gates - Product standard performance characteristics. CEN.
[3] EN 12978 (2003) Industrial, commercial and garage doors and gates - Safety devices for power operated doors and gates - Requirements and tes methods. CEN.
[4] EN 12635 (2002) Industrial, commercial and garage doors and gates - Installation and use. CEN.
[5] EN 12453 (2017) Industrial, commercial and garage doors and gates - Safety in use of power operated doors - Requirements and test methods. CEN.
[6] EN 16005 (2012) Power operated pedestrian doorsets - Safety in use - Requirements and test methods. CEN.
[7] ANSI/BHMA A156.10 (1999) Power Operated Pedestrian Doors. ANSI/BHMA.
[8] Semerdzhiev, S. et al. (1992) English-Bulgarian Technical Dictionary. 2nd ed. Sofia: Technica Publishing House.
[9] Artobolevskiy, I. et al. (1979) Polytechnical dictionary. Sofia: Technica Publishing House.
[10] Hornby, A. (2000) Oxford Advanced Learner's Dictionary of Current English. 6th ed. s.l.:Oxford University Press
[11] Khine, K. (2018) Simulation of Automatic Door Sliding Control System. International Journal of Scientific Engineering and Technology Research, 7(12), pp. 2000-2006.
[12] Haese, A. (2018) Revolving entrance doors: Machines or structural elements?. Glass Structures nad Engineering. DOI: 10.1007/s40940-018-0081-x

[13] Dormakaba. (2021a) Swing door operators ED100/250 Technical Product Brochure. Dormakaba.
[14] Allegion. (2020) LCN 4000 Series Catalog. Allegion.
[15] Schweiss. (2022, February 22) Bifold Doors - Hangar Doors, Hydraulic Doors, Schweiss Doors. Retrieved from Bifold.com, HydraulicDoors.com, Schweiss Doors: https://www.bifold.com/index.php
[16] Myakhor, D.A., Dimitrov, L.V. (2020) Development of a control board for walking mechatronic systems: Principle of operation and implementation. 2020 1st International Conference Problems of Informatics, Electronics, and Radio Engineering, PIERE 2020 09314657, pp. 257-260
[17] Chivarov N., Stoev P., Yovkov S., Stoimenov N., Chivarov S. (2021) Specialized Gripper-Dispenser for Dosing Laboratory Mill With Grinding Bodies, 20th IFAC Conference on Technology, Culture and International Stability, TECIS 2021, Volume 52, Issue 25, September 14-17, IFAC-PapersOnLine, Publisher: IFAC Secretariat, ISSN:2405-8963, pp. 539-542
[18] Dichev, D., Zhelezarov, I., Karadzhov, T., Madzharov, N., Diakov, D. (2019) Method for measuring motion parameters of moving objects. Vide. Tehnologija. Resursi - Environment, Technology, Resources, 3, pp. 27-31
[19] Medjugorac, M.; Susac, S.; Maric, P. & Filipovic, T. (2020) Importance of Energy Management in Buildings, Proceedings of the 31st DAAAM International Symposium, pp.0891-0898, B. Katalinic (Ed.), Published by DAAAM International, ISBN 978-3-902734-29-7, ISSN 1726-9679, Vienna, Austria
[20] Georgi Todorov, Konstantin Kamberov, and Adrian Ivanov (2018) Decreasing power loss through control improvement of kinetic UPS system, AIP Conference Proceedings 2048, 020017 https://doi.org/10.1063/1.5082035
[21] Dormakaba. (2021b) TS 97 A new closer creation for enhanced architectural elegance. Dormakaba.
[22] Dormakaba. (2020) KTV 3/KTV 4 automatic revolving doors. Dormakaba
[23] Alumil. (2022a) Alumil Suprime SD 115 PIVOT. Alumil.
[24] Alumil. (2022b) S650 e-MOTION. Alumil.
[25] Reynaers. (2022) CP 155/CP 155-LS. Reynaers.
[26] Dormakaba. (2018a) FFT Flex Green. Dormakaba.
[27] Hörmann. (2020) Garage doors RollMatic. Hörmann.
[28] Hörmann. (2021a) Sectional garage doors. Hörmann.
[29] Hörmann. (2021b) Up-and-over garage doors. Hörmann.
[30] Volkswagen. (2022, February 23) Volkswagen vans Bulgaria. Retrieved from https://www.vw-lekotovarni.bg/
[31] Doorson. (2022) 400 Timber Wood Sliding Door. Doorson.
[32] Kömmerling. (2019) KOMADOOR. Kömmerling.
[33] Inovastil. (2022, March 4) InovaStil Home. Retrieved from https://inovastil-bg.com/
[34] Euroglass. (2022, March 4) Euroglass Homepage. Retrieved from https://euroglass-bg.com/
[35] Artistico Systems. (2022, March 4) Artistico Systems Homepage. Retrieved from http://www.artistico-systems.bg/
[36] Larco. (2022, February 16) Automatic door mats. Retrieved from https://www.larco.com/door-access-controls/automatic-door-mats
[37] Hörmann. (2019) Garage door and entrance gate operators. Hörmann.
[38] Geze. (2022) Geze competence in bank buildings. Geze.
[39] Dormakaba. (2022) Meeting all your access needs. Dormakaba.
[40] Optex. (2017) A new sensing solution. Image-recognition automatic door sensor VVS-1(BL). Optex.
[41] Dormakaba. (2018b) ST TST automatic sliding door systems. Dormakaba.

DOI: 10.2507/33rd.daaam.proceedings.018

SUSTAINABILITY ASSESSMENT OF SMALL HYDROPOWER PLANTS

Sanda Midžić Kurtagić, Sadbera Trožić Borovac,
Branislav Bijelić & Halima Hadžiahmetović

This Publication has to be referred as: Midzic Kurtagic, S[anda]; Trozic Borovac, S[adbera]; Bijelic, B[ranislav] & Hadziahmetovic, H[alima] (2022). Sustainability Assessment of Small Hydropower Plants, Proceedings of the 33rd DAAAM International Symposium, pp.0123-0132, B. Katalinic (Ed.), Published by DAAAM International, ISBN 978-3-902734-36-5, ISSN 1726-9679, Vienna, Austria
DOI: 10.2507/33rd.daaam.proceedings.018

Abstract

The paper presents a set of criteria for assessing the sustainability of Small Hydropower Plants in Bosnia and Herzegovina developed on the basis of the catalogue of criteria for the use of hydropower in the Austrian province of Tyrol as well as the application of the criteria in a case study. Bearing in mind that the criteria used in the Tyrolean catalogue refer to all types of hydropower plants without capacity limitations, and the catalogue for SHP applies to plants with a capacity of less than 10 MW, it was necessary to choose a methodology for adapting the Tyrolean catalogue. The SHP catalogue includes five interdisciplinary areas, as well as the Tyrolean catalogue, namely: Energy, Water Management, Spatial Planning, Water Ecology, Nature Protection. The evaluation of the project by all criteria is done on the basis of a point scale from 0 to 5, and weighting factors are taken into account for the final evaluation by area. Ultimately, the values obtained according to the criteria are classified into ranges of values marked with red, yellow and green colours. Criteria whose values fall into the red range are not acceptable, while the yellow colour suggests the possibility of improving the project.

Keywords: Small hydropower plant; Multicriteria analyses; Sustainability assessment; Scoring intervals.

1. Introduction

Small Hydropower Plants (SHPs) are highly represented in Bosnia and Herzegovina (BiH). By 2020, 113 have been built, and another 340 are planned. Out of total 2373 MW of renewable energy installed in BiH in 2020 (2,42% below the targeted 40% of renewables share for 2020), 172 MW are produced in SHPs [1]. Decision-making on the construction of hydropower plants and feasibility assessment ware mainly based on the criterion of maximum utilization of water potential, in fact there are no procedures for mitigating the conflicting goals between energy, water and the environment. Investors often plan the construction of SHP in protected natural areas, and the competent authorities issue a concession regardless of the protection objectives. Several problems related to the development of hydropower has been identified in the country: over-ambitious assessment of usable hydro potential, subsequent planning without involvement and communication between different sectors in charge of managing and protecting water resources, while corruption is especially emphasized mainly due to lack of bureaucratic transparency. [2]

This leads to conflicts between local communities and different interest groups. The SHP projects are often considered as an alternative to large hydroelectric power plants due to the assumed lower environmental impact [3]. Although studies of the impact of SHP on the biological communities show different levels of impact, from very small [4] to significant [5], impacts related to SHP cannot be neglected, so there is a need for a comprehensive set of indicators allowing complete and transparent assessment of the sustainability of SHP projects, from various perspectives. With the growing interest in maintaining and improving the quality of the environment, companies and investors turning with increased rate their attention to the potential impacts of their activities on the environment. Interest in achieving and proving good environmental performance through managing the impacts of one's activities is a necessary direction for environmental protection and achieving the goals of sustainable development. [6] In the context of the development of hydropower projects, the environmental protection policy should be based on the Application of criteria for sustainability assessment of hydropower projects in the early stage of the project development, i.e. on the level of the project design.

Number of authors contributed to development of a sustainability criteria for hydropower assessment and multicriteria evaluation methods. Following the results of the CHOICE project [7], the authors [8] developed a concise criteria matrix that was used to evaluate impacts of the project and its alternatives during the construction phase and during the operational phase of hydro power plants (HPPs). They included quality elements of the river ecosystem into consideration, such are: phytobenthos, macrophytes, fish fauna, benthic, invertebrates, morphological, conditions, water quality, terrestrial and habitat. For selection within a set of alternative HPPs, i.e. for ranking, they took construction costs, installed capacity (in MW) and production capacity (GW h/y) using Vikor method as multicriteria decision aid. However, the tested method did not take social criteria into account. A chronological list of research studies along with the reported technical, ecological, economic and social parameters has been provided and analysed by [9]. Authors pointed on the complexity in quantifying social impacts, as one of the most challenging aspects of hydropower development. Under the reported parameters, only few authors consider limited number of technical criteria such are: installed capacity, development period, transmission, etc. Further development of environmental criteria has been presented by [10], who developed a comprehensive checklist of river function indicators for hydropower ecological assessment has been developed, but no social or economic criteria has been included. Authors [11], tested innovative multicriteria assessment model, i.e Mesohabitat Simulation Model, on hydropower projects in Aosta Valley, as a good representative of Alpine region. They replaced the EU Water Framework Directive (EU WFD)[12], biological indicators with the MesoHABSIM based river Habitat Integrity Index, amending he list of indicators with "energy index", that quantifies the production losses due to the flow releases; and to economy related indicators: , "producer income" and "local community income". this method also includes the indicator of "landscape protection level", assessing to which extent the landscape changes due to the changes in the natural flow. The analysed research papers show that the challenge of defining an appropriate set of sustainability indicators that covers social, economic, technical and environmental indicators is still present and that most of the papers refer to hydropower plants in general. Some authors have investigated the criteria for SHP analysis, but the research has mostly focused on aquatic ecology perspective [13], [14].

The transposition and implementation of the EU WFD (2000/60/EC) in BiH imposed a new imperative and the need to define criteria more precisely, bearing in mind that the directive requires the achievement of a good qualitative and quantitative state of waters, assuming the "principle of non-deterioration" of the existing ecological status. On the other hand, the national goals of increasing the share of renewable energy and the interest of investors in the construction of new renewable capacities as SHP, represent additional pressure on water bodies, which required the urgent development of a set of evaluation criteria that will enable the assessment of the sustainability of SHP, taking into account the efficiency in energy production, efficiency in the use of water potential, impact on water ecology, biological and landscape diversity, population and space, enabling decision makers to make an informed decision. The paper presents a set of criteria for assessing the sustainability of SHPs in Bosnia and Herzegovina as well as the application of the criteria in a case study.

2. Method and adaptation results

A set of criteria for assessing the sustainability of Small Hydropower Plants (SHP) in BiH is developed on the basis of the catalogue of criteria for the use of hydropower in the Austrian province of Tyrol [15]. The Tyrolean Catalogue has been prepared as a basis for regional planning in the field of hydropower supply, and it includes an assessment of hydropower projects in the early stages of planning. The Tyrolean Catalogue contains five thematic areas as follows: energy management, water management, spatial planning, water ecology and nature protection. The criteria used in the Tyrolean Catalogue refer to all types of hydropower plants without capacity restrictions. The catalogue of criteria for BiH refers only to small hydropower plants the capacity of which is less than 10 MW. Therefore, a first step in adaptation approach was to eliminate criteria not applicable on SHP with capacity less than 10 MW. A second step was elimination of criteria for which BiH lack of data and information. Third step was to adapt the names of the adopted criteria to the professional terminology used in the territory of BiH. This resulted in adoption of 33 out of total 55 criteria (6 for Energy management, 9 for Water management, 10 for Spatial planning, 22 for Water ecology 8 for Nature protection) from Tyrolean catalogue (Table1). In addition to the 33 criteria we introduced category of 5 eliminatory criteria that would suspend a project in case of a negative evaluation (Table 2). 4 of them are introduced due to the legally prescribed restrictions while existence of valid hydrological study is introduced with the aim to improve the quality of the studies accompanying the project documentation, i.e. the reliability of the input data based on which the hydropower facility is planned and designed.

Energy management	**Water management**	**Spatial planning**	**Water ecology**	**Nature protection**
Specific investment	Degree of utilization of hydropower potential	Spatial planning documents	Ecological status	Protection of species
Contribution to climate protection	Water management related characteristics of the SHP, using the following indicators: • Degree of Utilization • Ratio of Head and Length of Diversion • Duration of spillover	Direct use at watercourses	Hydromorphology	Protection of natural habitat
Connection to the distribution network	Efficiency of Utilization of the River Stretch	Infrastructure	Watershed area	Ecosystem
Additional Effects / Synergies.	Change of Hazard Potential	Agriculture	Presence of natural fish spawning site	Landscape-recreational values
	Effects on Immission Situation	Forestry	Special types of water bodies	Natural values of water courses
	Effects on Groundwater	Tourism	Thermal pollution	Sensitive water bodies
		Local Economy	Free flow pathways	Sensitive/ unique water bodies
			Reservoir size	Protected areas

Table 1. List of selected criteria for BiH Catalogue

No.	Subject area	Criterion	Reason for introduction
1	Water Management	Existence of valid hydrological study	Ensuring reliability of the input data
2	Spatial Planning	Cultural property	Legal
3	Water Ecology	Ecological Flow	Legal
4	Water Ecology	Reference sites	Legal
5.	Nature protection	Protected areas	Legal

Table 2. Eliminatory criteria for BiH Catalogue

For most of the criteria the Tyrolean Catalogue adopts a unique scoring method with a scale of 0-5 and is established in a way that score 0 corresponds to the lowest and score 5 to the highest score. For quantified type of indicators (energy management and water management) specific scoring intervals are associated to each score. Since the criteria in the field of Energy and Water management depend primarily on the technical and economic parameters of the hydropower plant, and the Tyrolean Catalogue covers all types and sizes of power plants, there was a need to adjust the range of values used in scoring. We used data from 2019, when 83 SHPs with an installed capacity of up to 10 MW were built. The following information was collected and analysed necessary to adjust scaling in scoring system for selected energy management and water management set of criteria:

1. Mean annual flow rate of river stretch
2. Length of pipeline
3. Installed flow rate of HPP
4. Installed rated capacity
5. Average annual electricity production
6. Gross head
7. Length of water course section under the influence of HPP
8. Full load hours - calculated based on input data
9. Length of connection to the distribution network (km)
10. Total investment costs –calculated based on input data,
11. Type of SHP

After classifying the collected data, we completed quantitative data for 30 SHPs. Hence, analyses per criterion were made by using all available data. The approach to adjusting the scoring system is illustrated on the example of the indicator "Energy related efficiency" for which the scoring intervals are not changed after analyses and the indicator "Aspects of the grid economy" for which the scoring intervals are changed after analyses. "Energy related efficiency" is an indicator of the ratio of the length of the waterbody (m) section affected by the plant and average annual production in a normal year (GWh/year). This shows the extent to which the hydropower plant uses the watercourse in relation to the production. The length of the waterbody under the influence of the power plant should be minimized, and annual production maximized. This criterion for BiH SHPs ranges from a very good 114 m / GWh to a very unfavourable 8344 m / GWh (Fig1).

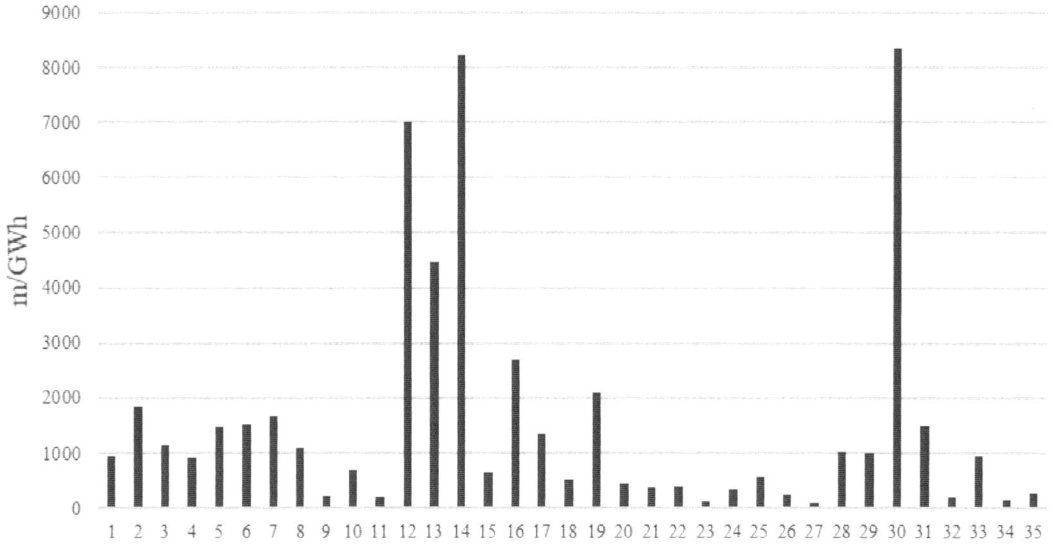

Fig. 1. Values of the criterion "Efficiency of Utilization of the River Stretch - energy efficiency" for existing SHP in BiH

Applying the Tyrolean Criteria to the SHPs in BiH shows that only in 2 SHPs, the waterbody is efficiently used (4 or 5 points), 11 show medium scoring (2 and 3 points) while 19 have 0 points, since the amount of the produced energy is small in relation to the length of the waterbody under the influence of a power plant (Table 3).

Score	Length of waterbody section/yearly production (m/GWh)	Number of existing SHP
0	800 < EE	19
1	500 < EE ≤ 800	4
2	250 < EE ≤ 500	6
3	125 < EE≤ 250	5
4	80 < EE ≤ 125	2
5	EE ≤ 80	0

Table 3. Number of SHP built in BIH falling into the scoring intervals of the "Energy related efficiency" as set in the Tyrolean Catalogue

Small power plants built in BiH have values from almost all intervals foreseen by the Tyrolean Catalogue, and as the criterion in terms of scope and purpose should be minimized, the scoring scale from the Tyrolean Catalogue is adopted for BiH without changes. Connection to electrical power transmission grid of new hydropower plants can require lower or higher cost, depending on the distance from a suitable network hub as well as the way it connects to the network (cables or transmission lines). On the other hand, new hydropower plants may have positive or negative impact on the electricity grid which is needed to transfer the electricity that can be produced.

With a decentralized method of production, it could be possible to avoid the need for further network development measures and support weaker areas of the network. However, additional production may make it necessary for the network to be further developed, in cases where there are no consumers in the immediate vicinity or when the existing network is too weak. The grid related aspects are determined based on the distance of the power plant from the point of connection to the public grid and the average annual electricity production. Additionally, for the assessment in the Tyrolean catalogue, the voltage level at the grid connection point is relevant. The result is expressed in km / GWh /per year. In the Criteria Catalogue for BiH, this criterion will be applied under the name "Connection to the Distribution network" (Fig. 2).

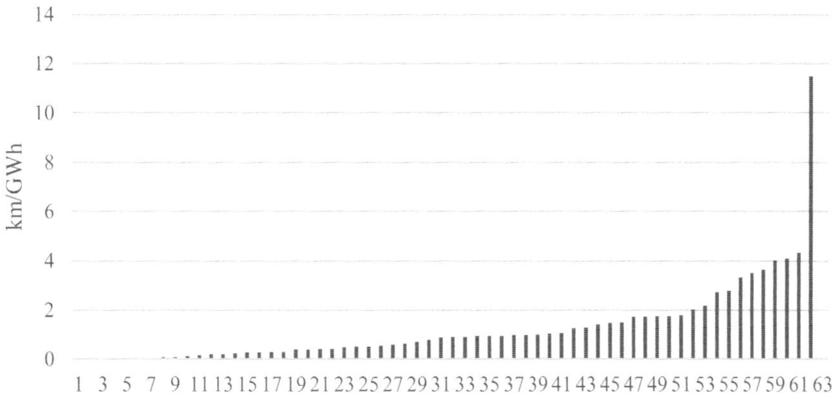

Fig. 2. Value of Criterion "Connection to the distribution network for SHPs built in BiH"

When applying the Tyrolean Catalogue scoring intervals on the selected BiH SHPs, out of 63 SHPs 41 have values of the criterion "Connection to distribution network" above 0.45, thus scored with 0. Bearing in mind that the electricity distribution network is quite underdeveloped, and that better scoring is not realistically possible to achieve in Bosnia and Herzegovina, the scoring intervals have been adjusted. (Table 4)

Score	Scoring intervals Aspects of Grid Economy - Tyrolean Catalogue e_{Net}[km / GWh/ann.]	Number of SHP built in BiH falling into the Tyrolean catalogue range	Adopted scoring intervals and criterion Connection to distribution network- BiH e_{Net}[km / GWh/ann.]	Number of SHP built in BIH falling into the BiH catalogue range
0	$e_{Net} \geq 0.45$	41	$e_{Net} \geq 1.50$	18
1	$0.35 \leq e_{Net} < 0.45$	4	$1.00 \leq e_{Net} < 1.50$	6
2	$0.25 \leq e_{Net} < 0.35$	3	$0.75 \leq e_{Net} < 1.00$	10
3	$0.15 \leq e_{Net} < 0.25$	5	$0.50 \leq e_{Net} < 0.75$	6
4	$0.05 \leq e_{Net} < 0.15$	3	$0.10 \leq e_{Net} < 0.50$	15
5	$e_{Ne} \leq 0.05$	7	$e_{Net} < 0.10$	9

Table 4. Analysis of the criterion "Connection to distribution network" for selected hydropower plants trough Tyrolian Catalogue scoring

For the criterion "Additional Effects / Synergies", under the Energy management thematic area the following country specific additional effects /synergies are selected as subcriteria: i) Impact on the operation of other hydropower plants, ii)Synergy with parts of existing plants; iii) Additional use as reversible plant, iv) Output regulation, v) Security of supply, vi) Contributions to local / regional energy programs. If the criteria are positively evaluated it will gain 1 or 2 points. In case of decrease in production in other hydropower plants the criterion "Impact on the operation of other hydropower plants" will receive negative points:

- (-1) in case of 5 - 10% decreased production in other hydropower plants
- (-2) in case of ≥10% decreased production in other hydropower plants

For the overall assessment of the "Additional Effects / Synergy" criterion, the points of individual subcriteria are added up, where the sum cannot exceed 5 and be less than 0. The criteria for the area of spatial planning have been adapted taking into account the availability of spatial planning documentation and the criteria and provisions of national legislation for the protection of natural resources included in the set of spatial planning area. The criterion "spatial planning at the local level" in the Tyrolean catalogue is impact analyses type of criterion based on two elements of planning: i) effects on the Local Concept of Spatial Planning, ii) compatibility with the categories of land use from the Land Use Local Development Plan. Data were collected on the existence of spatial planning documentation, which showed that out of 135 local self-government units in BiH that are obliged to draw up spatial plans, 29 do not have a plan, while in 21 plans are under preparation. These 50 units of local self-government make up 41.66% of the total territory with 26.75% of the population of BiH. Among them are those with significant hydropower potential, such as Foča, Kalinovik, Višegrad, Srebrenica, Kneževo, Šipovo and Drvar. In the past period, due to the lack of detailed spatial planning documents for the majority of realized SHPs, the institutions responsible for spatial planning prepared expert opinions on the suitability of the location for the construction of SHP, which created the basis for issuing location permits. When preparing expert opinions, criteria from the fields of Water management, Water ecology, and Nature protection were not taken into account. Therefore, instead of assessing the impact on spatial planning concept or land use, the criterion is based on the assessment of the existence of spatial planning documentation. Thus, a project whose spatial assessment is based on expert opinion gets 1 point, and a project that is specified in a detailed spatial planning document gets 5 points. For the selected 8 criteria from the field of Water ecology, the scoring approach remained the same as in the Tyrol Catalogue, but the scoring criterion were adapted. A first step is the sensitivity analyses of water bodies, which are classified as "very sensitive" and "sensitive". The baseline data for the sensitivity analyses could be found in water management plans [16], [17], [18], [19] in which the status of surface bodies of water in the River Sava and Adriatic Sea basin districts is determined in accordance with the EU WFD [12] and regulations in field of water management in BiH [20], [21], [22], [23],. As example, for the sensitivity analyses for Hydromorphology criterion we used data on assessments of the general hydromorphological status of the water bodies. High sensitivity classification requires a minimum length of 1 km (except for water bodies shorter than 1 km, in which case the entire length is used), as prescribed in the hydromorphology standard defined in the water district management plans in BiH. [24] Sensitivity assessment of a body of water under this criterion is based on the total length of a type-specific body of water in BiH, which is subject to changes in morphology. Types of bodies of water deemed sensitive are those where <20% of the length of sections of water body of a specific type in the territory of BiH is subject to morphological changes.

Assessment	Hydromorphological status
Very sensitive	Quality of the hydromorphological structure of the body of water Class 1 in the length of at least 1 km (or total length for water bodies <1 km)
Sensitive	Quality of the hydromorphological structure Class 1 - natural state, in the length of only 500 m; Quality of the hydromorphological structure Class 2 - slightly modified state, in the length >500 m
Low to moderately sensitive	Quality of the hydromorphological structure Class 3-5 (moderately modified to highly modified) in the length >0.5 km

Table 5. Sensitivity assessment for the Hydromorphology criterion (BiH Catalogue)

For all of 8 criteria under the Water ecology , the project is scored by combining two parameters: sensitivity and weight, whereby the highest score (5 points) is awarded to projects where none of the criteria were evaluated as highly sensitive and the lowest score (0 points) to projects with three or more "highly sensitive " criteria or where the ecological status of the water body is high. For the Hydromorphology criterion assessment weights are determined as follows:

*** High status sections of the water body cover <20% of the total length of a type-specific body of water in the territory of BiH

** High status sections of the water body cover >20% of the total length of a type-specific body of water in the territory of BiH

For the criteria from the field of Nature Protection, adjustment of the internal scoring within each individual criterion was done on the basis of the BiH laws on nature protection and the laws on freshwater fisheries, which regulate protection issues. Laws prescribe measures for the preservation of biological diversity through the preservation of habitats, ecological networks and ecologically significant areas. As example, evaluation of the criterion "Protection of species" is based on the assessment of the impact of a certain SHP project on the species listed on the BiH Red lists [25], [26] and species protected by laws on nature protection [27], [28]. The assessment is based on scoring from 0 to 5. The project receives 0 points when directly impacts vulnerable, critically endangered and endangered species from the Red Lists while 5 is awarded to projects having neither impact on endangered and/or protected species, nor on population/subpopulation, even at the level of individual specimens. (Table 6).

Score	Estimation - Range Definition
0	Threatens individuals, parts of the population or populations of critically endangered wild species (CR, RE) or endangered (EN) or vulnerable (VU)
1	Threatens population or part of populations of endangered (EN) or vulnerable (VU) species
2	Threatens population or parts of the populations of nearly threatened (NT) or data deficient (DD) species
3	Negative impact to individuals of least concern (LC) species and/or those that receive 2 points in the case of negative impact on population or parts of populations
4	Negative impact to individuals of non-threatened species
5	No impact on endangered and/or protected species, nor on any population/ subpopulation, even on the individual specimens

Table 6. Scoring for the Protection of species criterion (BiH Catalogue)

For the overall evaluation of the criteria under the Nature protection thematic area, a dominantly qualitative approach is used in the evaluation, with a scale from 0 to 5. If none of the criteria receives points, the Nature Protection score is calculated as mean value of individual criterion scoring. The criterion "Sensitive Water Types" and "Sensitive / Unique Water Bodies" will be awarded 0 points if the project is planned on watercourses that belong to this category (sensitive types and sensitive and unique water bodies). (Table 7) Depending on the number of criteria evaluated with 0 points, the Nature Protection overall score will receive 0 (more than 2 criteria), 0.5 (2 criteria) or 1 point (1 criteria).

Criterion	Score
Species Protection	0 to 5
Habitat Protection	0 to 5
Natural Value of Watercourses	0 to 5
Ecosystem	1 to 5
Landscape/Recreational Values	1to 5
Sensitive Water Types	0
Sensitive / Unique Water Bodies	0
Protected Areas	0 to 5

Table 7. Scoring of the Nature Protection criteria (BiH Catalogue)

The consolidated scoring of the criteria within the thematic areas is carried out in such a way that the results of the evaluation of individual criterion within the thematic area are assigned with their shares (weights) according to their importance in the overall evaluation. Determination of weight shares, as well as the entire process of analysis, selection and adaptation of criteria was done through consultations with a stakeholder working group composed of staff members of the ministries of environment, water management and energy management, regulatory renewable energy agency, agencies for granting concession, water management agencies, as well as non-governmental organizations, and other representatives of hydropower industry. Finally scoring results are summarised in spider graphic form. To facilitate the interpretation of scores, we followed the Tyrolean Catalogue approach and divide scores into three ranges (Fig. 2.):

- "Red" range - the score is between zero and the yellow-red boundary: If the thematic area score is in the red range, it has to be significantly improved or the project will be abandoned,
- "Yellow" range - the score is between the red-yellow and yellow-green boundary: If the thematic area score is in the yellow range , this indicates that the project can be approved only on the ground of favourable assessment of public interests unrelated to thematic area. Alternatively, assessment score in the yellow range can be interpreted as a recommendation to improve the project to bring the project score as close as possible to the green area,
- "Green" range - the score exceeds the yellow-green boundary: This indicates that the project is eligible for approval with respect to considered thematic area.

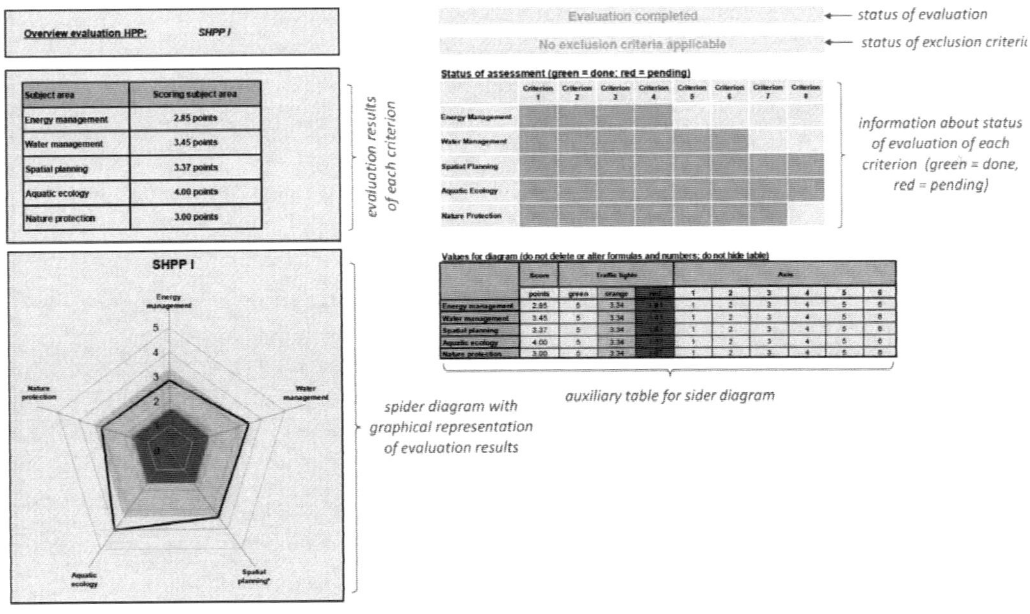

Fig. 3. Example of consolidated scoring for the SHP project presented in the spider diagram form

3. Results and discussion

To illustrate the applicability of the selected criteria to the assessment of the sustainability of the SHP, a diversion power plant with the installed capacity of 249 kW, annual production 965 MWh/a, total investment 0.924 mill. BAM, Head 66.9 m, Grid connection 0.7 km, mean annual flow 0,26 m3/s, installed flow 0,48 m3/s, length of pipline1,6 km. The project does not meet any of the eliminatory criteria, so an evaluation for all groups of criteria followed. With these technical characteristics, the project in the field of energy management has a favourable rating only for a specific investment.(Table 8).

Criterion	Score	Weight	Weighting score
1. Specific investment	4	30%	1.2
2. Climate protection	1	30%	0.3
3. Grid connection	2	20%	0.4
4.Additional aspects/synergies	1	20%	0.2
Total score			2.1

Table 8. Scoring for the thematic area Energy management

The location of the SHP is unfavourable as the available potential is utilised to a minor degree and poses major obstacles to future utilisation of the remaining hydropower potential. The characteristics of the hydropower plant are also unfavourable due to the fact that the degree of utilization of SHPP is 1,85 m3/s (1 point), duration of spillover during one-year period is 37 days (1 point) and head to length ratio is 4,2% (1 point). When applying weighting factors for this subcriterion as 40%, 40%, 20% we get total score of 1 point for the criterion "Characteristics of the hydropower plant" (Table 9). The aspect of impact on society is incorporated into the "Change of hazard potential" criterion, and for this pilot SHP was rated with a favourable score of 4 because the SHP is located in an unpopulated area without the presence of other users of the water body, there are no other buildings in the risk area. Efficiency of utilization of river stretch is lowest as possible as it is 1.337 m/GWh what gives 0 points on the scoring scale for this criterion (wEff > 800), while the 5 points gets 80≥ wEff >0.

Criterion	Score	Weighting factor	Weighting score
1. Degree of utilization of hydropower potentials	1	20%	0,20
2. Characteristics of the hydropower plant	1	25%	0,25
3.Efficiency of utilization of river stretch	0	20%	0,00
4. Change of hazard potential	4	10%	0,40
5. Effects on emission situation	0	10%	0,00
6. Impact on groundwater	1	15%	0,15
Total score		100%	0,20

Table 9. Scoring for the thematic area Water management

There is no agricultural land around the location of SHP while the forest land is solely of lowest cadastral class V – VIII. Since the location is outside urbanised area it is expected that the construction of the SHP will contribute to infrastructure development. There are also no other water users along the water course, thus the awarded score for the subcriterion "Direct use at watercourses" is 5 points. (Table 10). The only unfavourable scoring is given to the subcriterion "local economy" as the level of the annual concession fee is below 5,000 BAM/a. The best score (5) is given when the concession fee is over 20,000 BAM/a. Since two criteria are not applied the weighting is adjusted to give the total score of 100%. The total scoring is converted from a scoring range 1-5 to the scoring range 0-5 points that is used in all other thematic areas.

	Score	Adjusted weighting factor	Weighting score	Weighting factor
1. Spatial planning documentation	4	25%	1,00	20%
2. Direct use at watercourses	5	25%	1,25	20%
3. Infrastructure	5	13%	0,63	10%
4. Agriculture	5	13%	0,63	10%
5. Forestry	4	13%	0,50	10%
6. Cultural property	n/a			10%
7. Tourism	n/a			10%
8. Local economy	1	13%	0,13	10%
Total score (scope 1-5)		100%	4,13	100%
Total score (Score 0-5)			3,91	

Table 10. Scoring for the thematic area Spatial planning

Hydro-mophology	Ecological status	Basin size	Special types of w.c.	Special features by w.c type	Presence of spawning sites	Free flow pathways	Thermal pollution	Reservoir size
Very sensitive	Very sensitive	Sensitive	Low to moderately sensitive	Low to moderately sensitive	Low to moderately sensitive	Sensitive	Low to moderately sensitive	Low to moderately sensitive
**	***	***	**	*	***	***	*	***

Table 11. Scoring for the thematic area Water ecology

The water course (w.c) is of 1290 m length with where >20% of the total length of the water has a "High status" and has preserved natural features in length of 1000 m, i.e. the quality of hydromorphological structure is Class 1. Due to this fact the Hydromorphology criterion is assessed as "Very sensitive". Given that the ecological status is very high, the total score for the Water Ecology criterion will be 0, regardless of the scores of other criteria.

Protection of species	4
Protection of natural habitat	5
Ecosystem	3
Landscape and recreational values	n/a
Natural value of the water course	4
Sensitive water types	0
Sensitive and unique water bodies	0

Table 12. Scoring of the subject area Nature protection

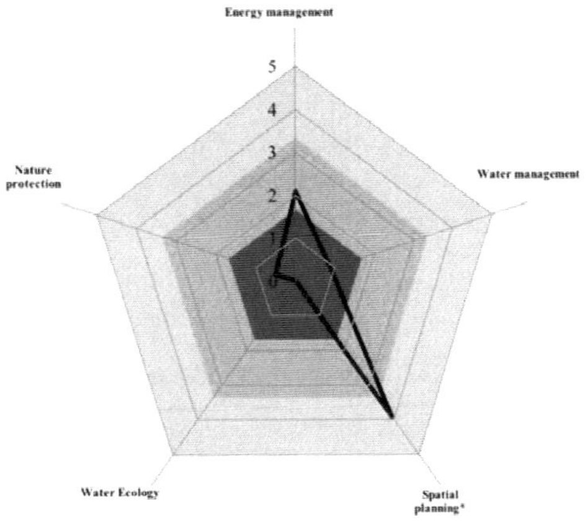

Fig. 4. Consolidated scoring for the SHP pilot project presented in a spider diagram form

The SHP is planned to be built on sensitive water type and that the water body is sensitive and unique those two criteria are scored with 0. Since the number of criteria scored with 0 is 2 the Nature protection thematic area is granted with 0 points. Bearing in mind that the three key criteria for the protection of natural assets, Water management, Water ecology and Nature protection, fall into the "red range" (Fig. 4.), this project would not be able to obtain construction permits. The fact that the project has favourable spatial planning characteristics, even that the location in the detailed local spatial plan is reserved for the SHP, indicates that these criteria were not taken into account during spatial planning.

4. Conclusion

Due to the increase in interest in the construction of SHP, the introduction of criteria for evaluating their sustainability deemed necessary. The criteria must cover all five areas (energy management, water management, spatial planning, water ecology and protection of natural resources), and issues of social protection should be mainstreamed in all the mentioned areas. The adaptation of the Tyrolean Catalogue to the context of Bosnia and Herzegovina proved to be a good choice, especially because the interested parties involved, who are at the same time future users of the criteria for decision-making, showed a high level of understanding regarding the purpose of criteria, calculation and scoring method. The adaptation methodology, which was based on the principles of applicability to the SHP context and national context, compliance with national laws and data availability, was a legitim choice, as it resulted in a comprehensive set of indicators was chosen in all five areas that decision makers can evaluate by themselves. Progress made in implementation of European directives in Bosnia and Herzegovina in particular: The Water Directive (status assessment and risk assessment for surface water bodies) and the Habitat Directive (red list of habitats and species) was a key for the data availability for thematic areas of Water Ecology and Nature Protection. The need to engage an external expert arises only in the field of Nature Protection, because it is the only field for which there is not enough data and an expert opinion based on field investigation is needed. Testing of the BiH catalogue on the pilot SHP showed the need to apply the criteria at the earliest stage, already during the creation of spatial planning documentation. The fact is that the SHP pilot was planned in the spatial planning documentation, but the evaluation of the criteria showed that the project could not be approved. Consideration of the sustainability of waterpower utilization at the spatial plan level should be preceded by an extensive study on water potential, to which the above criteria would be applied. That study would serve as a background study for the spatial plan.

5. Acknowledgments

The authors would to thank to the Dautche Gesellcfat fur Internationale Zusammenarbeit (GIZ) for their overall support provided within the process of development of the Methodology for the Adaptation of the Tyrolean Criteria and the Criteria Catalogue for the Development of a Sustainable SHPs in BiH, as well as to the working group members in obtaining data and information and their interaction and valuable comments provided during this process.

6. References

[1] Bosnia and Herzegovina- Annual Implementation Report, 1 November 2021, Energy Community Secretariat

[2] Özge Can D., Jonas Ø. N., (2019), Is the hydropower boom actually taking place? A case study of a South East European country, Bosnia and Herzegovina,, Renewable and Sustainable Energy Reviews, Volume 110, Pages 278-289, ISSN 1364-0321, DOI: 10.1016/j.rser.2019.04.077

[3] Himanshu N. Goel,V., (2020). Sustainability assessment of hydropower projects, Journal of Cleaner Production, 265, 121661. DOI: 10.1016/j.jclepro.2020.121661

[4] Vaikasas, S., Bastiene, N., Pliuraite, V. (2015), Impact of small hydropower plants on physicochemical and biotic environments in flatland riverbeds of Lithuania. J. Water Sec. 2015, 1. ISSN 2345-0363, DOI: 10.15544/jws.2015.001

[5] Premalatha,M.; Abbasi, T.; Abbasi, S.A. (2014), A critical view on the eco-friendliness of small hydroelectric installations. Sci. Total Environ. 2014, 481, 638–643, DOI: 10.1016/j.scitotenv.2013.11.047

[6] Rusko, M., Paulova, I., Vanova, J., Kralikova, R. (2017), Connectivity of Quality Management Systems and Environmental Management Systems, Proceedings of the 28th DAAAM International Symposium, pp.1213-1220, B. Katalinic (Ed.), Published by DAAAM International, ISBN 978- 3-902734-11-2, ISSN 1726-9679, Vienna, Austria , DOI: 10.2507/28th.daaam.proceedings.169

[7] Goltara A., Boz B.,at all. (2008),The guidelines for hydro power plants licensing for decision makers, CH2OICE project task 3.2 report, supported by The Intelligent Energy – Europe (IEE) Programme

[8] Vučijak, B., Kupusović, T. Midžić-Kurtagić, S., Ćerić, A. (2013), Applicability of multicriteria decision aid to sustainable hydropower, Applied Energy, Volume 101, Pages 261-267, ISSN 0306-2619, DOI:10.1016/j.apenergy.2012.05.024.

[9] Tahseen,S, Karney W., B., (2017), Reviewing and critiquing published approaches to the sustainability assessment of hydropower, Renewable and Sustainable Energy Reviews, Volume 67, Pages 225-234, ISSN 1364-0321, DOI:10.1016/j.rser.2016.09.031.

[10] Pracheil, B.M., McManamay, R.A., at all, (2019). A checklist of river function indicators for hydropower ecological assessment. Sci. Total Environ. 687, 1245e1260. Volume 687, Pages 1245-1260, ISSN 0048-9697, DOI: /10.1016/j.scitotenv.2019.06.049.

[11] Vassoney E., Mochet A.M., at all (2019) Integrating meso-scale habitat modelling in the Multicriteria Analysis (MCA) process for the assessment of hydropower sustainability, Water (Switzerland), 11 (4), art. no. 640, DOI: 10.3390/w11040640

[12] EC Directive 2000/60/EC of the European Parliament and the Council of 23 October 2000 Establishing a Framework for Community Action in the Field of Water Policy; Official J. EC: Brussels, Belgium, 22 December 2000; L 327; pp. 1–72.

[13] Česonienė L., Dapkienė M., Punys P., (2021), Assessment of the Impact of Small Hydropower Plants on the, Ecological Status Indicators of Water Bodies: A Case Study in Lithuania, 2021. Water 13, no. 4: 433. DOI:10.3390/w13040433

[14] Tomczyk, P., Wiatkowski, M. (2021) Impact of a small hydropower plant on water quality dynamics in a diversion and natural river channel. J Environ Qual.;50:1156–1170. DOI: 10.1002/jeq2.20274

[15] Wasserkraft in Tirol, Kriterienkatalog- Kriterien für die weitere Nutzung der Wasserkraft in Tirol, Version 3.0

[16] Water Management plan for the water district of the Adriatic Sea FBiH (2016-2021),

[17] Water Management plan for the water district of the Sava Basin of FBiH (2016-2021),

[18] Water Management plan for the water district of the Adriatic Sea of RS (2017-2022),

[19] Mater Management plan for the water district of the Sava Basin of RS (2017-2022),

[20] Law on Water of FBIH, (S.N FBiH 70/06)

[21] Law on Water of RS, (S.G RS, br. 50/06 i 92/09)

[22] Decision on the characterisation of surface waters and groundwaters, reference conditions and parameters for assessment of water status and monitoring of waters, (O.G. FBIH 1-14)

[23] Regulation on the Classification of Water and Categorisation of water courses, (O.G. RS 42/01)

[24] BAS EN 15843: 2011 Water quality - Guidance standard on determining the degree of modification of river hydromorphology

[25] Red list of wild species and subspecies of plants, animals and fungi (O.G F BiH 7/2014)

[26] Red list of wild species and subspecies of plants, animals and fungi (O.G RS 20/2014)

[27] Law on Nature Protection FBiH (O.G FBiH 66/13)

[28] Law on Nature Protection RS (O.G. RS 20/14)

DOI: 10.2507/33rd.daaam.proceedings.019

IMPROVEMENT OF AIRPORT'S ENERGY EFFICIENCY, CASE STUDY OF SARAJEVO AIRPORT

Sanda Midžić Kurtagić & Džana Kadrić

This Publication has to be referred as: Midzic Kurtagic, S[anda] & Kadric, D[žana] (2022). Improvement of Airports Energy Efficiency, Case Study of Airport Sarajevo, Proceedings of the 33rd DAAAM International Symposium, pp.0133-0142, B. Katalinic (Ed.), Published by DAAAM International, ISBN 978-3-902734-36-5, ISSN 1726-9679, Vienna, Austria
DOI: 10.2507/33rd.daaam.proceedings.019

Abstract

The aviation sector generates approximately 2,5 % of the world's carbon dioxide emissions. The majority of it is associated to the air traffic, while 5% goes to airports. With the expected increase in the number of airports worldwide and intensity of the air traffic, the carbon footprint of the aviation sector is expected to increase. Modern airports strive to rationalize resources and the energy consumption per passenger served, to increase their competitiveness and reduce energy consumption and carbon dioxide emissions. The paper presents an analysis of the energy efficiency of the selected regional airport, performed using benchmarking and energy audit methods. Historical data on energy consumption are used to calculate the company's carbon footprint and key energy performance indicators. Several improvement measures are analysed to reduce energy consumption of heating, the ventilation and air conditioning and lighting systems, improve the energy management system and to implement systems based on renewable energy. The implementation of the selected measures should result in the reduction of all relevant indicators as follows; reduction of specific carbon dioxide emissions by 43%, and reduction of specific energy consumption by 28%. Strengthening the airport's energy management system should ensure continuous evaluation and modification of the airport's resource and energy efficiency, which also contributes to the facility's energy savings.

Keywords: Airport efficiency; Energy efficiency; Carbon footprint; Performance indicators.

1. Introduction

In terms of climate change impact, the aviation sector, including passenger, cargo and military transport, accounts for 1.9% of global greenhouse gas emissions (including all greenhouse gases, not just the carbon dioxide), and 2.5% of the global carbon dioxide (CO_2) emissions [1]. Greenhouse gas (GHG) emissions from airports are caused by the consumption of fuel (gasoline and diesel) for airport vehicles and ground support equipment, the consumption of fossil fuels for the production of electricity and heat, and the consumption of kerosene for auxiliary power units that power airplanes from the airport gates and other sources [2]. According to Statista 2020 [3], in 2019, Dubai and London Heathrow airports were among the largest CO_2 emitting international airports globally, with more than 30 million metric tons of CO_2 emissions. According to the Airport Carbon Accreditation Annual Report 2010-2011, out of a total 2.5% of CO_2 emissions from air traffic, 5% refferes to airports [4], i.e. 0.1% of the global carbon emissions.

Airports, like many other resources that consume economic activities, need to fulfil international obligations to combat climate change but also to reduce their own operational costs. Modern airports are committed to rationalizing resources and energy consumption per passenger served, in order to increase their competitiveness and reduce energy consumption, carbon dioxide emissions and waste. In line with their commitments, some airports have made significant improvements and reported a reduction of airport CO_2 emissions to 2%-3% of the total aviation industry CO_2 emissions (equivalent to 15-20 million tonnes of CO_2 per year), suggesting that there is scope for further reductions [5]. This reduction is certainly a consequence of the developed methods and accreditation programs [6] for improving resource efficiency in airports as well as development and implementation of more resource efficient technologies and technologies based on renewable energy sources. Both methods and programs imply the measurement of environmental sustainability of an airport using indicators and the application of benchmarking methods [7], [8]. The benchmarking method allows airports to perform self-measurement (internally) to monitor progress towards a set of goals and/or to carry out a peer-benchmarking (externally) thus comparing their performance with other airports. In [7], 3 sets of resource related indicators are introduced, namely: i) Static power consumption- Fossil-fuelled electricity consumption, kWh (monthly, yearly), Fossil-fuelled gas consumption (kWh monthly, yearly) Wind, solar or bio-generated electricity consumption, (kWh monthly, yearly), ii) Water consumption- Monthly volume consumed (m^3), iii) Waste generation- Monthly volume arising (kg), Monthly volume recycled or re-used (not incinerated/sent to incineration). In [9], the authors proposed 77 indicators, organized into ten performance dimensions, including resource efficiency related indicators: water, waste and energy consumption reduction. The consumption of resources in airports is related either to the surface area of terminals (heating and cooling, lighting) and other buildings or to the number of passengers. In [10] and [8], normalized indicators expressed per reference unit are introduced (m^2 or passenger): i) energy per square meter of the terminal, ii) water consumption per passenger.

According to the Greenhouse Gas Protocol initiative [11], the calculation method for the CO_2 footprint depends on the extent of organizational and operational boundaries. In terms of organizational boundaries, airport sites contain variety of ground operations, some of which are not related to flights serviced. In [7], the authors emphasize that airports differ in terms of the area they occupy, transport access modalities, presence of commercial tenants, and other. Airport Carbon Accreditation Guidance Document [12], suggests the application of a so-called control approach, stating that the company accounts for 100 percent of GHG emissions from operations over which it has control. In terms of operational boundaries, the calculation may vary from taking into account direct and indirect emissions that are related to the direct airport operations but may also take into account all other indirect emissions, which are a consequence of the airport's activities but arise from sources not owned and/or controlled by the company (e.g., aircraft movements, vehicles and equipment operated by third parties, off-site waste management, etc.) [12]. When benchmarking, it is very important to take a sample with a similar scope of organizational and operational boundaries, so as not to draw wrong conclusions.

European airports are very active in the Airport Carbon Accreditation programme. According to data from 2022 [13], those registered in the program cover 73.1% of the total air passenger traffic in Europe. 76 airports have mapped their carbon footprints, 45 airports have actively reduced their CO_2 emissions, 22 airports have reduced their CO_2 emissions & have engaged others to do so, 39 airports are carbon neutral, 10 airports have aligned their CO_2 management with global climate goals, 19 airports have achieved transformation and have further compensated for residual emissions. Only four airports from the Western Balkan countries are included in the program, namely: Belgrade (Serbia), Pristina (Kosovo), Skopje (North Macedonia) and Tirana (Albania). Bosnia and Herzegovina (B&H) has 4 active airports, however none of them is included in any of the carbon reduction programs. Also, according to [14], [15] the industrial sector in B&H has a large potential for energy savings and carbon reduction, therefore the authors were motivated to conduct this research and apply methods for evaluating airport's energy performance while identifying opportunities for improvement. The authors selected an airport that has no previous practice of data monitoring and performance analysis. Sarajevo Airport, which is the largest airport in B&H, voluntarily agreed to participate in the author's research study. The results of the Study enable a detailed insight into the total and normalized values of energy consumption, energy breakdown and carbon footprint and allow for their benchmarking and analysis of the impact of various energy saving measures on the airport's carbon footprint. It is expected that the study data will be of great use to the Sarajevo Airport in strenghtening the airport's energy management system and reducing its carbon footprint.

2. Methods

A standard methodology for conducting energy audits is applied [16], [17] starting with an introductory meeting with top management and the core engineering team, followed by identification of the main energy consumers and consumption levels, energy related challenges, discussion on potential for energy efficiency improvements. The second phase included measurements and detailed engineering analyses of the proposed measures. Both phases implied full participation of the airport employees, both in the phase of data collection and during identification of improvement measures. The organizational boundaries included the terminal buildings (terminals A and B), the administration buildings, the cargo warehouse, the fire station building, the two-story hangar, the control tower, the runway, and the parking lot. Sarajevo Airport has no other services or commercial tenants. Operational boundaries included direct and indirect emissions for which the airport is directly responsible (heating and cooling, lighting, electrical consumers located in the buildings including electrical vehicles and internal transport fuel consumption). This corresponds to Scope 1 and Scope 2 of the CO_2 emission calculations as presented in the Airport Carbon Accreditation Guidance Document [12].

Data on monthly consumption of electrical energy and fuel consumption for the period 2017-2019 are provided. The energy derivied from a fuel is calculated by multiplying the fuel consumed and the lower heating value of the fuel. The total CO_2 emission from fuel and electrical energy is calculated by multiplying the value of energy derivied from a fuel (or electric energy) and the related fuel (or electric energy) conversion factor, regulated at the national level [18]. CO_2 emission from internal transport (mobile sources within the Scope 1) is calculated taking into account the vehicle type, its average annual distance travelled, and CO_2 emission coefficient [12].

Energy metering is not enabled for particular energy users or airport buildings. Therefore, the energy consumption breakdown was performed using the indirect method. Annual electrical energy consumption was calculated for each category of electrical energy users taking into account the installed power and working hours. There are no heating meters installed in the substations, therefore the approximate values of energy consumption for building heating are derived, based on the installed power, and working hours of the system. During the audit, thermal imaging of subsystems and facilities was also performed, as well as measurement of boiler combustion efficiency, using the indirect method.

For the benchmarking purposes, the authors selected 4 airports from the Western Balkans region that belong to the same climate area and have similarities in the scope of organisation boundaries, as well as in their respective national CO_2 calculation factors: Zagreb, Ljubljana, Priština and Belgrade. Edinburgh Airport represents large airports with more than 10 million passengers per year, while Inverness Airport is included due to similar annual number of passengers. Data on CO_2 footprint for sample airports were taken from their annual carbon footprint reports covering the same period 2017-2019 [19], [20]. The benchmarking exercise included the total carbon footprint (tCO_2/ann.), the normalised carbon footprint (tCO_2/passenger), as well as the annual number of passengers.

These methods enable identification of systems with excessive energy consumption and weaknesses in the energy management system, aiming to determine and select adequate improvement measures. The evaluation of the proposed energy saving measures included: i) investment's costs, ii) energy and iii) CO_2 saving potentials, as well as the effects of potential measures expressed through the iv) simple payback period (SPP) and v) reduction of the baseline indicator value in absolute value and in percentage.

3. Audit and benchmarking results

3.1 Energy consumption and carbon footprint indicators

The airport uses natural gas and fuel oil to produce energy for heating. Electricity is supplied from the public distribution network, and diesel fuel is used for the production of electricity in case of an interruption in the supply of electricity, and for vehicles. The facility does not use any renewable energy sources in the energy supply system. The total annual energy consumption of the facility amounts to 6.285 GWh/ann. Natural gas has the largest share in total energy consumption (51%), followed by electrical energy (46 %) (Figure 1). Diesel and fuel oil have a negligible percentage. The annual CO_2 emission of the facility amounts to 2.845 t/ann. Due to the high conversion factor of electrical energy (0,744 kgCO2/kWh), the CO_2 emission resulting from consumption of electrical energy is dominant, accounting for 76 % of the total CO_2 emissions of the facility (Figure 1).

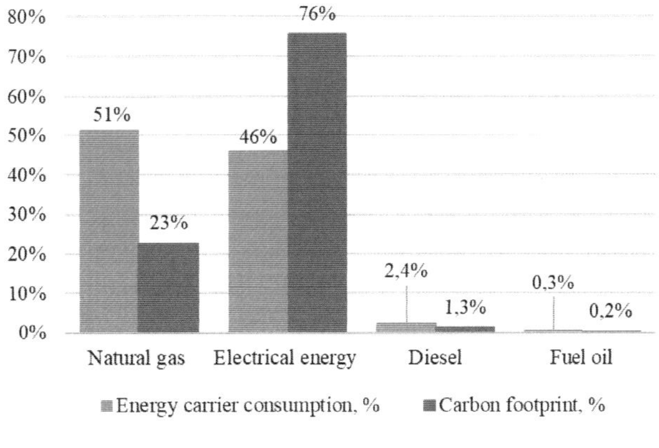

Fig. 1. Percentage of energy sources in total facility energy consumption and share of CO_2 for 2017-2019

3.2 Energy consumption breakdown

Energy for heating is supplied to the facility via two boilers with a nominal power of 2.900 kW each. The boilers efficiency is 94,4 and 93,8 %, and they work alternately every other day. Hot water from the boiler is supplied to seven heating substations, with a total nominal heating power of 2.743 kW. The terminal B has the largest share in energy consumption for heating (Figure 2.) as well as in electricity consumption (Figure 3).

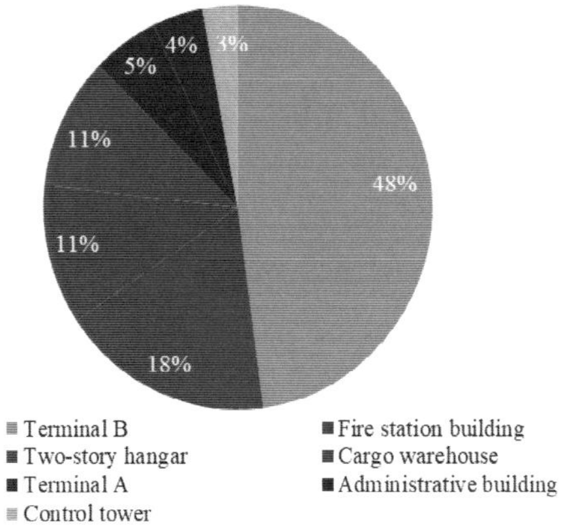

Fig. 2. Breakdown of heating energy consumption per buildings

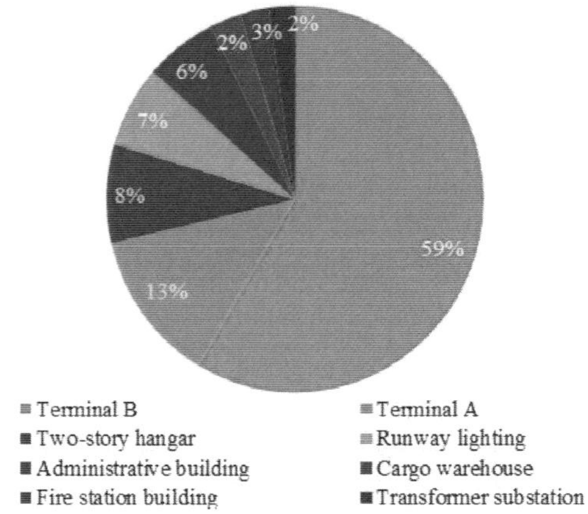

Fig. 3. Breakdown of electricity consumption per buildings

Electrical energy consumers can be classified into the following categories: heating, cooling, ventilation (HVAC), lighting, office equipment, kitchen equipment, workshops equipment and other consumers. The categories with the largest share in the total electrical energy consumption are HVAC, lighting, and office equipment (Figure 5.), therefore the analyses of energy saving measures is focused on the aforesaid systems.

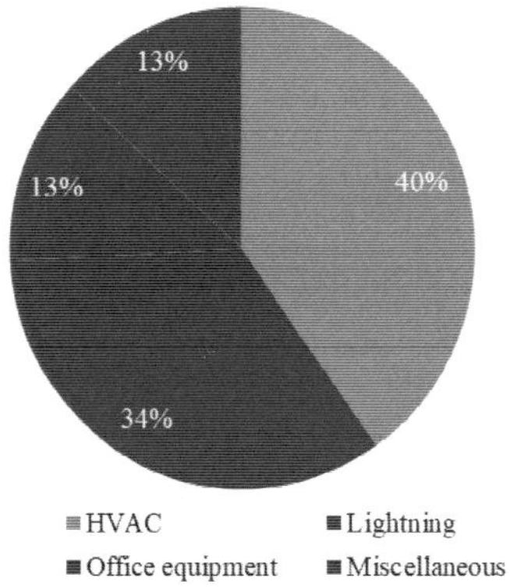

Fig. 4. Breakdown of electricity consumption by user category

Terminal B (Figure 5.) serves as a passenger terminal with a nett surface area of 9.233 m^2. Considering that it is in use throughout the year, with a complex HVAC system installed for heating and cooling, Terminal B represents a building with the largest share of the facility's thermal energy consumption (48 %) as well as electrical energy consumption (59 %). This is common in airport facilities worldwide [17].

Fig. 5. Terminal B exterior

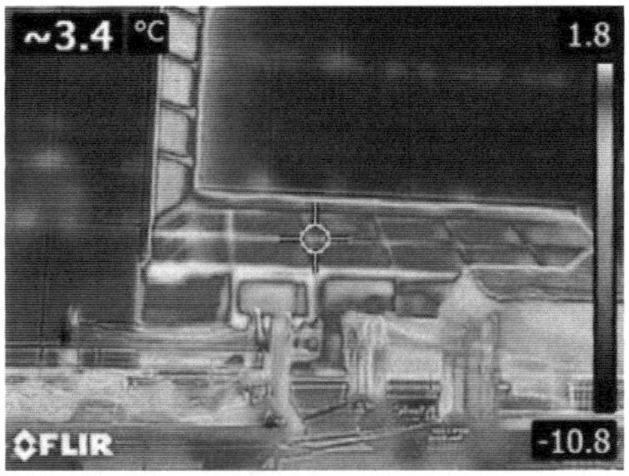

Fig. 6. Thermal image of terminal B

A central cooling system is installed in terminal B and the administration building, while the other building uses local split systems. The cooling system in terminal B accounts for approximately 11,7 % of the facility's electrical energy consumption. One of the major issues regarding the building envelope is large transparent surfaces that do not meet the regulations relevant to the heat transfer coefficient [18], glazed with large Solar Heat Gain Coefficient (SHGC), (Figure 6.). Air-conditioned areas, including terminal B, are overheated in winter with temperatures set at approx. 24 °C, which is much higher in comparison to the designed temperatures [21]. This is a common problem for other similar facilities worldwide [22], and results in excess energy consumption.

3.3 Benchmarking of energy and carbon footprint indicators

Energy related indicators and carbon footprint indicators are calculated using monthly data on energy consumption and monthly data on passengers served. (Figure 7.). The average values of the indicators are listed, with an average specific energy consumption of 5,99 kWh/pass, and a substantial seasonal variation of the indicator (the lowest value of the indicator is 1,8 kWh/pass and the highest is 19,9 kWh/pass). This is the result of an increase in energy consumption in the winter period (70,8 % of the total energy consumption is related to the winter period), with a decrease in the number of passengers (36,3 % of the total number of passengers). A similar trend is observed in the carbon footprint indicator (Figure 8.), where the average value of the indicator is 2,71 $kgCO_2$/pass., the lowest value is 1,3 $kgCO_2$/pass., and the highest value is 7,0 $kgCO_2$/pass.

Compared to the selected regional airports Zagreb and Ljubljana, the Sarajevo Airport has lower total annual CO_2 emission, but its average carbon footprint per passenger (2,71 $kgCO_2$/pass) is higher due to the lower number of passengers (Zagreb 1,57 and Ljubljana 1,85 $kgCO_2$/pass). The same can been seen in Edinburgh were the annual number of passengers of over 14 million has significantly reduced the normalised carbon footprint per passenger. However, a comparison with the Belgrade airport, which is much larger and with a higher number of passengers, having both indicators lower than Sarajevo, as well as with the Inverness Airport (Scotland), which is similar in both size and the number of passengers but with a 26% lower carbon footprint per passenger, indicates that Sarajevo has potential to improve its energy efficiency performance. Airport in Priština has the worst energy efficiency performance.

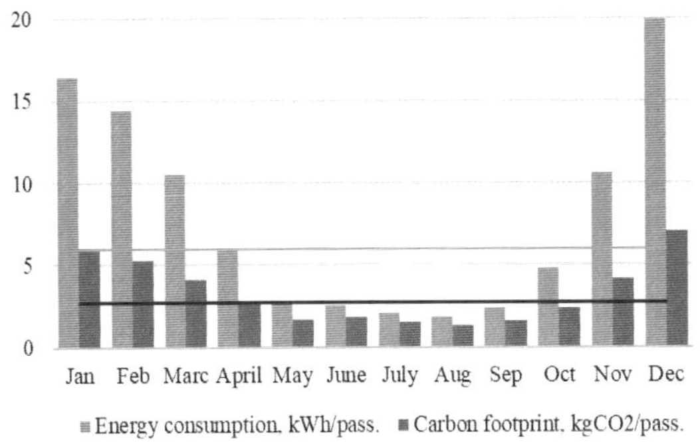

Fig. 7. Average monthly specific energy consumption and carbon footprint indicators for 2017-2019

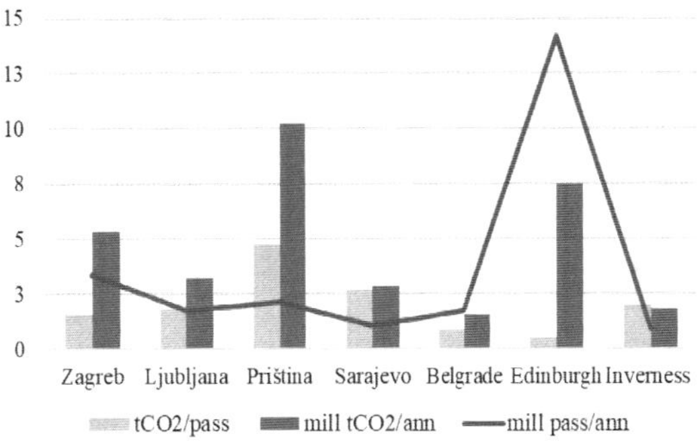

Fig. 8. Benchmarking of carbon footprint for selected airports

4. Key observations and proposal for improvement of energy efficiency

Considering the facility's energy consumption, the characteristics of the installed equipment and its operating regimes, the most important conclusions related to excessive energy consumption are summarized and presented in Table 1.

User category	Conclusion
HVAC system	Sarajevo Airport does not use any renewable energy sources in the energy supply system. Air-conditioned space is overheated during winter. The heating substation of terminal B is equipped with circulation pumps with a large number of annual operating hours. The pumps do not have frequency regulation. The transparent envelope of the terminal B building is a source of excessive heat loss in winter and heat gain in summer.
Lighting system	The lighting bulbs in terminal B are outdated, mostly based on fluorescent tubes with a magnetic starter, high power, low efficiency and manually controlled. In the parking lot in front of terminal B, 40 reflectors with a power of 150W and low efficiency are installed for advertising purposes.

	The lighting on advertising panels/billboards is outdated and consumes 30% more electricity in comparison to the new LED technology for the same lighting.
Office equipment	In the administration building, one printer is installed for each employee (total number of printers is 58). Laser printers remain on during the night and on non-working days. Much of the installed computer equipment is not actively used during working hours. The largest number of installed telephones is found in terminal B and in the administration building (261 units).

Table 1. Observations related to energy consumer categories

Based on a detailed analysis of collected data and observations, a large number of energy saving measures are listed and shown in Table 2.

Measure	Description
Integration of renewable energy sources in the energy supply system	
M1. Installation of photovoltaic solar panels in the parking lot to produce electrical energy. Parking lot has a surface area of 9.036 m^2.	Covering of the parking area would provide protection from the sun in summer and snow in winter, and would provide a base for installation of the photovoltaic solar panels. The number of panels is 3.780, the nominal installed power is 1,3 MW, the electricity produced is approx. 1.290,9 MWh/ann.
M2. Installation of a water-to-water heat pump for the heating and cooling of the administration building. The current system is connected to the heating substation and the central cooling system.	The thermal properties of the building envelope are in line with current regulation, and the specific energy needs of the building are low. The characteristics of the heat pump are Seasonal Coefficient of Performance (SCOP) – 5,17 and seasonal energy efficiency ratio (SEER) – 4,95, which is higher than the SEER of the existing system.
M3. Installation of a thin film of photovoltaic cells on the glazed surfaces of terminal B to produce electrical energy The total surface area of windows is 2.966,93 m^2, with glazing surface of 2.520,00 m^2	The installation of a thin film of photovoltaic cells, a thin film of amorphous silicon with low opacity, will enable the production of electrical energy [23] and the reduction of heat gain in the summer period. The nominal installed power is 113 kW, the generated electrical energy is 84,14 MWh/ann., and the reduction of electricity consumption for system cooling is up to 10 %.
HVAC system	
M4. Replacement of 18 pumps with variable frequency regulated pumps in heat substation of terminal B. The heating substation is equipped with circulation pumps, with no frequency regulation, and with nominal power of 0,55, 1,1 and 2 kW, with large number of annual operating hours.	By replacing 18 pumps in the terminal B substation with frequency-regulated pumps, the consumption of electrical energy will be reduced, and the overall efficiency of the heating system will be increased. Electrical energy savings of more than 30 % can be achieved [24], depending on the number, size and operating hours of pumps.
M5. Placing of low-E window film on all transparent surfaces of terminal B. The total glazing surface is 2.520,00 m^2, double glazed 8/16/8 mm, Low-E filled with argon, metal framed with non-insulated thermal bridges.	Thermal insulating UV films are placed on the inner surface of the windows. The implementation of this measure results in a reduction of inner temperature by 1 °C during summer, and can result in reduction of the cooling load and cooling energy consumption by up to 10 % [25].
M6. Night cooling during the summer period. The operating regime of the cooling system is completely related to the daily regime, and the advantages of the night cooling regime are not exploited.	A mechanical ventilation system can be used to automatically ventilate the building at night to cool the building's thermal mass and reduce the need for compressor cooling the next day. It is shown that energy consumption for cooling can be reduced up to 15 % [26].
M7. Hydraulic balancing and regulation of water supply temperature for terminal B. Installation of heat meters in each thermal substation.	The adjustment of flow rate, water temperature and pressure drop for terminal B, will result in a decrease of energy consumption of up to 9 % [27]. By installing thermal energy meters in thermal substations, a detailed insight into the energy consumption of each building will be provided and the basis for the implementation of the energy management system will be created.
Lighting system	
M8. Replacement of lighting bulbs in terminal B. A total of 781 standard fluorescent tubes (T12) with a power of 36 W are installed in terminal B.	The energy consumption of the lighting system accounts for 34% of the facility energy consumption, which is in line with the consumption share of other similar plants [28]. Replacement of lighting bulbs in terminal B with LED lightings, will result in a

Measure	Description
The annual operating hours of the lighting system are substantial.	reduction of electrical energy consumption of lighting system by up to 25 % [28].
M9. Installation of occupancy detection sensors and occupancy-driven lighting control.	By installing occupancy detection sensors and occupancy driven lighting control in the administration building, terminal A, terminal B and the fire station building, it is possible to save up to 25 % of electricity [29] .
M10. Replacement of reflectors from advertising panels/billboards in the parking lot. The panels are outdated, with a power of 150 W (40 psc.) with a large number of annual operating hours and poor characteristics such as illuminance, efficiency etc.	By replacing 40 old reflectors with a power of 150W each, with low-voltage tungsten halogen bulbs or with metal halogen HID bulbs, it is expected that energy savings for this category of users will be up to 50 %.
M11. Replacement of neon tubes from the advertising panel/billboards in the parking lot.	Replacement of 40 pcs. of 54W neon tubes from the advertising panel, with fluo-tubes of 36W. Energy savings will amount to 33 %.
Office equipment	
M12. Centralization of 58 printers in the administration building. The power of the printer during operation is 650W, and the power of the printer in stand-by mode is 8.5W.	Replacement of 58 local printers with 4 centralized printers. This will ensure a reduction in electrical energy consumption and cartridge consumption.
M13. Shutting down all laser printers at night and on weekends. The laser printer consumes 20W in stand-by mode, and 1W when turned off.	A total of 200 printers with a power of 350W are installed in the facility. By turning off laser printers at night and on weekends, electricity consumption will be reduced by up to 75% on an annual basis.
M14. Setting computer equipment in energy saving mode. In total, 407 computer units with monitors are installed in the facility.	Setting computer equipment to the "hibernate" energy saving option (computer operation mode), thereby saving up to 90 % of electrical energy consumption annually.
M15. Reduction in the number of telephones.	The total number of telephones can be reduced up to 50 %. As a result, the electrical energy consumption will decrease.

Table 2. Energy and saving measures

The implementation of all measures will reduce the annual consumption of electricity and natural gas in the facility, as well as the company's carbon footprint. Financial and energy related indicators for all analysed measures are presented in Table 3. Energy savings are estimated as very conservative with a smaller percentage than found in the literature, as indicated in Table 2, for individual measures, to ensure that the indicators were correctly calculated and not overestimated. The implementation of all selected measures should result in reduction of CO_2 emissions. Only for measure M2 (the installation of heat pump in administrative building), the electrical energy consumption will increase after implementation of the measure, as it will be in use during winter and summer, contrary to the existing system (heating from the boiler room and cooling from the central system with low SEER).

No.	Investment	Electrical energy saving	Natural gas saving	Savings	SPP	CO_2 reduction	
	[Eur]	[kWh/ann.]	[kWh/ann.]	[Eur/ann.]	[-]	[t/ann.]	[%.]
		Integration of renewable energy sources in energy supply system					
M1.	1.119.652	1.290.900	0	99.993	11,2	961	33,79
M2.	101.400	-20.726	128.832	7.145	14,2	10	0,37
M3.	152.265	104.572	0	8.100	18,8	78	2,74
		HVAC system					
M4.	16.112	22.083	0	1.711	9,4	16	0,58
M5.	87.937	23.834	0	1.846	47,6	18	0,62
M6.	0	17.024	0	1.319	0,0	13	0,45
M7.	18.247	0	60.873	4.135	4,4	12	0,43
		Lighting system					
M8.	3.594	50.016	0	3.874	0,9	37	1,31
M9.	6.519	37.512	0	2.906	2,2	28	0,98
M10.	5.113	8.760	0	679	7,5	7	0,23

No.	Investment	Electrical energy saving	Natural gas saving	Savings	SPP	CO_2 reduction	
	[Eur]	[kWh/ann.]	[kWh/ann.]	[Eur/ann.]	[-]	[t/ann.]	[%.]
M11.	123	2.628	0	204	0,6	2	0,07
			Office equipment				
M12.	4.090	10.456	0	4.900	0,8	8	0,27
M13.	0	9.850	0	763	0,0	7	0,26
M14.	0	6.417	0	497	0,0	5	0,17
M15.	0	20.577	0	1.594	0,0	15	0,54
Total	**1.515.052**	**1.583.901**	**189.704**	**139.665**	**10,8**	**1.217**	**42,80**

Table 3. Evaluation of energy and CO_2 saving measures

5. Discussion

A large number of energy-saving measures are analysed; ranging from organisational measures, measures for reduction of electrical energy consumption for lightings, HVAC system and integration of renewable energy sources into the facility's energy supply system. All measures have been shown to resultin a reduction in the facility's carbon footprint and, when combined, can result in a 43% reduction in the facility's carbon footprint and a 28 % redusction in energy consumption with favourable SPP. Reducing the specific carbon footprint will ensure that Sarajevo Airport ranks among the best-ranked airports in the region. The energy supply system of the Sarajevo Airport is based on fossil fuels, resulting in high CO_2 emissions, in both absolute value and normalized according to the number of passengers, therefore, there is a need to reduce the facility's carbon footprint.

Several improvement measures were analysed to reduce the energy consumption of HVAC and lighting systems, improve the energy management system and to implement systems based on renewable energy sources. The implementation of all measures will result in a substantial reduction of energy and energy costs, with an average Simple Payback Period (SPP) of 10,8 years. Also, the reduction of all relevant indicators will be achieved as follows; reduction of specific carbon dioxide emissions by 43% and reduction of specific energy consumption by 28% with a substantial amount of energy provided from renewable energy sources.

Parameter	Current	With measures	Reduction
Electrical energy, kWh/ann.	2.893.253	1.309.352	55%
Fossil fuels, kWh/ann.	3.391.452	3.201.748	6%
Renewable energy, kWh/ann.	0	1.374.745	-
Total energy consumption, kWh/ann.	6.284.705	4.511.100	28%
Total carbon footprint, tCO_2/ann.	2.845	1.627	43%
Specific energy consumption, kWh/pass.	5,99	4,30	28%
Specific carbon footprint, $kgCO_2$/pass.	2,71	1,55	43%

Table 4. Effect of energy and CO_2 saving measures

6. Conclusion

The paper presents the application of airport energy performance evaluation methods along with the identification of opportunities for improvement at an airport with no previous practice of data monitoring and performance analysis. The authors combined the performance benchmarking method and the energy audit method. Given that the airport has no practice of measuring, collecting and analysing data, certain parameters were obtained indirectly, such as breakdown of energy consumption. The energy related indicators and the carbon footprint indicators were calculated using monthly data on energy consumption and monthly data on passengers served.

The study showed that benchmarking the normalized value of the carbon footprint per passenger can lead to the conclusion that the t environmental impact of airport is small, which would not be the case if only the total annual carbon footprint was analyzed. Namely, Sarajevo Airport has a significantly lower total annual CO_2 emission in comparison to, for example, the airport in Zagreb, but its average carbon footprint per passenger (2.71 $kgCO_2$/pass) is higher due to the smaller number of passengers (Zagreb 1.57 $kgCO_2$/pass). The Edinburgh Airport has larger CO_2 emissions and larger negative impact on the environment in comparison to airports in Sarajevo and Zagreb, due to its high value of the carbon footprint, while its normalized value is the lowest among the analysed airports due to the large number of passengers (14 million). In order to get a complete picture of the overall environmental impact and the company's performance, it is necessary to perform an analysis of both the total energy consumption and the carbon footprint and their normalized values.

The analysis of only normalized values can also lead to a wrong conclusion about the overall impact of airport operation on the environment and climate change because, due to the large number of passengers, the normalized value becomes significantly lower than in airports with lower annual CO_2 emissions and a small number of passengers.

Carbon footprint benchmarking should be combined with the benchmarking of the total annual energy consumption, and the analysis of the monthly consumption, as well as consumption by energy sources. The share of renewable energy sources reduces the carbon footprint, but not the total energy consumption, therefore it can be wrongly concluded that the airport does not need to improve its energy efficiency. Seasonal fluctuations in monthly consumption may indicate the impact of heating and cooling on the overall result and the need to work on potential improvements in their energy efficiency.

The combined application of the performance benchmarking methods and energy audit method proved to be successful, as the airport identified areas of energy inefficiency that can be improved by implementing several improvement measures with the aim of reducing total and specific carbon dioxide emissions by 43%, and reducing total and specific energy consumption by 28%.

Further research into the efficiency and sustainability of airports should, in addition to analyses of water efficiency and further development of on-site renewable energy technologies, include waste flow analyses and technology solutions for the application of circular economy in airports.

7. Acknowledgments

Authors would like to thank to the Airport of Sarajevo for providing all the data for conducting this research. The authors would like to thank Sarajevo Airport for the data provided for purposes of conducting this research.

8. References

[1] D. S. Lee et al. (2021). "The contribution of global aviation to anthropogenic climate forcing for 2000 to 2018," Atmos. Environ., vol. 244, no. September 2020, doi: 10.1016/j.atmosenv.2020.117834.

[2] United States Department of Transportation. (2022). "Airport Carbon Emissions Reduction," https://www.faa.gov/airports/environmental/air_quality/carbon_emissions_reduction.

[3] S. R. Department. (2022). "Top departure airports with the largest passenger CO2 emissions from international operations in 2019," https://www.statista.com/statistics/1189557/carbon-dioxide-emissions-international-airports-worldwide/.

[4] Airport Carbon Accreditation. (2011). "Airport Carbon Accreditation Annual Report 2010-2011,"

[5] Airport Carbon Accreditation. (2019). "Airport Carbon Accreditation Annual Report 2018-2019," [Online]. Available: http://www.theemiratesgroup.com/system/aspx/download.aspx?id=tcm:409-2358317.

[6] Airport Carbon Accreditation. (2022). "Airport Carbon Accreditation," https://www.airportcarbonaccreditation.org/.

[7] P. J. Upham and J. N. Mills. (2005). "Environmental and operational sustainability of airports: Core indicators and stakeholder communication," Benchmarking, vol. 12, no. 2, pp. 166–179, doi: 10.1108/14635770510593103.

[8] M. E. Baltazar and J. Silva. (2020). "Spanish airports performance and efficiency benchmark. A PESA-AGB study," J. Air Transp. Manag., vol. 89, no. August, p. 101889, doi: 10.1016/j.jairtraman.2020.101889.

[9] G. C. L. Bezerra and C. F. Gomes. (2018). "Performance measurement practices in airports: Multidimensionality and utilization patterns," J. Air Transp. Manag., vol. 70, no. November 2017, pp. 113–125, doi: 10.1016/j.jairtraman.2018.05.006.

[10] Oliver Wyman. (2012)., "Guide to Airport performance measures,"

[11] S. Schmitz, B. Dawson, M. Spannagle, F. Thomson, J. Koch, and R. Eaton. (2004). "The Greenhouse Gas Protocol - A Corporate Accounting and Reporting Standard, Revised Edition," GHG Protoc. Corp. Account. Report. Stand., vol. 9, no. March, p. 116, [Online]. Available: http://www.ghgprotocol.org/files/ghg-protocol-revised.pdf.

[12] Airport Carbon Accreditation. (2016). "Airport Carbon Accreditation Guidance Document Issue 10," [Online]. Available: https://www.aca-pplication.org/media/guidance_documents/ACA_Guidance_Document_Issue_10_Sept_2016.pdf.

[13] Airport Carbon Accreditation. (2022). "Airports addressing their CO2 emissions,", https://airportco2.org/airports-across-the-world.html.

[14] D. Kadric, N. Delalic, S. Midzic-Kurtagic, B. Delalic, and K. Medar. (2017). "Energy efficiency assessment and improvement measures for furniture factory," Ann. DAAAM Proc. Int. DAAAM Symp., pp. 315–321, doi: 10.2507/28th.daaam.proceedings.043.

[15] D. Kadric, N. Delalic, B. Delalic-Gurda, M. Kotur, and A. Skulj. (2021). "Reduction of Energy use in Industrial Facility," J. Therm. Eng., vol. 7, no. 1, pp. 54–65, doi: 10.18186/thermal.840063.

[16] M. Z. Tahir, M. N. M. Nawi, and M. F. Baharum. (2015). "Energy management study: A proposed case of government building," AIP Conf. Proc., vol. 1660, doi: 10.1063/1.4915884.

[17] S. H. Abdallah, A. Makram, and M. Abdel-Azim Nayel. (2021). "Energy audit and evaluation of indoor environment condition inside Assiut International Airport terminal building, Egypt," Ain Shams Eng. J., vol. 12, no. 3, pp. 3241–3253, doi: 10.1016/j.asej.2021.03.003.

[18] N. 81/19. Official Gazette of Federation of Bosnia and Herzegovina. (2019). "Rulebook on minimum requirements for energy performance of buildings," [Online]. Available: http://www.ee-infos.ba/Content/docs/pravilnikominimalnimzahjtevima.pdf.
[19] Ricardo Energy & Environment. (2020). "Airport Carbon Footprint 2020 Scope 1, 2 & 3; Report for Edinburgh Airport Limited,"
[20] Fraport Slovenija. (2020)., "Fraport Slovenija Sustainability Report 2020,"
[21] M. A. Humphreys and J. F. Nicol. (2006). Environmental criteria for design. In environmental design: CIBSE Guide A, no. CIBSE Publications. Norwich, Norfolk: CIBSE Publications
[22] Z. Wang, H. Zhao, B. Lin, Y. Zhu, Q. Ouyang, and J. Yu. (2015). "Investigation of indoor environment quality of Chinese large-hub airport terminal buildings through longitudinal field measurement and subjective survey," Build. Environ., vol. 94, pp. 593–605, doi: 10.1016/j.buildenv.2015.10.014.
[23] S. A. Kalogiru. (2018). McEvoy's handbook of photovoltaics, Fundamentals and Applications, Third., vol. 1999, no. December. London: Academic Press Inc.
[24] H. Wang, H. Wang, and T. Zhu. (2017)., "A new hydraulic regulation method on district heating system with distributed variable-speed pumps," Energy Convers. Manag., vol. 147, pp. 174–189, doi: 10.1016/j.enconman.2017.03.059.
[25] Sedaghat et al. (2021). "Climate change and thermo-solar patterns of office buildings with/without window films in extreme hot-arid climate of Kuwait," Sol. Energy, vol. 217, no. October 2020, pp. 354–374, doi: 10.1016/j.solener.2021.02.051.
[26] X. Li, K. Huang, G. Feng, W. Li, and J. Wei. (2022). "Night ventilation scheme optimization for an Ultra-low energy consumption building in Shenyang , China," Energy Reports, vol. 8, pp. 8426–8436, doi: 10.1016/j.egyr.2022.06.059.
[27] X. Gao, M. Jia, S. Cao, and C. Qi. (2022). "Study of an Integrated Control Method for Heating Substations Based on Prediction of Water-Supply Temperature and Indoor Temperature," Buildings, vol. 12, no. 3, doi: 10.3390/buildings12030351.
[28] J. Wilding et al. (2010). "Airport Energy Efficiency and Cost Reduction,"
[29] T. Labeodan, C. De Bakker, A. Rosemann, and W. Zeiler. (2016). "On the application of wireless sensors and actuators network in existing buildings for occupancy detection and occupancy-driven lighting control," Energy Build., vol. 127, pp. 75–83, doi: 10.1016/j.enbuild.2016.05.077.

DOI: 10.2507/33rd.daaam.proceedings.020

INVENTORY OF REFRIGERANTS IN USE FOR COMMERCIAL PURPOSES IN BOSNIA AND HERZEGOVINA

Sanda Midžić Kurtagić, Džana Kadrić, Mirela Alispahić,
Rejhana Blažević & Halima Hadžiahmetović

This Publication has to be referred as: Midzic Kurtagic, S[anda]; Kadric, D[žana]; Alispahic, M[irela]; Blazevic, R[ejhana] & Hadziahmetovic, H[alima] (2022). Inventory of Refrigerants in Use for Comercial Purposes in BiH, Proceedings of the 33rd DAAAM International Symposium, pp.0143-0150, B. Katalinic (Ed.), Published by DAAAM International, ISBN 978-3-902734-36-5, ISSN 1726-9679, Vienna, Austria
DOI: 10.2507/33rd.daaam.proceedings.020

Abstract

The paper presents results of the inventory of refrigerants in use in Bosnia and Herzegovina and a discussion on their impact on the environment. Data on types, quantities and numbers of refrigerants in use, a type of a user and a purpose of the use were collected through the survey. The research sample included 297 facilities with 745 samples of refrigeration equipment covering the facilities from different sectors such as: banks, industry, shopping centres and stores, hotels and hospitals. Findings of this research are important for analysis of a level of the application of environmentally friendly refrigerants. The research showed that in Bosnia and Herzegovina HFC coolants are mostly used with a share of 62%, followed by 37% of natural refrigerants. The use of other types of refrigerants is negligible. Refrigerants that have the ozone depletion potential equal to zero are mainly used. Natural refrigerant R-717 has the largest share of 36.3%, followed by HFC R-134A with share of 24.9%. Refrigerant R-404A has the highest global warming potential (GWP) of 4210 and share in the total mass of 16.8%. Refrigerant R-410A has a share of 16.5% in the total mass and GWP of 2080. The work identifies promising alternative refrigerants with the lower GWP as potential environmental substitute for the refrigerants under control R -404A, R-410A and R-134.

Keywords: refrigerants; environmentally friendly gasses; global warming potential; ozone depletion potential

1. Introduction

Refrigerants are used in cooling and air conditioning processes, and they might have a negative impact on the environment in terms of ozone damage and global warming caused by the emission of refrigerants. Scientific findings about their harmfulness provoked the signing of international agreements to limit or phase out the use of refrigerants with the high ozone depletion potential (ODP) and the high global warming potential (GWP). The first agreement was the Vienna Convention for the Protection of the Ozone Layer (1985), followed by the Montreal Protocol on Substances that Deplete the Ozone Layer (1987) and series of subsequent amendments agreed to control the consumption and production of the ozone-depleting substances (ODS).

Since 2010, the use of chlorofluorocarbons (CFCs), halons and carbon tetrachloride - gases that are mostly used in refrigeration and air conditioning devices, fire protection systems and laboratories - has been completely prohibited. Ban on CFCs, that are the second generation of refrigerants, stimulated development of substitutes, first with hydrochlorofluorocarbons (HCFC) and then hydrofluorocarbons (HFC). HFCs have a minor effect on the stratospheric ozone, but some of them have significant GWP. Those were the third generation of refrigerants. The first generation of refrigerants (1830 and 1930) had high flammability, toxicity and reactivity. The second generation refrigerants were based on the low toxicity and flammability, while a focus of the third generation refrigerants was the protection of the ozone layer. (Fig1) [1]

However, the concentrations of HFCs in the atmosphere, of the most currently measured HFCs, are increasing in the global atmosphere. HFC emissions increased by 23% from 2012 to 2016 and currently amount to about 1.5% of the total emissions from all long-lived greenhouse gases as carbon dioxide-equivalent emissions (GtCO2-eq) [2]. This required the introduction of additional control measures, so the Montreal Protocol has been amended with the ratification and entry into force of the Kigali Amendment (2019), which includes control measures to reduce hydrofluorocarbons (HFCs). Since 2010, the fourth generation of refrigerants with low GWP, such as hydrofluoroolefins (HFO) and hydrochlorofluoroolefin (HCFO), has been developed. The fourth generation of refrigerants possess characteristics such as the low GWP, non-toxic, non-flammable, and zero-ozone depletion potential (ODP). (Fig1) [1],[3]. Refrigerants are also expected to have excellent thermodynamic and thermophysical properties. [3]

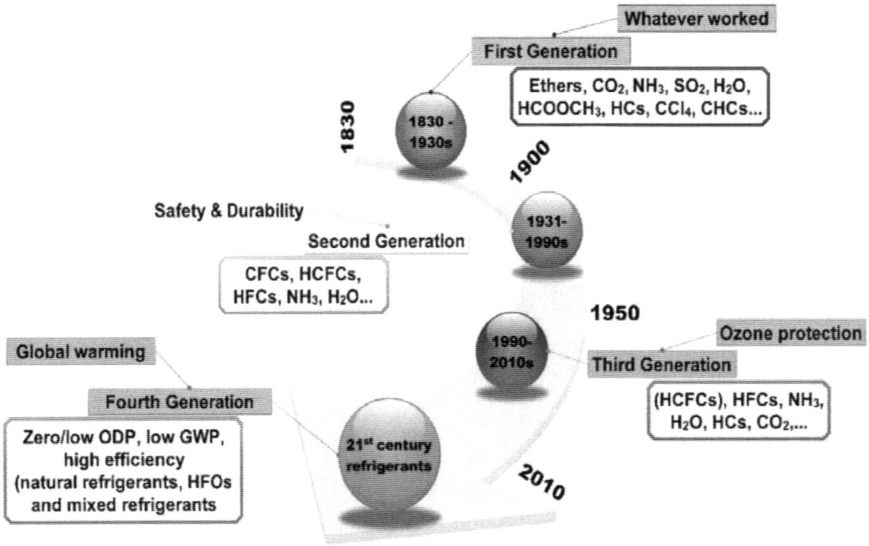

Fig. 1. Progression in the refrigerants over time [3]

Bosnia and Herzegovina has ratified the Vienna Convention and the Montreal Protocol, which created prerequisites for the implementation of activities on phasing down the consumption of substances that damage the ozone layer. Bosnia and Herzegovina, as a developing country and a country mentioned under Article 5 of the Montreal Protocol (consumption of the controlled substances does not exceed an annual calculated level of consumption of 0.3 kilograms per capita), is obliged to introduce HFC control measures as follows:

- Determination of base consumption for the period from 2020 to 2022;
- Freezing consumption at the level of basic consumption in 2024;
- Reduction of consumption by 10% until 2029;
- Reduction of consumption by 80% by 2045.

Under this background the objective of this research is to diagnose the use of refrigerants in Bosnia and Herzegovina and to determine the baseline as an input for policymakers in designing effective policies in order to comply with the Vienna Convention, the Montreal protocol and its amendments. The main contribution of this study is to understand which refrigerants are in use, in which type of a refrigeration system, what the purpose of a refrigeration equipment is and a sector in which an equipment is used.

2. Method

The research sample for analysis of refrigerants and their impact on the environment included 297 facilities located in 10 major cities and 64 municipalities in Bosnia and Herzegovina. (Fig. 2) .

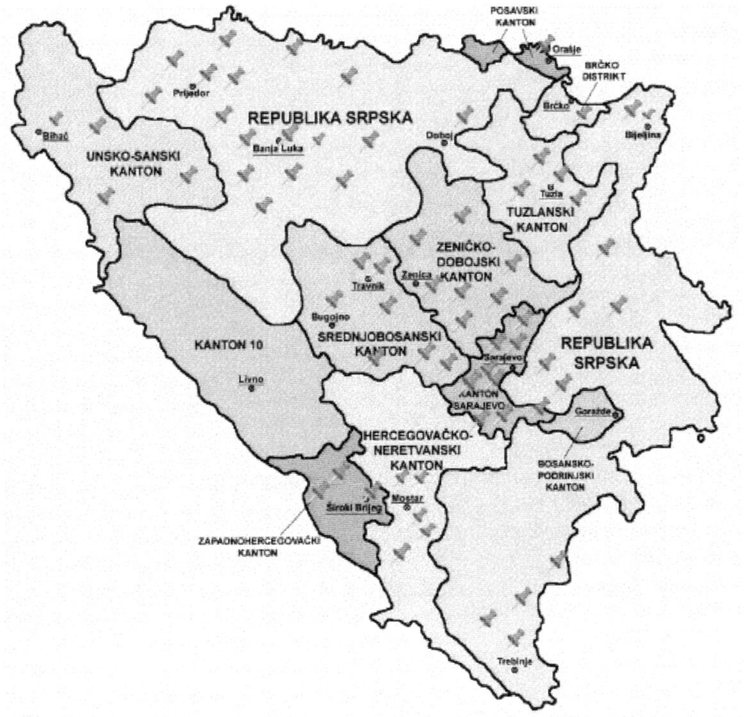

Fig. 2. Overview of places where refrigeration equipment was registered in BiH

The authors established a database that contains information on a type of refrigeration system, type of refrigerants used, a purpose of refrigeration equipment and a sector in which an equipment is used. Refrigeration devices and refrigerants are recorded in the facilities of various uses such as: banks, shopping canters and markets, hotels, hospitals, bakeries, processing and canning of fruits and vegetables, meat and dairy industry, etc. Data are grouped into 4 sectors: i) Trade sector, ii) Industry and manufacturing, iii) Tourism, iv) Health care/administrative buildings. Analysis of environmental impacts of refrigerants was done on the basis of the global worming potential and the ozone depletion potential.

3. Results and discussion

Shopping canters and stores (commercial sector) have the largest number of registered facilities, the share of which is 68%, while the tourism and hospitality sector has the smallest share, around 3%. (

Fig. 3). The most used are hydrofluorocarbon (HFC) refrigerants with a share of 62%. Natural refrigerants are presented with 37%, while the use of other types of refrigerants is minor.(

Fig. 4).

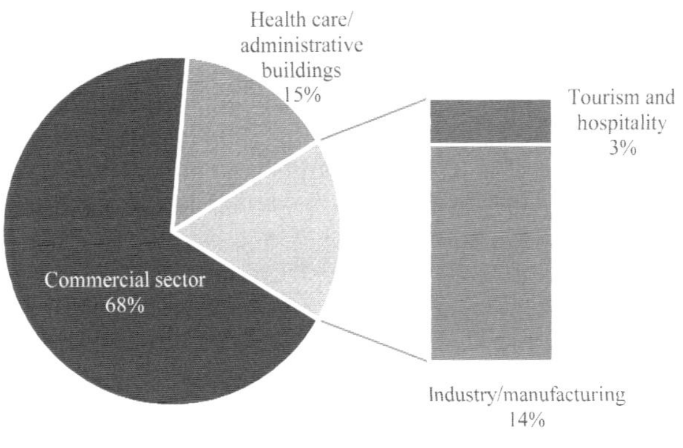

Fig. 3. Sector representation in the researched sample (%)

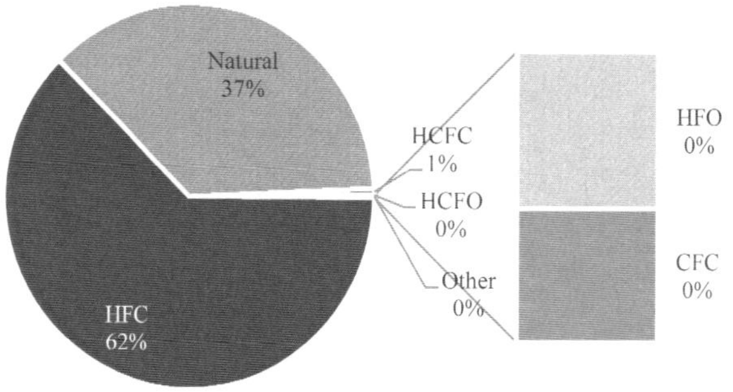

Fig. 4. Shares of use of refrigerants (%)

The total mass of refrigerants registered amounted to 6.7 t. The total of 17 types of refrigerants has been registered having the mass between 8,7 kg to 2,4 tones. Ten of them are HFC, while there is also banned CFC still in use.

No	Refrigerants	Type	Mass (kg)	GWP at 100-yr time horizon [2]	ODP
1	R-12	CFC	50,00	10300	1
2	R-407C	Blend HFC	2328,90	1730	0
3	R-404A	Blend HFC	10955,83	4210	0
4	R-410A	Blend HFC	11111,01	2080	0
5	R-417A	Blend HFC	10,00	2346	0
6	R-407F	Blend HFC	47,60	1790	0
7	R-407A	Blend HFC	70,00	2070	0
8	R-134A	HFC	16470,00	1360	0
9	R-32	HFC	36,00	705	0
10	R-143A	HFC	303,00	5080	0
11	R-449A	Blend HFO/HFC	8,70	1370	0
12	R-22	HCFC	360,90	1780	0,05
13	R-717	Natural (Ammonia NH_3)	24072,00	0	0
14	R-744	Natural (Carbon dioxide CO_2)	380,00	1	0
15	1233zd (E)	HCFO	25,00	3,7	0
16	1234yf	HFO	25,00	<1	0
17	1234ze (E)	HFO	25,00	<1	0

Table 1. List of registered refrigerants in BiH

CFC refrigerant R-12 has share of only 0, 02%, but it is ODPs. According to the Montreal Protocol, Bosnia and Herzegovina was required, as a developing country, to phase out CFCs by January 1, 2010. Besides its ozone depletion potential, it is refrigerant with the highest GWP of 10300, among those registered in Bosnia and Herzegovina. Refrigerant R-717 (ammonia) has the largest share in relation to the total mass and is 36.3%. GWP of this refrigerant is 0, with atmospheric lifetime of 0,019 years. This is a natural refrigerant, with widespread use in medium and large food, beverage and preservation industry and has growing potential in HVAC chillers, thermal storage systems, process cooling and air-conditioning, district cooling systems, supermarkets, and convenience stores. R-717 at concentrations above 300 ppm have toxic effects and is "moderately flammable" in the air when its concentration ranges between 16% and 28% weight. [4]

R-22 is HCFC that is, like R-12, regulated by the Montreal Protocol. The protocol identified CFC refrigerants and HCFC as substances possessing exceedingly high ODP. For developing countries, the complete phase out of HCFC is scheduled for January 1st, 2030. As for other cooling agents, the following have a significant share in relation to the total mass (fig. 5):

- R-404A with a share of 16.5% in the total mass, whose GWP is 4210, and ODP is zero;
- R-410A with a share of 16.8% in the total mass, whose GWP is 2080 and ODP is zero;
- R-134A with a share of 24.9% in the total mass, whose GWP is 1360 and ODP is zero.

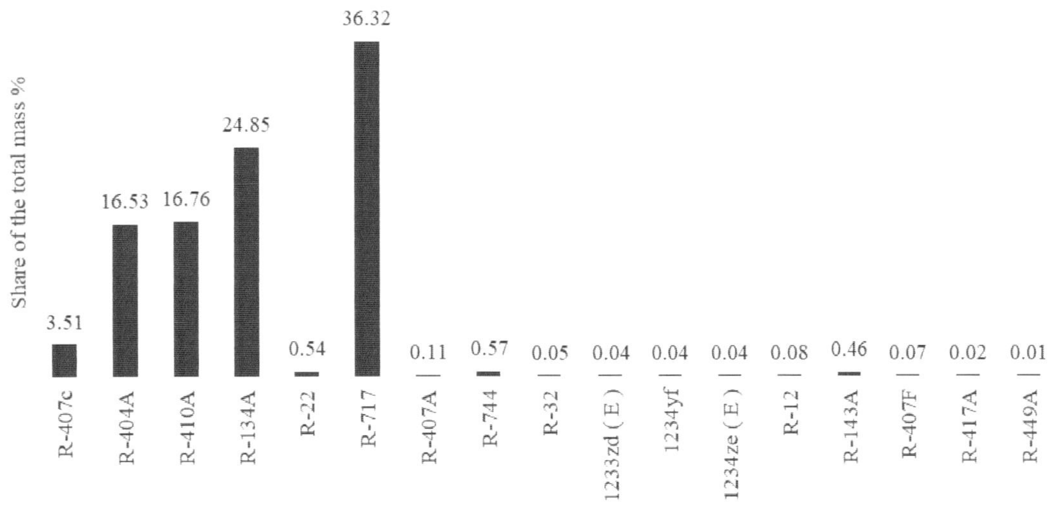

Fig. 5. Share of refrigerants in the total mass (%)

Taking into account GWP and the total mass, the most harmful refrigerant from the list, which is widely used in Bosnia and Herzegovina, is blend HFC R-404A with GWP of 4210. R-404A is composed of R-125/R-143a/R-134a with the share of 44/52/4 % respectively. It has a low toxicity, and it is not flammable; its boiling point allows it to work in low temperature refrigeration applications. R-404 A was used as a successful replacement for R-22 in refrigerated transport, which was possible even while maintaining the existing equipment. Keeping existing equipment in refrigerated transport has proven to be a cheap solution and a wise option in times of crisis. R-22 and R-404A have similar thermodynamic characteristics. [5]

The EU F-gas regulation of 2015 set restrictions on the use of HFC refrigerants with GWP >2500. The regulation aims to reduce supply of HFCs from 100% in 2015 to just 21% of that total in 2030, and prohibits the use of refrigerants with GWP > 2500 in new stationary refrigeration equipment, except for those intended for cooling the product to temperatures below – 50 °C. This restriction applies to R-404 A as well as to R-12 and R-143A, both having GWP above 2500. [6] Scientists have analysed gases R-448A, R-449A, R-454C, R-455A, R-442A, R-407H and R-452A as substitutes for R-404A and they have noticed a lot of differences between them. R-455A and R-465A represent the maximum reduction of the coefficient of performance (COP), while R-442A, R-449A and R-407H can represent an energy benefit. [7] The R-404 A is also widely used for the plug-in type refrigerating or freezing equipment in restaurants, health institutions, markets, etc. Due to provisions of the EU F-Gas regulation [6], which, from January 1st 2022, only allows the use of refrigerants with GWP < 150 for hermetic refrigerators/freezers of small capacity, alternative refrigerants to replace R404-A have been investigated. Researchers suggested that refrigerants R-454C, R-455A, R-457A and R-465A can be used as alternatives to R404A in all types of plug-in equipment such as both freezers and chillers and it is possible to achieve reduction in GWP about by 95% as a result of using those alternatives. [8]

R-410A is a blend refrigerant made up of 50% R-32 and 50% R-125 and was developed as an R-22 replacement for air conditioning systems. The EU F Gas regulation [6], also effects R-410A, as the split air conditioning systems, are limited to operating with refrigerants of which the GWP is below 150. Various researchers tested potential R-410 low GWP alternatives against main parameters such are: increases or decreases in cooling/heating capacity, COP, and compressor discharge temperature. Some studies addressed also the Total Equivalent Warming Impact (TEWI) and the Life Cycle Climate Performance (LCCP) analyses provide a measure of total CO_2 emissions over the equipment lifespan. A comprehensive analysis of different studies related to comparison between R-410A and alternative refrigerants is presented in [9]. Based on the analyses performed the author recommends R-452B and R-454B, both available on the market as suitable alternatives. Both refrigerants presented for the most of reviewed studies that the COPs and the cooling (or heating) capacities are similar to systems operating with R-410A, as well as that they have lower TEWI and LCCP values. R-452 provides overall performance closer to R-410A, resulting in minimal equipment redesign and cost, while R-454b is the most environmentally friendly refrigerant. [9].

Although with GWP lower than 2500, HFC-134a remains the most abundant HFC in the global atmosphere and has the largest annual growth rate and emission. The global annual mean mole fraction reached 89.5 ppt in 2016, up from 67.7 ppt in 2012. Global emissions have increased by the average of 10 Gg yr−1 since 2008, faster than emissions of any other HFC. [2]. Like other HFCs, it is regulated by the F-Gas Regulation, which determines deadlines, in relation to the type of equipment in which it is used, after which the gas can no longer be in use (Table 2).

Products and equipment		Date of prohibition
Refrigerators and freezers for commercial use (hermetically sealed equipment)	that contain HFCs with GWP of 2500 or more	1 January 2020
	that contain HFCs with GWP of 150 or more	1 January 2022
Stationary refrigeration equipment, that contains, or which functioning relies upon, HFCs with GWP of 2500 or more except equipment intended for the application designed to cool products to temperatures below –50 °C		1 January 2020
Multipack centralized refrigeration systems for commercial use with a rated capacity of 40 kW or more that contain, or which functioning relies upon, fluorinated greenhouse gases with GWP of 150 or more, except in the primary refrigerant circuit of cascade systems where fluorinated greenhouse gases with a GWP of less than 1500 may be used		1 January 2022
Movable room air-conditioning equipment (hermetically sealed equipment which is movable between rooms by the end user) that contains HFCs with GWP of 150 or more		1 January 2020
Single split air-conditioning systems containing less than 3 kg of fluorinated greenhouse gases, that contain, or which functioning relies upon, fluorinated greenhouse gases with GWP of 750 or more		1 January 2025

Table 2. F-Gas regulation restrictions against HFCs [6]

The fourth generation of refrigerants, with zero ODP and GWP below ten, namely 1233zd (E), 1234yf, 1234ze (E), have minor share of 0,12% in total mass. Those HFOs are the commercially important short-lived replacement compounds that have been developed as a replacement for the controlled HFC refrigerants [2]. In recent years, scientists have devoted much attention to studying various aspects of HFO refrigerants such as thermodynamic and transport properties, flammability and oil compatibility, heat transfer performance during boiling and condensation, and their performance in real vapor compression refrigeration systems. There are still some open areas to be further investigated: flammability of HFO refrigerants while considering, effects of temperature and relative humidity on lower flammability limit (LFL) and upper flammability limit (UFL), oil compatibility of high boiling HFO refrigerants at various temperature and pressure, identification of a suitable compressor oil for each and every HFO refrigerant, boiling and condensation heat transfer performance investigations of HFO/oil mixtures at various oil concentrations, etc. [10]

Registered refrigeration equipment has several areas of use:

- Refrigeration of storage;
- Refrigeration of premises;
- Deep freezing of goods;
- Heating of premises;
- Heat pumps;
- Ice production;
- Industrial processes cooling.

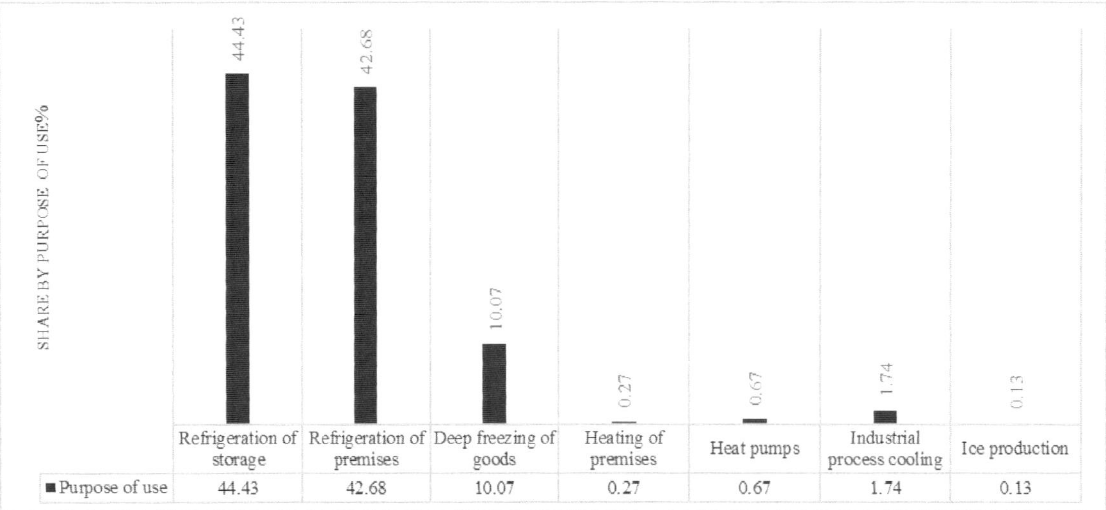

Fig. 5. Share by purpose of refrigeration equipment (%)

Refrigeration equipment has the largest application in the area of refrigeration of storage, which accounts for 44.4%, and refrigeration of premises, which accounts for 42.7%. Freezing of goods has the share of 10%, industrial process cooling 1,74%, while the share of other areas of use amounted to less than 1%. (Fig. 5)

R-717 is represented in Bosnia and Herzegovina in the industry and production sector. (Fig. 6) The largest application of R-134A is in the administration and healthcare sector, the share of which is 40.6%, and similar level of the application is in the commercial sector, which share is 37.5% (Fig. 7). Refrigerant 404 A has the largest application in the commercial sector, which share is 78.5%. Refrigerant R-410A is most widely used in the commercial sector, with the share of 60.6%, and this refrigerant is also widely used in the administration and healthcare sector, with the share of 25.7%. (Fig. 8) R-407C is most widely used in the commercial sector, with the share of 44.4%. Also, this refrigerant is widely used in the administration and health sector, which share is 42.8%. (Fig. 9)

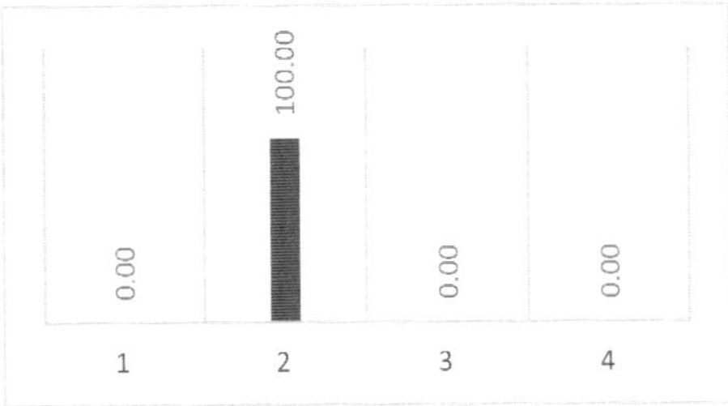

Fig. 6. Share of R-717 per sector (%)

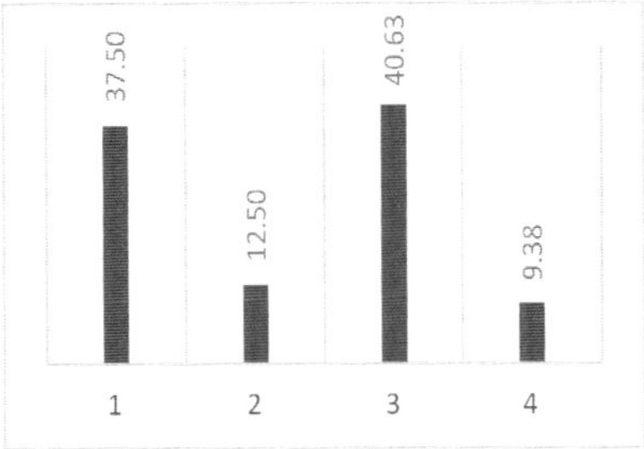

Fig. 7. Share of R-134A per sector (%)

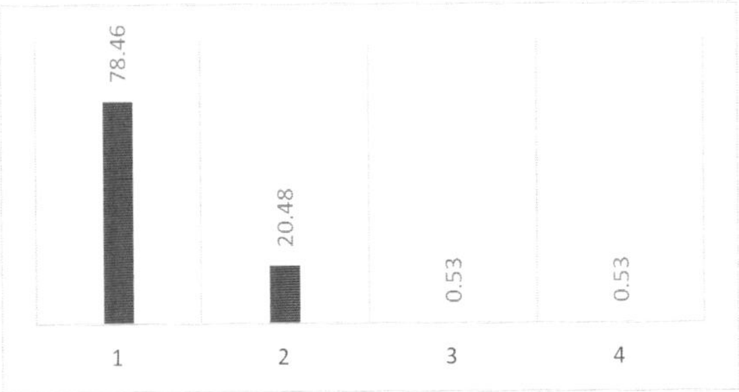

Fig. 8. Share of R-404A per sector (%)

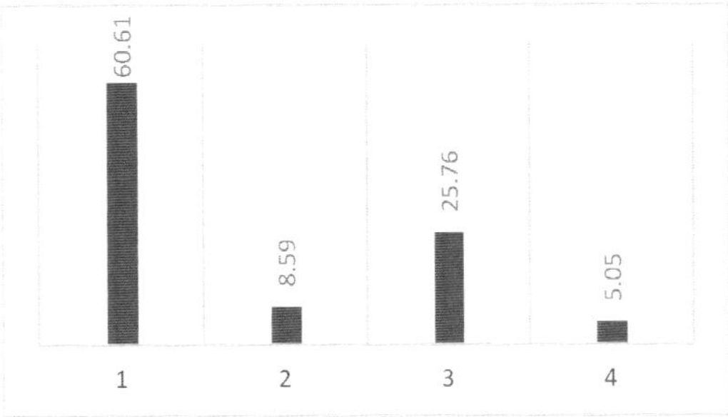

Fig. 9. Share of R-410A per sector (%)

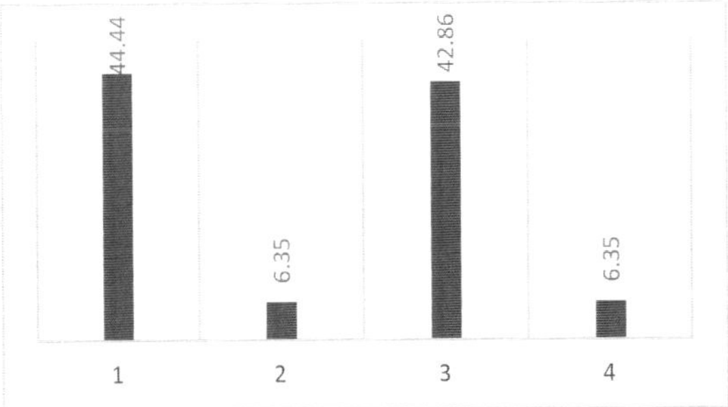

Fig. 10. Share of R-407C per sector (%)

Legend:
1-Commercial sector
2-Industry and manufacturing
3-Administrative and health care buildings
4-Tourism and hospitality

4. Conclusion

The results of this study provide an approximation of the way in which coolants are used in different sectors in Bosnia and Herzegovina and enable an assessment of the current situation in the context of the level of use of environmentally acceptable coolants. The results should be taken with a little reserve because not all facilities and equipment are registered, but they give a clear picture of types, dominant representation of refrigerants and sectors in which they are used. Main obstacles to registration of the refrigeration equipment are lack of data on number of hotels, motels, restaurants, supermarkets, etc., as well as a lack of legal obligation for equipment owners to submit data to competent authorities.

Out of the total registered mass of 6,7 t , the most commonly used refrigerant gases are HFC with share of 62%, followed by environmentally friendly natural gas R-717. Environmentally friendly gases such as the fourth generation gases HFO/HFCO are not frequently used. Refrigerants having ODP equal to zero are mostly in use, except for banned CFC R-12 and HCFC R-22 which has to be phased out until January 1st, 2030.

Although both of controlled refrigerants are present in the research sample with the very small percentage of 0,62 %, this indicates that better control on import, use and service and maintenance of equipment is needed. Refrigerants R-143a, R-404A have a share of approx. 17%, their GWP is greater than 2500 and should be phased out until January 1st, 2030. HFC gases with the high GWP: R-407C, R-407F,R-417A and HFC/HFO R-449A are represented with only 0,21%.

Although the EU F-regulation does not refer to non-EU countries, Bosnia and Herzegovina as an EU accession country should speed up phasing down of R-134 A and R-410-A, with gases having GWP below 150. 78% of the total R-134A is used in administration & healthcare and in a commercial sector. With a share of 78.5%, R-404 A has the largest application in the commercial sector. R-410A is widely used by administration & healthcare and a commercial sector. Industry and production sector mostly use natural refrigerant R-717 followed by R-134 A.

The commercial sector has become aware of the importance of changing their refrigerants for more environmentally friendly gases, such are 1233zd (E), 1234yf, 1234ze (E), but their share in the total registered mass of 6,7 t is just 0,12%. The recent scientific research has shown that there are environmentally friendly suitable alternatives for gases that need to be phase-down, and some of which are already present on the market as a commercial product. Further studies should include refrigeration and air conditioning equipment and refrigerants in residential and transport sectors, the mobile air conditioning (MAC) sector and other sectors (foam, aerosol, solvent and fire suppression).

5. Acknowledgments

The study was made as part of the Environmentally Sound Management of Persistent Organic Pollutants (POPs) in the Industrial and Hazardous Waste Sectors in Bosnia and Herzegovina project, which is supported by the UNDP. The results of the research would help to improve the existing legislation and policies in Bosnia and Herzegovina and make further steps towards fulfilling the country's obligations in the implementation of the Montreal Protocol on ODSs and their alternatives.

6. References

[1] Pinklesh A., Ajay K.T., Seshadri G., (2018) Fourth Generation Refrigerant: HFO 1234 yf, Current Science 115(08):1497, DOI: 10.18520/cs/v115/i8/1497-1503

[2] World Meteorological Organization (2018), Scientific Assessment of Ozone Depletion, Global Ozone Research and Monitoring Project—Report No. 58

[3] Vuppaladadiyam A.K, Antunes E., Vuppaladadiyam S.V, Baig Z.,T., Subiantoro A, Lei G, Leu S., Y., Sarmah A., J., Duan H., (2022) Progress in the development and use of refrigerants and unintended environmental consequences, Science of The Total Environment, Volume 823, 2022, 153670, ISSN 0048-9697, DOI:10.1016/j.scitotenv.2022.153670.

[4] Ayub , Z.,H, (2010) Status of Enhanced Heat Transfer in Systems With Natural Refrigerants, Journal of Thermal Science and Engineering Applications 2(4):044001, DOI:10.1115/1.4003343

[5] Mitu, D. E, Feiza Sucuran M. (2011), Considerations on adopting R 404a instead of R22 on reefers, Annals of DAAAM for 2011 & Proceedings of the 22nd International DAAAM Symposium, Volume 22, No. 1, ISSN 1726-9679, ISBN 978-3-901509-83-4, Editor B. Katalinic, Published by DAAAM International, Vienna, Austria, EU, 2011

[6] Regulation (EU) No 517/2014 of the European Parliament and of the Council of 16 April 2014 on fluorinated greenhouse gases and repealing Regulation (EC) No 842/2006, Official Journal of the European Union

[7] Heredia-Aricapa, Y., Belman-Flores, J.M., Mota-Babiloni, A., Serrano-Arellano, J., García-Pabón, Juan J., (2020), Overview of low GWP mixtures for the replacement of HFC refrigerants: R134a, R404A and R410A, International Journal of Refrigeration, Volume 111, 2020, Pages 113-123, ISSN 0140-7007, DOI: 10.1016/j.ijrefrig.2019.11.012.

[8] Devecioğlu A., Oruç V., (2022), Drop-in assessment of plug-in R404A refrigeration equipment using low-global warming potential mixtures, International Journal of Low-Carbon Technologies, Volume 17, 2022, Pages 991–999, DOI: 10.1093/ijlct/ctac078

[9] Guilherme I., F., Pico D., F., M., Dall'Onder dos Santos, D., Filho, E., P., B., (2022), A review on the performance and environmental assessment of R-410A alternative refrigerants, Journal of Building Engineering, Volume 47, 2022, 103847, ISSN 2352-7102, DOI: 10.1016/j.jobe.2021.103847.

[10] Vipin N., (2021) HFO refrigerants: A review of present status and future prospects, International Journal of Refrigeration, Volume 122, Pages 156-170, ISSN 0140-7007, DOI: 10.1016/j.ijrefrig.2020.10.039.

[11] Pardo Martínez C., Cotte Poveda A., Torralba Barreto D.,R., (2021), Diagnosis and baseline refrigerant use in the Colombian meat, dairy, and fruit and vegetable industries, International Journal of Refrigeration, Volume 131, 2021,Pages 448-458, ISSN 0140-7007, DOI: 10.1016/j.ijrefrig.2021.08.017.

DOI: 10.2507/33rd.daaam.proceedings.021

LINUX SECURITY IN PHYSICAL, VIRTUAL, AND CLOUD ENVIRONMENTS

Vedran Dakic, Karlo Jakobovic & Luka Zgrablic

This Publication has to be referred to as: Dakic, V[edran]; Jakobovic, K[arlo] & Zgrablic, L[uka] (2022). Linux Security in Physical, Virtual, and Cloud Environments, Proceedings of the 33rd DAAAM International Symposium, pp.0151-0160, B. Katalinic (Ed.), Published by DAAAM International, ISBN 978-3-902734-36-5, ISSN 1726-9679, Vienna, Austria
DOI: 10.2507/33rd.daaam.proceedings.021

Abstract

This paper will cover some of the available Linux methodologies and tools that can be used to enhance Linux security. It gives the reader some orientation in the world of Linux security. We will define what are the components of security and risk management, followed by device encryption and decryption using LUKS (Linux Unified Key Setup) and NBDE (Network Bound Disk Encryption) and restricting USB access by using USBGuard. The paper then shifts its focus to authentication methods utilizing PAM (Pluggable Authentication Modules), followed by firewalld, system auditing and monitoring file system changes with Audit and AIDE. We also go through ACL and SELinux concepts, ending with resource usage management with cgroups and OpenSCAP for compliance management. In the last part, we talk about an overall approach to putting all of these principles into operation, significantly impacting security footprint of our Linux-based resources. We do all this with the purpose being to have all of the necessary details for top-to-bottom security approach, as well to discuss future frameworks and standards development.

Keywords: NBDE; PAM; SELinux; OpenSCAP; ACL

1. Introduction

In this day and age, security is more important than ever before. By having in mind that most of the servers used in production are Linux/Unix[1], their security becomes the imperative of OS security[2]. This paper will provide a comprehensive overview of the necessary security components of Linux OS hardening. First, we will dive into the definitions of risk and security with their respective components. After explaining these concepts, we will transition to device encryption and decryption by using LUKS and NBDE. Then, we will discuss restricting USB access to Linux/Unix servers by using USBGuard. Authentication methods for users and applications are explained by using the Linux native PAM system. When it comes to auditing and monitoring changes, Audit toolkit for system auditing and AIDE for tracking file system changes are the norm. With that, let's start our discussion with core concepts – security and risk management.

2. Managing security and risk

A risk is any event that presents the possibility of loss, whether it be financial or productivity loss or delays to services. Continuous risk management is a process of taking a proactive approach to discovering potential risks, assessing facts, and taking action based on the facts to resolve those risks. Monitoring risk indicators and the actions taken to resolve them is important to succeed in risk management. Communicating to stakeholders what those risk indicators are, the action taken to resolve them, and the results, all promote confidence and stability in the process. The following diagram illustrates a process for continuously managing security risks by maintaining constant attention to potential security vulnerabilities and taking a proactive approach to maintaining a secure computing environment:

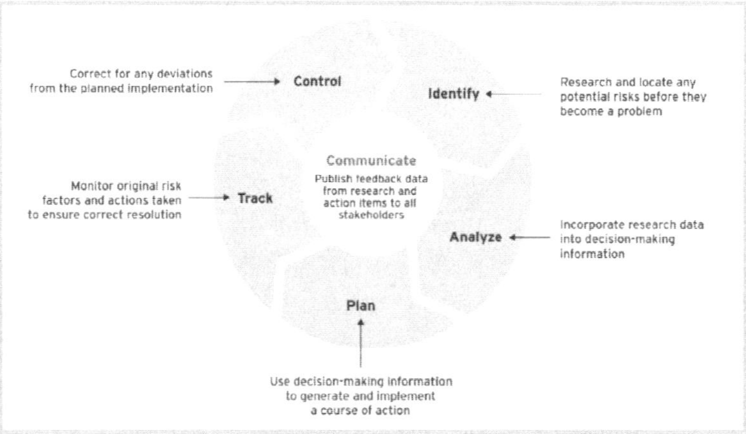

Fig. 1. Continuous risk management life cycle [3]

Managing security in our computing environments is a collection of continuous activities. Red Hat offers solutions that enable security in each phase of the continuous security life cycle. Security must be both proactive and reactive. It must be considered at every stage of our application and infrastructure life cycle. To do this effectively, we need to integrate security experts into our application, deployment, and infrastructure teams. The following diagram illustrates a continuous security life cycle that incorporates the risk management life cycle:

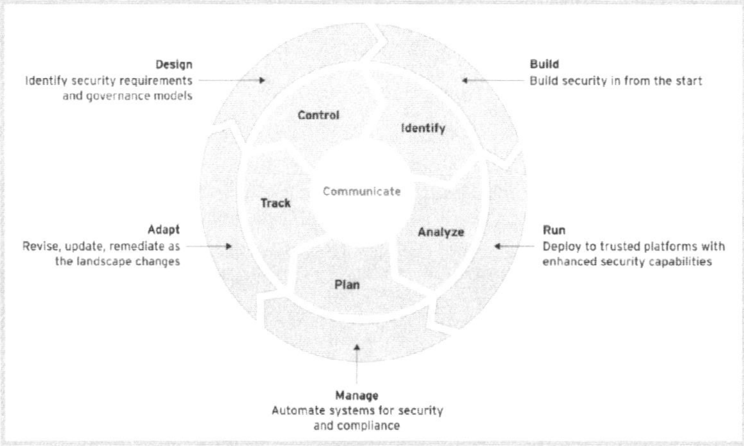

Fig. 2. Continuous security and risk management life cycle [3]

When using Linux, that will surely mean using a layered approach, as there are many technologies involved, and none of them is top-to-bottom. Let's discuss some of them.

3. Protecting data with LUKS and NBDE

The risk of a system's physical compromise puts sensitive data in its storage devices at risk of compromise. This is particularly true for mobile systems such as laptops or removable media, but servers may also have security requirements that the data they store is protected at rest. Encrypting this data helps mitigate the risk of its exposure if the system is lost. Red Hat Enterprise Linux supports block device encryption with Linux Unified Key Setup (LUKS) technology. Encrypting a block device [4] (such as a disk partition or LVM physical volume) is easiest at installation time, but LUKS can also be configured after installation.

However, setting up LUKS encryption requires that the file systems on the device be reformatted [5]. We can use the *cryptsetup luksDump* command post-installation to verify the encryption information for an encrypted device. This command displays information such as the LUKS header information[6], and the key slots (each of which may contain a valid passphrase) in use by the LUKS-encrypted device. The default encryption cipher is *aes-xts-plain64* [7].

One of the pitfalls of LUKS encryption is that we have to enter a passphrase manually to decrypt a disk at boot time. With a large number of servers, this requirement becomes a problem. Network-bound Disk Encryption (NBDE) automates the decryption of those encrypted disks without manually entering any passphrase at boot time in a secure way, by ensuring certain criteria are met. We can use NBDE to automatically decrypt LUKS storage devices containing root and non-root file systems. We must edit the */etc/crypttab* file to specify the encrypted block devices to open (decrypted and mapped) at boot time. This file contains an entry for each encrypted block device. We can use */etc/crypttab* on its own to configure LUKS to decrypt devices at boot time, but this requires either those passwords to be entered manually on the console or that they are stored in plain text on a decrypted device on the system. This is inadequate for use cases where servers need to boot unattended and still keep the decryption password secure [8].

NBDE provides a mechanism to automatically decrypt LUKS devices at boot time in a secure manner. This mechanism is based on two key components. Clevis runs on the system and decrypts devices and defines the policy whose conditions must be met for the data to be decrypted. Tang [9] is a very simple server that Clevis clients contact to determine if they are booting with access to a secure network and therefore decrypt their LUKS devices. We can use this to stop the automatic decryption of a device that has been removed from the data center [10].

The architecture of NBDE uses two key components: the Clevis framework, and the Tang server. The Clevis framework is a pluggable framework that supports NBDE on the client side and automates the unlocking of LUKS-encrypted block devices. The Tang server supports NBDE on the server side and makes data available on a system when the system can reach the Tang server over a specific secure network. The Clevis framework supports plug-ins to interact with Tang servers (PINs). Both the Clevis framework and the Tang server use the JOSE framework as their back end for encryption and decryption.

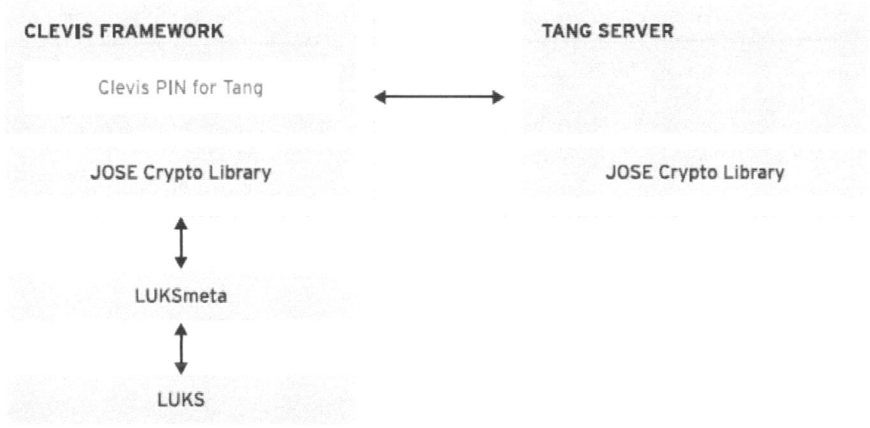

Fig. 3. NBDE architecture with Clevis and Tang [8]

Clevis gets a list of the keys for each Tang server and uses one of those keys to generate an encryption key. The Clevis PIN then uses the encryption key to encrypt the data. The encryption generates some state information, which is stored in the LUKS header. That state information is accessible with the *luksmeta show* command [11].

4. Restricting USB device access

USBGuard is a software framework that protects our systems against rogue USB devices by implementing basic whitelisting and blacklisting capabilities based on device attributes. The Linux kernel's USB device authorization system is used to enforce the settings configured by USBGuard. This allows us to control access to USB devices. For example, we can define what kind of USB devices are authorized and how a USB device may interact with our system. The USBGuard framework provides the following components:

- The daemon component has an inter-process communication (IPC) interface for dynamic interaction and policy enforcement
- The command-line interface used to interact with a running USBGuard instance
- The rule language used to write USB device authorization policies
- The C++ API used to interact with the daemon component, implemented in a shared library

The *usbguard* daemon determines whether or not to authorize a USB device based on a policy defined by a set of rules[12]. When a USB device is inserted into the system, the daemon scans the existing rules sequentially. When a matching rule is found[13], it either authorizes (allows), does not authorize (blocks), or removes (rejects) the device, based on the rule target. If no matching rule is found, the decision is based on an implicit default target. This implicit default is to block the device until a decision is made by the user [14].

5. Controlling authentication with Pluggable Authentication Modules (PAM)

The Pluggable Authentication Modules system, or PAM, provides a generic way for applications to implement support for authentication and authorization. Initially, an application that wanted to authenticate users used the local */etc/passwd* and */etc/shadow* files. Eventually, other authentication mechanisms were invented. For example, our company might decide to implement Kerberos, in which case the application must verify the correctness of the passwords against the KDC (*Key Distribution Center*). Alternatively, we may choose to deploy an LDAP (*Lightweight Directory Access Protocol*) infrastructure, and this time the application must contact the LDAP server for authentication. This means that every time a new authentication method came up, applications had to be rewritten to support it. Each application was handled separately, there was a lot of duplication of effort, and many different implementations had to be audited for correctness.

PAM provides a generic way for applications to implement support for authentication and authorization. A PAM-enabled application calls the PAM library, *libpam*, to perform all authentication tasks on its behalf and return a pass or fail response to the application. The PAM modules implement different authentication methods [15]. For example, the *pam_krb5* module performs authentication against Kerberos. The *pam_ldap* module authenticates against an LDAP server. The *pam_unix* module uses the standard local files, */etc/passwd* and */etc/shadow*. The PAM-enabled applications can directly use these authentication methods without having to be modified or recompiled [16].

The majority of the PAM configuration files are in */etc/pam.d/*, although some modules use additional configuration files in */etc/security*. PAM stores its binary modules responsible for evaluating criteria in the */usr/lib64/security* directory. Each PAM-enabled application has its own PAM configuration file in */etc/pam.d/*, and generally this file has the same name as the application: */etc/pam.d/login* for the login program, or */etc/pam.d/sshd* for the SSH daemon etc. If the service file for an application is missing, PAM uses */etc/pam.d/other* by default [17].

These configuration files contain rules that specify the modules to call for authentication and authorization. This way administrators can configure different authentication methods per application. In practice, we usually want all of our applications to use the same authentication method. To achieve that, most of the configuration files use the include directives to include the *password-auth* or *system-auth* files, also in the */etc/pam.d/* directory.

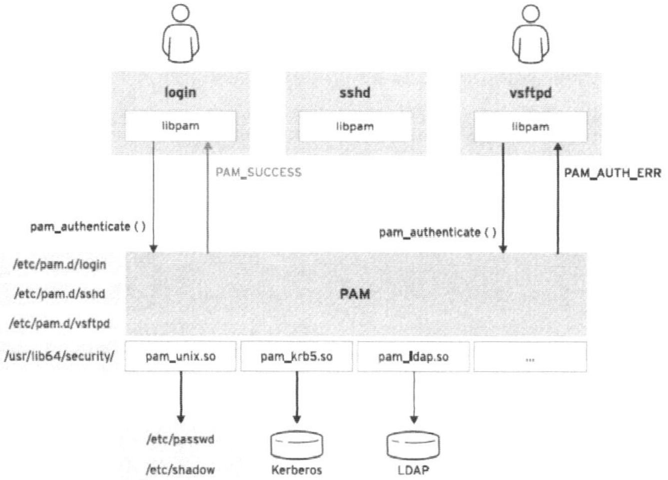

Fig. 4. PAM authentication [18]

In this figure, the login program contacts PAM for authentication. PAM reads the */etc/pam.d/login* configuration file to retrieve the list of modules to use for authentication. PAM calls the modules, stored in /usr/lib64/security/, and these modules perform the authentication. PAM returns to the login program a successful authentication status. The *vsftpd* FTP server also uses PAM to authenticate FTP users, but in the figure, the user provides incorrect credentials, and PAM returns a failure code to vsftpd [19]. The *System Security Services Daemon* (SSSD), introduced in Red Hat Enterprise Linux 6.0 allows user authentication against remote directories and authentication services such as LDAP, Kerberos, Active Directory, and Red Hat Identity Management. When a remote user authenticates on the local system, SSSD stores their credentials and authentication parameters in a local cache.

This way, even if the remote authentication server is not reachable, the remote user can still log in to the local system. This feature is particularly useful for laptops disconnected from the company network but also helps reduce the load on remote authentication services [20].

6. Managing network security with firewall (firewalld, iptables, ufw, ...)

Firewalls are an essential part of modern system security and Linux systems are no different. *Firewalld* is the default firewall in modern Red Hat-based Linux distributions. It allows users to shape the network traffic using the concept of zones and services. Zones are a set of rules that apply to a certain network interface [21]. For example, some of the default firewalld zones are trusted and external. As the names may suggest, trusted zone has a default set of rules that allow almost all traffic, while external zone won't allow any external traffic. Zones can be applied to an interface using firewall-cmd command.

Services are a set of rules that can be applied to a zone that would allow a certain network service to function properly. For example, service ssh would allow traffic on port 22 once applied to a zone. Their purpose is to simplify firewall configuration for the end user. Rather than manually allowing and configuring each port needed for a given network service, the concept of firewalld service allows the user to configure it using a single command [21]. Services can be configured for a given zone using the firewall-cmd command. Firewalld architecture is described in the following figure:

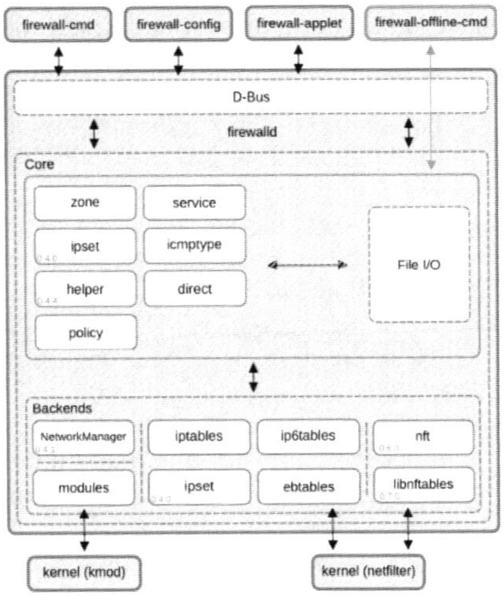

Fig. 5. Firewalld architecture [21]

Firewalld management is done using *firewall-cmd* command and the graphical counterpart *firewall-config*, which support a variety of options and configure the firewall while it's running, meaning there is no need to restart the machine or the service for the new rules to take effect. Rather than being a standalone firewall, firewalld is a frontend management tool for *nftables*, an IP packet filtering framework. Nftables was designed as a replacement for the widely used *iptables*. While their command and rule syntax differ, both use netfilter, the packet filtering component of the Linux kernel. Nftables provides a userspace configuration tool named *nft*. This tool is used by firewalld to configure nftables behind the scenes. All of the configuration done via firewalld is translated into nftables commands [22].

7. Recording system events with audit

The Linux Audit system is a mechanism in the kernel that provides a way to track security-related events and information on our systems. We define a set of rules that we load into the kernel to specify which events it should record in the audit log. The *auditd* system daemon writes the logged events to the local disk or forwards them to remote log servers. We can then use this information to discover and investigate the cause of violations or attempted violations of the system's security policy [23]. We can configure Audit to include or exclude events based on user identity, security contexts, or other labels. Some of the information that the Audit system can collect about events includes:

- Date and time, type, and outcome of an event
- Sensitivity labels of subjects and objects
- Association of an event with the identity of the user who triggered the event
- All modifications to the Audit configuration and attempts to access Audit log files
- All uses of authentication mechanisms, such as SSH, Kerberos, and others
- Changes to any trusted database, such as /etc/passwd
- Attempts to import or export information into or out of the system

We may need to use Audit system to meet the compliance requirements of particular security certifications. Audit has been evaluated by National Information Assurance Partnership (NIAP) and Best Security Industries (BSI). It has been certified to LSPP/CAPP/RSBAC/EAL4+ on RHEL5 and OSPP/EAL4+ on RHEL6. Audit is designed to meet or exceed the requirements of the CAPP, LSPP, RSBAC, NISPOM, FISMA, PCI-DSS, and STIG certifications or compliance guides [24].

Auditd is the user-space component of the Linux auditing subsystem. The system unit file that starts auditd normally loads any persistent audit rules when the daemon is started. After audit rules have been loaded into the kernel, the kernel uses them to record events of interest. The kernel sends information about those events to auditd. The main role of auditd is to collect the audit event messages recorded by the kernel and save them to a log file. When auditd is running, the log file configured for auditd (usually */var/log/audit/audit.log*) collects the audit messages sent by the kernel. If auditd is not running for any reason, *rsyslog* receives the kernel audit messages [25].

Without any extra configuration, only a limited number of messages pass through the audit system, mainly authentication and authorization messages (users logging in, sudo being used, etc.) and SELinux messages. Administrators can add audit rules that control the audit system, watch files, or record information about any system call with the *auditctl* command [26].

8. Monitoring filesystem changes with Advanced Intrusion Detection Environment (AIDE)

On an operational server, it is normal for files to be added, removed, and modified on its file systems. However, unexpected changes to certain files, such as executable programs and configuration files, might indicate unauthorized modifications or other security issues. Therefore, it is important to monitor those files for changes to their content, permissions, or other characteristics. Red Hat Enterprise Linux provides Advanced Intrusion Detection Environment (AIDE), a user-space utility that can help with this monitoring. AIDE monitors files for a variety of changes including permission or ownership changes, timestamp changes (modification or access time stamps), or content changes.

The configuration file for AIDE is */etc/aide.conf*. This file controls which files AIDE monitors for changes, and what characteristics are monitored for each file. For example, by default AIDE monitors most files in the */etc* directory for permission changes only, but specific files are monitored more closely. As the security administrator, we might want to tune exactly what AIDE monitors for different parts of our computer's file system. Each line in the */etc/aide.conf* file is a directive. The file contains three types of lines: configuration lines, selection lines, and macro lines. Any line that starts with a hash (#) is a comment and has no effect [27].

The configuration lines adjust the configuration parameters of AIDE. These lines either tune the functional behavior of AIDE globally or set a group definition [28]. Group definitions are used by selection lines to specify what characteristics of a file AIDE should monitor when detecting file system changes [29]. The selection lines specify the files and directories that AIDE monitors, and the changes for which AIDE will watch. Selection lines can be regular, equals, or negative. A regular selection line is a regular expression matching the absolute path to a file or directory, followed by the name of a group definition. Files and directories matching the regular expression are added to the AIDE database, with checks performed as specified by the line's group definition. This effectively means that if the regular expression is /etc, then the regex will also recursively match all files and directories in the /etc directory. An equals selection line starts with an equals (=) sign followed by a regex and a group definition. AIDE records the files that match the regular expression, considering all the checks the line's group definition mentions. However, an equals selection line will only match the children of directories if the regex ends with a forward slash (/) character. The children of subdirectories will not be recursively matched. A negative selection line starts with a bang (!) character followed by a regular expression matching the absolute path to a file or directory. AIDE does not monitor files or directories that match a negative selection line. The macro lines set or clear variables that are useful for referring to lengthy URLs or file system paths in multiple occurrences throughout the AIDE configuration file [30].

9. Mitigating risks with Access Control Lists (ACL) and Security Enhanced Linux (SELinux)

Every file and directory on a Linux/Unix system has a set of permissions targeted at three categories of users: the user who owns the file/directory, the group of users who own the file/directory, and all other users on the system. The types of access that can be defined using these permissions are: read permission, write permission and execute permission. While this offers some amount of control over which user can access which object in the file system, it does not offer specific control over user permissions because it's limited to defining permissions for one user, one group, and all other users [31].

Access Control Lists (ACLs) are used to control user permissions in a more granular way. Using ACLs, each file or directory can have an unlimited number of permissions set on it, and each user on the system can have his own set of permission on a given file or directory. This allows users and administrators to strictly define user permissions, allowing for a greater level of security on the system and preventing unwanted access to specified filesystem objects [32].

Unlike regular filesystem permissions which are managed using ls, chown, and chmod commands, ACLs are managed using a different pair of commands: *getfacl* and *setfacl*. As their names imply, *getfacl* is used to read permissions currently set on a filesystem object, while *setfacl* is used to modify these permissions. For example, to give a user named "testuser" read and write permissions on the "*/testdir*" directory, "*setfacl -m testuser:rw- /testdir*" command would be used.

Security Enhanced Linux (SELinux) is an additional layer of system security. The primary goal of SELinux is to protect user data from system services that have been compromised. SELinux confines programs, mainly system services to the minimum amount of privilege they require to do their jobs successfully. For that purpose, a set of security rules determine which processes can access which files, directories, and ports. If there is no explicit rule for access, SELinux denies the operation. Every file, process, directory, and port has a particular security label called the SELinux context. A context is used by SELinux rules to control access. For files and directories, SELinux stores the contexts as file system extended attributes. For processes and ports, the kernel maintains the contexts in memory. Administrators can view the context of processes and files by adding the -Z option to the ps or ls commands. For ports, use the semanage port -l command.

In Red Hat Enterprise Linux, SELinux bases most of its rules on the third element: the context type. The context type for the web server is *httpd_t*. Context types for processes are also known as domains. The context type for files and directories in */var/www/html* is *httpd_sys_content_t*. The context type for web server ports is *http_port_t*.

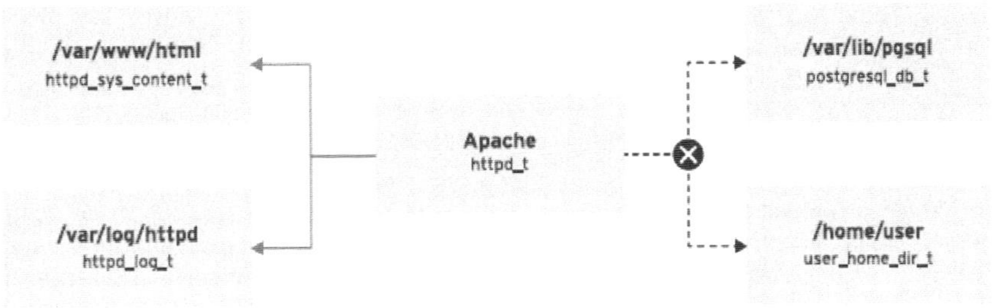

Fig. 6. Apache service with SELinux enabled [18]

The previous figure shows that an SELinux rule allows Apache, whose domain is *httpd_t*, to access files and directories with the *httpd_sys_content_t* context type. Another rule allows Apache to access its log files under /var/log/httpd. The rule allows the *httpd_t* domain to access files and directories with the *httpd_log_t* type.

There is no rule to allow the httpd_t domain to access files and directories with the user_home_dir_t type. Access is, therefore, denied [33]. If a malicious user manages to exploit a security flaw in Apache, they cannot access the contents of user home directories, even if the Linux access rights permit the access. This way, SELinux mitigates the risk while the security flaw is fixed [34].

10. Restricting resource usage via Control Groups (cgroups)

Control groups (*cgroups*) are a Linux kernel feature that allows users to group a set of processes on the system and allocate a certain amount of resources to them. Resources that can be allocated include CPU time, memory, network bandwidth, disk I/O etc. These groups are ordered hierarchically, meaning one group can be a subgroup of an already existing group, creating a tree. However, unlike with the regular Linux process model, in which only a single process tree exists, starting with the init process, multiple cgroup trees can exist. These cgroups are created and managed in the /sys/fs/cgroup/ directory. Cgroups are a vital component of containers, which use cgroups in combination with Linux namespaces as a basis of their architecture [35].

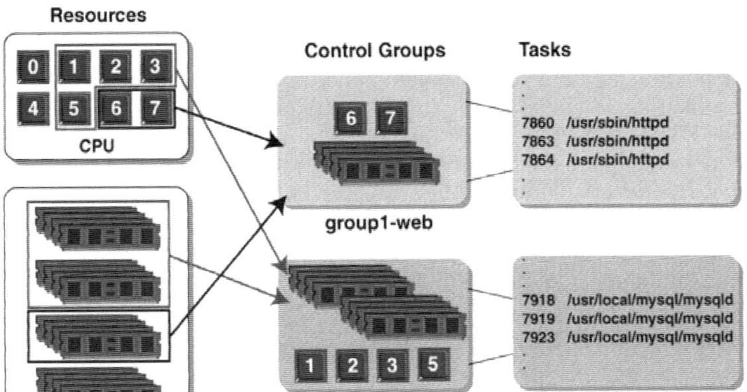

Fig. 7. Control group relationship with resources [37]

Cgroup resources are managed by a kernel component named resource controller (also named control group subsystem). These resource controllers represent a resource in question. For example, a resource controller named memory takes care of memory allocation for a given cgroup, while blkio controller takes care of I/O access [36]. Cgroups can be created and managed by manually creating and editing files and directories in the aforementioned /sys/fs/cgroup/ directory.

However, *systemd* can be used to create and manage cgroups in modern Linux systems. By default, every systemd service, scope, and slice (systemd units) map to objects in a cgroup tree, and new units can be created on the fly. Service units contain one or more processes that are started and stopped at the same time. These units are defined using a unit configuration file. In comparison, scope units do not create their processes, but manage those created externally. The third type of unit, slice, does not contain any processes on its own but is instead used to group services and scopes hierarchically [38].

To create a systemd unit, *systemd-run --unit=unit_name* command can be used to start command under a unit. This also creates a cgroup and starts that command within it. Furthermore, systemd provides a way to manage units/cgroups resources on the fly, using the systemctl command. For example, command *systemctl set-property unitname MemoryLimit=500M* sets a memory limit of 500 MiB for a unit named "*unitname*". These changes are persistent after reboots unless they're run with the *--runtime* parameter.

11. Managing compliance with OpenSCAP

Enterprise computing environments may consist of hundreds or thousands of interconnected computer systems, running numerous applications and services, and accessed by a large and diverse set of users and applications. To maintain control over the security of this vast environment, a standard way to scan systems for compliance with security policies is needed. The National Institute of Standards and Technology (NIST), in conjunction with other authorities, developed a standard compliance system called Security Content Automation Protocol (SCAP).

SCAP [39] can be described as a framework of security specifications. It supports automated configuration, vulnerability, and patch-checking measurement. OpenSCAP is an open-source project that develops tools for implementing and enforcing security policies using the SCAP standard. We need to perform a compliance audit to verify that a given object follows a rule in a compliance policy. Usually, the compliance policy varies across organizations. The needs and risk profiles of each organization are different and often require the ability to customize the compliance policy checklist. The OpenSCAP project provides several predefined and customizable compliance policies in SCAP format for use with OpenSCAP tools [40].

The SCAP Security Guide is a collection of security policies for Linux systems, in the form of SCAP documents. It consists of rules with detailed descriptions and proven remediation scripts and Ansible playbooks. The SCAP Security Guide can be used with OpenSCAP tools to automate the auditing of a Linux system. For ease of use, all the available security policies are broken into profiles. A profile can be defined as a grouping of security settings that correlate to a known policy. The SCAP Security Guide provides profiles for verifying system compliance against standards established by governmental or other organizations. For example, the PCI-DSS profile tests compliance against the rules mandated by the Payment Card Industry Security Standards Council.

However, in the real-world situations vary. Most profiles in the SCAP Security Guide are meant as a catalogue, not a checklist, and satisfaction of every item may not be sensible or even possible in many operational scenarios. Our organization, auditors, or other stakeholders may not require our systems to comply with every item specified in a particular profile, but instead, require our systems to comply with a specific subset of the profile. Some rules may not apply in our environment, and others, not in the profile, may be required. With SCAP Workbench, we can create custom profiles. As a starting point, we select an existing profile that we adjust by selecting and clearing compliance rules. We then save our new profile in an XML tailoring file that we can copy to the systems to scan [41].

Having this in mind, we can see that SCAP is a tool that can be used for a variety of different scenarios – albeit at the expense of time. But that also leads us to a larger discussion that needs to be further researched in the future - for example, problems with the correct detection of the identity of someone who's attacking our infrastructure [42]. That's also a chance to expand our research into different frameworks, like GDPR. Although it could be improved [43], it's still a valid framework for working with private data, and all of these technologies that we mentioned in this paper are underlying to that in a technical sense. That also means introducing new frameworks to govern security infrastructure, and that can also lead to some changes in the way we approach network infrastructure design [44]. Putting all of this together will be an area of intensive research in the future, and, as time goes on, we expect new methodologies and standards to be developed.

12. Conclusion

We have shown that managing security in our computing environments is a collection of continuous activities and layering different technologies. LUKS and NBDE provide us with a solid framework for managing encryption and decryption of our block devices, paired with USBGuard for defending USB access to our systems. We are provided a generic authentication mechanism with PAM, it enables us to authenticate users and applications on our systems. With Audit, we can discover and investigate the cause of violations or attempted violations of the system's security policy. On an operating server, it is normal for files to be added, removed, and modified on its file systems.

However, unexpected changes to certain files, such as executable programs and configuration files, might indicate unauthorized modifications or other security issues. Therefore, it is important to monitor those files for changes to their content, permissions, or other characteristics; AIDE can help us with these operations. In addition to standard POSIX permissions, we have introduced SELinux which confines programs, mainly system services to the minimum amount of privilege they require to do their jobs successfully. Lastly, we discussed OpenSCAP, the standardized compliance system. With OpenSCAP we can benchmark our system against predefined security policies and guidelines and then remediate the non-compliant systems. These tools and their overall methodology need to be a part of larger discussion and research that could lead to some new standard or framework being adopter for a more global approach to Linux security management. This is also going to be our future research topic.

13. References

[1] https://w3techs.com/technologies/details/os-unix, (2018). Q-Success Web-based Services, Accessed on: 2022-07-01

[2] https://w3techs.com/technologies/overview/operating_system, (2018). Q-Success Web-based Services, Accessed on: 2022-07-01

[3] https://docs.fedoraproject.org/en-US/quick-docs/getting-started-with-selinux, (2019). RedHat, Accessed on: 2022-08-15

[4] https://www.redhat.com/sysadmin/disk-encryption-luks, (2019). RedHat, Accessed on: 2022-07-01

[5] https://en.wikipedia.org/wiki/Linux_Unified_Key_Setup, (2022). Wikipedia, Accessed on: 2022-07-01

[6] Danut, A. & Simion E. (2018). Linux Unified Key Setup (LUKS) – The Good, the Bad, the Ugly, Proceedings of ECAI 2018 International Conference, Lasi, Romania, ISBN: 978-1-5386-4901-5, doi: 10.1109/ECAI.2018.8678978

[7] https://www.thegeeksearch.com/beginners-guide-to-luks-disk-encryption-in-linux, (2016). The Geek Search, Accessed on: 2022-07-02

[8] https://access.redhat.com/documentation/en-us/red_hat_enterprise_linux/8/html/security_hardening/configuring-automated-unlocking-of-encrypted-volumes-using-policy-based-decryption_security-hardening, (2019). RedHat, Accessed on: 2022-07-02

[9] https://www.redhat.com/en/blog/network-bound-disk-encryption-improvements-rhel-8, (2021). RedHat, Accessed on: 2022-07-02

[10] https://docs.openshift.com/container-platform/4.9/security/network_bound_disk_encryption/nbde-about-disk-encryption-technology.html, (2022). RedHat, Accessed on: 2022-07-02

[11] https://www.cybersecurity-insiders.com/network-bound-disk-encryption-in-red-hat-linux-7-2, (2017). Cybersecurity Insiders, Accessed on: 2022-07-02

[12] https://www.redhat.com/en/blog/built-protection-against-usb-security-attacks-usbguard, (2017). RedHat, Accessed on: 2022-07-01

[13] https://access.redhat.com/documentation/en-us/red_hat_enterprise_linux/7/html/security_guide/sec-using-usbguard, (2015). RedHat, Accessed on: 2022-07-01

[14] https://www.systutorials.com/docs/linux/man/5-usbguard-rules.conf, (2017). Systutorials, Accessed on: 2022-07-1

[15] https://www.redhat.com/sysadmin/pluggable-authentication-modules-pam, (2020). RedHat, Accessed on: 2022-07-02

[16] https://access.redhat.com/documentation/en-us/red_hat_enterprise_linux/6/html/managing_smart_cards/pluggable_authentication_modules, (2012). RedHat, Accessed on: 2022-07-03

[17] https://www.ibm.com/docs/en/aix/7.2?topic=system-pluggable-authentication-modules, (2022). IBM, Accessed on: 2022-07-03

[18] Vazquez, A.; Glogowski,A.; Costea, V.; Quatremain, H.; Paul, S. & Karmakar, S. (2018): Red Hat Security: Linux in Physical, Virtual, and Cloud, Bonneville, S.; O'Brien, D.; Kenlon, S.; Sacco, D. & Charles, H., 1-534, Red Hat, Raleigh, NC27601, USA

[19] https://en.wikipedia.org/wiki/Pluggable_authentication_module, (2022). Wikipedia, Accessed on: 2022-07-03

[20] http://web.mit.edu/rhel-doc/5/RHEL-5-manual/Deployment_Guide-en-US/ch-pam.html, (2007). MIT, Accessed on: 2022-07-01

[21] https://access.redhat.com/documentation/en-us/red_hat_enterprise_linux/9/html/configuring_firewalls_and_packet_filters/using-and-configuring-firewalld_firewall-packet-filters#when-to-use-firewalld-nftables-or-iptables_getting-started-with-firewalld, (2022). RedHat, Accessed on: 2022-07-02

[22] https://firewalld.org/2018/07/nftables-backend, (2018). FirewallD, Accessed on: 2022-07-02

[23] https://access.redhat.com/documentation/en-us/red_hat_enterprise_linux/6/html/security_guide/chap-system_auditing, (2011). RedHat, Accessed on: 2022-07-04

[24] https://www.redhat.com/sysadmin/configure-linux-auditing-auditd, (2021). RedHat, Accessed on: 2022-07-04

[25] https://linux.die.net/man/8/auditd, (2008). Die.net, Accessed on: 2022-07-04

[26] https://documentation.suse.com/sles/12-SP4/html/SLES-all/cha-audit-setup.html, (2009). SuSe, Accessed on: 2022-07-04

[27] https://en.wikipedia.org/wiki/Advanced_Intrusion_Detection_Environment, (2022). Wikipedia, Accessed on: 2022-07-04
[28] https://access.redhat.com/documentation/en-us/red_hat_enterprise_linux/7/html/security_guide/sec-using-aide, (2015). RedHat, Accessed on: 2022-07-03
[29] https://www.redhat.com/sysadmin/linux-security-aide, (2021). RedHat, Accessed on: 2022-07-03
[30] https://documentation.suse.com/sles/15-SP1/html/SLES-all/cha-aide.html, (2019). SuSe, Accessed on: 2022-07-03
[31] https://www.redhat.com/sysadmin/linux-access-control-lists, (2020). RedHat, Accessed on: 2022-07-02
[32] https://access.redhat.com/documentation/en-us/red_hat_enterprise_linux/7/html/system_administrators_guide/ch-access_control_lists, (2015). RedHat, Accessed on: 2022-07-05
[33] https://www.redhat.com/en/topics/linux/what-is-selinux, (2019). RedHat, Accessed on: 2022-07-06
[34] https://wiki.centos.org/HowTos/SELinux, (2020). RedHat, Accessed on: 2022-07-06
[35] https://opensource.com/article/21/8/container-linux-technology, (2021). RedHat, Accessed on: 2022-07-26
[36] https://access.redhat.com/documentation/en-us/red_hat_enterprise_linux/9/html/managing_monitoring_and_updating_the_kernel/setting-limits-for-applications_managing-monitoring-and-updating-the-kernel#what-kernel-resource-controllers-are_setting-limits-for-applications, (2022). RedHat, Accessed on: 2022-07-26
[37] https://www.oracle.com/technical-resources/articles/linux/resource-controllers-linux.html, (2012). Oracle, Accessed on: 2022-08-01
[38] https://access.redhat.com/documentation/en-us/red_hat_enterprise_linux/7/html/resource_management_guide/sec-default_cgroup_hierarchies, (2015). RedHat, Accessed on: 2022-08-15
[39] https://access.redhat.com/documentation/en-us/red_hat_enterprise_linux/6/html/security_guide/chap-compliance_and_vulnerability_scanning, (2012). RedHat, Accessed on: 2022-08-15
[40] https://access.redhat.com/documentation/en-us/red_hat_enterprise_linux/6/html/security_guide/sect-using_openscap_to_remediate_the_system, (2012). RedHat, Accessed on: 2022-08-15
[41] https://www.redhat.com/en/blog/getting-started-red-hat-insights-and-openscap-compliance-reporting, (2021). RedHat, Accessed on: 2022-08-15
[42] Zaharova, A.; Elin, V. & Panfilov, P. (2018): Technological and legal issues of identifying a person on the internet to ensure information security, Proceedings of DAAAM., at Zadar, Croatia, ISSN 1726-9679, DOI: 10.2507/29th.daaam.proceedings.069
[43] Dakic, V. & Ribaric, S. (2020): Judicial and technical improvement of General Data Protection Regulation, Proceedings of DAAAM., at Mostar, Bosnia and Herzegovina, ISSN 1726-9679, DOI: 10.2507/31st.daaam.proceedings.025
[44] Reithner, I.; Papa, M.; Lueger, B.; Cato, M.; Hollerer, S. & Seemann, D. (2020): Development and implementation of a secure production network, Proceedings of DAAAM., at Mostar, Bosnia and Herzegovina, ISSN 1726-9679, DOI: 10.2507/31st.daaam.proceedings.102

DOI: 10.2507/33rd.daaam.proceedings.022

CI/CD TOOLSET SECURITY

Vedran Dakic, Jasmin Redzepagic & Matej Basic

This Publication has to be referred as: Dakic, V[edran]; Redzepagic, J[asmin] & Basic, M[atej] (2022). CI/CD Toolset Security, Proceedings of the 33rd DAAAM International Symposium, pp.0161-0164, B. Katalinic (Ed.), Published by DAAAM International, ISBN 978-3-902734-36-5, ISSN 1726-9679, Vienna, Austria
DOI: 10.2507/33rd.daaam.proceedings.022

Abstract

Modern application development has redefined the way teams develop their solutions. Instead of using their workstations for building code, a lot of teams have resorted to using microservices for CI/CD systems. CI stands for continuous integration while CD denotes continuous delivery. Combined these two things mean that using a CI/CD system code goes into building and testing cycle as soon as the developer submits it. Since the whole system is inevitably complex, almost all the systems are using a combination of technologies to manage both the building and delivery part as well as underlying services that make building possible. In this space technologies such as Kubernetes or OpenShift are becoming a norm. From the security perspective this creates a whole new problem since such a system has to be deeply integrated into the core of the business network, and any potential threat to the CI/CD infrastructure is immediately a threat to the whole internal infrastructure. CI/CD tools need to have advanced privileges, they have to be able to access code repositories, user directories, complete development environments and even bare metal servers in order to optimize the delivery process. This means that attack surface in such a system is enormous and exploiting it means gaining access to large part of the business infrastructure. Securing such a heterogenous system is a big task and, in this paper, we address most important challenges.

Keywords: CI/CD; Kubernetes; Openshift; Gitlab; Jenkins; ArgoCD

1. Introduction

During the middle of 2010 decade there was a large shift when it comes to application development, especially in the web application space. Established monolithic building architecture involving dedicated servers and large server pools slowly became too complicated and too financially wasteful so a new solution was needed. Old model has an inherent problem of underutilisation since a lot of resources are allocated on servers that perform either a single or small number of steps in the build process. Since this creates a large serial pipeline, this creates a lot of wasted time spent waiting for a task to finish. Virtual machines are a partial solution to this but only when containers and orchestration frameworks become available was a true solution possible. [1] Containers enabled creating environments that contain only the objects and resources needed for a particular task and running a lot of tasks in parallel on same hosts. This meant that the hardware utilization went up while ability to manage software differences across platforms was simplified. Continuous Integration (CI) and Continuous Delivery (CD) systems have an inherent problem with dependencies, every attempt to support many architectures and platforms means that at any given time there are a lot of similar but subtly different environments being used [2].

At the same time updating the environments in which the build process is happening is almost constant and, in this task, containerization enables probably the quickest way to update or redeploy whole environments. Security in the context of a CI/CD system should therefore cover not only least privilege set, access control, monitoring and all the other best practices for the system itself, but we also need to track all of the components and their internals. This involves tracking changes in the source code, containers, all the tools and all the dependencies our system uses to produce the resulting application.

For all of the stated reasons CI/CD systems are a big attack surface from the security perspective and any improvement in the way security is designed in the build system itself has huge ramifications on the security of the final product. Our aim is to provide essential guidelines on how to both secure and restructure a typical CI/CD system to make it more robust.

2. Typical deployment scenario

How this all works in a typical development environment is relatively simple when abstracted enough. A developer usually manually develops and updates application code. In reality, this is mostly manual work, but sometimes rapid development tools assist in this part of the process. After a particular piece of code is finished it is submitted for review by the team. Review is usually done by a senior developer. Sometimes this step also involves automatic review by different tools that try to enforce standards in code syntax to keep the code uniform across the whole code base. After review code is merged with the appropriate code tree at which point CI/CD system identifies code changes and starts build deployment. Depending on the particular configuration this may start a partial of complete rebuild of the finished product.

Next step in CI/CD is usually some sort of compliance and functional testing done automatically to ensure that most obvious errors are spotted. Once all this is done without errors code is packaged into test environment to be further tested by QA (Quality Assurance) team. In this process CI/CD is the most important part of the infrastructure since it can access entire codebase, can introduce changes to the code, can deploy and run arbitrary code on any given part of the infrastructure and in the end can sign and deploy this code once application is assembled as a package.

All this makes CI/CD a prime target for any sort of attack since a successful attacker can disrupt the build process but also can directly inject code into the application and create even greater problems if this code injection ends up released as a signed and finished application. This kind of attack is sometimes called supply-chain attack since it tries to influence not the finished product but the production of the said product itself. If successfully done this kind of attack is complicated to detect and requires a lot of resources on response team side of things.

3. Best practices for CI/CD

CI/CD systems tend to be heterogeneous and comprise of different subsystems that are by themselves quite complex. Usually, we are talking about some sort of a code repository or code storage solution, Build / Deployment solution and some sort of orchestration tool as the core of the system. Rest of the system are different tools that are specific for each build. [4]

Name of the tool	Bamboo	Gitlab	Jenkins
Price (per agent)	Paid (from 10$)	Paid (from 4$/month)	Free
Type of deployment	OnPrem/Cloud	OnPrem/Cloud	OnPrem/Cloud
Free version exists?	Yes	Yes	Yes
Platforms supported	Windows, macOS, Linux, Solaris	Linux (Ubuntu, Debian, CentOS, Oracle)	Windows, macOS, Linux, Unix-alike operating systems

Table 1. Common CI/CD tools

Problem this poses from the security perspective is that we not only need to keep all the subsystems and tools updated, but we also need to keep inventory of their deployment and place in the whole system. Frequent updates and making sure we are not using outdated tools is the first thing that needs to be done. At the same time, we will sometimes HAVE to use tools that have known vulnerabilities due to their behaviour on a particular platform – a lot of applications we are developing can use older dependencies on older platforms and sometimes updates are no longer available. This is where inventory comes into play, every such dependency needs to be documented for our security team to apply additional measures to ensure safety.

Next thing to do is privilege reduction. Although the system needs to have broad permissions, every module has its own permissions that we can apply just to that module. For example, if there is a need for a module to read something from the database only read permissions must be given to that module ensuring nothing will end up being changed if there is a breach in that module. Network connections are another part of the security. Every connection is a potential risk, but we can be quite strict in limiting access. For example, we can be sure which workstations belong to developers and assign them special access to the system. Same goes for containers that build and deploy the solution.

IF a permission is not needed it shouldn't be given in the first place. Every access should be logged, and a monitoring solution used to ensure we are able to track all the actions and changes done to the system. [5] This monitoring and access also should protect from inside attack attempts and should keep information compartmentalized if at all possible. Next thing to monitor and pay special attention to are the credentials and the identification keys used across the system. A particular CI/CD can access not only the database but also other resources like cloud, both internal and external, remote shares, API endpoints, SSH endpoints, certificates and other sensitive data. Every piece of information should be encrypted, and keys automatically rotated to ensure security of data-at-rest.

Source repository is part of the CI/CD system but also has other uses inside the team. Since it will be accessed by the team directly, sometimes outside of the CI/CD scope, access should be monitored and limited. Last thing that can be considered a best practice is making sure we track possible vulnerabilities to all of our tools and to be ready to apply updates as needed while also keeping track on possible interactions between subsystems that can be broken by updates. CI/CD systems tend to have a lot of custom scripts and a lot of knowledge is needed to make sure system runs after updates.

4. Practical examples of security

Automated testing should not only be limited to testing pure functionality of the code, but also should include security[8], [10]. Static code analysis is one such solution that enables us to detect potential security problems in the code. There are solutions available on the market that can deal with a lot of specialized vulnerabilities, for example there is a library called Selenium that enables us to automate SQL injection checks in the code.[6] Solutions like this enable "shift-left" security concept, trying to implement security as early as possible in the application lifecycle.

Developers are often using special in-code tags to mark certain parts of the code to be executed, updated, or disabled. If we can implement these tags into our checks this can help us identify parts of the code that needs to be checked or even disabled. This also includes checks for obsolete dependencies. We mentioned that these can be part of the development process due to compatibility reasons, but we should make sure that usage scope for such parts of code must be limited to a minimum.

Rollback procedures are also a big part of the security planning. Sometimes a vulnerability will pass all the checks and end up in the production application. We must have a feasible plan on how to restore to a previous working state without the vulnerability. This includes CI/CD system having older versions of application available if needed. We mentioned the need to automate monitoring and logging of the CI/CD system. Production environments these days usually already have some sort of SIEM system implemented. Scope and reach of such system should be extended not only to application behaviour but also to the behaviour of the whole CI/CD system.

Our goal in all of this is to have complete control over CI/CD so we can spot not only changes in the code but also changes in the system that builds the code itself.[9] Even having implemented all the recommendations outlined here we are just scratching the surface of security since we are talking about a complex, heterogenous system. In this paper we concentrated only on the most important aspects of securing the CI/CD environment, reader should understand that the security of the CI/CD system also depends on all the underlying components.

5. CI/CD security in Jenkins

When it comes to One of the most popular and often used CI/CD tools is called Jenkins [7],[3]. This tool is free and open source, so it is both highly transparent for the user but also a frequent attack vector. Jenkins tends to be primary target of a lot of attacks that then use other tools to spread horizontally across resources. Being such a complicated and exposed tool, Jenkins has a robust security built in. A lot of the defaults are already strict out of the box, and update system is integrated into the tool itself. Jenkins also supports plugins, and they are in the scope of the update system. Unfortunately, Jenkins has seen hundreds of vulnerabilities published as public information so updating the system has to be frequent and complete.

Jenkins is structured in two parts – Jenkins server that enables all the functionalities of the system and Jenkins agents that are installed on build nodes, virtual machines or containers. Security of the server is usually in the hands of the administrator but since we are talking about a distributed system with high privileges a lot of functionality, especially scripting, is done by the users. Users are usually responsible for node configuration, script coding and deployment and everything else involving turning source into application itself. Since all this is executed by the agent there is a huge potential for security compromise.

Having agents also means that we can limit user interaction with the system – agents can be designated to specific projects so compromise of a single project can be isolated from other projects or the system itself.

Environment variables are one more thing we need to have in mind. Variables are used to pass different pieces of information between parts of the build process, most commonly build scripts. Variables can be divided by the phase of the build process they belong to (test, QA, production). This means we can keep sensitive data separate from the end product, but also that we need to keep track of variables at the same time.

Jenkins is a popular tool and a very powerful one, but it requires a lot of attention, and a dedicated team to keep it running. Main challenges in running Jenkins are updating the system while ensuring that the entire pipeline works after each update.

6. Conclusion

Modern development influences CI/CD systems and tools in making them more complex and advanced than ever before. Placing CI/CD in the core of the infrastructure and giving it control over almost every part of the system makes CI/CD security more important than ever. For this exact reason security has to be implemented from the moment we create any CI/CD infrastructure including not only the source control and build systems but also Kubernetes, OpenShift or whatever our system uses. Since CI/CD directly controls build process any intrusion not only compromises the system that we are working on but also the resulting application. All this means that security of the CI/CD process is probably the mose Security has to be layered, and implement all the standard security controls together with all the security practices for individual products that are used in the build system. Since systems tend to be heterogeneous and span different vendors, applying all the updates can prove problematic due to the way they can affect the system itself. Solution to this is to try and find the balance between using latest updates and making sure our system behaves as desired, which may also include unit tests that are performed while using the system on the system itself.

Human factor is also very important here since users are almost exclusively developers themselves and it can be difficult having them conform to different security challenge. But even when taking all the factors into account, in this particular case security has to be implemented since CI/CD is a core system that can not only compromise the work environment but also potentially can compromise the entire end product of the said system.

7. References

[1] Smart, J.F., Jenkins: The Definitive Guide: Continuous Integration for the Masses, , O'Reilly Media, 2011 ISBN: 9781449313654

[2] S. Garg and S. Garg, "Automated Cloud Infrastructure, Continuous Integration and Continuous Delivery using Docker with Robust Container Security," 2019 IEEE Conference on Multimedia Information Processing and Retrieval (MIPR), 2019, pp. 467-470, doi: 10.1109/MIPR.2019.00094.

[3] V. Armenise, "Continuous Delivery with Jenkins: Jenkins Solutions to Implement Continuous Delivery," 2015 IEEE/ACM 3rd International Workshop on Release Engineering, 2015, pp. 24-27, doi: 10.1109/RELENG.2015.19.

[4] Seremet, Z[eljko] & Rakic, K[resimir] (2021). Best Approach to Security in Azure Devops, Chapter 18 in DAAAM International Scientific Book 2021, pp.223-230, B. Katalinic (Ed.), Published by DAAAM International, ISBN 978-3-902734-31-0, ISSN 1726-9687, Vienna, Austria DOI: 10.2507/daaam.scibook.2021.18

[5] Kafol, C[iril] & Bregar, A[ndrej] (2017). Cyber Security – Building a Sustainable Protection, Chapter 07 in DAAAM International Scientific Book 2017, pp.081-090, B. Katalinic (Ed.), Published by DAAAM International, ISBN 978-3-902734-12-9, ISSN 1726-9687, Vienna, Austria DOI: 10.2507/daaam.scibook.2017.07

[6] https://owasp.org/www-community/attacks/SQL_Injection, 2022, Open Web Application Security Project, OWASP Foundation, Inc. Accessed: 2022-6-30

[7] https://www.jenkins.io/doc/book/security/, Jenkins project, Software in the Public Interest, Inc, Accessed: 2022-6-30

[8] Securing Devops: Safe Services in the Cloud (Paperback or Softback); Vehent, Julien; 2018; Published by Manning Publications; ISBN: 1617294136

[9] The Phoenix Project: A Novel about IT, DevOps, and Helping Your Business Win; IT Revolution Press; 2013; 1st edition ; ISBN 0988262592

[10]Agile Application Security: Enabling Security in a Continuous Delivery Pipeline; Laura Bell, Michael Brunton-Spall, Rich Smith, Jim Bird; 2017; O'Reilly Media; 1st edition; ISBN: 1491938846

DOI: 10.2507/33rd.daaam.proceedings.023

RUST AND WEBASSEMBLY FOR FAST, SECURE, AND RELIABLE SOFTWARE

Zlatan Moric, Loïc Branstett & Robert Petrunic

This Publication has to be referred as: Moric, Z[latan]; Branstett, L[oic] & Petrunic, R[obert] (2022). Rust and Webassembly for Fast, Secure and Reliable Software, Proceedings of the 33rd DAAAM International Symposium, pp.0165-0171, B. Katalinic (Ed.), Published by DAAAM International, ISBN 978-3-902734-36-5, ISSN 1726-9679, Vienna, Austria
DOI: 10.2507/33rd.daaam.proceedings.023

Abstract

Rust is a new systems programming language that promises to overcome the seemingly fundamental tradeoff between high-level safety guarantees and low-level control over resource management. WebAssembly (Wasm) is a new binary instruction format for a stack-based virtual machine, it is designed as a portable compilation target for programming languages, enabling deployment on the web for client and server applications. This paper explores these two new technologies and how they interact with each other. Our findings showed that the combination of those two technologies could revolutionize the way the entire industry design fast and secure software.

Keywords: Rust; WebAssembly; C++; memory security; reliable software.

1. Introduction

Systems programming languages like C and C++ give programmers low-level control over resource management at the expense of safety, whereas most other modern languages give programmers safe, high-level abstractions at the expense of control. It has long been a holy grail of programming languages research to overcome this seemingly fundamental trade-off and design a language that offers programmers both high-level safety and low-level control. Rust comes closer to achieving this holy grail than any other industrially supported programming language to date, compared to mainstream safe languages, Rust offers both lower-level control and stronger safety guarantees [1]. Some safe languages like Java or C# use a virtual machine but unlike WebAssembly, which is an abstraction over modern hardware, making it language-, hardware-, and platform-independent, with use cases beyond just the Web. WebAssembly has been designed with formal semantics from the start making it a strong contender for secure and reliable software [2].

From the perspective of a web browser, WebAssembly is a type of code that web browsers can execute directly – with its compact binary format we can expect to have near-native performance while using different programming languages as compilation targets. While it's debatable how useful that is, WebAssembly is also capable of running within a web page that already contains JavaScript code, so mixed use of these technologies is possible. In this paper, we are going to focus on Rust as a source for WebAssembly, as this seems to be a very popular choice for developers.

We also consider it to be a sensible approach, as Rust offers advantages over many existing programming languages in terms of security and semantics. So, let's explore Rust first with a brief overview, then move on to WebAssembly and finally check how they would partner for a high-performance web application.

2. Rust

In this section, we give a brief overview of some of the central features of the Rust programming language. As a programming language, Rust has a different focus than many other programming languages – it's very efficient with memory, very fast, and prioritizes memory security which seems to be a big problem for applications and computer architecture in the past ten or so years with a variety of different memory and CPU-related issues hampering application development and speed. It's well-documented and recently made it to the Linux kernel as a development platform (version 6.1).

A hugely popular feature of Rust is its capability to work with JavaScript and publish to the npm registry, which makes it a very attractive platform for web developers – so it's not meant to be used for non-web applications only. Also, with the proliferation of IoT devices and the general growth of the IoT concept, its capability to run decently on devices that are low on resources seems like a reasonable value proposition to use it. Let's now look at some basic capabilities and advantages of Rust.

2.1. Ownership

Ownership is Rust's most unique feature and has deep implications for the rest of the language. It enables Rust to make memory safety guarantees without needing a garbage collector. The ownership rules are quite simple:

- Data is assigned to a variable.
- The variable becomes the 'owner' of the data.
- There can only be one owner at a time.
- When the owner goes out of scope, the data will be dropped.

Primitive types are popped from stack memory automatically when they go out of scope (e.g., when a function block ends), while complex types must implement a drop function that Rust will call when out of scope (to explicitly deallocate the heap memory).

2.2. Borrowing

The concept of borrowing is designed to make dealing with ownership changes easier. It does this by avoiding the moving of owners, and by letting your program provide a 'reference' to the data. This means the receiver of the reference (e.g., a function, struct field, or a variable) can use the value temporarily without taking ownership of it. There are two main rules to the concept of borrowing:

- At any given time, you can have either (but not both of) one mutable reference or any number of immutable references.
- References must always be valid.

Those rules prevent concurrent access to the same underlying memory location and ensure that the underline memory location is safe to access at any time. These effectively make data races impossible [3].

2.3. Lifetimes

Lifetimes are tightly coupled to references. They are a way for the compiler to know how long a reference lives, and to be sure that any reference that is currently active doesn't refer to data that no longer exists (i.e., a 'dangling pointer'). At the language level lifetimes are just an annotation that has a specific naming convention: '<T> where <T> is a letter like 'a or 'b. The letters don't mean anything special, they're just a way for the user to differentiate them. [4] The below code example highlights how defining a single lifetime called 'a and assigning it to both arguments (and to the return value) allows the compiler to track these references and ensure they both live long enough to prevent any errors at runtime.

```
fn longest<'a>(x: &'a str, y: &'a str) -> &'a str {
   if x.len() > y.len() {
      x
   } else {
      y
   }
}
```

Code 1. Example of usage of lifetimes in a Rust code

2.4. Interior Mutability

Interior mutability is a design pattern in Rust that allows you to mutate data even when there are immutable references to that data; normally, this action is disallowed by the borrowing rules. To mutate data, the pattern uses unsafe code inside a data structure to bend Rust's usual rules that govern mutation and borrowing. The basic idea behind Interior Mutability is UnsafeCell [5] a basic primitive in Rust that allows checks normally done by the borrow checker to be deferred (not bypassed) to the library code. RefCell [6] is a type that is based on UnsafeCell and does the checks at runtime. This still provides the same safety guarantees as the borrow checker would give at compile-time. This allows developers to still be safe even when outside of the standard borrowing rules. Rust has many features to make the code safe but how do these features ensure memory safety? Let's look at a few examples of programs that should be memory unsafe and see how Rust identifies their errors.

2.5. Memory safety – dangling pointers

Let's take a look at the following code example:

```
fn foo() -> &i32 {
   let n = 0;
   &n
}
```

, and result from its compilation:

```
error[E0106]: missing lifetime specifier
--> test.rs:1:13
|
1 | fn foo() -> &i32 {
| ^ expected lifetime parameter
|
= help: this function's return type contains a borrowed value, but there is no value
  for it to be borrowed from
= help: consider giving it a 'static lifetime
```

If we attempt to return a pointer to a value owned by n, where the pointer would nominally live longer than the owner. Rust identifies that it's impossible to return a pointer to something without a "value for it to be borrowed from."

2.6. Memory safety – use-after-free

Let's take a look at the following code example to create a variable and pass it to the drop function:

```
use std::mem::drop; // equivalent to free()
fn main() {
let x = "Hello".to_string();
drop(x);
println!("{}", x);
}
```

, and result from its compilation:

```
error[E0382]: use of moved value: `x`
--> test.rs:6:18
|
5 | drop(x);
| - value moved here
6 | println!("{}", x);
| ^ value used here after move
|
= note: move occurs because `x` has type `std::string::String`, which does not implement
  the `Copy` trait
```

Due to Rust's ownership semantics, when we free a value, we relinquish ownership of it, which means subsequent attempts to use the value are no longer valid. This also protects against double frees, since two calls to drop would encounter a similar ownership type error.

2.7. Development safety

Rustc (the Rust official compiler) won't compile programs that attempt unsafe memory usage. Most memory errors are discovered when a program is running. Rust's syntax and language metaphors ensure that those common memory-related problems in other languages null or dangling pointers, data races, and so on—never make it into production. This makes the developers more confident about the safety and reliability of their code. [7] Let's now move on to WebAssembly - as a near-native-performance platform that's able to use various programming languages as sources, WebAssembly might play a significant role in the future of application development, especially when talking about web applications.

3. WebAssembly

WebAssembly is based on the principle of a stack-based virtual machine. It's an abstraction of a computer, that emulates a real machine. In this case, the emulation is based on a stack pointer that follows the instructions of the code. Its capability to execute binary code on the web while being independent of the programming language makes it a very attractive proposition for running fast, near-native performance web applications. WebAssembly has modules that are binary compiled by the browser (or a virtual machine). It has its own ISA (Instruction Set Architecture), and – at its core – it's not a project that wants to replace the ever-present JavaScript, as it tries to work alongside it to make the best of both worlds. Let's now discuss some WebAssembly specifics that make it such a good platform for web application developers.

3.1. Modules

A Wasm program is designed to be a separate module containing collections of various Wasm-defined values and program-type definitions. Every module is separated from the other modules. This provides native isolation between the code, meaning that the code from module A cannot access anything from module B that isn't explicitly exported/allowed by module B. This is particularly useful in our modern world where malware tries to access the memory of a program to give them privileges.

3.2. Linear memory

The main storage of a WebAssembly program is a large array of bytes referred to as linear memory or simply memory. This technique is fairly standard and has some advantages as the CPU can directly (and linearly) address all of the available memory locations without having to resort to any sort of memory segmentation or paging schemes [8].

3.3. Security

The linear memory is disjoint from code space, the execution stack, and the engine's data structures; therefore, compiled programs cannot corrupt their execution environment, jump to arbitrary locations, or perform other undefined behavior. At worst, a buggy or exploited WebAssembly program can make a mess of the data in its memory [9]. This means that even untrusted modules can be safely executed in the same address space as other code. It also allows a WebAssembly engine to be embedded into any other managed language runtime without violating memory safety, as well as enabling programs with many independent instances with their memory to exist in the same process.

3.4. Determinism

The design of WebAssembly has sought to provide a portable target for low-level code without sacrificing performance. Where hardware behavior differs it usually is corner cases such as out-of-range shifts, integer divide by zero, overflow or underflow in floating point conversion, and alignment. WebAssembly gives deterministic semantics to all of these across all hardware with only minimal execution overhead. However, there remain three sources of implementation-dependent behavior that can be viewed as non-determinism [10]:

- NaN Payloads: WebAssembly follows the IEEE-754 standard for floating-point arithmetic [11]. However, this standard doesn't define the representation of the NaN payload and instead only puts a rule for what it cannot be. This leads to CPUs handling and representing NaN payload differently.
- Resource Exhaustion: Available resources are always finite and differ wildly across devices. An "out of memory" error is unpredictable and therefore nondeterministic.
- Threads: Executing concurrently some code is highly dependent on the CPU Scheduler and subject to the difference between runs as the CPU has a different schedule profile [12].

Nevertheless, these nondeterminisms are limited and local, meaning that WebAssembly is as deterministic or even better than every other program.

3.5. Control-flow integrity

Within WebAssembly, Control-flow integrity (CFI) is enforced by the Memory Safety API. CFI protects against code-reuse attacks, such as Return Oriented Programming (ROP) [13]. The efficacy of control-flow integrity can be measured based on its completion. Generally, three external control-flow transitions need to be safeguarded because the caller may not be trusted. Those are direct function calls, indirect function calls, and returns.

3.6. Memory safety

To ensure memory safety, WebAssembly will use an expected control flow graph when it is compiled. Vectors of computer-readable instructions can be made safe by inserting runtime instrumentation at every call site to verify that the transition is safe.

3.7. Memory isolation

The Memory Isolation API provides facilities to protect against shared memory writes between two modules with different permissions (e.g., one module can read and change data while another cannot) [14]. This protects against vulnerabilities such as dangling pointers where an attacker might use knowledge of a pointer location from one context to attack it in another. The Memory Isolation API takes advantage of type system features available only when compiling to WebAssembly, such as linear types, which provide transparent compile-time enforcement at runtime.

3.8. Development safety

As for WebAssembly, its determinism of it makes developers also more confident and more productive as they don't have to worry about non-determinism and platform-specific methods/conversions or APIs. And, because WebAssembly methods are typed misuse of them (e.g., passing a short instead of an int) is impossible reaffirming the safety to the developers.

4. Speed of execution

As stated, previous WebAssembly is based on the principle of a stack-based virtual machine. This technique isn't new and is used in many implementations like the JVM (Java Virtual Machine) or the CLR (.NET Common Language). These implementations of a stack-based virtual machine prove to be extremely high-performant with only a small performance penalty [15]. However, unlike most JVM or CLR implementations many WebAssembly interpreter has also a compiler mode to compile WebAssembly to the native binary format saving the cost of the interpretation and allowing many optimizations making these binary close to native speed.

Rust is based on the now robust LLVM project a backend for compiler optimization, its most famous frontend is Clang, a C/C++ compiler. LLVM caries many optimizations accumulated over more than 2 decades. Studies as found that in general the same C/C++ program compiled with GCC, and Clang only has a +5% percent difference in performance [16]. This greatly helps Rust achieve the same or better performance than an equivalent C++ program compiled with Clang. Note that there is ongoing work to provide GCC as an alternative backend for the Rust compiler. This could potentially make Rust programs even faster than with LLVM [17]. For comparison, we executed a series of number operation counter code written in JavaScript and compare it with Rust code.

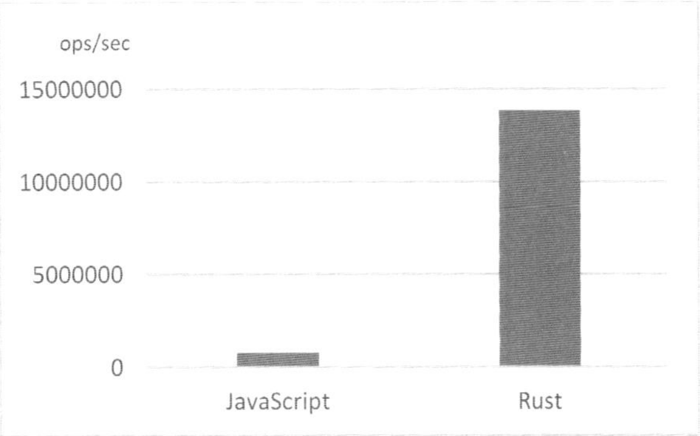

Fig. 1. JavaScript vs Rust speed comparison on simple counting loop

In this case (Fig. 1.), Rust outruns JavaScript by far. Next, we tried a similar test but now with strings to see how the technologies behave with a more complex set of data. Again, in both technologies, we create functions that accept two strings and check if the second string is part of the first.

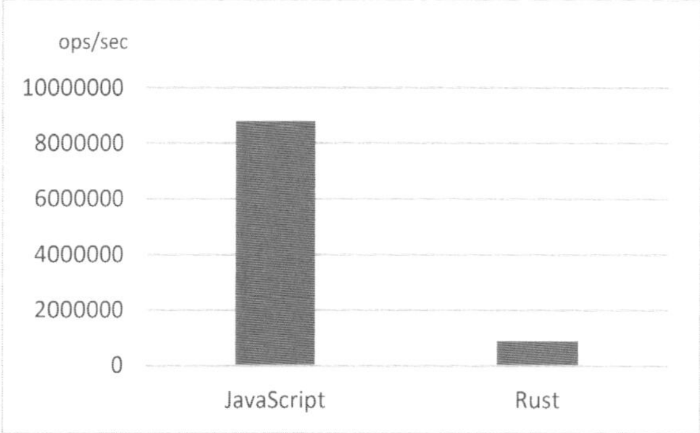

Fig. 2. JavaScript vs Rust speed comparison with strings

It was strange that JavaScript is so faster than Rust (Fig. 2). The raw execution of an algorithm in WASM is almost always faster than in JavaScript. However, when writing data into the WASM module's memory, the cost can be so high that it negates the benefit of using WASM. To verify this assumption, we conduct another set of experiments. Instead of just calling a method with parameters, we will call a method with static parameters, that are already in memory, and loop 1000 times.

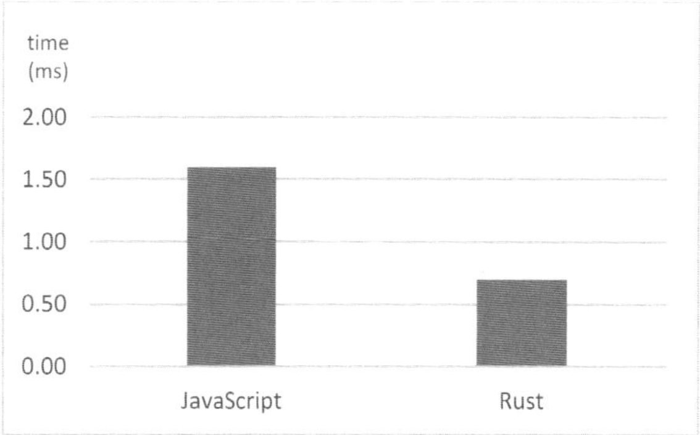

Fig. 3. JavaScript vs Rust speed comparison with static parameters on 1000 iterations

Rust-based WASM alternative outruns JavaScript by far (Fig. 3.). Writing complicated data structures into WASM memory is expensive. For example, to give a basic string to a WASM module, you must first encode the string, then find some free memory, and finally, write the encoded string into that memory. This can become slow depending on the memory management strategy being used. Additionally, the operation runs slower the longer the string.

5. Conclusion

Rust and WebAssembly, both have been designed with security and reliability in mind. This makes the combination of them through features like borrow checking and ownership for Rust and modularization and determinism in WebAssembly even more (data races, invalidation, ...) secure than any other alternative like C/C++ and JVM or CLR.NET even though most of them can be compiled to WebAssembly. The speed of WebAssembly and Rust also greatly improve their possible massive adoption as a secure combination for fast, secure, and reliable software. The capability of Rust/WebAssembly combination to work alongside JavaScript seems like a very good adoption point as it creates a non-destructive environment where WebAssembly applications can be seamlessly integrated into a rich ecosystem of JavaScript applications without development wars raging all over the world. Also, if this co-existence was to be a standardized way of deploying future web applications, it would mean that we would also get the best of both worlds in terms of functionality, security, and speed, which should be the point of developing web applications. At the end of the day, we do live in an age where almost anything needs to be a web application or have a web-application frontend.

Related to performance, WebAssembly is a good choice either if you are only handling numeric data or if your algorithm is complex enough for the cost of passing arguments to the WebAssembly module to become negligible. Another advantage is related to the obfuscation of code in browsers which increases the complexity of reverse engineering. In the future, we expect to see a massive increase in the usage of Rust and WebAssembly. In our future research, we will focus on trying to determine if the security of these technologies is a decisive factor in favor of adoption or if there are other key points besides the non-destructive nature of WebAssembly that could lead to an even better adoption rate.

6. References

[1] Jung, R.; Jourdan J.; Krebbers, R. & Dreyer D. (2018). RustBelt: Securing the Foundations of the Rust, Proc. ACM Program. Lang. 2, POPL, Article 66 (January 2018), 34 pages. DOI:10.1145/3158154 https://people.mpi-sws.org/~dreyer/papers/rustbelt/paper.pdf

[2] https://webassembly.org/docs/non-web (2022). The WebAssembly Contributors, Accessed on: 2022-06-30

[3] https://doc.Rust-lang.org/book/ch04-02-references-and-borrowing.html#mutable-references (2022). References and Borrowing, The Rust Contributors, Accessed on: 2022-06-30

[4] https://doc.Rust-lang.org/book/ch10-03-lifetime-syntax.html (2022). Lifetime Syntax, The Rust Contributors, Accessed on: 2022-06-30

[5] https://doc.Rust-lang.org/core/cell/struct.UnsafeCell.html (2022). UnsafeCell, The Rust Contributors, Accessed on: 2022-06-30

[6] https://doc.Rust-lang.org/std/cell/struct.RefCell.html (2022). RefCell, The Rust Contributors, Accessed on: 2022-06-30

[7] Jedlicka, M., et al. (2009). Reliability assessment using UML models, DAAAM International Scientific Book, pp. 53+

[8] Matei, R. (2021). A practical guide to WebAssembly memory, https://radu-matei.com/blog/practical-guide-to-wasm-memory, Accessed on: 2022-06-30

[9] Lewycky, N. (2021). Why is WebAssembly safe and what is linear memory model, https://stackoverflow.com/a/65933986, Accessed on: 2022-06-30

[10] Richey J. (2017). Nondeterminism in WebAssembly, https://github.com/WebAssembly/design/blob/main/Nondeterminism.md, Accessed on: 2022-06-30

[11] Hough D., et al. (2019). IEEE Standard for Floating-Point Arithmetic, IEEE, New York

[12] https://www.iitk.ac.in/esc101/05Aug/tutorial/essential/threads/priority.html (2011). Thread Scheduling, Sun Microsystems, Accessed on: 2022-06-30

[13] https://webassembly.org/docs/security/#control-flow-integrity (2021). Control Flow Integrity, Accessed on: 2022-06-30

[14] https://webassembly.org/docs/security/ (2020). WebAssembly - Security, Accessed on: 2022-06-30

[15] Lippert, E. (2011). Why have a stack?, https://web.archive.org/web/20120328082950/http://blogs.msdn.com/b/ericlippert/archive/2011/11/28/why-have-a-stack.aspx, Accessed on: 2022-06-30

[16] Kornelski, D. (2021). Speed of Rust vs C, https://kornel.ski/Rust-c-speed, Accessed on: 2022-06-30

[17] Antoyo (2020). Rustc_codegen_gcc, https://github.com/Rust-lang/Rustc_codegen_gcc, Accessed on: 2022-06-3

DOI: 10.2507/33rd.daaam.proceedings.024

STATIC-ANALYSIS TECHNIQUES OF MALWARE REVERSE ENGINEERING

Zlatan Moric, Loïc Branstett & Robert Petrunic

This Publication has to be referred as: Moric, Z[latan]; Branstett, L[oic] & Petrunic, R[obert] (2022). Static-Analysis Techniques of Malware Reverse Engineering, Proceedings of the 33rd DAAAM International Symposium, pp.0172-0179, B. Katalinic (Ed.), Published by DAAAM International, ISBN 978-3-902734-36-5, ISSN 1726-9679, Vienna, Austria DOI: 10.2507/33rd.daaam.proceedings.024

Abstract

Network and system security are critical issues of overall Internet security. Scientific papers and popular literature are full of new security issues being published and analysed daily. Due to the rapid proliferation of various types of malware tools that can be used both to create security attacks and to influence their results, traditional analysis methods struggle with the size and scope of samples needed to do proper analysis. For example, a well-disguised malware attack can easily penetrate an environment which is protected by a load balancer and/or Web Application Filter (WAF). In this paper, we will discuss different static-analysis techniques used to reverse engineer executable code to determine if the code is in fact malware.

Keywords: static analysis; malware; reverse engineering; artificial intelligence; dynamic analysis.

1. Introduction

Malware threats assessed by IT security organizations have been increasing in number by more than ten thousand per day. Symantec Internet Security Threat Report reveals that the total number of unique variants of malware in the world in 2011 amounts to around 403 million compared to 286 million variants in 2010. The usage of avoidance techniques such as self-defending code, packing, anti-debugging, and anti-virtualization techniques has led to advanced malware capable of passing the researchers detection barriers and anti-malware. This has increased the potential breach of corporate assets and individual computers. From the aspect of antivirus software companies and researchers, the most challenging are the threats that occur in computer applications because of the unknown vulnerability treat, also known as a zero-day attack.

These attacks take advantage of an application vulnerability unknown to anyone, so antivirus software cannot detect it either. That is why majority of the solutions used today try to go deeper and do the behavioural analysis that is the evolution of antimalware software we have the luxury of using. A reverse engineering approach can be used to detect and analyse malware by extracting data and finding out how it works on infected system. This study aims to compare different reverse engineering techniques to malware analysis. Reverse engineering has been shown to be highly successful in analysing and detecting of malware. However, it causes high complexity. Static analysis should be far more efficient and secure because the characteristics are extracted without actually executing the application [1].

2. Analysis methods

To defend against malware, analysts use different types of analysis: Static-analysis, Dynamic-analysis, and Hybrid-analysis. Each of these analysis types is unique and powerful, but the best way to use it is to combine them together. When combined, they offer more robust and more resilient solution to identify modern malware strains. Static analysis on its own is used to identify malware code by partially translating machine language into a human readable code, while the dynamic analysis tries to understand the malware behaviour while running it in the isolated environment called sandbox. Hybrid analysis is trying to bring the best from both worlds.

2.1. Static analysis

In static analysis, malware characteristics can be collected without execution of the code itself. For Portable Executable (PE) files, which are native to Windows, analysts often use tools that can present tabular views of PE header information, disassemble machine language, extract printable strings, file hashes, etc. Some features commonly investigated as part of static analysis include file hashes, strings, opcode sequences, DLL imports, API calls, and other metadata found in the PE header. In malware visualization, researchers propose to view the contents of software in raw numerical forms, such as binary, decimal, or hexadecimal, and convert these strings of numbers into images.

2.2. Dynamic analysis

In contrast to static analysis, dynamic analysis involves the execution of the malicious code. While this is much harder to scale than static analysis, monitoring how a binary interacts with an infected system can lead to more insights. Analysts are often interested in recording API and system calls, processes, modifications to system files and registries, network communication, etc.

2.3. Hybrid analysis

Both static and dynamic analysis have their limitations when conducted individually. For example, packers that compress software can be used as an obfuscation tool to obscure the contents of an executable. This will often necessitate the manual efforts of a malware analyst to conduct further static analysis on machine code. Dynamic analysis is not always successful since some types of malwares require a certain duration to pass or a specific event to trigger its execution. Hybrid analysis, the combination of both static and dynamic analysis, can lead to more comprehensive views of malicious programs and enable researchers to gather a larger set of features for classification.

3. Technology background

As contrast to the data file that needs to be processed (interpreted) by a program in order to be meaningful, an executable program, executable file, or executable piece of code permits a computer to "perform indicated tasks according to encoded instructions" [2]. Machine code tries to defend itself in a variety of ways, which makes it challenging to analyse.

3.1. Machine code and Assembly language

Machine code, also known as machine language, is the elemental language of computers. It is read by the computer's central processing unit (CPU), it is composed of digital binary numbers, and looks like a very long sequence of zeros and ones.

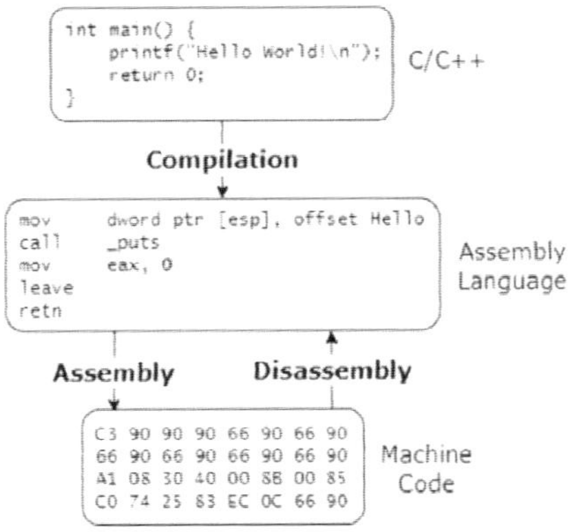

Fig. 1. Compilation and assembly, disassembly [4]

Ultimately, the source code of every human-readable programming language must be translated to machine language by a compiler or an interpreter because binary code is the only language that computer hardware can understand [3]. This transformation is generally called compilation, as it transforms source code to the assembly language understood by the CPU, this assembly language is then transformed into a series of machine instructions (aka machine code). However, static analysis generally requires a higher level of understanding of the code and uses a technique known as disassembly; the process of transforming machine code to a control flow graph of machine instructions, which can be transformed to a very low language like C.

3.2. Code obfuscation and encryption

Since executable files can be analysed by disassembly, new techniques called anti-reverse engineering have been invented to obstruct the process. Not only malware authors but also software companies use them to protect commercial software from being cracked or pirated. Code obfuscation [5] and encryption are two commonly used methods. One common technique used in code obfuscation is Control flow obfuscation. It synthesizes conditional, branching, and iterative constructs that produce valid executable logic, but yield non-deterministic semantic results when de-compiled.

In a managed executable, all strings are discoverable and readable. Even when methods and variables are renamed, strings can be used to locate critical code sections by looking for string references inside the binary. This includes messages (especially error messages) that are displayed to the user. To provide an effective barrier against this type of attack, string encryption can be used to hide strings in the executable and only restores their original value when needed.

```
Invalid Authentication - Try Again => !ù$àç_èè-'("'=):'")'
```

A more extreme version of string obfuscation is complete code encryption. Unlike obfuscation, code encryption packs and encrypts executable files on the disk. The code will be decrypted during the execution. It means that it is nearly impossible to analyse the code just by static disassembly so, we have to rely on execution and reviewing of system logs also.

3.3. Executable format

Executable formats encapsulate system-related information, such as API export and import tables, resources (icons, images, audio, etc.), and the distribution of data and code. This information is critical for malware analysis. Data and executable code are stored in different sections after the headers, depending on their functions.

3.4. Portable executable

Portable Executable (PE) [6] format is a file format for executables, object code, DLLs, and the like used in 32-bit and 64-bit versions of Windows operating systems.

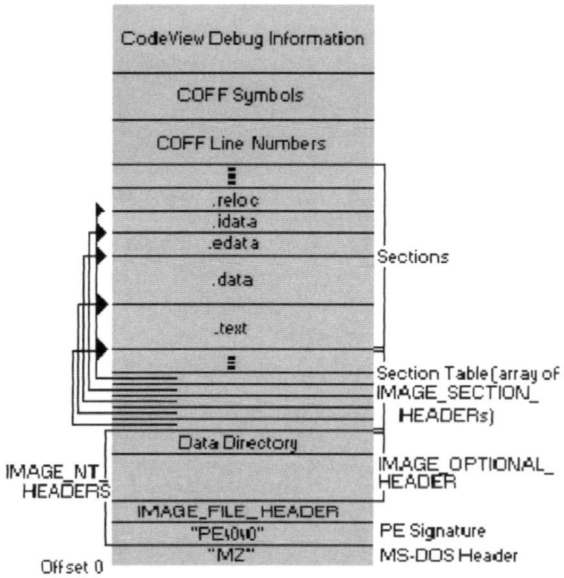

Fig. 2. The PE file format [6]

Executable file on a disk is very similar to what the module will look like after Windows loads it. The Windows loader does not need to work extremely hard to create a process based on the disk file, it just maps the necessary file segments into the virtual address space using the memory-mapped file technique.

3.5. Executable and Linkable Format

Executable and Linkable Format (ELF) format is flexible, extensible, and cross-platform file format. For instance, it supports different endianness and address sizes, so it does not exclude any particular central processing unit (CPU), or instruction set architecture. This has allowed it to be adopted by many different operating systems on many different hardware platforms.

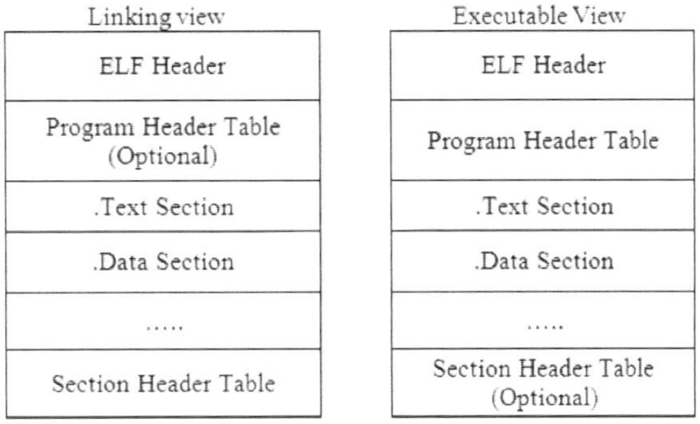

Fig. 3. Executable and linkable format (ELF) structural view [6]

Figure 3 illustrates the ELF file's structure. Every file has an ELF header at the top that describes how the file should be organized. A section header table is required, and a program header table is optional for relocatable files. A program header table and a section header table are both required in the case of an in-executable object file.

4. Analysis procedures

Analysis is the complex process that is trying to get to the conclusion of the sample behaviour, with sample being the malware analysed. Manual analysis is usually slow and heavily dependent on the analyst skills, while automated analysis allows for quick and reproducible results. There are problems with both approaches that can lead to an incomplete analysis when being done manually, or to an overseen parts of the malware code when full automation is used. That is the reason for mixed approach in the same way as explained with malware analysis types, meaning that human interaction is still required to some extent. Machine learning is closing that gap and it is used both by malware writers to hide an evolve malware, and malware analysts to identify the malicious samples and behaviour [6].

4.1. Manual analysis

At this stage, analysts reverse-engineer code using debuggers, disassemblers, compilers, and specialized tools to decode encrypted data, determine the logic behind the malware algorithm, and understand any hidden capabilities that the malware has not yet exhibited. Code reversing is a rare skill set and executing code reversals takes a great deal of time and knowledge. For these reasons, malware investigations often skip this step and therefore miss out on a lot of valuable insights into the nature of the malware.

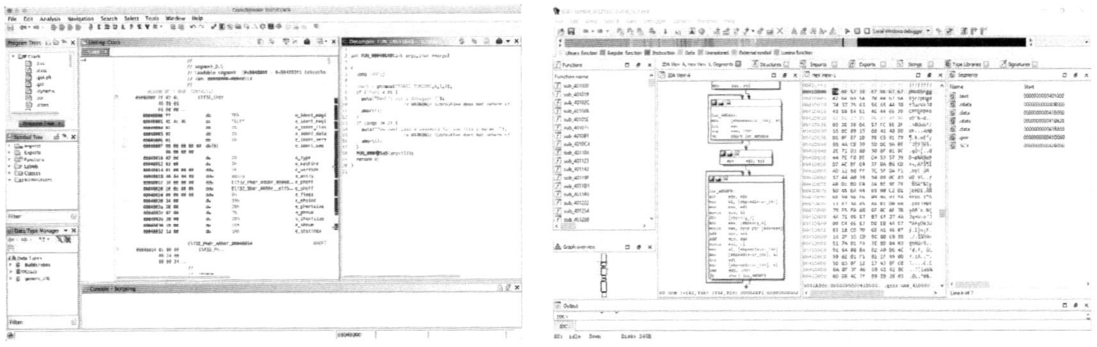

Fig. 4. Sample debugging tools (Ghidra and IDA Pro)

To manually reverse the code, malware analysis tools such as a debugger and disassembler are needed. The skills needed to complete manual code reversing is very important, but also difficult to find. Sometime forensic tools are also used to find hidden artifacts. [8]

4.2. Semi-automated analysis

Semi-automated analysis is a type of analysis that leaves only a small portion of the analysis to humans. The researcher presented an architecture that enables collaborative malware analysis by sharing resources (scripts, executable, ...) and leaving to human analysts only a small and manageable part of the whole effort. [9],

4.3. Automated analysis

The emergence of various malware tactics to get through network-based and host-based security defence's has led to an evolution in malware analysis and detection approaches during the past ten years. [10] Forensics investigators found it extremely challenging to respond promptly due to the rapidly increasing variety and number of malware species. To automate various processes of static and dynamic malware analysis, Machine Learning (ML) aided malware analysis become essential. [11]

One of the simplest ways to assess a suspicious program is to scan it with fully automated tools, which can quickly assess what malware is capable of if it infiltrated the system. This analysis can produce a detailed report regarding the network traffic, file activity, malware access and modified registry keys. Even though a fully automated analysis does not provide as much information as an analyst, it is still the fastest method to sift through large quantities of malware.

4.4. Using Equality

The "equality" technique uses information such as strings, and hashes, ... and compares them to a pre-existing database. Due to its simplicity, it is one of the mostly used techniques on a client, like the anti-malware solution. [12]

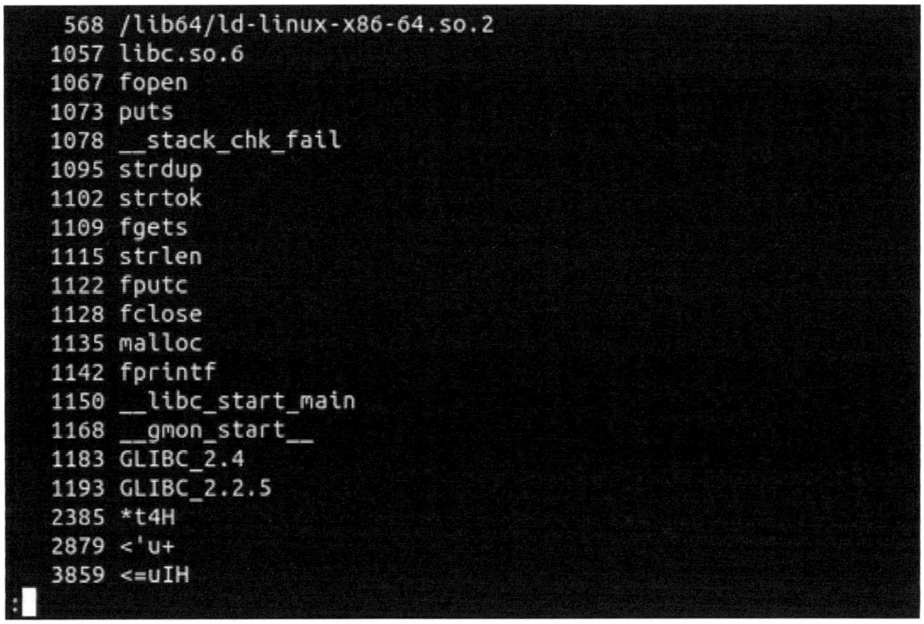

Fig. 5. Symbols of an executable

Most of the programs we have tested had triggers that were equality tests, however malware that checks value ranges may have many more trigger conditions.

4.5. Using Machine Learning

The semi-automated analysis is a type of analysis that leaves only a small portion of the analysis to humans. We can use tools like Android-COCO [13], a supervised approach for detecting Android malware.

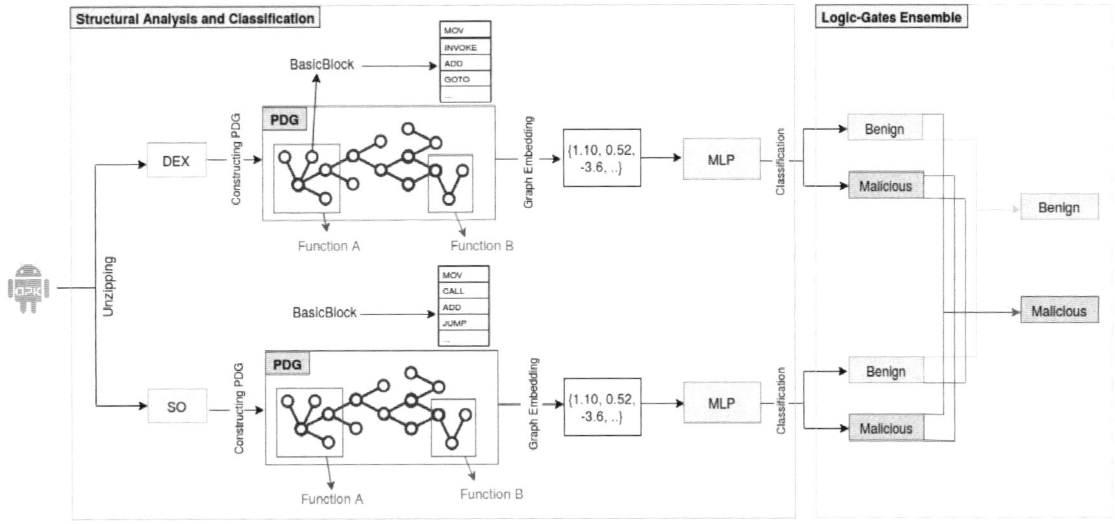

Fig. 6. Android-COCO architecture [13]

It works by automatically transforming the byte and native code to the program dependency graphs (PDG) and uses the embedding of these graphs to classify applications. Large-scale experiments on 100,113 samples (35,113 malware and 65,000 benign) shows that this approach can detect malware applications with an accuracy of 99.86%, significantly outperforming state-of-the-art solutions. [13] We can also use the N-GRAM approach [4]. This novel technique uses an N-GRAM neuronal network with different inputs:

- File Size
- API
- Op Codes
- Dependencies
- Section Size
- Section Permissions
- Content complexity

The new approach features together with a classical machine learning algorithm (Random Forest) present very good accuracy at 99.40%. Higher than the previous 4-grams than only achieve 57.96% accuracy and require a more complex model.

5. Fighting limitations of static analysis

As already mentioned, static analysis is not perfect and it requires a lot of analyst's time and effort, and especially knowledge in the debugging and assembly code analysis. The knowledge of architecture where the application is running is also one of the prerequisites the analyst is supposed to have while working on the static analysis process.

5.1. Lazy loading

Lazy initialization/loading [14] is one of those design patterns which is in use in almost all programming languages. Its goal is to move the object's construction forward in time. It is especially handy when the creation of the object is expensive, and you want to defer it as late as possible, or even skip it entirely. This implies that the object may not be precomputed in the data section like other data. Fighting this during the malware analysis is very hard and close to impossible. The only method that works reliably is to execute the TLS (Thread Local Storage) in a sand-boxed environment, but this defeats the goal of static analysis.

5.2. Name Mangling

Name mangling is a mechanism used by compilers to add additional characters to functions with the same name (function overloading). The goal of name mangling is to avoid any confusion when executing the program and calling a function that may have the same name as another one. [15]

This c++ code:

```
void add(int a, int b){cout << a + b << endl;}
```

, would result in these symbols inside the compiled library or binary:

```
$ nm a.out
[..]
00401382 T __Z3adddd
00401350 T __Z3addii
[..]
```

If name mangling were made the same across compilers, your applications would link to libraries given by other compilers but crash when run because of differences in how objects are put out, multiple inheritance is implemented, virtual function calls are handled, and other factors. Because of this, ARM urges compiler developers to differentiate their name from other compilers for the same platform. Then, rather than at run time, incompatible libraries are found at the time of linking. [16] Name mangling can be minimized. De-mangling identifiers could reveal the function's prototype. This could be used to discover special, undocumented API functions that the company would want to keep as a secret.

5.3. Jump tables

A jump table can be either an array of pointers to functions or an array of machine code jump instructions. If you have a relatively static set of functions (such as system calls or virtual functions for a class), you can create this table once and call the functions using a simple index into the array. This would mean retrieving the pointer and calling a function or jumping to the machine code depending on the type of table used. Recognizing the target addresses of conditional branches implemented via a jump table in a machine-independent manner is one of the core issues with static analysis of binary (executable) code. Without these addresses, it is impossible to fully decode the machine instructions for a specific procedure, which results in inaccurate code analysis. [17]

6. Conclusion

In this paper we have explored different techniques for malware analysis, from manually looking at the assembly code of a program to the usage of artificial intelligence to provide an analysis of a potentially malicious program. One part of the problem was identified as the inefficiency and inability to scale when the traditional methods such as static or dynamic analysis are used. This is partially solved by using the combination of these, and automation or semi-automation on top of it. Even then, the efficiency is still not at the desired level because modern malware is evolving faster than the methods and used techniques. The other part of the problem was identified as the process complexity, and the knowledge the analyst must have for the traditional malware analysis methods. This is again, only partially solved by automation and semi-automation techniques. Based on all of that, we believe that the future of malware analysis will be based on automatic analysis with some degree of machine learning embedded in both static and dynamic analysis techniques.

7. References

[1] Baskaran, B. and Ralescu, A. (2016). A study of android malware detection techniques and machine learning. In: MAICS, pp 15–23.

[2] M.-W. (n.d.), Executable Definition & Meaning - Merriam-Webster. Available from: https://www.merriam-webster.com/dictionary/executable. Accessed: 2022-10-10

[3] Kinder, J. (2010). Static Analysis of x86 Executables Statische Analyse von Programmen in x86 Maschinensprache, Ph.D. Thesis, Technische Universität, Darmstadt

[4] Chen, Z.; Brophy, E. & Ward, T. (2021). Malware Classification Using Static Disassembly and Machine Learning, The 29th Irish Conference on Artificial Intelligence and Cognitive Science 2021, Dublin

[5] Madou, M.; Van Put, L.& De Bosschere, K. (2006). Understanding obfuscated code, IEEE Int. Conf. Progr. Compr., vol. 2006, pp. 268–271

[6] Pietrek, M. (1994). Peering Inside the PE: A Tour of the Win32 Portable Executable File Format, Microsoft Syst. Journal-US Ed., vol. 9, no. 3, pp. 15–38

[7] Ucci, D.; Aniello L.; Baldoni R. (2019). Survey of machine learning techniques for malware amalysis, ScienceDirect, vol. 81, pp. 123–147

8] Moric, Z.; Redzepagic, J. & Gatti, F. (2021). Enterprise Tools for Data Forensics, Ann. DAAAM Proc. Int. DAAAM Symp., vol. 32, no. 1, pp. 98–105, Wiena

[9] Laurenza, G.; Ucci, D.; Aniello, L. & Baldoni, R. (2016). An Architecture for Semi-Automatic Collaborative Malware Analysis for CIs, Proc. - 46th Annu. IEEE/IFIP Int. Conf. Dependable Syst. Networks, DSN-W 2016, pp. 137–142

[10] Svoboda, J. & Lukas, L. (2019). Sources of Threats and Threats in the Cyber Security, DAAAM Int. Sci. Book, pp. 321–330, Wiena

[11] Shalaginov, A.; Banin, S.; Dehghantanha, A. & Franke, K. (2018). Machine learning aided static malware analysis: A survey and tutorial, Cyber threat intelligence, vol. 70, pp. 7–45, Springer

[12] Sharif, M.; Lanzi, A.; Giffin, J. & Lee, W. (2008). Impeding Malware Analysis Using Conditional Code Obfuscation, Proc. 15th Netw. Distrib. Syst. Secur. Symp. - NDSS, pp. 321–333, Citeseer

[13] Xu, P. & Khairi, A. E. (2021). Android-COCO: Android Malware Detection with Graph Neural Network for Byte- and Native-Code, arXiv:2112.10038v2

[14] Park, H. & Kim, C.(2019). Design of Adaptive Web and Lazy Loading Components for Web Application Development, Journal of the Korea Institute of Information and Communication Engineering, vol. 23, no. 5, pp. 516–522, The Korea Institute of Information and Commucation Engineering

[15] Kefallonitis, F. (2007). Name mangling demystified, pp. 1–21, Accessed: 2022-10-10

[16] Ellis, M. A. & Stroustrup, B. (1990). The Annotated {C++} Reference Manual, Addison-Wesley Longman Publishing

[17] Cifuentes, C. & Van Emmerik, M. (2001). Recovery of jump table case statements from binary code, Science of Computer Programming, vol. 40, no. 2–3, pp. 171–188, Elsevier

DOI: 10.2507/33rd.daaam.proceedings.025

THE EFFECT OF CUTTING EDGE RADIUS SIZES ON TOOL LIFE IN MACHINING NICKEL ALLOY INCONEL 718

Tomas Vopat*, Robert Straka, Marcel Kuruc & Jozef Peterka

This Publication has to be referred as: Vopat, T[omas]; Straka, R[obert]; Kuruc, M[arcel] & Peterka, J[ozef] (2022). The Effect of Cutting Edge Radius Sizes on Tool Life in Machining Nickel Alloy Inconel 718, Proceedings of the 33rd DAAAM International Symposium, pp.0180-0187, B. Katalinic (Ed.), Published by DAAAM International, ISBN 978-3-902734-36-5, ISSN 1726-9679, Vienna, Austria
DOI: 10.2507/33rd.daaam.proceedings.025

Abstract

This article deals with the edge preparation of cemented carbide cutting inserts. Several cutting edge radius sizes were manufactured by drag-finishing. The influence of cutting edge radius sizes on the tool life was investigated. Cutting tests were carried out during the machining of a difficult-to-cut material, in particular, a nickel alloy Inconel 718. The cutting parameters were selected with respect to the semi-finishing turning operation. The longest tool life was reached with the cutting edge radius of 35 μm which was the middle value between the three radii tested. The dependence of tool life on the cutting edge radius size was determined. The results can be applied to increase the efficiency of the Inconel 718 the turning process of nickel alloy Inconel 718

Keywords: Edge preparation; Tool life; Tool wear; Surface finishing; Drag-finishing

1. Introduction

The tool life of the cutting tool and the roughness of the machined surface are the two most important parameters as a result of the cutting process. The ever-increasing efficiency and productivity in the production process require constant new research into the manufacturing processes and the tools used in production. Therefore, there is a need to increase the life of the tool, which can be done by changing parameters such as the material of the tool, the geometry of the tool [1], and the coating [2]. There are various ways to describe the tool geometry accurately, the simplest being to describe it using the cutting edge radius, but this description only works with symmetrical cutting edges. The so-called K factor was created to describe more complex cutting edges, which uses the parameters S_γ, S_α, Δr, and the ratio $K = S_\gamma / S_\alpha$, to describe such geometries [3]. To achieve those geometries on the cutting edge of the tool, one of the preparation methods must be applied to the tool after it has been manufactured. The most commonly used are drag-finishing and brushing [4], [5], [6]. In addition to those, several other methods are used for preparation, such as abrasive jet machining [7], grinding, laser machining, magneto abrasive machining [8], or using plasma discharges in electrolyte [9]. The machining process with a geometrically undefined cutting edge, where the workpiece is inserted into a container with free abrasive grains, is called drag finishing. Abrasive grains can consist of ceramic, plastic, silicon, carbide, or walnut shells, corundum, and quartz.

The material is removed inside the container due to the rotation of the workpiece and moves through abrasive particles [8]. When parts need to have good surface quality, this technology is used because of its advantage, which is the short time needed to obtain the required results. Increasing the relative speed between the media and surface and media pressure in the workpiece, makes drag finishing a faster and more efficient process when compared to others. It can also be used for the preparation of cutting tools by modifying some process parameters such as the feed rate, the processing time, or the type of grain [4], [10]. From cutting theory, we know four physical causes of the formation of the machined surface [11]. Theoretical roughness of the machined surface, the so-called geometric roughness, we can calculate, simulate [12] and predict. Other causes, such as the existing vibration [13] of the machine-tool-workpiece system and the occurrence of the increase, cannot be unambiguously calculated by analytical equations. In these cases, we help ourselves with experiments [14], [15], [16]. Zhang and Zhuang researched the effect of cutting edge microgeometry on surface roughness in the turning of AISI 52100 steel. They found that increasing the chamfer width can reduce the surface roughness, but a reasonable chamfer geometry must be selected because the increase in a chamfer angle leads to an extreme change in the surface roughness value. They also found out that the enhancement of chamfer length and angle can contribute to the formation of the white layer [17]. Ventura et al. found that asymmetric geometry increases tool life.

They demonstrated the application of prepared cutting inserts and proved that the use of those microgeometries is adequate in hard turning [18]. Uhlmann et al. compared brush polishing, polish blasting, magnet finishing, and immersed tumbling as preparation methods for micro-milling tools. They found that magnet finishing, and immersion tumbling improved the tool performance in a matter of resulting forces F_z by 7.5 %, and 14 %, respectively, compared to brush polishing and polish blasting. Tool wear was reduced by 9 % compared to brush polishing and by 13.2 % compared to polish blasting [19]. The effect of cutting temperature on the tool life of cemented carbide drills was examined in the publication [20]. The effect of cutting edge rounding on tool life was determined in the publication [21]. The influence of cutting edge radius sizes on cutting forces was investigated in the publication [22]. The authors in the paper [23] determined the accuracy of the solid cemented carbide cutting tools manufactured on the grinding machine and determined the influence of drag finishing on the macro geometry of the solid cemented carbide cutting tools.

The purpose of this paper is to examine the difference between three cutting edge radii of 10, 35, and 50 μm and their influence on cutting inserts in terms of tool life in turning difficult-to-cut material, in particular Inconel 718 equivalent to DIN 2.4668, EN NiCr19FeNbMo or UNS N07718.

2. Materials and Methods

CNMG 120408-SM coated cutting inserts were used in the experiment (Fig. 1). They were made of WC-Co cemented carbide. The hardened steel has the designation PCLNL 2020K16 (Fig. 1). The cemented carbide cutting inserts were made by the tool manufacturer Dormer Pramet and the toolholder was made by the tool producer Seco. These cutting inserts are suitable for roughing and semi-finishing turning operations. The used cutting inserts were prepared by drag-finishing. The grade of cemented carbide and coatings are not specified further due to ongoing research. The prepared cutting edge radii were 10 μm, 35 μm, and 50 μm. The example of measurement of the cutting edge radius is shown in Fig. 2. It was carried out by the Alicona Infinite Focus SL measurement device. Table 1 shows the parameters of the cutting tool. The optical three-dimensional measurement principle was used to measure the radius of the cutting edge. The objective magnification was 50x, where the working distance is 10.1 mm, the vertical resolution is 20 nm, and the minimum measurable radius is 2 μm. In the experiment, DMG CTX Alpha 500 multi-axis turning centre was used. The parameters of the machine tool are described in Table 2. The longitudinal turning was used during cutting tests.

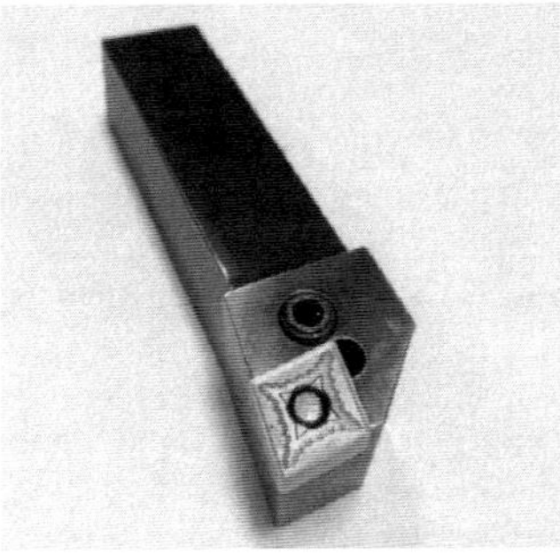

Fig. 1. Toolholder with cemented carbide cutting inserts

Parameter	Value	Parameter	Value
Length [mm]	12.90	Recommended f [mm]	0.15 – 0.45
Thickness [mm]	4.76	Recommended a_p [mm]	0.80 – 4.00
Nose radius [mm]	0.80	Cutting edge angel [°]	80.00

Table 1. Parameters of the CNMG 120408-SM cutting insert

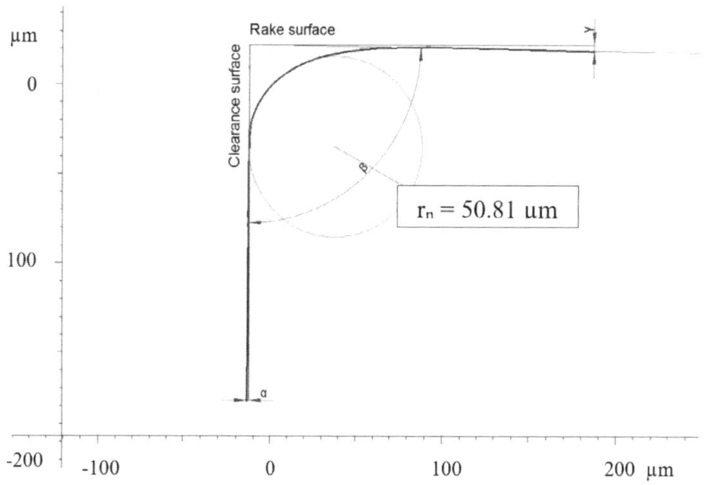

Fig. 2 Measurement of the cutting edge radius r_n

Travel distance	
X [mm]	190
Y [mm]	± 40
Z [mm]	525
Main spindle	
Drive power [kW]	20
Max. rotation frequency [min^{-1}]	6000
Tool revolver head	
Number of tool positions [-]	12
Drive power [kW]	5.4
Max. rotation frequency [min^{-1}]	5000
Operating software	Heidenhain Plus IT

Table 2. Parameters of DMG CTX Alpha 500 multi-axis turning centre

The cutting parameters were kept constant during the entire machining. The cutting speed of $v_c = 40$ m·min^{-1}, the feed of f = 0.15 mm, and the depth of cut of $a_p = 1$ mm were selected as process parameters. They were used to control the development of wear at the cutting edges. Inconel 718 (equivalent to DIN 2.4668, EN NiCr19FeNbMo or UNS N07718) was used as tested workpiece material. The chemical composition of Inconel 718 is in Table 3. It was an annealed nickel-chromium-based alloy. The tested workpiece material was a rounded bar with a diameter of 101.6mm that was produced by cold working.

C	Si	Mn	N	Cr	Mo	Ni	Cu	Co	Ti	Al	Nb	Fe
0.023	0.070	0.11	0.006	17.780	2.990	53.6	0.06	0.35	0.92	0.49	5.16	18.29

Table 3. Chemical composition of Inconel 718 (wt. %)

The chosen value of the tool wear criterion on the main flank was $VB_K = 0.22$ mm. After the tool wear reached this value on the cutting edge, the tool wear started to increase significantly, or edge chipping was observed on the cutting insert. The flank wear, in areas namely VB_N and VB_B, was measured by a Dino-Lite Edge AM73915MZT Dino-Lite Edge 3.0 microscope with a resolution of 2560 × 1920 pixels. This microscope uses magnification in the range of 10-220x. The frame rate was 15 fps at 5M pixels. Before the experiment, the cutting inserts were cleaned in an Elmasonic P ultrasonic cleaning device due to possible impurities on the cutting edges. Three cutting inserts with the same cutting edge radius were used to machine the workpiece material that was clamped into the three-jaw hydraulic chuck (SMW Autoblok 210BB) of the main spindle. The length of the machined surface was 60 mm. The top layer of the workpiece was strain hardened by the cold work process during the manufacturing the workpiece. This layer has a thickness of less than 0.3 mm. This layer was machined with another cutting insert so as not to affect the results of the experiment. The diameter of the tested workpiece material after the top layer was 101 mm. The cutting length of the first cutting insert was 0 to 20 mm. For the second cutting insert, the cutting length ranged from 20 to 40 mm, and for the last one it was from 40 to 60 mm. After the first minute of machining, the cutting inserts were cleaned by cleaning plasticine.

Then, they were placed in the special holder that was manufactured on a 3D printer. The microscope was focused on the cutting edge of the cutting insert. Then it was captured in the photo so that the cutting insert could not be moved by external influences, which could distort the measurements of tool wear. Subsequently, the value of the flank wear was measured in the image. After the first measurement, tool wear was measured every 2 minutes of machining until tool wear criterion VB_K was reached. The workplace of the machine tool is shown in Fig. 3.

Fig. 3. Workplace of machine tool
1 - hydraulic chuck, 2 - machined material, 3 – cemented carbide cutting insert, 4 – toolholder of cutting insert

3. Results and Discussion

Fig. 4 shows the development of the tool wear on the flank face in time for 2 different radii. The figure shows the cutting edges before the machining and then after exceeding the tool wear criterion. As can be seen in Fig. 4, the largest flank wear values were observed in the area VB_N. It is a typical tool wear type for machining hard materials or materials that are strain hardened after machining. The time dependence of tool wear (VB_N) with respect to the cutting edge radius is shown in Fig. 5. There are tool wear curves of average values of individual flank wear. For the experiment, three cutting inserts were used for each cutting edge radius r_n. The calculated average values of tool lives and standard deviations for cutting inserts with different cutting edge radii can be seen in Table 4.

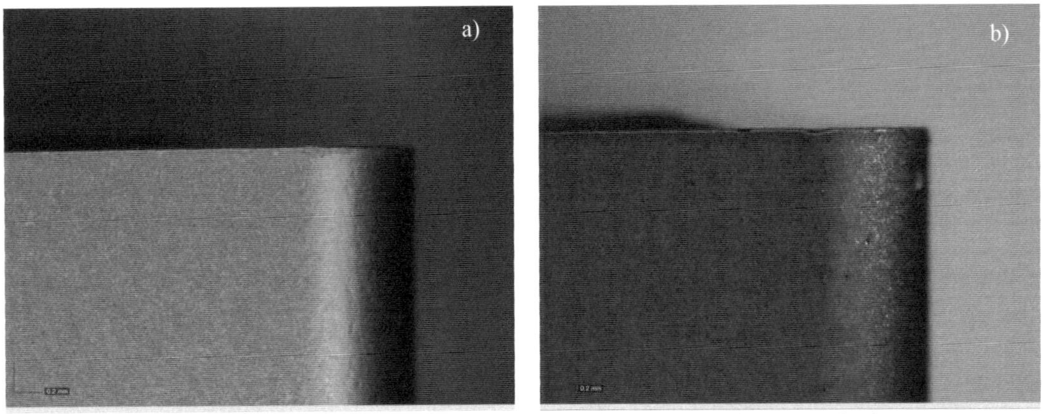

r_n = 10 μm, t = 0 min r_n = 35 μm, t = 0 min

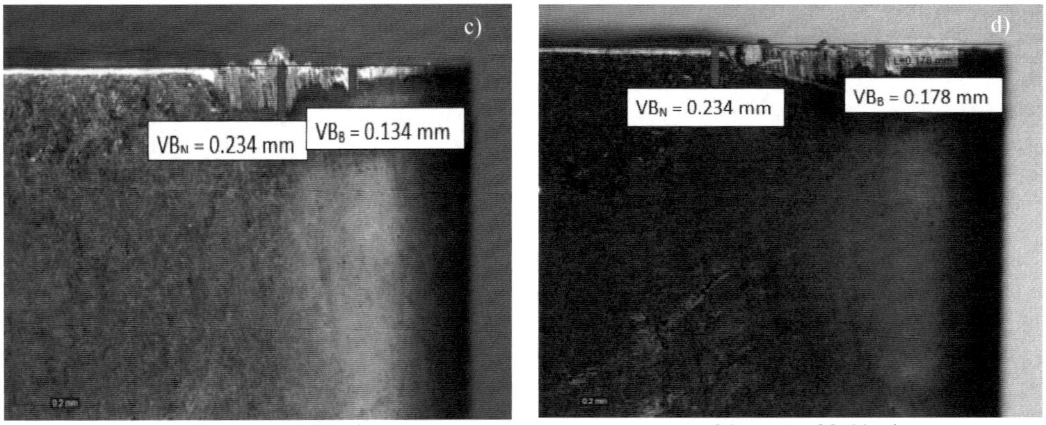

r_n = 10 µm, t = 11.84 min r_n = 35 µm, t = 20.64 min

Fig. 4. Development of the tool wear on the flank for
a) r_n = 10 µm before the experiment and b) r_n = 35 µm before the experiment
c) r_n = 10 µm after 11.84 minutes of machining d) r_n = 35 µm after 20.64 minutes of machining

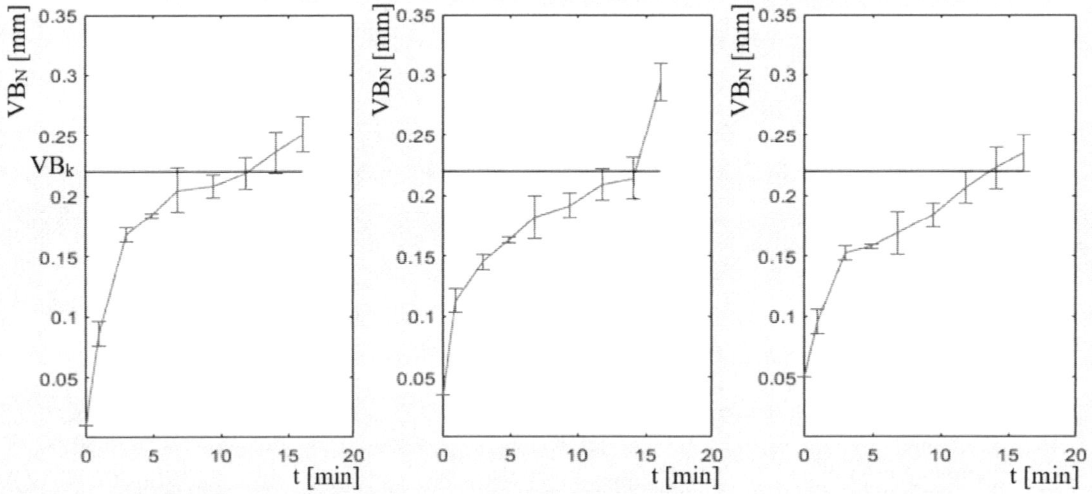

r_n = 10µm r_n = 35 µm r_n = 50µm

Fig. 5. Graphs of the time dependence of the VBN width of flank wear value

Cutting insert with	Average tool life	Standard deviations
r_n = 10 µm	12.00	2.08
r_n = 35 µm	15.13	4.08
r_n = 50 µm	13.6	1.81

Table 4. Average tool life of cutting inserts

The longest tool life was achieved by cutting the radius of 35 µm. The tool life of the cutting insert with r_n = 35 µm was increased by 25.83% compared to the cutting edge radius of 10 µm. The tool life of the cutting insert with r_n = 50 µm was increased by 13.33% compared to the cutting edge radius of 10 µm. The increased tool life of the cutting insert can be explained by the fact that the cutting edge with r_n = 10 µm is not as strong as with the larger radii of rounding of the cutting edge. For this reason, the cutting edge with r_n = 10 µm is more prone to tool breakage and fatal failure. On the other hand, the usage of too large cutting edge radii leads to higher friction and thermal load of the cutting edge due to the larger contact area of the cutting tool with the workpiece [24], [25]. When machining difficult-to-cut material such as a nickel alloy, the development of tool wear is difficult to predict. Therefore, it sometimes happens that the cutting tool fails sooner than expected. For this reason, the standard deviation values are higher in some cases.

The average values of the tool life were determined from the individual graphs of the time dependence of the VBN of the width of flank wear value for every sample. These values were evaluated by one-way ANOVA (Table 5). After verifying the normal (Gaussian) distribution of the measured values, the data were tested for equality of variances, and it was confirmed that the variances were not statistically different, therefore, the Tukey method of pairwise comparison of the average values of tool lives was used (Fig. 6). The α significance level was 0.05. Based on the results of the one-way ANOVA, it can be concluded ($p > 0.05$) that the durability values for the individual radii of the rounding of the cutting edges are not statistically significant.

Cutting insert with r_n	**DF**	**F-value**	**p-value**
r_n = 10 μm and r_n = 35 μm	1	1.4	0.302
r_n = 10 μm and r_n = 50 μm	1	1.01	0.372
r_n = 35 μm and r_n = 50 μm	1	0.35	0.584

Table 5. One-way ANOVA

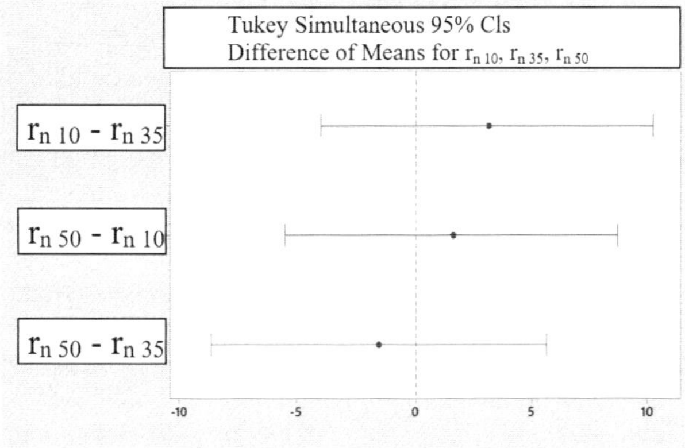

Fig. 6. Tukey method of pairwise comparison of the average values of tool lives

The graphical dependence of the tool life on the cutting edge radius is shown in Fig. 7. The equation in Fig. 7 is a mathematical expression of this dependence, where y is the tool life and x is the size of the cutting edge radius. It shows that the longest tool life was achieved by a cutting edge radius of 35 μm. The cutting edge radius of 35 μm seems to be the best option. However, the results obtained from the research are limited by adjusted variables. It means that the results could be properly applied only for turning nickel alloy Inconel 718. And they are valid for cutting parameters: cutting speed 40 $m \cdot min^{-1}$, the feed 0.15 mm, and the depth of cut of 1 mm. If the workpiece material or cutting parameters were changed, the results would be different (but probably similar). Moreover, different cutting insert geometry could also lead to different results. The presented results could be used to increase the productivity of nickel alloy machining, and it can also serve as a background for experiments for increasing the productivity of machining other metals.

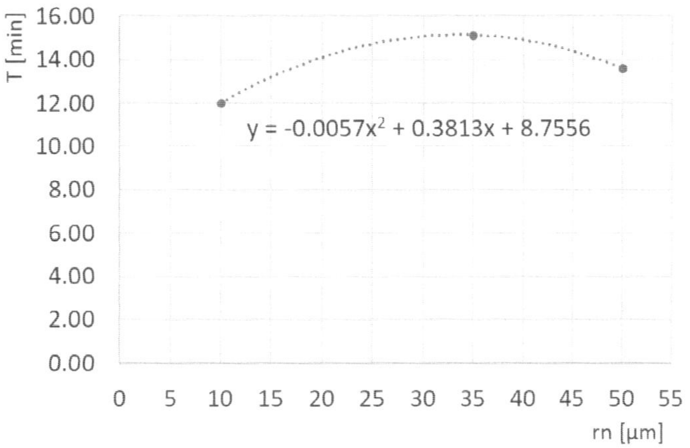

Fig. 7. Graph of the dependence of tool life on the cutting edge radius sizes r_n

4. Conclusion

This article deals with the edge preparation of cemented carbide cutting inserts. The influence of cutting edge radius sizes on the tool life was investigated when machining the difficult-to-cut material, in particular a nickel alloy Inconel 718. The cutting inserts were prepared by drag-finishing. The cutting parameters were selected with respect to the semi-finishing turning operation. This article presents the new finding on the impact of cutting edge preparation on selected aspects of the Inconel 718 machining of nickel alloy Inconel 718. The longest tool life was achieved with a cutting edge radius of 35 μm. The average tool life for this cutting edge radius was 15.13 minutes. The tool life of the cutting insert with r_n = 35 μm was increased by 25.83% compared to the cutting edge radius of 10 μm. The tool life of the cutting insert with r_n = 50 μm was increased by 13.33% compared to the cutting edge radius of 10 μm. Further research will examine the effect of cutting edge radius sizes prepared by various edge preparation methods on the tool life, cutting forces, and surface roughness during the milling and turning operations.

5. Acknowledgements

This research was funded by research projects APVV–16–0057 and APVV-21-0071. The authors are also grateful to the Dormer Pramet Tool Producer, who participated in this research.

6. References

[1] Peterka, J.; Hrbál, J.; Buranský, I. & Vozár, M. (2020). Design and manufacturing of cutting tools for milling. MM Science Journal March 2020, 3818-3821. ISSN 1805-0476.

[2] Denkena, B.; Michaelis, A.; Herrmann, M. & Pötschke, J. (2020). Influence of tool material properties on the wear behavior of cemented carbide tools with rounded cutting edges. Wear 456, 1-8. ISSN 0043-1648.

[3] Bergs, T.; Schneider, S. A. M.; Amara, M. & Ganser, P. (2020). Preparation of symmetrical and asymmetrical cutting edges on solid cutting tools using brushing tools with filament-integrated diamond grits. Procedia CIRP 93 (53), 873-878. ISSN 2212-8271.

[4] Peterka, J.; Pokorný P.; Václav, Š.; Pätoprstý, B. & Vozár, M. (2020). Modification of cutting tools by drag finishing. MM Science Journal March 2020, 3822-3825. ISSN 1805-0476.

[5] Peterka, J.; Vozár, M.; Vopát, T.; Pätoprstý, B. & Pokorný, P. (2020). Using multi-criteria analysis to evaluate the impact of drag-finishing technological parameters on the carbide tool radius. Materials Today: Proceedings 22, 205-211. ISSN 1876-6102.

[6] Pokorný, P.; Pätoprstý, B.; Vopát, T.; Peterka, J.; Vozár, M. & Šimna, V. (2020). Cutting edge radius preparation. Materials Today: Proceedings 22 (2), 212-218. ISSN 2214-7853.

[7] Wang, W.; Biermann, D.; Aßmuth, R.; Arif, A. F. M. & Veldhuis, S. C. (2020). Effect on tool performance of cutting edge prepared by pressurized air wet abrasive jet machining (PAWJAM). Journal of Materials Processing Technology 277, 116456. ISSN 0924-0136.

[8] Denkena, B. & Biermann, D. (2014). Cutting edge geometries. CIRP Annals - Manufacturing Technology 63, 631-653. ISSN 1726-0604.

[9] Vopát, T.; Podhorský, Š.; Sahul, M. & Haršáni, M. (2019). Cutting edge preparation of cutting tools using plasma discharge in the electrolyte. Journal of Manufacturing Processes 46, 234-240. ISSN 1526-6125.

[10] Malkorra, I.; Salvatore, F.; Arrayola, P. & Rech, J. (2020). The influence of the process parameters of drag finishing on the surface topography of aluminum samples. CIRP Journal of Manufacturing Science and Technology 31, 200-209. ISSN 1775-5817.

[11] Peterka, J. (2004). A new approach to calculating the arithmetical mean deviation of profile during copy milling. Strojniski Vestnik – Journal of Mechanical Engineering 50, 594-597. ISSN 0039-2480.

[12] Polakovic, M.; Buranský, I. & Peterka, J. (2008). Simulation concept for machined surface roughness and shape deviations prediction. Annals of DAAAM & Proceedings 19, 1089-1090. ISSN 1726-9679.

[13] Sentyakov, K.; Peterka, J.; Smirnov, V.; Božek, P. & Sviatskii, V. (2020). Modeling of boring mandrel working process with vibration damper. Materials 13(8). ISSN 1996-1944.

[14] Jurko, J.; Panda, A. & Behún, M. (2012). Prediction of a new form of the cutting tool according to achieve the desired surface quality. Applied Mechanics and Materials 268-270, 473-476. ISSN 1662-7482.

[15] Peterka, J. & Pokorný, P. (2014). Influence of the lead angle from the vertical axis milling on effective radius of the cutter. Precision Machining 581, 44-49. ISSN 1662-9795.

[16] Vopát, T.; Peterka, J.; Šimna, V. & Kuruc, M. (2015). The influence of different types of copy milling on the surface roughness and tool life of end mills. Procedia Engineering 100, 868-876. ISSN 1877-7058.

[17] Zhang, W. & Zhuang, K. (2020). Effect of cutting edge microgeometry on surface roughness and white layer in turning AISI 52100 steel. Procedia CIRP 87, 53-58. ISSN 2212-8271.

[18] Ventura, C. E. H.; Köhler, J. & Denkena, B. (2013). Cutting edge preparation of PCBN inserts by means of grinding and its application in hard turning. CIRP Journal of Manufacturing Science and Technology 6(4), 246-253. ISSN 1755-5817.

[19] Uhlmann, E.; Oberschmidt, D.; Kuche, Y.; Löwenstein, A. & Winker, I. (2016). Effect of different cutting edge preparation methods on micro milling performance. Procedia CIRP 46, 352-355. ISSN 2212-8271.

[20] Kolesnyk, V.; Kryvoruchko, D.; Hatala, M.; Mital, D.; Hutyrova, Z.; Duplak, J. & Alowa, M. (2015). The effect of cutting temperature on carbide drilling life in the process of CFRP/Steel stacks drilling. Manufacturing Technology, 15(3), 28-29. doi:10.21062/UJEP/X.2015/A/1213-2489/MT/15/3/357

[21] Pätoprstý, B.; Pokorný, P.; Vozár, M.; Vopát, T. & Buranský, I. (2019). The influence of grinding process and drag finishing on the milling tools macro geometry. Annals of DAAAM and Proceedings of the International DAAAM Symposium 2019; 30:1008–13. https://doi.org/10.2507/30TH.DAAAM.PROCEEDINGS.140.

[22] Vopát, T.; Kuruc, M.; Šimna, V.; Zaujec, R. & Peterka, J. (2017). Cutting edge microgeometry and preparation methods. Annals of DAAAM and Proceedings of the International DAAAM Symposium 2017:384–91. https://doi.org/10.2507/28TH.DAAAM.PROCEEDINGS.054.

[23] Šimna, V.; Jurina, F.; Vopát, T. & Kuruc M. (2019). Edge preparation of cutting tools and it impact on cutting forces during milling. Annals of DAAAM and Proceedings of the International DAAAM Symposium 2019; 30:764–9. https://doi.org/10.2507/30TH.DAAAM.PROCEEDINGS.105.

[24] Basset, E.; Köhler, J. & Denkena, B. (2012). On the honed cutting edge and its side effects during orthogonal turning operations of AISI1045 with coated WC-Co inserts. CIRP Journal of Manufacturing Science and Technology. Vol. 5, issue 2, 108-126. ISSN 1755-5817.

[25] Enders, W. J. & Kountanya, R. K. (2002). The Effects of Corner Radius and Edge Radius on Tool Flank Wear. Journal of Manufacturing Processes. Vol. 4, issue 2, 89-86. ISSN 1526-6125.

DOI: 10.2507/33rd.daaam.proceedings.026

CUTTING EDGE PREPARATION METHODS COMPARISON IN TERMS OF TOOL LIFE

Robert Straka & Tomas Vopat

This Publication has to be referred as: Straka, R[obert] & Vopat, T[omas] (2022). Cutting Edge Preparation Methods Comparison in Terms of Tool Life, Proceedings of the 33rd DAAAM International Symposium, pp.0188-0192, B. Katalinic (Ed.), Published by DAAAM International, ISBN 978-3-902734-36-5, ISSN 1726-9679, Vienna, Austria
DOI: 10.2507/33rd.daaam.proceedings.026

Abstract

The paper deals with three cutting edge preparation methods, namely brushing, drag finishing and abrasive jet machining. The cutting edge radius of 35 µm was prepared with each of these methods. The influence of used preparation methods on the tool life was investigated. Cutting tests were carried out during the turning of a difficult-to-cut nickel alloy Inconel 718. The chosen cutting parameters were considered for the semi-finishing turning operation. The longest tool life was achieved with drag finishing. The findings can be used to improve the efficiency of the turning process of nickel alloy Inconel 718.

Keywords: Cutting edge; Edge preparation; Tool life; Tool wear; Surface finishing

1. Introduction

The important aspect of the machinability of material, which was a topic of interest over the last 100 years is cutting edge geometry [1]. The tool life, coat adhesion and quality of machined surface are also influenced by it. Proper edge geometry stabilizes the cutting edge, which leads to the reduction of residual stresses in the cutting tool and improvement of the tool behaviour. The way to achieve such geometries on the cutting edge is preparation. It is usually carried out after the grinding of the tool and before coating [2], [3]. Denkena et al. showed that the tool life can be increase by 80% with an asymmetrical cutting edge rounding tilted to the rake face when machining C45E steel [4]. The research [5] found out that by enlarging the microgeometry of the cutting edge, the mechanical stress on the cutting tool is significantly reduced but it is also linked to an increase in the thermal load in the tool.

The article demonstrates a 70% reduction of the wear on the flank face when milling 42CRMo4-QT steel with milling tools with rounded cutting edges [6]. Ventura et al. used a tools with cutting edges with multiple chamfers and found out that more chamfers on the cutting edge increased tool life and compressive residual stresses on the finished surface but increases the cutting forces slightly [7]. The scanning laser interferometry was used in the article [8] to analyse, characterize, and verify the micro-geometry of the complex cutting edges when machining of Ti-6Al4V material. The research discovered an interaction between kinematic roughness and hone size of the cutting edge. It revealed that with larger hones the roughness increases or decreases depending on the kinematic roughness.

There are various cutting edge preparation methods, the most used ones in industry are brushing, drag finishing and abrasive jet machining. There are also new and progressive methods such as abrasive flow machining, magneto abrasive machining, laser machining, electrical discharge machining [9] or plasma discharge machining [10].

The abrasive jet machining is a process, where the material is removed trough erosion by an abrasive particle jet. A mixture of abrasive media is sprayed on the tool surface trough a jet nozzle. There are two types of this process. Dry abrasive jet machining uses air as a medium and the water is chosen for the wet abrasive jet machining. It is an effective method for creating cutting edges with reasonably accurate edge geometry [11]. Wyen discovered that the lack of thermally induced distortions on the machined surface is one of the most notable benefits of wet abrasive jet machining [12]. According to Jacob, the abrasive jet machining of the tool substrate may enhance coating adhesion. It could be used after coating to improve the hardness of the cutting edge. After the preparation procedure, the cutting tool had lower tool wear at a higher cutting speed [13]. The research of Bouzakis et al. showed that the abrasive jet machining may be used to precisely generate complicated cutting edge designs, while also enhancing the toughness and reliability of the cutting edge. Additionally, the procedure can induce compressive stress to the tool subsurface, improving the PVD coated tool film strength and tool performance [14]. The benefits of abrasive jet machining on cutting inserts for turning still require additional research due to a corresponding lack of studies.

Brushing allows to produce defined cutting edges with symmetrical and asymmetrical shape. For this reason, rotating circular brushes with filaments typically made of extruded polymer fibres containing scattered abrasive particles, such as silicon carbide or polycrystalline diamond, are frequently used [15]. It was first introduced as a novel 5-axis approach for the machining of cutting edges by Denkena et al. In their research, the fundamental effect of the brushing process parameters was examined. The cutting edges of carbide cutting inserts were machined using SiC 500 brushing tools, It has been demonstrated that the brushing process can be used to create customized microgeometries [16]. Using the same setup, they reported more research on the brushing process parameters in the subsequent study. It was demonstrated that the infeed and cutting speed had a considerable impact on the size of the cutting edge rounding. They also found out that the angle between cutting edge and the brushing tool affects the shape of the cutting edge. In the article it was shown that the tool life of the cutting inserts prepared by brushing could be increased by 100% compared to a tools without preparation [17].

Drag finishing is a machining process, where the workpiece is placed into a container filled with free abrasive grains with geometrically undefined cutting edges. This process is employed when parts need to have good surface quality, because it has the benefit of production of the desired results in a short time. It is quicker and more effective when compared to other processes, thanks to increase of the media pressure and relative speed between the media and surface of the workpiece. By changing specific process parameters, such as the feed rate, process time or the kind of grains, it can also be employed for the preparation of the cutting tools [18], [19]. Several researchers have studied the impact of media and their speed in the context of drag finishing. According to Barletta et al., the effectiveness of drag finishing is proportional to the both the impact forces and the kinematic energy of the medium [20]. Research of Kacaras et al. shows that the higher rotating speed result in greater impact forces and a quicker improvement of the surface [21].

The effect of cutting edge preparation on cutting forces was examined in publication [22]. Influence of cutting edge radius sizes on the tool life was determined in publication [23] The purpose of the article in [24] was to determine the accuracy of solid cemented carbide mills manufactured on grinding machine and determine the influence of drag finishing on macro geometry of solid cemented carbide mills. The purpose of this paper is to examine the difference between three cutting edge preparation methods, namely brushing, drag finishing and abrasive jet machining, and their influence on cutting inserts in terms of tool life in turning difficult-to-cut nickel alloy Inconel 718 equivalent to EN NiCr19FeNbMo.

2. Materials and methods

The manufacturer Dormer Pramet produced double-sided negative cutting inserts for the research. They were PVD coated and made of WC-Co cemented carbide. The designation of the cutting inserts is CNMG 120408-SM. Steel toolholder with designation PCLNL 2020k16 was used. These cutting inserts are suitable for machining steel, superalloys, hard materials, and for finishing or semi-roughing operations. The specifications of the cutting inserts are shown in Table 1. The cutting inserts were prepared by brushing (B), drag-finishing (DF) and abrasive jet machining (AJ). The cutting edge radius created with each of the preparation method was $r_n = 35$ μm.

Parameter	Value	Parameter	Value
Length [mm]	12.90	Recommended f [mm]	0.15 – 0.45
Thickness [mm]	4.76	Recommended a_p [mm]	0.80 – 4.00
Nose radius [mm]	0.80	Cutting edge angel [°]	80.00

Table 1. Cutting tool parameters

The measurement of the cutting edge radius was carried out by Alicona InfiniteFocusSL 3D measurement system. It is shown in Fig. 1 for the cutting edge prepared with brushing. The machine tool used in the experiment was a CTX Alpha 500 multi-axis turning centre. The machine tool parameters are in Table 2. During the cutting tests, a longitudinal turning strategy was used. Throughout the whole machining process, the cutting conditions remained unchanged.

The process parameters chosen were the cutting speed of v_c = 40 mm·min^{-1}, the feed of f = 0.15 mm, and the depth of cut of a_p = 1 mm. They were used to control the development of wear at the cutting edges.

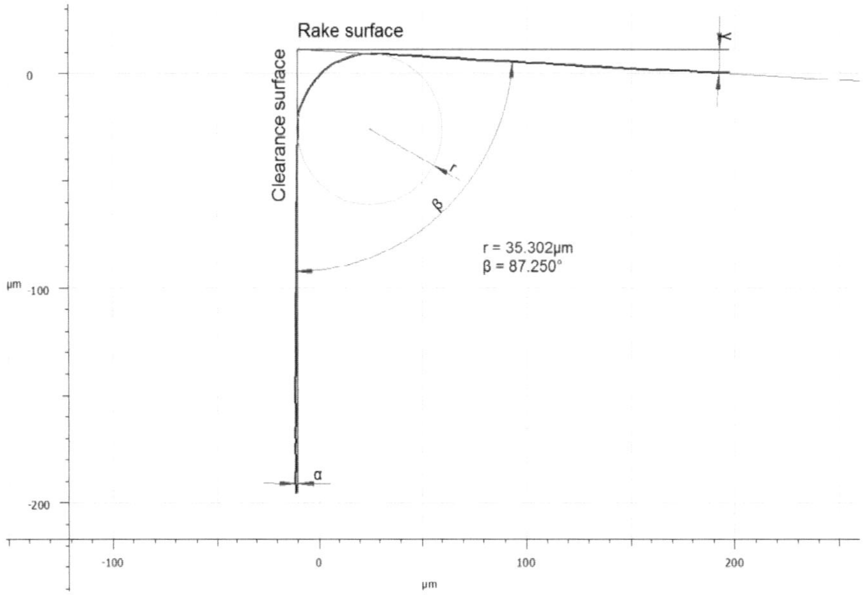

Fig. 1. Example of measurement of the r_n prepared by brushing

Travel distance		Tool revolver head	
X [mm]	190	Number of tool positions [-]	12
Y [mm]	±40	Drive power [kW]	5.4
Z [mm]	525	Max. rotation frequency [min-1]	5000
Main spindle		Control system	Heidenhain CNC Pilot 4290
Drive power [kW]	20		
Max. rotation frequency [min-1]	6000		

Table 2. Parameters of CTX Alpha 500 multi-axis turning centre

The material for the workpiece was Inconel 718 (equivalent to EN NiCr19FeNbMo). It is a high strength nickel-chromium alloy that has been annealed. Cold working was used to manufacture the workpiece. The chemical composition of the workpiece material is in Table 3.

C	Si	Mn	N	Cr	Mo	Ni	Cu	Co	Ti	Al	Nb	Fe
0.023	0.070	0.11	0.006	17.780	2.990	53.6	0.06	0.35	0.9	0.4	5.1	18.2

Table 3. Chemical composition of the Inconel 718 (wt. %)

The length of the wear area VB_N was measured by a Dino-Lite Edge microscope with a resolution of 1280 × 960 pixels. The cutting inserts were cleaned prior to experiment since the cutting edges had impurities. The workpiece material was clamped into the three-jaw chuck of the main spindle and machined using three cutting inserts prepared with the same method. The diameter of the workpiece was 101.6 mm. The machined surface was 60 mm in length. During the process of manufacturing the workpiece, the top layer become strain hardened. This layer is thinner than 0.3 mm. To avoid influencing the experiment's outcome, another cutting insert was used to machine this layer.

The first cutting insert had a cutting length range of 0 to 20 mm. The cutting length ranged from 20 to 40 mm for the second cutting insert and from the 40 to 60 mm for the final one. The cutting inserts were cleaned after the first minute of machining. Then, they were inserted into the customized holder that was created using 3D printer. The microscope was focused on the cutting edge. It was captured in the photo so that it could not be moved by outside sources that would have affected the measurement of tool wear. The flank wear was then measured in the image. The experiment continued after it was recorded, with cutting inserts being placed back to their positions in the turning centre. Tool wear was then measured every 2 minutes of machining until the tool wear started to increase significantly or edge chipping was observed on the cutting insert.

3. Results and discussions

The time dependence of the average tool wear with respect to the preparation method is shown in Fig. 3. For the experiment, three cutting inserts were used for each method with the prepared cutting edge radius of r_n = 35 μm with respect to statistical significance. The designation of the preparation methods in the Fig. 3 is as follow DF is for drag-finishing, AJ is for abrasive jet machining and the B is for brushing.

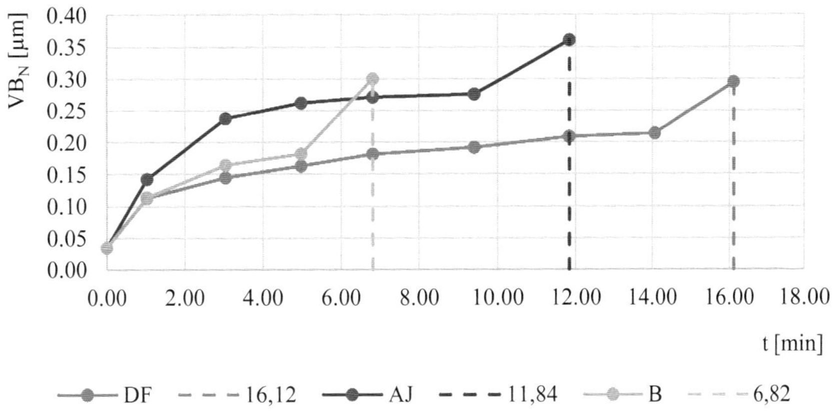

Fig. 2. Average wear development for cutting edges prepared with different methods

The longest tool life was achieved with the cutting inserts prepared with drag-finishing with a value of 16.12 minutes. The tool life increased by 36.15% compared to abrasive jet machining and by 136.36% compared to brushing. This increase in the tool life can be explained by the fact that the shape of the cutting edge is more uniform and that the adhesion of the coat to the substrate is better after the drag-finishing when compared to the other two used preparation methods. Tool wear can be difficult to predict when dealing with difficult-to-cut materials such as nickel alloys. As a result, it occasionally happens that the cutting tool breaks earlier than anticipated. This is also a reason why the shorter tool life occurred in some cases.

4. Conclusion

The topic of the paper deals with cutting edge preparation of cutting inserts made of cemented carbide. When cutting difficult-to-cut nickel alloy Inconel 718, the effect of 3 cutting edge preparation methods on the tool life was examined. The cutting edge radius prepared with each method was 35 μm. The cutting parameters were chosen in consideration of the semi-finishing turning operation. The tool life tests show that the longest tool life was reached with drag-finishing. The average tool life was 16.12 minutes. It was increased by 36% when compared to abrasive jet machining and by 136% compared to brushing.

Future research will examine the effect of more cutting edge radii within each method on the tool life and cutting forces during turning operations as well as their effect on the surface roughness parameters. The preparation methods will be compared with each other in these terms as well.

5. Acknowledgements

This research was funded by research projects APVV–16–0057 and APVV-21-0071. The authors are also grateful to Dormer Pramet tool producer, who participated in this research.

6. References

[1] Bernard, S. E.; Selvaganesh, R.; Khoshick, G. & Raj, S. (2021). A novel contact area based analysis to study the thermo-mechanical effect of cutting edge radius using numerical and multi-sensor experimental investigation in turning. Journal of Materials Processing Technology. Volume 293. ISSN 0924-0136.

[2] Zimmermann, M.; Kirsch, B.; Kang, Y.; Herrmann, T. & Aurich, J. C. (2020). Influence of the laser parameters on the cutting edge preparation and the performance of cemented carbide indexable inserts. Journal of Manufacturing Processes. Volume 58, Pages 845-856. ISSN 1526-6125.

[3] Denkena, B.; Krödel, A. & Hein, M. (2019). Innovative method for cutting edge preparation with flexible diamond tools. Procedia CIRP. Volume 86, Pages 121-125. ISSN 2212-8271.

[4] Bassett, E.; Köhler, J. & Denkena, B. (2012). On the Honed Cutting Edge and Its Effects During Orthogonal Turning Operations of AISI 1045 with Coated WC-Co Inserts. CIRP Journal of Manufacturing Science and Technology, 5/2: 108–126.

[5] Denkena, B.; Krödel, A. & Heckemeyer, A. (2021). Numerical and experimental analysis of thermal and mechanical tool load when turning AISI 52100 with ground cutting edge microgeometries. CIRP Journal of Manufacturing Science and Technology. Volume 35, Pages 494-501. ISSN 1755-5817.

[6] Denkena, B.; Köehler, J. & Rehe, M. (2012). Influence of the Honed Cutting Edge on Tool Wear and Surface Integrity in Slot Milling of 42CrMo4 Steel. Procedia CIRP. Volume 1, Pages 190 – 195.

[7] Ventura, C.E.H.; Köhler, J. & Denkena, B. (2015). Influence of cutting edge geometry on tool wear performance in interrupted hard turning. Journal of Manufacturing Processes. Volume 19, Pages 129-134. ISSN 1526-6125.

[8] Brown, I. & Schoop, J. (2020). The effect of cutting edge geometry, nose radius and feed on surface integrity in finish turning of Ti-6Al4V. Procedia CIRP. Volume 87, Pages 142-147. ISSN 2212-8271.

[9] Denkena, B. & Biermann, D. (2014). Cutting edge geometries. CIRP Annals - Manufacturing Technology 63, 631-653. ISSN 1726-0604.

[10] Vopát, T.; Podhorský, Š.; Sahul, M. & Haršáni, M. (2019). Cutting edge preparation of cutting tools using plasma discharge in the electrolyte. Journal of Manufacturing Processes 46, 234-240. ISSN 1526-6125.

[11] Wang, W.; Biermann, D.; Aßmuth, R.; Arif, A. F. M. & Veldhuis, S. C. (2020). Effects on tool performance of cutting edge prepared by pressurized air wet abrasive jet machining (PAWAJM). Journal of Materials Processing Technology. Vol. 277. ISSN 09240136.

[12] Wyen, C. F. (2011). Rounded cutting edges and their influence in machining titanium (Doctoral dissertation, ETH Zurich).

[13] Jacob, A. (2015). Effect of Micro-blasting on Characteristics and Machining Performance of PVD AlTiN Coated Cutting Tools (Doctoral dissertation).

[14] Bouzakis, K. D.; Gerardis, S.; Skordaris, G.; Katirtzoglou, G.; Makrimallakis, S.; Klocke, F. & Bouzakis, E. (2009). Effect of dry micro-blasting on PVD-film properties, cutting edge geometry and tool life in milling. Surface and Coatings Technology, 204(6-7), 1081-1086.

[15] Bergs, T.; Schneider, S.A.M. & Amara, M. & Ganser, P. (2020). Preparation of symmetrical and asymmetrical cutting edges on solid cutting tools using brushing tools with filament-integrated diamond grits. Procedia CIRP, Volume 93, Pages 873-878. ISSN 2212-8271.

[16] Denkena, B.; de Leon, L. & Bassett, E. (2008). Preparation of Designed Cutting Edge Microgeometries by Simultaneous 5-Axes Brushing. Proceedings of the 3rd Int. Conf. on Manuf. Eng. (ICMEN) and EUREKA Brokerage Event, Kallithea of Chalkidiki, Greece, 1–3 October, 117–123.

[17] Denkena B.; de Leon L.; Bassett E. & Rehe M. (2010). Cutting Edge Preparation by Means of Abrasive Brushing. 8th Int. Conf. ''THE Coatings'' in Manuf. Eng., Erlangen, Germany, April 14–15 Key Engineering Materials 438: 1–7.

[18] Peterka, J.; Pokorný P.; Václav, Š.; Pätoprstý, B. & Vozár, M. (2020b). Modification of cutting tools by drag finishing. MM Science Journal March 2020, 3822-3825. ISSN 1805-0476.

[19] Malkorra, I.; Salvatore, F.; Arrayola, P. & Rech, J. (2020). The influence of the process parameters of drag finishing on the surface topography of aluminum samples. CIRP Journal of Manufacturing Science and Technology 31, 200-209. ISSN 1775-5817.

[20] Barletta, M.; Giusario, A.; Venettacci, S. & Rubino, G. (2014). A comparative evaluation of fluidized bed assisted drag finishing and centrifugal disk dry finishing. Engineering Science and Technology, an International Journal, Volume 17, Issue 2, Pages 63-72. ISSN 2215-0986.

[21] Kacaras, A.; Gimbier, J.; Zanger, F. & Schulze, V. (2018). Influence of rotational speed on surface states after stream finishing. Procedia CIRP, Volume 71, Pages 221-226. ISSN 2212-827.

[22] Šimna, V.; Jurina, F.; Vopát, T. & Kuruc M. (2019). Edge preparation of cutting tools and it impact on cutting forces during milling. Annals of DAAAM and Proceedings of the International DAAAM Symposium 2019;30:764–9. https://doi.org/10.2507/30TH.DAAAM.PROCEEDINGS.105.

[23] Vopát, T.; Kuruc, M.; Šimna, V.; Zaujec, R. & Peterka J. (2017). Cutting edge microgeometry and preparation methods. Annals of DAAAM and Proceedings of the International DAAAM Symposium 2017:384–91. https://doi.org/10.2507/28TH.DAAAM.PROCEEDINGS.054.

[24] Pätoprstý, B.; Pokorný, P.; Vozár, M.; Vopát, T. & Buranský I. (2019). The influence of grinding process and drag finishing on the milling tools macro geometry. Annals of DAAAM and Proceedings of the International DAAAM Symposium 2019:1008-1113. DOI: 10.2507/30th.daaam.proceedings.140

DOI: 10.2507/33rd.daaam.proceedings.027

INCLUSIVE DESIGN IN ENGINEERING EDUCATION

Antoaneta Dobreva & Vasko Dobrev

This Publication has to be referred as: Dobreva, A[ntoaneta] & Dobrev, V[asko] (2022). Inclusive Design in Engineering Education, Proceedings of the 33rd DAAAM International Symposium, pp.0193-0197, B. Katalinic (Ed.), Published by DAAAM International, ISBN 978-3-902734-36-5, ISSN 1726-9679, Vienna, Austria
DOI: 10.2507/33rd.daaam.proceedings.027

Abstract

Various methods and approaches are to be applied in the University study process in order to activate new opportunities and potential for improving the quality of engineering education. The objective of the paper presented is to improve and adapt the established educational technology and to elaborate at a final stage a new educational methodology considering Machine Design subjects. The current study examines this methodology developed by the author's team aiming to ensure an inclusive design of Engineering Education. The inclusive design technology includes several stages: defining the problem of specifying the requirements for the "student - teacher - university environment" system; developing a system of solutions for creating personalized, tailor-made training and education, including preparation of special teaching materials containing sufficient information and literary sources that support the accelerated learning of the basic educational material; students' evaluation of the inclusive design educational technology suggested, in which important feedback comments are received on the efficiency of the new methodology applied. The results of the application of the inclusive design technology are presented and discussed. Conclusions are deduced.

Keywords: Inclusive Design; Innovative Training Technology; Machine Design subjects; Engineering Education.

1. Introduction

The contemporary level of industry development (Industry 4.0) calls for the integration of information and communication technologies with industrial achievements, aiming to transform production processes into more efficient and flexible ones. Employers from industry face barriers to implement this approach and the most significant of these barriers seems to be the lack of qualified staff. It is particularly important for the system of higher education that the qualifications of future university graduates, their competencies and professional skills, necessary to manage and control such modern industrial systems, have to be already achieved in the Bachelor and Master degree programs. Continuous monitoring of the academic and university environment is necessary in order to determine on time the necessary modernization of educational technology and curriculum contents that would ensure meeting the needs of modern industry, [1].

It is expected that in Europe, during the next 10 years, there will be almost one million new jobs in industry and a significant part of them should be directed to professionals with new profiles. The main features of these profiles are: creativity, strategic and very good organizational abilities [1], [2]. Therefore, the universities have to play a key role in the training and educating of such future professionals.

The main competences, which have to be developed within undergraduate and graduate degrees should include interdisciplinary thinking, decision-making, problem-solving abilities, intercultural skills and willingness for lifelong learning [3].

2. Theoretical background

According to the authors of [4], the training of future engineers has to be modified so that they should be prepared to confront the rapid changes in technology and adapt to entire new types of jobs. Future specialists will have to develop new technologies, to collaborate with colleagues at an international level and deal with the main challenges facing society, such as global environmental problems, [5]. The authors of [6] emphasize that it is necessary for future engineers to be trained with appropriate pedagogical strategies which are based on experiment-based, research-based, and active learning.

In addition to basic knowledge, skills and competencies in the field of innovation, the abilities to analyse data and solve problems play an increasingly important role. These abilities will allow the future graduates to analyze data sets from different sources, to optimize production processes and to make decisions in real time, [7].

The correct and appropriate application of educational technologies has the potential to enhance learning by maximizing the academic experience of students. Appropriate educational technology enables the delivery of face-to-face and/or distance education, improving learning outcomes by adapting assessment to learning style, facilitating access to up-to-date learning materials and to communication with famous experts in a certain scientific area, [8].

Based on the scientific literature sources analysed, the authors' team has focused on the following research objective: to improve, adapt and supplement the existing and established training technology and to create at a final stage a new educational methodology for inclusive design considering machine design subjects in engineering education. The methodology elaborated has been tested and validated in a real academic study process.

3. Research methods

The previous methodology according to the established curricula of the Machine design subjects, approved by the relevant Faculty Councils, includes two main stages. During the first stage, students receive assignments for course tasks and projects, consistent with the syllabus of the subject; lecturers control the current implementation of the assignments and deliver consultations, if necessary. At the end of the semester, the students fill out surveys about the quality of the study process.

A research methodology for ensuring an inclusive design of the educational technology in Machine design subjects within engineering training has been supplemented and improved. It should be characterized by the following features: education is carried out in both face-to-face and online mode; students are taught with a combination of such tools, methods, programs and aids, which are preferred, appropriate and most relevant for them; teaching and learning tools have to be adapted to each student's learning style and background; assessment is to be carried out on the basis of results, skills and competences achieved and less often – based on exams; information analysis and data interpretation represent a key competency that is additionally included in almost all Machine design subjects' syllabi according to the reccommendations of [8].

The methodology created by the author's team for ensuring an inclusive design of Engineering Education includes the following stages:

The first stage is related to defining the problem of specifying the requirements for the "student - teacher - university environment" system. Within this stage, the most important issue is the activity of the academic lecturer, who must identify the level of knowledge, skills and competences of each student, at the beginning of the semester, in order to prepare the corresponding personalized, tailor-made training.

The second stage involves developing a system of solutions defined by the student's entry level. For students with very good basic training, creative asssignments and projects in the field of Machine elements, Machine Science and Engineering Design are prepared and delivered. If some of these students manage to cope excellently with the organization and timing during elaboration of these assignments, it is recommended to switch to the mode of Independent research work, which ends with a scientific publication or a case study necessary for an industry enterprise.

The second group of students receives assignments for coursework and projects, consistent with the syllabus of the subject. The lecturer monitors and/or corrects the current performance of assignments by supervising some of the students (if necessary) in face-to-face or online mode.

In some cases, a third group of students with a lower level of creative activity and prior technical training literacy emerges. Special teachning and learning materials containing sufficient information, equations, standards and literature sources have been prepared for this part of the students, which support the accelerated study of the mandatory, basic study material.Concerning the preparation of drawings, face-to-face and/or online additional classes are provided, within which the lecturer delivers consultations concerning the preparation of drawings for each calculation, layout and design stage. Students from all groups are given the opportunity to participate in the "SMART in CAD" professional club, which is delivered every semester.

During the third stage, an evaluation of the inclusive design methods is implemented and feedback is received from the students concerning the efficiency and the quality of the new inclusive design educational technology applied.

The development approach of the new methododlogy presented takes into account several important limitations considering the research objective. The research has been implemented with students from engineering bachelor degree courses within the subjects from the area of Machine Elements, Design Projects, Machine Science and Engineering Design. These limitations have been defined by the authors' team due to the specific features of the subjects mentioned. They require relatively good prior education and training with STEM components (Science, Technology, Engineering and Mathematics). Besides, the machine design subjects require intensive work and efforts on behalf of the lecturers in order to facilitate and improve the students' creative activity and skills.

This new methodology of study process focused on Machine design subjects, has been applied within four semesters in a certain number of Engineering bachelor degree courses at the University of Ruse, Bulgaria. During this period of time, the previous, established methodology (currently approved by the relevant Faculty Councils) has been applied in the rest of the Engineering bachelor degree courses at the University of Ruse.

4. Research results and discussion

The period of implementation and validation of the newly created methodology for inclusive design continued four semesters. During that period, the following additional activities have been carried out:

A total number of eight scientific and methodological seminars have been organized on different topics, such as "Investigation of the possibilities of control and management of the work of technical systems and devices"; "Structure and organization of creating a scientific publication". Lecturers, Bachelor, PhD students and Incoming students from the Erasmus+ program participated in these seminars. Specialists from the industry also took part in some of the seminars as guest lecturers. The topics of the seminars have been determined by the academic staff of the Department of "Machine Science, Machine elements, Engineering Graphics and Physics" with the Transport Faculty at the University of Ruse and by the students according to their interests established during the first stage of the new methodology applied.

A total number of 12 questionnaires have been conducted among the bachelor degree students, concerning the feedback and assessment of the new educational technology applied. The students gave their opinion about the quality of lectures, tutorials, supervisors' consultations and additional activities. The contents of these questionnaires is elaborated by the academic staff of the Department of "Machine Science, Machine elements, Engineering Graphics and Physics".

Conducting this type of follow-up survey is an essential component of the feedback achieved. By expressing their opinion, the students have the opportunity to participate in the organization, management and control of the study process. The students evaluate this option extremely positively. Part of the results achieved concerning the diagnosis of the quality of the educational process are described and analyzed in [9], [10].

In order to facilitate this part of the students, who have a lower level of creative activity and prior technical literacy, two new special study teaching materals have been prepared, containing sufficient information and equations supporting the accelerated acquisition and assimilation of the obligatory, basic study material, [11], [12].

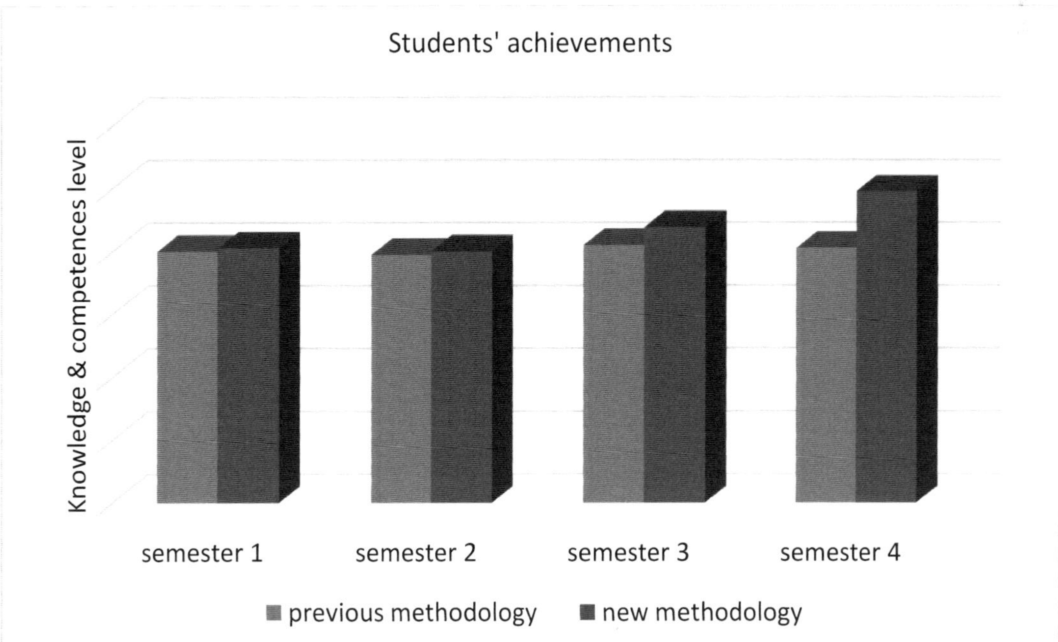

Fig. 1. Results of the tested and investigated new methodology for inclusive design example

The contributions of the department mentioned to delivering theoretical knowledge and practical skills in the field of computer-aided design are particularly significant. This addutional activity supports the work of bachelor, master and PhD students, who are members of the professional club "SMARTinCAD" headed by PhD Yuliyan Dimitrov. The achievements and results of this very important activity of this professional club have been described in details in [13], [14].

Considering that the authors' team and the academic staff of the Department of "Machine Science, Machine elements, Engineering Graphics and Physics" supervised more than 10 PhD students during the period of the research, the academic staff of the department has contributed significantly to the development of the creative thinking and the scientific skills of the students through several scientific activities. The most important of them are the active international activity improving the educational and internship opportunities of the students and the supervision of their relatively independent research work, [15], [16], [17], [18].

The main expected consequences of the new approach are directed towards the possibility that the methodology described can be applied by lecturers in other subjects in the same bachelor degree course through achieving horizontal and vertical integration. Horizontal integration can be achieved through active interaction between lecturers responsible for different subjects who could give joint independent assignments and projects (mainly for the students from the first and second groups described in research methods) within the same semester. Vertical integration can be realized if the academic staff responsible for the subjects during the last one or two academic years apply this new methodology and give students the opportunity to work on a topic of interest to them by setting new tasks and projects that deepen the knowledge and improve the skills of the students in the relevant professional area.

After applying the new methodology for inclusive design as educational technology, several very important groups of results have been obtained. The most important result is the significant increase of the level of knowledge and competences of students as is shown in Fig. 1. The students appreciated the tailor-made education technology and improved their interest in the study process and in the process of acquiring new knowledge and skills on their own. During each subsequent semester within the framework of the scientific research, the scientific and publication activity of the students increased by about 50 %. Some additional negative limitations are expected. Due to the relatively great work load of some lecturers, it is possible that they will not be able to apply the new methodology in its full scope.

5. Conclusions

The study presented examines the methodology developed by the author's team aiming to ensure an inclusive design of Engineering Education. Based upon the research implemented, several conclusions can be deduced. In the process of applying the new methodology, the level of knowledge, skills and competences increased significantly. Due to the introducing of this technology, the students' interest in the study process has increased. The students have begun to initiate their independent work to acquire new knowledge and technical skills.

The new methodology has also created prerequisites for achieving horizontal and vertical integration in the engineering training and education. This specific implication can be achieved through active interaction between lecturers in different subjects and giving the students joint and/or interdisciplinary tasks and projects. Some of these additional academic assignments can extend over more than one semester and deepen students' knowledge and skills in the relevant professional area.

The application of this methodology has created prerequisites for increasing students' motivation to participate actively in the study process and to improve the distribution of time necessary for preparing assignments and projects on their own. Based upon the conclusions deduced, it becomes clear that the objective of the research has been achieved. The students' motivation for learning has increased and the quality of the study process in Machine design subjects has been improved.

Future research plans include further development of inclusive design methodology. Enhanced work with companies based upon existing and new contacts with industry is envisaged.

6. Acknowledgments

The authors acknowledge the financial support from the Scientific Research Funds of the University of Ruse, Bulgaria. The research work done is realized in the framework of several scientific seminars at the Department of "Machine Science, Machine Elements, Engineering Graphics and Physics" with the Transport Faculty at the University of Ruse. The support is gratefully acknowledged.

7. References

[1] Hernandez-de-Menendez, M., Diaz, C. A. E. & Ruben Morales-Menendez, R. (2020). Engineering education for smart 4.0 technology: a review, International Journal on Interactive Design and Manufacturing (IJIDeM) 14: pp. 789–803, https://doi.org/10.1007/s12008-020-00672-x.

[2] Hecklau, F., Galeitzke, M., Flachs, S., Kohl, H. (2016). Holistic approach for human resource management in Industry 4.0. Procedia CIRP 54, pp. 1–6.

[3] Coskun, S., Kayıkcı, Y. & Genзay, E. (2019). Adapting engineering education to Industry 4.0 vision. Technologies 7, pp. 1–10.

[4] Jeganathan, L., Khan, A., Kannan Raju, J. & Narayanasamy, S. (2019). On a frame work of curriculum for engineering education 4.0. In: 2018 World Engineering Education Forum and the Global Engineering Deans Council. IEEE, pp. 1–6. https ://doi. org/10.1109/weef-gedc.2018.86296 29.

[5] Talikka, M. (2018). Recognizing required changes to higher education engineering programs' information literacy education as consequence of research problems becoming more complex. Lappeenranta University of Technology, Lappeenranta.

[6] Grodotzki, J., Ortelt, T.& Tekkaya, A. (2018). Remote and virtual labs for engineering education 4.0. Procedia Manuf. 26, pp. 1349–1360.

[7] Sackey, S.M. & Bester, A. (2016). Industrial engineering curriculum in Industry 4.0 in a South African context. S. Afr. J. Ind. Eng. 27, pp. 101–114.

[8] Mogos, R.I., Bodea, C.N., Dascalu, M.I., Safonkina, O., Lazarou, E., Trifan, E.L., et al. (2018). Technology enhanced learning for Industry 4.0 engineering education. Rev. Roum. Des. Sci. Tech. Ser. Electrotech. Energy 63, pp. 429–435.

[9] Ronkova, V. & Dimitrov, Y. (2019). Education in engineering – challenges and contributions. Annals of DAAAM and Proceedings of the International DAAAM Symposium, 30 (1), ISBN 978-3-902734-22-8, ISSN 1726-9679, DOI: 10.2507/30th.daaam.proceedings.098, pp. 717–723.

[10] Kamenov, K., Dobreva, A. & Ronkova, V. (2017). Advanced Engineering Methods in Design and Education, IOP Conference Series: Materials Science and Engineering, 252 (1), 012033, DOI 10.1088/1757-899X/252/1/012033, pp. 1-6.

[11] Dobreva, A. (2021). Initial project template. Contact stresses, 27 pp., ISBN 978-954-712-865-1.

[12] Dimitrov, Y. (2022). Screw jack. Buckling. 15 pp., ISBN 978-619-207-236-0.

[13] Dimitrova, Y. & Dimitrov, Y. (2018). The non-traditional practice of professional student club to connect the business with the engineering students of University of Ruse. Annals of DAAAM and Proceedings of the International DAAAM Symposium, 29 (1), ISBN 978-3-902734-20-4, ISSN 1726-9679, DOI: 10.2507/29th.daaam.proceedings.147, pp. 1032–1037.

[14] Dimitrov, Y. (2018). System for automatic distribution and working with technical documentation in small and micro-sized industrial enterprises. Annals of DAAAM and Proceedings of the International DAAAM Symposium, 29 (1), ISBN 978-3-902734-20-4, ISSN 1726-9679, DOI: 10.2507/29th.daaam.proceedings.140, pp. 976–982.

[15] Armas, A. (2021). Modular development of technical products. Proceedings of University of Ruse - 2021, Vol 60, book 4.3, BSc, MSc and PhD Students & Young Scientists, pp. 30-35, Supervisor: Prof. A. Dobreva. ISSN 2603-4123 (on-line).

[16] Dobreva, A. & Mollova, G. (2021). Research of driving systems - challenges and possible solutions. IN: Proceedings of University of Ruse, Vol 60, ISSN: 1311-3321, pp 36-40.

[17] Mollova, G. & Dobrev, V. (2021). Design methodology for investigating worm gear transmissions with significant dimensions. IN: Proceedings of University of Ruse, Vol 60, ISSN: 1311-3321, pp. 41-47.

[18] Dobreva, A. & Pavlov. P. (2021). Energy Efficiency of Worm Gear Drives, Annals of DAAAM & Proceedings. 2021, Vol. 10 Issue 2, ISBN 978-3-902734-33-4, ISSN 1726-9679, DOI 10.2507/32nd.daaam.proceedings.016, pp 110-116.

DOI: 10.2507/33rd.daaam.proceedings.028

PROBLEMS OF APPLICATION OF METHODS OF DYNAMIC REPROGRAMMING OF CONTROL SYSTEMS OF MOBILE ROBOT MODULES

Pavel Pletenev & Victor Andreev

This Publication has to be referred as: Pletenev, P[aul] & Andreev, V[ictor] (2022). Problems of Application of Methods of Dynamic Reprogramming of Control Systems of Mobile Robot Modules, Proceedings of the 33rd DAAAM International Symposium, pp.0198-0204, B. Katalinic (Ed.), Published by DAAAM International, ISBN 978-3-902734-36-5, ISSN 1726-9679, Vienna, Austria
DOI: 10.2507/33rd.daaam.proceedings.028

Abstract

This paper presents an analysis of existing and promising methods of dynamic reprogramming of embedded systems suitable for use in a mobile robot with a modular control system architecture. The functional-modular architecture of the mobile robot control system implementing distributed computing is briefly described. This control system architecture makes it possible to run a mobile robot in real time using functional modules control systems implemented on embedded systems – low performance microcontrollers. Remote dynamic reprogramming of the software aspects of the control systems of individual robot modules without stopping the functioning of the robotic complex is necessary to ensure an operational change in the configuration of the modular robot – reconfiguring the robot structure in order to adjust its functionality to the task that has arisen. The features of using 4 methods of dynamic reprogramming according to 6 criteria are considered: applicability on different embedded microprocessor systems, the amount of RAM and program memory required, the speed of calculations, the theoretical complexity of implementing the method, the theoretical complexity of using the method by the end user (customizer), the flexibility of solutions obtained using the method. As a result of the study, usage recommendations for the considered methods are given.

Keywords: mobile robot; modular robot; control system; communications channel; dynamic reprogramming; embedded systems.

1. Introduction

The experience of using robotic complexes (RTC) in extreme conditions (for example, the accident at the Chernobyl nuclear power plant in 1986) has shown that in order to accurately match the RTC functionality to the task when working in aggressive environments for humans (space, zones of chemical or radioactive contamination), not only robots in special design (extreme robotics) are required, but also it may be necessary to promptly change the composition and/or structure of the RTC directly at the work site.

The need for this functionality for extreme robotics was pointed out by such well-known Russian scientists as Academician of the Russian Academy of Sciences, Doctor of Technical Sciences Evgeny Pavlovich Popov [1], Doctor of Technical Sciences Evgeny Ivanovich Yurevich [2] and Doctor of Ph.D. Platonov Alexander Konstantinovich [3]. In other words, reconfigurable robots are needed for such applications. Reconfiguration - altering the configuration of a modular robot in order to achieve an intended change in the function of the modular robot [4].

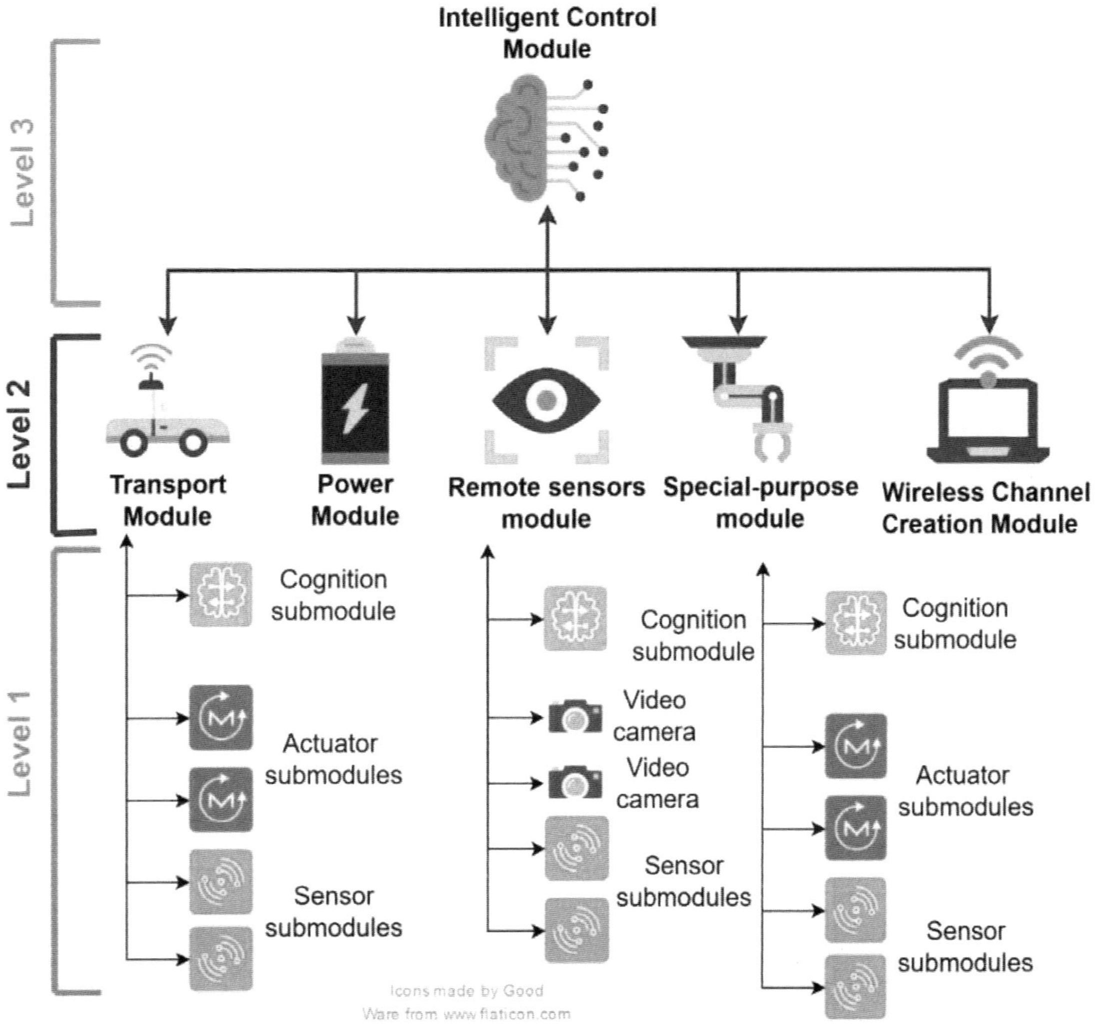

Fig. 1. Functional-modular architecture of a control system of a mobile robot

In fact, reconfiguration is a purposeful change in the configuration and/or technical characteristics of a mobile robot (MR) when changing the composition of its functional units without the need to reconfigure the software of its control system (CS). But such a reconfiguration is possible only with a modular solution of the architecture of the RTC control system. An example of such an architecture is shown in Fig.1 [5].

At the top of the architecture there is an intelligent control module responsible for planning and distributing tasks by function between the modules of the second level – system-wide management. In this architecture, modules at the third and second levels of the hierarchy are responsible for one enlarged function of the whole robot according to the principle of full functionality [6]: *each robot module should be able to perform its target function in any way convenient for it, using only its own means to execute commands from an external control system.* At the first level of the hierarchy there are submodules that perform the function of low-level control of actuators, for example, drives; in fact, this is a group of submodules (Worker group). The interaction of the second and first level modules is carried out according to the "Master-Worker group" principle. The control systems of most modern MR are built on the basis of a single computing device on which all computing operations are performed (both high-level and low-level control), which does not allow for the operational reconfiguration of a robotic system; the implementation of reconfiguration of such robots in practice means the creation of a new robot with a given new functionality.

The modular solution of the CS means splitting the RTC functionality into simpler subfunctions (see Fig.1), which are implemented by separate fully functional mechatronic modules with their own control system [5]. It should be particularly noted that in order to minimize the weight and size parameters and energy consumption of autonomous mobile robots, module control systems should be implemented on embedded systems – microcontrollers of low performance. To synchronize the operation of such a distributed structure, it is necessary to provide appropriate intermodule information interaction, which is implemented by combining control systems of all modules and submodules into a local area network (LAN) [7]. As a result, distributed computing is implemented by analogy with multi-agent systems [8].

In an MR with a modular architecture, the target task is solved due to the correct selection of functional nodes-modules and their automatic software integration into the robot's control system as a whole without processing the software aspects of both the general control system and the control systems of the modules themselves.

However, due to the uncertainty of the requirements for the parameters of the CS in cases where it is impossible to determine in advance the environmental conditions in which the robot will work (for example, reconnaissance MR on other planets or in areas of man–made disasters), there is a need to reconfigure the CS or even replace its software (software) - remote and externally controlled adaptation, settings and adjustments to the situation. There is a need for operational remote reprogramming of the software aspects of the control systems of individual robot modules without stopping the operation of the RTC, since under these conditions it turns out to be extremely costly or even impossible to deliver the robot to its developer or manufacturer. At the same time, if there is a connection with the robot, it is possible to remotely perform this work with the help of a software developer.

The purpose of this work is to analyze existing and promising methods of dynamic reprogramming suitable for use in a mobile robot with a modular control system architecture. The main problem of using dynamic reprogramming for a modular robot is the need to use embedded microcontroller system modules in the CS and update the software on the fly via the main data bus used by the microcontroller without disconnecting and dismantling the blocks. Since the control systems of modules and submodules are integrated into a LAN, the solution to this problem is based on the search for appropriate methods of intermodule communication (IMC) and network protocols that meet certain requirements.

Authors from different countries and scientific schools have proposed different methods, both for building modular robots [9], [10], [11] and methods of intermodular communication [12], [13], [14]. Our work [15] shows that the Cyphal specification [15] (formerly known as UAVCAN) and its OpenCyphal implementation are the most effective in terms of application as an IMC. Therefore, the various dynamic reprogramming mechanisms discussed in this article will use messages from the Cyphal specification as IMC.

2. Analysis methods

In the framework of this work, we will consider an embedded hardware and software system implemented on a microcontroller that controls some physical process. The control is performed by the "main algorithm". Any changes to the main algorithm are made by a "side algorithm".At the moment, there are many different and significantly different methods of dynamic reprogramming. Each of them has its own unique properties and differs in a different distribution of the complexity of tasks for changing the "main algorithm" between the human developer of the control system, the human adjuster (end user) of the control system, the software on the computer of the developer/adjuster of the embedded system and the direct control system of the embedded system. The differences also lie in the breadth of flexibility of solutions that can be created using each of these methods. The 4 main methods of dynamic reprogramming are discussed below:

- M1. Parametric reprogramming.
- M2. Replacing the microcontroller code.
- M3. Bytecode replacement.
- M4. Replacing a script in an interpreted programming language.

The analysis of each of the listed methods is performed according to the following criteria:

- K1. Applicability on various embedded systems – microcontrollers of low performance.
- K2. The amount of RAM and program memory required.
- K3. Computing speed (CPU time).
- K4. The theoretical complexity of creating an implementation of the method in the form of a basic control program.
- K5. The theoretical complexity of using the method to adapt to the operational situation of the basic control program by the end user.
- K6. Flexibility of the method – the depth and complexity of possible changes to the basic algorithm of the embedded system when using this method.

An important criterion of remote dynamic reprogramming methods is the possibility of ensuring information security, since remoteness is usually realized through the use of a radio channel. The methods discussed below enable Remote Code Execution, which is one of the largest vulnerabilities for any software package. But the authors are not specialists in the field of information security, so ensuring the information security of the methods presented below is a topic for separate consideration by relevant specialists.

In this case, it is assumed that all interactions occur inside the noise-proof internal bus of the robot, inaccessible to external influences. Also, the software of the internal network node that exchanges information with external, potentially open networks is written reliably enough that it does not allow downloading and executing code from unreliable and potentially malicious sources.

3. Analysis results

3.1. M1 – Parametric reprogramming

One of the first and simplest methods of dynamic reprogramming is parametric reprogramming – that is, the replacement of immutable (from the point of view of the main control algorithm) internal quantities that play an essential role in the operation of this algorithm. An example is the adjustment of the values of the proportional, integral and differential parts in the motor speed controller in the motion module's control system. The values (for example, default ones) for such parameters can be stored in the non-volatile memory of the computing device. This is the method of reprogramming provided by data types with access service (request-response): uavcan.register.Access and uavcan.register.List from the Cyphal specification. Both of these services provide information about named firmware parameters – the so-called registers. Each register has a name and a value. A list of register names can be obtained using consecutive calls to the uavcan.register.List service; one can get existing value or set a new one to a named register using the uavcan.register.Access command. The OpenCyphal implementation provides the possibility of mass installation of these parameters using a configuration file, which facilitates the configuration of the entire interacting system as a whole. It should be noted that the use of registers allows not only to set the values of the parameters of a separate algorithm, but also provides an opportunity to change the main algorithm itself on the fly by selecting one of the possible algorithms.

Evaluating the method by criterion K1, it can be noted that this method is suitable for the vast majority of embedded systems, since it does not imply modification of non-volatile memory in general or modification of only a small part of it. Many modern embedded microcontrollers and embedded systems either have the ability to write data directly to the non-volatile program memory, or contain a separate small non-volatile built-in memory area, which is designed specifically for storing parameters. If the microcontroller does not have any of these features, then such memory can be installed from the outside.

From the point of view of RAM and program memory consumption (criterion K2), this method requires an initial "investment" in the volume of implementations of the method of receiving and transmitting the messages listed earlier, the method of accessing non-volatile memory, as well as additional costs for storing both parameter names and parameters themselves. The Cyphal specification assumes a maximum message length per read and/or modification of one parameter of 515 bytes. Of these, 256 bytes are allocated to the name of the register, and 259 bytes are allocated to its value.

An additional burden also falls on the computational complexity of the most basic algorithm (criterion K3). If the parameters were set in the program text as constants known when creating machine codes, then there is scope for optimizations that are difficult or simply impossible to implement if these parameters can change during operation. This load is difficult to predict and needs to be measured in advance. On modern microcontrollers with a wide (32-bit) data bus, when using a limited number of parameters and a good optimizing compiler, we can assume a performance degradation of no more than a few percent.

From the point of view of both the developer of the control program (criterion K4) and the user (criterion K5), the use of this method turns out to be quite simple. It is enough for the developer to add another function that processes the corresponding messages. The user can use ready-made interfaces, both command-line and graphical, to configure a ready-made device without opening its source code.

The main disadvantage of this method is the lack of flexibility. It is impossible to adapt the module to a completely new scenario that was not described in the source code. For this method to work, there is no need to stop the operation of the microcontroller.

M2 – Replacing the microcontroller code

The following method of dynamic reprogramming is based on the replacement of machine code written in the program memory in the embedded system. Examples of the implementation of such an approach are discussed in [17], [18].

For this method to work with the Cyphal specification, it is necessary to implement message processing for the service of generalized command execution uavcan.node.ExecuteCommand.1.0. One of the generalized commands is COMMAND_BEGIN_SOFTWARE_UPDATE (launching program updates on an embedded system). The name of the file with the new program is specified in the request parameter for this command. In response to this command, the network node first makes a uavcan.file.GetInfo request to get information about a file with new machine codes, and then uacan.file.Read request to get this file in parts.

There are several implementations of this method:

M2.1. A separate bootloader program.
M2.2. Update inside the program.

3.2. A separate bootloader program (M2.1)

The simplest and at the same time the most difficult way to update the machine code is to use a separate loader program. This program is called by means of the embedded system itself before loading the main program and can perform basic configuration of the microcontroller, and then call the main program for execution. At the same time, the loader program has more extensive control capabilities of the embedded system, in particular, the ability to get from the outside and overwrite the control program. Not all embedded systems hardware support booting using a bootloader (criterion K1). On such devices, the work of the loader must either be replenished with software, or not to use the loader at all. Using the loader reduces the maximum amount of program memory by the size of the loader itself (criterion K2).

According to the K3 criterion, the use of the loader does not affect the operation of the main algorithm. A properly written loader completely gives control to the main subroutine, leaving no traces in RAM or setting up the entire peripherals. The main difficulty of loaders lies, from the point of view of the developer (criterion K4), in creating (a) a fairly reliable and (b) a small program. First, it is necessary that the embedded system remains online, despite errors during transmission and/or writing to the non–volatile part of the embedded system's program memory. Secondly, it is necessary that the loader can fit into a limited and small area of program memory.

From the user's point of view (criterion K5), the complexity of this method is quite high – you need to make edits directly to the source texts of programs for the embedded system, you need to know the programming language of the source texts, you need to have a programming environment and a sufficiently productive computer. One of the main problems here is creating a user–repeatable development environment with all the necessary tools.

This method provides maximum flexibility (criterion K6) in the modification and adaptation of software, but depends on the delivery of the source codes of the embedded system programs. Also a big disadvantage and advantage of this method is the need to stop and restart the microcontroller to update the program. The main algorithm should be able to respond adequately to requests for software updates.

3.3. Update inside the program (M2.2)

The method of updating inside the program and the next method – updating a part of the machine code – solve the problem of updating the software by merging the main and secondary algorithms. The update is performed in parallel with the operation of the main algorithm, the machine codes of the new program are written to a separate area of program memory, it is checked for errors and copied instead of the main program. After that, the control program is started again.

Unfortunately, not all embedded systems can use this method, since it requires the ability to change program memory during operation (criterion K1). This method uses program memory somewhat more efficiently (criterion K2) by using the same communication mechanisms as the main program. However, the main disadvantage of this method is the maximum amount of program memory that the program can use – it is reduced by half so that at any time the update algorithm can write another piece of update to the program memory.

The operation of the update algorithm does not take additional processor time (criterion K3) in standby mode, and in update mode it can work with low priority without interfering with the real-time mode of the main program. From the point of view of the developer (criterion K4), the implementation of this method is not particularly difficult – it is enough to create an additional command handler and a separate "thread" that performs the update. But, compared with a separate loader, the complexity is lower due to lower requirements for the amount of code and the ability to share the code of network interaction with the main program.

For the user (criterion K5), the use of this method is comparable in complexity to the complexity of the loader method – you also need to get a properly configured development environment, know and be able to use the programming language in which the source texts are written. However, due to the limitation on the size of the firmware, the user may also be forced to perform additional optimization of the program so that it can fit into the program memory area reduced by half. The flexibility of application (criterion K6) of this method is higher than that of the method with a loader due to the shorter downtime of the microcontroller - long operations for obtaining new firmware are integrated into the main algorithm, the main downtime occurs at the time of copying and restarting the control program.

3.4. M3 – Bytecode replacement

The bytecode replacement method is based on the fact that on the embedded system, instead of the main program, a program interpreter of codes other than the machine code of the embedded system itself, i.e. a bytecode program processor, is launched. At the moment, there are many different bytecode software processors with their own features. An example of such a software processor is given in [19].

This method can use intermodule communication methods according to the Cyphal specification by analogy with the two previous methods. In the case of parametric reprogramming, the bytecode of the control program is written to a parameter (register) of the "unstructured array" type. If a file update is used (discussed in the last section), then instead of updating machine codes, the program memory area responsible for storing the bytecode is updated.

In terms of performance, the bytecode program processor will always be a multiple of slower than similar code in machine codes. The software processor itself can take up significant amounts of program and data memory. A small improvement in the speed of work can be achieved by transferring complex computational operations to the machine code of the software processor. Because of this, this method may not be applicable on embedded systems with a narrow (8 and 16 bit) data bus and small (up to 1 kB) amounts of program and data memory (criteria K1 to 3).

From the point of view of the software developer (criterion K4), in this method, the main difficulty is the selection and transfer of an existing software processor to an embedded system, as well as the development of additional mechanisms for the interaction of this software processor with the peripheral devices of the microcontroller. Also, the developer must provide the user with a translator program of a high-level language in the bytecode of the program processor. If the bytecode of an existing interpreted programming language is used, then this task is significantly simplified. For the user (criterion K5), this method turns out to be easier than using machine code replacement (M2), but requires knowledge and use of some high-level language or even programming in bytecodes. However, the development cycle is significantly simplified, since a translator program from high-level languages can be much easier to set up and install than toolkits for working with the machine code of an embedded system.

The flexibility (criterion K6) of this method is significantly lower than that of the machine code replacement method due to the smaller volume of possible algorithms: a complex algorithm in bytecode may not meet the speed limit in real time, and the same algorithm converted into machine code and integrated into a software processor may not give that flexibility. adjustments that would give a complete replacement of machine codes. Because of this, the developer has to balance the algorithms of the main program (bytecode or machine code) at the place of implementation, and the user has to put up with this balance when adapting their algorithms.

3.5. M4 - Replacing a script in an interpreted programming language

This method is based on the bytecode replacement method and has similar characteristics. In this method, not a bytecode is sent to the embedded system, but the source code in a high-level language, which is then executed directly or using translation into bytecode. An example of such an interpreter is given in [19]. This method requires higher performance of the embedded system – it is necessary to perform the same or a larger amount of calculations compared to the bytecode program processor (criteria K1, K3). Due to the need to parse the text of the programming language for direct interpretation or conversion into bytecode, the volume of machine code and the amount of data increases. Also, due to the use of text as a program, the memory requirements for interpreted programs increase (criterion K2).

In comparison with the bytecode, the shoulders of the developer (criterion K4) also bear the transfer and optimization of the translator program on the embedded system. Also, this method increases the requirements for error handling and reporting mechanisms – the user must receive messages if any syntactic or logical errors are made in his program. From the user's point of view (criterion K5), this method is somewhat more complicated than parametric reprogramming, since it implies knowledge of a high-level programming language. However, this method is much simpler than the bytecode replacement method – there is no need to install a translator program. In terms of flexibility (criterion K6), this method is comparable to the bytecode replacement method. However, due to the need to work with the textual representation of commands, the developer has to search even harder for a balance between algorithms embedded in machine code and algorithms provided by the user.

4. Conclusion

At the moment, there are many different ways of dynamic reprogramming. Among those presented in the article, the easiest to use is the M1 method – parametric reprogramming. The method is implemented on any embedded systems, unlike other methods that require either hardware support for some functions (replacement of machine code – M2), and/or a large amount of program and data memory (replacement of bytecode (M3), or programs in an interpreted programming language (M4)), but it is the least flexible. The simplest methods for the user are M1 and M4, since they do not require special software or additional knowledge. The M2 method is ideal for the use of embedded systems with the smallest amount of program and data memory. The M3 method is good if the flexibility of the M4 method is needed, but the use of an embedded system that does not have the appropriate parameters does not allow using M4. In addition to the analysis of the listed methods according to six criteria, it is planned in the future to experimentally test the feasibility of each method on microcontrollers of the STM32F103, GD32VF103, ESP32 families and Raspberry Pi and Orange Pi microcomputers.

5. Acknowledgments

Research is supported by the Russian Foundation for Basic Research: Grant 19-07-00892a.

6. References

[1] Popov E.P. & Pismennyi G.V. (1990) Fundamentals of Robotics: Introduction to the specialty [Основы робототехники: введение в специальность], Higher School [Высшая школа] – 224 p.

[2] Lopota A.V. & Yurevich E.I. (2013) Stages and prospects of development of the modular principle of building robotic systems [Этапы и перспективы развития модульного принципа построения робототехнических систем], Scientific and technical bulletin of SPbSPU [Научно-технические ведомости СПбГПУ], Vol 1 2013 Computer science. Telecommunications. Management.[Информатика. Телекоммуникации. Управление.], pp. 98 - 103, ISSN 1994-2354.

[3] Platonov A.K. (2010) Robotics for a lunar base [Робототехника лунной базы], XXXIV Readings on cosmonautics KIAM RAS [XXXIV Чтения по космонавтике. ИПМ им. М.В. Келдыша РАН].

[4] ISO Central Secretary, (2021). Robotics – Modularity for service robots – Part 1: General requirements (Standard No. ISO/IEC TR 22166-1:2021). International Organization for Standardization, Geneva, CH.

[5] Andreev, V.; Kim, V. & Eprikov, S. ModRob: the Hardware-Software Framework for Modular Mobile Robots Prototyping // Proceedings of the 31st DAAAM International Symposium, B. Katalinic (Ed.), Published by DAAAM International, ISBN 978-3-902734-29-7, ISSN 1726-9679, Vienna, Austria, 2020, pp.0391-0402.

[6] Andreev, V.; Kim, V. & Pletenev, P. The principle of full functionality – the basis for rapid reconfiguration in heterogeneous modular mobile robots // Annals of DAAAM and Proceedings of the 28th International DAAAM Symposium on Intelligent Manufacturing and Automation, DAAAM 2017; Zadar; Croatia,B. Katalinic (Ed.), Published by DAAAM International, ISBN 9781510853270, ISSN 1726-9679, Vienna, Austria. Curran Associates, Inc. (Feb 2018), pp. 0023-0028.

[7] Andreev V. & Pletenev P. Organizing Intermodular Communication for Heterogeneous Modular Mobile Robot. Annals of DAAAM and Proceedings of the 28th International DAAAM Symposium on Intelligent Manufacturing and Automation, DAAAM 2017; Zadar; Croatia,B. Katalinic (Ed.), Published by DAAAM International, ISBN 978 1 510853 27 0, ISSN 1726-9679, Vienna, Austria. Curran Associates, Inc. (Feb 2018), pp. 0474-0480.

[8] Andreev V. Control System Mobile Robots with Modular Architecture as a Multi-Agent System with a Hierarchical Topology, Proceedings of the 30th DAAAM International Symposium, B. Katalinic (Ed.), Published by DAAAM International, ISBN 978-3-902734-22-8, ISSN 1726-9679, Vienna, Austria, 2019, pp.0010-0019.

[9] Herbrechtsmeier, S.; Korthals, T.; Schopping, T. & Ruckert, U. (2016). AMiRo: a modular & customizable open-source mini robot platform. 20th International Conference on System Theory, Control and Computing (ICSTCC), Sinaia. – 2016. – pp.687-692.

[10] Guifang Qiao; Guangming Song; Jun Zhang; Hongtao Sun; Weiguo Wang & Aiguo Song (2012). Design of Transmote: a Modular Self-Reconfigurable Robot with Versatile Transformation Capabilities, Proceedings of the 2012 IEEE International Conference on Robotics and Biomimetics, 2012. – pp.1331-1336.

[11] Baca J., Ferre M. & Aracil R. (2012) A heterogeneous modular robotic design for fast response to a diversity of tasks, Robotics and Autonomous Systems. vol. 60. no. 4. pp. 522–531.

[12] Bonarini, A.; Matteucci, M.; Migliavacca, M .& Rizzi, D. (2014). R2P: An open source hardware and software modular approach to robot prototyping. Robotics and Autonomous Systems. – 2014. – No.62. – pp.1073-1084.

[13] Losada D. P., Fernández J. L., Paz E. & Sanz R. (2017) Distributed and modular CAN-based architecture for hardware control and sensor data integration, Sensors. No.17. – pp.1013-1030.

[14] Peng, L.; Guan, F.; Perneel, L.; Fayyad-Kazan, H. & Timmerman M. (2016) EmSBoT: A lightweight modular software framework for networked robotic systems. 2016 3rd International Conference on Advances in Computational Tools for Engineering Applications (ACTEA), Beirut, 2016, pp.216-221. doi: 10.1109/ACTEA.2016.7560142.

[15] Andreev, V. & Pletenev, P. (2021) Problems of Choosing an Intermodule Information Interaction Protocol for Mobile Robots with Modular Control System Architecture, Proceedings of the 32nd DAAAM International Symposium, pp.0151-0157, B. Katalinic (Ed.), Published by DAAAM International, ISBN 978-3-902734-33-4, ISSN 1726-9679, Vienna, Austria.

[16] Kirienko, P.; Dixon, S. & others. (2022) OpenCyphal: Open technology for real-time intravehicular distributed computing and communication based on modern networking standards. URL: https://opencyphal.org/specification/Cyphal_Specification.pdf (accessed 10.09.22).

[17] Lobdell M. (2012) Robust over-the-air firmware updates using program flash memory swap on kinetis microcontrollers, Freescale Application Note, p. AN4533. 2012.

[18] Jaouhari S. E. & Bouvet E. (2022) Secure firmware Over-The-Air updates for IoT: Survey, challenges, and discussions, Internet of Things. 2022. (18). C. 100508.

[19] Zandberg K. & Baccelli E. (2020) Minimal virtual machines on IoT microcontrollers: The case of berkeley packet filters with rbpf, arXiv preprint arXiv:2011.12047. 2020.

DOI: 10.2507/33nd.daaam.proceedings.029

ANALYSIS OF ACUTE MYOCARDIAL INFARCTION USING ECG SIGNALS AND MACHINE LEARNING ALGORITHMS

Vyacheslav V. Potekhin & Ogul Unal
Peter the Great St. Petersburg Polytechnic University

This Publication has to be referred as: Potekhin, V[yacheslav] & Unal, O[gul] (2022). Analysis of Acute Myocardial Infarction using ECG Signals and Machine Learning Algorithms, Proceedings of the 33rd DAAAM International Symposium, pp.0205-0214, B. Katalinic (Ed.), Published by DAAAM International, ISBN 978-3-902734-36-5, ISSN 1726-9679, Vienna, Austria
DOI: 10.2507/33rd.daaam.proceedings.029

Abstract

Heart attack, alternatively known as myocardial infarction (MI), covers various conditions that impact the heart and is one of the most common causes of death worldwide. Electrocardiogram (ECG) can be used as a way to examine the functionality of the cardiovascular system. Researchers apply several machine learning techniques to analyze medical data, helping healthcare professionals to predict heart diseases. The focus of most studies has been on classifying heartbeats or classifying healthy ECG signals on a dataset. In this paper, we propose two methods based on machine learning algorithms for the classification of healthy subjects, subjects who have a myocardial infarction, and analysis of ST elevated (STEMI) or non-ST elevated MI (NSTEMI). We evaluated the proposed method on PhysionNet's PTB Diagnostics dataset. The suggested method can make predictions with accuracies of 95.29% and 82.93% on healthy-MI classification and STEMI-NSTEMI classification, respectively.

Keywords: Machine Learning; Supervised Learning; Electrocardiogram; Myocardial Infarction; Neural Networks

1. Introduction

MI is defined as myocardial cell death due to prolonged ischemia in pathology. Electrocardiography, according to Britannica is a method of graphic tracing of the electric current generated by the heart muscle during a heartbeat [3]. Global Burden of Disease Study 2017 published that the cardiovascular diseases were responsible for 31.8% of all deaths worldwide [21]. The mortality rate of Acute MI remains high, and most deaths occur outside of the hospital. Cardiovascular heart disease represents the leading cause of death in both men and women older than 65 years. Myocardial infarction (MI) can be recognized or defined as pathological by clinical features using electrocardiographic (ECG) findings, elevation of biochemical markers of myocardial necrosis, and imaging [10]. Machine Learning is a popular method to predict heart diseases, heartbeat classification or prediction of MI [19] [20]. The most challenging problem with the analysis of ECG signals, as with other time-series data, is the difficulty of classifying or detecting different morphology and waveforms in the signal. This is a time-consuming and challenging task that can cause errors. The medical data can be analyzed by different machine learning algorithms [23].

Datasets in various fields [22] such as in medicine can be used as training data therefore computers can learn to classify healthy subjects or subjects who have MI. In this research, we used PTB Diagnostic ECG Database. Machine learning is a method that a computer gradually learns what parameters or weights fit the data best. In this study, two classification models were developed for Acute MI and also for its subcategories. Results of the algorithms were compared. In this paper, we propose two different machine learning methods for ECG signals that can analyze ECG signal to classify the subcategories of MI (STEMI and NSTEMI). There are other heart diseases that can be detected utilizing ECG signals. In future work, we will further expand this research incorporating other machine learning techniques and heart diseases such as myocarditis or dysrhythmia and different features related to ECG.

2. Literature Review

2.1. Anatomy of the Heart

The heart is an organ that serves as a pump to circulate blood [1] situated between the two lungs. The heart consists of several layers of a muscle wall - the myocardium. A thin layer of tissue, the pericardium, covers the outer part, and another layer, the endocardium, lines the inner part. The heart cavity is divided in the middle into the right and left heart, which, in turn, are divided into two chambers. The upper chamber is called the atrium and the lower chamber is called the ventricle. The heart consists of four chambers: the atrium (right and left atria) and the ventricle (left and right ventricles) on the left and right. The mitral valve separates the left atrium and ventricle and the tricuspid valve separates the right atrium and ventricle [2].

The heartbeat is caused by changing contractions and relaxations of the myocardium. These contractions are stimulated by electrical impulses from a natural pacemaker, the sinoatrial, or SA, node located in the muscle of the right atrium [1]. The impulse from the SA node causes the two atria to contract, pushing blood into the ventricles. Ventricular contraction is controlled by impulses from the atrioventricular, or AV, node located at the junction of the two atria [1]. After contraction, the ventricles relax then the pressure inside the ventricles drops. The blood enters the atria again, and the pulse from S-A starts the cycle again. This process is called the cardiac cycle. The stimulus coming out of the sinus node spreads through both atria of the heart, along the conduction pathways specialized for this job, and with this stimulus, the atria contract and send the blood to the ventricles. Then the signal is sent to a certain area between the atria and the ventricles; comes to the atrioventricular (AV) node. The electrical message is held briefly in the atrioventricular node before being delivered to the ventricles. Thus, simultaneous contraction of the atria and ventricles is prevented.

The heart, by contracting and pumping the blood is caused by electrical currents. These currents are in the around millivolts [2]. These currents can be recorded with the help of an electrocardiogram or ECG by being amplified by specific devices.

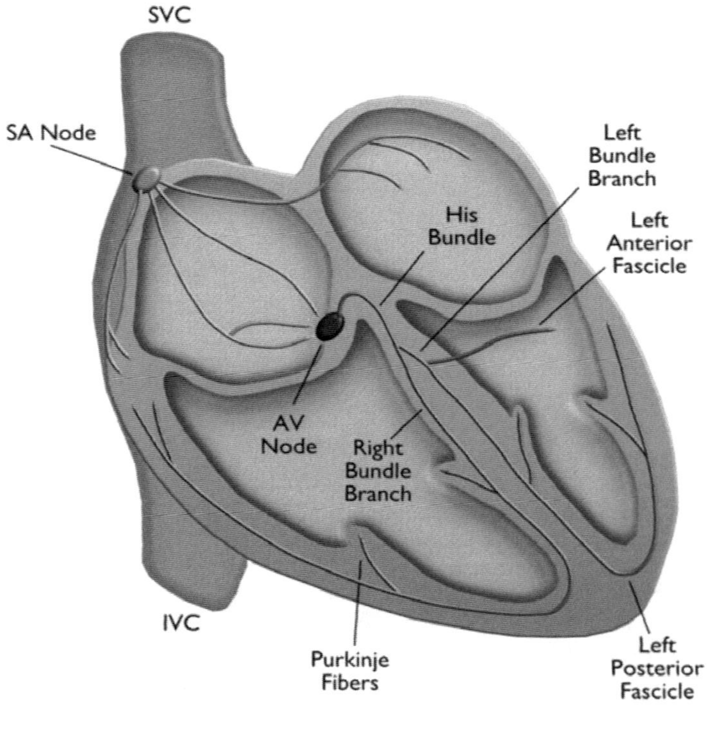

Fig. 1. Cardiac conduction system of heart

2.2. ECG

Dutch physiologist Willem Einthoven developed the first electrocardiogram in 1903, and for many years the tracing was called an EKG after the German Elektrokardiogramm [3]. ECG records of electrical changes that occur during the heartbeat can be used in the diagnosis of heart failure or detecting anomalies. Each heart cycle produces ECG waves called P, QRS, and T.

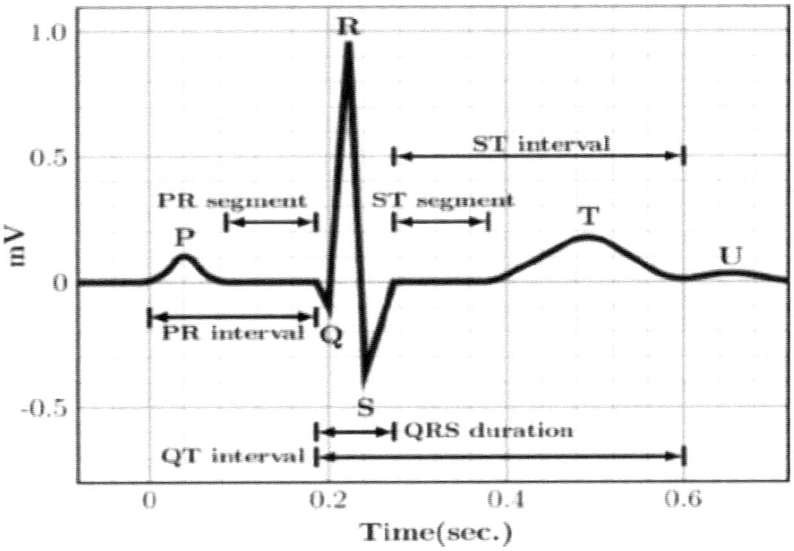

Fig. 2. Healthy ECG signal shown with PQRST complexes

Each wave or segment of the ECG corresponds to a certain event of the cardiac electrical cycle [4]. When the atria are full of blood, the SA node fires, and electrical signals spread throughout the atria and cause depolarization. This is represented by the P wave. Atrial contraction starts after the P wave begins. The PR interval is the time between the first deviation of the P wave and the first deviation of the QRS complex [5]. The PQ segment represents the time of signals travel from the SA node to the AV node. The QRS complex indicates the firing of the AV node and represents the ventricular depolarization:

- Q wave corresponds to depolarization of the interventricular septum.
- R wave is produced by depolarization of the main mass of the ventricles.
- S wave represents the last phase of ventricular depolarization at the base of the heart.

Atrial repolarization occurs during this time however the signal is not visible because of the large QRS complex. ST-segment reflects the plateau in the myocardial action potential [4]. This is when the ventricles contract and pump blood. The ST segment is the time interval between the end of the QRS complex and the start of the T wave. The T wave represents ventricular repolarization immediately before ventricular relaxation [6]. Thus the cycle repeats itself with every heartbeat.

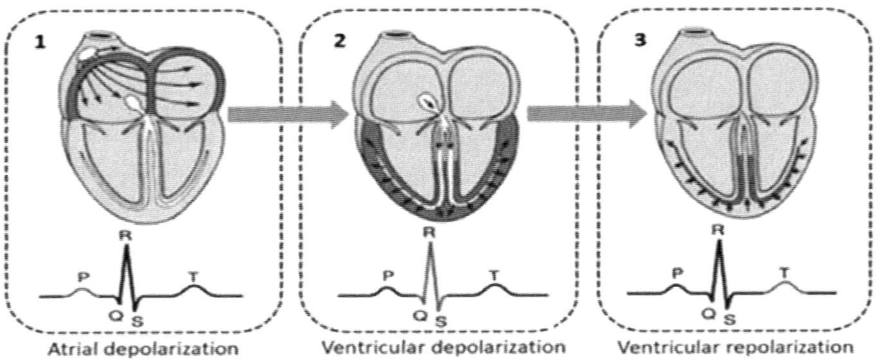

Fig. 3. Stages of depolarization and repolarization

The 12-lead ECG provides spatial information about the electrical activity of the heart in 3 orthogonal directions. Each of the leads represents a particular orientation in space [7].

- Bipolar limb leads are; Lead 1(Right Arm(-) to Left Arm(+)), Lead 2 (Right Arm(-) to Left Leg(+) and Lead 3 (Left Arm(-) to Left Leg(+)).
- Augmented unipolar limb leads are; Lead aVR (Right Arm (+) to (Left Arm & Left Leg(-)), Lead aVL (Left Arm(+) to (Right Arm & Left Leg(-)), Lead aVF (Left Leg to (Right Arm & Left Arm(-))
- Unipolar chest leads are; V1, V2, V3 (posterior anterior) and V4, V5, V6 (Right, Left or Lateral)

In a 12-lead ECG, there are 12 leads calculated using 10 electrodes. The last electrode is placed on the right leg which is earth. The 12-lead ECG provides more information from a larger surface area surrounding the heart [8].

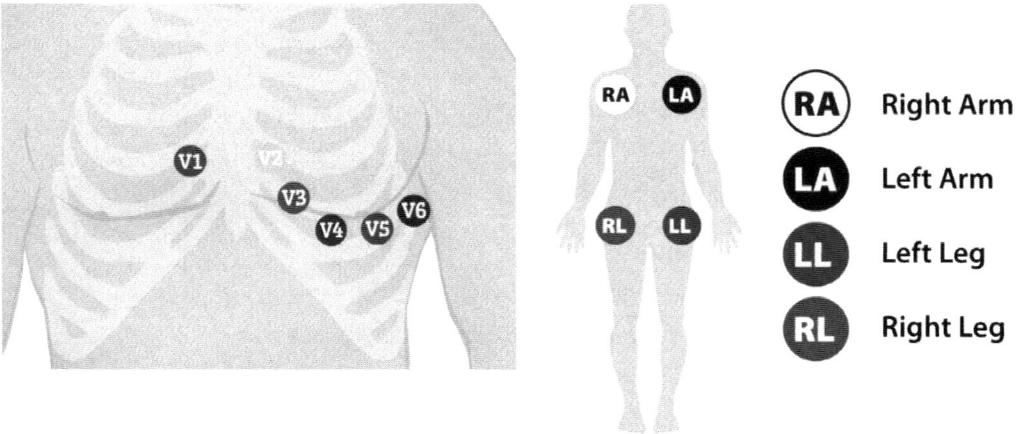

Fig. 4. Electrodes and their placement

2.3. Myocardial Infarction

The myocardium is heart muscle, myocardial infarction is defined pathologically as myocyte necrosis due to prolonged ischemia [9]. Ischaemia is the condition of restriction or reduction of blood flow in some part of the body. The heart muscle begins to die due to lack of oxygen caused by a blockage in the arteries (blood supply for the heart). Therefore it is crucial to restoring blood flow to prevent permanent damage to heart tissue [10]. Non-ST-segment elevation myocardial infarction(NSTEMI) is a type of MI that occurs on the inner layer of the ventricular wall. If the ECG Electrode sees the damaged area after the intact myocardial tissue in between, the ST segment collapses. Thus, ECG changes may include ST-segment depression or T segment inversion. ST-segment elevation myocardial infarction (STEMI) extends the entire thickness of the myocardium. ECG shows the ST elevated results because If the electrode directly sees the damaged area, the ST segment is elevated. In this research, Normal, STEMI, and NSTEMI (with ST depression) were used [18].

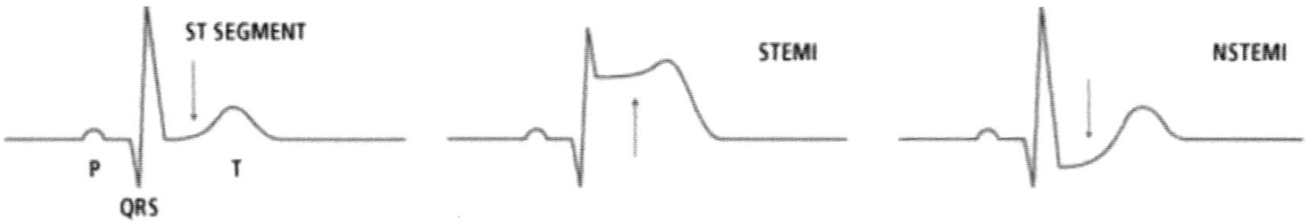

Fig. 5. Healthy, STEMI and NSTEMI ECG changes

3. Methods & Materials

3.1. PTB Dataset

The database contains 549 records from 290 subjects (aged 17 to 87) Each subject is represented by one to five records [11].The conventional 12 leads (i, ii, iii, aVR, aVL, aVF, v1, v2, v3, v4, v5, v6) together with the 3 Frank lead ECGs (vx, vy, vz). Each signal is digitized at 1000 samples per second, with a 16-bit resolution over a range of ± 16.384 mV [11]. The dataset was separated into two classes.

20 Healthy and 20 subjects with MI were selected. ECG signals with only acute MI were used in this research. The Myocardial Infarction class was separated into two sub-classes as STEMI and NSTEMI. ECG signals and leads chosen in this research have been approved by a medical expert to create correct dataset labels for classification algorithms. Leads 1, 2, aVF, v2, v3, v4 have been chosen to classify healthy or MI subjects. Afterward, leads v1, v2, v3, and v4 were used to classify STEMI and NSTEMI ECG signals. The dataset is balanced and contains six STEMI and seven NSTEMI ECG signals. Python waveform-database (WFDB) package was used to visualize results (4500 ms for visualization below).

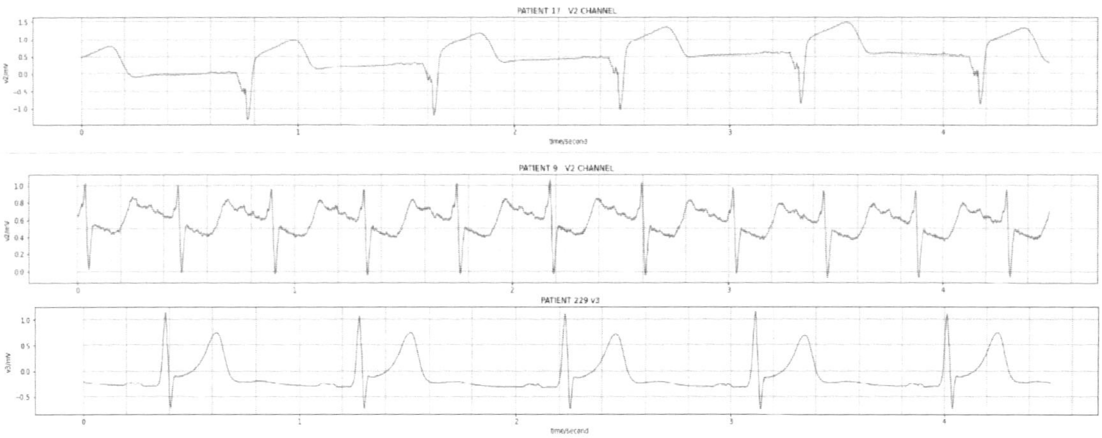

Fig. 6. ECG signals are shown as STEMI, NSTEMI, and non-MI respectively

3.2. Preprocessing

Raw ECG signals were preprocessed using delineation on signals. The wavelet algorithm was used to delineate the ECG dataset. The continuous-time signal is the "function of continuous-time variable that has an uncountable or infinite set of numbers in its sequence.

Wavelet transform is a mathematical approach widely used for signal processing applications. It can decompose special patterns hidden in the mass of data. Wavelet transform has the ability to simultaneously display functions and manifest their local characteristics in the time-frequency domain. CWT (Continuous Wavelet Transform) is where the scale and translation are set arbitrarily [12]. CWT (1) provides a description of the signal in the time-scale domain, allowing the representation of the temporal features of a signal at different resolutions; therefore, it is a suitable tool to analyze the ECG signal [13].

$$CWT(\alpha,\tau) = \int_{-\infty}^{\infty} f(x)\frac{1}{\alpha}\ \ \Psi*\left(\frac{x-\tau}{a}\right)dx \tag{1}$$

The wavelet transform is a mathematical technique that can decompose a signal into multiple lower resolution levels by controlling the scaling and shifting factors of a single wavelet function.

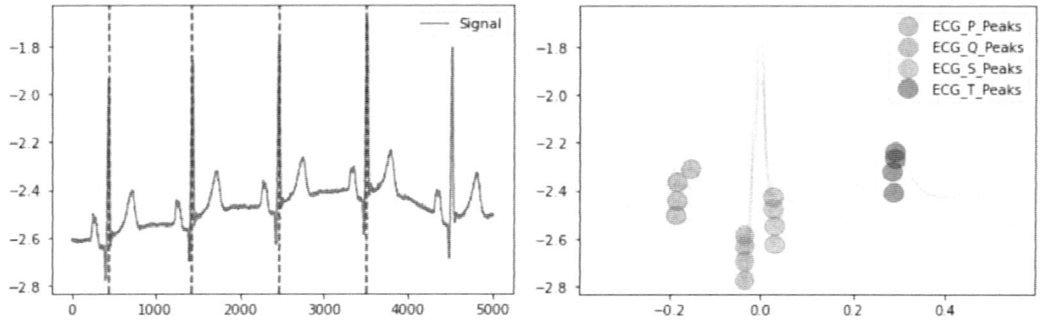

Fig. 7. Detected R peaks and delineated ECG signals using CWT (1000 sampling rate)

To identify MI S-T differences were detected and calculated. Extracted S-T onset and S-T-peak differences were used for related ECG leads. (1, 2, aVF, v2, v3, v4 for healthy-MI classification, v1, v2, v3, v4 for STEMI-NSTEMI).

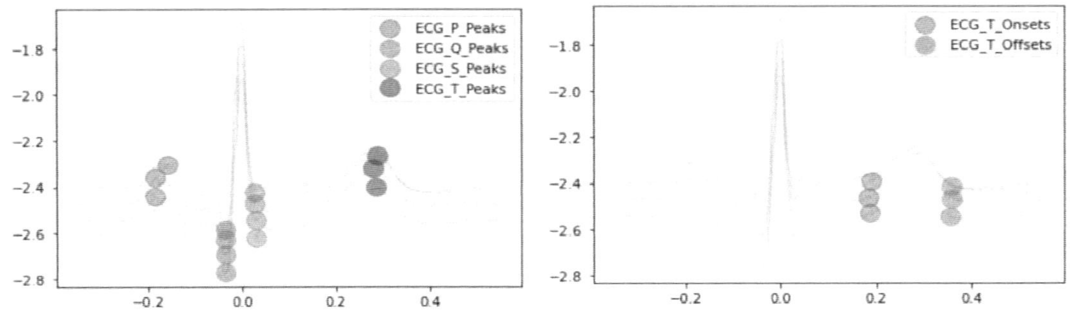

Fig. 8. S peaks, T peaks, and T onsets were extracted

After applying feature extraction new dataset that contains missing values was created, missing values in the dataset were filled with zeroes, and KNN (K nearest neighbor) imputer was used. The KNN is not robust to the input matrix with missing values, which may lead to the loss of effectiveness. [14] Therefore missing values were replaced with zeroes.

3.3. Algorithms

Machine learning is a process that which the computer gradually learns what parameters or weights fit the training data best. In this research, Support Vector Machine, Dense Neural Network using Keras and Torch, Random Forest, and XGBoost were used.

Support Vector Machine

SVMs have been used to solve many pattern recognition and classification problems and have been applied to the problems of prediction, forecasting, and implementing intelligent machines. The main goal of SVM is to find the optimal hyperplane for correct classification between data points of different classes (Figure 10). The dimension of the hyperplane is equal to the number of input objects minus one [15]. Data points on one side of the hyperplane will be classified for a certain class, while data points on the other side of the hyperplane will be classified for another class. The distance between the hyperplane and the first point (for all different classes) on both sides of the hyperplane is a measure of the algorithm's confidence in its classification decision. The greater the distance and the more confident we can be, SVM makes the right decision [15]. The data points closest to the hyperplane are called support vectors. The reference vectors determine the orientation and position of the hyperplane to maximize the classifier margin [15]. The number of support vectors that the SVM algorithm should use can be arbitrarily selected depending on the applications.

Neural Networks

Neural networks simulate the work of the human brain, allowing computer programs to find patterns and solve standard tasks in the field of AI, ML and deep learning. Neural Networks provide nonlinearity using activation functions (Relu and Sigmoid in this research). Keras provides a Python interface for artificial neural networks. Torch provides modules and classes torch.nn, torch.optim, Dataset, and DataLoader to create and train neural networks. Adam optimizer, Binary-Cross-Entrophy loss function, Dropout=0.2, L2 regularization have been used. Dataset was split into training (70%), testing (15%), and validation (15%) sets. Drop out and regularization was applied to prevent overfitting.

XGBoost

XGBoost is an implementation of decision trees with improved (boosting in ensemble learning) gradient, designed to improve performance and speed [16]. The model implementation supports the functions of the scikit-learn and R implementations with regularization. It is possible to overfit the model using XGBoost therefore, four different algorithms were used in this research.

In the beginning, hyperparameters were optimized using optuna library automatically. Afterward, they were manually adjusted to get better results. Chosen hyperparameters are: n_estimators: 200, max_depth: 5, reg_alpha: 3, reg_lambda: 2, min_child_weight: 1, gamma: 10, learning_rate: 0.01, colsample_bytree: 0.64

Random Forest

RF is an ensemble learning method, namely a random forest model consisting of a large number of small decision trees (estimators) that creates their own predictions. The RF combines the result to obtain accuracy (bagging in ensemble learning). A random forest builds trees in parallel, while boosting builts trees sequentially [17]. Hyperparameters chosen for RF algorithm: max_depth = 5, max_features=4, min_samples_leaf=2.

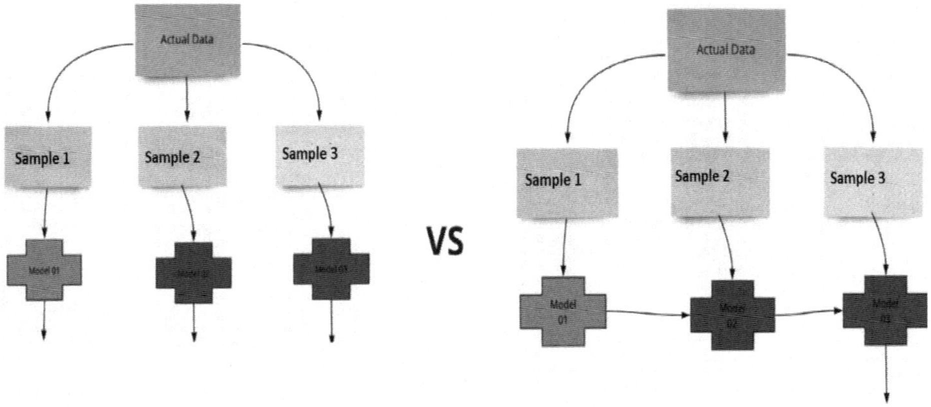

Fig. 9. Random Forest (bagging) and XGBoost (boosting)

4. Results

In this research, four different algorithms have been used. Same algorithms were applied to healthy-MI classification and STEMI-NSTEMI classification. For each algorithm, test accuracies and their confusion matrix are shown. Each figure has Predicted values for Healthy or MI binary classification on the left. On the right, classification results for STEMI or NSTEMI are shown. The XGBoost algorithm gave the highest accuracy for Healthy or MI classification (95.29). In the second classification (STEMI-NSTEMI). Random Forest gave the highest accuracy with 82.93%. Finally, SVM gave the worst results having 75.95% for the first and 70.03% accuracy for the second classification. Before using any algorithm, all ECG signals were preprocessed (delineated with CWT). S – T onset and S – T peaks were extracted and used as new features. For Healthy-MI classification, 12 features were used as; S-T differences of 1, 2, aVF, v2, v3, and v4. Eight features (S-T differences of v1, v2, v3, and v4 for STEMI-NSTEMI) were extracted. After preprocessing, missing values were replaced with zeroes.

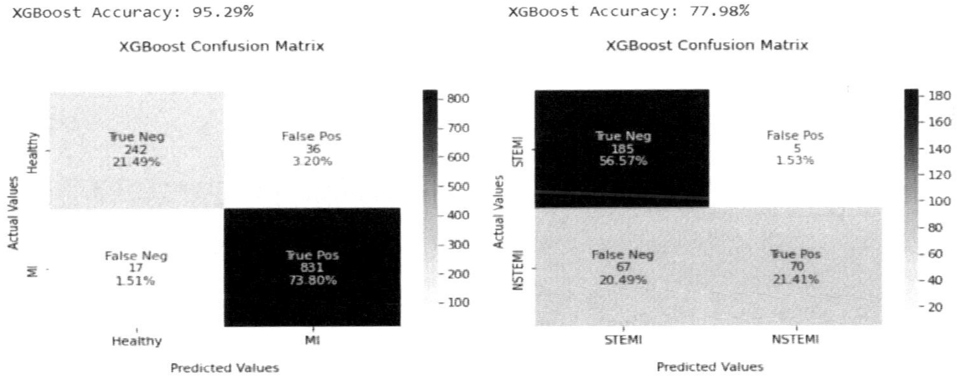

Fig. 10. Xgboost test accuracies and confusion matrix

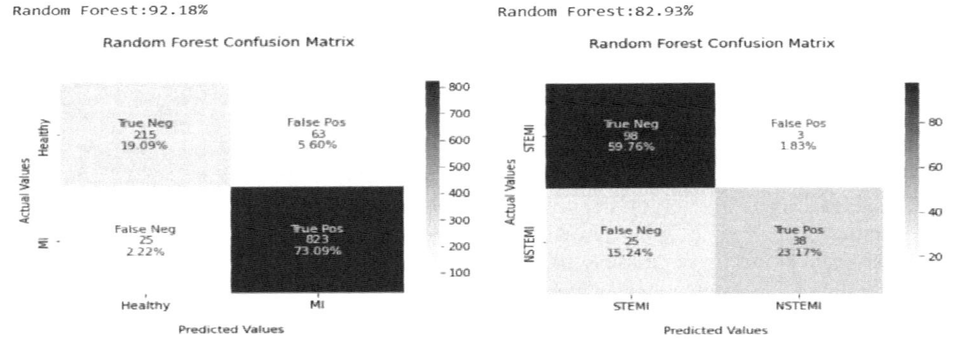

Fig. 11. Random Forest results

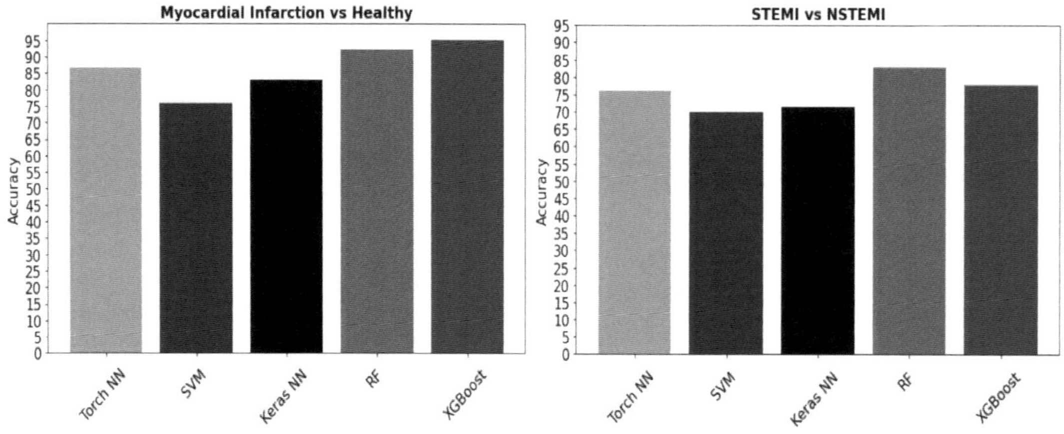

Fig. 12. Test accuracies for each algorithm

5. Discussion

In this research we analyzed Myocardial Infarction ECG signals to apply machine learning for classification. Data for MI represents Acute MI syndrome that occurred in the anterior part of the heart (formed mainly by the right atrium and ventricle). Thus Leads 1, 2, aVF, v2, v3, v4, and leads v1, v2, v3, v4 were chosen for both classification tasks. Chosen ECG signals were approved by a medical expert.

S-T differences differ in healthy subjects and subjects who have MI. Thus, extracted differences were used in the dataset for each chosen ECG lead. S-T differences might represent MI because electrical current will be able to bypass the necrosed area causing ST-segment to increase. Intact myocardial tissue in between the electrode damaged area causing the S-T segment to collapse. To classify STEMI and NSTEMI same approach was applied. In this case, S-T elevation or depression represents MI, as S-T elevation means STEMI and S-T depression means NSTEMI (sub-categories of MI).

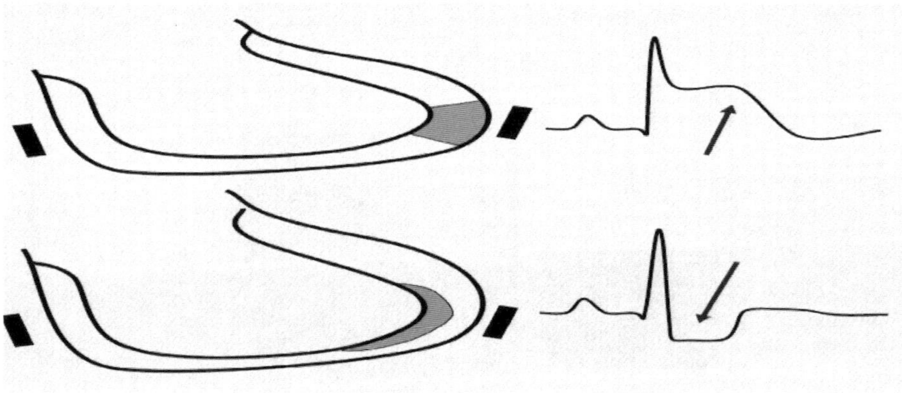

Fig. 13. Necroses (darker color on tissue) cause STEMI (above) and NSTEMI (below) ECG ST differences

The created dataset was in numeric format (contains missing values) and applied to several different machine learning algorithms. In both cases, the Neural network did not perform better than Xgboost and the Random forest algorithm due to having medium-sized data. Neural networks can perform better when utilizing bigger datasets and Ensemble learning can be more effective in this situation. To prevent overfitting, validation data (0.15) and a dropout value of 0.2 were chosen. Without overfitting, the neural network was able to train the model with around 80% accuracy, on the other hand, ensemble learning algorithms trained with 92% (shown in the previous section). All test accuracies drop approximately 10% when classifying STEMI-NSTEMI. The second classification has drawbacks as it is difficult to find appropriate ECG signals for the STEMI-NSTEMI classification. Model for healthy and MI ECG signals gave higher accuracies because S-T differences, in this case, were clearer in features. Another difficulty was during the training using ensemble learning algorithms, it is possible to overfit the model, and to prevent this, hyperparameters were tuned manually. The misuse of the max_depth parameter can cause overfitting even with shuffled validation sets. When the max_depth is greater than 5 for healthy-MI and more than 3 for STEMI-NSTEMI causing an overfitting model. As shown below, Log-loss was higher in the latter classification. As a result, lower ratios for colsample_bytree, training instances, lower values of max_depth, larger values of gamma, and lower values of learning rate were chosen to avoid over-fitting.

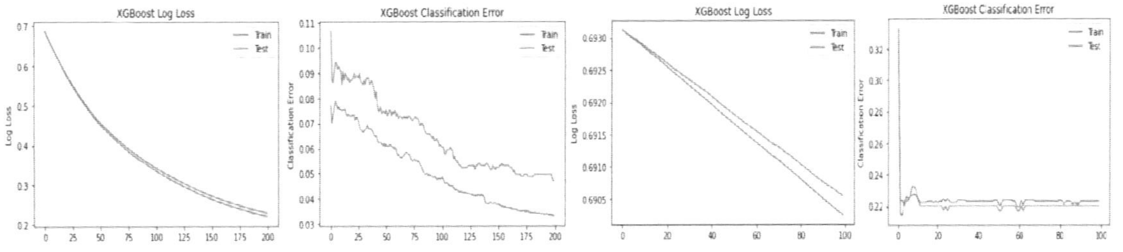

Fig. 14. Xgboost logloss and classification error for healthy-MI and STEMI-NSTEMI respectively

6. Conclusion

In this paper anatomy of the heart and characteristics of ECG signals were analyzed. Heart muscles work together to pump blood throughout our body. Every heartbeat can be recorded and compared to detect heart diseases. S-T segments in PQRST, represent either non-ST elevated or ST elevated Myocardial Infarction. Continuous wavelet transform were used to prepare new features for the dataset. S-T differences were extracted and used in two different ML models to classify healthy and MI subjects. Latter, S-T elevated and, Non-S-T elevated ECG signals were classified using ML algorithms. XGBoost and Random Forest gave high accuracies on test data. To evaluate the model, parameters were chosen manually and tested to prevent overfitting the ML models. Using ECG signals, it is possible to characterize signals because feature extraction can be performed with the help of PQRST locations and CWT. Binary classification models for healthy-MI gave 95.29%, model for STEMI-NSTEMI had 82.93% accuracy.

Many patients suffer from acute MI. Thus it can be helpful to detect it using a machine. Detecting S-T and Non-S-T elevated MI can be useful to classify different types of systemic heart diseases. The proposed methodology has achieved higher accuracy in healthy-MI classification. Considering the limitations of this study, there is a need to implement more complex models and use combination of datasets and features to increase the accuracy and classify other heart diseases.

7. Acknowledgments

Firstly, I would like to thank my father Dr. Turgay Ünal for his help in ECG studies and data preparation, also would like to thank Associate professor Vyacheslav Potekhin for all his help during my studies. Finally, I thank my mom Dr. Seher Gülçin Ünal for her assistance in giving all the support in my life. This research is being conducted at the Peter the Great St. Petersburg Polytechnic University and partially funded by the Ministry of Science and Higher Education of the Russian Federation as part of World-class Research Center programme: Advanced Digital Technologies (contract No. 075-15-2020-934 dated 17.11.2020).

8. References

[1] https://www.britannica.com/science/heart, (2021). Britannica, heart Accessed on: 2022-03-21

[2] Robert, W. (2006). Anatomy of the Heart, Medicine, Vol. 34, No. 5, pp. 163-165., doi: 10.1383/medc.2006.34.5.163

[3] https://www.britannica.com/science/electrocardiography, (2020), Britannica, electrocardiography Accessed on: 2022-04-03

[4] Tae W., Kee K. (2020). ECG PQRST complex detector and heart rate variability analysis using temporal characteristics of fiducial points, Biomedical Signal Processing and Control, Vol. 66, pp. 1-7., ISSN 1746-8094, doi: 10.1016/j.bspc.2020.102291

[5] Medina, V., Gonzalez, R., & Echeverria, C. (2005). Effect of noise sources on the averaged PQRST morphology, Computers in Cardiology, pp. 743-746., doi: 10.1109/CIC.2005.1588211

[6] Sabar, S., Djanggan, S., Setyawan, P. & Sakti, S. (2016). The peak of the PQRST and the trajectory path of each cycle of the ECG 12-lead wave, Indonesian Journal of Electrical Engineering and Computer Science, Vol. 4, pp. 169-175., doi: 10.11591/ijeecs. v4.i1.pp169-175

[7] Houghton, R.; Gray, D. (2012). Making Sense of the ECG, Hodder Education. ISBN 978-1-4441-6654-5

[8] Jin, B., Wulff, H., Widdicombe, J., & Zheng, J., Bers D., & Puglisi, J. (2012). A Simple Device to Illustrate the, Einthoven Triangle, Advances in physiology education, doi: 10.1152/advan.00029.2012

[9] Kristian, T., Hygesen, S., Joseph, A., Allan J., & Harvey, W. (2018). The universal definition of myocardial infarction, The ESC Textbook of Intensive and Acute Cardiovascular Care, pp. 356-364., doi: 10.1093/med/9780199687039.003.0041

[10] Kristian, T., Joseph, A., Allan J., Maarten, L., Simoons, R., Chaitman, & Harvey, W. (2012). Third universal definition of myocardial infarction, European Society of Cardiology, European Society of Cardiology, Vol. 126, pp. 2020-2035., doi: 10.1161/CIR.0b013e31826e1058

[11] Bousseljot, R., Kreiseler, D., & Schnabel, A. (1995). Nutzung der EKG-Signaldatenbank CARDIODAT der PTB. über das Internet, Biomedizinische Technik / Biomedical Engineering, Vol. 40, pp. 317-318., doi: 10.1515/bmte.1995.40.s1.317

[12] Najmi, A., Sadowsky, J., & Morlet, O. (1997). The Continuous Wavelet Transform and Variable Resolution Time Frequency Analysis, Johns Hopkins Apl Technical Digest, Vol. 18, pp. 134-139.

[13] Juan, M., Rute A., Salvador, O., Ana, R., & Pablo, L. (2004). A Wavelet-Based ECG Delineator: Evaluation on Standard Databases, IEEE Transactions On Biomedical Engineering, Vol. 51, pp. 570-581., doi: 10.1109/TBME.2003.821031

[14] Yang, F., Du, J., Lang, J., Lu, W., Liu, L., Jin, C. & Kang, Q. (2020). Missing Value Estimation Methods. Research for Arrhythmia Classification Using the Modified Kernel Difference-Weighted KNN Algorithms, Biomed Res Int, pp. 714-725., doi: 10.1155/2020/7141725

[15] Vladimir, V. (1995), Statistical Learning Theory, Vladimir, V., ISBN: 978-0471030034, AT & T Research

[16] Tianqi, C., Carlos, G. (2016). XGBoost: A Scalable Tree Boosting System, Cornell University, doi: 10.48550/arXiv.1603.02754

[17] Ali, J.,& Khan, R., Ahmad, N. & Maqsood, I. (2012). Random Forests and Decision Trees, International Journa of Computer Science Issues, Vol. 9, pp. 272-278

[18] Basit, H., Malik, A., Huecker, M. (2021). Non-ST Segment Elevation Myocardial Infarction, StatPearls Publishing, PMID: 30020600, National Library of Medicine

[19] Shah, D., Patel, S. & Bharti, S. (2020). Heart Disease Prediction using Machine Learning Techniques, SN Computer Science, Vol. 1, doi: 10.1007/s42979-020-00365-y

[20] Miquel, A., Miguel, C. & Silvia, O (2019). A Fast Machine Learning Model for ECG-Based Heartbeat Classification and Arrhythmia Detection, Frontiers in Physics, Vol. 7, doi: 10.3389/fphy.2019.00103

[21] Roth, G., Abate, D., Abate, K., Abay, M., Abbafati, C., Abbasi, N., Abbastabar, H., Abd-Allah, F., Abdela J. & Abdelalim, A. (2018). Global, regional, and national age-sex-specific mortality for 282 causes of death in 195 countries and territories, 1980–2017, A systematic analysis for the Global Burden of Disease Study, pp. 1736–, 1788., doi: 10.1016/S0140-6736(18)32203-7

[22] Potekhin, V.V., Selivanova, E.N., Katalinič, B. (2022). Development of a digital transformation model for industrial cyber-physical systems, AIP Conference Proceedings 2456, pp. 1-7., doi: 10.1063/5.0075038

[23] Potekhin, V.V., Unal, O. (2021). Analysis of emotions using EEG data and machine learning, Annals of DAAAM and Proceedings of the International DAAAM Symposium, 2021, 32(1), pp. 158–167., doi: 10.2507/32nd.daaam.proceedings.025

DOI: 10.2507/33rd.daaam.proceedings.030

NEW TEACHING APPROACHES IN CONDITIONS OF GLOBALISATION IN DIGITIZED HIGHER ENGINEERING EDUCATION

Oleg Ipatov, Darina Barinova*, Olga Ni, Sofia Saramaha & Maria Odinokaya
Peter the Great St.Petersburg Polytechnic University

This Publication has to be referred as: Ipatov, O[leg]; Barinova, D[arina]; Ni, O[lga]; Saramaha, S[ofia] & Odinokaya, M[aria] (2022). New Teaching Approaches in Conditions of Globalisation in Digitized Higher Engineering Education, Proceedings of the 33rd DAAAM International Symposium, pp.0215-0220, B. Katalinic (Ed.), Published by DAAAM International, ISBN 978-3-902734-36-5, ISSN 1726-9679, Vienna, Austria
DOI: 10.2507/33rd.daaam.proceedings.030

Abstract

The study was intended to explore the potential of new teaching approaches in digitized higher engineering education for conducting professional communication as it serves a significant competitive advantage for many specialties, and in a number of companies it is a mandatory item of the professionogram in the conditions of globalisation. This research aims to fill this void in its exploration of the effects of training engineering specialists in learning foreign languages in order to make them possess the knowledge, skills and abilities to successfully participate in foreign language professional communication. They also should be able to present the results of their professional and scientific activities in digital context, as well as possess the skills of academic writing. The conducted research shows that the main principles of teaching second foreign language for engineering specialities in the digitized context are consciousness and the principle of taking into account interference and transference. The frequency of occurrence of interference and transference phenomena depends on three factors, which are investigated in the research. The article serves the interest for those who research digital trends in education as well as those who are interested in introducing new teaching approaches into the educational process of engineers.

Keywords: Digitized contents; Multilingualism; Didactic triglossia; Engineering; Foreign language.

1. Introduction

Modern higher professional education, like science in general, cannot develop effectively without international relations in the field of research and practical testing and application of the results of scientific and educational activities. In the context of the global world and the digitalization of education, students get wide access to information resources, have the opportunity to listen to lectures, participate in webinars and scientific and educational events of various educational institutions, including foreign ones, thus setting their own individual learning path. At the same time, knowledge of a foreign language is certainly an important and necessary factor for conducting oral and written communication and for understanding and processing large volumes of textual material on professional topics [1], [2], [3].

The digitalization of modern higher professional education makes it possible to integrate digital technologies and innovative tools into the educational process, which increase the involvement and interest of students [4], [5]. Many educational institutions have tested and implemented the Learning Management Systems, LMS, which facilitates access to educational materials and makes the learning process more flexible and accessible for any learning format (blended, distance, traditional) [6]. Peter the Great Saint Petersburg Politechnic University (SPbPU) has long-standing partnerships with technical universities in Germany. Among them are such Technical Universities as Munich, Leibniz University Hanover, Wildau University, Ulm University, etc. Within the framework of partnership relations, SPbPU students are given the opportunity to participate in joint research projects, semester study programs, defense of master's and PhD theses. In addition, modern digital technologies, the possibility of online education allow a large number of students to participate in short-term, modular programs of partner universities, which makes education attractive, modern, and international [7]. The language of professional and scientific communication is most often English, but in order to work at an enterprise and participate in international joint project work, engineering specialists need knowledge of terminology and professional vocabulary in the language of the country (in our case, in German). Taking into account the fact that many students have been purposefully preparing to study or practice at partner universities since the first years of study at the university, it should be noted that many of the undergraduates already have sufficient knowledge of English [8], [9]. However, students have no or insignificant experience of professional communication in a foreign language and, as a result, students do not have professional vocabulary and terminological system. When teaching a foreign language at a university, the most important, in our opinion, is the development of foreign language professionally oriented communicative competence [10], [11]. The use of modern digital technologies in teaching a professionally oriented foreign language makes it possible to make classroom work more diverse and individualized, and independent work more meaningful and effective.

The main educational and methodological support in teaching a foreign language in professional activities is a professionally oriented or special text, which is not only of great informational value, but also equipped with terms and special vocabulary. The digital educational environment allows the use of various forms of work with POT (professionally oriented text) at various stages (search for equivalents, semantization, interpretation of professional vocabulary and terms, production phenomena) [12], [13]. When selecting them, it is necessary to take into account the direction of training of students, the availability of up-to-date information on the specialty, the presence of professional vocabulary and terms. Students are most interested in POTs from the original websites of equipment manufacturers, for example (*https://www.vw.com/en.html, https://www.ford.com/, https://info.daimler.com /en/, https://www.youtube.com/channel/UCxN-Csvy_9sveql5HJviDjA*), etc. Modern technologies make it easy to integrate text, audio and visual material of original sites into the educational process [14], [15].

Primary when working with POT is the understanding of information. Professionally-oriented, special texts are of great educational and developmental importance. When selecting them, it is necessary to take into account the direction of students' training, the availability of up-to-date information on the specialty, the availability of professional vocabulary and terms [16], [17]. When translating technical literature, students use background knowledge and compensatory skills, and when translating terms, they most often use methods of calculus, transcription and transliteration. To help students to memorize professional vocabulary faster and better, the teacher should build the process of semanticization based on the first foreign language or native language [18], [19].

A survey of students showed that more than 80% of respondents (1st-year master students in the direction of study: Energy, Construction engineering, Mechanical Engineering, Transport and Materials studies, Computer Science and Technology consider knowledge of one or more foreign languages to be an important component of professional success. 76% believe that it is necessary to study foreign languages in professional activities as part of master's studies, while almost 70% consider the number of classroom hours allocated by the program to be insufficient, they also believe that 4 or more hours per week are needed during the entire master's course and the organization of optional classes is preferable. Almost 100% of respondents use mobile applications and various digital technologies when studying a foreign language (FL). The survey involved 89 undergraduates of the 1st year in the areas of Energy, Construction engineering, Mechanical Engineering, Transport and Materials, Computer Science and Technology.

Knowledge of a foreign language(s) at the level that allows to establish professional contacts, negotiate, promote the product of the enterprise, present the results of collective or individual scientific research and practical activities is today one of the qualification requirements for specialists in any field of activity. This is due to integration processes, the impossibility of an isolated existence and the development of civilization. This is the reason for the increasing interest in learning a foreign language and the demand for specialists who speak a foreign language at a professional level. However, the rapid changes taking place in all areas of the economy and production dictated new requirements for foreign language proficiency. In the system of higher professional education, the discipline "foreign language" has always been mandatory and belonged to the federal component. According to Federal State Educational Standards (FSES):

- Proficiency in a foreign language is an integral part of the professional training of all specialists at the university.
- The foreign language course is multilevel and is developed in the context of continuing education.
- The study of a foreign language is based on an interdisciplinary integrative basis.
- Foreign language teaching is aimed at the comprehensive development of students' communicative, cognitive, informational, socio-cultural, professional and general cultural competencies.

At the moment, in the amended standards, much attention is paid to the formation of the language (foreign language) competence of the graduate, the practical level of proficiency should not be lower than B1+, in connection with which the discipline "Foreign language in professional communication" was introduced in many technical areas of training. Despite the fact that the objective difficulties for achieving the level stipulated by the Federal State Educational Standard are a limited number of academic hours and a break in learning a foreign language (as a rule, students study this discipline in 1-2 and in the 9th, 10th or 11th semesters), recently there has been an increase in the number of students who are aware of the role of foreign languages in their future professional career. The orientation of the Russian Federation to global development trends and the entry of our state into the world educational space have led to changes in the field of language education. Currently, language education in Russia is developing in two directions: on the one hand, an increasing number of students are choosing English as their first language, on the other hand, the desire to learn the second language is growing.

Over the past ten years, the study of several foreign languages has become a common phenomenon in modern general education schools. German, French and Spanish are most often used as the second language. "In accordance with the updated Federal State Educational Standard, since September 1, 2022, in all educational organizations for students mastering the educational program of basic general education, the study of the subject "Second foreign Language" is mandatory." (Letter of the Ministry of Education of the Russian Federation dated August 31, 2021 No. 03-1420 "On the study of the academic subject "Second foreign language").

Similar changes are taking place in the educational environment of higher education. The study of a second foreign language in the system of higher professional education of the Russian Federation in technical areas is becoming more and more in demand, due to modern educational trends: training of specialists capable of presenting and defending the results of innovative engineering activities in a foreign language; increasing academic mobility; internationalization of Russian higher education in general. Extensive international partnerships of Peter the Great St.Petersburg Polytechnic University allow students to participate in exchange programs, scientific conferences, industrial practice at a foreign enterprise, defend master's theses and continue research at partner universities. All these promising opportunities are a motivating factor for studying the second in a short time. Despite the processes of globalization and the dominant role of English as the language of international communication, knowledge of the language of the country of the proposed scientific internship is necessary for direct participation in professional communication. The unconditional factor of the success of cooperation is communication (personal and professional) in the partner's language.

Since teaching a second foreign language takes place in an organized educational process, we single out its didactic component as the main characteristic of the trilingualism under consideration, designating it with the terminological phrase "didactic triglossia" (DT) [20]. By DT we mean the coexistence of three languages (native and two foreign) in the speech-thinking sphere of an individual who learns these languages consistently and (with the exception of the native) in educational conditions, uses them in different communicative situations. In the situation of DT, the nature of the mutual influence of the three languages is determined primarily by the genetic relationship of the two foreign languages, German acts as the second foreign language, the study of which begins at the stage when students already speak English at an average level. Since English and German belong to the Germanic group, the similarity between them is much stronger than between German and the Russian language (RL), i.e. when studying FL2, students will primarily rely on FL1. Thus, the possession of RL is characterized by the greatest autonomy, while FL1 and FL2 interact in the student's mind. In the conditions of DT, this means the need to take into account this mutual influence when building the educational process, i.e. it is necessary to develop and apply such approaches in teaching FL2 that would rely on students' knowledge of the system and phenomena of FL1 and their native language [21]. In order to optimize the educational process, previous linguistic experience must be taken into account when studying FL2, where one of the main educational tools is the comparison of the phenomena of the languages that make up DT.

The comparison of linguistic elements of several language systems allows students to more easily and efficiently assimilate linguistic phenomena common to the two components of DT, as well as to avoid errors of negative transference (interference). The presence of regular comparison of linguistic phenomena of co-studied languages determines the expediency of the contrastive approach. Taking into account the importance of taking into account the mutual influence of the languages forming DT in the minds of students, we come to the conclusion that the main principles of teaching FL2 are consciousness (since direct methods do not provide for the analysis and comparison of phenomena of language systems of different languages) and the principle of taking into account interference and transference.

The educational process should be structured in such a way that the reliance on the language and the FL1 is carried out – then the prerequisites are created for the positive transfer of the use of phenomena of one language to similar phenomena of another by analogy. The frequency of occurrence of interference and transference phenomena depends on three factors:

1) the level of speech development in the native language and conscious proficiency in it;

2) the level of proficiency in FL1: the better the learner speaks FL1, the fewer interference phenomena he has and the more opportunities there are for positive transference. But it also means that a low level of proficiency in FL1 can have a retarding effect on the mastery of FL2;

3) the length of time that separates the study of FL2 from IY1: the smaller the interval, the greater the impact of FL1 on the mastery of IY2. Interference is usually based on differences in terms of expression, when the meanings present in both languages are transmitted by different linguistic means. However, the semantics of language units and their functioning even more provoke violations of the norms of FL (word order, tense coordination, specific meanings of verb forms, etc.). The work on the prevention of interference is based on the principles of consciousness and activity, the speech orientation of learning, taking into account the meaning of RL and FL1, i.e. FL2 training is based on theoretical information that reveals the features of the functioning of the phenomenon under study in comparative terms. At the same time, the constant availability of oral practice is ensured to activate knowledge and skills in order to solve communicative tasks [22]. Summarizing the above, we can conclude that when teaching IY2, one of the main tasks of the teacher and the compilers of educational and methodological complexes (EMC) is to consistently rely on the knowledge, skills and abilities already formed in students, similar in FL1 and FL2, and to prevent errors in the study of those phenomena where the possibility of such an analogy is only apparent.

So, undergraduates of the SPbPU Institutes of Energy, Construction engineering, Mechanical Engineering, Transport and Materials, Institute of Computer Science and Technology after a 2-3-year break, begin studying the discipline "Foreign language in professional communication". The study of this discipline takes 34-68 hours. In this situation, a contradiction arises: on the one hand, the desire and motivation of undergraduates to study the second language for use in professional activities (many undergraduates have sufficient knowledge of English), on the other – a limited number of study hours. In this regard, it is natural to teach the second language, in particular, to teach translation taking into account the principles of DT, i.e. conscious reliance on the native language and the first foreign language. For effective work, it is necessary to use the possibilities of optimal organization of classroom classes and independent work of students to prepare for professional communication, scientific discourse and academic writing. To achieve this goal, it is necessary to teach how to work with a professionally oriented text, a scientific text in the specialty, which includes the study of the basic terms of a certain term system in a foreign language, with the help of which they will be able to further understand special literature and, if possible, participate in foreign-language professional communication [23].

Russian	**English**	**German**
Авиационный двигатель	Aero-engine	Flugzeugtriebwerk, n.
Высшая математика	Higher mathematics	die höhere Mathematik
Гидроаэродинамика	Hydroaerodinamics	Hydroaerodynamik, f.
Гидротехника	Hydroengineering	Hydrotechnik, f. / Wassertechnik, f
Информационные технологии	Information technology (IT)	Informationstechnologien,
Квантовая электроника	Quantum electronics	Quantenelektronik, f.
Котельная установка	Boiler plant	Dampfkesselanläge, f.
Лазерные технологии	Laser technologies	Lasertechnologien, f. pl
Литейное производство	Casting production	Gießbetrieb, m.
Машиностроение	Engineering	Maschinenbau, m.
Парогенератор	Steam generator	Dampferzeuger, m.
Полупроводник	Semiconductor	Halbleiter, m
Прикладная механика	Applied mechanics	Angewandte Mechanik
Проводник	Conductor	Leiter, m.
Радиотехника	Radio engineering	Radiotechnik, f
Сопротивление материалов	Resistance of materials	Festigkeitslehre, f
Станкостроение	Machine-tool construction	Werkzeugmaschinenbau, m.
Теоретическая физика	Theoretical physics	Theoretische Physik
Физика плазмы	Plasma physics	Plasmaphysik, f
Холодильная техника	Refrigeration engineering	Kühltechnik, f.
Экспериментальная физика	Experimental physics	Experimentalphysik
Электрическая станция	Power plant	Elektrizitätswerk, n
Ядерная энергетика	Nuclear power	Kernenergetik

Table 1. Examples of using the calculus technique

The algorithm of work on a professionally-oriented text assumes reliance on the knowledge of the 1st century, i.e. when translating terms and professional vocabulary, first of all, the search for internationalisms and Anglicisms is conducted. Students cope with this type of work quite easily, because when translating professional vocabulary, the calculus technique is used more often than others. Calculus should be understood as one of the methods of translation, which consists in the fact that words and expressions of one language are translated into another language by means of exact reproduction by means of their morphemic or verbal structure (examples of using the calculus technique are presented in Table 1).

Due to the fact that students of technical specialties did not study theoretical linguistic disciplines, the task of the teacher is to explain word formation in German by the example of simple phrases and compound words (for example: Theoretische Physik, Angewandte Mechanik, die Kernenergetik, die Kühltechnik usw.) The next technique used in the translation of professional vocabulary is transliteration, i.e. the technique of translating the lexical unit of the original by recreating its graphic form using the letters of the target language (see Table 2). When using techniques such as transcription and borrowing, it is necessary to pay students' attention to the phonological side of words and phrases (for example: design – in English, design – in Russian and das Design – in German; manager, der Manager, manager; die Telekommunikation in German, which was gradually borrowed into English and Russian, i.e. "telecommunication" and "telecommunications" respectively). The primary data of the analysis of the use of methods of translation of professional vocabulary are presented in Table 1 and Figure 1.

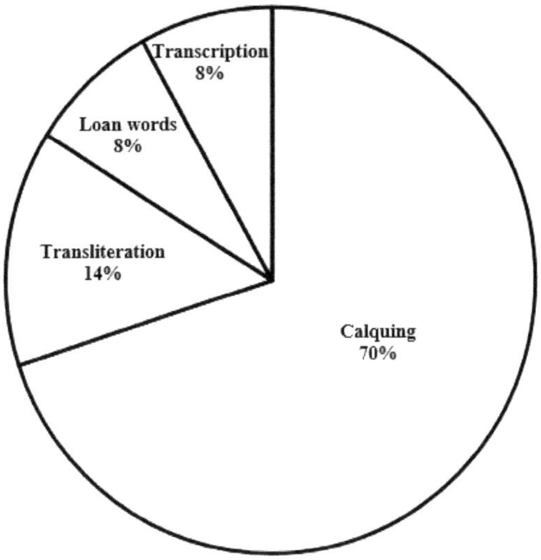

Fig. 1. The analysis of the results of the use of professional vocabulary translation techniques

2. Conclusion

As can be seen in the figure, the calculus technique was used more often than others (70%), i.e. it can be assumed that students use their linguistic experience when translating, but in order to achieve more effective results, given the limited time allotted by the program to study the discipline, the teacher should build classes using the full potential of the DT system.

In conclusion, it should be noted that the conducted research may be of interest to those who are engaged in the study of new trends in education. Thus, the modern educational environment is associated with the use of digital technologies, which provides the educational process with relevant information material and significantly optimizes teaching methods and obtaining the final result. When teaching students of technical areas of training, the resources of the original sites are used, as well as various forms of education (traditional, mixed and online). With the use of digital technologies, classroom and independent work can be more effective. In general, the problem investigated in the work presents great prospects for further research. Prospects for further research are seen in the need to continue working on further finding models for better improvement of the students' language skills in the foreign language learning process. If taken into account multilingualism can help to reduce individual and social risks of it and stimulate its positive impact on the individual and society there is the necessity of interdisciplinary interaction in research of multilingualism on various individual and social processes.

3. References

[1] Alenkina, T. (2020). English medium instruction in the Russian university. Vysshee Obrazovanie v Rossii, Vol. 29, No. 1, pp. 98–103., DOI:10.31992/0869-3617-2020-29-1-98-103

[2] Barinova, D., Ipatov, O., Odinokaya, M. (2019). Pedagogical assessment of general professional competences of technical engineers training. In: Annals of DAAAM and Proceedings of the International DAAAM Symposium, Vol. 30, No.1, pp. 0508–0512, DOI:10.2507/30th.daaam.proceedings.068

[3] Almazova, N., Krylova, E., Rubtsova, A., Odinokaya, M. (2020). Challenges and opportunities for Russian higher education amid covid-19: Teachers' perspective. Education Sciences, Vol. 10, No.12, pp. 1–11

[4] Anderson, K. D. (2004). The Nature of Teacher Leadership in Schools as Reciprocal Influences Between Teacher Leaders and Principals. School Effectiveness and School Improvement, No. 15, pp. 97–113., DOI:10.1076/sesi.15.1.97.27489

[5] Bekirogullari, Z., Bylieva, D., Lobatyuk, V., Nam, T. (2021). How Virtual Personal Assistants Influence Children's Communication. Lecture Notes in Networks and Systems, No. 184, pp. 112–124., DOI:10.1007/978-3-030-65857-1_12

[6] Botha, W. (2013): English-medium instruction at a university in Macau: Policy and realities. World Englishes, No. 32, pp. 461–475., DOI:10.1111/weng.12053

[7] Bylieva, D., Lobatyuk, V., Fedyukovsky, A. (2019). Ways of sociotechnical integration of scientists and volunteers in citizen science. IOP Conference Series Materials Science and Engineering, DOI:10.1088/1757-899X/940/1/012150

[8] Chappel, J. (2015). Teaching in English Is Not Necessarily the Teaching of English. International Education Studies, Vo. 8, No. 3, DOI:10.5539/ies.v8n3p1

[9] Chernysheva, M.A., Kozlova, A.O., Donova, E.V. (2019). Obuchenie akademicheskogo pis'ma na angliiskom yazyke russkoyazychnykh nauchno-pedagogicheskikh rabotnikov (iz opyta raboty). Vysshee obrazovanie v Rossii, Vol. 28, No .10, pp. 50-57., DOI:10.31992/0869-3617-2019-28-10-50-57

[10] Doiz, A., Lasagabaster D., Sierra, J.M. (Eds.). (2013). English-medium instruction at universities: Global challenges. Clevedon: Multilingual Matter, DOI:10.1093/elt/cct045

[11] http://www.education.ox.ac.uk/crdemi-oxford, (2021). EMI Oxford: Centre for Research and Development on English Medium Instruction, Accessed on: 2021-03-21.

[12] Hasan, Al-W. (2016). Role of Using English as medium of Instruction in Reshaping Bahraini Senior Teachers' Perceptions of Their Roles as Middle Leadership in their Schools. Journal of Language and Education No. 1, pp. 6-15., DOI:10.17323/2411-7390-2016-2-1-6-15

[13] Hollenbeck, A., Kalchman, M. (2012). Professional development for conceptual change: Extending the paradigm to teaching reading comprehension in US schools. Professional Development in Education, Vol. 39, No. 5, pp. 638-655., DOI:10.1080/19415257.2012.728535

[14] https://www.spbstu.ru/international-cooperation/international-educational-programs/international-graduate-programs/graduate-programs-in-english, (2021). International Master's Degree Programs, Accessed on: 2021-03-21

[15] Karpovich, I., Odinokaya, M. (2020). Interactive technology of pedagogical assistance as a means of adaptation of foreign first-year students. IOP Conference Series Materials Science and Engineering, DOI:940. 10.1088/1757-899X/940/1/012130

[16] Klochkova E., Volgina A., Dementyev S., Klochkov Y. (2016). Human FactorinQualityFunctionDeployment.SecondInternationalSymposiumon Stochastic Models in Reliability Engineering, Life Science and Operations Management (SMRLO), IEEE, 466-468, DOI:10.1109/SMRLO.2016.81

[17] Lukacs, K.S., Galluzzo, G.R. (2014). Beyond empty vessels and bridges: Toward defining teachers as the agents of school change. Teacher Development: An international journal of teachers' professional development, Vol. 18, No. 1, pp. 100-106., DOI:10.1080/13664530.2013.856338

[18] Lo, Y.Y. (2017). Development of the beliefs and language awareness of content subject teachers in CLIL: Does professional development help? International Journal of Bilingual Education and Bilingualism. Advance online publication, DOI:10.1080/13670050.2017.1318821.

[19] Macaro, E., Curle, S., Pun, J., An, J., Dearden, J. (2018). A systematic review of English medium instruction in higher education. Language Teaching, Vol. 51, No. 1, pp. 36–76., https:// https://dx.doi.org/10.1017/S0261444817000350

[20] O'Dowd, R. (2018). The training and accreditation of teachers for English medium instruction: an overview of practice in European universities. International Journal of Bilingual Education and Bilingualism, Vol.21, No. 5, pp. 553–563., Doi:10.1080/13670050.2018.1491945

[21] Rubtsova, A., Odinokaya, M., Krylova, E., Smolskaia, N. (2021). Problems of Mastering and Using Digital Learning Technology in the Context of a Pandemic. Lecture Notes in Networks and Systems, No. 184, pp. 324–337.

[22] Odinokaya, M., Krylova, E., Rubtsova, A. (2021). Using the discord application to facilitate EFL vocabulary acquisition. Education Sciences, Vol. 11, No.9, pp. 470, DOI:10.3390/educsci11109047'.

[23] Wiseman, A., Odell, A. (2021). Should non-English-speaking countries teach in English? Available from: https://www.britishcouncil.org/voices-magazine/should-nonenglish-speaking-countries-teach-in-english, Accessed: 2021-03-21.

DOI: 10.2507/33rd.daaam.proceedings.031

TENSILE MECHANICAL PROPERTIES COMPARATION OF PETG, ASA AND PLA-STRONGMAN FDM PRINTED MATERIALS WITH AND WITHOUT INFILL STRUCTURE

Adi Pandžić & Damir Hodžić

This Publication has to be referred as: Pandzic, A[di] & Hodzic, D[amir] (2022). Tensile Mechanical Properties Comparation of PETG, ASA and PLA-Strongman FDM Printed Materials With and Without Infill Structure, Proceedings of the 33rd DAAAM International Symposium, pp.0221-0230, B. Katalinic (Ed.), Published by DAAAM International, ISBN 978-3-902734-36-5, ISSN 1726-9679, Vienna, Austria
DOI: 10.2507/33rd.daaam.proceedings.031

Abstract

Additive technologies have certainly brought many novelties into today's industry. Also, the development of AM technologies directly depends on the development of AM materials themselves, so the examination of the mechanical properties of new AM materials is of great importance. The aim of this research is to examine the tensile mechanical properties of three different FDM 3D printed materials: PETG, ASA and PLA-Strongman. The advantage that certainly distinguishes AM technologies compared to classic production technologies is the production of materials with an infill structure. So, the tensile properties of the materials with and without the infill structure were also analysed. For all materials, tensile and yield strength, tensile strength at brake, elastic modulus and deformation at brake are analysed, compared, and presented in this paper.

Keywords: FDM material; Infill; PETG; ASA; PLA.

1. Introduction

Today, one of the most popular and fast-growing production technologies are surely additive technologies (AM). On the market, there is a huge interest in AM technologies, and certainly the key factor in the development of AM technologies is the development of AM materials. In addition to a large number of different technologies, one of the most frequently used is Fused Deposition Modeling (FDM) technology, also known as Fused Filament Fabrication (FFF), or material extrusion [1], [2]. Some of the reasons why FDM is such a popular technology today are certainly the simple application, safe to use at home, the relatively low cost of the 3D printers and materials themselves and a wide range of available materials [3], [4]. So currently, FDM 3D printers are so prevalent that we can find them in homes where children use them, in schools, creative hubs, universities, medicine, bioengineering, automotive and aviation industry, food industry and more [5].

FDM technology is based on the layer-by-layer production of products, using a previously created CAD 3D model. The process of 3D printing with FDM technology is presented on Fig. 1. The material, known as filament, comes in the form of wire wound on rolls, and is fed to the extrusion body (print head) via a feeder.

In the extrusion body, the material is heated until it turns into a semi-liquid state and further trough the nozzle it is transferred to the print bed. Then on the print bed product is 3D printed layer by layer along the Z axis [6], [7].

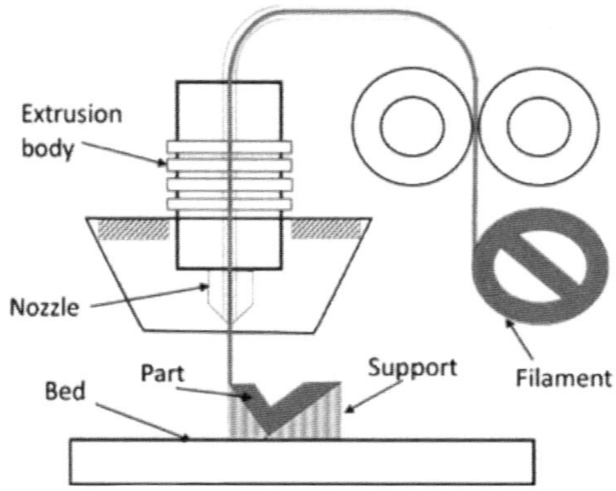

Fig. 1. The working principle of FDM technology [8]

As the engineering approach to designing a new product consists of a basic idea, adequate planning, material selection, all with the aim of optimizing the production process, surely the technology that can answer these challenges is FDM as well as other AM technologies [9]. Generally, AM technology provides many advantages, as well as FDM, and among some interesting (important) are rapid prototyping, production of products with complex geometries, savings in the production of individual products and small-series products, etc. [10].

FDM technology basically uses amorphous thermoplastic materials [11], but there is an increasing focus on the development of other materials such as composite materials with short and continuous fibers, metals, etc. Certainly, the most popular thermoplastic materials in FDM technology are PLA (polylactic acid) and ABS (acrylonitrile butadiene styrene) [12], but in addition to them, PETG (polyethylene terephthalate glycol), PC (polycarbonate), PP (polypropylene), ASA (acrylonitrile styrene acrylate), PA6 (polyamide, nylon), PEEK (polyetheretherketone), TPU (thermoplastic polyurethane), PEI (polyetherimide), PVA (polyvinyl alcohol) are widely used [3].

The material selection is essential when it comes to product manufacturing with FDM or other AM technologies, especially products for engineering applications. Knowing the mechanical and other material properties is of great importance when selecting material through an engineering approach [13]. This study aims to analyse the tensile mechanical properties of PETG, ASA and PLA-Strongman (PLAs) FDM printed materials. One of the goals is to compare the mechanical properties of these materials.

From literature review and analysing recent studies, it can be seen that the research of lightweight materials is of great importance [14]. FDM technology enables the production of materials with an infill structure, which achieves a smaller mass of the finished product as well as lower printing time and material consumption, which leads to lower overall costs presented in Fig. 2.

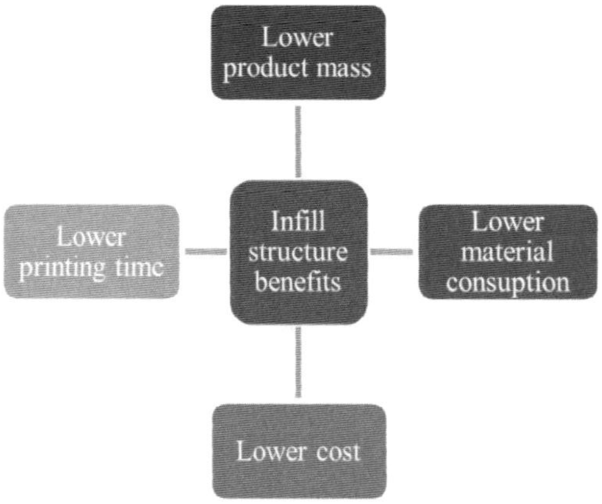

Fig. 2. Benefits of materials with infill structure

Due to the advantages of using the infill structure in FDM printed materials, the tensile mechanical properties of mentioned materials with the infill structure were also analysed.

2. Materials

In this paper, all materials (PETG, ASA and PLA-Strong) are from the same manufacturer, the company AzureFilm. PETG filament is a thermoplastic polymer, an environmentally friendly material that can also be recycled. It is a material that is relatively easy to 3D print and combines the properties of both ABS and PLA materials. Due to its good chemical stability, PETG material meets the European Parliament requirements for use in products that come into contact with food products and medicines. PETG is a material with good durability, solid and odourless. It is a material that has a combination of strength and durability, which is why it is also used to make functional products [6], [15]. Mechanical properties of AzureFilm PETG material are presented in Table 1.

Material properties	Value	Test method
Infill density	100 %	-
Colour	Transparent	-
Specific density	1,29 g/cm³	ASTM D-729
Tensile strength	51 MPa	ISO 527-2
Elastic modulus	2,98 GPa	ISO 527-2
Deformation at brake	29 %	ISO 527-2
Tensile testing speed is not mentioned.		

Table 1. PETG material – mechanical properties [15]

ASA is an amorphous thermoplastic polymer quite similar to ABS material when comparing mechanical properties and 3D printing conditions. With a 118.5 °C glass transition temperature, ASA material has good resistance to ultraviolet (UV) resistance and good resistance to weather conditions [16]. Products made of ASA material are very durable and impact resistant, also it can withstand higher temperatures and exposure in rain, cold and sea water [15]. Due to its good properties, ASA material is used in aerospace, automotive industry and other fields [16]. Mechanical properties of AzureFilm ASA material are presented in Table 2.

Material properties	Value	Test method
Infill density	20 %	-
Colour	-	-
Specific gravity	1,07	ASTM D-729
Tensile strength[1]	48 MPa	ASTM D638
Elastic modulus[2]	2,12 GPa	ASTM D638
Deformation at brake[1]	25 %	ASTM D638
[1]Tensile testing speed 50 mm/min [2]Tensile testing speed 1 mm/min		

Table 2. ASA material – mechanical properties [15]

PLA-Strongman is high performance polylactide with high impact toughness and good printability like standard PLA material. PLAs properties are similar to ABS material, but the essential difference is that PLAs is easier to print. Compared to standard PLA material it is much tougher and with better impact-resistant. PLAs is used for manufacturing drones parts, bearings and other products that require materials with increased strength and toughness [15]. Mechanical properties of AzureFilm PLAs material are presented in Table 3.

Material properties	Value	Test method
Infill density	20 %	-
Colour	-	-
Specific gravity	-	-
Tensile strength[1]	41,2 MPa	-
Elastic modulus[2]	2,9 GPa	-
Deformation at brake[1]	8,9 %	-
[1]Tensile testing speed 50 mm/min [2]Tensile testing speed 1 mm/min		

Table 3. PLAs material – mechanical properties [15]

3. Experimental methodology

This study is based on experimental research, where specimens for each material were FDM 3D printed with different infill density (20%, 40%, 60%, 80% and 100%), and then the tensile mechanical properties were examined, and results were statistically analysed. The experimental methodology is presented on the flow diagram in Fig. 3.

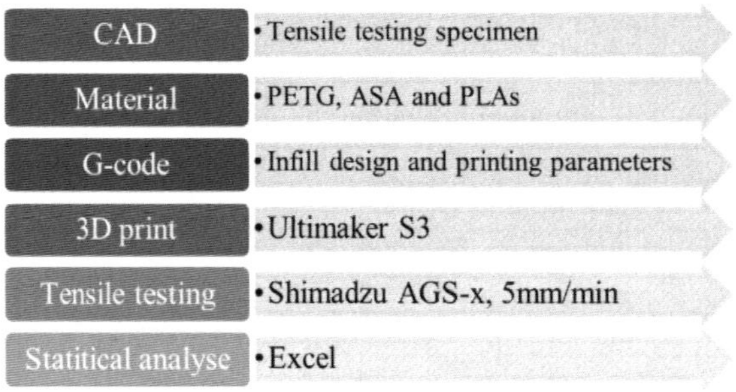

Fig. 3. Experimental methodology flow diagram

Using the Solidworks CAD tool, a 3D model of the test specimen was designed according to then ISO 527-2 standard. Technical drawing of test specimen is presented on Fig. 4.

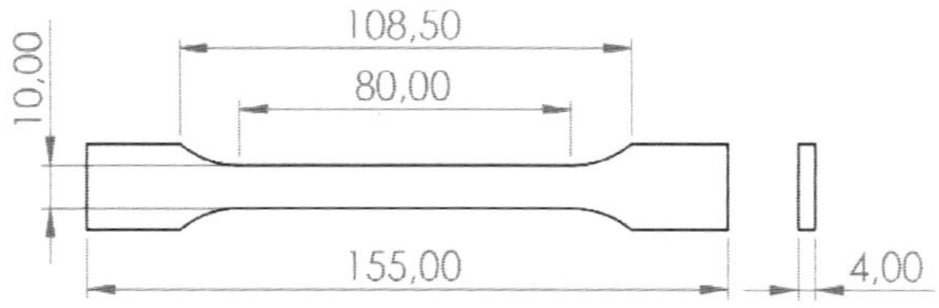

Fig. 4. Test specimen design according to ISO 527-2

The materials were in the form of wire (filament) with a diameter of 2,85 mm wounded on 1 kg rolls. As the colour of the material affects the mechanical properties, it is important to state the colours of the analysed materials, PETG and ASA materials were in black and PLAs in white.

G-code and 3D printing parameters were defined using Cura 3D printing software, so-called. slicer, from Ultimaker. The printing parameters are presented in Table 4. PETG material is 3D printed with predefined "generic PETG" parameters from the Cura software were used, for PLAs material "generic PLA" and for ASA material "generic ABS" with modified parameters.

Printing parameters	PETG	ASA	PLAs
Layer height	0,15 mm		
Printing temperature	240 °C	240 °C	200 °C
Build plate temperature	85 °C	80 °C	60 °C
Infill pattern	Triangles		
Infill density	20%, 40%, 60%, 80% and 100%		
Print speed	60 mm/s	60 mm/s	70 mm/s
Fan speed	20 %	10 %	100 %
Retraction distance	6,5 mm	3 mm	6,5 mm

Table 4. Test specimen printing parameters

As already mentioned, all materials were tested with and without the infill structure. In the case of samples with infill structure, a "triangle" infill pattern was used, with varying infill densities of 20%, 40%, 60%, 80% and 100%. It should be noted here that when the infill density is defined as 100%, the product is 3D printed in the same way, regardless of the defined infill pattern.

Tensile testing specimens are 3D printed on Ultimaker S3 professional desktop FDM 3D printer. Ultimaker S3 is composite-ready dual extrusion 3D printer, compatible with over 110 materials.

For each material and infill density five replicates and a total of 75 tensile test specimens were 3D printed and then tensile tested. Tensile testing was performed using Shimadzu AGS-X universal testing machine, with maximum load of 10 kN (Fig. 5). Tensile testing was performed according to ISO 527-1 with testing speed 5 mm/min.

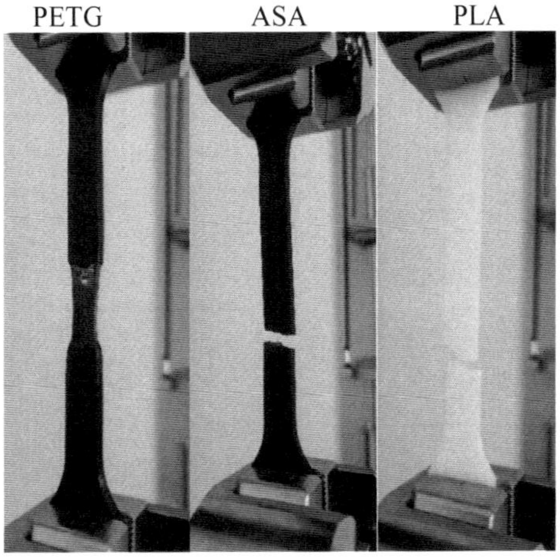

Fig. 5. Tensile testing of PETG, ASA and PLAs FDM printed materials

Mechanical properties data monitoring was performed using Trapezium-X software by Shimadzu. During the test, tensile mechanical properties such as tensile strength (R_m), yield strength (R_y), tensile strength at brake (R_b), elastic modulus (E) and deformation (strain) at brake (ε) were analysed. After testing, collected data was statistically analysed and results are presented trough stress-strain diagrams and histograms.

4. Results and discussion

The results obtained from testing the tensile mechanical properties of FDM printed PETG, ASA and PLA-Strongman materials are presented in this section. A total of seventy-five specimens were tested and analysed. Stress-strain diagram for PETG, ASA and PLAs materials is presented on Fig. 6.

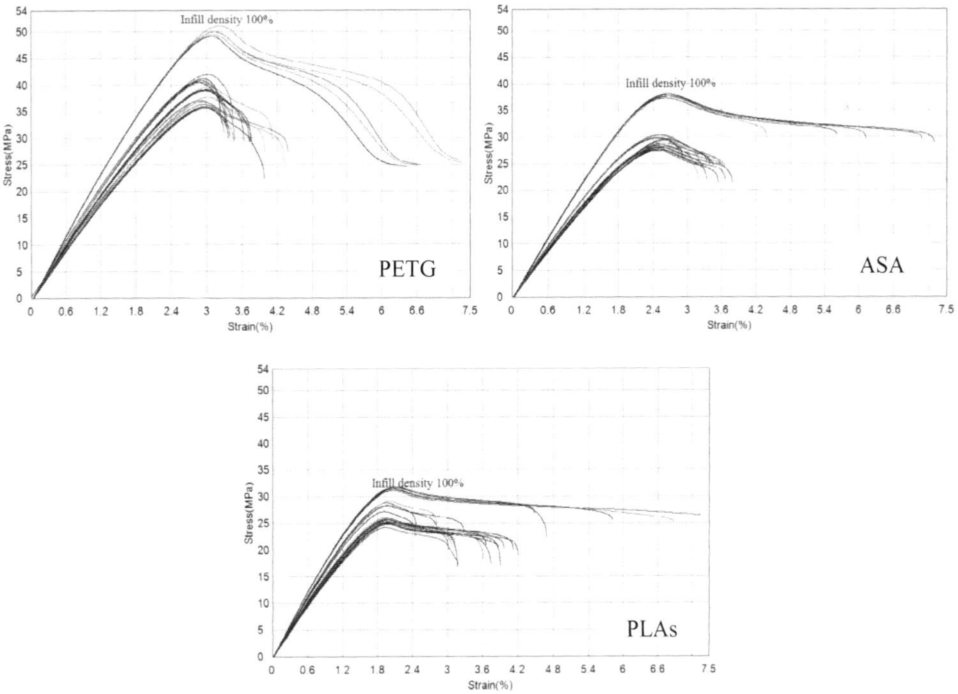

Fig. 6. Stress-strain diagram for PETG, ASA and PLAs material

The average values of the measured tensile mechanical properties, based on five replicas, for PETG, ASA and PLAs materials, are presented in Table 5.

Infill density	R_m	R_y	R_b	E	ε
PETG					
20%	35.88	27.86	28.53	1.56	3.52
40%	37.08	26.64	26.56	1.72	4.08
60%	39.08	30.86	29.53	1.72	3.74
80%	40.98	31.16	33.11	1.86	3.30
100%	49.84	38.80	25.13	2.10	6.88
ASA					
20%	29.56	24.82	24.93	1.50	3.48
40%	27.74	23.90	21.76	1.50	3.50
60%	28.40	23.44	24.84	1.54	3.38
80%	30.06	25.10	26.55	1.70	3.00
100%	37.70	31.34	29.70	1.98	6.10
PLAs					
20%	24.96	22.42	17.37	1.66	3.52
40%	25.46	22.58	20.44	1.70	3.78
60%	25.68	22.48	20.47	1.80	3.26
80%	28.34	24.68	23.73	1.94	2.78
100%	31.62	26.54	24.83	2.10	5.94

Table 5. PETG, ASA and PLAs materials – average tensile test results

An analysis of the stress-strain diagram for PETG and ASA materials FDM printed specimens without infill structure (100% infill density) showed significantly higher tensile strength and deformation compared to samples with infill structure (Fig. 7.). While the PLAs material with 100% infill density did not show a significant difference in tensile strength compared to the specimens with infill structure. While analysing the deformation at brake for the PLAs material (Fig. 10.), a significantly higher total deformation can be seen for the specimens without infill structure.

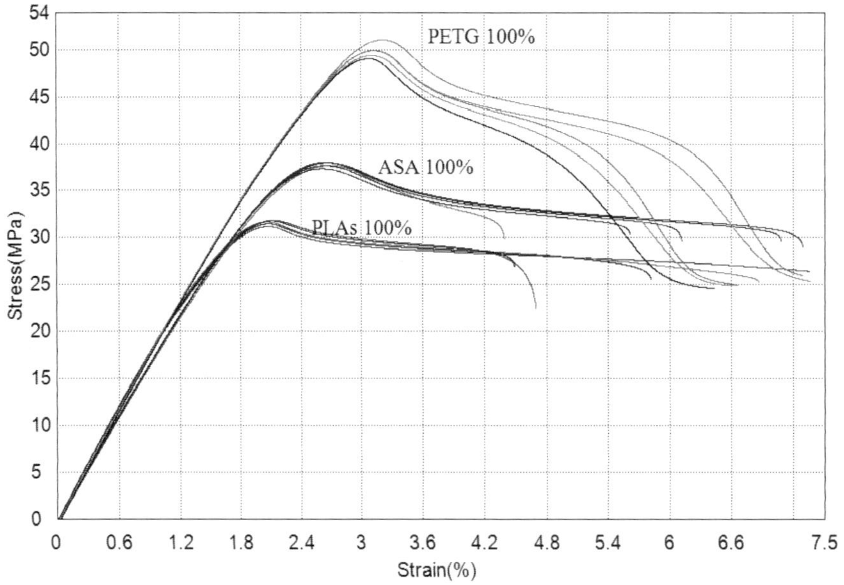

Fig. 7. Stress-strain diagram for PETG, ASA and PLAs material with 100% infill structure

Generally, PETG specimens without infill structure have highest tensile strength. It is not possible to see significant differences in elastic modulus and deformation at brake based on the curves from the stress-strain diagram for specimens with 100% infill. By using PETG material compared to ASA material, the tensile strength increases by 32%, while compared to PLAs material it increases up to 57% (Fig. 8a). Comparing the yield strength of the materials without infill structure, the highest values are observed for the PETG material (Fig. 8b). Yield strength of PETG material is 23% higher than ASA and 46% than PLAs material.

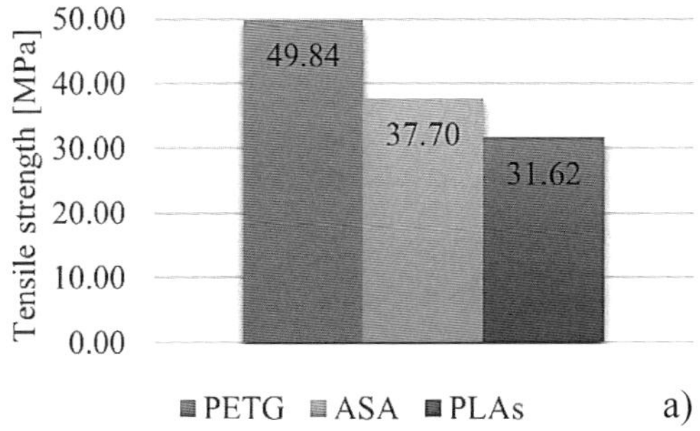

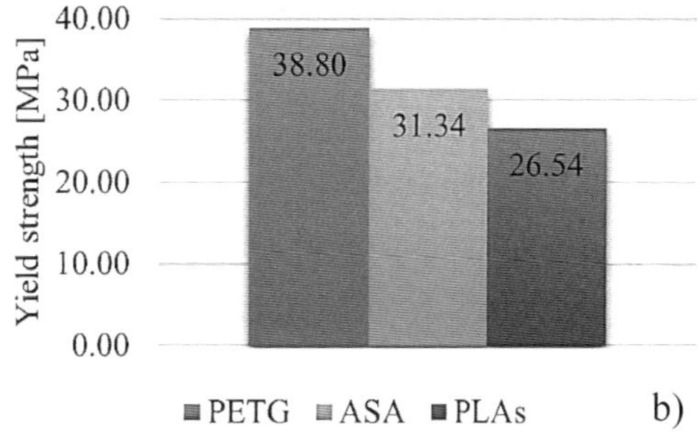

Fig. 8. PETG, ASA and PLAs tensile (a) and yield (b) strength comparison – 100% infill density

The highest values of tensile strength at brake were obtained with specimens made of ASA material (Fig. 9a). Comparing to PETG and PLAs materials, ASA material with infill density of 100% showed up to 20% increase in tensile strength at brake. Between PETG and PLAs materials, there is no significant difference in tensile strength at brake, not greater than 1%. Comparison of elastic modulus values for specimens with 100% infill density is shown in Fig. 9b. The elastic modulus of PETG and PLAs materials is 2,1 GPa, and compared to ASA material, it is higher by 6%.

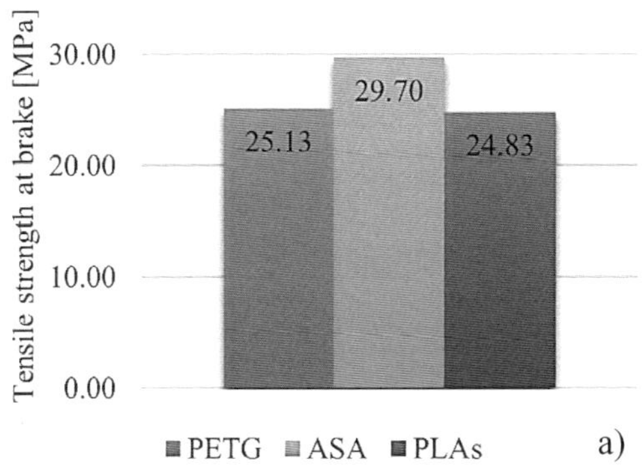

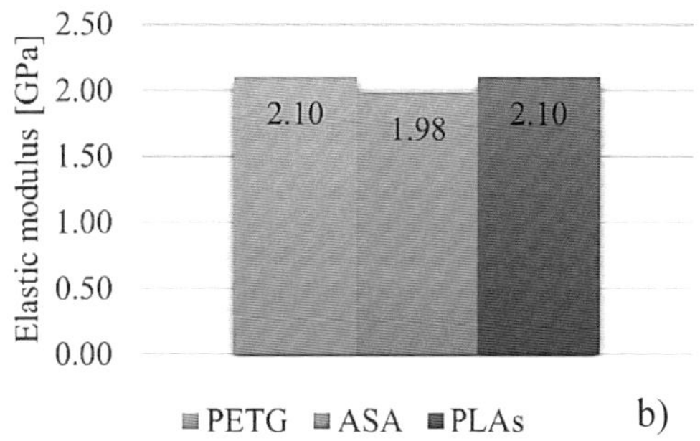

Fig. 9. PETG, ASA and PLAs tensile strength at brake and elastic modulus comparison – 100% infill density

Deformation at brake for PETG, ASA and PLAs materials without infill structure is presented on Fig. 10. The PETG material showed the highest deformation values, which was 13% higher than ASA and 16% higher than PLAs material.

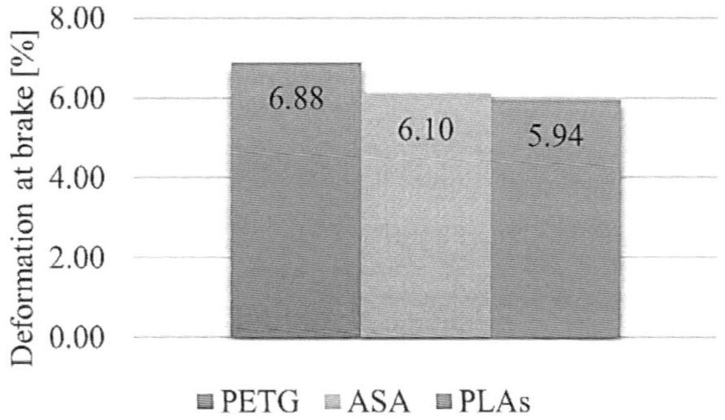

Fig. 10. PETG, ASA and PLAs deformation at brake comparison – 100% infill density

Analysis of PETG, ASA and PLAs material specimens with infill structure was also performed and results are compared with samples without infill structure. Different infill density for all materials was analysed and presented in further text.

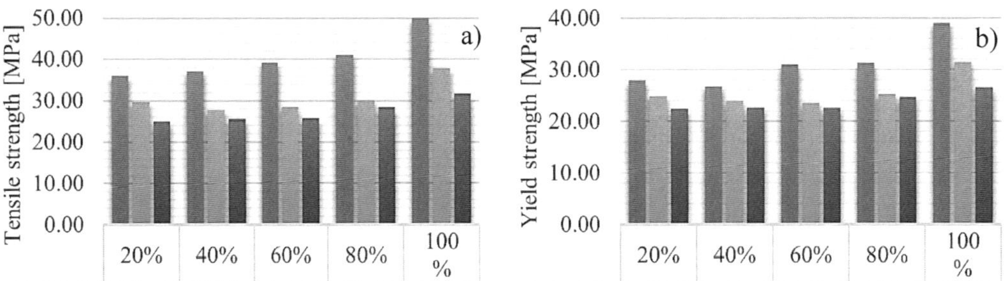

a)

	20%	40%	60%	80%	100 %
PETG	35.88	37.08	39.08	40.98	49.84
ASA	29.56	27.74	28.40	30.06	37.70
PLAs	24.96	25.46	25.68	28.34	31.62

b)

	20%	40%	60%	80%	100 %
PETG	27.86	26.64	30.86	31.16	38.80
ASA	24.82	23.90	23.44	25.10	31.34
PLAs	22.42	22.58	22.48	24.68	26.54

Fig. 11. PETG, ASA and PLAs tensile (a) and yield (b) strength comparison – different infill density

Tensile strength comparison of PETG, ASA and PLAs materials with different infill density is presented on Fig. 11a. In general, for all tested materials, the results showed a linear increase in tensile strength with increasing infill density. For all analysed infill densities, the highest tensile strength values were determined for PETG, while PLAs material showed the lowest tensile strength. By changing the infill density from 20% to 100%, for PETG material, the tensile strength can be increased up to 39%, i.e., it changes from 35,88 MPa to 49,84 MPa. Analysing only the results of samples with an infill structure (infill density from 20% to 80%), the tensile strength of PETG material can increase up to 14%, of ASA material up to 8%, and of PLAs material up to 14%, depending on the infill density.

What could also be noticed in the case of ASA material with an infill structure, is that for specimens with 20% infill density, the results showed higher tensile strength than specimens with 40% and 60% infill density, which was not expected. Although the difference is not significant, not over 6.5%, the test should be repeated for ASA material with 20% infill density.

Yield strength comparison for PETG, ASA and PLAs materials with different infill density is presented on Fig. 11b. The yield strength didn't show a linear increase with increasing infill density. For PETG and ASA materials, the yield strength for 20% infill density is slightly higher than for samples with infill density of 40%, after which it increased slightly with increasing infill density. While for PLAs material, for infill density from 20% to 60%, the yield strength did not change by more than 1%. Analysing the yield strength change for an increase in infill density from 20% to 100%, the results showed that yield strength increases up to 46% for PETG, up to 34% for ASA, and up to 27% for PLAs material.

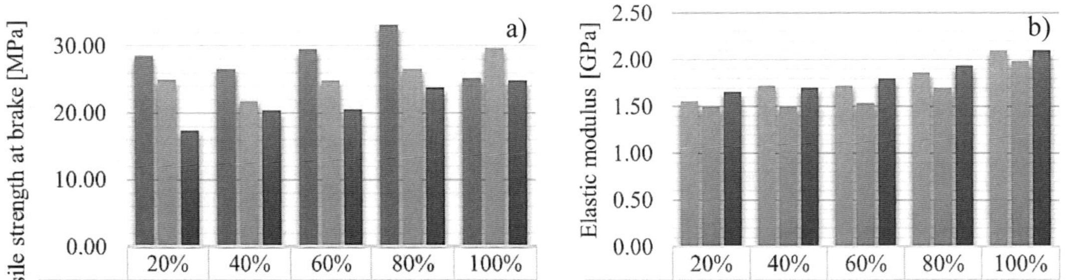

a)

	20%	40%	60%	80%	100%
PETG	28.53	26.56	29.53	33.11	25.13
ASA	24.93	21.76	24.84	26.55	29.70
PLAs	17.37	20.44	20.47	23.73	24.83

b)

	20%	40%	60%	80%	100%
PETG	1.56	1.72	1.72	1.86	2.10
ASA	1.50	1.50	1.54	1.70	1.98
PLAs	1.66	1.70	1.80	1.94	2.10

Fig. 12. PETG, ASA and PLAs tensile strength at brake (a) and elastic modulus (b) comparison – different infill density

Tensile strength at brake comparison for PETG, ASA and PLAs materials with different infill density is presented on Fig. 12a. As expected, tensile stress at brake is lower than tensile and yield strength for all analysed materials. For PLAs material, tensile strength at brake increased linearly from 17,37 MPa to 24,83 MPa (up to 9%) with increasing infill density. In the case of ASA material, the results show higher values of tensile strength at brake for samples with a infill density of 20% compared to samples with a density of 40%. While with the increase of infill density from 40% to 100%, the tensile strength at brake continued to increase linearly. Analysing the PETG material, it can be seen a similar behaviour of the material for tensile strength at brake, with the fact that the lowest values were measured at an infill density of 100%, which can also be seen from Fig. 7.

Elastic modulus comparison for PETG, ASA and PLAs materials with different infill density is presented on Fig. 12b. With the increase of infill density, the elastic modulus also increased for all analysed materials. By analysing the specimens with the infill structure, the PLAs material generally showed the highest values of measured elastic modulus. When increasing the infill density from 20% to 100%, the elastic modulus increases for PETG up to 35%, for ASA up to 32% and for PLAs material up to 26%.

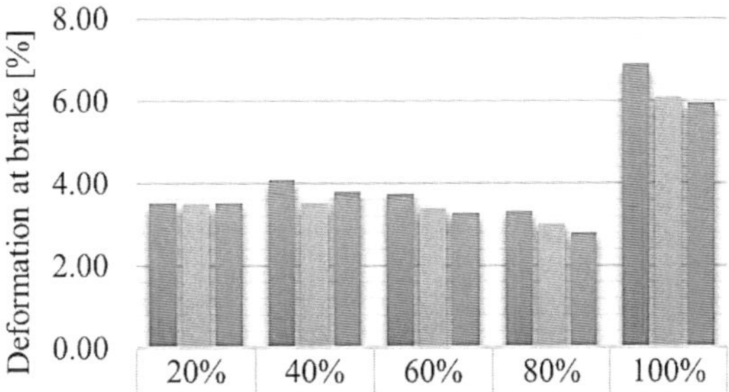

	20%	40%	60%	80%	100%
PETG	3.52	4.08	3.74	3.30	6.88
ASA	3.48	3.50	3.38	3.00	6.10
PLAs	3.52	3.78	3.26	2.78	5.94

Fig. 13. PETG, ASA and PLAs PLAs deformation at brake comparation – different infill density

Deformation at brake comparison for PETG, ASA and PLAs materials with different infill density is presented on Fig. 13. As expected, the highest values of deformation at break were obtained for samples with a infill density of 100%. For samples with an infill structure, as the infill density increases up to 80%, a slight decrease in the deformation at break can be seen. Materials with 100% infill density have significantly higher deformation at break compared to samples with a infill structure, and in the case of PETG material it is higher up to 108%, for ASA material up to 103% and for PLAs material up to 114%. Also, the lowest deformation at brake, for all tested materials, were for specimens with 80% infill density, which is not expected.

5. Conclusion

As expected, materials without infill structure (100% infill density) showed higher strength and deformation at brake compared to materials with infill structure. Analysing only the materials without infill structure, the results showed that the tensile strength is 32% higher when using PETG material compared to ASA and 57% compared to PLAs material. By comparing the yield strength, the results showed the highest values for PETG material, where the yield strength is 23% higher compared to ASA and 46% compared to PLAs material. ASA material showed highest values of tensile strength at brake, and it was 20% higher compared to PETG and PLAs materials. The same elastic modulus values were measured for PETG and PLAs materials, which are 6% higher than ASA material. By comparing the deformation at break, PETG material has higher values by 13% compared to ASA and 16% compared to PLAs material.

By analysing the tensile strength of materials with infill structure, a linear increase can be seen for all materials with increasing infill density. From 20% to 80% of infill density, PETG material showed the highest and PLAs material the lowest values of tensile strength. Also, by increasing the infill density from 20% to 80%, tensile strength goes up to 14% for PETG, up to 8% for ASA and up to 14% for PLAs material. Higher tensile strength values were measured for ASA material with 20% infill density compared to samples with 40% and 60% infill density, which was not expected, so the testing of these samples should be repeated.

In relation to the tensile strength, the yield strength did not increase linearly for PETG and ASA materials with infill structure. For PLAs material, the yield strength did not change more than 1% with increasing infill density from 20% to 60%. As expected, the tensile strength at brake was lower compared to tensile and yield strength for all materials with infill structure. A linear increase in tensile strength at brake is visible only for PLAs material. In contrast to ASA and PLAs materials, PETG material with 100% infill density had lower tensile strength at brake values compared to samples with 80% infill density.

The values of the elastic modulus increased linearly with the increase of infill density for all three materials. Generally, the PLA material with infill structure showed highest values of elastic modulus. With the increase of infill density from 20% to 100%, the elastic modulus values increase by 35% for PETG, by 32% for ASA and 26% for PLAs material.

Analysing the deformation at break, a slight drop in the value can be observed when the infill density increases from 20% to 80%, and a sudden jump for samples with an infill density of 100%. Materials without infill structure showed higher values of deformation at brake up to 108% for PETG, up to 103% for ASA and up to 114% for PLAs materials compared to the same materials with a infill structure.

Certainly, for future research, other mechanical properties of these materials should be analysed to better understand the behaviour of the PETG, ASA and PLAs materials under different loads, with the aim of their application for engineering and other applications.

6. Acknowledgments

The author gratefully acknowledges the support from AzureFilm company and University of Sarajevo – Faculty of Mechanical Engineering.

7. References

[1] Dezaki, M. L., Ariffin, M. K. A. M., Serjouei, A., Zolfagharian, A., Hatami, S., & Bodaghi, M. (2021). Influence of Infill Patterns Generated by CAD and FDM 3D Printer on Surface Roughness and Tensile Strength Properties. Applied Sciences (Switzerland), Vol. 11, No. 16, DOI: 10.3390/app11167272

[2] Pandzic, A., & Hodzic, D. (2021). Mechanical Properties Comparison of PLA, Tough PLA and PC 3D Printed Materials With Infill Structure – Influence of Infill Pattern on Tensile Mechanical Properties. IOP Conference Series: Materials Science and Engineering, Vol. 1208, DOI: 10.1088/1757-899x/1208/1/012019

[3] El Magri, A., Ouassil, S. E., & Vaudreuil, S. (2022). Effects of Printing Parameters on the Tensile Behavior of 3D-Printed Acrylonitrile Styrene Acrylate (ASA) Material in Z Direction. Polymer Engineering and Science, Vol. 62, No. 3, pp. 848–860, DOI: 10.1002/pen.25891

[4] Verma, P., Ubaid, J., Schiffer, A., Jain, A., Martínez-Pañeda, E., Kumar, & S., Kumar, S., & Org, K. O. (2021). Essential Work of Fracture Assessment of Acrylonitrile Butadiene Styrene (ABS) Processed Via Fused Filament Fabrication Additive Manufacturing. The International Journal of Advanced Manufacturing Technology, Vol. 113, pp. 771–784, DOI: 10.1007/s00170-020-06580-4

[5] Sodupe-Ortega, E., Sanz-Garcia, A., Pernia-Espinoza, A., & Escobedo-Lucea, C. (2018). Accurate Calibration in Multi-Material 3D Bioprinting for Tissue Engineering. Materials, Vol. 11, No. 8, DOI: 10.3390/ma11081402

[6] Algarni, M., & Ghazali, S. (2021). Comparative Study of the Sensitivity of PLA, ABS, PEEK, and PETG's Mechanical Properties to FDM Printing Process Parameters. Crystals, Vol. 11, No. 8, DOI: 10.3390/cryst11080995

[7] Pandzic, A., Hodzic, D., & Milovanovic, A. (2019). Effect of Infill Type and Density on Tensile Properties of PLA Material for FDM process. Proceedings of the 30th DAAAM International Symposium on Intelligent Manufacturing and Automation, Mostar, ISSN: 1726-9679, ISBN 978-3-902734-22-8, Katalinic, B. (Ed.), pp. 0545-0554, Published by DAAAM International, Vienna, DOI: 10.2507/30th.daaam.proceedings.074

[8] Martín, M. J., Auñón, J. A., & Martín, F. (2021). Influence of Infill Pattern on Mechanical Behavior of Polymeric and Composites Specimens Manufactured Using Fused Filament Fabrication Technology. Polymers, Vol. 13, No. 17, DOI: 10.3390/polym13172934

[9] Harpool, T. D., Alarifi, I. M., Alshammari, B. A., Aabid, A., Baig, M., Malik, R. A., Sayed, A. M., Asmatulu, R., & El-Bagory, T. M. A. A. (2021). Evaluation of the Infill Design on the Tensile Response of 3D Printed Polylactic Acid Polymer. Materials, Vol. 14, No. 9, pp. 1–20, DOI: 10.3390/ma14092195

[10] Hsueh, M. H., Lai, C. J., Wang, S. H., Zeng, Y. S., Hsieh, C. H., Pan, C. Y., & Huang, W. C. (2021). Effect of Printing Parameters on the Thermal and Mechanical Properties of 3D-Printed PLA and PETG, Using Fused Deposition Modeling. Polymers, Vol. 13, No. 11, DOI: 10.3390/polym13111758

[11] Schirmeister, C. G., Hees, T., Licht, E. H., & Mülhaupt, R. (2019). 3D Printing of High Density Polyethylene by Fused Filament Fabrication. Additive Manufacturing, Vol. 28, DOI: 10.1016/j.addma.2019.05.003

[12] Tanoto, Y. Y., Anggono, J., Budiman, W., & Philbert, K. V. (2020). Strength and Dimension Accuracy in Fused Deposition Modeling : A Comparative Study on Parts Making Using ABS and PLA Polymers. Jurnal Rekayasa Mesin, Vol. 11, No. 1, pp. 69–76, DOI: 10.21776/ub.jrm.2020.011.01.8

[13] Bonada, J., Pastor, M. M., & Buj-Corral, I. (2021). Influence of Infill Pattern on the Elastic Mechanical Properties of Fused Filament Fabrication (FFF) Parts Through Experimental Tests and Numerical Analyses. Materials, Vol. 14, No. 18, DOI: 10.3390/ma14185459

[14] Lesueur, M., Poulet, T., & Veveakis, M. (2021). Predicting the Yield Strength of a 3D Printed Porous Material From its Internal Geometry. Additive Manufacturing, Vol. 44 (May), DOI: 10.1016/j.addma.2021.102061

[15] AzureFilm 3D filaments. (2022). Material Technical Data Sheet, Available from: https://www.azurefilm.com/en/3d-filaments, Accessed: 2022-07-03

[16] Raam Kumar, S., Sridhar, S., Venkatraman, R., & Venkatesan, M. (2020). Polymer Additive Manufacturing of ASA Structure: Influence of Printing Parameters on Mechanical Properties. Materials Today: Proceedings, Vol. 39, No. 4, 1316–1319, DOI: 10.1016/j.matpr.2020.04.500

DOI: 10.2507/33rd.daaam.proceedings.032

INFLUENCE OF PRINTHEAD NOZZLE DIAMETER ON MECHANICAL PROPERTIES OF FDM PRINTED PLA MATERIAL AND PRINTING TIME

Damir Hodžić & Adi Pandžić

This Publication has to be referred as: Hodzic, D[amir] & Pandzic, A[di] (2022). Influence of Printhead Nozzle Diameter on Mechanical Properties of FDM Printed PLA Material and Printing Time, Proceedings of the 33rd DAAAM International Symposium, pp.0231-0237, B. Katalinic (Ed.), Published by DAAAM International, ISBN 978-3-902734-36-5, ISSN 1726-9679, Vienna, Austria
DOI: 10.2507/33rd.daaam.proceedings.032

Abstract

There are various 3D printing parameters that affect the mechanical properties of FDM printed materials as well as the final price of the finished product. One of the influential factors on printing time is certainly the diameter of the print head nozzle. In this paper influence of print head nozzle diameter on the mechanical properties of FDM printed PLA material was analysed. Tensile and flexural mechanical properties were examined, and tensile strength, elastic modulus, flexural strength, flexural modulus, and strain are presented in results. Also, the influence of the nozzle diameter on the total printing time was analysed because the printing time directly affects the final price of the product.

Keywords: FDM; nozzle diameter; PLA; material properties; printing time.

1. Introduction

Undoubtedly, one of the technologies that increasingly influences the course of industry development is additive technologies (AM). In the last few years, a huge focus is on development of AM technologies. There are many different AM technologies on the market today, but one of the most popular and used is fused deposition modeling, better known as FDM technology [1].

Compared to the conventional manufacturing technologies, reducing the time from initial concept design to the final product is one of the main advantages of AM technologies. With this technologies, designers or engineer are able to produce the prototype or final product in just a few days, without need for special tools or developing special manufacturing process. The rapid production of small series products using AM technologies has been best proven in the crisis period during COVID 19 for making different products in medical purposes [2].

FDM 3D printing technology uses thermoplastic polymeric material as a base material, also known as a filament. Process is based on the material extrusion through a nozzle mounted in the printhead (Fig. 1). The filament is fed from a coil into the printer head, where it is heated in heater element, and in molten phase material is forced out of the printhead nozzle on to build platform. With precisely defined movement of the printer head based on G-code, the finished product is made layer by layer on the build platform [3], [4].

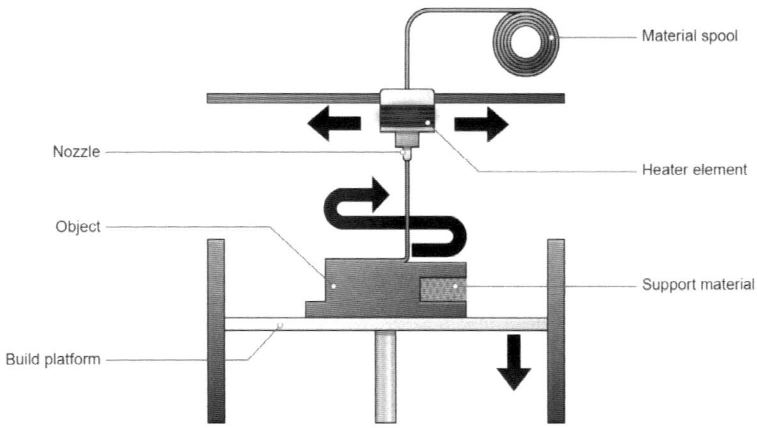

Fig. 1. A schematic of the FDM printing technology process [5]

Analysing the literature and previous research in this area, it can be concluded that many printing parameters affect the mechanical properties of FDM printed materials. Also, in addition to the mechanical properties of the printed material, the 3D printing parameters also affect other product characteristics such as surface quality, accuracy of the product geometry, porosity, total mass as well as the production time which directly affects the price of the product itself. Some of the printing parameters that affect the mechanical properties of the material as well as the total printing time are: layer height, print orientation, nozzle diameter, printing speed, infill design, printing temperature, etc. However, the choice of the right combination of printing parameters still needs further investigation because optimal process parameters defined for one material may not work well for another material [6], [7], [8].

One of the printing parameters that has an impact on the material mechanical properties, print quality, printing time and overall price of the printed product is the printhead nozzle diameter. In study [9], influence of nozzle diameter on the macrostructure and selected strength properties of PLA printed material was analysed. It was concluded that for nozzle diameter greater than 0,4 mm, an increase in the share of air voids between the print rasters occurred. It was stated that perhaps this phenomenon has an influence on the impact strength, because the results showed the highest impact strength values for samples with the largest voids between the rasters of the polymer material. It is also stated that the cause of this is that he voids in the sample structure may absorb some of the energy generated during the impact test. Results in this study showed highest value of impact strength when using a 0,8 mm nozzle diameter. For tensile strength, the highest values were obtained for 0,4 and 0,8 mm nozzle diameters. Also, in study [10] was concluded that higher pore size was obtained with higher nozzle diameter.

Influence of the printing nozzle diameter on tensile strength and printing time was analysed in study [11]. In this study, seven different nozzle diameters were varied, from 0.2 mm to 0.8 mm. Also, testing samples were printed with 50% and 100% infill density. Analysing the breaking force, results showed that for samples with 50% infill density breaking force was increasing with increasing the nozzle diameter. On the other hand, for samples with 100% infill density, breaking force was decreasing with the increase of the nozzle diameter. Also, printing time dependence on the nozzle diameter was presented, where for both infill densities (50% and 100%), printing time decreased with nozzle diameter increase.

In this study aim was to analyse influence of four different printhead nozzle diameters: 0,25 mm; 0,4 mm; 0,6 mm and 0,8 mm, on tensile and flexural mechanical properties of 3D printed PLA material. Also, influence of nozzle diameter on total printing time was analysed.

2. Materials and methods

In order to analyse the influence of the printhead nozzle diameter on the tensile and flexural mechanical properties of FDM printed PLA material, a detailed experimental study was carried out, presented through the flow diagram in Fig 2.

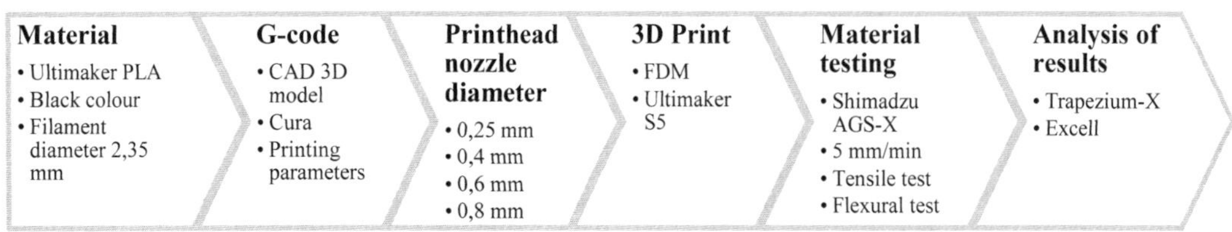

Fig. 2. Experiment flow diagram

The material used in the tests was thermoplastic polymer polylactic acid (PLA), produced by Ultimaker company, in black colour. Certainly, one of the most commonly used materials in FDM technology, which is also biodegradable material. Mechanical properties defined by manufacturer are presented in table 1.

Material properties	Test method	Typical value
Tensile strength [MPa]	ASTM D3039 (5 mm / min)	52,5 MPa
Tensile elastic modulus [GPa]	ASTM D3039 (1 mm / min)	3,25 GPa
Strain at brake [%]	ASTM D3039 (5 mm / min)	7,8 %
Flexural strength [MPa]	ISO 178 (5 mm / min)	96,8 MPa
Flexural modulus [GPa]	ISO 178 (1 mm / min)	3,02 GPa
Flexural strain at brake [%]	ISO 178 (5 mm / min)	4,8 %
Charpy impact strength (at 23 °C)	ISO 179-1 / 1eB (notched)	3,9 kj/m^2
Hardness	ISO 7619-1 (Durometer, Shore D)	84 Shore D

Table 1. FDM printed PLA material properties by Ultimaker manufacturer

Tensile and Flexural test specimens 3D models were designed using CAD software Solidworks 2022. Tensile test specimen were designed according to ISO 527, while the flexural test specimen was designed according to ISO 178, presented on Fig. 3.

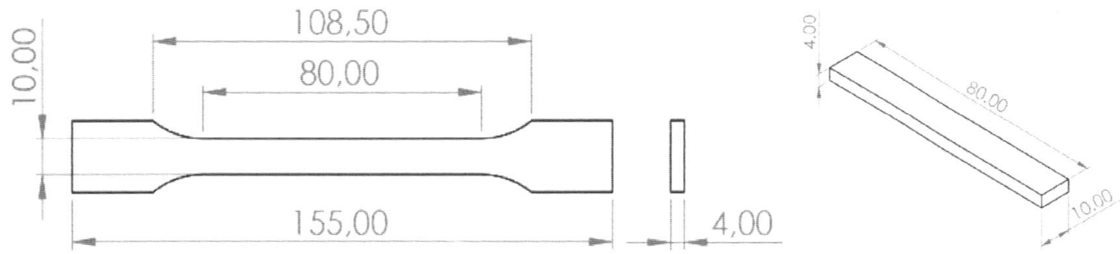

Fig. 3. Tensile test specimen according to ISO 527 (left) and flexural test specimen according to ISO 178 (right)

G-code with 3D printing parameters was prepared using Ultimaker Cura slicer. For all specimens, printing parameters are defined by default Ultimaker printing profile "Normal" for Ultimaker black PLA filament and for four different nozzle diameters: 0,25 mm; 0,4 mm; 0,6 mm; and 0,8 mm. Also, although the printhead nozzle diameter was changed, a layer height of 0,15 mm was used for printing all samples (Fig. 4).

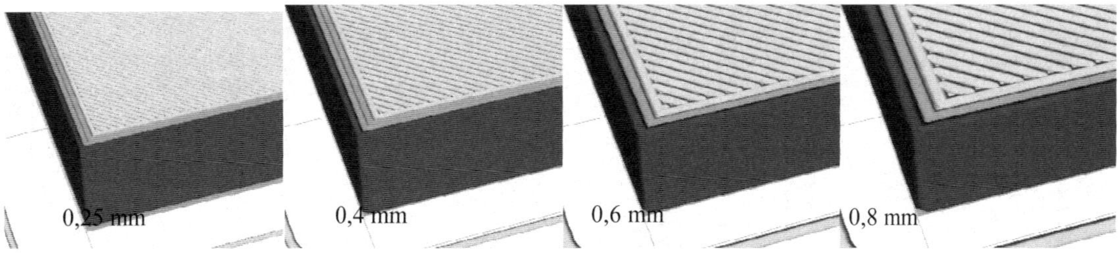

Fig. 4. Example of 3D printing product with different printhead nozzle diameter

After the g-code was defined, 3D printing was performed using the FDM "Ultimaker S5" desktop 3D printer. Ultimaker S5 is composite-ready dual extrusion professional desktop 3D printer with 330x240x300 mm build volume.

Tensile and flexural testing specimens were printed in five replicas and presented on Fig 5. All specimens are tested on Shimadzu AGS-X universal testing machine with capacity of 10 kN with testing speed of 5 mm/min. Data monitoring was performed using Shimadzu Trapezium-X software. After data collection, the results were processed using the Excel program and presented in further text.

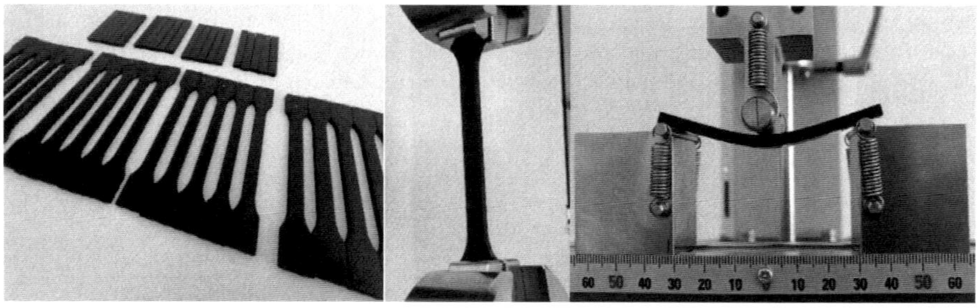

Fig. 5. Tensile and flexural FDM 3D printed specimens

3. Results and discussion

After the 3D printing of the test samples, the tensile and flexural mechanical properties were analysed. Stress-strain diagrams for tensile and flexural mechanical properties are presented on Fig. 6, where curves representing the mean values of five replicas of each sample.

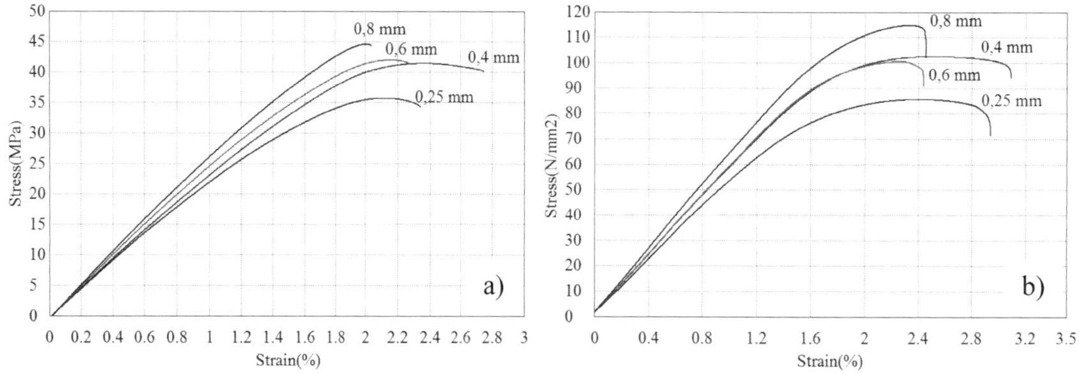

Fig. 6. Stress-strain diagrams for tensile (a) and flexural (b) mechanical properties of FDM printed PLA material with different printhead nozzle diameter

Influence of printhead nozzle diameter on tensile strength and elastic modulus is presented on the diagrams in Fig. 7. The diagrams show a linear increase in tensile strength and elastic modulus increasing printhead nozzle diameter. The highest tensile strength values (44,6 MPa) were measured for samples made with a 0,8 mm nozzle diameter. Samples 3D printed with a 0,25 mm nozzle diameter showed the lowest tensile strength values. For samples 3D printed with 0,4 mm and 0,6 mm nozzles diameter, tensile strength values were measured with a very small difference, no more than 0,3%. Analysing the elastic modulus, the highest values (2,8 GPa) can be seen for samples made with a 0,8 mm nozzle diameter, while the lowest values (2,4 GPa) were measured for samples printed with a nozzle diameter of 0,25 mm. Compared to the tensile strength, the elastic modulus has a larger difference (4%) in values for samples printed with a nozzle diameter of 0.4 and 0.6 mm.

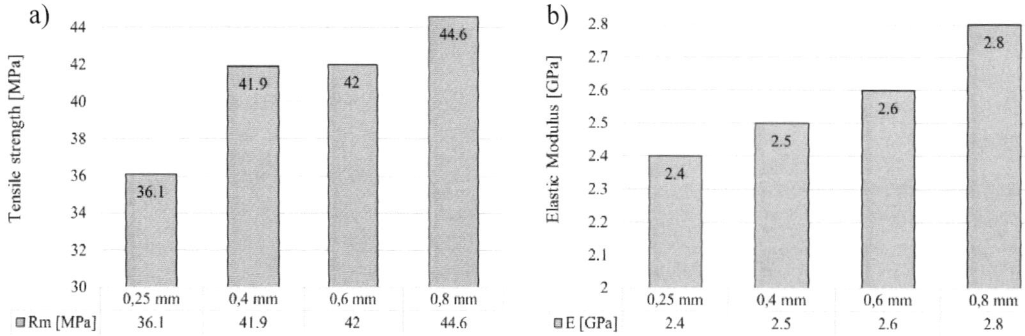

Fig. 7. Influence of printhead nozzle diameter on tensile strength (a) and elastic modulus (b)

Diagrams on Fig. 8 is presenting the influence of printhead nozzle diameter on flexural strength and flexural modulus. By examining the flexural mechanical properties, similar results were obtained as with the tensile tests results. The highest flexural strength values (114.9 MPa) were obtained for a 0,8 mm nozzle diameter, while the lowest values (85.7 MPa) were obtained for a 0,25 mm nozzle diameter.

The highest value of flexural modulus (6,3 GPa) was obtained for 0,8 mm nozzle diameter, while the lowest flexural modulus (5,2 GPa) was obtained for 0,25 mm nozzle diameter. Also, with an increase in the nozzle diameter from 0.25 mm to 0.8 mm, a linear increase in flexural strength and flexural modulus can be seen, which was previously shown in tensile mechanical properties results.

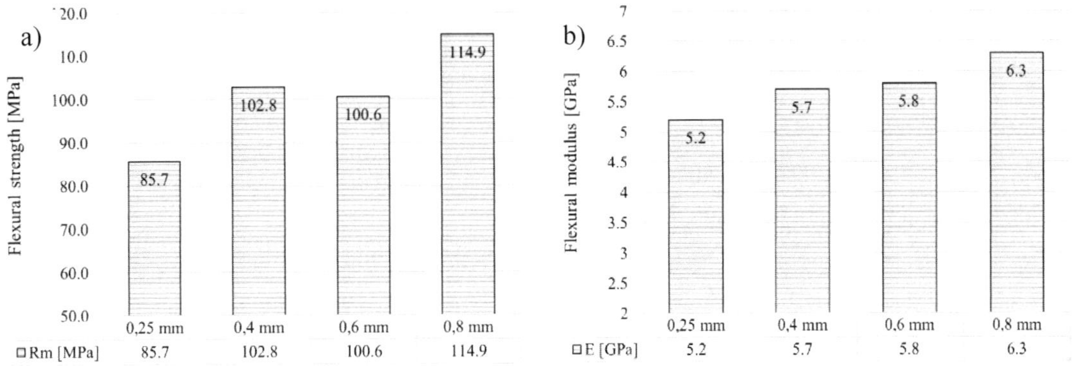

Fig. 8. Influence of printhead nozzle diameter on flexural strength (a) and flexural modulus (b)

In addition to the tensile and flexural strength, the tensile and flexural strain at brake was also analysed, and results are presented in diagrams on Fig 9. Similar changes of strain are observed when the nozzle diameter increases for both cases.

With an increase in nozzle diameter from 0,25 mm to 0,4 mm, an increase in strain is seen, then with an increase in diameter from 0,4 mm to 0,8 mm, a decrease in the value of strain can be seen. Also, it should be noted that the same flexural strain values were measured for nozzle diameters of 0.6 mm and 0.8 mm.

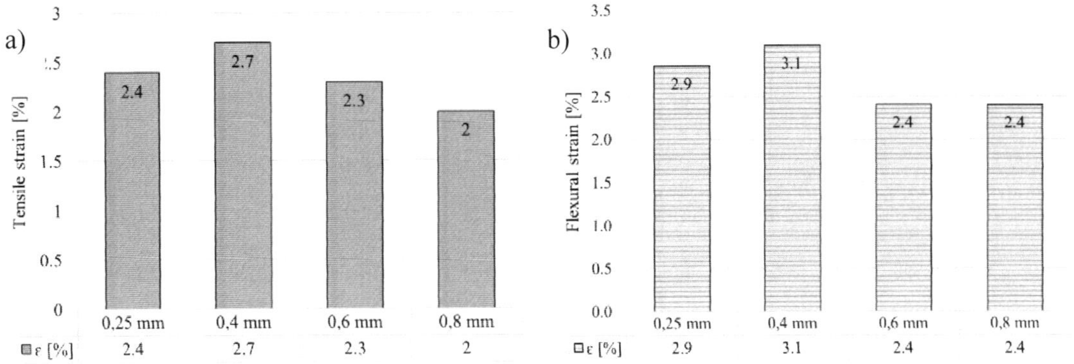

Fig. 9. Influence of printhead nozzle diameter on tensile (a) and flexural (b) strain

The effect of the printhead nozzle diameter on the total printing time was also analysed, whereby the total printing time refers to the 3D printing of one test sample for testing tensile mechanical properties. Influence of printhead nozzle diameter on total printing time is presented on Fig 10.

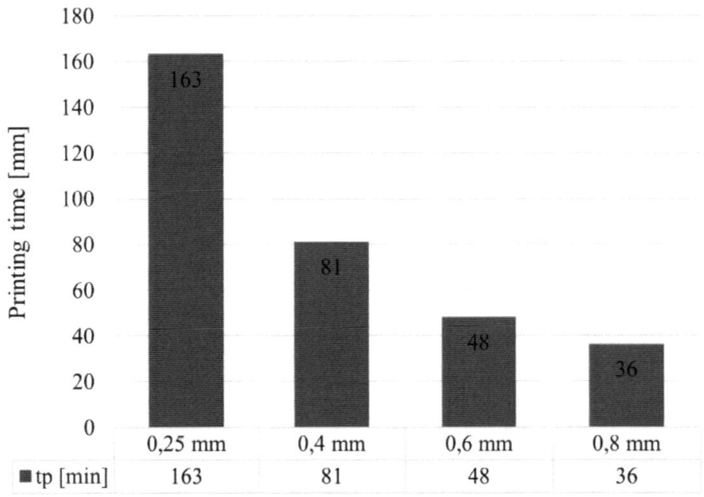

Fig. 10. Influence of printhead nozzle diameter on total printing time (tp)

As expected, with the nozzle diameter increase, the total printing time decreases. Significantly longer printing time with a nozzle diameter of 0.25 mm is visible compared to other nozzle diameters, where the difference is up to four times higher.

Table 2 presents the overall results of the nozzle diameter influence on the tensile and flexural mechanical properties of the FDM printed PLA material, as well as on the total printing time.

Nozzle diameter [mm]	**0,25**	**0,4**	**0,6**	**0,8**	**Ultimaker TDS***
Tensile strength [MPa]	36.10	41.90	42.00	44.60	52,5 MPa
Elastic modulus [GPa]	2.40	2.50	2.60	2.80	3,25 GPa
Tensile strain [%]	2.40	2.70	2.30	2.00	7,8 %
Flexural strength [MPa]	85.74	102.83	100.61	114.89	96,8 MPa
Flexural modulus [GPa]	5.20	5.70	5.80	6.30	3,02 GPa
Flexural strain [%]	2.86	3.09	2.41	2.40	4,8 %
Total printing time [min]	163.00	81.00	48.00	36.00	-
*Material mechanical characteristics from Ultimaker PLA material technical data sheet.					

Table 2. Overall results – Printhead nozzle diameter influence on tensile and flexural mechanical properties and total printing time

4. Conclusion

The research showed an influence of the printhead nozzle diameter on the tensile and flexural mechanical properties of FDM printed PLA material. Some of the main conclusions are:

- With the increase of the printhead nozzle diameter, the tensile strength, elastic modulus, flexural strength, and flexural modulus increased linearly.
- All analysed mechanical properties, tensile and flexural, change pretty much the same with the change in printhead nozzle diameter.
- The highest values of tensile strength and elastic modulus were measured for samples made with a 0,8 mm nozzle diameter, and the lowest values for 0,25 mm diameter.
- The highest values of flexural strength and modulus were measured for samples made with a 0,8 mm nozzle diameter, and the lowest values for 0,25 mm diameter.
- Maximum tensile and flexural strain at brake values were measured for 0,4 mm printhead nozzle diameter.
- Changing the printhead nozzle diameter it can be seen the same change for both the tensile and flexural measured mechanical properties.
- Total printing time was linearly decreasing with printhead nozzle diameter increase, which was expected.

Through future research, the same analysis should be done but for other materials, to see if the change in nozzle diameter shows the same effect as with FDM printed PLA material. Also, the influence of the nozzle diameter on other mechanical properties, as well as on other properties of FDM printed materials in general, should also be analysed.

5. References

[1] Pandzic, A. & Hodzic, D. (2021). Mechanical Properties Comparison of PLA, Tough PLA and PC 3D printed Materials with Infill Structure – Influence of Infill Pattern on Tensile Mechanical Properties, IOP Conf. Series: Materials Science and Engineering, 1208, DOI: doi:10.1088/1757-899X/1208/1/012019

[2] Muminović, A. J.; Pandzic, A.; Pervan, N. & Delic, M. (2022). Product Development and Design Inside Industry 4.0 From Rapid Prototyping to Additive Manufacturing, Special Editions ASAB&H CCII, DTS, Vol. 20, pp. 69-80, DOI: 10.5644/PI2022.202.21

[3] Khosravani, M. R.; Schuurmann, J.; Berto, F. & Reinicke, T. (2021). On the Post-Processing of 3D-Printed ABS Parts, Polymers, Vol. 13, No. 10, DOI: 10.3390/polym13101559

[4] Hodzic, D. & Pandzic, A. (2021). Influence of Infill Design on Compressive and Flexural Mechanical Properties of FDM Printed PLA Material, Proceedings of the 32nd DAAAM International Symposium, pp.0191-0199, B. Katalinic (Ed.), Published by DAAAM International, ISBN 978-3-902734-33-4, ISSN 1726-9679, Vienna, Austria, DOI: 10.2507/32nd.daaam.proceedings.029

[5] https://make.3dexperience.3ds.com/processes/3D-printing, (2018). 3D Printing – Additive: Introduction to 3D Printing – Additive processes, Accessed on: 2022-09-15

[6] Pandzic, A.; Hodzic, D.; Hajro, I. & Tasic, P. (2020). Strength Properties of PLA Material Obtained by Different Models of FDM 3D Printer, Proceedings of the 31st DAAAM International Symposium, pp.0315-0322, B. Katalinic (Ed.), Published by DAAAM International, ISBN 978-3-902734- 29-7, ISSN 1726-9679, Vienna, Austria, DOI: 10.2507/31st.daaam.proceedings.044

[7] Gebisa, A. W. & Lemu, H. G. (2018). Investigating Effects of Fused-Deposition Modeling (FDM) Processing Parameters on Flexural Properties of ULTEM 9085 using Designed Experiment, Materials, Vol. 11, No. 4, DOI: 10.3390/ma11040500

[8] Pandzic A.; Hodzic D. & Milovanovic A. (2019). Effect of Infill Type and Density on Tensile Properties of PLA Material for FDM Process, Proceedings of the 30th DAAAM International Symposium on Intelligent Manufacturing and Automation, Zadar, ISSN: 1726-9679, ISBN 978-3-902734-22-8, Katalinic, B. (Ed.), pp. 545-554, Published by DAAAM International, Vienna, DOI: 10.2507/30th.daaam.proceedings.074

[9] Czyzewski, P.; Marciniak, D.; Nowinka, B.; Borowiak, M. & Bielinski, M. (2022). Influence of Extruder's Nozzle Diameter on the Improvement of Functional Properties of 3D-Printed PLA Products, Polymers, Vol. 14, No. 2, 356, DOI: 10.3390/polym14020356

[10] Corral-Buj, I.; Bagheri, A.; Fernandez, A. D. & Lopez, C. R. (2019). Influence of Infill and Nozzle Diameter on Porosity of FDM Printed Parts with Rectilinear Grid Pattern, 8th Manufacturing Engineering Society International Conference, Vol. 41, pp. 288-295, DOI: 10.1016/j.promfg.2019.09.011

[11] Kinski, W. & Pietkiewicz, P. (2020). Influence of the Printing Nozzle Diameter on Tensile Strength of Produced 3D Models in FDM Technology, Agricultural Engineering, Vol. 24, No. 3, pp. 31-38, DOI: 10.1515/agriceng-2020-0024

DOI: 10.2507/33rd.daaam.proceedings.033

NON-INVASIVE STEAM FLOW MEASUREMENT AS AN ENERGY MEDIUM BY CLAMP-ON ULTRASONIC METHOD IN CONTEXT OF ENERGY MANAGEMENT CONCEPT IMPLEMENTATION, OPTIMIZATION AND SAFETY

Jan Ilko, Miroslav Rusko & Jaroslav Demko

This Publication has to be referred as: Ilko, J[an]; Rusko, M[iroslav] & Demko, J[aroslav] (2022). Non-Invasive Steam Flow Measurement as an Energy Medium by Clamp-on Ultrasonic Method in Context of Energy Management Concept Implementation, Optimization and Safety, Proceedings of the 33rd DAAAM International Symposium, pp.0238-0246, B. Katalinic (Ed.), Published by DAAAM International, ISBN 978-3-902734-36-5, ISSN 1726-9679, Vienna, Austria
DOI: 10.2507/33rd.daaam.proceedings.033

Abstract

The aim of this study is to present the results of test measurements of steam flow for the development of energy balancies and the design of optimization plans, provide application model for energy management implementation concepts and the safety aspects. Furthermore, to demonstrate innovative technology in field of industrial measurement and theoretical analysis of such technics as a part of modern way of technology improvement. The flow measurement was performed on a low-temperature steam branch. An ultrasonic flow meter from a leading manufacturer of ultrasonic measuring technology was used for the measurement, which supplies non-invasive ultrasonic devices in addition to other process media the clamp-on technology for measuring quantitative parameters of low-temperature and high-temperature steam. Another goal is to use procedures and collected data for further research and deployment of such technology in industrial practice to increase process efficiency and operational safety.

Keywords: Steam flow measurement; Ultrasonic method; Energy management; Operational safety.

1. Introduction

Steam is widely used mainly in industry, but it also finds its place in the energy, pharmaceutical, medical, and other sectors. To monitor consumption, optimize utilization, energy balances and reduce losses, it is necessary to measure its flow reliably and as accurately as possible. Until now, this has been performed by measuring devices that use conventional flow measurement methods, such as e.g., measuring orifice together with differential pressure gauge, etc. All these methods are invasive ones and the integration into the system, repair, or maintenance, require the medium to be redirected to bypass, or in the worst case, shut down of a production section or whole operation. Non-invasive ultrasonic flow meters do not interfere with the process and there is no need to change the operation in any way, which brings huge benefits as financial savings and significantly increases the reliability of the technological process and safety at work.

In case of energy management, precise flow and energy monitoring has a significant place in the whole schematics to provide reliable input data for further proceedings. Nowadays, the ISO 50 001 is a common concept to lead to energy efficiency in a proper and reliable way.

2. Energy management

Sustainable business development and companies market positioning require from companies to maximize added value with minimal resource utilization. The three main goals to optimize are to ensure safety, high efficiency, productivity and economy of „man-machine environment" system [1]. "Our Common Future", a report published by the World Commission for Environment and Development presided by Gro Harlem Brundtland, defines sustainable development as the kind of development that meets the needs of the present without compromising the ability of future generations to meet their own needs [2]. Growing load and deterioration of the environment can be interpreted as a result of some external effects interventions. While the positive externalities influence the positive productional and utilizational functions of other subjects, the negative externalities influence the negative ones. Both types of external effects can act as partial or global externalities. Linking of environmental issues to economy and finance is an important sphere [3]. In context of the current socio-economic activities of society, the solution of issues and problems related to environmental pollution is becoming more and more popular. Environmental degradation not only affects densely populated areas, heavy industrial pollution etc., but often affects areas that do not directly attack environmentally unsuitable activities [4]. Innovations refer to the actions required to create new ideas, processes or products. The implementation of new ideas, processes or products leads to positive effective change in a company. The positive effective change in a company is accomplished through savings. Thus, company makes savings in different types of resources via innovations, contribute to social goal or influence the development of technological solutions. In industry, savings via innovations are mostly reflected in redesigning products and processes to cut out unnecessary costs. That type of innovations are called frugal innovations. Frugal innovations represent innovations of existing solutions with accent of achieving resource savings. Nowadays, frugal innovations, which are used in industry, play a vital role in era of fourth industrial revolution, named Industry 4.0 [5]. In the year 2010, Germany initiated the "Industry 4.0" project, in order to promote its economic development and global competitiveness. China issued the "Made in China 2025" strategic plan, in order to transform the country from a manufacturing giant to a world manufacturing power. Japanese government introduced a new concept called "Society 5.0 (Super-smart society)" [6].

Technological progress leads to the implementation of artificial intelligence in the field of industry. Such systems require high reliability of partial blocks and sensors to ensure high reliability of the entire system controlled by artificial intelligence. Contactless systems show a lower failure rate and higher reliability as they are not often in contact with an aggressive process medium, and this makes them an ideal solution for securing input data for processing and controlling the process with artificial intelligence. Artificial intelligence (AI) is the most important general-purpose technology of our times [7]. It is getting used as an input by many downstream sectors enabling innovations in rapidly growing number of products, services and processes [8]. For some industries, AI is becoming a commonplace. The peak of the main wave of artificial intelligence (AI) tsunami is about to hit industries, governments and societies accelerating the pace of product and service innovation, and putting data in the focus of the business interest [9].

The rational use of energy and energy sources is also a growing challenge, which aims to preserve the environment. Industrial systems, primarily production systems, are the largest energy consumers. The need for this theme is caused by the pressure of European regulation on the implementation of energy saving measures as well as to have positive impact on the environment [10]. Energy Management Concept as part of Green Building Concept is focused to Improve Energy Efficiency Index (EEI) and Water Consumption Index (WCI). The Implementation Energy Management Concept in an office building is based on the management system model of continual improvement ISO 50 001. ISO 50 001 has been developed for an organization that is committed to resolving its impact, saving resources, and improving the result through efficient energy management. Designed to support organizations in all sectors, this ISO standard provides a practical way to improve energy use, through the development of an energy management system (EnMS) [11].

Accurate, fast, and stable electrical energy consumption forecasting plays a vital role in decision making, energy management, effective planning, reliable and secure power system operation. Inaccurate forecasting can lead to the electricity shortage, wastage of energy resources, power outage, and in the worst case, power grid collapse. Contrarily, accurate forecasting enables policymakers and public agencies to make real-time decisions imperative for the energy management and power system's secure and reliable operation [12].

Gryshchenko, I. (2021) published an interesting study on the implementation of ISO 50 001 in the context of current topic, the COVID-19 pandemic [13]. Several studies have been carried out on the implementation of ISO 50 001: 2018. The authors focused on various aspects of this issue in terms of buildings, production and organizational units and regions:

- Kurniawan, R. & A. Feinnudin (2021) - this study focuses on the public building in Indonesia that has implemented an energy management system compliant with ISO 50 001 standard [14].
- Siregar, M. et al. (2019) - implementation of Energy Management Concept and Energy Management System in High Rise Office Building [15].
- Cuanga, O. M. & G. C. Sánchez (2020) - study was carried out in National Bus Company (BEU) Cienfuegos, addressed to the aim of applying energy planning stage tools energy system management in correspondence with ISO 50 001 Standard [16].

- Dall'O, G. et al. (2020) - a case study on energy management for heating load reduction for a social building stock in Northern Italy [17].
- Quispe, E. et al. (2021) - this study presents the application of tools to develop a procedure to estimate the energy saving potentials in a small foundry company in Colombia [18].
- Rovňák, M. et al. (2021) - study focuses on the issue of energy efficiency management in the region of Prešov self-governing region (PSR), Slovakia, as the energy market is liberalized, and the behaviour of electricity consumers is influenced not only by conventional but also by alternative suppliers of this type of energy [19].

3. Steam production

The availability of this medium is essential for industrial production [20]. Steam generation is familiar process of boiling water to make steam. Thermodynamically, the heat energy used results in a change of phase from liquid to gaseous state, i.e., from water to steam. A steam generating system must provide a continuous and uninterrupted heat source for this conversion [21]. The effect of equipment failures on system operation performance and costs are analysed by system reliability, which is contributed by the system configuration, system redundancy, individual equipment failure and repair characteristics, operation mode and mode transfer [22].

4. Methods

For our test measurement a clamp-on ultrasonic portable flow meter was used. Clamp-on ultrasonic flow meters have the potential to be applied to existing pipes. Therefore, they have been widely used in industrial plants for measuring liquid and gas flow rates. Their principle is based on the time-of-flight (TOF), which changes with the flow velocity along the propagated ultrasonic pulses. Thus, flow rate is obtained based on the transit time difference of the ultrasonic pulses between upstream and downstream sensors. To measure the flow rate accurately, the transmitted ultrasonic signals must be detected. However, the measurement is difficult in the case of steam flow because ultrasonic attenuation in steam is higher than in liquid. Furthermore, the large difference in the acoustic impedance between the pipe material and steam results in a lower intensity of the transmitted ultrasonic signal for clamp-on ultrasonic flow meters. Thus, measuring liquid flow rates is easier than measuring gas flow rates using the clamp-on system [23].

5. Research

For our test we used portable unit Fluxus G601ST with ultrasonic sensors GRK1SC3. Before implementing of the ultrasonic sensors, the thermal insulation of the pipe must be dismounted in proper length. The surface of the pipe must be painted with special noise-damping varnish as shown by Fig. 1. Variofix mounting brackets were used for sensor fixation on the pipe. Both brackets must be mounted in precise opposite position as shown by Fig. 2.

Fig. 1. Noise-damping varnish on the steam pipe

Fig. 2. Mounted brackets in opposite position

The ultrasonic sensors need to have a good contact with the pipe surface, therefore the dumping varnish between the active size of the sensors and the pipe to be scratched, as shown by Fig. 3.

Fig. 3. Scratched noise dumping area under sensor`s active side

By transit time method the sensors need to be heated up on minimum 100 °C to get signal. This can take one hour until the sensors reach the limit. In our case, the sensors were preheated on the surface of another steam pipe on 70 °C to shorten the time of heating up. Before starting of the measurement, the control unit to be configured on proper process parameters. After the parameterisation, the ultrasonic flow meter unit software leads the operator through diagnostic parameters to ensure the reliable measurement. The fluid parameters were following:

- Outer diameter: 275.00 mm.
- Pipe wall thickness: 7.20 mm.
- Pipe wall material: Carbon steel.
- Roughness: 0.10 mm.
- Fluid: Low pressure steam.
- Fluid sound speed: 505.92 m/s.
- Fluid temperature: 187.00 °C.
- Fluid pressure: 9.50 bar(g).
- Unit of measurement: kg/h.
- Damping: 10 s.

Software supplied by the flow meter producer allows the operator to see the shape of the ultrasonic signals with the diagnostic values, simultaneously. Such screen is shown on the following picture.

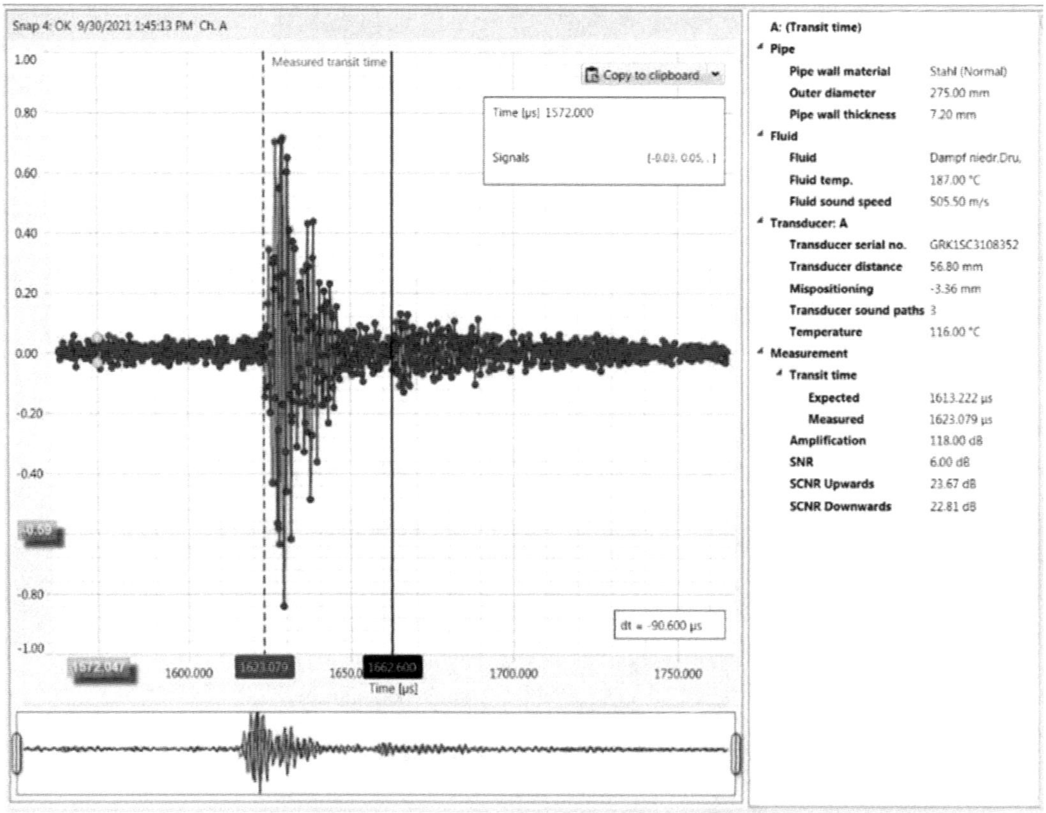

Fig. 4. Diagnostic screen of the steam flow measurement process using FluxDiag program

There is an internal data logger integrated in the flow meter unit. In our case, the data logger was set to store the data as average in ten minutes intervals. The measured data are shown in the Table 1. Graphical interpretation of the data from table 1 is in the Fig. 5.

Time hh:mm	Flow velocity [m/s]	Mass flow rate [kg/h]	Sound speed [m/s]	Volumetric flow rate [m³/h]	SCNR [dB]	SNR [dB]
5:07	1.31	1334.09	505.33	250.97	29	11

5:17	1.32	1343.28	505.21	252.7	30	11
5:27	1.35	1375.94	505.36	258.84	30	11
5:37	1.34	1365.74	505.21	256.92	30	11
5:47	1.37	1400.44	505.53	263.45	26	10
5:57	1.36	1392.28	505.39	261.91	28	10
6:07	1.35	1377.98	505.31	259.22	30	11
6:17	1.37	1398.4	505.36	263.06	30	11
6:27	1.31	1339.2	505.49	251.93	28	10
6:37	1.32	1348.38	505.28	253.66	27	10
6:47	1.41	1439.23	505.46	270.74	28	10
6:57	1.31	1339.2	505.29	251.93	28	10
7:07	1.27	1293.26	505.38	243.29	25	10
7:17	1.27	1296.33	505.53	243.86	25	11
7:27	1.21	1229.98	505.25	231.38	28	10
7:37	1.16	1187.11	505.45	223.32	30	11
7:47	1.07	1095.24	505.41	206.03	27	10
7:57	1.08	1098.31	505.44	206.61	28	10
8:07	1.06	1086.06	505.52	204.31	27	11
8:17	1.07	1088.1	505.35	204.69	30	10
8:27	1.06	1080.95	505.62	203.35	31	11
8:37	0,99	1013,58	505.26	190.67	30	11
8:47	0,94	955,4	505.5	179.73	31	10
8:57	0,98	999,29	505.63	187.99	31	11
9:07	0,91	932,95	505.39	175.5	29	11
9:17	0,87	889,06	505.53	167.25	31	11
9:27	0,85	871,7	505.45	163.98	27	10
9:37	0,87	885,99	505.43	166.67	31	11
9:47	0,86	880,89	505.64	165.71	28	11
9:57	0,78	793,11	505.48	149.2	29	11
10:07	0,81	827,81	505.61	155.73	30	11
10:17	0,73	741,05	505.46	139.4	30	11
10:27	0,8	814,54	505.48	153.23	30	11
10:37	0,75	770,65	505.37	144.97	31	11
10:47	0,74	759,42	505.51	142.86	31	11
10:57	0,65	660,41	505.65	124.24	31	11

Table 1. Logged data during steam flow measurement in eight minutes interval and ten seconds period

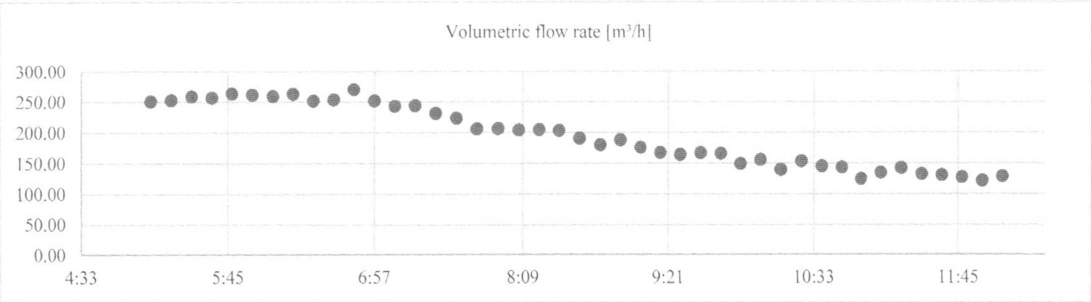

Fig. 5. Volumetric steam flow rate in eight minutes interval

Using the software supplied by the flow meter producer, there is possibility to generate a report with the uncertainty of the measurement. The uncertainties are shown in the Table 2. Graphical evaluation of the measurement uncertainty is shown in the Fig. 6.

Standard uncertainty component	**Source of uncertainty**	**Value of standard uncertainty**		**Sensitivity factor**	**Flow velocity**			
					0.10 m/s	**0.37 m/s**	**1.36 m/s**	**5.00 m/s**
					Transit time difference			
					$9.682e^{-8}$ s	**$3.582e^{-7}$ s**	**$1.317e^{-6}$ s**	**$4.841e^{-6}$ s**
$u(x_i)$		$u(x_i)$	$u_r(x_i)=u(x_i)/x_i$	$c_{r,i}=c_i*x_i/\acute{v}$	Contribution $\lvert c_{r,i}u_r(x_i)\rvert$			
$u(K_{Re})$	Flow profile		0.28 %	1	0.28 %	0.28 %	0.28 %	0.28 %
$u(A)$	Cross sect. area		0.08 %	1	0.08 %	0.08 %	0.08 %	0.08 %
$u(K_\alpha)$	Acoustical cal. factor		0.96 %	1	0.96 %	0.96 %	0.96 %	0.96 %
$u(\Delta t)$	Transit time difference	$6e^{-10}$		1	0.62 %	0.17 %	0.05 %	0.01 %
$u(t_o)$	Delay time		0.37 %	0.03	0.01 %	0.01 %	0.01 %	0.01 %
$u(t_{fr})$	Transit time		0.01 %	-1	0.01 %	0.01 %	0.01 %	0.01 %
Standard uncertainty $u_r(\acute{v})=u(\acute{v})/\acute{v}$					**1.18%**	**1.02%**	**1.00%**	**1.00%**
Expanded uncertainty					**2.36%**	**2.03%**	**2.00%**	**2.00%**

Table 2. Total uncertainty overflow velocity generated by the software FluxDiag

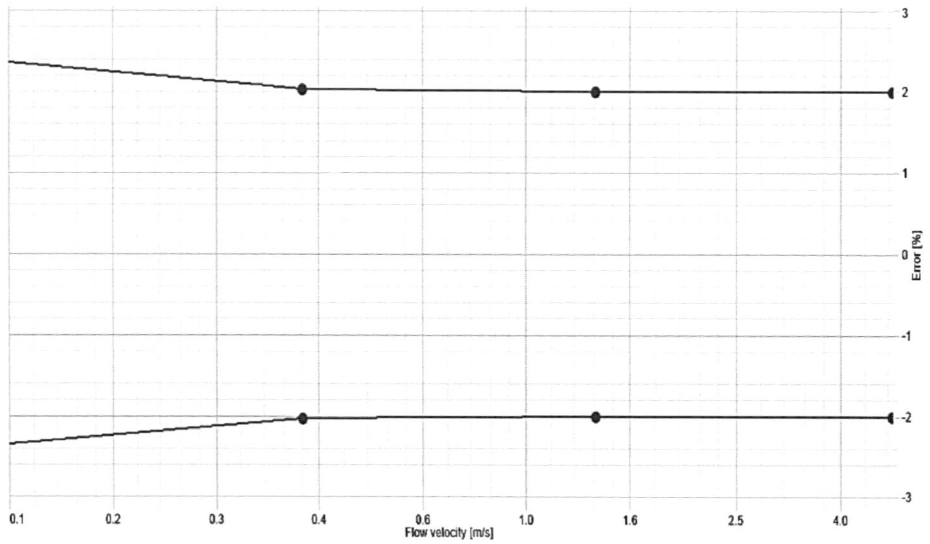

Fig. 6. Graphical evaluation of the steam flow measurement uncertainty generated by the software FluxDiag

6. Discussion

The measurement was done without any problems or data interruption. The signals SCNR and SNR are in stable and proper level, normally the SCNR should be kept higher than 25 dB to get the real and correct data. SNR is the signal provided by the electronics to be kept around 10 dB. One of the most important diagnostic parameters is the sound speed of the media shown in the Table 1. This value must correlate to the sound speed value awaited at the real media temperature. When it is correlating then the measurement is precise and provides correct measurement results. The data are compared to an existing older in situ flow meter and they are correlating what makes place for modernisation of the process control using ultrasonic technology. Such system provides very high precision of calibration and of course very high precision in the process. This system called transit time is nowadays suitable only for the steam parameters between 130 and 200 °C but there is a new principle invented by Flexim to be able to measure high temperature steam using non-invasive ultrasonic principle, so called CFM.

7. Conclusion

The goal of this measurement was to perform the steam flow measurement with innovative clamp-on ultrasonic measurement technology for balances and the design of optimization plans. Second target was to use collected data for further research and deployment of such technology in industrial practice to increase process efficiency and operational safety.

One of the big advantages of clamp-on technology is the possibility to change the measuring point without any process interruption. This type of flow meter has the possibility to also measure gases and liquids as well. These units are calibrated by the producer using a special kind of calibration, invented by the producer itself, where the fluid is in stable state and the sensors mounted on a moving system with very precise movement control are moving on the surface of the calibration fluid.

Results of this study were presented to specialists from energy management as a progressive measurement method of measurement to be implemented in energy strategy plans. Ultrasonic monitoring systems by specific configuration allow to monitor the energy consumption and have a potential for implementation in energy consumption reduction concepts in industry and facility management as well. Comparing to standard heat energy monitors is the price higher but precision and technical and mechanical properties do this kind of measurement devices high valued and comfortable mounting and operation.

8. References

[1] Kralikova, R., Dzunova, L., Pinosova, M., Wessely, E. & Koblasa, F. (2019). Man-Machine-Environment System Analyses and Impact of Environment Factors to Productivity and Health of Employees. - In: Proceedings of the 30th DAAAM International Symposium, pp. 0131-0138, B. Katalinic (Ed.), Published by DAAAM International, ISBN 978-3-902734-22-8, ISSN 1726-9679, Vienna, Austria DOI: 10.2507/30th.daaam.proceedings.017

[2] World Commission on Environment and Development (1987). Our Common Future. - Oxford: Oxford University Press. ISBN 019282080X

[3] Rusko, M., Sablik, J., Marková, P., Lach, M. & Friedrich, S.M. (2013). Sustainable development, quality management system and environmental management system in Slovak Republic. - In: Annals of DAAAM and Proceedings of DAAAM Symposium, 24th DAAAM Zadar 2013, pp. 418 - ISSN 2304-1382. - Procedia Engineering 69 (2014) 486 – 491, DOI: 10.1016/j.proeng.2014.03.016

[4] Ilko, J., Rusko, M., Halper, Ch., Majernik, M. & Majernik, S. (2020). Flow measurement on hot water lines at geothermal power plant using ultrasonic method. - In: Proceedings of the 31st DAAAM International Symposium, pp. 0341 / 0347, B. Katalinic (Ed.), Published by DAAAM International, ISBN 978-3-902734-29-7, ISSN 1726-9679, Vienna, Austria DOI: 10.2507/31st.daaam.proceedings.048

[5] Tesic, S., Bajic, B., Cosic, I. & Rikalovic, A. (2021). Frugal innovations in Industry 4.0: The Identification of Key Indicatiors. - In: Proceedings of the 32nd DAAAM International Symposium, pp. 0262-0269, B. Katalinic (Ed.), Published by DAAAM International, ISBN 978-3-902734-33-4, ISSN 1726-9679, Vienna, Austria DOI: 10.2507/32nd.daaam.proceedings.037

[6] Takakuwa, S., Veza, I. & Celar, S. (2018). „Industry 4.0" in Europe and East Asia. - In: Proceedings of the 29th DAAAM International Symposium, pp.0061-0069, B. Katalinic (Ed.), Published by DAAAM International, ISBN 978-3-902734-20-4, ISSN 1726-9679, Vienna, Austria, DOI: 10.2507/29th.daaam.proceedings.009

[7] Brynjolfsson E. & Mitchell, T. (2017). What can machine learning do? Workforce implications. - Science, vol. 358, No. 6370, pp. 1530–1534, 2017. DOI: 10.1126/science.aap80

[8] WIPO. (2019)."WIPO Technology Trends 2019: Artificial Intelligence." - Geneva, Switzerland

[9] Tekic, Z., Cosic, I. & Katalinic, B. (2019). Manufacturing and the Rise of Artificial Intelligence: Innovation Challenges. - In: Proceedings of the 30th DAAAM International Symposium, pp. 0192-0196, B. Katalinic (Ed.), Published by DAAAM International, ISBN 978-3-902734-22-8, ISSN 1726-9679, Vienna, Austria DOI: 10.2507/30th.daaam.proceedings.025

[10] Rajić, M.; Maksimović, R.; Milosavljevic, P.; Mančić, M.; Mančić, M. & D. Pavlovic (2020). Energy management model for environmental protection in industry. - In: Facta Universitatis, Series: Working and Living Environmental Protection. 12/2020, pp. 75 - 86. DOI: 10.22190/FUWLEP2002075R

[11] ISO50001 Energy management. – [on-line] Available on - URL: https://www.iso.org/iso-50001-energy-management.html

[12] Hafeez, G. & Alimgeer, K. & N. Javaid. (2021). Electrical Energy Consumption Forecasting for Efficient Energy Management in Smart Grid. - In: COMSATS University Islamabad, Islamabad – PakistanSpring. – [on-line] Available on - URL: https://www.researchgate.net/publication/351248870_Electrical_Energy_Consumption_Forecasting_for_Efficient_Energy_Management_in_Smart_Grid

[13] Gryshchenko, I. (2021). Using the energy efficiency hub in the university energy management system in the context of preventing a new outbreak of the COVID-19 pandemic. - In: Management, 1(33): 9–18. DOI: 33. 9-18. 10.30857/2415-3206.2021.1.1.

[14] Kurniawan, R. & Feinnudin, A. (2021). Assessing the Implementation of the Energy Management System in the First ISO 50001 Building in Indonesia. - In: Indonesian Journal of Energy 4(2):129-139. DOI: 10.33116/ije.v4i2.125 ISSN: 2775-5142

[15] Siregar, M. & Purbantoro, Firma & T. Nur (2019). Implementation of Energy Management Concept and Energy Management System in High Rise Office Building. - In: Jurnal TIARSIE. 16. 85. DOI: 10.32816/tiarsie.v16i3.55.

[16] Cuanga, O. M. & G. C. Sánchez (2020) Energy Management System implementation in National Bus Company (BEU) Cienfuegos. - In: Implementación de la etapa de planificación energética en la Empresa de Ómnibus Nacional (UEB) Cienfuegos. - Universidad Carlos Rafael Rodríguez de Cienfuegos DOI: 10.13140/RG.2.2.18798.74561

[17] Dall'O, G.; Ferrari, S.; Bruni, E. & L. Bramonti. (2020). Effective implementation of ISO 50001: A case study on energy management for heating load reduction for a social building stock in Northern Italy. - In: Energy and Buildings. 219. 110029. DOI: 10.1016/j.enbuild.2020.110029. ISSN: 3787788

[18] Quispe, E.; Medina, J.; Chaves, J.; Pérez, F. & D. Aguilar (2021). Estimation of the Energy Saving Potentials in Foundry Industries Based on ISO 50001. - In: IEEE Latin America Transactions. 18. 2155-2165. DOI: 10.1109/TLA.2020.9400444.

[19] Rovňák, M.; Tokarčík, A.; Štofejová, L.; Novotný, R.; Adamišin, P. & Bakoň, M. (2021). Design of the Model of Optimization of Energy Efficiency Management Processes at the Regional Level of Slovakia. - In: Energies 2021, 14, 6502. https://doi.org/10.3390/en14206502

[20] Flowserve. Průmyslová výroba páry. – [on-line] Available on - URL: https://www.sterlingsihi.com/cms/cz/CzechRepublic/home/trhy/prumysl/strojirenstvi-a-investicni-celky/prumyslova-vyroba-pary.html

[21] EnggCyclopedia. Steam generation fundamentals. – [on-line] Available on - URL: https://www.enggcyclopedia.com/2012/01/steam-generation-fundamentals/

[22] Sun, Li; Liu, Chang. (2015) Reliable and flexible steam and power system design. - In: Applied Thermal Engineering, 2015, 79: 184-191. ISSN: 13594311

[23] Murakawa, Hideki, et al. (2021) Measurement of steam flow rates using a clamp-on ultrasonic flowmeter with various wetness fractions. - In: Flow Measurement and Instrumentation, 2021, 101997. ISSN: 9555986

DOI: 10.2507/33rd.daaam.proceedings.034

OPTIMAL CONTROL OF THE DC MOTOR IN THE TUNING OF THE SPACE-BASED REFLECTOR OF THE RADAR-REFLECTING NETTING

Fedor Mitin & Alexey Krivushov

This Publication has to be referred as: Mitin, F[edor] & Krivushov, A[lexey] (2022). Optimal Control of the DC Motor in the Tuning of the Space-Based Reflector of the Radar-Reflecting Netting, Proceedings of the 33rd DAAAM International Symposium, pp.0247-0253, B. Katalinic (Ed.), Published by DAAAM International, ISBN 978-3-902734-36-5, ISSN 1726-9679, Vienna, Austria
DOI: 10.2507/33rd.daaam.proceedings.034

Abstract

This paper discusses the DC motor control applied to a large-sized transformable reflector when tuning a radar-reflecting netting. For technology used in outer space, energy efficiency is an important indicator. For large-sized structures, additional requirements are set to minimize the structural vibration. Fluctuations in the active surface of the reflector's netting affect the directivity diagram, so the radar-reflective form must be adjusted gradually. This paper investigates the possibility of minimizing the energy costs of motor control by applying real-time optimal control algorithms. Numerical simulation results showing the possibility of using the proposed algorithm are presented. The problem of exact fulfillment of terminal conditions is solved. The study results allow control optimization for different types of DC motors.

Keywords: DC motor; optimal control; energy costs; space-based reflector; radar-reflecting netting.

1. Introduction

Electric motors, in view of their purpose, namely the conversion of electrical energy into mechanical energy, have found their application in products from various industries: from industrial machines to the space industry [1], [2], [3], [4], [5]. When choosing of the electric motor for products of space products for use as a part of actuating devices of systems of space vehicles, for example, systems of adjustment of a form of a reflecting surface of a large-area reflector of the antenna or deployment mechanisms for a space telescope [6] a number of the requirements shown to the electric motor, such as: service life, efficiency, power consumption, torque, speed of rotation, model of external influencing factors, mass-size characteristics and others should be considered. With the optimal choice of electric motor, it is possible to ensure the reliable and efficient operation of both individual mechanical systems and the spacecraft as a whole.

Having analyzed the works [6] and [7], it was found that direct current motors (hereinafter – DCM) have proven to be a more versatile type of electric motors for space technology products due to their long service life, high torque and efficiency, relatively low heat generation, the ability to widely and smoothly regulate the speed of rotation, as well as small size and weight.

Let us consider the application of a DCM for tuning the radio-reflecting netting of a large-sized space-based reflector. Currently, research is actively underway to create these designs. [8], [9], [10], [11], [12], [13], [14], [15]. Fig. 1 shows the spoke 1, the two main radial cords of the rear 3 and front 2 networks and cables 4 going between them. At the right end of the spoke 1 is divided into two parts to set the desired shape of the radio-reflective surface. In each cable 4 is located actuator 5, which here is taken as a DC motor. It is necessary, by adjusting the length of cables 4 with the help of actuators 5 to set the shape of the radio-reflective netting defined by the front net 2 while minimizing energy costs.

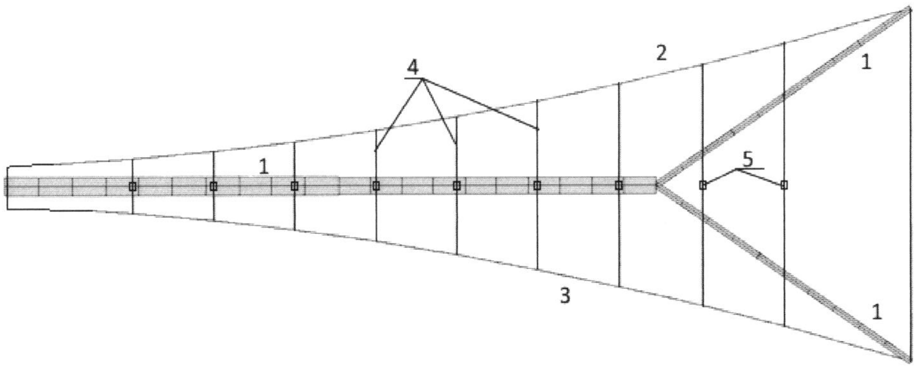

Fig. 1. Scheme of the space-based reflector spoke

Since the space-based reflector has a limited energy supply, the question of minimizing the energy cost to accomplish the tasks at hand arises separately. Currently, the control is performed by PID control structure, which allows achieving the required parameters in the terminal part. The application of optimal control algorithms will minimize energy costs. In addition, it is important to minimize the oscillation of the netting, this can be achieved by a smooth stop of the motor.

2. Building a mathematical model

The mathematical model must consist of electrical and mechanical parts [16], [17], [18], [19], described by equations to describe the DCM:

$$U = L \cdot \frac{dI(t)}{dt} + R \cdot I(t) + E(t), \tag{1}$$

$$M(t) - M_{ex} = J \cdot \frac{d\omega(t)}{dt}. \tag{2}$$

Equation (1) is a differential equation of electric equilibrium of the armature (rotor) motor circuit, where R is armature (rotor) active resistance, L is armature (rotor) winding inductance, $E(t)$ is counter EMF, $I(t)$ is motor current, U_a is armature (rotor) voltage. (1) does not consider the voltage drop (relatively small compared to the value of the rotor voltage), which nonlinearly depends on the rotor current, and assume that the armature (rotor) response is fully compensated, the excitation flux is constant, the active resistance of the armature circuit does not change during motor operation.

Equation (2) is the equation of mechanical equilibrium of the motor, where $M(t)$e is electromagnetic moment of the motor, M_{ex} is load resistance moment, J is rotor moment of inertia, $\omega(t)$ is angular velocity of the motor shaft. (2) does not consider the effect of frictional forces in the rotor rotation as they have a weak effect on the acceleration of the DCM shaft.

Let us describe each component (1) and (2) in more detail.

$$M(t) = k_M \cdot I(t), \tag{3}$$

where $M(t)$ is electromagnetic torque of the motor, k_M is motor torque constant, corresponding to the ratio between the torque developed by the motor and the current drawn, $I(t)$ is motor current.

$$E = k_E \cdot \omega(t), \tag{4}$$

where $E(t)$ is counter EMF, k_E is coefficient of proportionality, called back EMF constant, $\omega(t)$ is angular velocity of the motor shaft. From (1) and (2) one can see that the speed of rotation of the motor shaft $\omega(t)$ is controlled by changing the armature (rotor) voltage.

Taking into account the above, as well as the fact that when using the DCM in the system of adjusting the shape of the large-sized reflector of the spacecraft antenna, it is necessary to control the angle of rotation of the engine shaft, the final system of differential equations of the DCM will take the following form $X = (I\ \omega\ \varphi)^T$ or in a piecemeal form:

$$\frac{dI(t)}{dt} = \frac{U_a - R \cdot I(t) - k_E \cdot \omega(t)}{L}, \tag{5}$$

$$\frac{d\omega(t)}{dt} = \frac{k_M \cdot I(t) - M_{ex}}{J}, \tag{6}$$

$$\frac{d\varphi(t)}{dt} = \omega(t). \tag{7}$$

DCM actuators of the system for adjusting the shape of the reflecting surface of the large-sized transforming reflector of the spacecraft antenna are often used as part of the gearmotor. As a first approximation, we will assume that the gear motor uses an instrumental gearbox with a gear ratio of $k_g = 1/i_g$, where i_g is the gear ratio of the instrumental gearbox.

The control task is to drive the DCM shaft from the initial position $X(0)$ to a given end state $X(t_f)$ under the control constraint, which is taken as the external supply voltage $U_a = \pm 12\ V$ for a given time t_f This rotation angle φ_M will provide a change in the length of the cable.

3. Building a mathematical model

Various structures of proportional-integral-differential (PID)-controller, including those with control constraints, were considered as regulators.

The full vector of state variables X is available for measurement. Then the deviation of the vector from the set value $\Delta X = X_f - X(t)$, the integral $\int_0^t \Delta X dt$ and the derivative $\Delta \dot{X}$ are available for measurement and calculation. The PID controller generates control in the form of:

$$u = k_p \cdot \Delta X + k_i \cdot \int_0^t \Delta X dt + k_d \cdot \Delta \dot{X}, \tag{8}$$

where k_p, k_i, k_d are the proportional, integrating and differentiating gain components.

The application of the PID control structure does not consider the minimization of energy costs, which is important in conditions of energy limitation. Therefore, let us consider applications of the sequential optimization algorithm.

If for a system n of equations the optimization by a single objective functional solution leads to a two-point boundary value problem for a system of order $2n$, then for a hierarchy of two criteria this number is doubled [20]. It is possible to reduce the volume of calculations by using on two levels the functionals of the generalized work of A.A. Krasovsky [20] and [21]. The paper [20] shows the feasibility of a simplified version of the algorithm, when considering the second level of control of the first level is considered to be implemented. That is, there is a sequential optimization of controls by level. Studies for different systems have shown that the nature of the trajectories of controlled motion by the simplified algorithm does not change in comparison with the full solution [20]. The algorithm has been successfully used to control an aircraft, including the use of spiral prediction; to control a car, a nuclear reactor. Here the well-known advantages of applying the Krasovsky criterion remain, when instead of a two-point boundary value problem one has to solve two one-point Cauchy problems (in forward and backward time, respectively) [21].

Let us extend the system (5) – (7) by adding to it the equation $\dot{U}_a = u$. Assume $Y = U_a$, $x = (X^T\ Y)^T = (I\ \omega\ \varphi)^T$, $f = (F^T\ u)^T$. Consider a hierarchy of target functionals of the form

$$J_1 = V_{f1}(X, t_f), \tag{9}$$

$$J_2 = V_{f2}(X, t_f) + \int_{t_0}^{t_f} f_0(U_a, t)\, dt., \tag{10}$$

Here $V_f = 0{,}5\rho_1 \left[I_a(t_f) - I_{af}\right] + 0{,}5\rho_2 \left[\omega_M(t_f) - \omega_{Mf}\right] + 0{,}5\rho_3 \left[\varphi_M(t_f) - \varphi_{Mf}\right]$, $f_0 = 0{,}5\left(\frac{U_a}{R}\right)^2$, where t_0 is the initial time, t_f is the final time, R is the resistance of the armature winding, ρ_1, ρ_2 are the given coefficients. Since the task was to minimize the energy cost, the expression for the consumed supply voltage U_a over the entire simulation interval was taken as a function f_0.

A simplified version of the algorithm is appropriate to control the electric motor with end conditions for the angle φ at small finite values of its derivative ω.

The control is calculated in the form $u = u_1 + u_2$, where u_1 and u_2 minimize the quality criteria J_1 and J_2 respectively. At the first level, the predictive model is adjusted, for example, so that it is possible to take $u_1 = 0$, $U_m(t) = U_a(t) + \Delta U_a$.

Taking into account the assumptions made, the Hamiltonian of the system (5) – (7) will take the form $H = P^T f(X) + f_0$, where $P = [P_I\ P_\omega\ P_\varphi\ P_U]$ is the vector of conjugate variables, or

$$H = P_\varphi \omega + P_\omega \frac{k_m \cdot I - M_{ex}}{J} + P_I \frac{U_a - R_a \cdot I_a - k_e \cdot \omega_M}{L} + P_U u + 0{,}5\left(\frac{U_a}{R}\right)^2. \tag{11}$$

The equations of the predictive model at zero control [20] and [21] have the form

$$\dot{I} = \frac{U_a - R \cdot I(t) - k_E \cdot \omega(t)}{L}, \tag{12}$$

$$\dot{\omega} = \frac{k_M \cdot I(t) - M_{ex}}{J}, \tag{13}$$

$$\dot{\varphi} = \omega(t), \tag{14}$$

$$\dot{U}_a = 0. \tag{15}$$

Equations for conjugate variables are obtained by taking partial derivatives of equation (11) with a negative sign:

$$\dot{P}_I = -\frac{\partial H}{\partial I_a} = -P_\omega \frac{k_m}{J} + P_I \frac{R}{L}, \tag{16}$$

$$\dot{P}_\omega = -\frac{\partial H}{\partial \omega} = -P_\varphi + P_I \frac{k_e}{L}, \tag{17}$$

$$\dot{P}_\varphi = -\frac{\partial H}{\partial \varphi} = 0, \tag{18}$$

$$\dot{P}_U = -\frac{\partial H}{\partial U_a} = -P_I \frac{1}{L} - P_U - \frac{U_a}{R}. \tag{19}$$

The values of the conjugate variables at the final time t_f are determined from:

$$P_I(t_f) = \alpha_1[I(t_f) - I_f], \tag{20}$$

$$P_\omega(t_f) = \alpha_2[\omega(t_f) - \omega_f], \tag{21}$$

$$P_\varphi(t_f) = \alpha_3[\varphi(t_f) - \varphi_f], \tag{22}$$

$$P_U(t_f) = \alpha_4[U(t_f) - U_f]. \tag{23}$$

The value ΔU is chosen by iterating from the condition $V_{f1} = 0$ to the nearest small $\sigma > 0$ by integrating the equations of the first level model on the interval $[t, t_f]$. In this case, neglecting the friction moment on the predictive model, we can determine the control value required to minimize the first criterion in the model U_{md} from equation (6) in the form of $U_{md} = [\omega(t_f) - \omega_f/(t - t_f)/A_d]$. Here $A_d = \frac{k_d L}{k_m}$, $k_d$is the additive coefficient.

To determine the control u_2 is required to find pu_2 only from the system of equations of the fitted model and the system of equations for p_2 at $V_{f1} = 0$. That is, set the value $u_1 = -k_1^2 p_{1U}$ in the form of $-k_1^2 p_{1U} = \delta(t)\Delta U$, where $\Delta U = U_{md} - U(t)$, $\delta(t)$ is a delta function. Approximately, we can take $\delta(t) = \frac{1}{\Delta t}$, where Δt is the step of numerical integration of the system (5) – (7).

The control is found from the condition $\frac{\partial H}{\partial U_a} = 0$, considering the limitation on the value of the control:

$$U_a = 12 \cdot sign(U_a). \tag{24}$$

4. Simulation

The data for the 0816K012SR series 0816SR electric motor from FAULHABER [22]. Let us simulate the system (5) – (7) using the developed optimal control algorithm and using a PID controller.

Consider the control problem of bringing the shaft of a servo motor from an initial position φ_0= 0 rad, ω_0= 0 rad/s, I_{a_0}= 0 A to a given final state φ_f= 8π rad, ω_f= 0 rad/s, I_{a_f}= 0 A in the absence of an external load M_{ex}= 0 for a time t_f= 1 s.

The simulation was performed by the Euler method with integration step $h = 0.001$. Application of a PID controller with coefficients $k_p = 0.05$, $k_i = 0.002$, $k_d = 1.758$ and an optimal controller with coefficients $\alpha_1 = 1$, $\alpha_2 = 1$, $\alpha_3 = 14$, $\alpha_4 = 1$, $k_1 = 0.32$ showed identical results. Both algorithms successfully solved the problem. The error in fulfilling the terminal conditions for the angle of rotation of the motor rotor φ was not more than 0.0001 rad. The discrepancies in the total power input for control and the final shaft speed were insignificant and were caused by the choice of the integration method. When the final angle of rotation of the rotor shaft is changed to $\varphi_f = 160\pi$ rad (80 revolutions) and the coefficients are maintained, there is a deviation of the final speed ω from 0 when using PID controller (line 2 in Fig. 2a) in contrast to the optimal controller (line 1 in Fig. 2a). Due to the fact that in J_1 the condition of reaching $\omega = 0$ at the finite moment of time is introduced, the optimization algorithm quickly reduces ω and at the last step of slow control brings it to the required value. The deviation of the final φ in both cases did not exceed 0.3 rad (Fig. 2b)

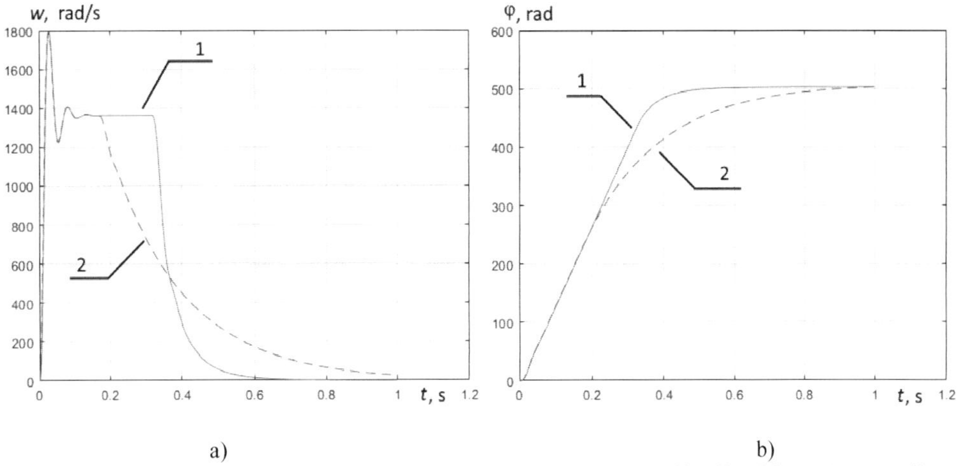

Fig. 2. Transients diagram: a) $\omega(t)$, b) $\varphi(t)$, line 1 is a optimal controller, line 2 is a PID controller

Thus a PID controller requires adjustment of the coefficients. When the final angle of rotation of the rotor shaft is increased to $\varphi_f = 250\pi$ rad (125 revolutions), adjustment of the coefficients is required for both types of regulators. A further increase in the final angle φ requires an increase in control time φ_f. Due to the absence of load on the rotor shaft, the power consumption does not differ significantly with different control options. The gain of the optimal regulator is ~1%. Consider the application of the algorithms when torque $M_{ex} = 0.61$ mN·m and when setting the final rotor shaft angle of rotation $\varphi_f = 200\pi$ rad (100 revolutions). Since the presence of torque leads to an increase in time, we assume $t_f = 1.5$ s. The application of the PID controller with coefficients $k_p = 1.1$, $k_i = 0.002$, $k_d = 1.3186$ and the optimal controller with coefficients $\alpha_1 = 1$, $\alpha_2 = 1$, $\alpha_3 = 965$, $\alpha_4 = 1$, $k_1 = 0.32$ showed identical results. Both algorithms successfully solved the problem. The change in angular velocity ω is similar to the considered version without load. The PID control structure provides a smoother and longer variation in shaft speed. The optimum controller reduces the speed sharply and then slowly brings it to the desired value. This reduces the power consumption by ~13% from 1.0741 W to 0.9345 W when the optimum controller is used. Fig. 3 shows graphs of the dependence $I(t)$ (line 1 is a optimal controller, line 2 is a PID controller). You can see that up to 1.1 s the control is at the limit ($U = 12$ V), then the current drops to 0 A. By changing the control of the optimum controller more quickly, the energy costs can be minimized.

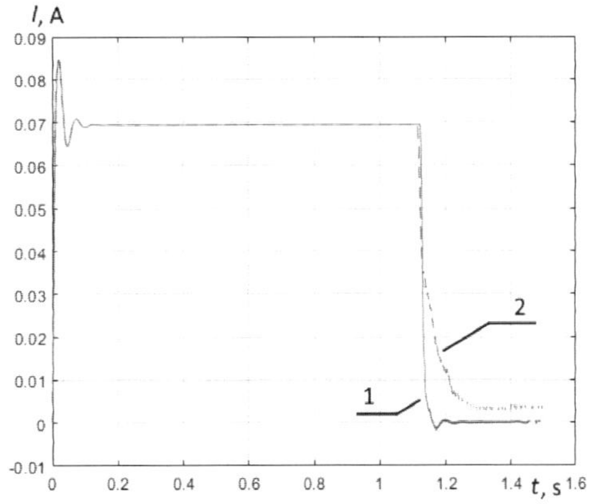

Fig. 3. Transients diagram $I(t)$, line 1 is a optimal controller, line 2 is a PID controller.

The selected coefficients allow us to successfully solve the problem in the range φ_f from 0 to 400π rad (200 revolutions). As experience with finite-time control shows [23], it is possible to achieve the necessary accuracy and quality of regulation of the system in a wide range of terminal conditions. The proposed algorithm allows a real-time smooth approach to a given motor shaft speed. Taking into account the perturbations and noise of the assumed sensors acting on the system under study leads to complex problems of measurement result processing and planning [24] and [25].

5. Conclusion

The use of the algorithm of sequential optimization according to the hierarchy of criteria does not give a significant advantage over the PID controller when the regulation time is long enough. The reduction in energy costs was ~1%. However, in some cases it is possible to reduce the electrical costs with the optimal controller by ~13%. The sequential optimization algorithm investigated in the article is reasonable to apply to more complex solutions using measurement processing and optimization of observation intervals. The results of these studies can be used to calculate the energy costs, the choice of actuator and the power plant of the space-based reflector.

In the future it is supposed to investigate the application of optimal control algorithms for a refined mathematical model of a direct current motor. Including an augmented gearbox model and a device that translates the angle of rotation of the motor shaft into changes in the length of the cable. It is also planned to consider the task of simultaneous control of several actuators for setting up and maintaining a given form of a radio-reflecting network drone, taking into account the optimization of finite time.

6. Acknowledgments

The study was supported by grant No. 22-79-10112 from the Russian Science Foundation, https://rscf.ru/en/project/22-79-10112/.

7. References

[1] Alekseev, K.B. & Bebenin, G.G. (1974). Control of space aircraft, - 2nd ed, Mashinostroenie, Moscow

[2] Dmitriev, V.S.; Kostyuchenko, T.G. & Gladyshev, G.N. (2013). Electromechanical executive bodies of spacecraft orientation systems, Part I, Stipend studies, Publishing house of Tomsk Polytechnic University, Tomsk

[3] https://ie.rs-online.com/web/generalDisplay.html?id=ideas-and-advice/dc-motors-guide, (2022). Radionics Ltd. Official site, Everything You Need To Know About DC Motors, Accessed on: 2022-09-03

[4] Kankam, M.D. & Elbuluk, M.E. (2001). A Survey of Power Electronics. Applications in Aerospace Technologies, NASA/TM-2001-211298 IECEC2001-AT-32, Available from: https://ntrs.nasa.gov/api/citations/20020013943/downloads/20020013943.pdf Accessed: 2022.09.03

[5] https://ntrs.nasa.gov/api/citations/19750007247/downloads/19750007247.pdf Brushless DC Motors, Application of Aerospace Technology, NASA CR-2506, Accessed: 2022.09.03

[6] Tran, A. N. (2016). James Webb Space Telescope Deployment Brushless. DC Motor Characteristics Analysis, Proceedings of the 43rd Aerospace Mechanisms Symposium, 4-6 May 2016, NASA Ames Research Center, Available from: https://ntrs.nasa.gov/api/citations/20160004038/downloads/20160004038.pdf Accessed: 2022.09.03

[7] https://extapps.ksc.nasa.gov/Reliability/Documents/Preferred_Practices/1229.pdf, Selection of Electric Motors for Space Applications, NASA Preferred Reliability Practices, Practice NO. PD-ED-1229, Accessed: 2022.09.03

[8] Yiqun, Z.; Na, L.; Guigeng, Y. & Wenrui, R. (2017). Dynamic analysis of the deployment for mesh reflector deployable antennas with the cable–net structure, Acta Astronautica, Vol. 131, pp 182–189, ISSN 0094-5765, DOI 10.1016/j.actaastro.2016.11.038

[9] Kabanov, S.A.; Krivushov, A.I. & Mitin, F.V. (2017). Modeling of joint deployment of units of the large-sized transformable reflector of space basing, SPIIRAS Proceedings, Vol.54, No 5, 2017, pp. 130-151, Print ISSN 2078-9181, Online ISSN 2078-9599, DOI: 10.15622/sp.54.6

[10] Mitin, F.; & Krivushov, A. (2017). Control deployment of mobile units of large-sized spacecraft, Proceedings of the 28th DAAAM International Symposium, pp. 0773-0779, B. Katalinic (Ed.), Published by DAAAM International, ISBN 978-3-902734-11-2, ISSN 1726-9679, Vienna, Austria, DOI: 10.2507/28th.daaam.proceedings109

[11] Ramachandran, S.; Neve, M. J. & Sowerby, K. W. (2018). Millimetre wave antenna deployment in a single room environment, IEEE Aerospace and Electronic Systems Magazine, Vol. 12., No. 15, pp. 2390-2394, Print ISSN 1751-8725, Online ISSN 1751-8733, DOI: 10.1049/iet-map.2018.5512

[12] Rahmat-Samii, Y.; Manohar, V.; Kovitz, J. M. R. Hodges, Freebury, G. & Perale, E. (2019). Development of Highly Constrained 1 m Ka-Band Mesh Deployable Offset Reflector Antenna for Next Generation CubeSat Radars, IEEE Aerospace and Electronic Systems Magazine, Vol. 67, No. 10, pp. 6254-6266, ISSN: 0885-8985, DOI: 10.1109/TAP.2019.2920223

[13] Tserodze, S.; Prowald, J. S.; Chkhikvadze, K.; Nikoladze, M. & Muchaidze, M. (2020). Latest modification of the deployable space reflector structure with V-folding bars, CEAS Space Journal: An Official Journal of the Council

of European Aerospace Societies, Vol. 12, No. 2, pp. 163-169, Print ISSN 1868-2510, Online ISSN 1868-2502, DOI: 10.1007/s12567-019-0281-9

[14] Kabanov, S.A. & Mitin, F.V. (2020). Optimization of the stages of deploying a large-sized space-based reflector, Acta Astronautica, Special Issue on 6th SFS 2019, Vol. 176, pp. 717-724, ISSN 0094-5765, DOI: 10.1016/j.actaastro.2020.04.066

[15] Kabanov, S.A. & Mitin, F.V. (2021). Optimization of the Processes of Deployment and Shape Generation for a Transformable Space-Based Reflector, Journal of Computer and Systems Sciences International, Vol. 60, No. 2, pp. 283–302, Print ISSN 1064-2307, Online ISSN 1555-6530, DOI: 10.1134/S1064230720060052

[16] Petrov, U.P. (1971). Optimum control of the electric drive taking into account the limitations on heating, Energy, Leningrad

[17] Emelyanov A.A.; Beskletkin V.V.; Agzamov I.M.; Zozulin M.S.; Onishchenko K.Y.; Zorin D.I.; Blinov E.K.; Bukhryakov I.F.; Dyatlov O.A.; Chumichev P.E. & Shamiev R.R. (2019). Mathematical modeling of a DC motor in the system of relative units in Matlab and C, Young Scientist, Vol. 11 (249), pp. 1-7, ISSN 2077-8295.

[18] Pal, D. (2016). Modeling, Analysis and Design of a DC Motor based on State Space Approach, International Journal of Engineering Research & Technology (IJERT), Vol. 5 Issue 02, pp. 293-296, ISSN 2278-0181, DOI: 10.17577/IJERTV5IS020332

[19] Mitin, F. & Krivushov, A. (2018). Application of Optimal Control Algorithm for DC Motor, Proceedings of the 29th DAAAM International Symposium, pp. 0762-0766, B. Katalinic (Ed.), Published by DAAAM International, ISBN 978-3-902734-20-2, ISSN 1726-9679, Vienna, Austria, DOI: 10.2507/29th.daaam.proceedings110

[20] Kabanov, S.A. (1997). Control systems on predictive models, Publishing house of St. Petersburg State University, ISBN: 5-288-01941-X, St. Petersburg

[21] Krasovsky, A. A. (1987). Handbook on the theory of automatic control, Nauka, Moscow.

[22] https://www.faulhaber.com/en/products/series/0816sr/, (2022). FAULHABER GROUP Official site, FAULHABER SR Series 0816 ... SR, Accessed on: 2022-09-03

[23] Kabanov, S. A. & Mitin, F. V. (2021). Solution of the filtration problem with the optimal adjustment of the radio-reflecting net of a transformable reflector, Siberian Aerospace Journal, Vol. 22, No. 4, pp. 577–588, ISSN 2712-8970, DOI: 10.31772/2712-8970-2021-22-4-577-588

[24] Malyshev, V.V.; Krasilshchikov, M.N.; Bobronnikov, V.T.; Nesterenko, O.P. & Fedorov, A.V. (2000). Satellite monitoring systems, V. Malyshev (Ed.), Publishing House of Moscow Aviation Institute, ISBN: 5-7035-2384-2, Moscow

[25] Kabanov, S. A.; Kabanov, D. S.; Nikulin, E. N. & Mitin, F. V. (2021). Optimal control of deployment of the spoke of a transformable reflector in the presence of disturbance, Siberian Aerospace Journal, Vol. 22, No. 4, pp. 649–659, ISSN 2712-8970, DOI: 10.31772/2712-8970-2021-22-4-649-659

DOI: 10.2507/33rd.daaam.proceedings.035

ELECTROMAGNETIC INDUCTION PUMP WITH ROTATING PERMANENT MAGNETS CAVITATION PROCESS INFLUENCE ON PRESSURE – FLOW-RATE CURVES

Buligins Leonids, Bucenieks Imants, Lacis Arturs,
Kravalis Kalvis & Mikanovskis Oskars

This Publication has to be referred as: Buligins, L[eonids]; Bucenieks, I[mants]; Lacic, A[rturs]; Kravalis, K[alvis] & Mikanovskis, O[skars] (2022). Electromagnetic Induction Pump With Rotating Permanent Magnets Cavitation Process Influence on Pressure – Flow-Rate Curves, Proceedings of the 33rd DAAAM International Symposium, pp.0254-0261, B. Katalinic (Ed.), Published by DAAAM International, ISBN 978-3-902734-36-5, ISSN 1726-9679, Vienna, Austria
DOI: 10.2507/33rd.daaam.proceedings.035

Abstract

The study aims to investigate the performance of the electromagnetic induction pump with rotating permanent magnets operating at low pump inlet pressure values to determine the cavitation process's influence on the pump's pressure–flow–rate characteristics. The experiments used the disc-type pump with rotating permanent magnets with a channel active cross-section of 20 x 70 mm and a mean diameter of 330 mm. The drive unit of the pump is a 22 kW 3-phase asynchronous electric motor. The pump's rotors have nine pole pairs of samarium-cobalt magnets. A series of experiments utilized the test loop operating with liquid sodium, equipped with two pumps and different pump interconnection possibilities, allowing series and parallel or standalone pump operation. Various values of pump inlet pressure and flow rate combinations were investigated. Pumps inlet pressure values vary from 0,01 to 3.2 bar absolute pressure. A severe flow rate decrease at low inlet pressures was observed alongside a boiling noise allowing us to judge the cavitation process occurrence in the pump. The flow rate differences at higher and lower inlet pressure values reach more than 30 %. Also, the pumps drive unit power consumption curves to identify the pump's cavitation process.

Keywords: cavitation; liquid sodium; electromagnetic pump; pressure-flow rate curves.

1. Introduction

Electromagnetic induction pumps for liquid metals, as every type of pump, face cavitation issues when operating at low inlet pressures. It leads to flow parameters deterioration and can cause damage to the pump channel. Electromagnetic induction pumps (EMP) with rotating permanent magnets (PM) gained broader application at the end of 1990ties [1] when permanent magnets with sufficiently high magnetic fields became commercially available [2]. The operation principle relies on the Lorentz force created by a movement of the rotating permanent magnets displaced near the pump's channel [3]. Permanent magnets are placed on each rotor with varying polarity, thus creating a traveling magnetic field.

The rotor is not touching the pump's channel resulting in a fully hermetical channel without any sealings. The EMPs with PM main advantages are the hermetic pump channel, relatively simple design and low costs, industrial drive unit use, and the magnetic rotor that can be easily disassembled from the pump channel [4]. Very little research has been done to investigate the electromagnetic induction pump with a rotating permanent magnet cavitation process. The study utilizes a purely experimental approach to determine the margins of the cavitation process for the EMP with rotating PM. The experimental investigation used the disc-type pump installed in the loop operating with liquid sodium to understand the limitations of cavitation occurrence. The cavitation process should not be neglected during the pump operation due to the danger of damage to the pump channel [5]. Test results reveal that the pump's performance is strongly influenced by inlet pressure, varying from 0,01 to 3,2 bar absolute pressure. The decrease in the flow rate reaches almost a factor of two.

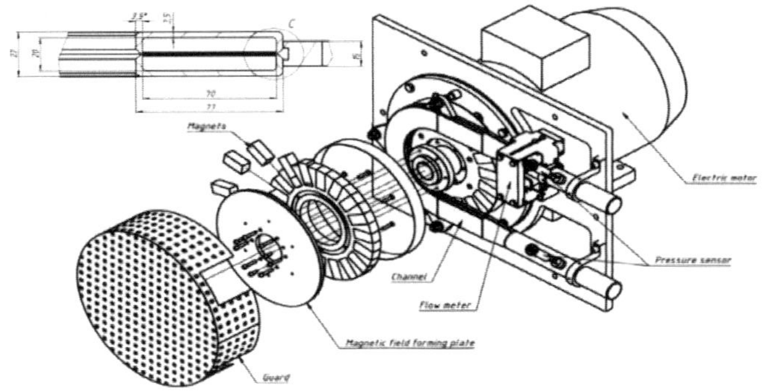

Fig. 1. Disc-type pump used for tests in the loop and its channel cross-section dimensions

2. Experimental setup and conditions

The investigated pump and its channel cross-section are shown in figure 1. Its inner cross-section is 20 x 70 mm. The loop represented in figure 2 is from AISI 316L stainless steel. It is a closed circulation loop, using a 60,3 mm outer diameter tube with a wall thickness of 3 mm to connect all loop elements. Each pump has a conduction flow meter for the flow rate measurements, and the loop has another two flow meters for integral flow rate measurements. Loops hydraulic resistance is measured at the outlet of the investigated pump by pressure sensor 3.2 and adjusted using the valveVmD1.

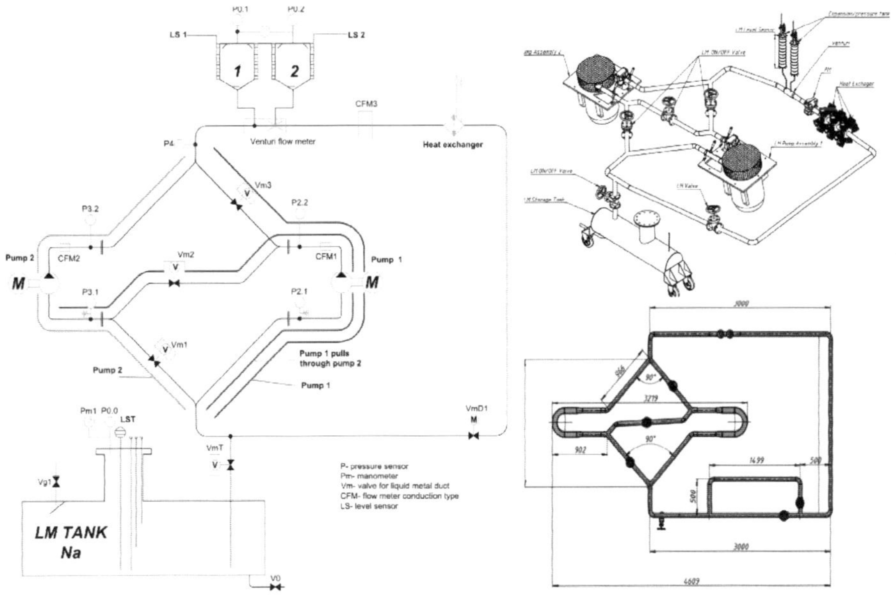

Fig. 2. Experimental loop principal scheme. The sodium flow path is drawn with a black line. The green alongside the sodium path marks the flow path "pump two." The brown line alongside the sodium path marks the flow path "pump one pulls through pump two." The blue line shows the flow path "pump one."

Pumps developed pressure is a difference between pressure measurements at sensors P2.1. and P2.2. further referred to as P_{in} and P_{out}. Loop also has two expansion tanks, 1. and 2. with continuous level meters LS1 and LS2. The pressure in the expansion tanks 1. and 2. forms the initial pressure P_{start}, measured by pressure sensors P01 and P02 when the pumps are entirely stopped. The pressure in the expansion tanks was measured using two Walfront LEPAZA60119 pressure sensors placed at the top of the expansion tanks. Four flow meters are installed in the loop. One of them is shown in figure 3, where the EMP with PM is delivered separately and within the sodium loop. Conduction flow meters are placed at each pump, and another for integral flow measurements. An additional Venturi flow meter is utilized for the reference flow measurement. Our team performed several measurements, and a counterclockwise flow direction was chosen as the most appropriate for these experiments. The measurements were carried out with a fully open valve VmD1 to get the lowest loop hydraulic resistance value and highest flow rate.

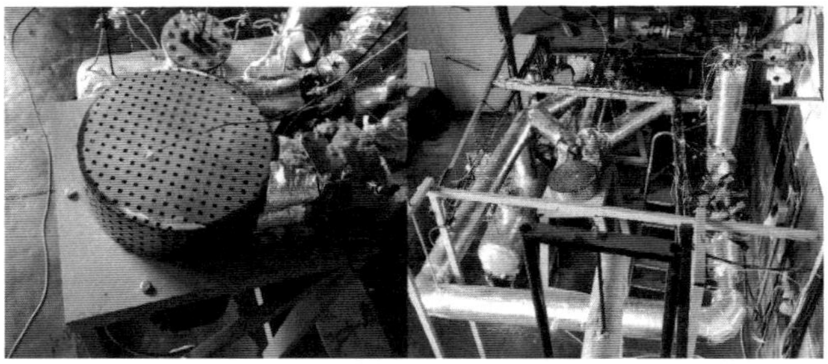

Fig. 3. Permanent magnet pump view from the top (on the left) and experimental loop.

The P_{start} value was set of 0,3;1; 1,5; 2; 2,5 and 3.0 bar absolute pressure. Pressure drop measurements along the flow pass in the loop utilizing BD DMP 331 pressure sensors P0.1.; P0.2.; P3.1.; P3.2. All pressure values are absolute pressure. The team investigated two LM loop flow configurations. The first is when pump one or pump two operates alone, through the route where the valve Vm3 is displaced, named "pump one" (or valve Vm1 for "pump two"). The other configuration has higher hydraulic resistance, where pump 2 pulls liquid metal (LM) through pump one and the leg with Vm2.

3. Measurement results and their interpretation.

Pressure–flow rate curves shown in figure 4 were obtained by varying the pump rotor's rotation speed. Measurements were carried out at a fully open valve VmD1. The flow rate values for the "pump one" or "pump two" flow path can reach 11.8 l/s at the pump magnetic rotor rotation rate of 735 rpm. The maximal flow rate at 735 rpm at P_{start} values of 2.5, 2.0, 1.5, and 1.0 bar follow the same path lowering to 11.4 l/s at 1.0 bar P_{start} value. The flow rate at the 0.3 bar P_{start} value drops to 5,9 l/s, caused by the cavitation observed by vibration measurements and sound spectrum recording. Also, the pump operation at low P_{start} values was accompanied by a strong "boiling" sound. The vibration and sound record analysis is not the scope of this paper.

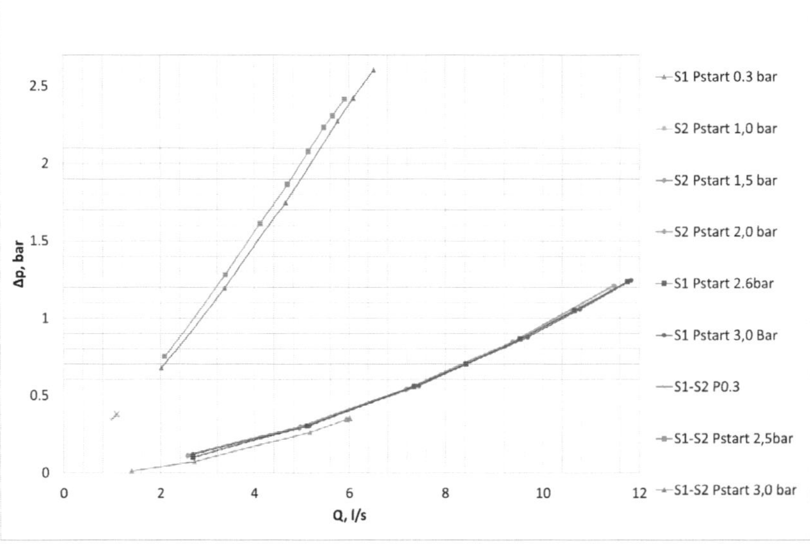

Fig. 4. Pumps developed pressure difference–flow rate (p-Q) measurement results at different flow path configurations.

For the flow path "pump one pulls through pump two," the pump parameter deterioration is even more dramatic. The flow rate reaches only one l/s at higher P_{start} values, a clear sign of cavitation at lower inlet pressures. The plot of the flow rate versus the pump rotor's rotation speed change is shown in figure 5. For the flow path "pump one" and "pump two," the flow rate increase is proportional to the pump magnetic rotor rotation speed increase at all initial pressure values, and the flow rate – rpm relation is a straight line. The exception is the P_{start} = 0.3 bar curve, where the line angle changes from the overall tendency at 294 rpm. The flow rate increase stops at 367,5 rpm. Further rotors rotation speed increase does not gain any flow rate increase, more so the flow rate slightly decreases at higher rotation rates. Referring to Brennen, we can consider this point a "cavitation breakdown" [6] point, where the critical combination of the inlet pressure and flow speed in the pump channel is achieved, and further magnetic field movement speed increase will not create an additional flow rate increase.

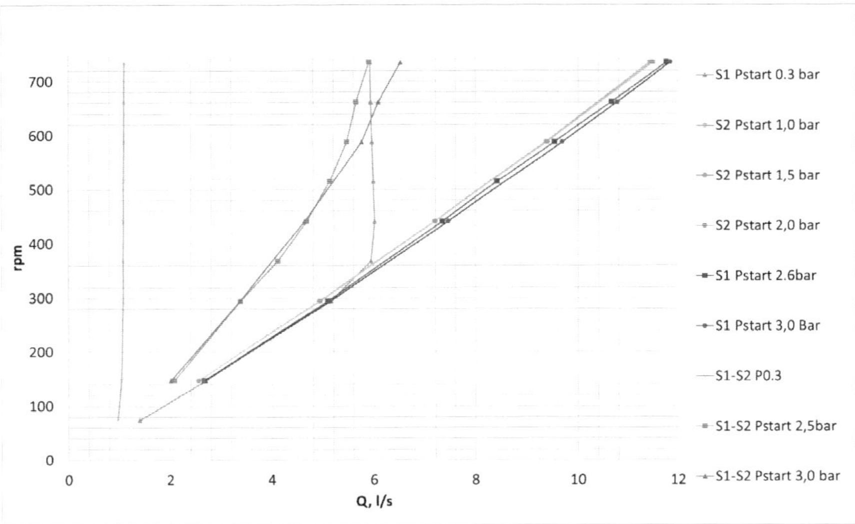

Fig. 5. Pumps developed pressure flow rates at different pump rotor rotation speed values.

For the case where "pump one pulls through pump two," the line angle is much steeper due to the higher hydraulic resistance of the loop. The deviation from the straight line starts at a 5.0 l/s flow rate even at P_{start} = 3.0 bar. At P_{start} = 2.5 bar the flow rate difference from the previous P_{start} = 3.0 bar begins from 514.5 rpm. At P_{start} = 0,3 bar, the severe cavitation starts from lower flow rates, not allowing to reach more than 1.1 l/s flow rate.

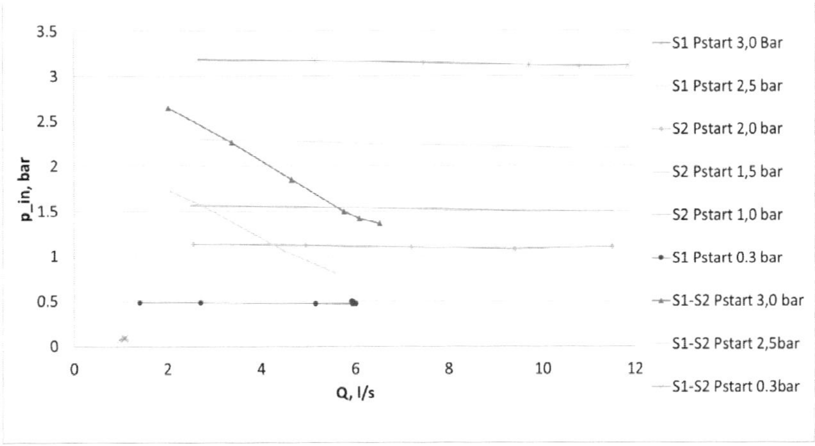

Fig. 6. Pumps developed pressure flow rates at different pump rotor rotation speed values.

Plotting the inlet pressures against the flow rate in figure 6, for the "pump one pulls through pump two" case, shows us that the inlet pressure is P2.2. drops down with the flow rate increase. This pressure drop is determined by the hydraulic resistance created by the Lorentz force interaction between the flowing sodium and rotor of pump two [3]. For the extreme cavitation case at P_{start} = 0,3 bar, the inlet pressure does not change much and is around 0,1 bar.

As it is uncertain if the cavitation process is happening already in the channel of pump two or pump one, the case will be viewed just for reference to compare the values with the single pump operation corresponding to "pump one" and "pump two" points. For "pump one" and "pump two" cases, the inlet pressure is P2.2. or P3.2. values do not change much from the P_{start} value and are little influenced by the flow speed as we know that the pump reaches the "cavitation breakdown" point at the inlet pressure around 0,5 bar, which is determined by the combination of the inlet pressure and flow speed. This relation regarding Brennen is the cavitation number, which evaluates the probability of cavitation [5]. The cavitation number was calculated using the following formulae:

$$\sigma=2(p_1-p_v)/\rho U^2 \quad (1)$$

, where p1 is the inlet pressure P_{in}, p_v – vapor pressure [6], and U – mean flow velocity measured by the loop's flow meter. Vapor pressure varied from 2.29733E-05 Pa at 130°C to 0.00906 Pa at 227°C. A particular value of σ corresponds to the inlet pressure, P_{in}, in our case, at which cavitation occurs as the pressure is decreased, which is called the cavitation inception number and is denoted by σ_i. Cavitation number σ_a, at which the head loss is 2, 3, or 5%, often is defined as a critical cavitation number. Further reduction in the cavitation number will lead to significant deterioration in the performance; the cavitation number at which this occurs is termed the breakdown cavitation number and is denoted by σ_b. [5]. As there were no acoustic measurements, it is impossible to determine cavitation inception and, thus, the corresponding number. Loop's tubing has an inner diameter of 54,3 mm and a cross-section of 2315,6 mm². Pumps channel cross-section is 1400 mm², which gives higher speeds at the cross-section and increases cavitation probability.
Cavitation numbers are plotted against flow speed in figure 7. For the "pump one" and "pump two" configurations, the cavitation numbers decrease with the flow speed increase for higher magnetic rotor rotation rates. At 735 rpm cavitation number varies from 9.6 at P_{start} value 3.0 bar to 3.65 at P_{start} value 1.0 bar. The P_{start} value 0,3 bar cavitation number values expectedly should follow the same path, but the situation is different. The cavitation numbers at low rotation rates drop with the flow speed increase, but the values are lower because of the lower initial inlet pressure. At 367.5 rpm, the decrease of the cavitation number stops, reaching 5.67. This s caused by the minor changes in the flow rate and, accordingly, to flow speed from 367,5 rpm to a maximal 735 rpm. Thus, we get a situation where the cavitation numbers for non-cavitating flow at P_{start} 1.0 and 1,5 bar are lower than the ones for the cavitating flow, accordingly 3.67 and 4.9 against 5.66 at P_{start} 0.3 bar at 735 rpm.

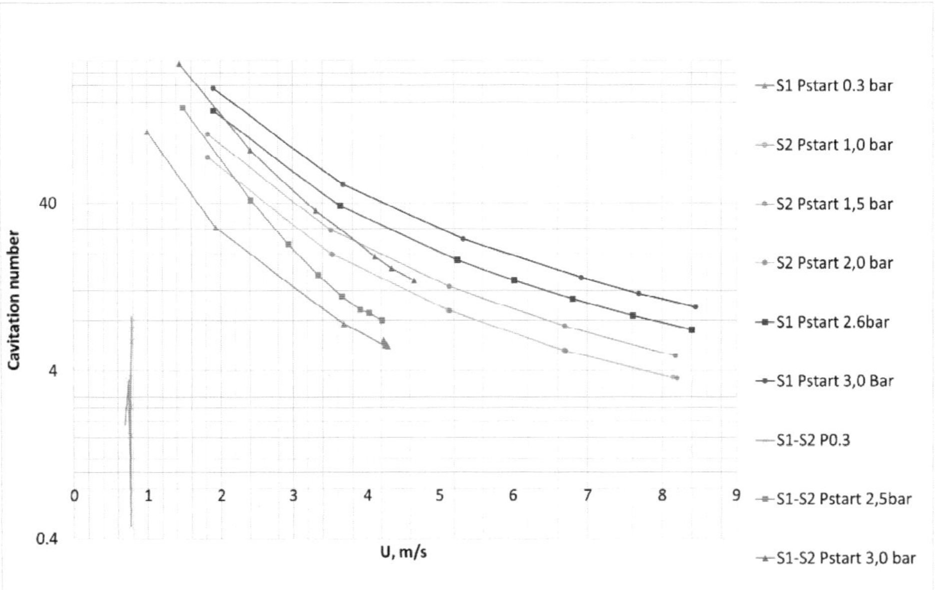

Fig. 7. Pumps flow speed at channel mean cross-section – cavitation number values at different flow path configurations.

This tendency is confirmed by plotting the cavitation number against the magnetic rotor rotation speed shown in figure 8. Here we can see the cavitation number decrease accordingly to the pump's rotor rotation speed increase for the "pump one" and "pump two" configurations for P_{start} values from 3.0 bar to 1.0 bar. The cavitation number curve for P_{start} 0.3 bar shows the change of the curve angle at 294 rpm and reaching the steady cavitation number values around 5.8 up to maximal 735 rpm.

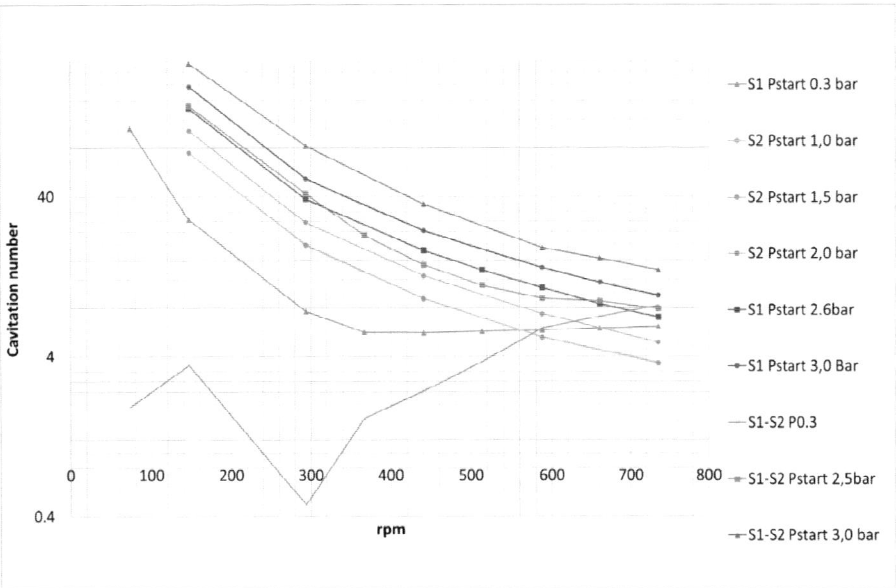

Fig. 8. Cavitation number – pumps magnetic rotor rotation rate.

The author suggests utilizing another approach for cavitation number calculation is necessary. The flow speed value is not an appropriate characteristic, which does not show the actual situation linking the electromagnetic force creating magnetic rotor movement to the conditions in the pump channel. For the centrifugal pump cavitation number calculation, the term "tip speed" is used [6]. Alternatively, the speed of the magnetic field movement would give more understanding of the "flow driving force" speed. The average radius of the pump's channel and rotation speed is used to calculate the linear speed of the magnetic rotor.

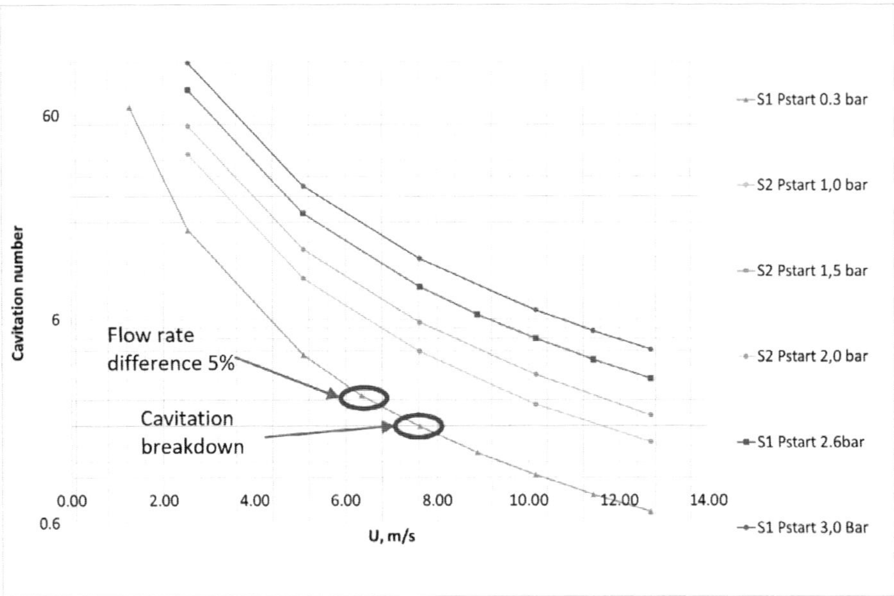

Fig. 9. Cavitation number – pumps magnetic rotor rotation rate.

The cavitation number plot against the pump's magnetic rotor rotation rate is shown in figure 9. The "pump one pulls through pump two" lines are excluded from the plot as the cavitation number values depend on the inlet pressure for P_{start} =0.3 bar giving expectedly low cavitation numbers varying from 0.57 to 0.03. For the P_{start} value, 2.5 bar cavitation numbers vary from 50.4 at 147 rpm to 0,88 at 735 rpm. For the P_{start} value, 3.0 bar cavitation numbers range from 87 at 147 rpm to 1,86 at 735 rpm.

The "pump one" and "pump two" configurations for P_{start} values from 3.0 bar till 1.0 bar give a stable decrease of the cavitation numbers from 38.7 till 1.5 at P_{start} 1.0 bar and corresponding values for P_{start} 3.0 bar from 108.6 till 4.3 at a magnetic rotor rotation rate from 147 till 735 rpm. The cavitation number curve for P_{start} 0.3 bar shows a similar tendency but with lower values varying from 16.4 to 0,68.

Those values gain meaning if we view them together with figure 5, where the flow rate and the pump magnetic rotor rotation rate are plotted. When the P_{start} 0.3 bar curve is compared to the P start 3.0 bar, we can see that flow rate differences of 5% start around 367,5 rpm value, corresponding to the cavitation number of 2.53. The "cavitation breakdown" for the P_{start} 0,3 bar starts at 441 rpm, corresponding to the cavitation number value of 1.8. Those numbers should also be considered considering the slip, which is always present in electromagnetic induction pumps, and for PEMDyn and Toshiba annular linear induction pumps, they are accordingly 0.172 and 0.22 [8].

Comparing the gained cavitation number values differs from literature sources for water pumps and turbines and the data on sodium pump investigation. In the work of K. Forster, for the sodium pump with a flow rate up to 150 m3/h, the cavitation inception number lies in the range of 2 to 3 for the water experiments, with slightly lower values for sodium experiments. [8]. The cavitation inception numbers for centrifugal pumps are between 0.3 and 1,4, as shown in the work of Gopalakrishnan. [10] Brennen finds the cavitation inception numbers lie around 0,8 - 1,2 and cavitation breakdown numbers for centrifugal water pumps between 0,4 and 0,5 [6]. The experimental investigation of the electromagnetic induction pump with rotating permanent magnets operating with liquid lead-bismuth eutectics (PbBi) shows the cavitation breakdown happened at the cavitation number values of 7 [11]. Although the high speed caused this in the pump inlet tubes, it is hard to compare both cases directly. The loop in the case mentioned above had a free surface, which differs from the hydraulic behavior of both loops.

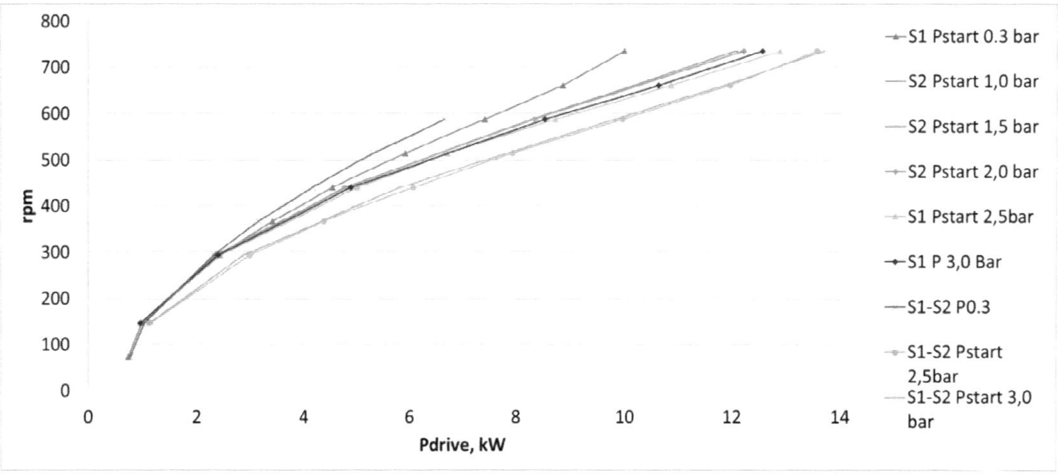

Fig. 10. Pumps drive unit consumed power – pumps magnetic rotor rotation rate for different P_{start} values.

The cavitation process can also be observed by viewing the drive unit's power consumption and plotting the pump's drive unit power consumption against the pump magnetic rotor rotation rate shown in figure 10. The 22 kW 3-phase asynchronous electric motor power supply uses an inverter with the possibility to measure the electrical power supplied to the motor. The highest power consumption rates are for the "pump one pulls through pump two" cases, reaching 13,7 kW for high P_{start} values due to increased pressure drop in the sodium loop. The case of the P_{start} 0,3 bar shows significantly less power consumption of a max of 8.88 kW, which also evidences considerable pump channel filling with cavitation bubbles.

The "pump one" and "pump two" configurations for P_{start} values from 3.0 bar to 1.0 bar curves follow a similar path with some differences at higher rotation rates reaching 12.2-12.5 kW. For the P_{start} 0.3 bar, the curve follows the way mentioned above. At higher pump rotor rotation rates, the power consumption curve becomes steeper, resulting in less consumed power of 10 kW, which links this curve to rpm – the flow rate curve plot.

4. Conclusions.

Electromagnetic induction pumps with rotating permanent magnets can suffer from cavitation when operating at low inlet pressure values. A series of experiments showed the relation between the pumps developed flow rate and the inlet pressure alongside with pump magnetic rotor rotation rate. We can use the cavitation number to characterize the pump operation conditions to determine the cavitation probability. The experiment allowed us to determine the cavitation number values corresponding to cavitation breakdown and flow rate increase rate change. However, we must be cautious regarding designing EMP with PM because there is very little data on cavitation processes in those pumps.

The data gained in the experiments described in the current paper does not give any information on cavitation inception but can provide rough initial details on the pump cavitation probability. Cavitation number values below $\sigma = 1{,}8$ can lead to flow increase breakdown. The pump's flow rate loss of 5% is achieved at cavitation number 2.5. Considering the slip, which is always present in electromagnetic induction pumps, those numbers should also be considered. The slip ratio should be regarded as in the cavitation number calculation, but further research is necessary. Further, the test loop should be modified and equipped with improved acoustic measurements for the cavitation inception point location.

5. Acknowledgements.

The publication is developed in the frame of Activity 1.1.1.2 "Post-doctoral Research Aid" contract No 1.1.1.2/VIAA/3/19/515.

6. References.

[1] I. E. Bucenieks. (2000). Perspectives of using rotating permanent magnets for electromagnetic induction pump design. Magnetohydrodynamics. 2000. T36. No2. pp.181-187.

[2] S. Constantinides. (2012). “The demand for rare earth materials in permanent magnets" (PDF). www.magmatllc.com. Available from: https://web.archive.org/web/20180329120806/https://www.magmatllc.com/PDF/Demand%20for%20rare%20earth%20materials%20in%20permanent%20magnets_Constantinides_120409.pdf. Accessed 2021.11.05.

[3] Dubovikova N., Kolesnikov Y. and Karcher Ch. (2015). Experimental study of an electromagnetic flow meter for liquid metals based on torque measurement during pumping process. Measurement Science and Technology, Volume 26, Number 11. Pp. 1-10. DOI:10.1088/0957-0233/26/11/115304.

[4] I. E. Bucenieks. (2003). High pressure and high flowrate induction pumps with permanent magnets. Magnetohydrodynamics 2003. 39, No. 4, pp.411-418, 2003. DOI: http://doi.org/10.22364/mhd.

[5] Kremljak, Z[vonko] & Petkovski, M[ilan] (2020). Cavitation - measurements for industrial valves, Chapter 01 in DAAAM International Scientific Book 2020, pp.001-014, B. Katalinic (Ed.), Published by DAAAM International, ISBN 978-3-902734-27-3, ISSN 1726-9687, Vienna, Austria DOI: 10.2507/daaam.scibook.2020.01.

[6] C. E. Brennen. Hydrodynamics of pumps. (1994). Concepts NREC and Oxford University Press .1994. Available from: https://resolver.caltech.edu/CaltechBOOK:1994.001. Accessed 2020.09.03.

[7] V. Sobolev. Database of thermophysical properties of liquid metal coolants for GEN-IV. Sodium, lead, lead-bismuth eutectic (and bismuth). Scientific report of the Belgian nuclear research centre. SCK•CEN-BLG-1069. 2011. - Coolants_GEN-IV. Available from: https://inis.iaea.org/collection/NCLCollectionStore/_Public/43/095/43095088.pdf?r=1. Accessed 2021.11.05.

[8] Martin-Lopez, Elena. (2018). Study of MHD instabilities in high flowrate induction electromagnetic pumps of annular linear design. Universite de Grenoble Alpes, 621 Avenue Centrale, 38400 Saint-Martin-d'Heres (France); CEA, DEN-DTN (France). PhD Thesis. 2018.

[9] Foerster, K. (1985). Acoustic surveillance as a detection method for cavitation in pumps. Specialist’s meeting on cavitation criteria for designing mechanisms working in sodium: Application to pumps. Summary report. International Atomic Energy Agency, International Working Group on Fast Reactors, Vienna (Austria); 289 p; 2002; p. 169-201; IAEA-IWGFR specialists' meeting on cavitation criteria for designing mechanisms working in sodium: Application to pumps; Bergisch Gladbach (Germany); 28-29 Oct 1985.

[10] Gopalakrishnan, S. (1985). Modern cavitation criteria for centrifugal pumps. Proceedings of the 2nd International Pump Symposium, Texas A&M University. Available from: https://www.911metallurgist.com/blog/wp-content/uploads/2016/01/Modern-Cavitation-Criteria-for-Centrifugal-Pumps.pdf. Accessed 2021.11.10.

[11] K. Kravalis, F. Boix, I. Bucenieks, et.al. (2022). Experimental cavitation investigation of the electromagnetic PbBi pump with rotating permanent magnets. Magnetohydrodynamics Vol. 58 (2022), No. 1-2, pp. 195–203 DOI: 10.22364/mhd.58.1-2.21.

DOI: 10.2507/33rd.daaam.proceedings.036

OPTIMIZATION FDM PROCESS PARAMETERS FOR FLEXURAL STRENGTH IMPROVEMENT OF CARBON FIBRES REINFORCED POLYAMIDE PARTS

Kenan Muhamedagić & Ahmet Čekić

This Publication has to be referred as: Muhamedagic, K[enan] & Cekic, A[hmet] (2022). Optimization FDM Process Parameters for Flexural Strength Improvement of Carbon Fibers Reinforcered Polyamide Parts, Proceedings of the 33rd DAAAM International Symposium, pp.0262-0270, B. Katalinic (Ed.), Published by DAAAM International, ISBN 978-3-902734-36-5, ISSN 1726-9679, Vienna, Austria
DOI: 10.2507/33rd.daaam.proceedings.036

Abstract

The FDM process was primarily developed only for certain types of polymer materials. Due to limited mechanical, thermal and electrical properties, these materials can't be used in industrial applications. By adding different nanoparticles and fibers to pure polymers, the aforementioned properties of pure polymers can be improved. However, optimization of process parameters is crucial to produce parts with optimal properties, such as mechanical strength. In this paper was presented the use of Taguchi method for optimization FDM process parameters (layer thickness, raster angle, wall thickness, printing speed, printing temperature and build plate temperature) to improve flexural strength of carbon fiber reinforced polyamide parts. The ranking of the most influential parameters was carried out using Taguchi analysis, and the statistical significance of the analyzed parameters was checked using ANOVA analysis. Based on the results of the Taguchi analysis and ANOVA analysis, it was concluded that the raster angle has a dominant influence on the flexural strength. Maximum flexural stress increases with the increase of the raster angle and maximum value is at a raster angle of 90°. Taguchi analysis showed that increasing build plate temperature and printing temperature are also important for improving the flexural strength.

Keywords: FDM; polymer composite; Taguchi method; flexural strength; optimization.

1. Introduction

The ISO/ASTM 52900 standard defines Additive Manufacturing (AM) as the the process of joining materials to make objects from three-dimensional (3D) model data, usually layer by layer, as opposed to subtractive manufacturing methodologies. According to the type of material used and the process physics, several of the most significant additive manufacturing procedures can be distinguished, such as fused deposition modeling (FDM), stereolithography (SLA), selective laser sintering (SLS), Ink jet modeling, etc. The FDM process is one of the most commonly used additive manufacturing processes, which is used for modeling, prototyping, but also for the production of fully functional products in various engineering applications.

The base material which is in the form of thermoplastic filament wound on a reel is delivered to the extruder of the device where it is heated and melted and then extruded to the build platform. Thermoplastic filament material such as polylactic acid (PLA), acrylonitrile butadiene styrene (ABS), poly carbonate (PC), polyamide (PA), polyethylene terephthalate (PET), and thermoplastic polyurethane (TPU) are used commonly in the FDM process [1][2][3]. However, the application of these materials, due to inadequate mechanical and physical properties, is limited only to conceptual modelling and prototyping due to the limited functionality of the manufactured parts. For this reason, in recent times, the application of various types of polymer composites for the FDM process is becoming more and more common. Different ways of reinforcement can be used for polymers, such as metal and ceramic particles, as well as continuous and short fibres of carbon, glass and Kevlar. Parts made using these composite materials have better mechanical, thermal and electrical properties, compared to the unreinforced polymer printed parts [4][5].

Ning et al. [6] presented influence of adding carbon fibres, with different content and length, on the mechanical properties of ABS printed parts including tensile properties (tensile strength, Young's modulus, toughness, yield strength, and ductility) and flexural properties (flexural stress, flexural modulus, flexural toughness, and flexural yield strength). It was concluded that the content and length of carbon fibres have a great influence on the mechanical properties of the material. Compared with the pure plastic specimen, CFRP composite specimen with 5 wt.% carbon fibre content had larger flexural strength, flexural modulus, and flexural toughness with an increase of 11.82%, 16.82%, and 21.86%, respectively. Also, by increasing the content of carbon fibres, the porosity of the material increases, which affected the deterioration of the mechanical properties for samples made with 10% carbon fibres. The effect of printing orientation on the mechanical behaviour of 3D printed carbon fibre reinforced polyamide composites was investigated in [7]. Dynamic mechanical analysis, flexural and tensile tests are performed to evaluate the mechanical behaviour of test specimens. The dynamic mechanical analysis test reveals higher storage and loss modulus in horizontal orientation as compared with angular and vertical orientation, mainly credited with better mechanical restraints imposed by carbon fibres on the matrix movement. Flexural properties of horizontally printed specimens are better as compared with the other configurations attributed to good stress transferability between the constituents. Influence of the percentage content of short carbon fibres in polyamide on the mechanical and thermal properties of PA/CF composite parts made by the FDM process was investigated in Liao et al. [8]. Compared to parts made of pure PA12, adding 10% carbon fibre to the PA12 matrix can achieve an increase in tensile strength, tensile modulus and flexural strength without significantly reducing the impact toughness of the material. The results showed that the tensile strength and flexural strength of polyamide (PA12) reinforced with 10% carbon fibre improved by 102.2% and 251.1%, respectively. Also, Spoerk et al. [9] were observed that the addition of short carbon fibres improves the flexural strength and modulus, compared to pure PP. The percentage content of fibres up to 10% enables a good material flow through the nozzle, printability, good surface quality and strong adhesion between the matrix and the fibres, compared to the fibre content of 20% and 30%.

Wang et al. [10] investigated the influence of the percentage content of short carbon fibres (CF) and glass fibres (GF) in the PEEK matrix on the mechanical properties, microstructure, surface quality and porosity of the printed samples. Test samples were made with three different fibre percentages (5%, 10% and 15%). The results show that the addition of CF/GF to PEEK can significantly increase the tensile and flexural strength at the expense of reducing the ductility. A lower percentage fibre content of 5% results in an increase in all analysed mechanical properties, improving surface quality and reducing porosity. It was observed that with GF/PEEK, a stronger adhesion between the fibres and the matrix is enabled than with CF/PEEK composites.

The aim of this paper was to determine the effect and optimization of process parameters (layer thickness, raster angle, wall thickness, printing speed, printing temperature and built plate temperature) to improve flexural strength of FDM printed carbon fibre reinforced polyamide parts using Taguchi method. In this research, a large number of input parameters were included, some of which have not been analysed in previous research when analysing the impact on flexural strength. Hence, this research provides a complete picture of the influence of all important FDM process parameters on the flexural strength of the carbon fibres reinforced polyamide parts.

2. Experimental procedure

Appropriate equipment and devices for the production of test samples, as well as appropriate equipment for testing mechanical properties, were used for the realization of experimental research. All test samples with process parameters defined by the experiment plan were produced on one material extrusion device (FDM device), Ultimaker S3 (Fig. 1.). The Ultimaker S3 3D printer enables the heating of the extruder up to 280 °C, and the use of special nozzles (CC nozzles) for the extrusion of composite polymer materials (Fig. 1a.). These nozzles have a diamond tip that has a high resistance to abrasion, which is very important when using composite materials. Another important feature of this device is the heated build plate (Fig. 1b.). The build plate of this device can be heated up to 140 °C, which is especially important when using certain materials that have weak adhesion to the surface and are sensitive to high temperature gradients, which can lead to deformations of parts as well as delamination due to insufficient cohesion between individual layers. This problem is even more pronounced when using polymer composite materials reinforced with short fibres where, in addition to the adhesion of the first layer and interlayer cohesion, it is also necessary to achieve good adhesion between the polymer matrix and the fibres. This is achieved by choosing the appropriate build plate temperature and extrusion temperature.

The material used in this device is a thermoplastic filament with a diameter of 2.85 mm wound on a reel and placed on a suitable material holder on the device (Fig. 1c.) and delivered to the extruder through special guides using a material feeder (Fig. 1d).

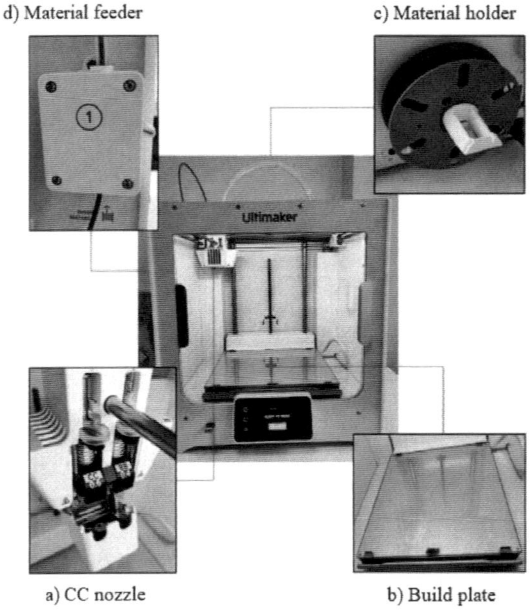

Fig. 1. Ultimaker S3 3D printer

A polymer composite material based on polyamide (PA6) reinforced with 10% short carbon fibres was used to make the test samples. One such commercially available filament material that can be used on the Ultimaker S3 printer is the "DSM Novamid® ID1030 CF10" material. The manufacturer of the material states that parts made by the FDM process from this material can have a strength close to that which can be achieved by injection moulding. The dimensions of the specimens were determined in accordance with ISO 178 standard (standard test method for flexural properties of plastics) (Fig. 2.). Specimens was designed using Solidworks software. Models of test specimens were saved in STL format which then were imported into 3D printing software „Ultimaker Cura". Regarding the software and printing configurations, the toolpath calculation (G-code) was made. The machine reads the G code and prints the specimens according to the instruction provided (Fig. 3.).

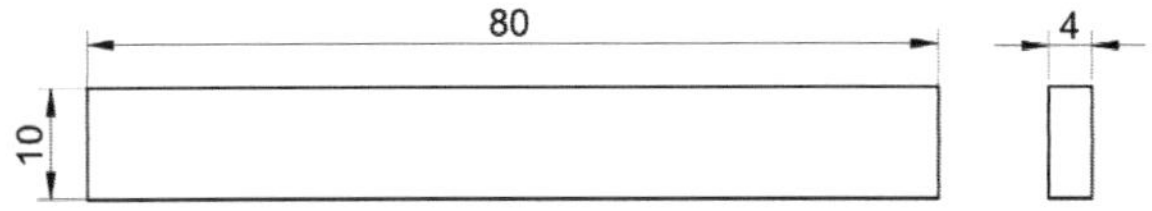

Fig. 2. Dimensions of the test specimen according to ISO 178

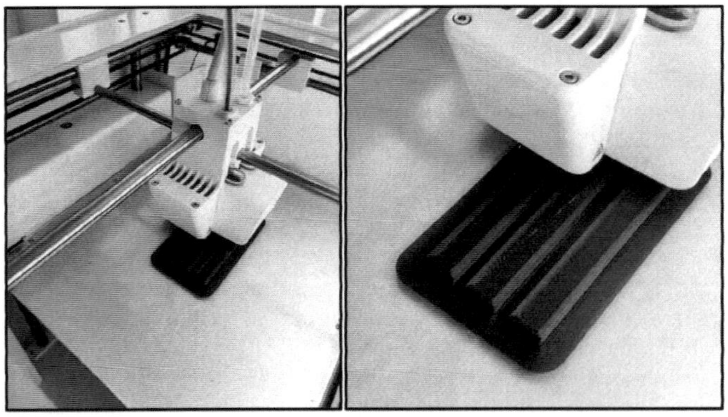

Fig. 3. Printing the specimens

After printing the samples, bending tests were performed with the aim of determining the flexural properties. For testing, the Schimadzu AGS-X universal testing machine was used, which achieves a maximum force of 10 kN. The test samples are placed between two bending measuring cells. The lower part of the measuring cells for bending has two cylindrical supports, and the upper measuring cell is movable and has one cylindrical edge that acts on the test sample (Fig. 4.). Distance between the supports is set to 64 mm and the test speed, in accordance with ISO 178, is 5 mm/min. All results and diagrams were tracked and saved using the "Trapezium" software (Fig. 4.).

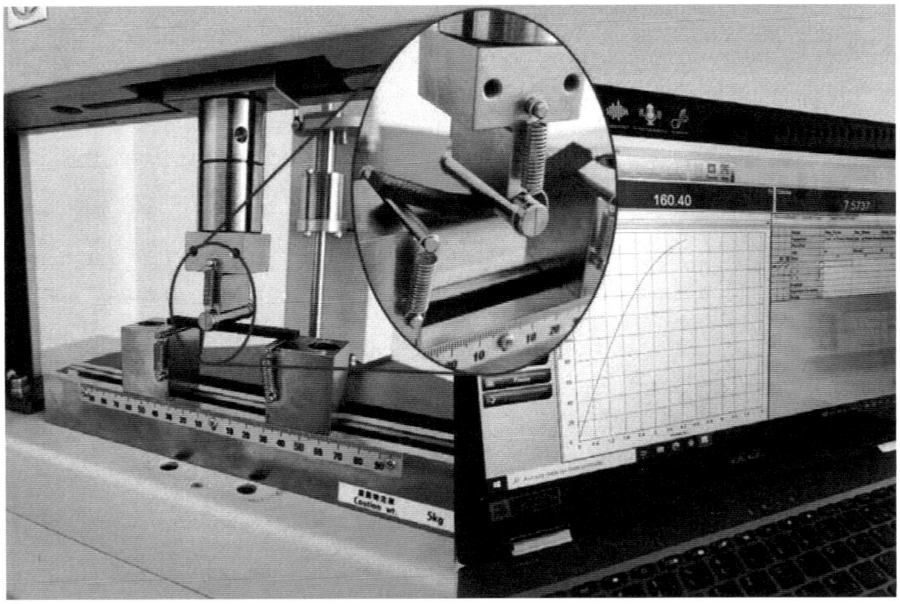

Fig. 4. Bending testing on Shimazu AGS-X

2.1 Design of experiments

To define the design of the experiment, a Taguchi robust plan was used which based on orthogonal arrays. The significance process parameters were tested by statistical analysis of the results. Selected input parameters (independent variables) and their variation levels are shown in table 1. Constant input parameters are given in table 2.

Factor	Symbol	Unit	Level		
			(1)	(2)	(3)
Layer thickness	A	[mm]	0,1	0,2	0,3
Wall thickness	B	[mm]	0,8	1,4	2
Raster angle	C	[°]	0	45	90
Printing speed	D	[mm/s]	40	60	80
Printing temperature	E	[°C]	240	260	280
Build plate temperature	F	[°C]	50	90	130

Table 1. Control factors and their levels

Infill	100 %
Orientation	Flat
Material	DSM Novamid® ID1030 CF10
Nozzle	Diamond nozzle (CC nozzle) with a diameter of 0.6 mm

Table 2. Constant process parameters

Based on the selected input factors and the defined levels of variation, an experimental matrix was formed in accordance with the used experimental methodology. According to the Taguchi method, for the analysis of the selected six numerical variables that vary on three levels, a total of 27 experimental tests are required based on the use of the standard orthogonal array L27 (3^6). The matrix of the experimental design according to the Taguchi design in coded form was given in table 3.

No.	Controlled input factors					
	A	B	C	D	E	G
1	1	1	1	1	1	1
2	1	1	1	1	2	2
3	1	1	1	1	3	3
4	1	2	2	2	1	1
5	1	2	2	2	2	2
6	1	2	2	2	3	3
7	1	3	3	3	1	1
8	1	3	3	3	2	2
9	1	3	3	3	3	3
10	2	1	2	3	1	2
11	2	1	2	3	2	3
12	2	1	2	3	3	1
13	2	2	3	1	1	2
14	2	2	3	1	2	3
15	2	2	3	1	3	1
16	2	3	1	2	1	2
17	2	3	1	2	2	3
18	2	3	1	2	3	1
19	3	1	3	2	1	3
20	3	1	3	2	2	1
21	3	1	3	2	3	2
22	3	2	1	3	1	3
23	3	2	1	3	2	1
24	3	2	1	3	3	2
25	3	3	2	1	1	3
26	3	3	2	1	2	1
27	3	3	2	1	3	2

Table 3. Experimental matrix in coded form

3. Results and Discussions

During the examination of the flexural mechanical properties of the test samples made of the material "Novamid® ID1030 CF10", it was not possible to unambiguously determine the character of the material's behaviour. Depending on the process parameters, some samples showed a brittle behaviour without a pronounced yield point, while some samples showed a plastic behaviour (Fig. 5.). In each case, the material broke and the maximum flexural stress could be clearly determined which occurred before break.

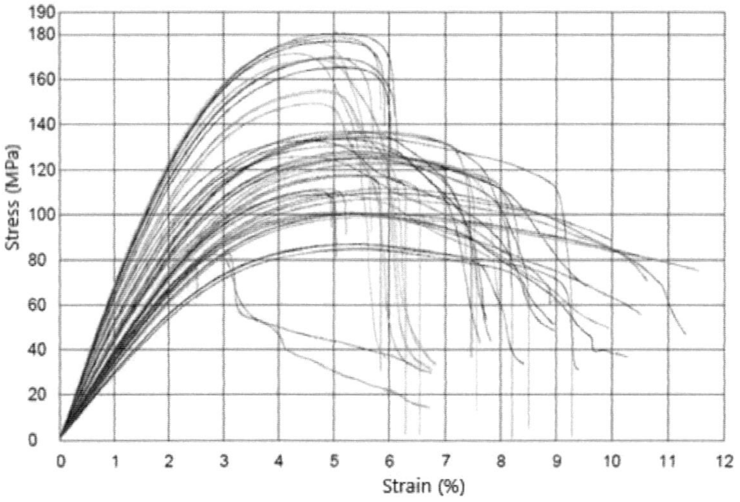

Fig. 5. The stress–strain curves of the test samples

The flexural strength test was carried out according to the previously defined matrix of the experiment plan. For each experimental test, three repeated measurements were performed, and the mean values of the maximum flexural stress σ_{fm}, as well as the values of the signal-to-noise ratio (*S/N*) are shown in table 4.

No.	Controlled input factors						Output	
	A	B	C	D	E	G	σ_{fm}	S/N
1	0,1	0,8	0	40	240	50	77,13	37,74
2	0,1	0,8	0	40	260	90	117,43	41,40
3	0,1	0,8	0	40	280	130	125,93	42,00
4	0,1	1,4	45	60	240	50	100,60	40,05
5	0,1	1,4	45	60	260	90	126,70	42,05
6	0,1	1,4	45	60	280	130	136,80	42,72
7	0,1	2	90	80	240	50	132,33	42,43
8	0,1	2	90	80	260	90	176,33	44,93
9	0,1	2	90	80	280	130	180,20	45,12
10	0,2	0,8	45	80	240	90	86,97	38,79
11	0,2	0,8	45	80	260	130	111,17	40,92
12	0,2	0,8	45	80	280	50	100,90	40,08
13	0,2	1,4	90	40	240	90	154,80	43,80
14	0,2	1,4	90	40	260	130	168,20	44,51
15	0,2	1,4	90	40	280	50	137,00	42,73
16	0,2	2	0	60	240	90	108,63	40,72
17	0,2	2	0	60	260	130	123,13	41,81
18	0,2	2	0	60	280	50	114,03	41,14
19	0,3	0,8	90	60	240	130	150,13	43,53
20	0,3	0,8	90	60	260	50	133,20	42,49
21	0,3	0,8	90	60	280	90	143,90	43,16
22	0,3	1,4	0	80	240	130	105,67	40,48
23	0,3	1,4	0	80	260	50	89,80	39,07
24	0,3	1,4	0	80	280	90	105,97	40,50
25	0,3	2	45	40	240	130	121,00	41,66
26	0,3	2	45	40	260	50	111,00	40,91
27	0,3	2	45	40	280	90	123,53	41,84

Table 4. The experimental data

Based on the data shown in table 4 and through the analysis of mean values, the character of the influence of FDM process parameters on the maximum flexural stress σ_{fm} was presented. The diagram of the partial influences of the analysed input parameters was shown in the form of a graphic in figure 6.

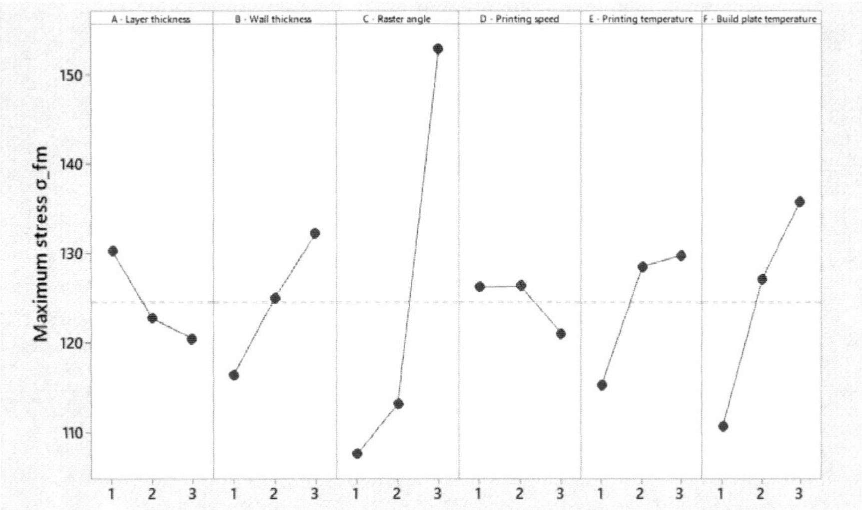

Fig. 6. Diagram of the partial influences of the analysed input parameters

Analysing the diagrams shown, it can be concluded that all analysed parameters have a clear influence on the flexural strength, where it can be seen that the raster angle has the greatest influence on the range of variability of the mean values of the maximum stress, i.e., that the raster angle has the greatest influence on the flexural strength. This influence can be explained with the direction of deposition of the material in the XY plane, which has the greatest influence on the anisotropic behaviour of the material in the XY plane. From the diagram it can be seen that the value of the maximum flexural stress increases with the increase of the raster angle and that the maximum stress is at a raster angle of 90°. After the raster angle, the build plate temperature had the greatest influence on the flexural strength. During flexural tests, shear stresses occur between individual layers of material. In order to prevent delamination between individual layers of material under shear stresses, a stronger cohesion between the layers of material is required. Increasing the printing temperature enables better melting of the material and stronger bonding to the previous layer. However, if the temperature of the previous layer is much lower than the temperature of the material that has just been applied, due to large temperature gradients, thermal stresses can occur within the material, which can reduce the interlayer strength. By increasing the build plate temperature, the temperature difference (gradient) between the temperature of the previous layer and the temperature of the newly applied material is reduced. This results in a stronger interlayer cohesion and, therefore, a higher total flexural strength. The layer thickness, walls thickness and printing speed had a smaller influence on the flexural strength than the other analysed parameters. It can be observed that by increasing layer thickness and printing speed, the flexural strength decreases.

Based on the individual values of the maximum stress σ_{fm}, the values of the signal-to-noise ratio (*S/N*) were calculated for each experimental test, based on which the significance of the influence of individual factors on the flexural strength was determined, and then the average values of the signal-to-noise ratio (S/N_{avg}) for individual levels of all six analysed input parameters were determined (Table 5.).

Level	A	B	C	D	E	F
1	42,05	41,12	40,54	41,84	41,02	40,74
2	41,61	41,77	41,00	41,96	42,01	41,91
3	41,51	42,28	43,63	41,37	42,14	42,53
Range	0,54	1,16	3,09	0,60	1,12	1,79
Rank	6	3	1	5	4	2

Table 5. Mean values of the signal-to-noise ratio

Based on the previous table, a ranking of the dominant input parameters can be performed, which confirms that the most dominant influence on the flexural strength has the raster angle, followed by the build plate temperature, the wall thickness and the printing temperature. Layer thickness and printing speed have the least influence. The previous table can also be displayed using a diagram of the partial influence of individual input parameters on the mean value of the signal-to-noise ratio (S/N_{avg}) (Fig. 7.). From these diagrams it is possible to determine the optimal parameters that achieve the maximum value of flexural strength.

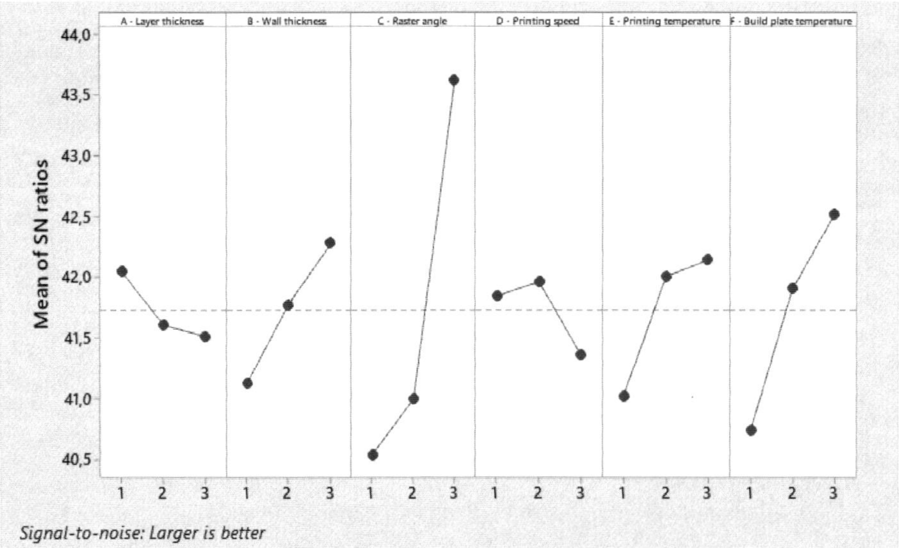

Fig. 7. Plot of mean values of the signal-to-noise ratio

Optimal values of process parameters were adopted for those levels at which S/N_{avg} is maximum. In this case, the optimal parameters for maximizing flexural strength would be:

- layer thickness: 0,1 mm,
- wall thickness: 2 mm,
- raster angle: 90°,
- printing speed: 60 mm/s,
- printing temperature: 280 °C,
- build plate temperature: 130 °C.

In order to determine the statistical significance of the analysed FDM parameters, data processing was performed through analysis of variance (ANOVA) for a significance level of 5%. By analysing the influence of individual parameters through ANOVA analysis, the percentage contribution of parameters on the experimental results can be obtained. The results of the ANOVA analysis of the signal-to-noise ratio *S/N* are shown in table 6.

Source	DF	Seq SS	Contribution	Adj SS	Adj MS	F-Value	P-Value
A	2	1,466	1,72%	1,466	0,7330	2,41	0,126
B	2	6,072	7,12%	6,072	3,0361	10,00	0,002
C	2	50,137	58,76%	50,137	25,0686	82,55	0,000
D	2	1,790	2,10%	1,790	0,8948	2,95	0,086
E	2	6,756	7,92%	6,756	3,3778	11,12	0,001
F	2	14,859	17,41%	14,859	7,4297	24,47	0,000
Error	14	4,252	4,98%	4,252	0,3037		
Total	26	85,332	100,00%				

Table 6. The results of the ANOVA analysis of the signal-to-noise ratio

According to the results presented in table 6, it can be determined which members are statistically significant. Based on the p-value, statistically significant parameters, for a significance level of 5%, are those parameters whose p-value does not exceed 0.05. Therefore, statistically significant parameters would be: wall thickness (*B*), raster angle (*C*), printing temperature (*E*) and build plate temperature (*F*). The influence of the analysed parameters can also be estimated based on the percentage contribution of each parameter in the total variation of the experimental results. This confirms that the raster angle has the greatest impact on flexural strength with a contribution of 58.76%. On second place is the build plate temperature with contribution of 17.41%. Although the parameters of the wall thickness and the printing temperature do not have a great percentage contribution in the total variation of the results, based on their p - values they are classified as statistically significant parameters. The layer thickness and printing speed, due to their p-value as well as the small percentage contribution in the total variation of the results, are statistically insignificant and their influence compared to the other analysed parameters is negligible. The error percentage of 4.98% is relatively small and it can be assumed that it arose as a result of external uncontrolled influences, i.e., noise.

4. Conclusion

The aim of this work was to optimize selected FDM process parameters (layer thickness, raster angle, wall thickness, printing speed, printing temperature and build plate temperature) to improve flexural strength of carbon fibre reinforced polyamide parts. Optimization of the selected parameters was carried out using Taguchi analysis, and the statistical significance of the analysed parameters was checked using ANOVA analysis Based on the conducted experimental research on the influence of selected FDM process parameters on the flexural strength of parts made of polyamide reinforced with short carbon fibres, the following conclusions can be drawn:

1. Using Taguchi analysis, optimal parameters were determined with the aim of maximizing flexural strength. The optimal parameters are: minimum value of layer thickness (0.1 mm), maximum value of wall thickness (2 mm), raster angle in the direction of load (90°), average value of printing speed (60 mm/s), maximum value of printing temperature (280 °C), maximum value of the build plate temperature (130 °C).
2. Based on the results of the ANOVA analysis of the *S/N* ratio, the dominant parameters with respect to flexural strength were selected and ranked. Four statistically significant parameters were singled out and ranked according to dominance: raster angle, build plate temperature, printing temperature and wall thickness. The parameters layer thickness and printing speed are marked as statistically insignificant.
3. The highest value of the flexural strength is achieved for a raster angle of 90°. With a contribution of 58.76% and p-value of 0.000, it represents the most dominant and statistically significant parameter in the analysis of the influence on flexural strength.
4. The influence of the build plate temperature is directly proportional to the maximum flexural stress. By increasing the build plate temperature, the temperature difference (gradient) between the temperature of the previous layer and the

temperature of the new deposited material is reduced. This results in a stronger interlayer cohesion and, therefore, a higher flexural strength.

5. The influence of the printing speed is inversely proportional to the value of the maximum flexural stress. At higher printing speed, an insufficient amount of material is deposited, which prevents a strong cohesion between the raster and the previously applied layers of material.
5. The influence of the wall thickness is directly proportional to the value of the maximum flexural stress. The highest value of flexural strength is achieved with a wall thickness of 2 mm.
6. Based on the results of the ANOVA analysis, it is concluded that the influences of layer thickness and printing speed are statistically insignificant parameters.

The future research is to investigate the effect of different polymer composite materials (type of polymer matrix, fibre type, fibre length and fibre content) on the mechanical properties of FDM printed parts.

5. Acknowledgments

This work was supported by Ministry of Education and Science Federation of Bosnia and Herzegovina and Faculty of Mechanical Engineering University of Sarajevo. The support is gratefully acknowledged.

6. References

[1] Muhamedagic, K., Cekic, A., & Begic-Hajdarevic, D. (2020). Experimental Investigation and Modelling of FDM Process Parameters for Tensile Strength Improvement Using RSM Method. In B. Katalinic (Ed.), DAAAM Proceedings (1st ed., Vol. 1, pp. 0490–0498). DAAAM International Vienna. https://doi.org/10.2507/31st.daaam.proceedings.069

[2] Pinero Vega, D., Batista Ponce, M., Valerga Puerta, A. P., Vazquez Martinez, J. M., & Fernandez, S. R. (2018). A Comparison of Macro and Microgeometrical Properties of Specimens Made With a FDM Commercial Printer and its Opensource Retrofit Version. In B. Katalinic (Ed.), DAAAM Proceedings (1st ed., Vol. 1, pp. 1108–1115). DAAAM International Vienna. https://doi.org/10.2507/29th.daaam.proceedings.158

[3] Cekic, A., Begic-Hajdarevic, D., Muhamedagic, K., & Guzanovic, N. (2018). Experimental Investigations of Process Parameters Influence on Dimensional Accuracy and Mechanical Properties of FDM Manufactured Parts. In B. Katalinic (Ed.), DAAAM Proceedings (1st ed., Vol. 1, pp. 0210–0214). DAAAM International Vienna. https://doi.org/10.2507/29th.daaam.proceedings.030

[4] Muhamedagic, K., Berus, L., Potočnik, D., Cekic, A., Begic-Hajdarevic, D., Cohodar Husic, M., & Ficko, M. (2022). Effect of Process Parameters on Tensile Strength of FDM Printed Carbon Fiber Reinforced Polyamide Parts. Applied Sciences, 12(12), 6028. https://doi.org/10.3390/app12126028

[5] Li, Zhao, Li, Yang, & Wang. (2019). Flexural Properties and Fracture Behavior of CF/PEEK in Orthogonal Building Orientation by FDM: Microstructure and Mechanism. Polymers, 11(4), 656. https://doi.org/10.3390/polym11040656

[6] Ning, F., Cong, W., Qiu, J., Wei, J., & Wang, S. (2015). Additive manufacturing of carbon fiber reinforced thermoplastic composites using fused deposition modeling. Composites Part B: Engineering, 80, 369–378. https://doi.org/10.1016/j.compositesb.2015.06.013

[7] Alarifi, I. M. (2022). Investigation of the dynamic mechanical analysis and mechanical response of 3D printed nylon carbon fiber composites with different build orientation. Polymer Composites, 43(8), 5353–5363. https://doi.org/10.1002/pc.26838

[8] Liao, G., Li, Z., Cheng, Y., Xu, D., Zhu, D., Jiang, S., Guo, J., Chen, X., Xu, G., & Zhu, Y. (2018). Properties of oriented carbon fiber/polyamide 12 composite parts fabricated by fused deposition modeling. Materials & Design, 139, 283–292. https://doi.org/10.1016/j.matdes.2017.11.027

[9] Spoerk, M., Savandaiah, C., Arbeiter, F., Traxler, G., Cardon, L., Holzer, C., & Sapkota, J. (2018). Anisotropic properties of oriented short carbon fibre filled polypropylene parts fabricated by extrusion-based additive manufacturing. Composites Part A: Applied Science and Manufacturing, 113, 95–104. https://doi.org/10.1016/j.compositesa.2018.06.018

[10] Wang, P., Zou, B., Ding, S., Huang, C., Shi, Z., Ma, Y., & Yao, P. (2020). Preparation of short CF/GF reinforced PEEK composite filaments and their comprehensive properties evaluation for FDM-3D printing. Composites Part B: Engineering, 198, 108175. https://doi.org/10.1016/j.compositesb.2020.108175

DOI: 10.2507/33rd.daaam.proceedings.037

TECHNOLOGICAL WEAR INFLUENCE ANALYSIS ON THE DECREASE IN THE EFFICIENCY OF A CLOSED LOOP CONTROL OF HEAT EXCHANGE EQUIPMENT

Michael Peshko, Petr Batrakov, Alexander Lasitsa & Vasilii Khomchenko

Omsk State Technical University, Mira avenue 11, Omsk, 644050, Russia Federation

This Publication has to be referred as: Peshko, M[ichael]; Batrakov, P[etr]; Lasitsa, A[lexander] & Khomchenko, V[asilii] (2022). Technological Wear Influence Analysis on the Decrease in the Efficiency of a Closed Loop Control of Heat Exchange Equipment, Proceedings of the 33rd DAAAM International Symposium, pp.0271-0278, B. Katalinic (Ed.), Published by DAAAM International, ISBN 978-3-902734-36-5, ISSN 1726-9679, Vienna, Austria
DOI: 10.2507/33rd.daaam.proceedings.037

Abstract

A physical model of heat and mass transfer processes in industrial equipment is presented in the work. The transfer functions of the model main signals taking into account the equipment's technological wear are obtained. The effect of deposits on the operation of a closed loop for temperature control has been studied. As a result of numerical experiments, a quantitative dependence has been established between the thickness of deposits on the inner working surface of heat exchange equipment and the quality indicators of the control system. Conclusions are drawn about the inefficiency of linear control systems in the conditions of technological wear during in real operation of heat exchange equipment.

Keywords: heat and mass transfer; heat exchange surface fouling; linear control systems, modelling control object; simulink matlab.

1. Introduction

Deposits on the surfaces of heat exchange and process equipment are a serious problem in the heat power industry, as well as in the chemical, food and other industries. These deposits impair the efficiency of heat transfer, which leads to an increase in specific energy consumption, reducing the effective flow area of equipment elements and pipelines, and increasing hydraulic resistance leads to an increase in energy consumption for transporting the coolant [1], [2], [3], [4], [5], [6], [7]. As a result, the chance of an accident at work due to the failure of the heat exchanger increases. The above factors in the framework of control theory can be classified as perturbations of a parametric nature [8]. The lack of compensation for parametric perturbation in combination with the most common linear PID controllers in the industry can lead to a decrease in the accuracy of maintaining the required process parameters and reduce the overall efficiency of the process.

To solve this problem, in most cases, traditional PID regulation is used with the use of special corrective links that ensure the adjustment of the linear control system to the changing parameters of the control loop [9], [10], [11], [12], [13]. However, such approaches require significant effort for the study of mathematical models of specific technological objects and the qualifications of service personnel. The construction of a universal analytical model of heat exchange equipment, taking into account polluting deposits on internal surfaces, can make it possible to assess the degree of influence of deposition on the operation of the control loop and design corrective actions to compensate for the operation of the regulator. The purpose of this study is to identify the influence of the deposits thickness of the closed-loop control systems performance for heat exchange equipment while maintaining the required temperature regime using linear control systems.

2. Mathematical model of the heat exchanger taking into account technological wear

The schematic diagram of the simulated heat exchanger displaying energy and mass flows is shown in Figure 1. The object of the simulation is the heating circuit of raw water (two heat exchangers T-3 and T-1, connected in series) for technological needs, where it is required to maintain the temperature within 20-30 °C, the permissible fluctuation is ± 1 °C per hour.

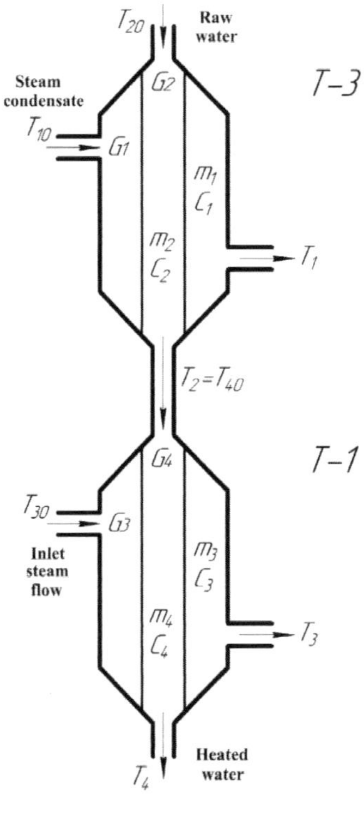

Fig. 1. Scheme of the simulation object

The value of this parameter is influenced by raw water temperature T_{20}, steam temperature T_{30}, condensate temperature T_{10} and other system parameters. The maximum water flow through the heat exchanger is assumed to be no more than 250 m3/h. The temperature of the heat carrier entering the unit can vary within different limits depending on technological factors, and in order to comply with the optimal conditions of the technological process, the temperature at the outlet of the heat exchanger must be maintained within the specified limits. Let's develop a mathematical model of the heat exchanger T-3. The heat carrier in it is water vapor condensate. The basis for creating a mathematical model of heat exchange processes is the equation of the law of conservation of energy in differential form. Equation 1 of the energy conservation law has the form [14], [15]:

$$mC\frac{dT}{dt} = \sum Q, \qquad (1)$$

where m is the mass [kg], C is the specific heat capacity [kJ/(kg °C)],
T – temperature [°C], Q – heat flow [kJ]

On the basis of equation 1, we will compose differential equations for the temperature of the coolant T_1(t)and the heated product T_2(t), characterizing the water heating model - equations 2 and 3, respectively:

$$m_1C_1\frac{dT_1(t)}{dt} = G_1C_1T_{10} - G_1C_1T_1(t) - k_{T-3}S_{T-3}\big(T_1(t) - T_2(t)\big), \tag{2}$$

where m1 is the mass of the heat carrier (condensate) [kg], C_1 is the specific heat capacity of the heat carrier [J/(kg °C)], $T_1(t)$ is the temperature of the heat carrier [°C], $T_2(t)$ is the temperature of the heated product [°C], G_1 is the mass coolant flow rate [kg/s], k_{T-3} – heat transfer coefficient through the wall T-3 [W/(m^2 °C)], S_{T-3} – heat exchange surface area T-3 [m2].

$$m_2C_2\frac{dT_2(t)}{dt} = G_2C_2T_{20} - G_2C_2T_2(t) + k_{T-3}S_{T-3}\big(T_1(t) - T_2(t)\big), \tag{3}$$

where m_2 is the mass of the product to be heated [kg], C_2 is the specific heat capacity of the product to be heated [J/(kg °C)], $T_2(t)$ is the temperature of the product to be heated [°C], G_2 is the mass flow rate of the product to be heated [kg/s], k_{T-3} – heat transfer coefficient through the wall T-3 [W/(m^2 °C)], S_{T-3} – heat exchange surface area T-3 [m^2]. Similarly, we obtain a mathematical model of the heat exchanger T-1 - equations 4 and 5. The heat carrier in it is steam.

$$m_3C_3\frac{dT_3(t)}{dt} = G_3C_{30}T_{30} - G_3C_3T_3(t) - k_{T-3}S_{T-3}\big(T_3(t) - T_4(t)\big), \tag{4}$$

where m_3 is the mass of the heat carrier (condensate) [kg], C_3 is the specific heat capacity of the heat carrier [J/(kg °C)], $T_3(t)$ is the temperature of the heat carrier [°C], $T_4(t)$ is the temperature of the heated product [°C], G_3 is the mass coolant flow rate [kg/s], k_{T-1} – heat transfer coefficient through the wall T-1 [W/(m^2 °C)], S_{T-1} – heat exchange surface area T-1 [m^2].

$$m_4C_4\frac{dT_4(t)}{dt} = G_4C_4T_{40} - G_4C_4T_4(T) + k_{T-1}S_{T-1}\big(T_3(t) - T_4(t)\big) + r\cdot G_3, \tag{5}$$

where m_4 is the mass of the heated product [kg], C_4 is the specific heat capacity of the heated product [J/(kg °C)], $T_4(t)$ is the temperature of the heated product [°C], G_3 is the mass flow rate of the heat carrier [kg/s], G_4 is mass flow rate of the heated product [kg/s], k_{T-1} – heat transfer coefficient through the wall T-1 [W/(m^2 °C)], S_{T-1} – heat exchange surface area T-1 [m^2], r – specific heat vaporization of water [J/kg]. The heat transfer coefficient k through a single-layer wall is defined as [16]:

$$k = \frac{1}{\frac{1}{\alpha_1} + \frac{\delta}{\lambda} + \frac{1}{\alpha_2}}, \tag{6}$$

where δ is the wall thickness [m], λ is the thermal conductivity of the wall material [W/m °C], α_1 is the heat transfer coefficient from the hot coolant to the wall [W/(m^2 °C)], α_2 is the heat transfer coefficient from the wall to the cold substance [W/(m^2 °C)]. The physical and technical parameters used for the simulation are given in tables 1 and 2.

Parameter name	Parameter designation	Meaning
Condensate temperature for heating (T-3)	T_{10}	80°C
Raw water temperature per unit	T_{20}	3°C
Steam temperature for heating (T-1)	T_{30}	280°C
Water temperature after T-3 for reheating in T-1	T_{40}	T_2
Mass flow of condensate for heating	G_1	$44\frac{kg}{s}$
Mass flow of raw water per plant	$G_2 = G_4$	$70\frac{kg}{s}$
Mass flow of steam for heating	G_3	$0 \ldots 2{,}7\frac{kg}{s}$
Mass of coolant (condensate) in T-3	m_1	680 kg
The mass of the heated product in T-3	m_2	550 kg
Mass of coolant (steam) in T-1	m_3	1,253 kg
The mass of the heated product in T-1	m_4	250 kg
Heat exchange surface area T-3	S_{T-3}	80 m^2
Heat exchange surface area T-1	S_{T-1}	48м^2
Wall thickness of the heat exchanger T-3 and T-1	δ	0,002м

Table 1 - Initial data

Calm water - metal wall	$450 \frac{W}{m^2 \cdot °C}$
Flowing water - metal wall	$450 + 2100\sqrt{v} \frac{W}{m^2 \cdot °C}$
Condensing water vapor	$10500 \frac{W}{m^2 \cdot °C}$
Specific heat of the heat carrier (water) in T-3, specific heat of the heated product (water) in T-3 and T-1	$4183 \frac{J}{kg \cdot °C}$
Specific heat capacity of the coolant in T-1 (steam)	$2200 \frac{J}{kg \cdot °C}$
Specific heat of vaporization of water	$2256 \frac{kJ}{kg}$
Thermal conductivity coefficient of heat exchanger material (steel «12X18H10T»)	$15 \frac{W}{m \cdot °C}$

Table 2 - Physical constants and dependencies [15]

The proposed model makes it possible to accurately consider the process of water heating in shell-and-tube heat exchangers T-3 and T-1. The model takes into account the thermal process of the phase transition when steam is used as a heat carrier, as well as the heat transfer coefficient through the walls of the heat exchanger. The use of such a model provides a mathematical tool for analyzing the behavior of control systems with changing parameters in a known range. The system of equations (2)-(5) will be used as a basis for obtaining a linear model of the control object in the form of transfer functions, taking into account the pollution of the working surfaces of heat exchange equipment.

3. Modelling of the functional elements of the temperature control loop by transfer functions

The block diagram of the temperature control loop is shown in Figure 2.

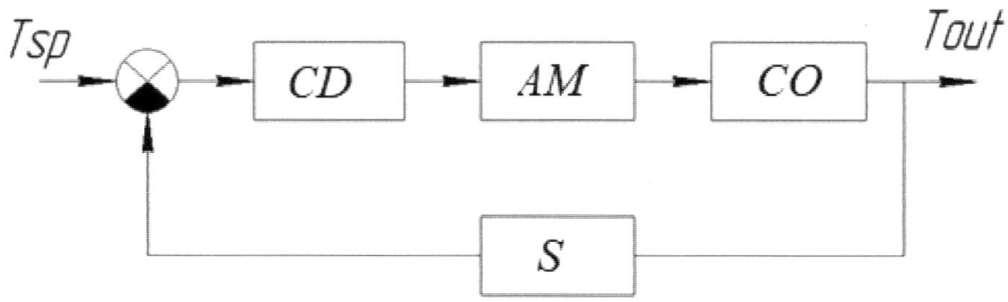

Fig. 2. Scheme of the simulation system

This scheme consists of the following elements:

- control device (CD) – PID controller;
- actuating mechanism (AM) – steam flow control valve at T-1;
- control object (CO) – T-3 and T-1 heat exchangers connected in series;
- sensor (S) – thermal converter.

For the synthesis of the control system, we obtain the transfer functions of the sensor (thermal converter) and the actuating mechanism (control valve) and the control object. Thermal resistance is a first-order aperiodic link, since the sensor converts signals of various physical nature: the input signal (temperature) into the output signal (resistance). The transfer function of the thermal resistance will be written as:

$$W_S(p) = \frac{k_S}{T_S p + 1}. \tag{7}$$

The dependence of the analog output signal of the sensor on temperature is linear, therefore, in order to simplify the model, we will accept the gain factor k_S=1. The time constant in the liquid, according to the reference data for the thermal converter, is T_S= 15 s.

Therefore, the transfer function of the temperature sensor has the form:

$$W_S(p) = \frac{1}{15p + 1}. \quad (8)$$

As the transfer function of the control valve, we use an first-order inertial link:

Time constant T_{am} is defined as the valve opening time with a small input action T_{am} = 0.03 s. We find the proportionality coefficient based on the maximum flow rate G_3=2.7 kg/s as the ratio $k_{am} = \frac{2{,}7\frac{kg}{s}}{100\%} = 0{,}027\frac{kg}{s \cdot \%}$.

The transfer function of the steam control valve has the form:

$$W_{am}(p) = \frac{0{,}027}{0{,}03p + 1}. \quad (9)$$

The heat exchanger described by equations (2)-(5), taking into account changing parameters, is a non-linear element of the model. To carry out the synthesis of the control system, the linearization of the control object was carried out and the transfer function was obtained for the control channel and the perturbation channel using the Linear Analysis Tool Simulink.

The calculated transfer function for the disturbance channel has the form:

$$W_{dist}(p) = \frac{0{,}427082p^2 + 10{,}0575p + 0{,}86435}{11{,}9832p^4 + 288{,}197p^3 + 147{,}394p^2 + 22{,}5285p + 1}. \quad (10)$$

The calculated transfer function for the control channel has the form:

$$W_{ctrl}(p) = \frac{0{,}00886758p + 0{,}264018}{0{,}00456621p^3 + 0{,}260913p^2 + 3{,}65342p + 1}. \quad (11)$$

The resulting transfer functions describe the control object behaviour dynamics, and are used for values close to some steady value.

4. Investigation of the influence heat exchange equipment pollution processes on decrease in the control quality of linear control systems

To perform a study of the influence of contamination processes of heat exchange equipment on reducing the quality of regulation, it is necessary to assess the quality of closed loop single-circuit regulation of the heat exchanger, taking into account the technological wear of the equipment and to simulate the change in its characteristics over time. The main study objective is to find out linear regulators are able to provide the required control quality of a non-stationary control object.

In the software environment Matlab/Simulink a complete model of the temperature control loop was developed (fig. 3). As the initial conditions, the parameters of the plant operation in winter were used: the temperature of heat exchanger T-3 inlet water in the range of 2-5 °C. Subsequent heating of water and temperature control at the outlet heat exchangers cascade is carried out by changing the mass flow rate of steam at T-1 with a set range of 0-2.7 kg/s. Control action on the system is G3 - the mass flow rate of steam at T1, and the disturbing action T20 is the temperature of raw water at the plant. The simulation results are shown in fig. 4.

From the simulation results, it can be concluded that the system reaches the required temperature value of 27 °C. According to the technological regulations, deviations in the outlet temperature should not exceed ±1 °C per hour. This condition is met even if the inlet water temperature deviation widely (1°C...11°C). At the second stage, modeling was carried out taking into account technological wear (2 mm fouling), which leads to a decrease thermal conductivity coefficient by 4 times. The simulation results are shown in fig 5. The graph shows deviation in the outlet temperature in the range of 26.6°C ... 27.3°C. These deviations meet the requirements (not exceed ±1 °C per hour).

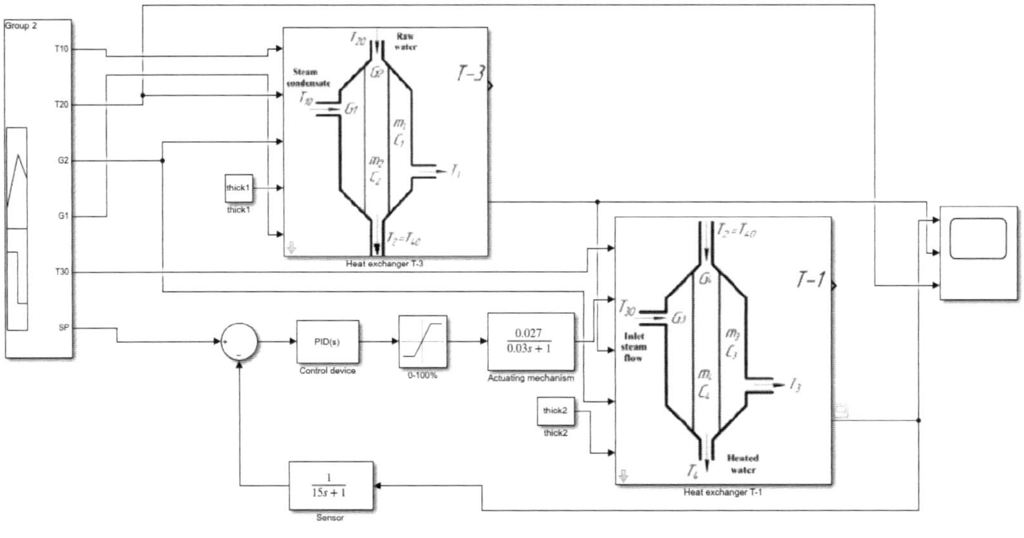

Fig. 3. System simulation diagram in Matlab/Simulink

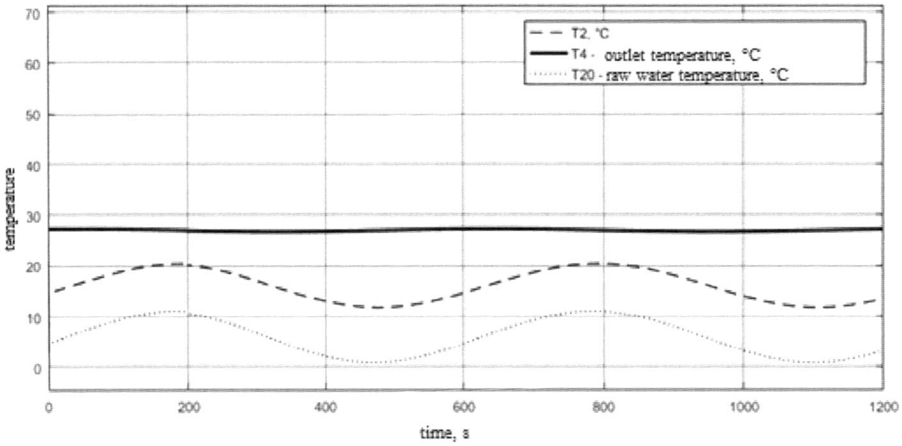

Fig. 4. System simulation results

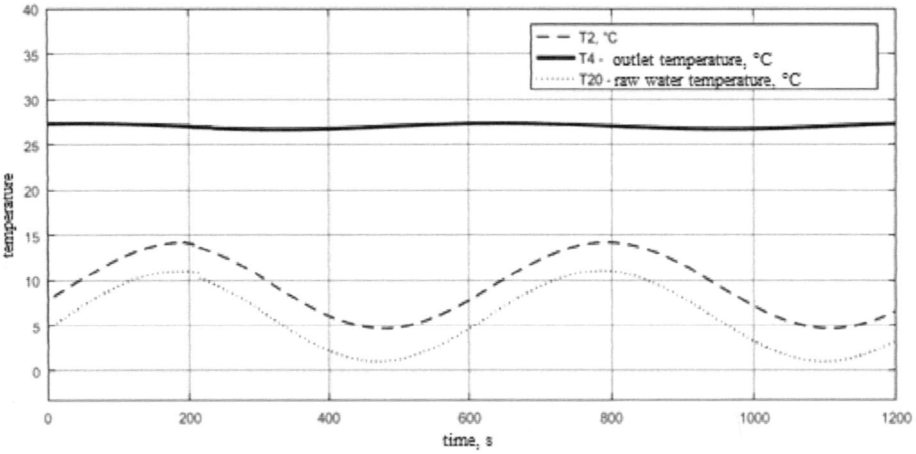

Fig. 5. System simulation results with technological wear

At the final stage, the operation of the automatic control system was simulated, taking into account scale deposits of 3 mm (the results are shown in fig. 6). In this case, temperature deviation is observed in the range of 25.9-29.8 °C, which does not meet the requirements of the water treatment process.

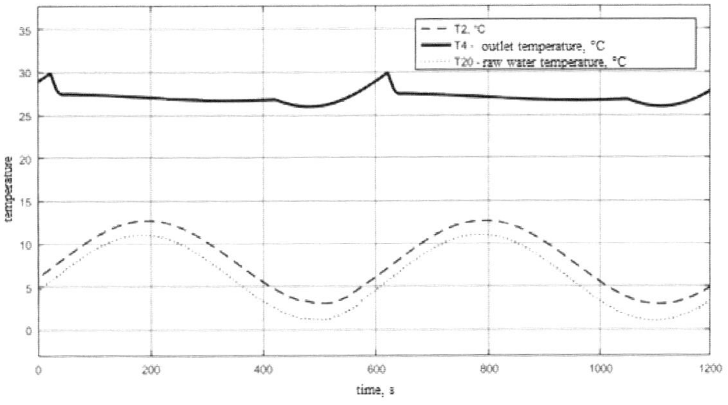

Fig. 6. System simulation results with strong technological wear

5. Conclusion

The main purpose of the work was to investigate the negative impact of the deposits thickness on the walls of heat exchange equipment on the linear control systems performance. A system consisting of two heat exchangers connected in series was chosen as the control object under study. The physical and mathematical model of this system took into account the deterioration of the thermal conductivity of the system with an increase in the thickness of deposits through the thermal conductivity coefficient. The resulting differential equations were used to build the model in the form of transfer functions of the entire outlet water temperature control system in the Matlab/Simulink environment. System studies were carried out taking into account winter operating conditions (2-5 °C of raw water) and various degrees of pollution.

The optimally tuned PID controller provided the required outlet temperature range, up to 3 mm of deposition on the inner walls of the heat exchanger. Further operation of this facility without human intervention is not advisable due to a decrease in economic and technological efficiency. The performed simulation of the temperature control loop operation with a PID controller, taking into account technological wear, shows that linear control systems are not able to provide the required quality of control of an object with characteristics that change due to strong wear. An analysis of the causes of this phenomenon requires additional research; as a hypothesis, it can be assumed that a decrease in the heat transfer coefficient due to scale layers formed during operation seriously affects both the time constants of the controlled system and other static coefficients. To compensate for these changes in real practice, the control system has to adjust the parameters or introduce special corrective links, which requires an individual approach to each installation and type of coolant. In special cases, ignoring such processes can lead to a loss of stability of the control system, followed by a violation of the technological process and a loss in the quality of temperature-dependent processes.

There are also positive aspects in the discovered phenomenon - since the difference in signals in heat exchange equipment is associated with deposits on its walls, then in the presence of an adequate physical and mathematical model, a comparison of these signals can make it possible to calculate the thickness of the degree of contamination. Thus, the linear control method can be indirectly used as a virtual analyzer of fouling in heat exchangers.

6. References

[1] Dytnersky Yu. I. (1995). Processes and apparatuses of chemical technology, Chemistry, Moscow, Russian Federation

[2] Cernecky, Jozef, Jan Koniar, and Zuzana Brodnianska. (2011). Impact of lower heat exchange plate's geometry on heat transfer in horizontal channel, Annals of DAAAM for 2011 & Proceedings of the 22nd International DAAAM Symposium, pp. 0245–0246, Published by DAAAM International Vienna, TU Vienna, Austria

[3] Galkovsky V.A., Chupova M.V. (2017). Analysis of the decrease in the heat transfer coefficient of heat exchangers due to surface contamination, Internet journal "NAUKOVEDENIE", Vol. 9, Issue 2.

[4] Bolotova Yu. V., Ruchkinova O. I. (2015). Corrosion of heat exchange equipment of petrochemical industries, Bulletin of the Perm National Research Polytechnic University. Mechanical engineering, materials science. Vol. 17, Issue 4. pp. 102–119.

[5] Matukhnov T.A., Matukhnova O.D., Yurkina M.Y. and Khomchenko N.V. (2020). Assessment of scale formation in heat exchangers, The Third Conference. Problems of Thermal Physics and Power Engineering. Journal of Physics: Conference Series, Ser. 1683 052031, DOI: 10.1088/1742-6596/1683/5/052031.

[6] Ren L., Cheng Y., Yang J., Wang Q. (2020). Study on Heat Transfer Performance and Anti-Fouling Mechanism of Ternary Ni-W-P Coating, Applied Sciences, Vol. 10(11), No. 3905. DOI: https://doi.org/10.3390/app10113905.

[7] Harche, R., Absi, R., Mouheb, A. (2014). Study of the fouling deposit in the heat exchangers of Algiers refinery, International Journal of Industrial Chemistry, Vol. 5. No. 16. DOI: 10.1007/s40090-014-0016-6.

[8] Sivakumar P., Prabhakaran D., Kannadasan T. (2012). Temperature Control of Shell and Tube Heat Exchanger by Using Intelligent Controllers-Case Study, International Journal of Computational Engineering Research (IJCER). Vol. 2, Issue 8. p. 285–291.

[9] Bespalov A. V., Kharitonov N. I. (2007). Chemical Process Control Systems, ICC "Akademkniga", Moscow, Russian Federation

[10] Shteinberg Sh.E., Zalutsky I.E., Serezhin L.P., Varlamov I.G. (2003). Adjustment and adaptation of automatic regulators. Tool set of programs, Industrial ACS and controllers, No. 10, p. 43–47.

[11] Skorospeshkin M.V., Tsapko G.P. Adaptive corrector of dynamic characteristics of automatic control systems, Scientific and technical conf. students and graduate students, MPEI, Vol. 1. p. 498–499.

[12] Khare Y.B., Singh Y. (2010). PID Control of Heat Exchanger System, International Journal of Computer Applications, Vol. 8 (6), p. 22-28, DOI: 10.5120/1213-1742.

[13] Shanthi K, Bhanumathi V., Singh Ajith. B., Manju P., Narendiran S. (2020). Investigation Of Intelligent Pid Controller Intended For Heat Exchanger, International Journal of Future Generation Communication and Networking, Vol. 13, No. 2. p. 1642–1654.

[14] Bukhmirova V.V., Sozinova T.E., Solnyshkov Yu.S. (2015). Calculation of heat transfer through impermeable walls, ed. ISPU, Ivanovo, Russian Federation

[15] Nitsche, M., & Gbadamosi, R. O. (2015). Heat exchanger design guide: a practical guide for planning, selecting and designing of shell and tube exchangers. Butterworth-Heinemann.

[16] Bilonoga, Y., & Maksysko, O. (2018). Specific features of heat exchangers calculation considering the laminar boundary layer, the transitional and turbulent thermal conductivity of heat carriers. International Journal of Heat and Technology, 36(1), p. 11-20.

DOI: 10.2507/33rd.daaam.proceedings.038

CREATION OF TECHNOLOGY FOR BUILDING AUTOMATED PRODUCTION WITH REMOTE CONTROL (INDUSTRY 4.0 CONCEPT)

Aleksander Aryskin & Valentin Pryanichnikov

This Publication has to be referred as: Aryskin, A[leksander] & Pryanichnikov, V[alentin] (2022). Creation of Technology for Building Automated Production With Remote Control (Industry 4.0 Concept), Proceedings of the 33rd DAAAM International Symposium, pp.0279-0284, B. Katalinic (Ed.), Published by DAAAM International, ISBN 978-3-902734-36-5, ISSN 1726-9679, Vienna, Austria
DOI: 10.2507/33rd.daaam.proceedings.038

Abstract

Production processes at technological enterprises often face many problems. One of the main problems is the failure of the mechanisms involved in the technological cycle, errors that occur during the operation of the system associated with data loss or incorrectly defined parameters, etc. To solve a limited number of problems arising at technological enterprises with automatic control, based on KIAM Russian Academy of Sciences, the International Laboratory "Sensorika" with the assistance of partners, a complex to produce experimental products was created. This complex allows not only to produce test samples of innovative materials, but also, in turn, allows conducting research to improve the reliability of the technological process. In a fairly short period of time, our specialists have developed and implemented control systems for the manufacturing, sorting and transportation of products. The developed dispatching systems allow intelligent analysis by means of distributed control of multiple controllers integrated into a single data exchange network. Each of the controllers allows you to interrogate dozens of sensor devices responsible for the processes, collecting positioning data, engines speeds, temperature, and pressure in tanks, etc. The obtained data is analyzed and allow to organize feedback from that technological equipment, in which it is not possible to conduct a direct survey of the state. The use of the distributed control principles opens opportunities for maximum scalability of all production processes. Processes in which a person plays an important role are equipped with an adaptive interface, that opens the possibility to conduct logical analysis and remote monitoring, at a higher level to determine failure situations in conditions of a large range of products, manufactured according to experimental recipes.

Keywords: Industry 4.0 concept; automation of technological processes; logical analysis; sensorics; PLC.

1. Introduction

When developing and creating industrial and civil complexes with an automated process control system, it is customary to use a standard architecture conditionally divided into three levels: information (upper level), control (middle level) and executive (lower level).

Each of the levels differs from each other in the functions, performed in the system architecture and is regulated by the normative document GOST standard 34.201-2020. The interaction between the levels takes place via fieldbus or local area networks. Often, when choosing the concept of building a system and architecture, it is impossible to conduct a complete examination of the condition of the executive elements of production equipment. Without receiving data related to the state of the actuators, mechanisms can fail, directly increasing the likelihood of defective products. To solve such problems based on KIAM Russian Academy of Sciences, the International Laboratory "Sensorika" with the assistance of partners, a complex to produce experimental products was created. During the development and creation of this complex, the approach of phased implementation and commissioning of production sectors was chosen. To date, the production complex includes: a suspension base preparation unit, a suspension synthesis unit, and a transport conveyor [1], [2],[3]. This complex allows not only to produce test samples and production of innovative materials, but also, in turn, allows conducting research to improve the reliability of the technological process.

2. Suspension-base preparation unit

The result of the study to improve the reliability of the technological process of suspension preparation was the development of special software, that allows you simultaneously monitor all control actions, coming from the operator, the technologist and has the control logic unit with the ability to analyze the presence of contradictions, thereby eliminating the occurrence of emergency[4],[5],[6] [7],[8],[9],[10].

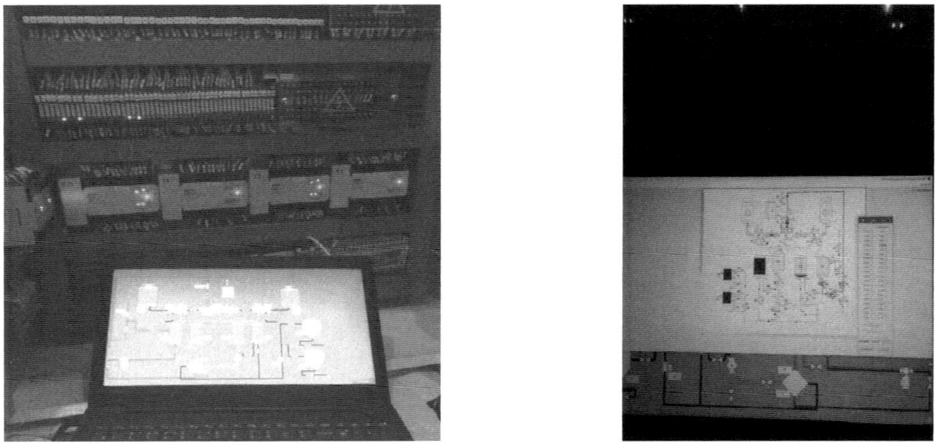

Fig. 1. Test of the algorithm for logical analysis of the technological process of manufacturing and casting finished raw materials into molds.

The software interface of the implemented program was created in the Delphi integrated development environment. The software, after analyzing based on the logic of branching time, visually displays more rational options for the implementation of the technological process, to achieve maximum efficiency of the selected recipe. Figure 2 shows the result of the logical analysis of the technological process by the system - interface displays in real time the operation process in the PLC (programmable logic controller), according to the specified parameters [11],[12],[13]. In that case, we find the optimal solution of the problem of filling the steam generator has been determined: it was necessary to control the corresponding valves.

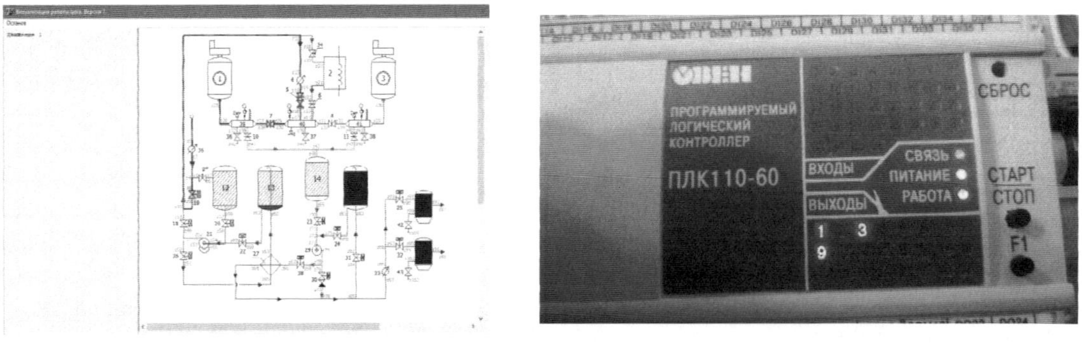

Fig. 2. Visualization of the algorithm for logical analysis of the technological process of manufacturing and casting of the finished raw materials into molds.

3. Suspension synthesis unit

This year, a new suspension synthesis plant started the operations. Several studies were conducted to determine the critical conditions of the equipment. As the result of research, we've developed an appropriate system architecture and assembled the electronic rack-box for sensors monitoring and control of power units up to 22 kW (Figure 3).

Fig. 3. Montage and testing of the electronic rack-box for monitoring and control of power units.

The conducted studies of power units have shown that the main reason for the failure of power units is the jamming of moving parts during operation, which leads to a complete shutdown of the system and the failure of several engines [14],[15],[16],[17]. The way out of this situation was to modernize the some working elements of the engines, inject sensors, because of it, became possible to control the engine speed and compare them with the reference ones for the corresponding unit. Thus, when the parameters deviate from the nominal ones, the system analyzes, shows what happened, and generates signals about it.

At large production facilities a vibration monitoring system for aggregates was used for the same purpose, but the accuracy of the measured parameters, in this way, was not optimally accurate and can lead to a false alarm, when working near the power unit. Figure 3 (on the right) shows the testing of a system with a connected power unit standing next to it [18],[19]. Also, the result of receiving feedback from power units was the ability to control the volume of bulk additives using indirect data with an accuracy of up to a hundred grams without using additional weighing equipment, i.e., by receiving readings from a single actuator and performing calculations from temperature, pressure, etc. The use of such sensors, make possible to evaluate the concentrations and properties of the suspension without resorting to chemical analyses of specialized laboratories.

4. Transport conveyor

To date, a transport conveyor is being prepared for launching into test operation. Engine control algorithms (about 100 units) are being investigated in accordance with the law of trapezoidal change of speed [20],[21],[22]. Due to the need to consider many different parameters of mobile trolleys/platforms on conveyor rails (loading, wear of drive parts, etc.), it is necessary to pay special attention to the calculations of the moments of approach to the acceleration and braking sections to maintain a constant speed, as well as a smooth stop and start of movement.

Earlier, we developed a corresponding application package with a simulation model of the process and integrated it into the system. Thanks to this package, it becomes possible to analyze the state of each mechatronic modules in the complex and adjust of the production scenarios. In addition, a line of executive consoles has been assembled for testing the movement of trolleys/platforms in real production conditions (Figure 4).

Fig. 4. Remote control for testing of industrial vehicles motion

The upper level of visualization of the entire complex is based on the generally accepted SCADA system interface for industrial enterprises. The SCADA system includes a set of control settings for selecting recipes and making changes to them [23],[24],[25]. The operator could control part of the actuators, depending on the state of the technological process. Each keystroke requires confirmation, and an event log is kept not only of the use of the operator's interface, but also all control posts on the territory of the industrial cluster. The use of the event logs of the state of equipment and actuators, as well as the presence of a video surveillance system, play one of the main roles in the industrial safety of the enterprise [26],[27],[28],[29]. The administrative staff has remote access to the video surveillance system and the enterprise monitoring system, using the cloud services implemented at the enterprise, Figure 5 [30],[31],[32],[33].

Fig. 5. Interface of the system for industrial production monitoring and control.

5. Conclusion

The created software and hardware implement typical processes, that take place in a significant part of the control systems of smart factories, built based on the concept of Industry 4.0. Thus, the data obtained from the implemented software and hardware are a good example of creating a technological base, a constructive approach to the organization of intelligent industries and other similar industrial applications. The system in automatic mode successfully calculates the results of the control algorithms and issues recommendations to the operator and technologist, based on which the automated process can work flawlessly and produce high-quality products.

Further horizons of planning the development of the system consist in the complete exclusion of a person from the production cycle of production. The main idea is a web interface on the consumer side, where a potential buyer can configure the required number of products himself, as well as make his wishes, the logical analysis system will adjust the production process itself. At the exit, you can get a finished product, packaged and ready for further transportation.

6. References

[1] Aryskin, A[leksander] A[.]; Bogdanovich, A[lena]; Davydov, O[leg]; Khelemendik, R[oman]; Petrakov, M[axim] &Pryanichnikov, V[alentin] (2019). Control and Interaction Algorithms for Industrial and Service Robots, Chapter 33 in DAAAM International Scientific Book 2019, pp.377-384, B. Katalinic (Ed.), Published by DAAAM International, ISBN 978-3- 902734-24-2, ISSN 1726-9687, Vienna, Austria DOI: 10.2507/daaam.scibook.2019.33

[2] Pryanichnikov V. E., Chernyshev V. V., Arakantsev V. V., Aryskin A. A., Ksenzenko , A. Ya., Petrakov M. S., Treveskyn A. S., Eprikov S. R. / / Intelligent adaptive robots (Special issue of the journal - Information-measuring and control systems, vol. 16, no. 12), Moscow: Radiotechnika, vol. 14, no. 1-2, 2019, ISSN 2070-0814, pp. 49-56, DOI 10.18127 / j20700814-201812-05

[3] The principle and methodological foundations of constructing software systems for logical control of technological equipment: abstract dis. doctor of technical Sciences: 05.13.06 / Nezhmetdinov Ramil Amirovich; [place of defense: MSTU "STANKIN"]. Moscow, 2020, 39 p. (in Russian).

[4] GOST R IEC 61131-3-2016 programmable Controllers. Part 3. Programming languages (IEC 61131-3: 2013, IDT). – M.: STANDARTINFORM, 2016, 230 p. (in Russian).

[5] GOST R IEC 61508-1-2007 Functional safety of electrical, electronic, programmable electronic systems related to safety.:STANDARTINFORM, 2008, 58 p. (in Russian).

[6] GOST 27.301-95. Reliability in technology. Calculation of reliability. Basic provisions-Moscow: publishing-houseofstandards, 1995-19 p.

[7] GOST 27.310-95. Reliability in technology. Analysis of the types, consequences, and criticality of failures. Basic provisions-Moscow: publishinghouseofstandards, 1997g7.– 12 p.

[8] GOST 26.013-81. Measuring and automation tools. Electrical signals with a discrete change of parameters input and output. - M.: Standardinform, -7 pages.

[9] GOST 15971-90 System of information processing. Terms and definitions. Moscow: publishing house of standards, 1990, 14 p.

[10] RD 03-418-01. Guidelines for conducting risk analysis of hazardous production facilities. Normative documents of Gosgortehnadzor of Russia-Moscow: publishing house of standards, 2001-10 p.

[11] Aryskin A.A., KsenzenkoA.Ya., MarzanovYu.S., Prysev E.A., Pryanichnikov V.E., Khelemendik R.V., Eprikov S. R. Industrial automation with remote access and automatic resolution of logical contradictions for Industry 4.0, ekstremal'najarobototekhnika [Extreme robotics]. // Collection of abstracts of the International scientific and technical conference. - Saint Petersburg: Publishing and printing complex "Gangut", 2017. - 272 p., With.190-193. URL: http://er.rtc.ru/images/docs/Sbornik_tezisov_ER_2017.pdf

[12] Khelemendik R. V. (2000). Applications of linear and branching time logics. Algorithm for testing satisfiability// Bulletin of the Nizhny Novgorod state University. Mathematical modeling and optimal control. Nizhny Novgorod: Publishing house of the Nizhny Novgorod University, 2000, Vol. 1 (22), Pp. 176-191.

[13] Khelemendik R.V. (2006). An algorithm for determining satisfiability formulae of computation tree logic and an effective algorithm for constructing derivations of valid formulas from axioms. Mathematical Issues of Cybernetics, no. 15, A collection of papers, ed. by O.B.Lupanov. Moscow: Fizmatlit, 2006, pp. 217-266.

[14] Aryskin A.A., Kozak N.V. Construction of components of a distributed control system based on a single-Board raspberry PI2 computer. Materials of the student scientific and practical conference "automation and information technologies (AIT-2016)". First round. Collection ofreports, Moscow: MSTU STANKIN, 2016.

[15] Aryskin A.A., Kozak N.V. Building a development environment applications of automation single board computers. Materials of the student scientific and practical conference "automation and information technologies (AIT-2015)". First round. Collection ofreports, Moscow: MSTU STANKIN, 2015, Pp. 18-21.

[16] Pryanichnikov V.E., Aryskin A.A., Eprikov S.R., Kirsanov K.B., Khelemendik R.V., KsenzenkoA.Ya., Prysev E.A., Travushkin A. S. (2017). Technology of Multi-Agent Control for Industrial Automation with Logical Processing of Contradictions // Proceedings of the 28th DAAAM International Symposium, 2017. — DAAAM International, Vienna, Austria, 2017. — P.1202-1207. DOI: 10.2507/28th.daaam.proceedings.167.

[17] V.E.Pryanichnikov, A.Ya.Ksenzenko, S.V.Kuvshinov, Yu.S.Marzanov, Yu.V.Poduraev, E.A.Pryses, R.V.Khelemendik, A.A.Ariskin, O.O.Belyaev, P.A.Bryzgalov, D.V.Davydov, D.A.Kuzovkin, P.F.Pletenev, S.R.Eprikov. The developmentof a networkofrobotariumand mobile robots "Amur" for innovative learning intelligent robotronica. C.141-148. // Proceedings of the International multiconference of Network partnership in science, industry and education. Saint Petersburg, Russia 4-6 July 2016. "Politehnika-Service", 2016, Peter the Great Saint Petersburg Polytechnic University, 2016. ISBN 978-5-906841-54-4.

[18] Aryskin, A[lexander]; Grigorev, A[leksei]; Khelemendik, R[oman]; Petrakov, M[axim]; Pryanichnikov, V[alentin] & Tarasov, R[adomir] (2020). System for Monitoring and Control in Industrial Technological Processes, Proceed-

ings of the 31st DAAAM International Symposium, pp.0644-0649, B. Katalinic (Ed.), Published by DAAAM International, ISBN 978-3-902734-29-7, ISSN 1726-9679, Vienna, Austria DOI: 10.2507/31st.daaam.proceedings.089.

[19] Aryskin, A[lexander]; Grigoriev, A[leksey]; Petrakov, M[axim]; Pryanichnikov, V[alentin]; Travushkina, A[nastasia] & Travushkin, A[leksandr] (2021). Industry 4 0 - Automation in the Conditions of Reconfiguration of Production Lines, Proceedings of the 32nd DAAAM International Symposium, pp.0660-0665, B. Katalinic (Ed.), Published by DAAAM International, ISBN 978-3-902734-33-4, ISSN 1726-9679, Vienna, Austria DOI: 10.2507/32nd.daaam.proceedings.093.

[20] The official Russian-language site of the company "CoDeSys" [electronic resource]: the Official Russian-language site of the company "CoDeSys" - access Mode: http://www.codesys.ru/ (accessed 01.01.2020).

[21] The official website of the software product CoDeSys of the company 3S-Smart Software Solution GmbH [electronic resource]: the official website of the software product CoDeSys of the company 3S-Smart Software Solution GmbH-access Mode: https://www.codesys.com/(accessed 01.01.2020).

[22] Official site of the company "ARIES" [electronic resource]: Official site of the company "ARIES" - access Mode: http://www.owen.ru/(accessed 10.01.2020)

[23] N. V. Gusev, S.V.Lyapushkin, M. V. Kovalenko Automation of technological complexes and systems in industry: a textbook / N. V. Gusev, S. V.Lyapushkin, M. V. Kovalenko; Tomsk Polytechnic University. – Tomsk: Publishing house of Tomsk Polytechnic University, 2011 – p. 198

[24] Denisenko V.V. Computer control of technological process, experiment, equipment / V. V. Denisenko, Moscow: Hotline-Telecom, 2009. -608 p. 47.

[25] Denisenko V.V. Computer control of technological process, experiment, equipment / V. V. Denisenko, Textbook. Stavropol: AGRUS publ., 2009. 100 p. 48.

[26] Minaev, I. G. Programmable logic controllers. Practical guide for a novice engineer / I. G. Minaev, V. V. Samoylenko-Stavropol: AGRUS, 2009. - 100 p.

[27] Official site of the company "Siemens" [electronic resource]: Official site of the company "Siemens" - access Mode: https://www.siemens.com/global/en/home.html(accessed 08.12.2019).

[28] Official site of the company "National Instruments" [electronic resource]: Official site of the company "National Instruments" - access Mode: http://www.ni.com/ru-ru.html(accessed 07.12.2019).

[29] The official Russian-language site of the software product ISaGRAF of Rockwell Automation [electronic resource]: the Official Russian-language site of the software product ISaGRAF of Rockwell Automation-access Mode: http://www.isagraf.ru/(accessed 10.01.2020).

[30] Official portal of the Modbusdevelopers Association [electronic resource]: Official portal of the Modbusdevelopers Association-access Mode: http://www.modbus.org/(accessed 07.12.2019).

[31] Poduraev Yu. V. Fundamentals of mechatronics / Yu. V.Poduraev, M.: IC MSTU "Stankin", 2000. -103 p.

[32] Egorov O. D., Poduraev Y. V., Bubnov M. A. Robotic mechatronic systems / O. D. Egorov, Yu. V. Poduraev, M. A. Bubnov, M.: MSTU "STANKIN", 2016. - 326 p. (in Russian).

[33] Pryanichnikov V. E. The Combination of ultrasonic data filtering and linear drive software control / Pryanichnikov V. E.// Devices+automation. -2008. - №12, 22-29-c.

DOI: 10.2507/33rd.daaam.proceedings.039

ANALYSIS OF THE PISTOL BULLET PENETRATION THROUGH A WOODEN BARRIER

Stefan Turopoljac, Zvonko Trzun & Milan Vrdoljak

This Publication has to be referred as: Turopoljac, S[tefan]; Trzun, Z[vonko] & Vrdoljak, M[ilan] (2022). Analysis of the Pistol Bullet Penetration Through a Wooden Barrier, Proceedings of the 33rd DAAAM International Symposium, pp.0285-0294, B. Katalinic (Ed.), Published by DAAAM International, ISBN 978-3-902734-36-5, ISSN 1726-9679, Vienna, Austria

DOI: 10.2507/33rd.daaam.proceedings.039

Abstract

Effects of handgun (pistol) bullets on wooden barriers, and the ballistic gelatine placed behind these barriers, are compared. In one comprehensive analysis, the penetration of three frequently used handgun bullets through wooden barriers of different hardness and thickness is compared, and at the same time, the effect on the replacement biological tissue behind such barriers is analysed. The ballistic gelatine is prepared according to the NATO standard and therefore results are directly comparable with results of earlier tests for other bullet calibres. A significant difference in the efficiency of the tested bullets is discovered, based on which recommendations can be made on the most effective calibre for use in police and military for urban operations.

Keywords: terminal ballistics; ballistic gelatine penetration; pistol bullets; wooden barrier penetration.

1. Problem Statement and Overview of Previous Works

By reviewing previous works, it is clear that primary focus is maintained on rifle bullets where [1] and [2] analyse the effect on ballistic gelatine, whereas [3] and [4] analyse the effect of rifle bullet when hitting barriers made of stronger materials (respectively shock-absorbing concrete and ceramic composite armour).

On the other side, pistol bullets and their ability of penetration is less examined, and if so it is mainly focuses on penetration and terminal capabilities on actual or replacement biological tissue [5], [6]. This is understandable since the mentioned analyses aim for the military purposes, and the primary infantry weapon is the rifle. Handguns are mainly used for police force but also have purpose in military environment (secondary weapon, special operations, close-quarters combat).

Capabilities of wooden barriers penetration become more and more interesting for police and armed forces lately mainly due to increased number of combat operations in urban environment (Military Operations in Urban Terrain – MOUT). Because of this it is very important to know penetration capabilities of wooden barriers with individual cartridge, enabling soldier or policeman to survive in urban environment. Wooden barriers can be frequently seen in urban environment (doors, furniture, walls). In this paper, testing of multiple pistol cartridges on different types of wood will be presented.

Wood is the second most common material in the household so it can be expected that it will be necessary to penetrate wooden barrier which protects the opponent. Two most common types of furniture wood were examined in this test – fir and oak. Aside from wooden barriers, pistol cartridges which would be tested are 9mm Luger, .40 S&W and .45 ACP. These cartridges are the three most used calibres today in military and police worldwide. From recent research on this topic, papers that should be singled out are [7] and [8] although they analyse the effectiveness of only one calibre. The paper will demonstrate the advantages of individual calibres when penetrating through wooden obstacles. It will also be proven how effective such obstacles can be as a protection against the analysed calibres.

2. Wood as a Building Material

Wood has been used as construction material since ancient times. Nowadays materials like polymers, alloys or composites have found greater application thus reduced the share of wood in construction and everyday application. Despite this revolution of new materials wood is still the second most used material in construction, and therefore there is still a good chance of enemy taking cover behind wooden barrier during the close-quarter combat. Anisotropy is the greatest problem when performing examination connected with wood behaviour. Namely, physical properties of wood are different in all directions (longitudinal, radial and tangential) [9]. During the experiment in this work penetration of bullet was examined on the radial axis, i.e. perpendicular to tree rings. Mechanical properties which may be of interest for this test are toughness, torsional, compressive, tensile and shear strength, Poisson's ratio, wear resistance and more. Hardness and density of wood are the most important parameters for determining projectile penetration. These two characteristics are in strong mutual correlation, which means that denser wood usually also has a higher hardness. Hardness is usually tested by Brinell test by pressing the steel ball into the wood surface. Diameter of cavity is measured after the application of force, and according to that diameter hardness of material can be detected. Hardness according to Brinell is then:

$$HB_D = \frac{2F_B}{\pi D_B \sqrt{D_B^2 - d_B^2}} \tag{1}$$

where F_B presents applied force [N], D_B diameter of test ball [mm] and d_B diameter of imprinted trace in the wood [mm]. In our analysis two most common types of European construction wood were used: fir and oak. Except for being two most used wood types in construction applications, these wood types were also used due to different characteristics. Fir is very soft and according to Brinell $HB_D = 1{,}81 \pm 0{,}12$ (depending on dryness), while on the other hand oak is pretty hard with hardness $HB_D = 3{,}56 \pm 0{,}2$ [10].

3. Pistol Cartridges

For the penetration analysis, three most popular pistol calibres were used: 9x19 Parabellum, .40 S&W and .45 ACP. All three are presented od Figure 1.

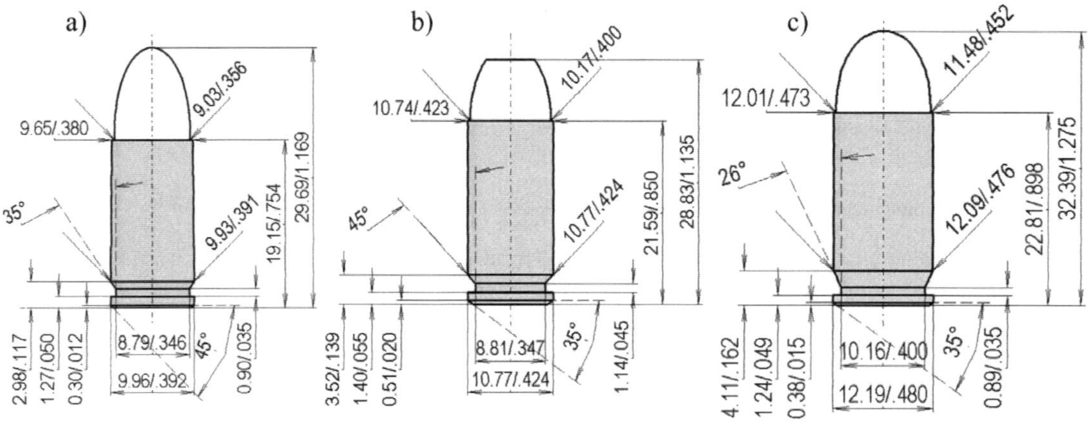

Fig. 1. Analysed cartridges: a) 9x19 PARA, b) .40 S&W, c) .45 ACP [14]

According to STANAG 4090 9x19 Parabellum characteristics are: calibre 9 mm, case length 19 mm, bullet mass between 7 g and 8,3 g. The bullet should inflict deadly wound on a target which stands at a distance of 23 meters, even when the target has basic level of head and body protection. According to NATO STANAG 4090 from 1982 group on a target should be less than 3 inches wide with standard barrel length.

This experiment was performed with Sellier&Bellot's 9x19 Nontox TFMJ with cartridge mass of 12.7 g, and bullet mass of 8 g. According to manufacturer this bullet achieves average velocity of 360 m/s from 150 mm long barrel witch comes to 518 J of kinetic energy on the muzzle exit. This part of the test was executed with HS Produkt's SF19 4.5" with barrel length of 116,5 mm which gives a slightly lower muzzle velocity and energy compared to the manufacturer's statement. 9x19 Parabellum is appreciated due to its low mass and recoil during firing process, even though there were some speculations about low efficiency of this cartridge in the past. That was the reason for a constant improvement of this cartridge during the past and today there is a wide variety of 9 mm ammunition. At 25 meters from the muzzle bullet velocity decreases to 326 m/s providing 425 J of energy and on 50 m from muzzle velocity decreases to 301 m/s with 362 J of energy which is enough for instantaneous neutralization of target [11].

Bullet trajectory keeps flat trajectory for the first 30 m, and on 50 m from the weapon there is a slight bullet drop that amounts only 37 mm. Manufacturer's data are used for the purposes of this paper, however if higher data accuracy is required it is possible to calculate the bullet trajectory with flight models like MMPM (*Modified Mass Point Model*) or 6DOF (*Six Degrees of Freedom*) model [12], [13]. Figure 2 shows bullet velocity decrease with distance and Figure 3 presents decrease in kinetic energy up to a distance of 100 m for all cartridges used in the experiment.

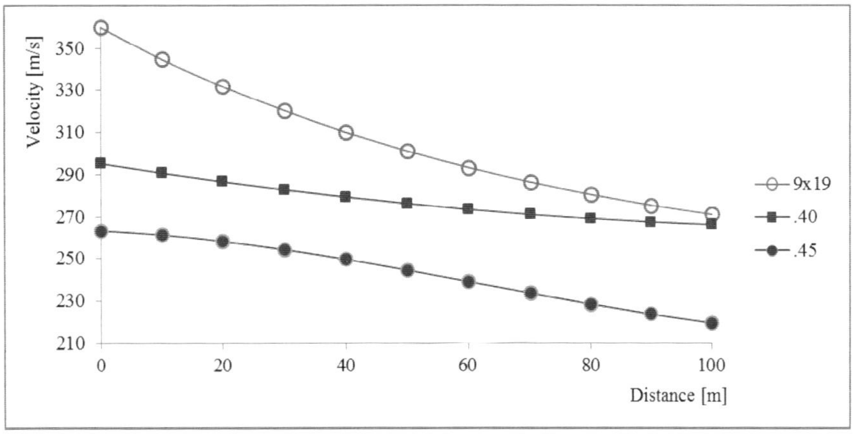

Fig. 2. Bullet velocity decrease for 9x19 mm, .40 S&W and .45 ACP; according to [11]

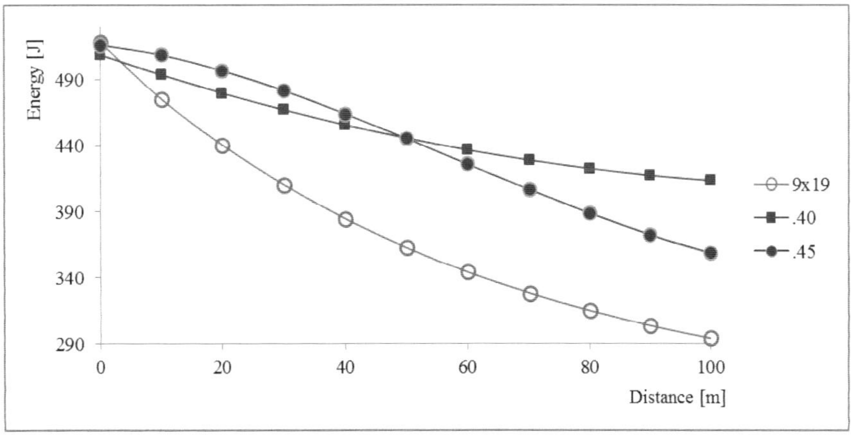

Fig. 3. Kinetic energy decrease for cartridges 9x19 mm, .40 S&W and .45 ACP; according to [11]

Cartridge .40 S&W (Smith&Wesson) is the second most used option for police and armed forces around the globe. It was introduced in 1990 as a replacement for 9x19 which was considered as to be of inadequate mass and stopping power. Weight of the cartridge (S&B TFMJ Nontox) was 16.7 g, and bullet mass was 11.7 g. HS Product's XDM 40 4.5" was used for the experiment, giving the bullet 295 m/s and 509 J of energy at the barrel muzzle. Although the bullet is heavier compared to the 9x19, due to the lower initial velocity it has less kinetic energy (509 J), which is why it failed to replace the 9x19 calibre. Quite the opposite, in the last several years more and more agencies and armies leave .40 S&W and return to the 9x19 Parabellum. The velocity decrease due to the aerodynamic force is not so dramatic for .40 S&W compared to 9x19 Parabellum, meaning that the .40 S&W bullet keeps its kinetic energy better (Figures 2 and 3).

However, since in the real battle conditions pistol cartridges are mostly used at a distances from 10 to 20 meters, the velocity preservation is not of great importance. On the other hand, due to its higher mass, the recoil is stronger, and the trajectory drop is more accentuated (55 mm at the 50 m distance) than one of the 9x19 bullets.

The .45 ACP (*Automatic Colt Pistol*) is the largest and heaviest of all three cartridges included in this experiment. Weight of the cartridge (S&B TFMJ Nontox) is 20.9 g, and the bullet weight is 14.9 g. For the experiment purposes HS Product's XDM 45 4.5“ was used. With the bullet velocity 263 m/s at the muzzle it gives 516 J of kinetic energy. Figures 2 and 3 show a decrease in bullet velocity and kinetic energy; it loses both faster than the .40 bullet (due, among other things, to a higher coefficient of resistance), but not as drastic as the 9x19 bullet. As for the trajectory curvature, the .45 bullet drop is the largest of the three calibres analysed (76 mm at a distance of 50 m).

The large mass potentially suggests that the penetrability of this bullet will be the highest, but since the aeroballistic coefficient is very low it could cause a large negative acceleration and difficult pass through the surrounding medium (including both the air and the set barrier). The aeroballistic coefficient is calculated according to the following expression:

$$C_b = \frac{m_p}{C_D A} = \frac{\rho l}{C_D} \tag{2}$$

where C_b presents aeroballistic coefficient, m_p bullet mass, A is cross-sectional area of the bullet, ρ is surrounding media density, while l presents characteristic bullet length. C_D is drag coefficient which directly determines the value of aeroballistic drag according to expression:

$$D = \frac{\rho v^2}{2} A C_D \tag{3}$$

where the velocity of the bullet v is also introduced. Equation (2) shows inversely proportional relationship between aeroballistic coefficient and drag coefficient, meaning that the bullet with higher drag coefficient C_D will also have lower aeroballistic coefficient C_b. The second part of (2) indicates that a shorter bullet (and therefore of shorter characteristic length l) would have a lower coefficient C_b.

Ammunition manufacturer (Sellier&Bellot (2022)) indicates that the .45 ACP has the lowest aeroballistic coefficient $(C_b = 0{,}068)$, 9x19 Parabellum has slightly higher coefficient $(C_b = 0{,}09)$ while the most favourable aeroballistic coefficient has the .40 S&W bullet $(C_b = 0{,}157)$.

4. Anticipated Bullet Penetration

Penetration of the bullet is affected by its kinetic energy at the point of impact, construction of the bullet, and material properties for both the bullet and the target. Kinetic energy depends on mass and velocity of the bullet ($E_k = m_p v^2 / 2$). Kinetic energy grows linearly with increasing mass of the bullet m_p, and with a square of increasing bullet velocity v. During the flight heavier bullets lose their velocity more slowly. Deceleration is also affected by the drag coefficient. Handgun bullets usually have a higher drag coefficient due to construction limitations of the weapon itself: bullet and cartridge should not be too long because the magazine should fit into the handgun grip, and the only way to increase the kinetic energy of a short bullet is to increase its calibre. Therefore, pistol bullets are more spherical than protruded (Figure 1) which leads to a higher drag coefficient compared to rifle bullets.

When it comes to construction, the tip of the bullet is the primary factor determining bullet penetrability and other terminal ballistics aspects. Some bullets are intended for fragmentation upon impact (e.g. 5.56x45 NATO) with less penetration into the target. Our experiment was performed with lead core bullets coated with CuZn30 alloy. This alloy (69-71% copper, small shares of iron, aluminium and nickel, and the rest of alloy makes of zinc) is characterized by great tensile strength which prevents fragmentation upon impact. Therefore, the bullet penetrates into the target, and if there is enough kinetic energy left, the bullet makes a complete breakthrough. But this is not a necessary condition for pistol bullets with rounded tips (Figure 1); they primarily aim for enemy neutralization by transferring as much energy as possible, wherein the bullet stays in the target. This suggests that pistol bullets could have problems with penetrating wooden barriers – which will be verified during the experiment.

For preliminary calculating of bullet penetration capabilities, two expressions were used. The first one is Journee's equation:

$$X = c_p d v_C^2 \tag{4}$$

where X is depth of bullet penetration [mm], c_p is penetrability coefficient, d bullet calibre [mm], while v_C embraces bullet velocity upon impact in wooden barrier [m/s] [15]. The penetrability coefficient for fir is $c_{pJ} = 0{,}000093$ while for oak the coefficient is $c_{pH} = 0{,}000076$. Journee's expression takes into account calibre and velocity of the bullet. Poncelet's expression is better because it is more complex and takes more elements in consideration [8]:

$$m_p \frac{dv}{dt} = -\beta - \alpha v^2 \tag{5}$$

where β presents material resistance-dependent parameter while α is contribution of inertial stress. By integrating (5) follows the expression for penetration depth X in a function of bullet velocity upon impact v_c:

$$X = \frac{m_p}{2\alpha} \ln\left(1 + \frac{\alpha v_c^2}{\beta} \right) \tag{6}$$

For the case of uniform deceleration $\alpha = 0$, (6) transfers into:

$$X = \frac{m_p v_c^2}{2\alpha} \tag{7}$$

which is actually the Robins-Euler's model. Bullet penetration into living tissue can be calculated according to:

$$-m_p \frac{dv}{dt} = \frac{1}{2} \rho v^2 A C_D + \frac{1}{2} \rho (aU)^2 A C_D \tag{8}$$

or as a function of the travelled distance x:

$$-m_p v \frac{dv}{dx} = \frac{1}{2} \rho A C_D \left[v^2 + (aU)^2 \right] \tag{9}$$

where a present modification of C_D due to tissue tearing, while U presents characteristic tissue velocity [16]. Due to the small number of wood and bullet samples, the penetration calculation was not done in advance, and the experimental results were subsequently analysed.

5. Description and Course of Experiments

Two main variables were combined in experiments, pistol calibre and selection of wooden barrier. Twelve separate experiments were analysed:

- Experiments No. 1-3: shooting into the ballistic gelatine with each pistol calibre (3 bullets per calibre) with a 50 mm thick barrier of fir wood
- Experiments No. 4-6: shooting into the ballistic gelatine with each pistol calibre (3 bullets per calibre) with a 50 mm thick barrier of oak wood
- Experiments No. 7-9: shooting into a 100 mm thick barrier made of fir wood
- Experiments No. 10-12: shooting into a 100 mm thick barrier made of oak wood

The shooting was performed at the HS Produkt's closed shooting range which eliminated the influence of other variables (including atmospheric conditions). Wooden barriers were at a distance of 10 m and the ballistic gelatine was placed on a stand right behind those barriers. The shooting range was equipped with a laser system that measures the velocity of the bullets. Due to high-tech equipment at the shooting range impact velocity v_c was measured and not approximated according to ballistic charts.

The ballistic gelatine can be 10-percent or 20-percent. The percentage indicates the proportion of beef gelatine in the water. 10 kg heavy block of 20% ballistic gelatine is made of 8 litters of water and 2 kg of beef gelatine while the 10 kg heavy block of 10% ballistic gelatine is made of 9 litters of water and 1 kg of beef gelatine. According to available online sources, the FBI uses the 10% ballistic gelatine while the 20% ballistic gelatine is in experiments performed by NATO. However, this is not so important as long as the ballistic gelatine is properly validated, and the results can be extrapolated onto living tissue. On the other hand, the same type of ballistic gelatine can be used while examining the effects of different calibres, and these results can be directly compared. Demonstrations of the ballistic gelatine use and gelatin preparation techniques can be seen in [17], while in [18] is analysed how different concentrations, temperatures, and curing times affect calibration results affect the experiment results.

Four steps are required to make a ballistic gel. The first step involves the preparation of a hand and electric mixer as well as beef gelatine. In the first step, 56 litters of water are poured into the caldron and heated to the appropriate temperature. This procedure required a water dosimeter as well as a thermometer sensor. The second step started with adding 14 kg of beef gelatine (20% ballistic gel) into the caldron. Mixing of the mixture immediately followed and a hand mixer was used. Using the hand mixer eliminates the creation of foam which can be caused by an electric mixer. The execution of the second step can be seen in Figure 4.

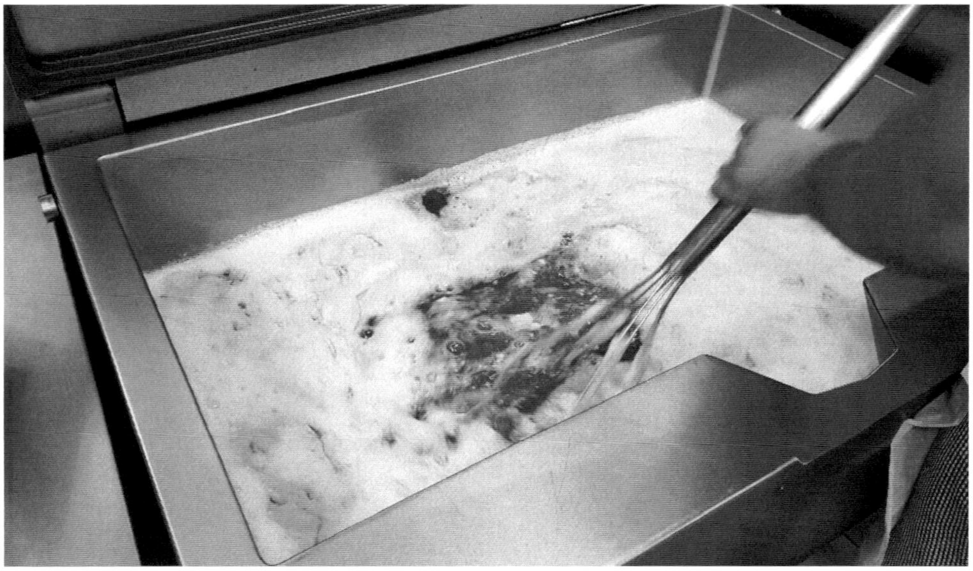

Fig. 4. Water and beef gelatine mixing procedure

An electric mixer was used at the end of this step to crush greater clumps of beef gelatine. This procedure created a small amount of foam which was subsequently removed. Pouring the ballistic gelatine mixture into moulds (with dimensions of 150x150 mm and 400 mm in length) was the third step of the procedure. The last step is cooling. The moulds were left at room temperature for half an hour. After the initial cooling process, they were placed in the refrigerator. This cooling process lasted for 38 hours thus achieving the required level of gelatine strength.

The main part of this experiment was executed in HS Produkt's indoor shooting range in the town of Karlovac. The shooting was performed by HS Produkt's authorized handgun shooter under the supervision of an engineer. As already mentioned, shooting range was equipped with HPI's laser system for measuring bullet velocity. At the distances of 10 m (ballistic gelatine and wooden barriers were placed at that distance during the experiment) following bullet velocities were measured:

1) 9 x 19 mm Nontox TFMJ with a bullet weight of 8 g achieved an average speed of 331,27 m/s;
2) .40 S&W Nontox TFMJ with a bullet weight of 11,7 g achieved an average speed of 290 m/s;
3) .45 ACP Nontox TFMJ with a bullet weight of 14,9 g achieved an average speed of 240,7 m/s.

Bullet velocity variations are acceptable for both 9x19 Parabellum and .40 S&W while for .45 ACP it was to high: a difference between the maximum and minimum velocity was almost 19 m/s, while for the test group of 10 bullets a very large velocity dispersion was recorded (standard deviation was 7,7 m/s).

After evaluating bullet velocities, the main part of the experiment started consisting of 12 separate experiments. In all experiments, the gelatine was situated 6 to 8 cm behind a wooden barrier which can be compared with a real situation when someone takes cover behind an object.

The experiment No. 1 consisted of shooting 9x19 mm Parabellum through a 50 mm thick barrier made of fir. Considering high bullet velocity and lower drag due to the small impact area, good penetration capabilities were presupposed. This assumption proved to be correct because the bullet made a complete breakthrough in the wooden barrier, and after that made a deep penetration into the gel. The average gelatine penetration depth was 36 cm. One of the bullets hit the table on which the gelatine was situated and lost a part of its kinetic energy; it is therefore possible that the gelatine penetration could be even higher (Figure 5).

According to the FBI protocol, bullet penetration of 25 to 30 cm into the ballistic gelatine is enough to neutralize the opponent. Hit into vital organs leads to instant neutralization and a fatal outcome. Therefore, a barrier made of 50 mm thick fir is insufficient to be used as a cover.

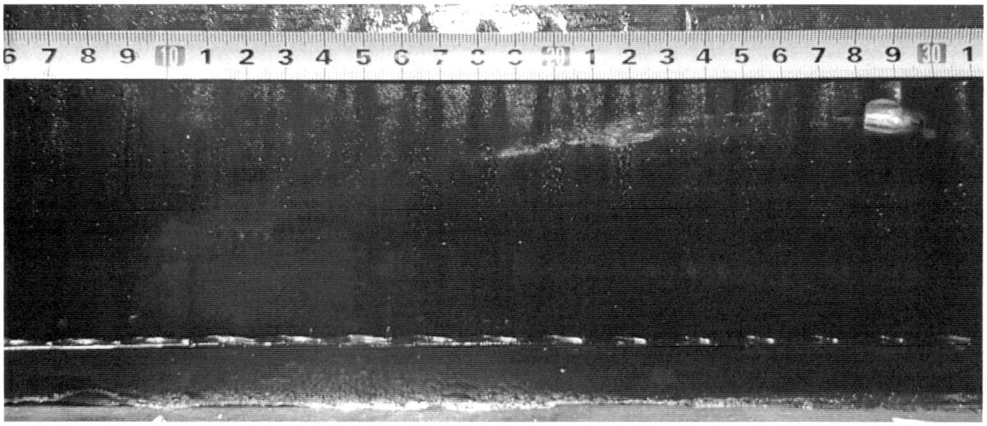

Fig. 5. Bullet 9x19 Parabellum, penetration of one bullet into the ballistic gelatine behind a fir barrier

The experiment No. 2 involved shooting at a ballistic gelatine with a .40 S&W projectile behind a 50 mm fir wooden barrier. The assumption was that this projectile would have a similar effect as 9 mm because of similar kinetic energy, although having a slightly larger cross-section. The assumption proved to be correct, and an average penetration of 36.5 cm was recorded. Similar to the 9x19 bullet, the barrier made of fir proved to be inadequate. As for the permanent cavity in the gel, it is substantial and it can be assumed that such a hit in the body would result in heavy internal bleeding.

The experiment No. 3 involved shooting with a .45 calibre ACP bullet, again through 50 mm of fir. Preliminary measurements have shown that this bullet, despite its large mass, has a lot of problems when trying to break through rigid wooden barriers. During the experiments, a mean penetration of only 29 cm was recorded. This still means that the opponent will be neutralized. Interestingly, the .45 ACP calibre does not create a larger permanent cavity (as the .40 bullet S&W does), despite the larger calibre and higher mass. One example of permanent cavity is clearly visible in Figure 6.

Fig. 6. Bullet 9x19 Parabellum, penetration into the ballistic gelatine behind a fir barrier

Next three experiment (4-6) involved the same procedures as the first three, using a barrier of the same thickness (50 mm), but this time made of oak wood. Due to the higher density and hardness of the oak, the main assumption was that the penetration would be lower compared to the experiments in which fir wood was used.

The experiment No. 4 included the same procedures as the first experiment, but with a barrier made of oak. The 9x19mm bullet made its way through the wooden barrier, but the penetration into the gelatine was not so deep (23,5 cm, which is some 35% less compared to the first experiment).

Because of the more shallow penetration, neutralization of the opponent would not be instant (depending on the hit position); instant neutralization may still occur if a vital organ is hit. In any case, the opponent would need urgent medical treatment.

The experiment No. 5 is best compared to the fourth one. The 40. S&W bullet proved similar terminal ballistic properties as a 9x19 Parabellum. A mean penetration was 24 cm, with one ejected sample - that was a bullet that achieved an unrealistically deep penetration of 34 cm because the bullet passed through the hole in the wooden barrier made by the previous bullet. The permanent cavity was greater compared to the one made by the 9 x 19 mm Parabellum bullet, leading to a higher probability of fast neutralization.

The experiment No. 6 included a test with the slowest bullet in the .45 ACP calibre. The lack of kinetic energy resulted in a mean penetration value of just 13 cm which is noticeably less compared to the other two calibres. During the experiment, bullets hit an area very close to the wooden knot, and that is exactly the place where anisotropy of wood can be present. According to that, these results cannot be taken with a sufficient level of confidence. What this experiment proved is that high calibre bullets with low velocity and high drag coefficients suddenly lose velocity upon impact. Due to the velocity loss, bullets tumble in and shortly after the barrier, causing them to hit the gelatine upside down. This increases drag and reduces the depth of penetration. But this is a desirable feature for this calibre, because it is not meant to break through the target but to stay trapped inside, consequently transferring the entire kinetic energy on the target. What lacked in this experiment was a tumble of a bullet in gel, because without tumbling it cannot make the large temporary cavity. As for the permanent cavity, it is not larger compared to the cavity made by the .40 S&W bullet - a conclusion that yet has to be confirmed by the additional experiments.

After conducting the first six experiments where the main focus was the depth of penetration into a gel, as well as the size of the permanent cavity, six more experiments were performed. These experiments were focused on the penetration through the wooden barriers (fir or oak) for each calibre. Two connected 50 mm thick wooden planks formed a single 100 mm wooden barrier. That thickness is chosen because it represents the upper limit for some practical barriers (a thick table, a door frame, or a multi-compartment closet).

In experiments No. 7-9, the possibility of penetration for each calibre through a 100 mm thick fir barrier was examined. Since fir is a wood of low density and low hardness, each calibre completely went through the barrier. A decisive conclusion can be drawn that the fir block (even when it is 100 mm thick) cannot be used as an adequate cover. Figure 7 shows the front of the wooden barrier (bullet inlet holes - left) and the rear side (exit holes - right).

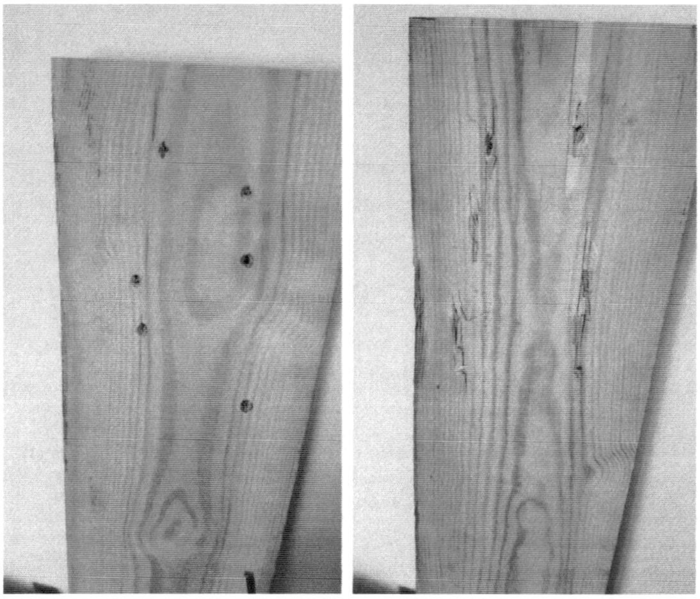

Fig. 7. Bullet inlet holes (left) and exit holes (right) for the fir barrier

Experiments No. 10-12 provided information on the suitability of oak when used as a protection. Due to the high density of this wood, no calibre has achieved a complete breakthrough of the 100 mm thick oak barrier. The average penetration depth of 9x19 Parabellum was 6.85 cm, for .40 S&W it was 5.85 cm, and the lowest average penetration was achieved with a .45 ACP bullet (5.1 cm). This confirms that oak indeed can be used as protection during armed conflict in urban conditions. The summary results for all experiments are shown in Table 1.

Experiment No.	Wooden Barrier		Bullet Calibre	Penetration	
	Type	Thickness [cm]		Through Wood [cm]	Into gelatine [cm]
1	Fir	5	9x19 mm	Breaks Through	36,0
2	Fir	5	.40 S&W	Breaks Through	36,5
3	Fir	5	.45 ACP	Breaks Through	29,0
4	Oak	5	9x19 mm	Breaks Through	23,5
5	Oak	5	.40 S&W	Breaks Through	24,0
6	Oak	5	.45 ACP	Breaks Through	13,0*
7	Fir	10	9x19 mm	Breaks Through	Not Measured
8	Fir	10	.40 S&W	Breaks Through	Not Measured
9	Fir	10	.45 ACP	Breaks Through	Not Measured
10	Oak	10	9x19 mm	6,85	-
11	Oak	10	.40 S&W	5,85	-
12	Oak	10	.45 ACP	5,10	-

* *During experiment No. 6 there was a hit in a wooden knot*

Table 1. Summary results for all experiments

The graphical representation of results is shown in Figure 8:

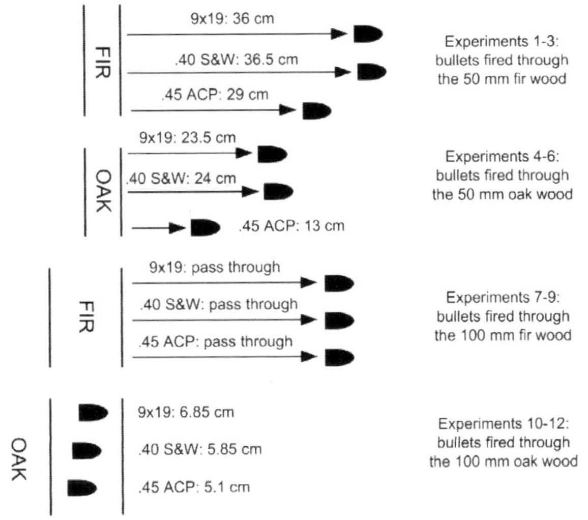

Fig. 8. Graphical representation of each calibre penetration capability, barriers of different wood and thickness

6. Results and Discussion

Experiments have shown that both the 9x19 Parabellum and the .40 S&W have good penetration capabilities, with the .40 S&W bulled showing slightly better terminal ballistic properties and creating a larger permanent cavity. This test showed why there is still no single decision by armies and police forces on the bullet to use. Both the 9x19 Parabellum and the .40 S&W have a number of strong arguments, both with their advantages and disadvantages. The analysis presented here, therefore, represents a contribution to the debate that is still ongoing. The 9 x 19 Parabellum bullet still seems to prevail over the .40 S&W bullet, and the main reason is the fact that the 9 x 19 Parabellum is standardized and widely accepted in NATO's defence systems. Other arguments in favour of this calibre are a long tradition, wider choice of pistols, easier supply of ammunition, and price. There is also a weaker recoil, factor important for improved control over the handgun.

Bullet.45 ACP, the largest of the three analysed, shows good characteristics when it comes to gelatine penetration, but due to its low flight speed and large cross-section, it is less successful in penetrating a wooden barrier. Therefore, this bullet relies primarily on its large mass, and then on the tumbling in the body of the opponent, which results in a large permanent cavity. As this bullet stops faster in the body, it is more likely to transfer complete kinetic energy to the target. The 9x19 mm and .40 S&W bullets can easily pierce the target if they hit a part of a small depth such as the extremities, after which the bullet continues its movement without transferring the remaining kinetic energy to the target. Significant drawbacks of the .45 ACP calibre are its low ballistic coefficient, low penetration ability, and a large bullet drop: in a collision with a hardwood barrier, this bullet gets completely stopped. As for the wooden barrier, the fir does not provide adequate protection - even a slow and cumbersome .45 ACP bullet will achieve a complete breakthrough.

In contrast, dense and hard oak wood provides much better protection, so a barrier of sufficient thickness (> 50 mm) manages to stop all three bullets. The surprisingly poor results shown by the largest analysed calibre (.45 ACP) show how important it is to properly test the behaviour of bullets in a particular scenario. Choosing the wrong calibre, even if it is widely accepted by the public as "the most powerful" (as is the case with the .45 ACP round), could have fatal consequences for soldiers and police officers. Perhaps such analyses will encourage improvements to the .45 bullet, just as the 9x19 mm bullet was improved after criticism for its poor efficiency. As for the limitations of the analysis, they were primarily related to the small number of samples. Therefore, the results should be confirmed in the future with an extended experiment. We also plan to examine the penetration of other pistol calibres, and the penetration of frequently used rifle bullets (for example 5.56 × 45 mm NATO, or 7.62x39 mm). In this way, an even wider range of potential scenarios related to military action in urban areas could be examined.

7. Acknowledgments

Authors kindly thank HS Produkt, Croatia, for assistance in the preparation and implementation of the experiment.

8. References

[1] Wen Y, Xu C, Jin Y, Batra RC. (2017). Rifle bullet penetration into ballistic gelatin. J Mech Behav Biomed Mater. Vol 67, March 2017, pp. 40–50. ISSN: 1751-6161.

[2] Park NR, Kim KH, Mo JS, Yoon GH. (2017). An experimental study on the effects of the head angle and bullet diameter on the penetration of a gelatin block. Int J Impact Eng. Vol 106, August 2017, pp 73–85.

[3] Fabian GL, O'Donnell RH, Tom JG, Malone PG. (1996). Use of Shock-Absorbing Concrete (SACON) as an Environmentally Compatible Bullet-Trapping Medium on Small-Arms Training Ranges. In: Enhancing Readiness Through Environmental Quality Technology. Hershey, PA, USA: Army Environmental Center Aberdeen Proving Ground MD.

[4] Liu W, Chen Z, Cheng X, Wang Y, Amankwa AR, Xu J. (2016). Design and ballistic penetration of the ceramic composite armor. Compos Part B Eng. Vol.84, January 2016, pp 33–40. ISSN: 1359-8368.

[5] Kerkhoff W, Mattijssen EJAT, Riva F. (2020). Influence of bullet type and muzzle-to-target distance on trajectory deflection through a soft tissue simulant. Forensic Sci Int.Vol.311, June 2020,110289.

[6] Hub J, Komenda J, Racek F. (2013). The Analysis of Terminal-Ballistic Behaviour of a Pistol Bullet Penetrating a Block of Substitute Biological Material. Probl Mechatroniki Uzbroj lotnictwo, inżynieria bezpieczeństwa. Vol.4, No.1, pp. 19–35. ISSN: 2081-5891.

[7] Koene B. (2017). Bullet penetration into wooden targets. In: Chocron S, Walker J, editors. Proceedings of the 30th International Symposium on Ballistics. Long Beach,CA,USA: DEStech Publications. DOI: 10.12783/ballistics2019/33172.

[8] Koene B, Koene L, Broekhuis FR. (2019). Bullet penetration into medium density fibreboard targets. In: Saraswat VK, Reddy GS, Woodley C, editors. Proceedings of the 31st International Symposium on Ballistics. Hyderabad, India: DEStech Publications.

[9] Green DW, Winandy JE, Kretschmann DE. (1999). Mechanical properties of wood. Vol. 113, Wood handbook : wood as an engineering material. Madison, WI : USDA Forest Service, Forest Products Laboratory. General technical report FPL ; GTR-113. Pages 4.1-4.45.

[10] Sydor M, Pinkowski G, Jasińska A. (2020). The Brinell Method for Determining Hardness of Wood Flooring Materials. Forests.Vol.11, No.8. ISSN: 1999-4907.

[11] Sellier&Bellot. Nontox cartridges. (2022). Available from: https://www.sellier-bellot.cz/en/products/pistol-and-revolver-ammunition/nontox-cartridges/list/. Accessed: 2022-02-11.

[12] Trzun Z, Vrdoljak M. Monte Carlo Simulation of Missile Trajectories Dispersion due to Imperfectly Manufactured Warhead. (2020). In: Katalinic B, editor. Proceedings of the 31st DAAAM International Symposium. Mostar, BiH. pp. 0574-0583. DOI: 10.2507/31st.daaam.proceedings.079

[13] Trzun Z, Vrdoljak M, Cajner H. (2021). The Effect of Manufacturing Quality on Rocket Precision. Aerospace. Vol.8, No.6. ISSN: 2226-4310.

[14] C.I.P. Pistol / Revolver Cartridges; Commission internationale permanente pour l'épreuve des armes à feu portatives. (2022). Available from: https://www.xxl-reloading.com/Calibers/Pistol-Revolver-Cartridges/?order=name-asc&p=1. Accessed 2022-02-14.

[15] Kršić N. (1968). Konstruisanje projektila i upaljača - Dejstvo projektila na cilj. Zagreb: Tehnički školski centar JNA.

[16] Carlucci DE, Jacobson SS. (2018). Ballistics Theory and Design of Guns and Ammunition, Third Edition. 6th ed. ISBN: 978-1420066180. Boca Raton: CRC Press.

[17] Jussila J. (2004). Preparing ballistic gelatine—review and proposal for a standard method. Forensic Sci Int. Vol.141, No.2-3. pp 91–98. ISSN: 0379-0738.

[18] Maiden NR, Fisk W, Wachsberger C, Byard RW. (2015). Ballistics ordnance gelatine - How different concentrations, temperatures and curing times affect calibration results. J Forensic Leg Med. August 2015, Vol.34. pp 145–150. ISSN: 1752-928X.

DOI: 10.2507/33rd.daaam.proceedings.040

ANALYSIS OF CLAMPING EFFECT IN HMF CONNECTOR MEASUREMENT

Nikola Skřivanová, Martin Melichar & Kateřina Kašová

This Publication has to be referred as: Skrivanova, N[ikola]; Melichar, M[artin] & Kasova, K[aterina] (2022). Analysis of Clamping Effect in HMF Connector Measurement, Proceedings of the 33rd DAAAM International Symposium, pp.0295-0301, B. Katalinic (Ed.), Published by DAAAM International, ISBN 978-3-902734-36-5, ISSN 1726-9679, Vienna, Austria

DOI: 10.2507/33rd.daaam.proceedings.040

Abstract

This paper focuses on the influence of HFM connector clamping quantification in evaluating the concentricity of internal contact against the external rosette in Automotive. The experiment was conducted in an accredited laboratory of a company that manufactures HFM connectors. The measuring devices used in the experiment were a VERTEX 311UM (3D coordinate measuring machine) and a Keyence VHX-5000 (digital microscope). This kind of concentricity check is implemented very often due to frequent customer complaints. Due to capacity reasons, the measurement is carried out by two different devices with two different operators. In order to determine the capability of each process and to evaluate it correctly, it is also necessary to know the effect of clamping and the measurement uncertainty. Determining the measurement uncertainty is also a guarantee that the measurement result is reliable. Through this processing, the influence of unstable connector clamping on the measurement result using the Keyence VHX-5000 was revealed and subsequently eliminated. This benefit is best demonstrated in the data obtained.

Keywords: measurement uncertainty; clamping; CMM ; digital microscope

1. Introduction

The automotive industry is one of the most competitive sectors in the engineering industry. Companies are constantly striving to prove their good reputation. For this reason, progressive technologies, product features and customer service are constantly being developed. While the information and telecommunication industry deals with life cycles of 0.5 to 3 years on average, the automotive industry designs its products with a life cycle of 6 to 10 years. Accurate measurement is an essential prerequisite for improving the quality of production in the automotive industry and is one of the fundamental ways of obtaining quantitative information about the variable under study. It has become an essential prerequisite for trust in the exchange of goods and is also one of the important conditions for efficient production. The experiment was carried out in an accredited laboratory. [1], [2], [3]. The entire research was carried out in an accredited laboratory of a company producing cable harnesses for the automotive industry. The company is certified according to IATF 16 949. The IATF 16 949 standard for the automotive industry clearly states that if a customer requires a manufacturer to take a measurement of one of the final products, the manufacturer is obliged to take the measurement.

If the conditions are not adequate, the manufacturer is obliged to provide a suitable alternative. The requirements for product accuracy are constantly increasing, especially in the automotive industry. It is therefore becoming essential not only to ensure the production of accurate dimensions, but also to control them by ensuring a reliable and competent measurement process. In practice, however, not only this fact but also time plays a role. The customer has clear requirements for product quality. Within the framework of IATF 16 949, he can obtain, and in most cases requires, proof of the dimensional accuracy of the required product. Due to frequent complaints, the company's customers have requested a concentricity check of the HFM connector. This parameter affects the assembly of the harness into the car. Failure to comply with the required concentricity may result in incorrect assembly or damage to the counterpart inside the car. The company has been forced to ensure regular inspection of the supplies. Therefore, there was a need to ensure the correct procedure, measurement and clamping strategy for this product. [3], [4], [5]

Two different measuring devices are used to measure HFM alignment (Micro-VU's VERTEX 311 UM coordinate measuring machine, KEYENCE VHX digital measuring microscope). Each device has its own measurement procedure, which will be briefly introduced in the following sections. To ensure the correct procedure and strategy, repeated measurements were performed and the measurement uncertainty was calculated. However, the results varied from device to device. Why? It was necessary to determine the cause of the unstable measurements.

2. Product characteristics

The HFM connector is part of a bundle of high-frequency cables used for high-speed data transmission in automobiles. HFM is capable of transmitting large amounts of data in a relatively short time. These harnesses are used for communication between, for example, sensors and the car control unit. These harnesses are produced in different variants and lengths according to the application and the needs of the end customers. The HFM connector itself consists of an internal two-segment contact, an insulator and an external rosette. The two segment contact is connected to the inner conductors of the cable by a crimp joint. Then the outer rosette together with the insulator is fitted onto the inner contact and crimped onto the outer sheath of the cable. Both components (inner contact and outer rosette) are made of spring bronze. The coefficient of thermal expansion of this material is $\alpha = 0{,}000018$mm.

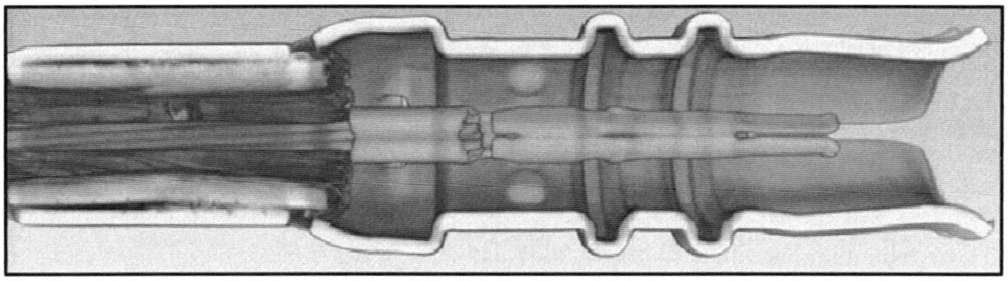

Fig. 1. Ilustrative example CT scan - longitudinal cut through the HFM connector

3. CMM – Vertex 311UM

The Vertex 311 UM coordinate measuring system is used for dimensional inspection of products. The principle of the Vertex 311 UM is to scan the part to be measured with a CCD camera and then measure it on the captured image. The machine contains a precision electromechanical table that sends X, Y, Z coordinates to the measuring software. If the table moves, the linear encoder sends the position change information to the computer. [6]

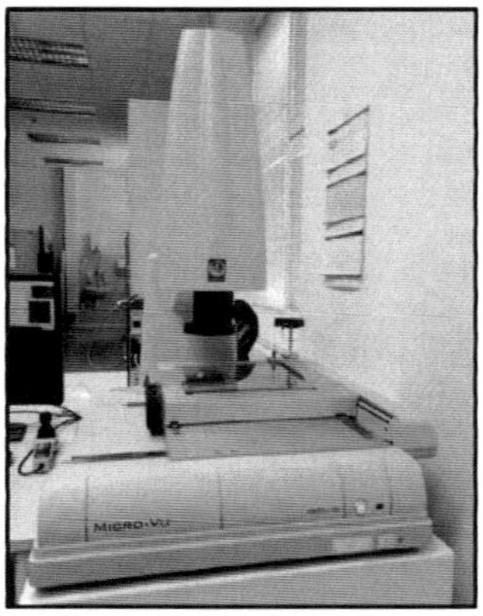

Fig. 2. CMM - Vertex 311UM

3.1. The measuring procedure

The clamping of the FHM connector is carried out with a steel vise with prismatic jaws supported by steel gauges in the case of measurements with the Vertex 311 UM measuring instrument to prevent the sample from being affected by the bending of the cable. Tightening of the sample must be gentle to avoid damaging the crimp joint and consequently damaging the sample.

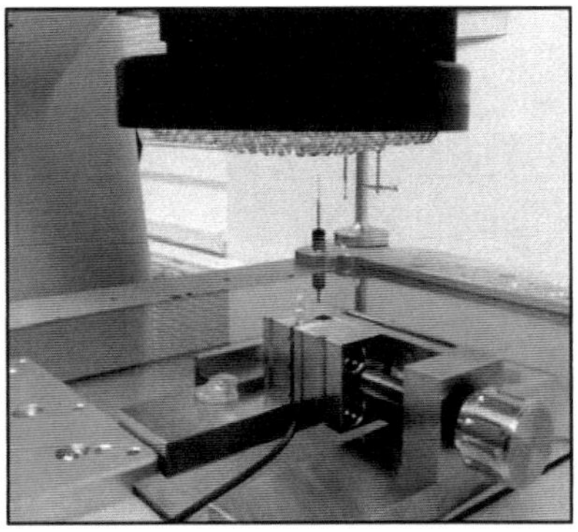

Fig. 3. Clamping of the connector

The actual measurement is performed using the corresponding program created with the InSpec software. The first step of the program is to enter a starting point, in this case a circle, which is taken in manual mode. This circle defines the zero point of the coordinate system for the X and Y axes. Once the starting circle is removed, the machine switches to CNC mode. The next step is to scan the plane at the top of the outer rosette to determine the Z-axis zero point. The coordinate system is calculated from all scanned features. Next, each edge of the outer reference circle is scanned to evaluate concentricity (see Figure 4).

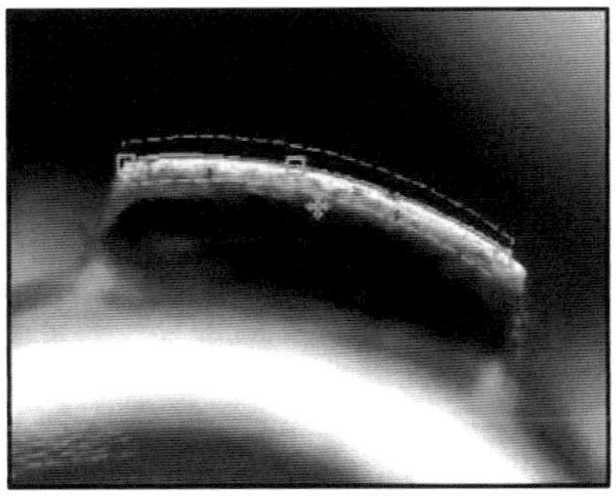

Fig. 4. Segment of the reference circle

The last step is to photograph the leading edges of each segment of the internal contact (Figure 5). A circle is constructed from these segments and then the concentricity to the reference circle is calculated.

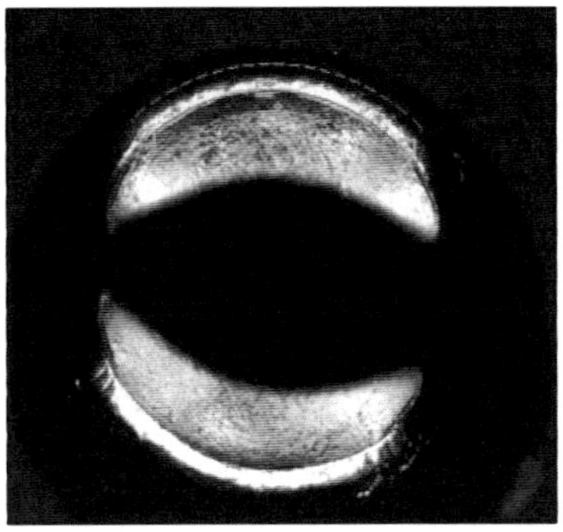

Fig. 5. Segment of the inner circle of contact

3.2. Determination of measurement uncertainty for CMM - Vertex 311UM

The part was repeatedly clamped at each measurement. From these measurements, the uncertainty of the type A measurement was calculated. Other uncertainty components were determined: temperature variation and MPE of the measuring device. The input data and calculations are presented in the tables.

Measured parameter - concentricity - tolerance	0,23mm
MPE (2,6+L/170) µm	(2,6+0,23/170) = 2,6 µm = 0,0026mm
Temperature in the laboratory	23°C
Temperature fluctuations	2°
Coefficient of thermal expansion α spring bronze	0,000018mm

Table 1. Input data

1	**2**	**3**	**4**	**5**	**6**	**7**	**8**	**9**	**10**
0,113mm	0,119mm	0,103mm	0,110mm	0,110mm	0,110mm	0,112mm	0,112mm	0,118mm	0,118mm

Table 2. Measured values (10 repetitions)

Systematic error at 23°C	$\Delta L_{20°C} = \propto \cdot \Delta T$	L=0,000018*3	0,00001242mm
Fluctuation 2°C	$\Delta L_{kol.} = \propto \cdot \Delta T$	L=0,000018*2	0,00000828mm

Table 3. Calculation of the effect of temperature change

Average of measured values	=PRŮMĚR()	0,113mm
Selection directional deviation	=SMODCH.VÝBĚR.S()	0,0048591mm
Average correction to 20°C	$X\ korigovaný = \bar{X} - L_{20°C}$	0,112mm
u_A	$u_A = k_{UA} \cdot \frac{s}{\sqrt{n}}$	0,00154mm

Table 4. Calculation of Type A measurement uncertainty based on repeated measurements

	Zmax	χ	u_{bi}	u_{bi}
MPE Vertex 311 UM	0,0026mm	1,732051	$u_{bi} = \frac{Zmax}{\chi}$	0,001501111
Fluctuation 2°C	0,00000828mm	1,732051	$u_{bi} = \frac{Zmax}{\chi}$	0,000000478
Calculation of the standard combined uncertainty				
u_y	=ODMOCNINA(SUMA.ČTVERCŮ($u_A; u_{bivertex}; u_{bikol}$))			0,002148131
Calculation of the expanded combined uncertainty				
U	$=2 \cdot u_y$			0,005mm

Table 5. Calculation of type B measurement uncertainty and expanded combined uncertainty U [8]

The measurement result is with a specified expanded combined uncertainty = 0.112 mm +/- 0.005 mm.

4. Keyence VHX-5000

A digital measuring microscope is a device composed of several components. One of the main parts of the microscope is a fully motorized stage in the X and Y axes. This stage is fully controlled by the microscope operator software. Another integral part of the microscope is the vertically mounted objective. The objectives can be changed directly on the microscope. In this case, a variable magnification objective with a total magnification range of 20x - 200x is mounted on the instrument. Above this interchangeable objective is a high-speed camera with a high resolution of 1600x1200 pixels. This camera transmits the image to the imaging unit, which is also an integral part of the complete measuring microscope set. The entire microscope, including its components, is controlled partly manually and partly digitally via a software interface supplied by the manufacturer. [7]

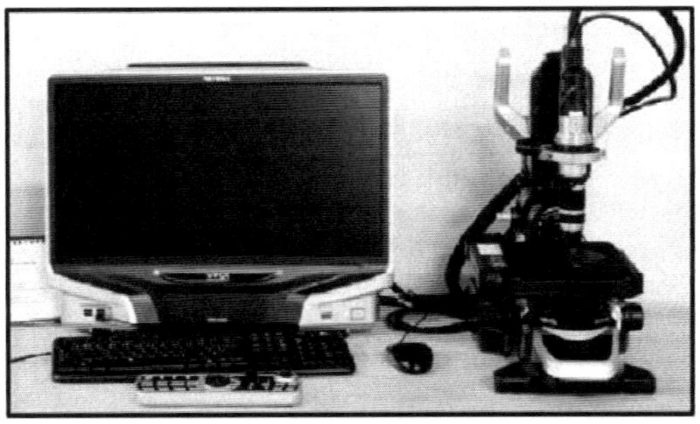

Fig. 6. Keyence VHX-5000

4.1. The measuring procedure

Measurement of the HFM connector is performed at 50x magnification. The part is clamped with a vise and then placed under the microscope objective. The position of the part should be as close as possible to the centre of the field of view of the objective and perpendicular to the objective.

Fig. 7. Clamping of the connector

After the object to be measured has been located and photographed, a reference circle on the outer edge of the connector and an inner circle on the leading edge of the contact are captured with the measuring tool. The circles are captured manually using three points. The KEYENCE VHX-5000 cannot automatically evaluate the geometric concentricity tolerance, but only the distance of the centers of the circles. It is therefore necessary to convert the center distance to concentricity (concentricity = 2* measured center distance of the circles). A graphical representation of the measurement is shown below (Figure 8).

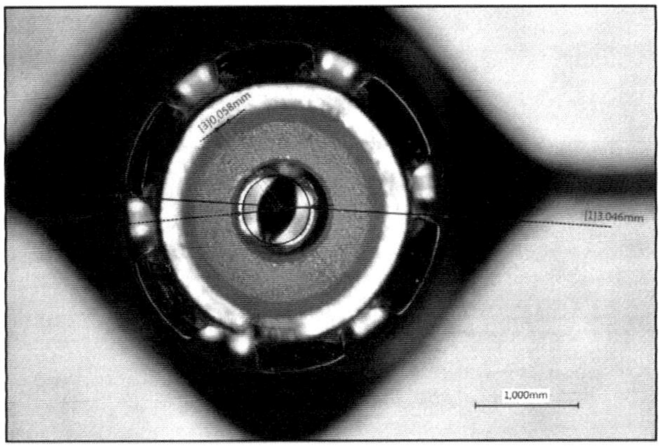

Fig. 8. Graphical representation

4.2. Determination of measurement uncertainty for Keyence VHX-5000

The part was repeatedly clamped at each measurement. From these measurements, the uncertainty of the type A measurement was calculated. Other uncertainty components were determined: temperature variation and MPE of the measuring device. The input data and calculations are presented in the following tables. Since this equipment can be operated by more than one laboratory personnel, two personnel were selected to make the specified measurements independently and the uncertainty was then calculated for each of them so that they could be compared with each other.

Measured parameter - concentricity - tolerance	0,23mm
MPE (2,6+L/170) μm	0,0038mm
Temperature in the laboratory	23°C
Temperature fluctuations	2°
Coefficient of thermal expansion α spring bronze	0,000018mm

Table 6. Input data

Systematic error at 23°C	$\Delta L_{20°C} = \propto \cdot \Delta T$	L=0,000018*3	0,00001242mm
Fluctuation 2°C	$\Delta L_{kol.} = \propto \cdot \Delta T$	L=0,000018*2	0,00000828mm

Table 7. Calculation of the effect of temperature change

1	**2**	**3**	**4**	**5**	**6**	**7**	**8**	**9**	**10**
0,126mm	0,108mm	0,116mm	0,102mm	0,118mm	0,110mm	0,106mm	0,120mm	0,106mm	0,114mm

Table 8. Measured values (10 repetitions) - worker A

1	**2**	**3**	**4**	**5**	**6**	**7**	**8**	**9**	**10**
0,142mm	0,112mm	0,102mm	0,134mm	0,116mm	0,130mm	0,124mm	0,128mm	0,142mm	0,104mm

Table 9. Measured values (10 repetitions) - worker B

Average of measured values	=PRŮMĚR()	0,113mm
Selection directional deviation	=SMODCH.VÝBĚR.S()	0,0074863mm
Average correction to 20°C	$X\ korigovaný = \bar{X} - L_{20°C}$	0,113mm
u_A	$u_A = k_{UA} \cdot \frac{s}{\sqrt{n}}$	0,00237mm

Table 10. Calculation of Type A measurement uncertainty based on repeated measurements - Worker A

Average of measured values	=PRŮMĚR()	0,123mm
Selection directional deviation	=SMODCH.VÝBĚR.S()	0,0144852mm
Average correction to 20°C	$X\ korigovaný = \bar{X} - L_{20°C}$	0,123mm
u_A	$u_A = k_{UA} \cdot \frac{s}{\sqrt{n}}$	0,00458mm

Table 11. Calculation of Type A measurement uncertainty based on repeated measurements - Worker B

	Zmax	χ	u_{bi}	u_{bi}
MPE Keyence VHX	0,0038mm	1,732051	$u_{bi} = \frac{Zmax}{\chi}$	0,002193931
Temperature fluctuation 2°C	0,00000828mm	1,732051	$u_{bi} = \frac{Zmax}{\chi}$	0,000000478
Calculation of the standard combined uncertainty				
u_y	=ODMOCNINA(SUMA.ČTVERCŮ($u_A; u_{bivertex;}\ u_{bikol}$))			0,003227662
Calculation of the expanded combined uncertainty				
U	=2· u_y			0,007mm

Table 12. Calculation of type B measurement uncertainty and expanded combined uncertainty U - worker A [8]

	Zmax	χ	u_{bi}	u_{bi}
MPE Keyence VHX	0,0038mm	1,732051	$u_{bi} = \frac{Zmax}{\chi}$	0,002193931
Temperature fluctuation 2°C	0,00000828mm	1,732051	$u_{bi} = \frac{Zmax}{\chi}$	0,000000478
Calculation of the standard combined uncertainty				
u_y	=ODMOCNINA(SUMA.ČTVERCŮ($u_A; u_{bivertex;} u_{bikol}$))			0,005078935
Calculation of the expanded combined uncertainty				
U	=2· u_y			0,011mm

Table 13. Calculation of type B measurement uncertainty and expanded combined uncertainty U - worker B [8]

The measurement result of operator A with the specified expanded combined uncertainty =0.113mm +/- 0.007mm. The measurement result of operator B with the specified expanded combined uncertainty is 0.123mm +/- 0.011mm.

5. Conclusion

This paper focused on determining the effect of clamping quantification for concentricity measurements of HFM connectors in an accredited laboratory using non-sensitive measurements. This type of inspection is common due to the nature of manufacturing and the proliferation of customer complaints. Therefore, the requirement to verify the measurement system using. With the Vertex 311UM measuring device, the result obtained was expected, however the difference in the measurement results obtained with the Keyence VHX-5000 when changing the operator is already apparent at a glance. The result obtained with operator A is 0.113 mm +/- 0.007 mm. The result obtained with operator B is 0.123 mm +/- 0.011 mm. Analysis of the measuring system due to differences between operators A and B revealed a significant proportion of unstable part clamping increasing the measurement uncertainty. This analysis including clamping optimization and subsequent verification will be part of further research.

6. Acknowledgments

The article contribution has been prepared under project SGS-2022-007 - Research and Development for Innovation in Engineering Technology - Machining Technology IV.

7. References

[1] Skrivanova, N[ikola] & Melichar, M[artin] (2019). The Comparison of Destructive and Non-Destructive Forms of Measurement in the Automotive Industry, Proceedings of the 30th DAAAM International Symposium, pp.0995-1002, B. Katalinic (Ed.), Published by DAAAM International, ISBN 978-3-90273422-8, ISSN 1726-9679, Vienna, Austria

[2] Bicova K. Kutlwaser J., Sklenicka J.: (2016). Issue of High Precision Manufacturing Analysis in Automotive Industry, Proceedings of the 27th DAAAM International Symposium, ISBN 978-3-902734-08-2, ISSN 1726-9679, Vienna, Austria DOI: 10.2507/27th.daaam.proceedings.052

[3] Skrivanova, N[ikola] & Melichar, M[artin] (2019). Comparing the TimeConsumption of Destructive and Non-Destructive Forms of Measurement in the Automotive Industry, Proceedings of the 30th DAAAM International Symposium, pp.1003-1007, B. Katalinic (Ed.), Published by DAAAM International, ISBN 978-3-902734-22-8, ISSN 1726-9679, Vienna, Austria

[4] Melichar M., Kubátová D., Kutlwaše J.: (2018). The Influence of humidity on ABS plastic measurement results, Proceedings of the 29th DAAAM International Symposium, ISBN 978-3-902734-20-4,ISSN 1726-9679, Vienna, Austria DOI: 10.2507/29th.daaam.proceedings.065

[5] Płowucha W.. Jakubiec W.. Wojtyła M.: Possibilities of CMM Software to Support Proper Geometrical Product Verification. Procedia CIRP. Volume 43. 2016. Pages 303-308. ISSN 2212-8271. http://dx.doi.org/10.1016/j.procir.2016.02.124

[6] Digitální mikroskop - VHX-5000 | KEYENCE International Belgium(Čeština). KEYENCE INTERNATIONAL (BELGIUM) NV/SA [online]. available from: https://www.keyence.eu/cscz/products/microscope/digital-microscope/vhx-5000/models/vhx-5000/, Accessed on: 2022-10-10

[7] Automatický měřicí přístroj VERTEX 311 UM. TZUS [online]. Copyright © 2022 TZÚS Praha, s.p. [cit. 10.10.2022]. available from: https://www.tzus.cz/kontakty/pobocky-a-odstepny-zavod/oz-zkusebni-ustav-lehkeho-prumyslu-cb/automaticky-merici-pristroj-vertex-311-um, Accessed on: 2022-10-10

[8] NĚMEČEK, Pavel. Nejistoty měření. 1. vydání 2008. Praha: česká společnost pro jakost,o.s., 2008. 96 s. ISBN 978-80-02-02089-9.

DOI: 10.2507/33rd.daaam.proceedings.041

ANALYSIS OF CLAMPING OPTIMIZATION FOR HFM CONNECTOR DURING MEASUREMENT

Nikola Skřivanová, Martin Melichar & Kateřina Kašová

This Publication has to be referred as: Skrivanova, N[ikola]; Melichar, M[artin] & Kasova, K[aterina] (2022). Analysis of Clamping Optimization for HFM Connector During Measurement, Proceedings of the 33rd DAAAM International Symposium, pp.0302-0306, B. Katalinic (Ed.), Published by DAAAM International, ISBN 978-3-902734-36-5, ISSN 1726-9679, Vienna, Austria
DOI: 10.2507/33rd.daaam.proceedings.041

Abstract

This paper focuses on the quantification of the influence and optimization of HFM connector clamping in the evaluation of the concentricity of internal contact against the external rosette in the automotive industry. It is related to the article Analysis of clamping effect in HFM connector measurement. In the previous experiment, which was carried out in an accredited laboratory of an automotive company, the negative effect of HFM connector clamping in concentricity measurements using a Keyence VHX-5000 digital microscope was found. Now the previous findings have been verified and a new way of clamping the parts has been devised. The new clamping method eliminated this negative effect. This benefit is best demonstrated by the data obtained.

Keywords: uncertainty of measurement; clamping; CMM; digital microscope

1. Introduction

This article is a follow-up to the article Analysis of the effect of clamping in HMF connector measurements, in which the effect of negative clamping of the HFM connector in concentricity measurements using the Keyence VHX-5000 digital microscope was determined. Thus, the problem and its cause were found in the previous experiment. Subsequently, the question arose, how to achieve optimal clamping without a negative contribution to the measurement result? Therefore, the next experiment consisted in verifying the consistency of the measurement using the EN parameter and quantifying the new clamping. With the new type of clamping, this effect was eliminated. This result is represented by the obtained measurement data and verified by calculations. Thus, we can conclude that the problem that the accredited laboratory had has been solved.

2. Determination of measurement uncertainty for Keyence VHX-5000

The part was repeatedly clamped at each measurement. From these measurements, the uncertainty of the type A measurement was calculated. Other uncertainty components were determined: temperature variation and MPE of the measuring device. The input data and calculations are presented in the following tables.

Since this equipment can be operated by more than one laboratory personnel, two personnel were selected to make the specified measurements independently and the uncertainty was then calculated for each of them so that they could be compared with each other.

Measured parameter - concentricity - tolerance	0,23mm
MPE (2,6+L/170) µm	0,0038mm
Temperature in the laboratory	23°C
Temperature fluctuations	2°
Coefficient of thermal expansion α spring bronze	0,000018mm

Table 1. Input data

Systematic error at 23°C	$\Delta L_{20°C} = \propto \cdot \Delta T$	L=0,000018*3	0,00001242mm
Fluctuation 2°C	$\Delta L_{kol.} = \propto \cdot \Delta T$	L=0,000018*2	0,00000828mm

Table 2. Calculation of the effect of temperature change

1	**2**	**3**	**4**	**5**	**6**	**7**	**8**	**9**	**10**
0,126mm	0,108mm	0,116mm	0,102mm	0,118mm	0,110mm	0,106mm	0,120mm	0,106mm	0,114mm

Table 3. Measured values (10 repetitions) - worker A

1	**2**	**3**	**4**	**5**	**6**	**7**	**8**	**9**	**10**
0,142mm	0,112mm	0,102mm	0,134mm	0,116mm	0,130mm	0,124mm	0,128mm	0,142mm	0,104mm

Table 4. Measured values (10 repetitions) - worker B

Average of measured values	=PRŮMĚR()	0,113mm
Selection directional deviation	=SMODCH.VÝBĚR.S()	0,0074863mm
Average correction to 20°C	$X\ korigovaný = \bar{X} - L_{20°C}$	0,113mm
$\boldsymbol{u_A}$	$u_A = k_{UA} \cdot \frac{s}{\sqrt{n}}$	0,00237mm

Table 5. Calculation of Type A measurement uncertainty based on repeated measurements - Worker A [7], [8]

Average of measured values	=PRŮMĚR()	0,123mm
Selection directional deviation	=SMODCH.VÝBĚR.S()	0,0144852mm
Average correction to 20°C	$X\ korigovaný = \bar{X} - L_{20°C}$	0,123mm
$\boldsymbol{u_A}$	$u_A = k_{UA} \cdot \frac{s}{\sqrt{n}}$	0,00458mm

Table 6. Calculation of Type A measurement uncertainty based on repeated measurements - Worker B [7], [8]

	Zmax	**χ**	$\boldsymbol{u_{bi}}$	$\boldsymbol{u_{bi}}$
MPE Keyence VHX	0,0038mm	1,732051	$u_{bi} = \frac{Zmax}{\chi}$	0,002193931
Temperature fluctuation 2°C	0,00000828mm	1,732051	$u_{bi} = \frac{Zmax}{\chi}$	0,000000478
Calculation of the standard combined uncertainty				
u_y	=ODMOCNINA(SUMA.ČTVERCŮ(u_A; $u_{bivertex;}$ u_{bikol}))			0,003227662
Calculation of the expanded combined uncertainty				
U	=2· u_y			0,007mm

Table 7. Calculation of type B measurement uncertainty and expanded combined uncertainty U - worker A [7], [8]

	Zmax	χ	u_{bi}	u_{bi}
MPE Keyence VHX	0,0038mm	1,732051	$u_{bi} = \frac{Zmax}{\chi}$	0,002193931
Temperature fluctuation 2°C	0,00000828mm	1,732051	$u_{bi} = \frac{Zmax}{\chi}$	0,000000478
Calculation of the standard combined uncertainty				
u_y	=ODMOCNINA(SUMA.ČTVERCŮ($u_A; u_{bivertex}; u_{bikol}$))			0,005078935
Calculation of the expanded combined uncertainty				
U	=2· u_y			0,011mm

Table 8. Calculation of type B measurement uncertainty and expanded combined uncertainty U - worker B [7], [8]

The measurement result of operator A with the specified expanded combined uncertainty =0.113mm +/- 0.007mm. The measurement result of operator B with the specified expanded combined uncertainty is 0.123mm +/- 0.011mm.

3. Assessment of the measurement results of operators A and B

At first glance, the difference in the measurement results of the Keyence VHX-5000 obtained when changing the operator is noticeable. The result obtained with operator A is 0.113 mm +/- 0.007 mm. The result obtained with operator B is 0.123 mm +/- 0.011 mm. The EN parameter can be used to assess the agreement between the results of the two operators under the condition that EN ≤ 1.

$$E_n = \frac{\bar{x}_{lab} - \bar{x}_{pilot}}{\sqrt{U_{lab}^2 + U_{pilot}^2}} = |E_n| \leq 1,$$

$\overline{X_{lab}}$ = the average of the measured values of the operator, in our case operator B,
$\overline{X_{pilot}}$ = the average of the measured values of the operator - pilot, in our case operator A,
U_{lab} = the extended combined uncertainty of the operator, in our case operator B,
U_{pilot} = the expanded combined uncertainty of the operator, in our case operator A.

After substituting the values into the formula, we obtain the value EN = 0.77. This value satisfies the condition |EN|≤1. Based on the EN calculation, the resulting value is within tolerance, although it is close to the warning value of 0.80. When comparing the sampling standard deviations and calculated uncertainties of the type A measurements between operator A and operator B, it is clear that not all random effects entering the measurement system have been sufficiently removed. The resulting values are:
the sampling standard deviation of operator A = 0,0074863mm,
the sampling standard deviation of operator B = 0,0144852mm,
u_A operator A = 0,00237mm,
u_A operator B = 0,00458mm.

Analysis of the measuring system due to differences between operators A and B revealed a significant contribution of unstable part clamping, which increases the measurement uncertainty. This contribution was reduced by constructing a more suitable and stable part clamping under the microscope objective.

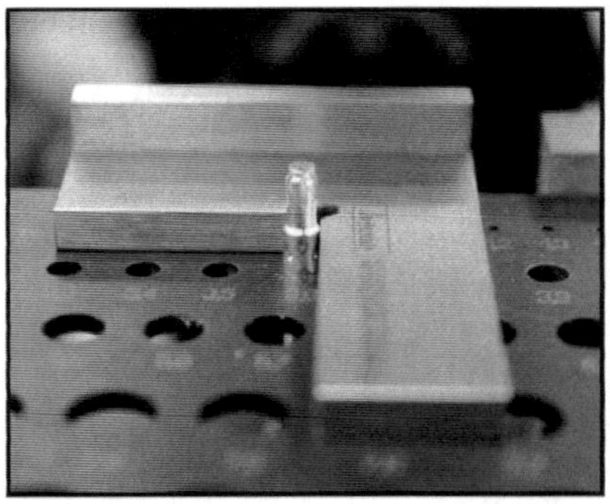

Fig. 1. More stable clamping of the HFM connector

4. Recalculation of the measurement uncertainty after clamping optimization in the measurement system

After setting up the new clamping system and sensitizing the operator, the whole procedure of determining the measurement uncertainties was repeated. The repeat measurement was performed under the same input conditions with the same operators A and B as in section 4.2.

1	2	3	4	5	6	7	8	9	10
0,115mm	0,111mm	0,114mm	0,118mm	0,108mm	0,111mm	0,120mm	0,109mm	0,113mm	0,111mm

Table 9. Measured values (10 repetitions) - worker A [7], [8]

1	2	3	4	5	6	7	8	9	10
0,124mm	0,108mm	0,116mm	0,104mm	0,126mm	0,114mm	0,126mm	0,122mm	0,106mm	0,106mm

Table 10. Measured values (10 repetitions) - worker B [7], [8

Average of measured values	=PRŮMĚR()	0,113mm
Selection directional deviation	=SMODCH.VÝBĚR.S()	0,0039001mm
Average correction to 20°C	$\overline{X}\ korigovaný = \overline{X} - L_{20°C}$	0,113mm
u_A	$u_A = k_{UA} \cdot \frac{s}{\sqrt{n}}$	0,00123mm

Table 11. Calculation of Type A measurement uncertainty based on repeated measurements - Worker A

Average of measured values	=PRŮMĚR()	0,115mm
Selection directional deviation	=SMODCH.VÝBĚR.S()	0,0088544mm
Average correction to 20°C	$\overline{X}\ korigovaný = \overline{X} - L_{20°C}$	0,115mm
u_A	$u_A = k_{UA} \cdot \frac{s}{\sqrt{n}}$	0,00280mm

Table 12. Calculation of Type A measurement uncertainty based on repeated measurements - Worker B

	Zmax	χ	u_{bi}	u_{bi}
MPE Keyence VHX	0,0038mm	1,732051	$u_{bi} = \frac{Zmax}{\chi}$	0,002193931
Temperature fluctuation 2°C	0,00000828mm	1,732051	$u_{bi} = \frac{Zmax}{\chi}$	0,000000478
Calculation of the standard combined uncertainty				
u_y	=ODMOCNINA(SUMA.ČTVERCŮ($u_A; u_{bivertex}; u_{bikol}$))			0,002516837
Calculation of the expanded combined uncertainty				
U	=2· u_y			0,006mm

Table 13. Calculation of type B measurement uncertainty and expanded combined uncertainty U - worker A [7], [8]

	Zmax	χ	u_{bi}	u_{bi}
MPE Keyence VHX	0,0038mm	1,732051	$u_{bi} = \frac{Zmax}{\chi}$	0,002193931
Temperature fluctuation 2°C	0,00000828mm	1,732051	$u_{bi} = \frac{Zmax}{\chi}$	0,000000478
Calculation of the standard combined uncertainty				
u_y	=ODMOCNINA(SUMA.ČTVERCŮ($u_A; u_{bivertex}; u_{bikol}$))			0,003557156
Calculation of the expanded combined uncertainty				
U	=2· u_y			0,008mm

Table 14. Calculation of type B measurement uncertainty and expanded combined uncertainty U - worker B [7], [8]

The measurement result of operator A with the specified expanded combined uncertainty is 0.113mm +/- 0.006mm. The measurement result of operator B with the specified expanded combined uncertainty is 0.115mm +/- 0.008mm.

5. Conclusion

This paper focuses on optimizing the clamping of HFM connectors for concentricity measurements in an accredited automotive laboratory. The research is related to the article Analysis of the effect of clamping in HMF connector measurements, where a negative effect on the measurement result due to unstable clamping of this connector was found. The experiment was performed using a Keyence VHX-5000 digital microscope with two different operators. By assessing the agreement using the EN parameter, this negative effect was demonstrated. Subsequently, the clamping optimization and subsequent recalculation were performed. This experiment was of practical benefit and helped to increase the stability of the HFM connectors. This benefit can best be demonstrated by the data obtained.

Before clamping optimization:
operator A = 0,113mm +/- 0,007mm /// operator B = 0,123mm +/- 0,011mm,
the sampling standard deviation of operator A=0,0074863mm,
the sampling standard deviation of operator B= 0,0144852mm,
u_A Operator A = 0,00237mm /// u_A Operator B = 0,00458mm,
parameter EN = 0,77.

After clamping optimization:
operator A = 0,113mm +/- 0,006mm /// operator B = 0,115mm +/- 0,008mm,
the sampling standard deviation of operator A = 0,0039001mm,
the sampling standard deviation of operator B = 0,0088544mm,
u_A operator A = 0,00123mm /// u_A operator B = 0,00280mm,
parameter EN = 0,20.

6. Acknowledgments

The article contribution has been prepared under project SGS-2022-007 - Research and Development for Innovation in Engineering Technology - Machining Technology IV.

7. References

[1] Skrivanova, N[ikola] & Melichar, M[artin] (2019). The Comparison of Destructive and Non-Destructive Forms of Measurement in the Automotive Industry, Proceedings of the 30th DAAAM International Symposium, pp.0995-1002, B. Katalinic (Ed.), Published by DAAAM International, ISBN 978-3-90273422-8, ISSN 1726-9679, Vienna, Austria

[2] Bicova K. Kutlwaser J., Sklenicka J.: (2016). Issue of High Precision Manufacturing Analysis in Automotive Industry, Proceedings of the 27th DAAAM International Symposium, ISBN 978-3-902734-08-2, ISSN 1726-9679, Vienna, Austria DOI: 10.2507/27th.daaam.proceedings.052

[3] Skrivanova, N[ikola] & Melichar, M[artin] (2019). Comparing the TimeConsumption of Destructive and Non-Destructive Forms of Measurement in the Automotive Industry, Proceedings of the 30th DAAAM International Symposium, pp.1003-1007, B. Katalinic (Ed.), Published by DAAAM International, ISBN 978-3-902734-22-8, ISSN 1726-9679, Vienna, Austria

[4] Melichar M., Kubátová D., Kutlwaše J.: (2018). The Influence of humidity on ABS plastic measurement results, Proceedings of the 29th DAAAM International Symposium, ISBN 978-3-902734-20-4,ISSN 1726-9679, Vienna, Austria DOI: 10.2507/29th.daaam.proceedings.065

[5] Płowucha W.. Jakubiec W.. Wojtyła M.: Possibilities of CMM Software to Support Proper Geometrical Product Verification. Procedia CIRP. Volume 43. 2016. Pages 303-308. ISSN 2212-8271. http://dx.doi.org/10.1016/j.procir.2016.02.124

[6] Digitální mikroskop - VHX-5000 | KEYENCE International Belgium(Čeština). KEYENCE INTERNATIONAL (BELGIUM) NV/SA [online]. available from: https://www.keyence.eu/cscz/products/microscope/digital-microscope/vhx-5000/models/vhx-5000/, Accessed on: 2022-10-10

[7] ČSN P ENV 13005: Pokyn pro vyjádrení nejistoty merení, Praha: Ceský normalizacní institut, 2005.

[8] NĚMEČEK, Pavel. Nejistoty měření. 1. vydání 2008. Praha: česká společnost pro jakost,o.s., 2008. 96 s. ISBN 978-80-02-02089-9.

DOI: 10.2507/33rd.daaam.proceedings.042

MACHINABILITY OF THERMAL SPRAY STELLITE 6 WITH CHEMICAL AND MECHANICAL ANALYSIS

Tomas Zatloukal, Ondrej Marsalek & Simon Syrovatka

This Publication has to be referred as: Zatloukal, T[omas]; Marsalek, O[ndrej] & Syrovatka, S[imon] (2022). Machinability of Thermal Spray Stellite 6 with Chemical and Mechanical Analysis, Proceedings of the 33rd DAAAM International Symposium, pp.0307-0312, B. Katalinic (Ed.), Published by DAAAM International, ISBN 978-3-902734-36-5, ISSN 1726-9679, Vienna, Austria
DOI: 10.2507/33rd.daaam.proceedings.042

Abstract

This article deals with the machinability of Stellite 6 thermal spray, which is widely represented in many industries today, where it is used as an additive to increase resistance to wear, corrosion and oxidation. The aim of the research was to achieve the same or better quality of the Stellite 6 thermal spray surface with a different technology than grinding technology. The main idea of this change is to shorten production times and thereby increase productivity during thermal spray machining. After the experiment, the chemical analysis of the processed thermal spray is evaluated, during which the microhardness, microstructure, chemical analysis and thickness of the thermal spray were evaluated. Subsequently, an evaluation of the surface roughness and the amount of insert wear was carried out, from which the optimal cutting conditions were determined.

Keywords: Stellite 6; Machinability; Thermal spray; Metallography; Chemical analysis

1. Introduction

Currently, the requirements for materials used in the engineering industry are increasing rapidly. With the development of new materials, emphasis is also placed on surface integrity. In certain cases, it is necessary for the manufactured component to be able to withstand very demanding conditions in operation. This can be achieved in several ways. Among the most well-known variants chemical and heat treatment can be considered, which includes, for example, hardening, tempering, cementation, nitriding, etc. Another possibility that is at the forefront of interest is the use of thermal spraying of very resistant materials. Thermal spraying is practically an irreplaceable technology in many industries. Their biggest advantages are increased resistance to mechanical wear, chemical environment, corrosion and oxidation. They also withstand high temperatures very well. There are many areas in which it is possible to encounter thermal sprays. Take, for example, general engineering, where these sprays are applied to pumping and hydraulic technology to increase resistance to wear and corrosion. It is also used in the printing industry, where sprays are applied to the surfaces of printing cylinders. In the energy industry, they are used as a thermal barrier, protection against cavitation, erosion and hot corrosion. Last but not least, thermal sprays are widely used in the automotive and aerospace industries, where they are applied to internal combustion engine parts to increase resistance to wear and high temperature. We can list here, for example, pistons, valves, cylinder heads, pins, crankshafts or cams. [1] [3] [5]

It is known in the engineering industry that with every machining technology used, residual stresses are introduced into the machined material. These can be divided into tensile and compressive residual stresses. They arise as the machined material tries to return to an equilibrium state. A typical operation for the formation of tensile residual stress is grinding technology. During grinding, a large amount of heat is generated between the grinding wheel and the machined surface, which penetrates into the workpiece, and begins to cool after the process is finished. Tensile residual stresses start to form in the material. Tensile stresses can be considered a negative phenomenon in most cases. When a crack occurs, they help its propagation and thus shortens the lifetime of the component. During grinding, the residual stresses can be partially reduced by using coolant. The second type is compressive residual stresses. These arise during technologies where the surface layer of the machined material is plastically deformed by means of mechanical loading. These are technologies where tool has a defined cutting edge geometry, such as milling, turning, or drilling. In the event of a crack, on the other hand, compressive residual stresses tend to close the crack and thus extend the already mentioned service life even in cases where intensive mechanical wear occurs. [1] [4]

The publication titled "Machinability of Stellite-6 Coatings with Ceramic Inserts and Tungsten Carbide Tools" [2] also dealt with the issue of machinability of thermal sprays. This study focused not only on the cutting conditions but also on the difference in cutting materials and their effect on the resulting surface quality. The Stellite alloy was applied to the base material of austenitic stainless steel with the commercial designation AISI 304. Two types of cutting materials were tested for the thermal spray of Stellite 6. In the first case, it was a ceramic cutting plate fixed to the body of the turning knife. The ceramic plate from Sandvik has the designation RNGN 120700 T0 1020 and consists of the base material Al_2O_3, which is reinforced with so-called whiskers. According to the authors of the study, this type of insert is suitable for high-speed machining of alloys resistant to high temperatures. The holder was designed in such a way that the inserted plate ensures a negative face angle with a value of -6°. [2]

In the second case, a one-piece turning knife with a brazed cutting edge made of grade K20 tungsten carbide, which was uncoated, was used. Compared to the ceramic plate, the soldered edge has a positive face geometry, the value of which is not stated. The process fluid was used only for the variant with a brazed cutting edge, and it was an oil emulsion. In the case of machining with a ceramic plate, the entire process took place without the use of any process liquid. Cutting conditions took place in combinations of changes in cutting speed and feed per revolution. A detailed elaboration of the entire experiment plan with achieved roughness values are shown in table 1. [2]

Insert type	Cutting speed (m/min)	Feed rate (mm/rev)	Average surface roughness (μm)	Tool type	Cutting speed (m/min)	Feed rate (mm/rev)	Average surface roughness (μm)
	V	*f*	Ra		*V*	*f*	Ra
Whisker-reinforced ceramic CC670	30	0.25	2.48	Tungsten carbide	30	0.1	2.6
		0.3	3.44			0.15	3.8
		0.35	4.74			0.20	5.1
	50	0.25	2.42		40	0.1	2.1
		0.3	3.29			0.15	3.1
		0.35	4.63			0.20	4.7
	70	0.25	2.30		50	0.1	1.8
		0.3	3.19			0.15	2.5
		0.35	4.54			0.20	4.0
	90	0.25	2.21				
		0.3	3.10				
		0.35	4.40				

Table 1. Experiment plan with measured roughness [2]

In the case of machining with a ceramic insert, it can be noted that the best achieved roughness Ra = 2.21 μm was achieved at the highest cutting speed v_c = 90 m/min and at the smallest feed value f = 0.25 mm/rev. When comparing the individual variants in more detail, however, it is evident that the cutting speed does not have such a fundamental effect on the roughness values, in contrast to the displacement values. The measured roughness are very similar at the selected feed values, with the fact that they improve slightly with increasing cutting speed. A similar trend was also observed when machining with a tungsten carbide soldered plate. The best value of the roughness of the machined surface Ra = 1.8 μm was recorded at the highest cutting speed v_c = 50 m/min and the smallest feed f = 0.1 mm/rev. Compared to the previous case, it is clear from the table at first glance that both the feed parameter and the cutting speed value have an effect on the roughness value. However, the overall process is identical to the previous version with a ceramic cutting edge. [2]

In conclusion, it can be said that similar results of surface quality were achieved in both variants, with the fact that the variant with a tungsten carbide solder plate turned out slightly better. This could have been caused both by significantly smaller displacement values and also by the use of an operating fluid in the form of an oil emulsion. This article again confirms the claim that better surface roughness values are achieved at higher cutting speeds and smaller feed values.

2. Description and preparation of the experiment

Stellite 6 alloy contains between 45-55% cobalt (Co) and is referred to as the base. The second most represented element is chromium (Cr), the amount of which ranges from 25 to 35%. Tungsten (W) also has a non-negligible presence in this alloy, from 5 to 25%. Another very important element is carbon (C), which creates individual carbides and is contained here in the range of 0.1 - 1.5%. These alloys also contain other elements in small quantities, such as molybdenum (Mo), nickel (Ni), niobium (Nb), tantalum (Ta), silicon (Si), manganese (Mn) and iron (Fe). According to the given values, it is evident that Stellite alloys can have a very different chemical composition and thus the resulting properties. The exact chemical composition is described in Table 2. The listed contents of individual chemical elements are available from the Deloro and Kennametal catalog. [10] Chromium and tungsten forming carbides with carbon provide the required hardness and strength. The hardness is in the range of 36 – 45 HRC and the tensile strength limit is given in the range of 800 – 1200 MPa. In addition, chromium carbides increase resistance to corrosion and oxidation even at higher temperatures. The melting point is determined by the combination and content of the individual elements and ranges from 1200 to 1400 °C. [7] [8]

Compo-sition in wt %	Co	Cr	W	C	Ni	Fe	Si	Mo	Mn	Other
	45 - 60	27 - 32	4 - 6	0,9 - 1,4	max. 3	max. 3	max. 1,5	max. 1	max. 1	max. 1

Table 2. Chemical composition of Stellite 6 alloy in weight percent [6]

Stellite 6 is popular mainly due to the balanced ratio between hardness and toughness. It contains a large number of hard carbides of chromium and tungsten. Carbides contribute to a very good ability to resist many types of wear, even at elevated temperatures. At room temperature, the hardness value is 410 HV. When the temperature increases, the alloy still retains the appropriate hardness value of 400-300 HV up to 500 °C. As a cutting tool, a curved carving knife made of material marked as 42CrMo4 with a tensile strength of 1000 MPa, which is commonly used for the production of lathe holders, was used. As it is planned to change the angle of the face during the experiment (in the values of -7°, 0°, +7°), it is necessary to produce 3 variants of turning knives. The preparation of the machining knife with IC20 cemented carbide insert can be seen in the picture (Figure 1). [7], [9]

Fig. 1. Machining tool with Iscar IC20 insert

In the experiment, the cutting speed was varied in the range $v_c = 20 – 55$ m/min. Furthermore, the depth of the cut was changed in two variants of 0.1 and 0.3 mm. A cemented carbide insert from the company Iscar marked IC20 is attached to the tool. The chosen technology was a simplified imaging process known as orthogonal free cutting. Steel with dimensions of 400 x 40 mm with a width of 3 mm from material 11 373 according to ČSN was chosen as a semi-finished product. This sheet was treated from the top with Stellite 6 thermal spray produced by the HVOF method with a thickness of 500 μm (Figure 2). High-velocity flame spraying known as HVOF (High-Velocity Oxygen Fuel) belongs to the most modern methods in the field of thermal sprays.

Fig. 2. Manufactured workpieces with Stellite 6 coating

3. Evaluation and result of the experiment

Rockwell, Brinell or Vickers tests are generally used to assess hardness. The main difference between the above tests is the use of different indentors. The Vickers test was chosen to measure the thermal spray. The test according to Vickers consists of perpendicularly pressing a four-sided diamond pyramid into the material being examined. The pyramid has a given wall angle of 136°. According to the magnitude of the load, Vickers is further divided into hardness and microhardness tests. For microhardness testing, the load values are a maximum of 0.2 kg. Vickers is denoted by the abbreviation HV (Vickers Hardness) and a number indicating the value of the load in kilograms, e.g. $HV_{0.2}$. Vickers is used to measure the hardness of both soft and hard materials and is limited only by the surface being measured, which must be flat and polished. The hardness was measured on the workpiece at two distances (Figure 3). The values are around 700 HV. At a distance of 0.5 mm, there is a significant decrease in hardness (Table 3), due to the transition of the thermal spray to the base material, which is steel 11 373 according to ČSN.

Depth [mm]	0,05	0,1	0,15	0,2	0,25	0,3	0,35	0,4	0,45	0,5	0,55	0,6	0,65	0,7
Hardness $HV_{0,1}$	634	793	736	650	623	627	731	680	741	646	284	182	164	148

Table 3. Measured dependence of hardness on the depth of thermal spray

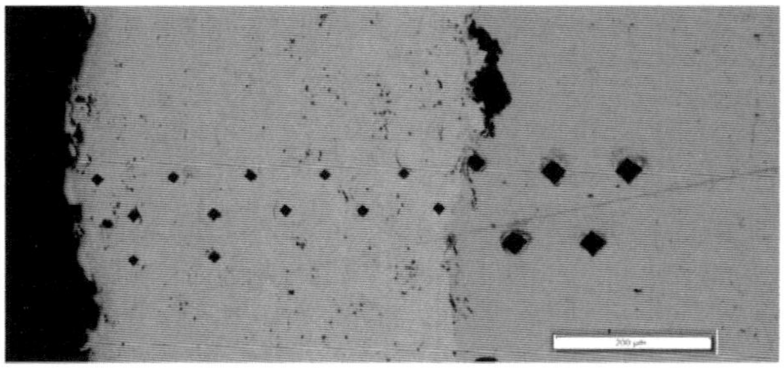

Fig. 3. Vickers hardness test (Olympus BX61, magnification 100x)

Chemical analysis of thermal spray were carried out in order to find out how the chemical composition of the used powder differs from the chemical composition of the already applied coating. This difference is mainly caused by the temperature during the injection process, which causes certain structural changes due to the high temperature. Sometimes there may even be partial or complete evaporation of some elements. Unfortunately, the chemical composition of the applied powder was not known, so it was decided to compare only the measured chemical composition with the table values of the Stellite 6 spray. The measurement was carried out in several places with very similar chemical composition results. When comparing the measured values from Figure 4 with values from Table 2 for the Stellite 6 alloy, it follows that the chemical composition of the measured coating corresponds to the table values in percentage.

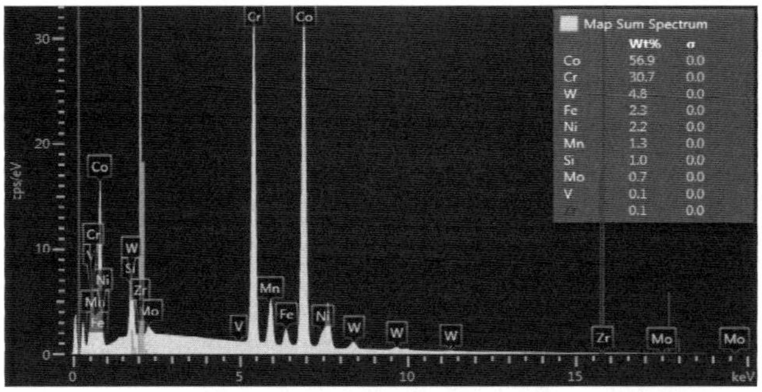

Fig. 4. The measured chemical composition of Stellite 6

The metallographic analysis was carried out from the metallographic specimen. Images are taken with an Olympus BX61 microscope at 50x – 1000x magnification. Light microscopy was performed in the etched and unetched states. 3% Nital was chosen as the etchant. Figure 5 shows the lamellar structure typical pro thermal spraying. There are inclusions and larger impurities at the interface. The coating also contains a certain percentage of pores and probably oxidized parts, which, however, cannot be determined using a light microscope. The thickness of the thermal spray is declared by the manufacturer to be in the range of 0.4 – 0.5 mm. After evaluation, the thickness in the centre of the cross-section is in the range of 0.485 – 0.532 mm.

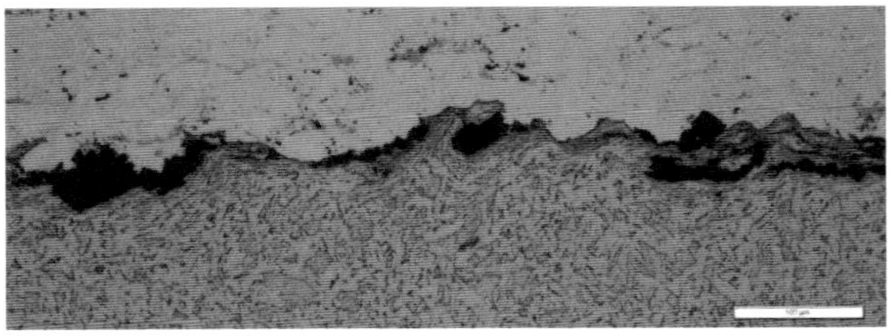

Fig. 5. Light microscopy (Olympus BX61, etched, magnification 200x)

The roughness of the unprocessed thermal spray is given around Ra = 6 μm. This value is influenced by the structure itself, which consists of fused or partially fused deformed particles (splats). For most mechanical applications, the sprayed coating needs to be machined to reduce its surface roughness. In our case, profile roughness was measured on a Hommel Etamic T8000 device. According to the DIN EN ISO 4288:1998 standard, a tip with a size of 2 μm was selected. The measurement took place at 7 places of the semi-finished product with a constant measured length of 4.8 mm. The experimental plan was set so that each cut was made by an insert with an unworn cutting edge. In this way, the effect of the wear of the cutting tool on the quality of the machined coating was eliminated. At the same time, the type and amount of wear was investigated depending on the changed parameters, because the results can be helpful in continuing to solve this issue. After the cut, the plates were marked with the same number as the numbered workpiece. The wear of these plates was measured on a Multicheck PC500 optical microscope, during which the cutting edge, the size of the notch and the increase were examined. The achieved results from the experiment are shown in the following Graph 1.

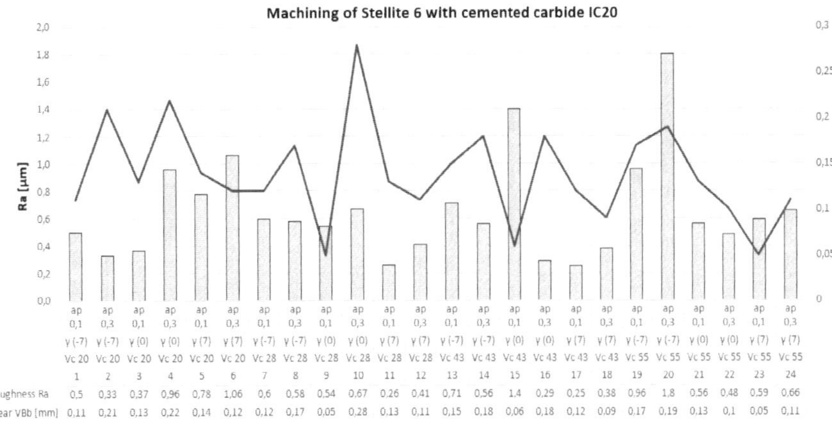

Graph 1. Dependence of cutting conditions on the roughness of the machined surface

4. Conclusion

The main objective of this research was to achieve the same or better surface quality of Stellite 6 thermal spray, which was machined with a defined cutting edge geometry when compared to grinding technology. The main idea behind this change is to shorten production times and thereby increase productivity during thermal spray machining. In the theoretical part, the conditions of the experiment are defined, such as the cutting conditions, the machine and the tool.

The experiment is followed by the evaluation section, where in the first part a chemical analysis of the machined thermal spray was carried out, during which the microhardness, metallography, chemical analysis and thickness of the thermal spray were evaluated. The highest measured microhardness was found to be $HV_{0.1}$ = 623, which is three times higher than steel 11 373 according to ČSN. When comparing the results of the chemical composition with the table values, we did not notice significant differences. After evaluation, the thickness in the centre of the cross-section is in the range of 0.485 – 0.532 mm. At the interface, there are several inclusions and impurities in the metallographic specimen. The coating also contains a certain percentage of pores and probably oxidized particles, which, however, cannot be clearly determined using a light microscope. These impurities and pores can have a major effect on the service life.

From the obtained machining results, a better surface roughness Ra < 0.4 μm was achieved for several samples. The depth of cut did not depend on the achieved roughness or the amount of wear. For each variant of the cutting speed, the setting of the face angle had the greatest effect and is very different for each variant. The best result appears to be sample 18 with positive face geometry $\gamma = 7°$, where the roughness value Ra = 0.38 μm and the size of the back wear VB_b = 0.09 mm. From the wear results, it can be seen that uncoated cemented carbide is not suitable for machining these hard coatings. To continue the research, it is necessary to choose harder insert materials and follow up on the results from this experiment.

5. Acknowledgments

The article contribution has been prepared under project SGS-2022-007 - Research and Development for Innovation in Engineering Technology - Machining Technology IV.

6. References

[1] Hoesel, D.; Wertheim, R.; Haenel, T. & Thurner, S. (2011). Milling of Cast Material and PTA-Hard-Facing Layers of the Cobalt-Base-Alloy Stellite 12, Annals of DAAAM for 2011 & Proceedings of the 22th International DAAAM Symposium, Published by DAAAM International, Vienna, 2011, ISBN 978-3-901509-83-4

[2] Ozturk, S. (2014) Machinability of Stellite-6 Coatings with Ceramic Inserts and Tungsten Carbide Tools. Arabian Journal for Science and Engineering DOI: 10.1007/s13369-014-1343-9.

[3] www.stellite.com, (2019). Kennametal Stellite. Hardfacing Alloys. Accessed on: 2022-06-06. Available from: http://www.stellite.com/content/dam/kennametal/stellite/literature/Brochures/B-16-05107_KMT_Stellite_Hardfacing Alloys _Brochure_ EN_LR.pdf

[4] Farsky, J.; Baksa, T. & Zetek, M. (2018).Influence of Grinding Parameters on Forces when Grinding Maraging Steel X3Nicomo 18 9 5 with SiC Grinding Wheels; Proceedings of the 29th International DAAAM Symposium 2018. DAAAM International Vienna, 2018, s. 0767-0771. DOI: 10.2507/29th.daaam.proceedings.111.

[5] Cesanek, Z. (2015). Thermally Sprayed Coatings Used in Corrosive Environments, Ph.D. Dissertation, ZCU, FST, Pilsen, Czech republic

[6] www.exocor.com (2017). Technical publications. Accessed on: 2022-06-06. Available from: http://exocor.com/downloads/product-datasheets/Stellite-6-Datasheet.pdf

[7] Moravek, M. (2016). The influence of the input parameters of the micro and macro structure of the selected type of coating on its machinability with tools with a defined cutting edge geometry, Thesis, ZCU, FST, Pilsen, Czech rep.

[8] www.mmspektrum.com, (2005). MM Industrial Spectrum. Innovation – Surface treatment. Accessed on: 2022-06-06. Available from: http://www.mmspektrum.com /clanek/zarove-strikane-povlaky-moderni-technologie-tvorby-povrchovych-ochra.html

[9] Hajdarevic, D. B.; Cekic, A. & Kulenovic, M. (2014). Experimental Study on the High Speed Machining of Hardened Steel, Annals of DAAAM for 2014 & Proceedings of the 25th International DAAAM Symposium, Published by DAAAM International, Vienna, 2014, pp. 291 – 295.

DOI: 10.2507/33rd.daaam.proceedings.043

APPLICATION OF LINEAR CUTTING EDGE TECHNOLOGY FOR MACHINING NICRBSI THERMAL SPRAY

Tomas Zatloukal

This Publication has to be referred as: Zatloukal, T[omas] (2022). Application of Linear Cutting Edge Technology for Machining NiCrBSi Thermal Spray, Proceedings of the 33rd DAAAM International Symposium, pp.0313-0318, B. Katalinic (Ed.), Published by DAAAM International, ISBN 978-3-902734-36-5, ISSN 1726-9679, Vienna, Austria
DOI: 10.2507/33rd.daaam.proceedings.043

Abstract

New trends in the field of mechanical engineering place ever greater demands on the properties of machine components, especially in the field of material engineering. With this fact, the application of special types of applications, such as thermal spraying, which increases resistance to mechanical wear, corrosion and oxidation, is growing. The functionality of the sprayed surface is not suitable due to the high roughness of the surface, so subsequent machining is needed. For very hard coatings such as NiCrBSi, grinding technology is used to reduce the roughness. This article deals with the idea of machining this unique material using turning technology. The machining process use a linear cutting edge under specified cutting conditions was analysis. Square IC20 carbide inserts and special tool were used for machining. The result of the experiment was the knowledge of the non-standard machining process depending on the roughness of the surface and the amount of insert wear.

Keywords: Machining; Thermal Spray; NiCrBSi; Linear Cutting Edge; HVOF

1. Introduction

The uniqueness of thermal spraying technology lies in the ability to apply coatings to all structural materials and thus increase the functional properties of the coated parts, in the order of hundreds of percent compared to classic surface treatment technologies. There is a continuous expansion of the possibilities of technology applications in the entire range of engineering production. In many cases, it is the only technology for producing, repairing or renovating special parts. Increasing work productivity in this area is of indisputable economic importance, as it is mainly about applications for expensive machine parts with a high added value. This can be achieved by reducing the thickness of the coating, while guaranteeing the properties of the functional surfaces of the parts and/or by more efficiently machining the coatings to final dimensions with defined surface quality. [3] [6] [10]

Applied highly resistant coatings are generally very difficult to machine, as a rule only grinding with special discs is used. Machining productivity is therefore low and, on the contrary, production costs are high. Therefore, it is necessary to investigate the possibilities of their machinability and to increase the cutting capacity of tools, especially with more productive approaches and technologies. This also requires optimizing the preparation parameters of these coatings, considering the increased homogeneity of the coating structure and thereby achieving higher integrity of the machined surface and reliability of the cutting process. [3] [6] [10]

Thermal spraying technology is, from today's point of view, a very important technology affecting several industries. Thermal sprays are most often divided according to the heat energy source used. From this point of view, today there are many variants that create specific coatings with regard to the required functionality and the difficulty of production. From a historical point of view, thermal spraying technology cannot be considered a novelty. The first to come up with this technology was Dr. Ing. Max Ulrich Schoop as early as 1909. The idea itself arose quite by accident that same year when looking at a lead shot that had been fired against a wall. The lead shot flattened not only due to the high kinetic energy when it hit the wall, but also stuck to the mentioned wall due to plastic deformation. In the same year, the first experiment in the form of this technology was tried, simply by pouring melt from a casting pan in front of a nozzle generating compressed air. After filing a patent for the newly discovered technology, Dr. Schoop didn't get much credit and the whole process was likened to applying melted chocolate. [1] [3] [6]

The very principle of creating a thermal spray is described in a simplified manner in the previous text. It can be understood as the application of a continuous layer of additional material on the pre-prepared surface of the sprayed part. In the first point, it is necessary to have a suitable additive material with a certain chemical composition. This material is fed into a special device, where the individual particles of the additive material are melted. The melted particles are subsequently accelerated by the air stream and directed toward the surface by the exit nozzle. After the impact on the base material, due to the high kinetic energy, the grains are deformed and create a continuous surface with a typical lamellar structure. [1] [3] [4]

The structure of the created thermal spray is very specific in terms of construction and appearance. It follows from the very principle of the technology that the melted grains of the new material are significantly deformed into individual disk-shaped lamellae after impacting the base material. Grains deformed in this way are called "splats" in the field of thermal spraying. Due to the wide range of process parameters and the many methods of thermal spraying, it is logical that the resulting structure will not always be the same in terms of appearance and quality. The graphic appearance of the structure of the thermal spray can be seen in Figure 1. In addition to the fused disc-shaped grains, there are also grains in the structure that were not completely fused. These grains are only partially deformed after hitting the surface and are therefore more globular in nature. Due to the purity of the additive material and the application process, there are also various inclusions and impurities in the structure. Due to the technology, the coating also contains a certain percentage of porosity. All these imperfections have a major impact on the quality of the coating. [3] [4] [7] [9]

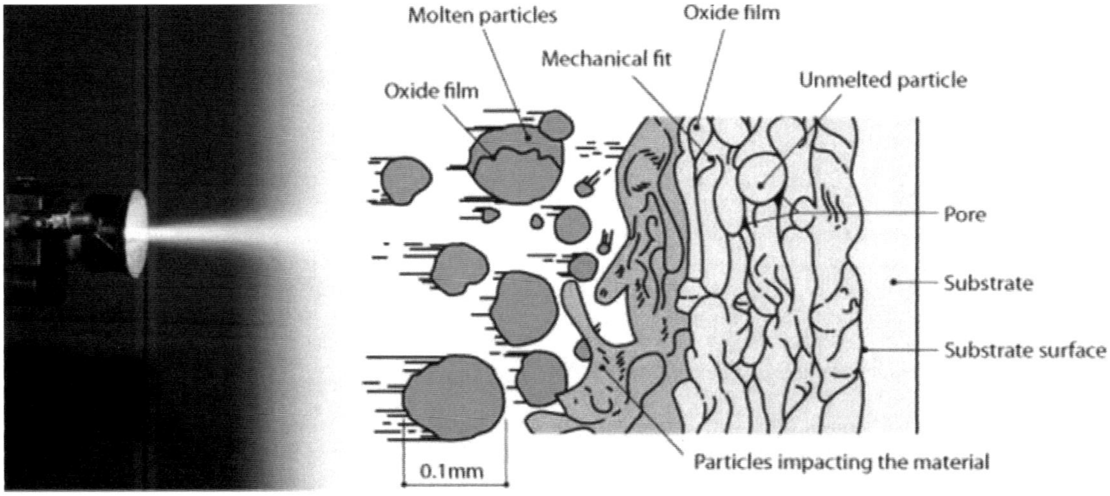

Fig. 1. Description of the HVOF spray application principle [9]

The technology of this coating is referred to in the English literature as HVOF (High-Velocity Oxygen Fuel). This method can be described as the most modern way of creating thermal spray. HVOF technology follows on from the detonation method of spraying, with the difference that the HVOF process takes place continuously. In the combustion chamber, a mixture of fuel and oxygen is burned, which is supplied continuously, and this ratio has a major influence on the melting temperature, which ranges from 2800 to 5000 °C. The most commonly used burning fuel is kerosene, but propylene, acetylene or hydrogen are also used to a lesser extent. [3] [4] [7]

In the combustion chamber, after the fuel is mixed with oxygen, ignition occurs with the help of a candle. During combustion, the additional material is melted in the form of a powder, which is brought to the combustion chamber by a carrier gas, usually nitrogen. The molten particles are subsequently accelerated in a special nozzle, which is convergent-divergent in shape. The molten particles are accelerated up to supersonic values reaching over 1000 m/s. To ensure a stable application process, the temperature and pressure in the combustion chamber are monitored on the one hand, and the entire structure is also equipped with internal cooling for better process regulation, see Figure 2. [3] [4] [7]

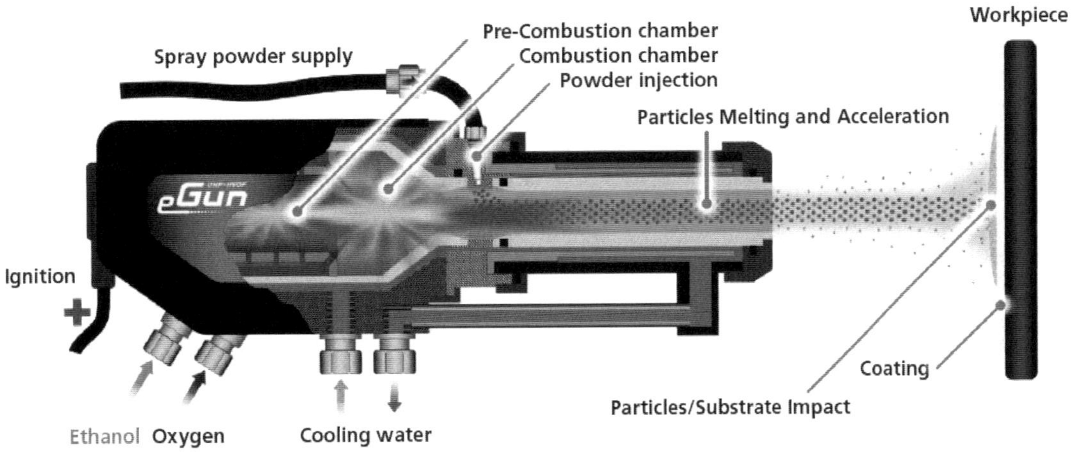

Fig. 2. Scheme of equipment for creating a coating by HVOF technology [4]

Compared to other thermal spraying technologies, HVOF achieves the highest speed of molten parts. Mainly for this reason, a very high-quality coating with adhesion reaching up to 90 MPa is created. Also, in terms of structure, the created coating is of very high quality with a low oxide content and a small porosity of up to 0.5%. However, when compared to spraying using plasma, the HVOF method has a lower melting temperature, therefore it is more suitable for materials with a medium and low melting temperature, such as nickel, chromium, cobalt, tungsten carbide, etc. The advantage of a low melting temperature is, on the contrary, the preservation of the original properties of the additive material, as there are no significant phase changes here. The HVOF method is mainly used in conjunction with the Stellite material (an alloy based on Co and Cr) and the NiCrBSi thermal spray we investigated. [2] [7] [10]

2. Materials used and methodology of the experiment

Additive material NiCrBSi in the form of powder belongs to the cheaper group of metal additive powder materials. Equipment from the American company was used to optimize the spraying process. Praxair. This device belongs to the systems in which the range of setting individual technological parameters is very detailed, therefore the implementation of complex optimization is very important. The only parameter on the burner is the possibility of using barrels of different lengths for different additional materials. The selected technological parameters were: the amount of oxygen, the amount of kerosene, the flow of transport gas, the amount of additional material, the peripheral and transverse speed of the substrate and the distance of spraying. The results achieved in the individual stages confirmed the predicted properties, and by combining the results achieved in the individual stages, coatings were created with properties meeting the standards stated in the foreign professional literature. The coating created from this additive material is significantly more homogeneous in terms of structure and exhibits significantly better structural characteristics and mechanical, physical and chemical properties compared to coatings of the same chemical composition created by all remaining methods. Significantly better mechanical properties were achieved by optimizing technological and movement parameters, see Table 1. [2] [5] [8]

Structural cracks	0
Delamination of the structure	0
Deadhesion	0
Contamination of the interfacial interface	0,50 %
Measurement of the thickness of the functional coating / interlayer	540 μm
Porosity measurement	do 0,25 %
Hardness measurement	$HV_{0,3}$ 641,4
Adhesion measurement	53,65 MPa
Process efficiency	57,80 %

Table 1. Measured properties of HVOF sprayed NiCrBSi coating [11]

The NiCrBSi thermal spray machining test was carried out on the CTX Beta 1250 TC machine, which is located in the laboratories of the Regional Institute of Technology in Pilsen. The first experiment was carried out in order to know the machining process with a linear cutting edge. In particular, the correct position of the exchangeable insert in the section, the correct orientation of the tool and the testing of the program were investigated. The first (trial) experiment was followed by a series of several experiments, which are mainly devoted to learning the NiCrBSi thermal spray machining process. Finally, the best cutting conditions are selected, with which long-term cutting tests are carried out. From long-term tests, the durability value for selected cutting conditions and types of inserts is determined. Durability is determined based on the results of edge wear and surface quality achieved.

For the first pre-experiment, an IC20 uncoated cemented carbide insert was selected for the purpose of stopping in the cut and monitoring the correct position of the cutting edge. The first idea is to find the correct position of the cutting edge and use the entire length of the cutting edge. The Y axis is used to set the displacement value at the first contact of the plate with the workpiece and at the same time the change of the Y axis after the given distance has been moved. The insert is not focused on the tip of the cutting edge as with conventional tools but on the centre of the cutting edge. The movement of the y-axis during machining is therefore set from a positive value to a negative value ($y = -3$; $y = +3$). When setting a positive y-axis correction, the tool moves relative to the workpiece axis more to the tip of the insert, on the contrary, with a negative value, towards the bottom of the insert. With this setting, it is possible to use the entire length of the cutting edge in one pass with uniform wear on the cutting edge. Figure 3 shows the position of the y-axis during initial contact with the workpiece and the position of the cutting edge after a defined distance of 100 mm.

Fig. 3. Test machining of special tool with IC20 insert

3. Evaluation and result of the experiment

According to the previous pre-experiment, the cutting conditions for machining an uncoated carbide insert with the designation SCMW 120408 were selected. The material to be machined, with its composition and properties, is classified as a difficult-to-machine material, therefore worse wear and surface quality results can be expected. The aim of this experiment is to test the fundamental behaviour of the tool with different variants of cooling and movement of the Y axis at a defined distance of 100 mm see Figure 4. Cutting speed, feed, depth of cut and blade angle settings are constant. From the conducted experiment, a table was created with a description of the cutting conditions, the achieved surface roughness, the resulting wear on the VBD edge and the overall appearance of the machined surface. A Keyence VHX 6000 optical microscope was used to measure and evaluate the resulting wear. The wear of the inserts was measured on the face (KB value) and the back (V_{Bmax} value). A Marsurf M300 roughness meter was used to measure the roughness.

Fig. 4. Workpiece with a machined layer of NiCrBSi coating

From the results achieved (Table 2) when machining with a cemented carbide insert, it appears at first glance that machining with this material clearly does not meet the defined requirements for the machined surface. During the experiment, the effect of stiffness during machining and the effect of cooling on the resulting machined surface was investigated. From the experiment, some basic information about this method of machining was found. It can be seen from the table that the inclusion of the movement of the Y axis during machining has a negative effect on the appearance of the machined surface due to large vibrations. During machining, vibrations did not occur along the entire machined length, but only in approximately 1/3 of the surface. When machining sample 2.1, the appearance of the surface was unified without audible vibrations, but at the same time, the roughness of the machined surface Ra and Rz turned out to be significantly worse than in the other variants. Of course, worse roughness Ra and Rz were also found in samples with Y-axis movement. On the other hand, the uniform appearance of the surface came out well when machining with a constant value of Y. The best-achieved roughness Ra = 1.97 μm was obtained for sample 7.1 on surface number 1 using cooling. However, when comparing sample number 1 and 5, it follows that the inclusion of liquid during machining does not have a significant effect on the roughness or appearance of the surface.

Insert	No.	Roughness	Ra [μm]	Edge wear [mm]		Cutting conditions						Surface appearance
						Vc [m/min]	f [mm/ot]	ap[mm]	λ_s	Y axis	Fluid	
SK IC20	1.1	Ra [μm]	2,34	VBmax	0,27	80	0,6	0,15	70°	cons.	no	Homogeneous
		Rz [μm]	12,80	KB	0,24							Without vibration
SK IC20	2.1	Ra [μm]	**2,94**	VBmax	0,26	80	0,6	0,15	70°	move	yes	Homogeneous
		Rz [μm]	**16,63**	KB	0,26							Without vibration
SK IC20	3.1	Ra [μm]	2,48	VBmax	0,28	80	0,6	0,15	70°	move	no	**Heterogeneous**
		Rz [μm]	13,66	KB	0,25							**Vibration**
SK IC20	4.1	Ra [μm]	**3,00**	VBmax	0,16	80	0,6	0,15	70°	move	air	**Heterogeneous**
		Rz [μm]	**15,41**	KB	0,23							**Vibration**
SK IC20	5.1	Ra [μm]	2,32	VBmax	0,29	80	0,6	0,15	70°	cons.	yes	Homogeneous
		Rz [μm]	14,29	KB	0,22							Without vibration
SK IC20	6.1	Ra [μm]	2,26	VBmax	0,29	80	0,6	0,15	70°	cons.	yes	Homogeneous
		Rz [μm]	13,00	KB	0,24							Without vibration
SK IC20	7.1	Ra [μm]	**1,97**	VBmax	0,11	80	0,6	0,15	70°	cons.	yes	Homogeneous
		Rz [μm]	**11,41**	KB	0,25							Without vibration

Table 2. Summary and measured results during spray machining with the IC20 insert

4. Conclusion

From today's point of view, technology using thermal spraying is a very important technology affecting many important industries. Thermal sprays are basically divided according to the selected heat energy source. The technology referred to as HVOF (High-Velocity Oxygen Fuel) can be described as the most modern way of creating coating. The applied structure of the thermal spray is very specific in terms of construction and appearance. It follows from the principle of creation that the molten grains of the sprayed material is significantly deformed into specific disk-shaped lamellae after hitting the base material.

The NiCrBSi thermal spray machining test was carried out on the CTX Beta 1250 TC machine, which is located in the indoor laboratories of RTI at the University of West Bohemia. The trial experiment was conducted to obtain basic information about the linear cutting-edge machining process. The correct position of the exchangeable insert in the section, the correct orientation of the tool and the testing of the program were examined. According to this experiment, the cutting conditions were selected for machining an uncoated cemented carbide insert marked SCMW 120408.

The aim of the main experiment was to test the basic behaviour of the tool during the variation of cutting conditions, cooling and movement of the Y axis when crossing a surface with a length of 100 mm. The achieved surface roughness, the occurrence and extent of wear on the cutting edge and the resulting appearance of the machined surface were monitored. The cutting speed, feed, depth of cut and blade angle setting were set constant. A table 2 was created from the experiment with a description of the cutting conditions and the results achieved.

The best-measured surface roughness Ra = 1.97 μm was achieved for sample 7.1 on surface number 1 using cooling. However, when comparing sample number 1 and 5, it follows that the use of process fluid during machining does not have a significant effect on the quality or appearance of the surface. From the measured wear results, a strong abrasion on the edge was found in a very short time. The results revealed basic information about the machining process with the conclusion to continue this issue with more progressive types of cutting material and coatings.

5. Acknowledgments

The article contribution has been prepared under project SGS-2022-007 - Research and Development for Innovation in Engineering Technology - Machining Technology IV.

6. References

[1] Seigmann, S. & Abert, CH. (2020). About the inventor Max Ulrich Schoop. Surface and Coatings Technology 220:3-13, 100 years of thermal spray, DOI: 10.1016/j.surfcoat.2012.10.034

[2] Povolny, M.; Zatloukal, T. & Rehor, J. (2021). Construction of Tool For Machining Hard Materials, Annals of DAAAM for 2021 & Proceedings of the 32nd International DAAAM Symposium, Published by DAAAM International, Vienna, 2021, ISBN 978-3-902734-33-4

[3] https://www.triboing.cz/zarove-nastriky.html, (2018). Thermal sprays method, Tribo Engineering, Accessed: 2022-09-09

[4] https://www.ep-coatings.us/processes/hvof/ (2021). Thermal spraying HVOF. Engineered Performance Coatings, Accessed: 2022-09-09

[5] Hajdarevic, D. B.; Cekic, A. & Kulenovic, M. (2014). Experimental Study on the High Speed Machining of Hardened Steel, Annals of DAAAM for 2014 & Proceedings of the 25th International DAAAM Symposium, Published by DAAAM International, Vienna, 2014, pp. 291 – 295.

[6] https://www.plasmametal.cz/pouzite-technologie, (2019). Thermal spraying HVOF. Plasmametal, Accessed: 2022-09-09

[7] Cesanek, Z. (2015). Thermally Sprayed Coatings Used in Corrosive Environments, Ph.D. Dissertation, ZCU, FST, Pilsen, Czech republic

[8] Sykora, J.; Povolny, M.; Kroft, L. & Syrovatka, S. (2021). Study of The Influence of Cutting Tool Geometry During Turning of Submillimetre Nickel Parts, Annals of DAAAM for 2021 & Proceedings of the 32nd International DAAAM Symposium, Published by DAAAM International, Vienna, 2021, ISBN 978-3-902734-33-4

[9] Verdian, M. M. (2017). Finishing and Post-Treatment of Thermal Spray Coatings, Comprehensive Materials Finishing, Published by Elsevier, Iran, 2017, pp. 191 – 206. ISBN 9780128032497

[10] Zatloukal, T.; Rehor, J.; Fulemova, J. & Povolny, M. (2020). Optimization of The Machining Process of NiCrBSi Thermal Spraying, Annals of DAAAM for 2020 & Proceedings of the 31st International DAAAM Symposium, Published by DAAAM International, Vienna, 2020, ISBN 978-3-902734- 29-7

[11] REHOR, J. Theory of Machining. ZCU in Pilsen [Internal documentation], [cit. 2022-09-09]

DOI: 10.2507/33rd.daaam.proceedings.044

VIBRATION MEASUREMENT DURING MILLING BY ACOUSTIC EMISSION

Michal Povolny, Lubos Kroft & Jindrich Sykora

This Publication has to be referred as: Povolny, M[ichal]; Kroft, L[ubos] & Sykora, J[indrich] (2022). Vibration Measurement During Milling by Acoustic Emission, Proceedings of the 33rd DAAAM International Symposium, pp.0319-0325, B. Katalinic (Ed.), Published by DAAAM International, ISBN 978-3-902734-36-5, ISSN 1726-9679, Vienna, Austria
DOI: 10.2507/33rd.daaam.proceedings.044

Abstract

Production processes are nowadays very much challenged by the continuous improvement of productivity, the reduction of energy and ecological burden and the pursuit of the highest possible economy of the whole process in general. In order to optimize any process, it is essential that the process quality can be measured, and its behaviour adjusted ideally "just in time". In the case of machining, vibration is a major problem. This phenomenon, which always accompanies machining and especially milling, has an essentially negative impact. The aim of this article is to present the possibility of measuring vibration both with special active tool grooves from SCHUNK and indirectly with the acoustic emission sensing method. The article also deals with the nature of vibration itself as well as the comparison of both methods and the validation of the results.

Keywords: Vibration; Vibration measurement; Acoustic emissions; Machining

1. Introduction

Nowadays, the machining industry is focusing on the use of multifunctional machining centres, where, together with a suitable monitoring system, it is possible to achieve a significant improvement in production efficiency. Machining processes are associated with deformation of the cutting tool and workpiece, which is caused by cutting forces, thermal effects and vibrations. Modern monitoring systems are used to monitor the measured characteristics of the machining process, collect and evaluate the data. This enables the machining process to be controlled retrospectively by adjusting the cutting conditions, e.g. cutting speed, feed rate or cooling medium supply. The monitoring system must therefore identify the actual cutting conditions during the production process via sensors. Reliable sensing systems coupled with signal processing techniques are required to monitor processes and extract useful information from the machining process. Based on the information obtained from the sensors, the adaptive control system must be able to handle a wide range of modifications to the machining process, from simple stopping of machining to advanced control. Vibration measurement is now quite common and can also be transferred to the machines. However, the whole device is very costly and requires specialized software and hardware. One simple method of measuring phenomena in machining, but also in other processes, is acoustic emission. This signal can be sensed relatively easily, even at greater distances from the cutting point, and it is possible to measure certain behaviours of the system with this signal.

The aim of this paper is to give answers to whether it would be possible to use the acoustic emission signal to evaluate vibrations during the machining process. Then to compare these signals with a commonly available active tool clamp which is used to measure vibration. This should make this monitoring of the machining process simpler and cheaper in the future. [2] [3] [10].

2. Acoustic emissions in the machining process

Acoustic emission (AE) is a physical phenomenon in which stored energy is released in a material in the form of transient stress waves due to dynamic processes. These dynamic processes are stimulated by internal or external forces. The waves propagate in the material from the source towards the surface. Due to the passage of the stress wave, part of the energy is converted into heat, part of the energy is returned to the material and part of the energy induces a so-called Rayleigh wave at the surface. The waves that are on the surface are then suitable for capture by a sensor and converted into an electrical signal, which we call an emission signal. The resulting amplitude is on the order of micrometres. The waves generated by acoustic emission range in frequency from the lowest (0,1-100 Hz) to high frequencies (up to 10 MHz), which can be observed in lattice faults or fracture propagation due to mechanical breaking. Figure 1 shows the principle of measuring and evaluating acoustic emission [2] [3] [7].

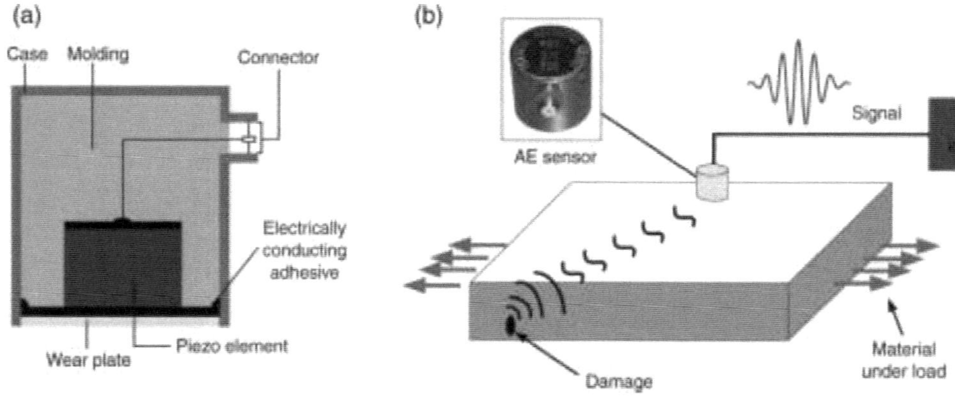

Fig. 1. The principle of acoustic emission [1]

In the machining process, AE is mainly used for comprehensive tool wear monitoring. Sometimes as a warning of complete tool destruction by micro cracks in the tool that cause AE signals. This is used for quality monitoring of tool wear or process control. Tool wear changes the contact area between the tool and the workpiece and thus the intensity of AE. When tool wear or catastrophic tool failure occurs, the AE signal changes during cutting or, in the case of tool breakage, shows dramatic peaks from the energy released during breakage [2] [3] [8].

The benefits of acoustic emission include:

- High sensitivity
- Early detection of defects, cracks, breaks, etc.
- Real-time monitoring
- Cost reduction
- No need to scan the entire structural surface.

The disadvantage of acoustic emission is that we do not know the exact cause of the acoustic wave, as the energy released is influenced by several factors such as the shape and surface of the body, the transmission path of the wave (acoustic wave propagation function) given by the structure and homogeneity of the material [2] [3] [8]. The AE signal obtained usually characterizes the number of memory waves in a certain amplitude range released per unit time. AE signals are distinguished based on their sources into continuous (continuous) and impulsive (discontinuous). Continuous signals are associated with the area in the primary region where chip shearing occurs and then at the wear face and tool back. In contrast, impulsive signals are caused by cracks in the material, chip breakage or impact, or breakage of the tool edge [2] [8].

2.1. Sources of AE in the machining process

During machining, acoustic emission (AE) is mainly caused by plastic-mechanical processes in the form of cracks, chip evacuation and surface friction. These sources are illustrated in the following figure [3] [4].

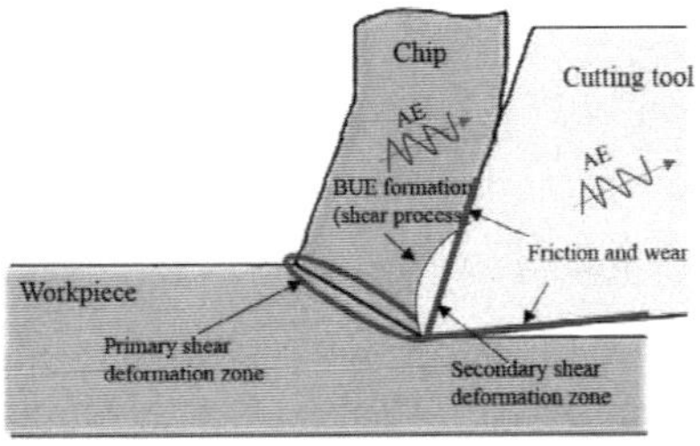

Fig. 2. AE sources in the Machining Process [11]

Each of these sources is manifested in a specific frequency band and has a specific intensity. The AE signal arising from plastic deformation is continuous in nature, while in other cases the signal is discontinuous and more energetic. One of the phenomena that limits the use of AE is the attenuation properties of the material. This means that the same phenomenon in different materials can result in different character of acoustic emission. It is therefore necessary to make a sufficient number of comparative measurements for the accuracy of the measurements to be meaningful. AE is therefore particularly suitable for series production where a large amount of measurement data can be used to determine exactly what a given AE characteristic means. Another disadvantage is also the scattering and reflection of the voltage waves, which is given by the size of the measured body [3] [4].

2.2. Acoustic emission measurement

Multiple sensors and multi-channel measurement systems are used to measure acoustic emissions. The acoustic emission sensors, which are distributed on the structure, form a measurement network. This network allows the location of the origin of the emission activity to be located. The localisation is based on the time difference of the arrival of elastic waves of acoustic emission at the different sensors of the network [5].

Linear localization – the arrival of elastic waves at two sensors is used to determine the location of the emission event. The source lies at the junction of these sensors [5].

Area localization – requires the arrival of elastic waves to at least three sensors. The location of the emission event can be determined if we know the time differences of the arrival of the waves at each sensor and the speed of propagation of the acoustic waves [5].

The design of the sensors is adapted to their use, so there are many types and types of sensors. In practice, piezoelectric sensors are the most used. In the design of piezoelectric sensors, the piezoelectric effect is used, where inside some crystalline dielectrics, due to mechanical deformations, an electrical polarization is created, which causes apparent charges on the surface, which can bind or release real charges in the attached electrodes. If the mechanical stresses disappear, the dielectric returns to its original state [6][10].

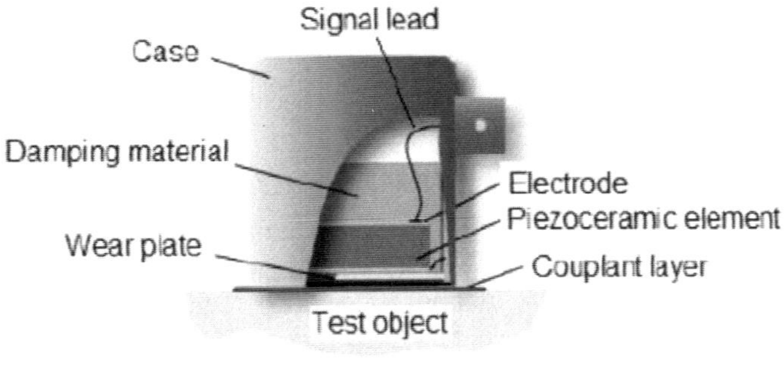

Fig. 3. Piezoelectric AE sensor [6]

Acoustic emission sensors can be divided into:

- according to the frequency range of the measured signals:
- low frequency sensors 0 Hz ... 1 kHz
- mid-frequency sensors 1 kHz ... 100 kHz
- acoustic emission sensors 100 kHz ... 2 000 kHz

According to the permissible operating temperatures of the sensors:
- standard temperature range
- extended operating temperature range
- high temperature sensors

According to the connection method to the measured object:
- direct bonding of the sensor to the surface
- by magnetic holder
- screw connection (external or internal thread on the sensor)
- by means of a welded rod waveguide
- belt clip (on tubes)

Due to the nature of the AE monitoring process, it is necessary to place the sensors as close as possible to the machining point. For operations such as turning, where the main rotational motion is performed by the workpiece, it is necessary to place the AE sensors directly on the tool body or tool clamp. In contrast, for operations where the tool performs the main rotary motion (e.g., milling, drilling), the encoders are placed on the workpiece and the toolholder. The sensors must therefore be resistant to coolant, dirt, flying chips as well as mechanical, electromagnetic and thermal effects. They must not restrict the working space and cutting parameters [5] [7] [8].

2.3. AE signal modification and evaluation

When measuring AE under industrial conditions, acoustic noise/noise present at lower frequencies (below 30-50 kHz) must also be considered. It is therefore necessary to ensure that it is filtered out. This can be done in two ways. Either directly at the time of measurement by appropriate sensor selection, using a filter on the preamplifier and on the AE measuring unit, or by digital processing and filtering of the measured data [3] [4] [6]. The aim of acoustic emission evaluation is to detect the source of these emissions during the measurement. The basic principle is to evaluate the acoustic sources when loading the measured objects at or above the nominal load value, finding and characterizing the impulses, their temporal localization, oscillation frequency, amplitude, and phase. This information is indicative of the sudden emission of elastic energy generated within the material. Obtaining such physical parameters from AE signals is one of the most common problems in its processing. This is since these signals are non-stationary and often contain overlapping transitions, and their waveforms and the time at which they occur are unknown and involve variations in terms of time and frequency. Often such events are partially overlapping in time or are subject to noise [3] [4] [6] [9]

The basic parameters in acoustic signal evaluation include: [9]
- *Reference count*: the number of times the amplitude of the signal crosses a threshold line,
- *AE events*: the micro-structural displacement that leads to the generation of pro-elastic waves in the material under load and stress,
- *Rise time*: the time taken to reach the maximum amplitude from the first time the signal crosses the boundary,
- *RMS voltage*: the voltage that defines the intensity of the AE signal

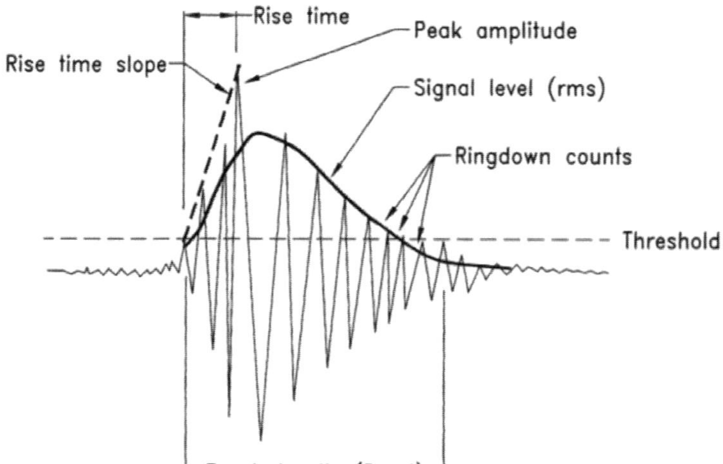

Fig. 4. Characteristics of AE signal [9]

3. Experiment

Acoustic emission has interesting potential for measuring various processes during the machining process. For this reason, an experiment was carried out to experimentally verify the possibility of measuring vibrations in the machining process using AE. This experiment was carried out on the milling of aluminium alloy AW EN 7075, where only sideways machining of the tool occurred. In order to evaluate whether the AE signal could be used in any way for vibration measurement, it was necessary to perform a retention measurement with an already validated device. For this reason, an active toolholder with integrated axis vibration measurement from SCHUNK, called iTENDO^2, was used. This fixture allows vibration to be measured directly on the tool, during the machining process. This signal is transmitted to a display device where the signal can be evaluated and processed retrospectively.

Fig. 5. Setting up the aperture in the machine

Thanks to this, it was possible to compare the acquired signals and evaluate with what accuracy it is possible to monitor vibrations as a very important phenomenon of the machining process using AE. The experiment was carried out under the following conditions, see Table 1.

Sample number	Cutting speed [m/min]	Feed per tooth [mm/tooth]	Sidestep [mm]	Depth of cut [mm]
1	250	0,01	0,5	15
2	450	0,073	5	15
3	425	0,065	0,2	15
4	425	0,065	0,2	15
5	450	0,073	0,2	15

Table 1. Conditions of experiment

The measurements were carried out according to the scheme under variations of different conditions, as shown in Table 1. In the case of AE, the recording was realized by means of an amplifier and a measuring card into a measuring PC and evaluated directly there. In the case of the active iTENDO^2 fixture, the recording was transmitted wirelessly to the connected tablet and then exported from there. The data recording from the measuring PC to the active iTENDO^2 fixture can be seen in Figure 10.

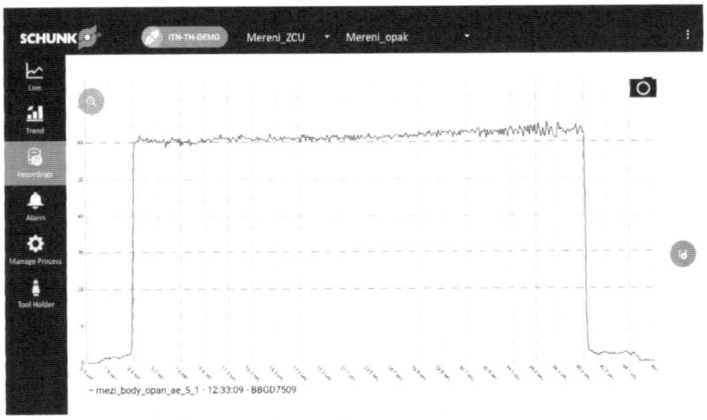

Fig. 6. Data recording from the measuring software from the active tool holder iTENDO^2

The result of this experiment is a significant match between the AE outputs and the vibrations sensed by the iTENDO² active fixture. Although the two outputs are in different units, it is evident that the nature of the data is very similar, both at lower cutting conditions and at higher conditions.

The individual outputs can be seen in the figures below. Here, especially at the beginning of the measurement and at the end, there is still a slight difference in the progress of the two functions, but this difference is due to the different starting point for the two systems. If this effect is filtered out the agreement of the resulting functions is very good.

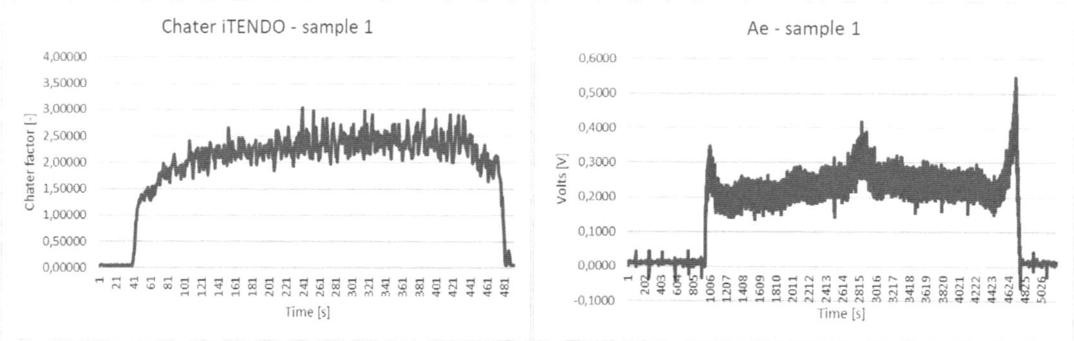

Fig. 7. Comparison of the resulting signal functions – sample 1

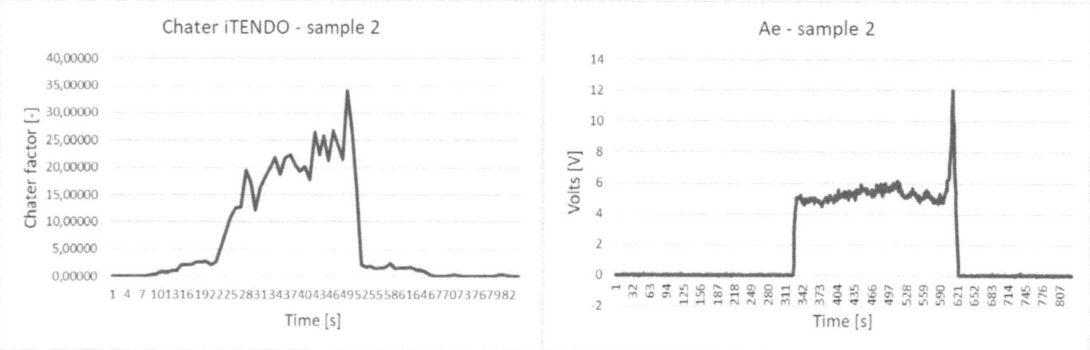

Fig. 8. Comparison of the resulting signal functions – sample 2

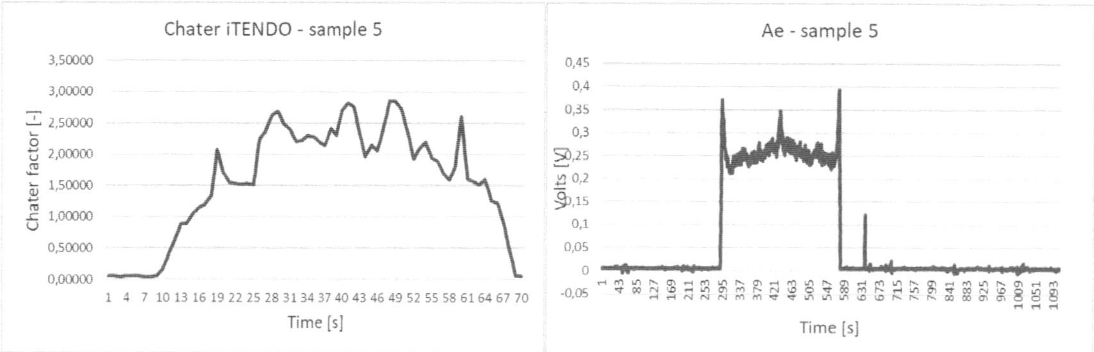

Fig. 9. Comparison of the resulting signal functions – sample 5

4. Conclusion

As a result of the whole experiment, it is found that vibration monitoring by acoustic emission can be used for this type of machining operation. The AE signal corresponds very well with the measured data from the active toolholder.

This experimental result is very interesting both from an academic point of view, where it gives further scope for a much greater validation of the results obtained, both on a variety of operations, and with a much more thorough evaluation methodology, such as DoE. And at the same time, it is a very good result for deployment in a real machining process. Here, the AE measurement system is very suitable both in terms of installation and sensor size, which allows deployment in many practical applications. Here, however, further measurements are first required, especially with respect to vibration attenuation versus the position of the encoder in the machine space and other variables.

Overall, the whole concept can be evaluated very positively as it has fulfilled expectations and at the same time opened further possibilities for research into the subject, which will be continued subsequently. In the future it will be advisable to include other variables monitored during the machining process in the evaluation and to try to find correlations between other phenomena.

5. Acknowledgments

The article contribution has been prepared under project SGS-2022-007 - Research and Development for Innovation in Engineering Technology – Machining Technology IV.

6. References

[1] Wang, M, L.; Lynch, J, P.; Ssohn, H. (ed.). (2014) Sensor technologies for civil infrastructures.

[2] Rubio,E.M.; Teti,R.; Baciu,I.L. (2006) Advanced signal processing in acoustic emission monitoring systems for machining technology. Cardiff University, Manufacturing Engineering Centre

[3] Žižka, J, (2003). Použití akustické emise ke sledování stavu řezného nástroje, Liberrec, Technická univerzita. 106s. ISBN 8070836881

[4] Li, X. (2001) A brief review:acoustic emission method for tool wear monitoring during turning, School of Electric Engineering, Yanshuan University, Qinhuandao, P.R. China,

[5] Introduction to Acoustic Emission Testing [online]. [cit. 2017-02-12]. Available from: http://www.nde-ed.org

[6] Svecko, R.; Kusi´c, D.; Kek, T.; Sarjaš, A.; Han˘ci˘c, A.; Grum, J. (2013) Acoustic Emission Detection of Macro-Cracks on Engraving ToolSteel Inserts during the Injection Molding Cycle Using PZT Sensors. Sensors 2013, 13, 6365–6379

[7] Cekic, A; Muhamedagic, K; Cohodar, M; Begic-Hajdarevic, D & Biogradlija, S. (2019). Experimental Investigation of Effect of Overhang Tool Length on Tool Vibration and Surface Roughness, Proceedings of the 30th DAAAM International Symposium, pp.0184-0191, B. Katalinic (Ed.), Published by DAAAM International, ISBN 978-3-902734-22-8, ISSN 1726-9679, Vienna, Austria DOI: 10.2507/30th.daaam.proceedings.024

[8] Kaufnerova, A& Kutlwaser, J (2019). Measuring Stability of the Cutting Process when Turning Using TXF Metalmax Software, Proceedings of the 30th DAAAM International Symposium, pp.0982-0988, B. Katalinic (Ed.), Published by DAAAM International, ISBN 978-3-902734-22-8, ISSN 1726-9679, Vienna, Austria DOI: 10.2507/30th.daaam.proceedings.136

[9] Jayakumar, T. (2001) NDT Techniques: Acoustic Emission: Acoustic Emission. In: Oxford: Elsevier, s. 6001-6003. ISBN 978-0-08-043152-9.

[10] Ahmed, Y, S.; (2019) Md. Shafiul alam, a. f. m. arif a s. c. veldhuis. Use of acoustic emission and cutting force signals to monitor built-up edge formation in stainless steel turning. The International Journal of Advanced Manufacturing Technology, 103(5), 2257-2276. ISSN 1433-3015.

DOI: 10.2507/33rd.daaam.proceedings.045

INDUSTRIAL CYBER SECURITY ASPECTS IN ICS APPLICATIONS

Mladen Babić, Miloš Stanojević, Gordana Ostojić,
Srđan Tegeltija & Stevan Stankovski

This Publication has to be referred as: Babic, M[laden]; Stanojevic, M[ilos]; Ostojic, G[ordana]; Tegeltija, S[rdjan] & Stankovski, S[tevan] (2022). Industrial Cyber Security Aspects in ICS Applications, Proceedings of the 33rd DAAAM International Symposium, pp.0326-0331, B. Katalinic (Ed.), Published by DAAAM International, ISBN 978-3-902734-36-5, ISSN 1726-9679, Vienna, Austria
DOI: 10.2507/33rd.daaam.proceedings.045

Abstract

Regarding already growing technological progress and the new industrial revolution, Industry 4.0, every day there is a bigger need of using security systems in Industrial Control Systems. Using closed or open architectures, which are connected locally or over the Internet, is much more vulnerable in terms of outside or inside attacks in terms of gathering pieces of information or just sabotage. In this paper, we make a survey of possible concepts which are used to eliminate the possibility of cyber-attacks on Industrial Systems. We will also make a classification of CPS Domains, Attacks, Defenses, Research-trends, Network-security, Security level implementation, and Computational Strategies.

Keywords: CPS; ICS; Security; Vulnerabilities; Industry 4.0;

1. Introduction

In the last decade, the development of various types of Cyber-Physical Systems (CPS) are growing exponentially which is resulting in their wider application. These systems can be found in daily life as a part of transport systems, oil, gas and electrical energy distribution, medicine, household appliances, and a lot of other systems. Many of these systems are used in processes that are endangering humanity and the environment and at the same time in processes that are improving the quality of life. For all these reasons, these systems are considered critical and they must be resilient to failures and all kind of attacks which is proven to be very challenging to achieve.

As there is not yet an official definition for Cyber-Physical Systems (CPS), it can be described as a system that is used for monitoring and controlling the outside world. They are considered a new generation of embedded control systems that are interconnected [1]. Also, CPS can be defined as IT systems, integrated into the physical world applications [2]. More and more industries, such as energy, military, auto industry, and medical are starting to depend on and to leverage CPS. CPS can be named differently based on an area of application. One of the most known CPS is Supervisory Control and Data Acquisition (SCADA) systems, which are used for example in Smart Grid and Industrial Control Systems (ICS). In the wide spectrum of systems in which CPS are present, in this paper, the accent will be on Industrial Control Systems (ICS). This paper contains an analysis of potential attacks on these systems, systems vulnerabilities, threat mitigation, and the proposal for securing these systems.

2. Industrial Control System

The main responsibility of Industrial Control Systems (ICS) is the acquisition, control, and monitoring of several types of industries such as the auto industry, pharmaceutical, nuclear power plants, water systems, sewerage, and many others. ICS is composed of several control units with different capabilities that work together to achieve a common goal. The most used control unit is the Programmable Logic Controller (PLC). PLC is nothing more than a microprocessor designed for continuous operation in difficult working conditions [3]. This controller is connected to the real world through sensors and actuators. Such controllers may have wireless and/or wired communications, depending on the system to be controlled. They can also be connected to desktop computers located in the control center, which serves to monitor and control the entire system. These systems are expected to be resistant to all types of attacks, whether those attacks are external or performed by insiders. Lack of or weak security in ICS systems can lead to catastrophic consequences depending on the area of application. For example, if there was a problem in the security of the ICS used in a nuclear power plant, it would pose a great threat both to the broader and narrower environment. Therefore, ICS are not yet ready to be connected to the Internet [4]. In order to better understand the system and create better protection, it is possible to divide the entire system into three parts: cyber, cyber-physical, and physical [5]. The physical part of the system represents all components that are in direct contact with the real world, such as sensors and actuators. Cyber and cyber-physical parts of the system have no connection with the real world and they represent software, communication as well as monitoring systems. These two mentioned parts of the system share some common functionalities, but the key difference between them is the way of interaction with the physical part of the system. The cyber-physical part communicates directly with the physical part, while the cyber part does not communicate directly. The cyber-physical part is the mediator between the cyber and physical parts. Cyber, cyber-physical and physical parts of ICS are shown in figure 1.

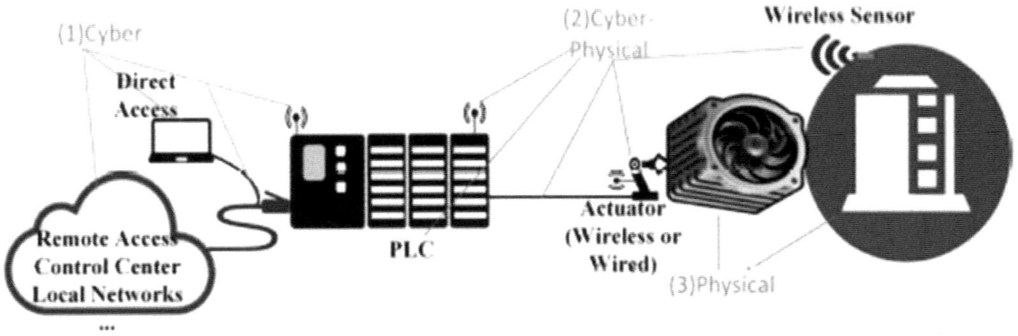

Fig. 1. Possible division of ICS [5]

3. Security threats

In order to create a secure system, first it is necessary to understand potential threats to that system [6]. System is considered safe if it meets the following three requirements: confidentiality, integrity and availability. The first step is to understand what is threat. One of the possible threat definition is that the threat is „a set of circumstances that has the potential to cause loss or harm"[7]. Each threat can have five different factors.

- First is the threat source, it represents the attacker. Attacks can be intentionally performed by an individual or a group of people, they can be accidental, for example through legitimate ICS components, or they can be environmental threats that include natural disasters (floods, earthquakes), disasters caused by humans (fires, explosions) or infrastructure failures (power or telecommunications failure).
- Second factor is the goal that is gained by performing the attack. Targets can be ICS applications and their components or users.
- The third factor is motive. Attackers usually have a number of different motives such as espionage, criminal, terrorism or perhaps political.
- Method of attack is the fourth factor. The attack can be based on one or more of the following four mechanisms: interception, interruption, modification or fabrication [8].
- The fifth factor represents the consequences of a possible attack. Consequences can be for confidentiality, integrity, availability, privacy, or safety.

In most cases, insiders participate in the attacks, because they are fully aware of the system (the source) and have the knowledge to attack the system and harm its normal functioning. Another attack example can be a capable customer who aims to reduce the utility bill by tampering with physical equipment or entering false information to deceive the utility company and thereby cause financial damage.

Also, hostile organizations could cause a cyberwar against a nation by remotely attacking critical infrastructures such as nuclear power plants, gas pipelines or other facilities with the help of malware and thus causing power utility shutdowns, sabotage or environmental pollution. Another example of an attack would be to tamper with the data that the sensor measured and sent to the control center.

4. System vulnerabilities

The most frequent method in the last few decades for implementing security in almost all ICS was "security by obscurity". The focus was on creating reliable systems, while security was not so important. The reason for this is that these systems were isolated from the outside world and were therefore considered secure. Perhaps the best example of such systems is the auto industry. ICS is demanding more and more connection to the Internet than ever before and almost every part of today's systems relies on communication technology. For example, an ICS that controls the operation of an industrial utility is connected to a control center that is connected to the internet. Until 2001, most attacks on ICS were from the inside, but with the beginning of the Internet era, attacks started coming from the outside world as well [9]. Also, to speed up data transfer, many field devices are directly connected to the Internet [10], [11], [12], [13], [14], [15], [16], [17] and [18]. System vulnerabilities can also be observed as three types: cyber, cyber-physical, and physical.

4.1. Cyber vulnerabilities

The first area of interest needed investigation for determining cyber vulnerabilities is communication between ICS components. ICS uses standard protocols such as TCP/IP and ICCP. As these protocols are vulnerable, it makes ICS vulnerable. TCP/IP communication, although it has been studied for many years, still has some shortcomings in terms of security [19]. Another protocol used is Remote Procedure Call (RPC), which also has data security problems, which was exploited by the well-known Stuxnet attack. Wired communication in ICS uses fiber-optic and Ethernet. Local networks are usually based on Ethernet and therefore have problems with message interceptions and man-in-the-middle attacks [20]. For example, an inside attacker could tamper messages with false data or reveal confidential information [21] and [22]. Security problems can also occur if the ICS uses short-range wireless communications. If an employee connects his possibly insecure device to the network, there is a possibility to expose the system to threats, so that an attacker can use his device to attack the system through the employee's device. Another area needing investigation when cyber vulnerabilities are the subject is software. One of the most popular attacks is the attack on databases and all the records in them. Databases containing important data can be directly or indirectly connected to ICS servers, which is the reason why this vulnerability exists. Also, emails can be one of the most common ways of attack. Via email, it is possible to send malware that will later spread through the network. In the end, we should pay additional attention to the devices that are connected to the internet as well as to the local network, such as servers in control centers, laptops and mobile phones of employees, through which malware can be inserted into systems [23].

4.2. Cyber-physical vulnerabilities

The area of observation is the connectivity between the cyber and physical components of ICS. The most used communications are Modbus and DNP3 protocols for monitoring and sending commands from the control center to sensors and actuators. As Modbus has shortcomings in terms of data transmission security, it exposes ICS to various types of attacks. Because it has no encryption capability, the system is exposed to espionage [24]. Also due to the lack of data validation, the integrity of the data is questioned [25]. The authenticity of the sent data is not taken into account, so manipulation of the transmitted data can occur on the source. As a consequence, the actuators can receive wrong data which can lead to their misbehavior, or the sensors can send wrong data to the controller [26]. It is similar to the DNP3 protocol, it does not have any mechanized encryption or authentication. Direct access to remote devices such as PLCs used in smart grids can also be problematic. Furthermore, in today's systems more and more PLCs are directly connected to the internet. Another area of vulnerabilities refers to Operating System (OS) vulnerabilities. Applications used to manage and monitor field devices run on a general-purpose OS. If the OS or the specified application has flaws, they can be a potential risk for an attack on field devices. The well-known Stuxnet attack used two flaws in Windows OS, Print Spooler Service and Windows Server Services for the attack. The third and final area for vulnerabilities refers to Software vulnerabilities. Programs running on a general-purpose OS for controlling and monitoring controllers can be an opportunity for attack. Simatic WinCC is an example of such a program, which is a supervisory control and data acquisition (SCADA) and human-machine interface (HMI) system developed by Siemens. During the Stuxnet attack, the first step was to find a vulnerable computer on which WinCC software is running [27].

4.3. Physical vulnerabilities

Physical exposure of many ICS components is a problem for system security. A soil moisture measurement system can be observed as an example. It can consist of multiple sensors at different locations powered by solar panels so that they can measure the desired values and send the measured values to the control center. If the solar panels are stolen, the sensors no longer have the necessary power and the control center is left without the necessary data.

5. Improvements of security in ICS

One of the main directions in improving ICS security is the improvement of security in communications protocols. Even if protocols are used to connect the system to the Internet or if they are used to connect system hardware locally, some standard protocols are Modbus, DNP3, TCP/IP. The main directions in improving security in ICS communication protocols are: Cyber, Cyber-Physical and Physical.

5.1. Cyber

The best mechanism to prevent attacks is to use an encryption mechanism in all ICS networks. One of the problems with encryption is the delay in data transmission that occurs due to data encryption, which can be a big problem in real-time systems. The delay problem that occurs during data encryption can be solved by using an ICS-specific key management solution [27]. In addition, there are studies that say it is possible to use a layered approach to protect important ICS data [28]. This study is based on Hash Chains to create layered protection by dividing it into two zones: high and low-security zones and a lightweight key management mechanism. Thanks to this division, if an attacker gets full access to the low-security level, he will not be able to access the data that is in the high-security level. The next step in security is the constant improvement of OS security.

Windows improved its security after the Stuxnet attack, so that this type of attack is no longer a security threat [29]. In addition to the OS and applications used for management and monitoring, ICS must also follow trends and constantly improve their protection measures. Another type of improvement in security protection is standardization. By following existing security standards when designing an ICS, it can contribute to a greater degree of security. One of the studies provides guidelines for technical controls such as firewalls, IDS and access control, which can influence the increase of ICS security [30].

5.2. Cyber-Physical

The best solution would be to create a new security solution designed exclusively for ICS. That new solution should take into account both the system's communication with the cyber part and the system's communication with the real world. Until now, the solutions have mainly aimed to ensure the reliable functioning of the solution in case of some unintentional problems [6]. Although reliability is a very important feature, malicious cyber-attacks are more possible today than ever, and must be taken into account when creating new ICS security solutions. It is also necessary to increase the security of communication protocols. Secure Modbus framework is one of the solutions [31]. This method of communication ensures authentication, non-repudiation, and obstructs replayed packets. Another problem is related to remote access to field devices. One solution is that only authorized persons can access remote devices [32]. This could be achieved by using one particular computer and through a VPN.

5.3. Physical

National Institute of Standards and Technology provides a list of defence-in-depth physical controls [30]. Examples of which include the protection of physical access, locations, etc.

6. Conclusion

This paper showed the main problems of security in ICS applications. We presented threats, vulnerabilities and possible improvements separated into the cyber, cyber-physical and physical aspects. For each ICS application, it is mandatory to isolate threats, in the last years the main solution for security in ICS was to isolate the system from the outside world, and not to implement security solutions because it was considered safe from an outside attack. Today, more and more industrial systems are connected to the Internet and they are no more isolated from the outside world, this brings a very big risk of attacks and the possibility to interfere in the system process. There are possible improvements that are proposed as the solution to diminish risk in ICS and they are considered in the cyber, cyber-physical and physical aspects.

Cyber aspect solutions should implement an encryption mechanism in all ICS networks. Cyber-Physical aspect solution implements a new security framework for Modbus communication protocol which is used for communication between system hardware, and implements security in terms of authorized remote access using VPN to control the ICS. In physical it is only mentioned that NIST provides standardized solutions for lowering the risk of vulnerabilities of ICS. With the increasing implementation of Industry 4.0 and IoT, there is a need to connect ICS applications to the outside world and this increases the risk of outside attacks on ICS applications which in many ways are real-time applications that need to be isolated and secured from outside attacks. On basis of described aspects we intend to implement every aspect on existing ICS application, and to validate implemented solutions and gather results on which we will make distinction and show results of which solution is easiest to implement and which solution gives us best protection against outside attacks.

7. Acknowledgment

This research was funded by the Ministry of Education, Science and Technological Development of the Republic of Serbia through project no. 451-03-68/2022-14/200156 and the Provincial Secretariat for Higher Education and Scientific Research of Autonomous Province of Vojvodina, Republic of Serbia, through project no. 142-451-2004/2022-01.

8. References

[1] Amin, S.; Schwartz, G. A. & Hussain, A. (2013). In quest of benchmarking security risks to cyber-physical systems. IEEE Network, 27(1), 19-24.

[2] Amin, S.; Litrico, X.; Sastry, S. S. & Bayen, A. M. (2010, April). Stealthy deception attacks on water SCADA systems. In Proceedings of the 13th ACM international conference on Hybrid systems: computation and control (pp. 161-170).

[3] Anderson, R. & Fuloria, S. (2010, October). Who controls the off switch?. In 2010 First IEEE International Conference on Smart Grid Communications (pp. 96-101). IEEE.

[4] Francia III, G. A., Thornton, D., & Dawson, J. (2012). Security best practices and risk assessment of SCADA and industrial control systems. In Proceedings of the international conference on security and management (SAM) (p. 1). The Steering Committee of The World Congress in Computer Science, Computer Engineering and Applied Computing (WorldComp).

[5] Humayed, A.; Lin, J., Li, F. & Luo, B. (2017). Cyber-physical systems security—A survey. IEEE Internet of Things Journal, 4(6), 1802-1831.

[6] Cardenas, A.; Amin, S.; Sinopoli, B.; Giani, A.; Perrig, A. & Sastry, S. (2009, July). Challenges for securing cyber physical systems. In Workshop on future directions in cyber-physical systems security (Vol. 5, No. 1).

[7] Ankaralı, Z. E.; Demir, A. F.; Qaraqe, M.; Abbasi, Q. H.; Serpedin, E.; Arslan, H. & Gitlin, R. D. (2015, September). Physical layer security for wireless implantable medical devices. In 2015 IEEE 20th International Workshop on Computer Aided Modelling and Design of Communication Links and Networks (CAMAD) (pp. 144-147). IEEE.

[8] Cardenas, A.; Amin, S.; Sinopoli, B.; Giani, A.; Perrig, A. & Sastry, S. (2009, July). Challenges for securing cyber physical systems. In Workshop on future directions in cyber-physical systems security (Vol. 5, No. 1).

[9] Byres, E. & Lowe, J. (2004, October). The myths and facts behind cyber security risks for industrial control systems. In Proceedings of the VDE Kongress (Vol. 116, pp. 213-218).

[10] Leverett, É. & Wightman, R. (2013, November). Vulnerability inheritance programmable logic controllers. In Proceedings of the Second International Symposium on Research in Grey-Hat Hacking. GreHack 2013 Grenoble, France.

[11] Slay, J., & Miller, M. (2007, March). Lessons learned from the Maroochy water breach. In International conference on critical infrastructure protection (pp. 73-82). Springer, Boston, MA.

[12] Stankovski, S., Ostojić, G., Baranovski, I., Babić, M.& Stanojević, M. (2020). The Impact of Edge Computing on Industrial Automation, 2020 19th International Symposium INFOTEH-JAHORINA (INFOTEH), East Sarajevo, Bosnia and Herzegovina, 2020, pp. 1-4.

[13] Klašnja, N.; Sremčev, N.; Vukelić, Đ.; Simeunović, N. & Lazarević, M. (2019). Optimization of Cable Harness AssemblySystems Based on Lean Concept Application. International Journal of Industrial Engineering and Management, 10(1), 115.

[14] Anišić, Z.; Veža, I.; Suzić, N.; Sremčev, N. & Orčik, A. (2013). Improving product design with ips-dfx methodology incorporated in PLM software. Tehnicki vjesnik/Technical Gazette, 20(1).

[15] Ostojic, G., Lazarevic, M., Stankovski, S., Cosic, I. & Radosavljavic, Z. (2008). Radio frequency identification technology application in disassembly systems, Strojniski Vestnik/Journal of Mechanical Engineering, 54 (11), 759-767.

[16] Tegeltija, S.., Lazarević, M., Stankovski, S., Ćosić, I., Todorović, V. & Ostojić, G. (2016). Heating circulation pump disassembly process improved with augmented reality, Thermal Science, 20, 611-622.

[17] Stankovski, S., Ostojić, G., Šaponjić, M., Stanojević, M.& Babić, M., (2020). Using micro/mini PLC/PAC in the Edge Computing Architecture, 19th International Symposium INFOTEH-JAHORINA (INFOTEH), East Sarajevo, Bosnia and Herzegovina, 2020, 1-4.

[18] Ostojic, G., Stankovski, S., Vukelic, D., Lazarevic, M., Hodolic, J., Tadic, B. & Odri, S. (2011). Implementation of automatic identification technology in a process of fixture assembly/disassembly, Strojniski Vestnik/Journal of Mechanical Engineering, 57 (11), 819-825.

[19] Harris, B. & Hunt, R. (1999). TCP/IP security threats and attack methods. Computer communications, 22(10), 885-897.

[20] Francia III, G.; Thornton, D. & Brookshire, T. (2012, June). Cyberattacks on SCADA systems. In Proc. 16th Colloquium Inf. Syst. Security Educ (pp. 9-14).

[21] Paukatong, T. (2005, August). SCADA security: A new concerning issue of an in-house EGAT-SCADA. In 2005 IEEE/PES Transmission & Distribution Conference & Exposition: Asia and Pacific (pp. 1-5). IEEE.

[22] Wang, W. & Lu, Z. (2013). Cyber security in the smart grid: Survey and challenges. Computer networks, 57(5), 1344-1371.

[23] Cardenas, A. A.; Roosta, T. & Sastry, S. (2009). Rethinking security properties, threat models, and the design space in sensor networks: A case study in SCADA systems. Ad Hoc Networks, 7(8), 1434-1447.

[24] Alcaraz, C. & Zeadally, S. (2013). Critical control system protection in the 21st century. Computer, 46(10), 74-83.

[25] Fovino, I. N.; Carcano, A.; Masera, M. & Trombetta, A. (2009). An experimental investigation of malware attacks on SCADA systems. International Journal of Critical Infrastructure Protection, 2(4), 139-145.

[26] Zhu, B.; Joseph, A. & Sastry, S. (2011, October). A taxonomy of cyber attacks on SCADA systems. In 2011 International conference on internet of things and 4th international conference on cyber, physical and social computing (pp. 380-388). IEEE.

[27] Chen, T. M.,& Abu-Nimeh, S. (2011). Lessons from stuxnet. Computer, 44(4), 91-93.

[28] Choi, D.; Kim, H.; Won, D. & Kim, S. (2009). Advanced key-management architecture for secure SCADA communications. IEEE transactions on Power delivery, 24(3), 1154-1163.

[29] Cao, H.; Zhu, P.; Lu, X. & Gurtov, A. (2013). A layered encryption mechanism for networked critical infrastructures. IEEE Network, 27(1), 12-18.

[30] Langner, R. (2011). Stuxnet: Dissecting a cyberwarfare weapon. IEEE Security & Privacy, 9(3), 49-51.

[31] Fovino, I. N.; Carcano, A.; Masera, M. & Trombetta, A. (2009, March). Design and implementation of a secure modbus protocol. In International conference on critical infrastructure protection (pp. 83-96). Springer, Berlin, Heidelberg.

[32] Fernandez, J. D. & Fernandez, A. E. (2005). SCADA systems: vulnerabilities and remediation. Journal of Computing Sciences in Colleges, 20(4), 160-168.

DOI: 10.2507/33rd.daaam.proceedings.046

SECURITY CLEARANCE AND POLYGRAPH EXAMINATION

Lucie Sousedikova & Milan Adamek

This Publication has to be referred as: Sousedikova, L[ucie] & Adamek, M[ilan] (2022). Security Clearance and Polygraph Examination, Proceedings of the 33rd DAAAM International Symposium, pp.0332-0335, B. Katalinic (Ed.), Published by DAAAM International, ISBN 978-3-902734-36-5, ISSN 1726-9679, Vienna, Austria
DOI: 10.2507/33rd.daaam.proceedings.046

Abstract

This research provides the insight into the problematics of the security clearance process that is very important in the field of national security. The security clearance must be passed by everyone who has the access to sensitive information. One of the steps in this process is polygraph examination that verify the trustworthiness of the applicant. The purpose of this paper is to show that the polygraph results are valuable although there are still doubts about its accuracy in scientific community.

Keywords: Security Engineering; Security Clearance Process; Polygraph Examination; Detection of Deception.

1. Introduction

The security clearance, as a verification process of the person determining the trustworthiness of an individual, dates back to the early 1950s. Whether the person applies for employment as a border patrol or FBI agent, he or she has to complete successfully the security clearance process. A security clearance must hold anyone whose job requires access to classified national information or restricted areas. There are a lot of cleared careers – federal, military, and government positions or jobs in intelligence agencies. Although polygraph is mainly used in law enforcement and criminal investigation [1], it finds its use al so in this type of investigation and represents an important step in the security clearance process that has to be passed by candidates.

It has evolved since the 1920s and till now it is the world's best-known and most efficient tool for detecting deception and for profiling people of interest. Despite that, there are doubts about the polygraph measurement results occurring in some sources or scientific community and sometimes the polygraph examination is considered controversial. That is why this research focuses on the polygraph screening process with the use of the security clearance questions and evaluates the polygraph measurement results [2].

2. Security Clearance Process

When receiving a conditional job offer, the candidate must go through a security clearance process to obtain a final job offer from a government agency that provides it. Figure 1. shows the entire process which takes a few months depending on the required level of security clearance.

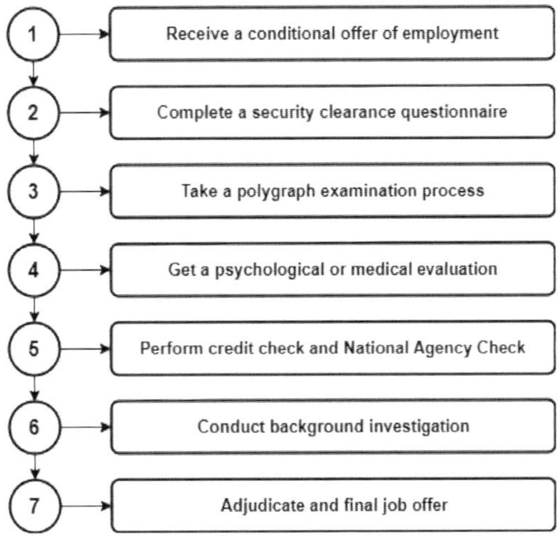

Fig. 1. Security clearance process

There are three levels of security clearance related to the sensitivity of the information to which the person will have access:

- Confidential, the access to information that could cause damage to national security when it was unauthorized uncovered.
- Secret, the access to information that could cause serious damage to national security if it were unauthorized uncovered.
- Top Secret, the access to information that could cause exceptionally grave damage to national security when it was unauthorized uncovered.

The security clearance determines the eligibility against:

- Criminal conduct
- Foreign influence
- Financial considerations
- Allegiance to a given state
- Handling protected information
- Use of information technology systems
- Drug involvement and alcohol consumption
- Psychological conditions and personal behaviour.

The security clearance investigation has two types of polygraph examination - lifestyle and counterintelligence. The first one is based on the security clearance questionnaire and it focuses on personal questions regarding to illegal activity, drug use, sexual misconduct, mental health issues, and family relationships, or falsification of the information on the questionnaire provided. In general, it determines if a potential employee is trustworthy. The second type of examination deals with the questions related to spying, including the areas of espionage, sabotage, and terrorism [2], [3].

3. Polygraph Measurement

Detection of deception by polygraph is based on analysis of physiological changes in the body that occur during a stressful situation. A such stressful situation may be the asked questions during the polygraph examination process. The physiological changes can be changes in the heart rate, pulse rate, blood pressure, respiration rate, or skin conductivity. In addition, the minor and major movements are measured [4].

The polygraph system is composed of the following sensors that are connected to the data acquisition subsystem and simultaneously measure physiological values:

- Blood Pressure Cuff.
- Pneumograph tubes tied around the torso to measure breathing.
- Electrodermal (EDA) Electrodes attached to fingers to detect the increased skin conductivity caused by the activation of the sympathetic nervous system.
- Activity Sensor Seat, Arm, and Foot Pads in the form of pouches place under the seat, arms, and feet and detect every movement.
- Photoelectric Plethysmograph recording the rapidly occurring changes in pulse blood volume [5], [6].

In this research, the first four sensors were used for the measurement – blood pressure cuff, pneumograph tubes, EDA electrodes, and activity sensor seat pad. These sensors are depicted in Figure 2. Those are the polygraph components of a state-of-the-art computerized polygraph system known as LX6. With the use of the special LXSoftware and the data acquisition subsystem, the signals from all sensors can be recorded, stored, and analysed.

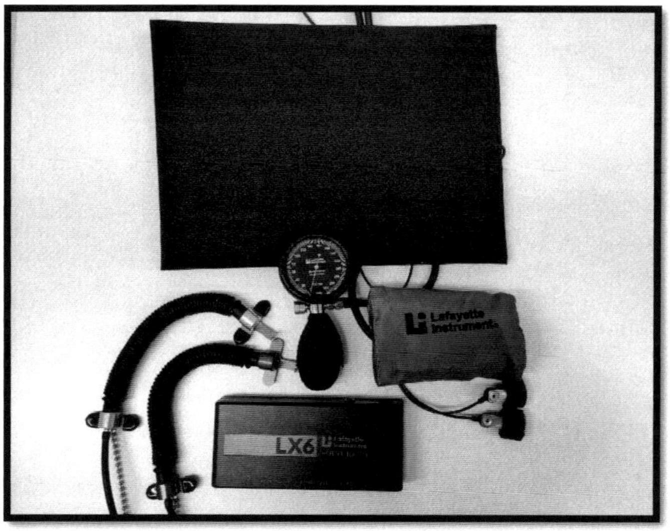

Fig. 2. LX6 Polygraph components used for measurement

For evaluation of data was used modified variation of 3-position scoring [7], [8]. Values of -1, 0 and 1 were assigned to changes after each question for each measured channel (pneumograph, blood pressure cuffs, movement sensor and electrodermal). Negative value was assigned to greater reactions to question compared to benchmark question, positive value to smaller reactions and 0 was assigned for measurements without any noticeable difference [9]. For implying of detection of potential deception sum of values from all channels had to be negative. Values of scores from all channels separately were also used for final evaluation of usefulness of each sensor.

4. Results

Measured dataset has total of 140 answers with 14 sets of 10 questions that were asked repeatedly. Total number of lies reported by examinee was 36 and out of this number just 3 (8,4%) remained undetected. This means that 91.6% lies were successfully detected. Number of true answers scored as potential lies (false positives) were 29 out of 114 (25,4%). From this evaluation It can be implied that there is high probability of raising question marks on lies, and if the answers are evaluated as true, there is low probability of deception that remained unnoticed. In the Figure 3. is the example of detected lie that was clearly visible on multiple channels during questioning.

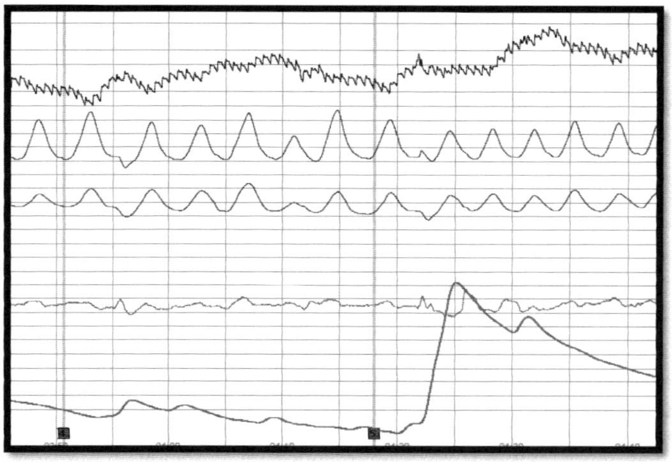

Fig. 3. Polygraph chart - example of detected lie on polygraph (5. question)

On the other hand, answers that are marked as potential lies must be inspected in more depth because there is small but still significant chance of false positive evaluation. For more accurate measurements, important questions can be asked again with slightly different wording, and this should confirm lies and reduce chance of false positives.

5. Conclusion

The use of the polygraph in security clearance investigations has a long and controversial history but this analysis shown that polygraph is effective tool which has its place in lie detection. The best evaluations from measurements were from blood pressure cuffs, their evaluation detected 16 lies and just 7 false positives. Most changes were detected on electrodermal with 30 lies and 23 false positives. Both movement sensor and pneumograph had similar scores, movement sensor detected 10 lies and 7 false positives and pneumograph 13 lies and 9 false positives so even these two sensors with worst results are still showing useful.

Data also showed that answers on questions asked 1 or 2 questions after lie are less likely to be evaluated as false positives. There were just 2 false positives for questions right after the measured lie and 3 false positives for two questions after a lie. This is probably because changes in measured values are more distinct, easier to be interpreted and change of physiological response is more noticeable after the examinee calms down when the moment of his lying pass. In the future, research could use the other polygraph sensors for measurement and the Virtual Reality for better profiling of the people. In addition, it could be tested on a larger sample of people.

6. Acknowledgments

This publication was supported by the Internal Grant Agency of Tomas Bata University in Zlin, the Department of Security Engineering, Faculty of Applied Informatics, under the project No. IGA/FAI/2022/006

7. References

[1] Sousedikova, Lucie, et al. The Role of Lie Detection Based System in Controlling Borders. Annals of DAAAM & Proceedings, annual 2021, pp. 384+.
[2] Christensen, M. D.; Kaiser F. M. (2014). Security Clearance Process: Answers to Frequently Asked Questions (January 2014) pp. 153-168.
[3] Cohen I. S. (2000). Security Clearances and the Protection of National Security Information: Law and Procedures, Defense Personnel Security Research Center, Technical report 00-4, pp. 62-68
[4] Rastogi, P.; Kanchan, T.; Das, S.; Daya, P. (2008). The Truth behind Lie Detector, Journal of Indian Academy of Forensic Medicine, Vol. 30, No. 3, pp. 161-163, ISSN 0971-0973
[5] National Research Council. (2003). The Polygraph and Lie Detection, The National Academies Press, ISBN 978-0-309-13313-5, Washington DC
[6] Krapohl, D. J.; Shaw P. K. (2015) Fundamentals of Polygraph Practise, 1st ed., Academic Press, ISBN 978-0-12-802924-4
[7] Gordon, N. J. (2017) Essentials of Polygraph and Polygraph Testing, 1st ed., CRC Press, ISBN 978-1-4987-5771-3
[8] Senter, S. M. (2003). Modified general question test decision rule exploration, Polygraph, Vol. 32, pp. 251-263.
[9] Backster, C. (1962). Technique tips and polygraph chart interpretation, Newsletter of the Academy for Scientific Interrogation, (September-October 1962), p. 4.

DOI: 10.2507/33rd.daaam.proceedings.047

INFLUENCE OF TOOL GEOMETRY ON THE PROCESS OF TURNING TI-6AL-4V

Jindrich Sykora & Simon Syrovatka

This Publication has to be referred as: Sykora, J[indrich] & Syrovatka, S[imon] (2022). Influence of Tool Geometry on the Process of Turning Ti-6Al-4V, Proceedings of the 33rd DAAAM International Symposium, pp.0336-0343, B. Katalinic (Ed.), Published by DAAAM International, ISBN 978-3-902734-36-5, ISSN 1726-9679, Vienna, Austria DOI: 10.2507/33rd.daaam.proceedings.047

Abstract

Titanium alloys are widely used in the aerospace and medical industry and these industries generally require parts of high precision and good surface finish. The ratio between Young's modulus of elasticity and the ultimate tensile strength of titanium alloys is approximately half in comparison with C45QT steel. This manifests itself in the lower resistance of titanium workpieces against deformation induced by cutting forces. It can be especially detrimental when workpieces of a high length-to-diameter ratio are turned. This work is focused on the analysis of the process of turning Ti-6Al-4V under various cutting conditions when machining cylindrical and conical geometry. Cutting forces were measured and evaluated to find requirements for minimizing the Fp component of the total cutting force. This will result in a decrease in workpiece deformation and therefore vibration and an increase in the surface quality and dimensional accuracy. Positive inserts of ISO C and D shapes with 0,2 mm nose radius are evaluated and the depth of cut is varied.

Keywords: turning; cutting forces; depth of cut; Ti-6Al-4V

1. Introduction

Titanium as a pure metal is characterized by its corrosion resistance, high tensile strength and high biocompatibility. This type of metal is often used for the production of alloys, with which its very good properties are further enhanced. The most widely used titanium alloy is Ti-6Al-4V. This alloy is widely used, for example, in the aviation and space industry or in medicine [1]. However, high-speed machining of this alloy is very problematic precisely because of its excellent mechanical properties, which include low modulus of elasticity or low thermal conductivity. Poor machinability of this alloy can lead to surface defects or poor surface integrity, which can affect the service life of components made from this alloy. Regarding to the turning process, the final surface quality is influenced by several factors, including the macro and micro geometry of the tool, the cutting parameters of the manufacturing process or the type of cooling medium [2]. During the machining method of turning, three basic working parameters are used to influence the process. These parameters are cutting speed (v_c), feed (*f*) and depth of cut (*Ap)*.

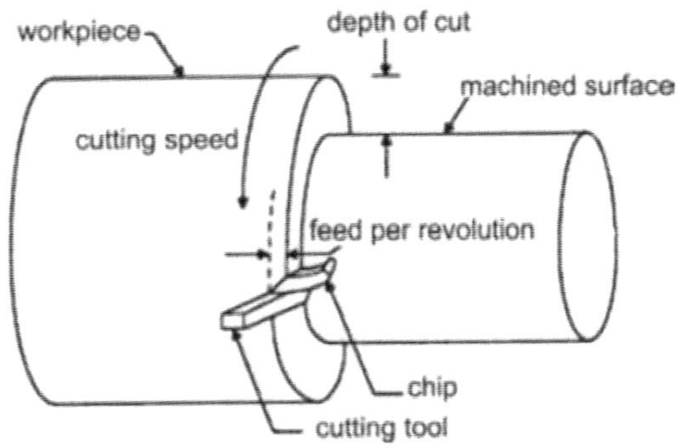

Fig. 1. Scheme of turning process [3]

Determining the correctness of the operating parameters can be done in several ways. These include, for example, measuring the surface roughness of the final surface, measuring cutting tool wear, the efficiency of the machining process or measuring the magnitude of the cutting forces generated. The main cutting force F in the turning process consists of three forces, namely the cutting force F_c which is usually the largest, then the feed force F_f and the passive force F_p which is perpendicular to the machined surface [4].

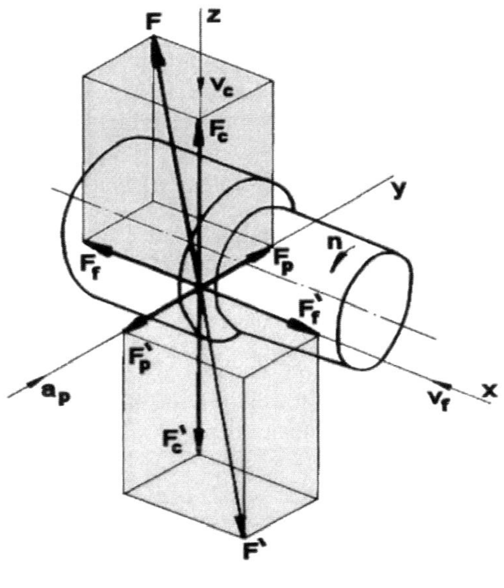

Fig. 2. Cutting forces [5]

The main cutting force F can be calculated using the following equation

$$F = \sqrt{F_c^2 + F_f^2 + F_p^2} \tag{1}$$

In the case that the correctness of working parameters is decided based on the size of the main cutting force F it can be said that smaller values of the main cutting force F indicate the choice of more suitable working parameters. For to measure the cutting forces generated during the turning process three-component dynamometers are very often used [6]. These devices can be divided according to the method of measuring (mechanical, hydraulic, pneumatic, electrical (piezoelectric), optical) [7]. The magnitude and the cutting force ratio are primarily influenced by the tool's geometry, cutting parameters and the workpiece material. Although things like tool wear, the material of the tool and the surface finish of the tool can also play a major role. This work is focused on the influence of the side cutting edge angle (SCEA) and end cutting edge angle (ECEA).

The importance of the side cutting edge angle was shown by Saglam. In this work, AISI 1040 steel was turned and the cutting forces were measured at varying side cutting edge angles (45°, 60°, 75°, 90°) and rake face angles of 0°, 6° and 12°. Fig. 3 clearly shows the significant influence of the side cutting edge angle (entering angle). Feed force increases with the growth of SCEA while the thrust force decreases [8]. Trust force and main cutting force (Fc) in combination contribute to workpiece deformations during the turning process. This can lead to dimensional and geometrical inaccuracies, and it is, therefore, advantageous to minimize these two components of the main cutting force. As stated above, this can be achieved through changes in geometry and SCEA plays a significant role. One of the ways the main cutting force can be lowered is by decreasing an uncut chip cross section area (UCCA). This can be achieved by reducing the feed rate and depth of cut. This work is focused on the influence of depth of cut, which should always be related to nose radius. Manufacturers generally advise not to use the deep of cut (DOC) lower than 2/3 of the nose radius. The main reason is that with low DOC nose radius changes the effective SCEA angle and makes it lower. Consequently, the ratio of the thrust force increases leading to undesirable deformations during the machining process.

The main cutting force is also largely influenced by the tool rake angle and edge radius. With an increase in the rake face angle, the main cutting force lowers. The same is true for a decrease in the edge radius [9]. However, these two parameters also reduce the edge stability, and their modification can lead to rapid tool wear. In this work highly positive inserts are used, for this reason, other methods, such as reduction in DOC are investigated.

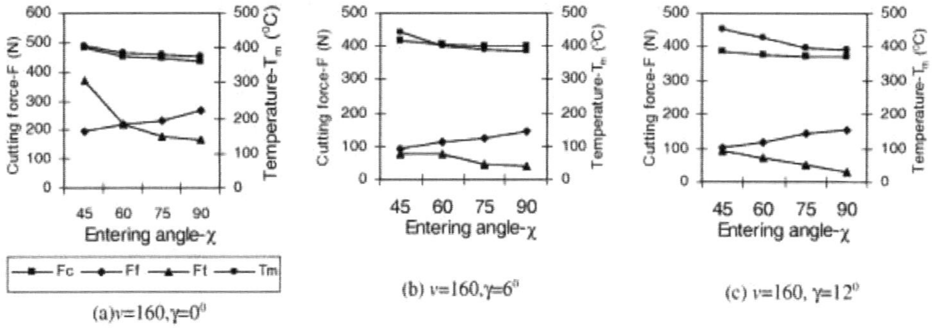

Fig. 3. Influence of SCEA on the cutting forces

One of the unique characteristics of titanium alloys is the ratio between tensile strength and modulus of elasticity. This ratio is approximately double in comparison to C45 steel (see Table 1) [10], [11]. This fact must be considered for machining setups. This also is the reason why turning slender workpieces from titanium alloys can be especially challenging. The low modulus of elasticity causes the part to deflect much more severely due to the cutting forces. In this case, the role of Fp is especially important. Deformation of the workpiece caused by this force component causes geometrical and dimensional inaccuracies of the part. Furthermore, the stability of the process may be decreased and lead to vibrations and compromised surface finish [12]. This study, therefore, investigates how various conditions influence the cutting forces and how these results can be used when turning Ti6Al4V parts on non-swiss type lathes.

Material	Rm [MPa]	E [GPa]	Thermal conductivity [W/m.K]
Ti6Al4V – Grade 5 (annealed)	950	113.8	6.5-7.5
ISO C45QT	630-780	210	40-45
ISO 7075-T6	572	71.7	130

Table 1. Comparison of mechanical properties largely influences the machinability [10], [11], [13]

2. Materials and methods

Cylindrical workpieces of Ti6Al4V Grade 5 alloy were used with an initial diameter of 30 mm. The cutting tests were performed using Emco Maxxturn 25 with 6.5 kW and 8000 RPM spindle. There were two types of tools used in the experiment, both manufactured by Iscar. Firstly, it was the SCLCL 1212F-09 insert holder with CCGT 09T302 – AS IC520 inserts which provide the side cutting edge angle of 95° and ECEA of 5°. These inserts have highly positive geometry with an 8 μm cutting edge radius and rake angle of 20° and are PVD coated with a combination of TiCN and TiN. Secondly, the SDJCL 1212F-11 insert holder was used with an SCEA of 93° and an ECEA of 32°. The used insert was DCGT 11T302 – AS IC320, it is again a highly positive insert with a cutting edge radius of 6 μm and a rake angle of 20°. Most of the geometrical parameters of both these tools are very similar and the only significant difference is ECEA. The cutting forces were measured using a Kistler MiniDyn Multi-component dynamometer. To make the results of these tests relevant for industrial purposes and minimize tool wear, an external coolant line was used to deliver the coolant medium to the cutting edge.

Fig. 4. Tool and dynamometer set up in the machine tool

3. Experimental setup

The experiment was designed to compare the influence of tool geometry (ECEA), workpiece geometry and depth of cut on the cutting forces. Therefore, the rest of the parameters were kept constant. The cutting speed was fixed at 60 m/min which is usually at the bottom of the cutting speed range recommended by tool manufacturers for this material and carbide inserts. This somewhat low cutting speed was used to minimize the influence of tool wear on the results of the experiment. Because this study aims to explore conditions for finishing operation, the feed rate was set to 0.1 mm/rev. It is a value in the middle of the range recommended by the tool manufacturer for this nose radius and should be suitable for general finishing operations.

There were two workpiece geometries machined, a cylindrical section with a constant length of 30 mm and a taper with an included angle of 30° and an initial length of 17.5 mm. The cutting time decreased with each successive pass for the cylindrical specimen and increased for the tapered specimen, however, it should not severely influence the cutting forces. The tool wear VB was measured for the CC insert after the first 4 min in cut and at the end of the cutting trials. The difference is 8 µm, therefore, its influence on the measured data is not very significant and was not evaluated. The DOC was varied on four levels, it was 0.5, 0.3, 0.2 and 0.1 mm. This way the relationship between the nose radius to DOC ratio and cutting forces could be evaluated. The goal of this paper is to provide data that can be applied to the machining of titanium workpieces of high length to diameter ratios. One of the main concerns in these situations is the deformation induced by cutting forces. The most significant role is played by Fp and Fc force. When a cylindrical part is turned the Fp force corresponds to the direction of an X axis of a lathe and the Fc force corresponds with the direction of a Y axis. However, for a tapered workpiece, the direction of the Fp would be at an angle corresponding with the taper. Although, for this study forces in the directions of the machine coordinate system were evaluated. This approach allows for better comparison between the workpiece geometries. The uncut chip cross section area is one of the main factors that influence the magnitude of the cutting force. For the cylindrical workpiece and DOC of 0.5 mm with SCLCL insert is the value of this parameter 0.0498 mm^2 the difference in this factor between the insert types and workpiece geometries is within 0.05 %, therefore, it is extremely small and is not considered in the evaluation.

Fig. 5. Workpiece geometries

4. Results and discussion

In total 16 individual tests were performed. As predicted, the main cutting force decreases with the reduction of DOC. The uncut chip cross section area certainly plays a major role and a reduction in this value will play a major role in total cutting forces. Fig. 5 compares the main cutting forces, the CCGT insert generated a significantly higher main cutting force when DOC 0.5 and 0.3 mm were used. An even higher difference can be observed in figure X where Fx forces are compared. These two results are exceptions in the data set and point toward an unknown variable in the process. For this reason, the hardness of the material was measured on the original surface and a diameter of 17 mm. The original surface shows a hardness of 393 HB whereas the middle of the specimen shows only a hardness of 105 HB. This demonstrates a significantly higher value of ultimate tensile strength on the surface which would lead to significantly higher cutting forces. As the test with 0.5 mm and 0.3 mm DOC were performed as the first ones, they were probably influenced by the described mechanical properties of the specimen. Towards the end of a cut (near the surface), forces are much higher. Unfortunately, this fact was realised during the evaluation phase and the first two trials could not be repeated, for this reason, the results are excluded from further evaluation.

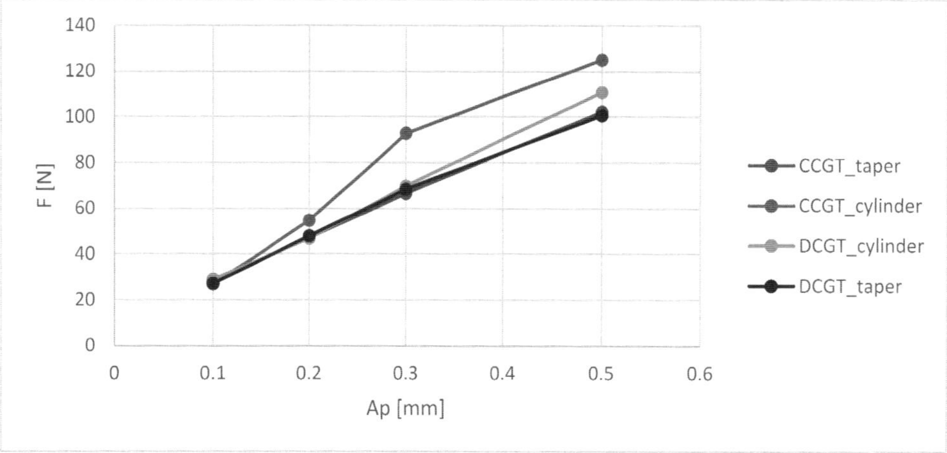

Fig. 6. Dependence between DOC and main cutting force

The first of the evaluated variables is the workpiece geometry. The direction of an Fp and Ff forces obviously changes with the specimen geometry. However, this study is concerned with force directions Fx and Fz which match the machine coordinate system. The main cutting force varies by less than 15 % and there is not a significant difference between tapered and cylindrical geometry. Nonetheless, figures 8 and 10 show a clear difference in Fx and Fz forces where Fx forces are reduced when the taper is machined, and the opposite is true for Fz forces. In the case of the cylindrical geometry, the cutting force Fx is on average 21 % of the main cutting force but when tapered geometry is machined it is only 15 % of the main cutting force.

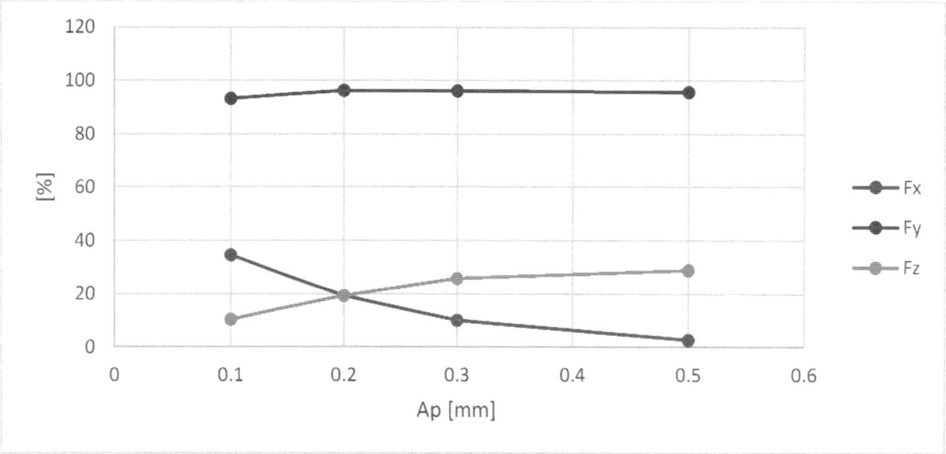

Fig. 7. The influence of DOC on the ratio of force components and main cutting force – turning cylinder with DCGT insert

The Fp and Ff forces are very similar for both machined geometries, yet the Fx force which induces significant deformations in the workpiece is lower when the taper is machined. The tool geometry was a second evaluated factor and the main differences between the tools were SCEA and ECEA. The results show that higher Fx force is generated when CCGT inserts are used for both cylinder and taper. For a tapered workpiece, the Fx force is on average only 4 % higher for the CCGT insert. In terms of a cylindrical workpiece, the difference is 17 %. This result confirms the benefit of higher ECEA when the Fx force needs to be reduced. The tapered workpiece geometry essentially increases the effective ECEA making it 20° for the CCGT insert. Therefore, the difference between the inserts is less noticeable in this case.

Lastly, the influence of DOC was investigated. It is generally advised to use DOC values higher than 2/3 of the nose radius. Lower values may lead to vibrations and inferior chip breaking. One of the reasons for this recommendation is the growth of the Fx cutting force. The results clearly show the exponential growth of this force component with a decrease in DOC. As can be seen in Fig. 8 and Fig. 9 there is growth in the magnitude of this force as well as its ratio to the main cutting force with a decrease of DOC. From these results can therefore be argued that if the goal is to maximize the stability of the process, then DOC higher than nose radius should be used. A decrease in DOC significantly lowers the main cutting force, however, when minimizing the deformation of the workpiece in the X axis is also required then inserts with a nose radius lower than DOC should be used.

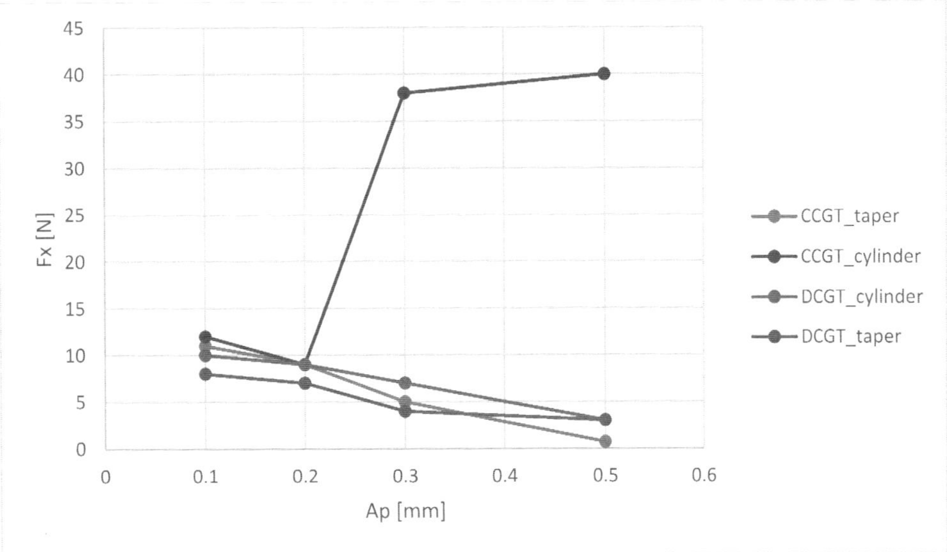

Fig. 8. Dependence between DOC and cutting force in the X axis direction

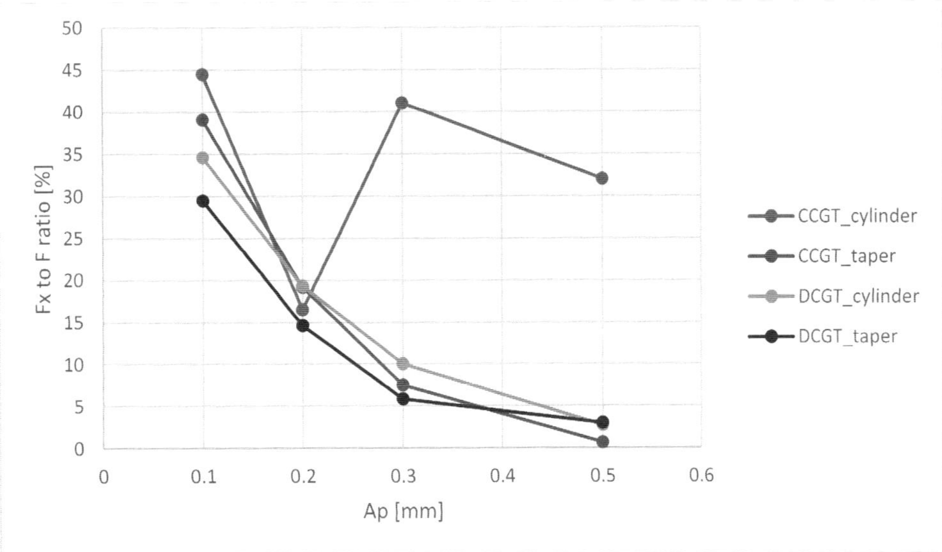

Fig. 9. The influence of DOC on the ratio of Fx and main cutting force

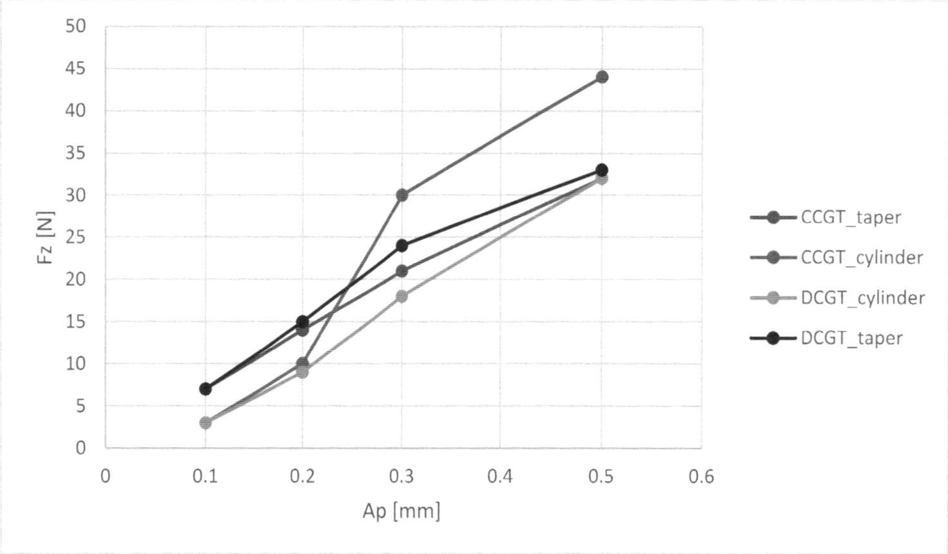

Fig. 10. Dependence between DOC and force in the Z axis direction

5. Conclusion

This study aims to investigate the influence of several variables on the cutting forces generated during the turning of titanium alloy. The focus was given to the Fx component of the main cutting force which plays a role in the stability of the turning process and the final dimensional and geometrical accuracy of the parts. The magnitude of the Fx component is generally significantly lower than the Fy (Fc) component, however, under certain conditions, Fx can become substantial. Force components coincident with the coordinate system of the machine tool (Fx, Fy, Fz) were used in the study together with the more common Fp, Fc, Ff, because Fx, Fy, Fz are not dependent on the geometry of a workpiece. The following conclusions can be drawn from the results:

- Machining tapered geometry (30° included angle) causes a shift in the direction of Fp and Ff forces, reducing the Fx force and increasing the Fz force.
- The ECEA can play a significant role in the Fx force component, although there seems to be a value above which its influence becomes less significant. This can be observed by comparing both inserts and machined geometries. When cylindrical geometry is machined, the CCGT (effective ECEA is 5°) insert shows on average a 17 % higher Fx force than the DCGT (effective ECEA is 32°) insert. In the case of tapered geometry, the difference in Fx force is on average only 4 % and the effective ECEA is 20° for the CCGT and 47° for the DCGT insert.
- With a decrease in DOC, the exponential growth of the Fx to F ratio can be observed. Lowering of DOC leads to increased contribution of Fx at the expense of Fz, therefore, possible increase in radial deformation of a workpiece.

It can be concluded that a reduction in the Fx force can be achieved by eider increase in DOC or a reduction of nose radius. Furthermore, if an ECEA is under a certain value its increase can also play a significant role. Finally, an increase in effective SCEA will also reduce Fx force. The force measurements that were influenced by the difference in mechanical properties in some sections of the workpiece will be repeated in further research. Additionally, tools with different nose radii will be investigated in the future as well as higher values of DOC. Lastly, other workpiece materials with significantly dissimilar mechanical properties will be tested to explore the effect of this variable.

6. Acknowledgements

The article contribution has been prepared under project SGS-2022-007 - Research and Development for Innovation in Engineering Technology - Machining Technology IV.

7. References

[1] D. Singha, A. Mandal, and S. Kar, "Simulation, analysis and comparison of cutting parameter during turning of Ti-6Al-4V alloy with assorted industrial coated carbide," Materials Today: Proceedings, vol. 62, pp. 6136–6142, 2022, doi: 10.1016/j.matpr.2022.05.025.

[2] G. Hou, A. Li, X. Song, H. Sun, and J. Zhao, "Effect of cutting parameters on surface quality in multi-step turning of Ti-6Al-4V titanium alloy," Int J Adv Manuf Technol, vol. 98, no. 5–8, pp. 1355–1365, Sep. 2018, doi: 10.1007/s00170-018-2317-y.
[3] B. Buldum, U. Eşme, M. Kemal Külekci, A. Şik, and Y. Kazançoğlu, "Use of Grey-Taguchi Method for the Optimization of Oblique Turning Process of AZ91D Magnesium Alloy," Materials Testing, vol. 54, no. 11–12, pp. 779–785, Nov. 2012, doi: 10.3139/120.110392.
[4] E. Netolická, "Parametry řezného procesu," Czech technical university in Prague, Prague, 2016.
[5] R. Nur, N. M. Yusof, I. Sudin, F. M. Nor, and D. Kurniawan, "Determination of Energy Consumption during Turning of Hardened Stainless Steel Using Resultant Cutting Force," Metals, vol. 11, no. 4, p. 565, Mar. 2021, doi: 10.3390/met11040565.
[6] S. Yaldız and F. Ünsaçar, "A dynamometer design for measurement the cutting forces on turning," Measurement, vol. 39, no. 1, pp. 80–89, Jan. 2006, doi: 10.1016/j.measurement.2005.07.008.
[7] R. Čep and J. Petrů, "Experimentální metody v obrábění," p. 146.
[8] H. Saglam, S. Yaldiz, and F. Unsacar, "The effect of tool geometry and cutting speed on main cutting force and tool tip temperature," Materials & Design, vol. 28, no. 1, pp. 101–111, Jan. 2007, doi: 10.1016/j.matdes.2005.05.015.
[9] D.-A. Coroni and S.-M. Croitoru, "Prediction of Cutting Forces at 2D Titanium Machining,". Annals of DAAAM for 2013 & Proceedings of the 24th International DAAAM Symposium, Published by DAAAM International, Vienna, 2013, doi: 10.1016/j.proeng.2014.02.206.
[10] Ovako, "C45." https://steelnavigator.ovako.com/steel-grades/c45/ (accessed Apr. 20, 2022).
[11] Aerospace specification metals Inc., "Titanium Ti-6Al-4V (Grade 5), Annealed." https://asm.matweb.com/search/SpecificMaterial.asp?bassnum=MTP641 (accessed Apr. 20, 2022).
[12] A. Pramanik, "Problems and solutions in machining of titanium alloys," Int J Adv Manuf Technol, vol. 70, no. 5–8, pp. 919–928, Feb. 2014, doi: 10.1007/s00170-013-5326-x.
[13] Aerospace specification metals Inc., "Aluminum 7075-T6; 7075-T651." https://asm.matweb.com/search/SpecificMaterial.asp?bassnum=ma7075t6 (accessed Jun. 30, 2022).

DOI: 10.2507/33rd.daaam.proceedings.048

ANALYSIS AND COMPARATION OF EMERGENCIES IN THE ZLIN REGION

Martin Dzermansky, Nikola Cajkova & Tereza Sanderova

This Publication has to be referred as: Dzermansky, M[artin]; Cajkova, N[ikola] & Sanderova, T[ereza] (2022). Analysis and Comparision of Emergencies in the Zlin Region, Proceedings of the 33rd DAAAM International Symposium, pp.0344-0351, B. Katalinic (Ed.), Published by DAAAM International, ISBN 978-3-902734-36-5, ISSN 1726-9679, Vienna, Austria
DOI: 10.2507/33rd.daaam.proceedings.048

Abstract

Every day, people are surrounded by emergencies that affect not only their health, lives, and property but also the environment. These events are a manifestation of the harmful effects of forces and phenomena, either on the environment or on the human activity itself. To improve the preparation not only of crisis management bodies and the protection of the population but also of the general public, it is necessary to carry out more detailed analyzes of threats and the impact of individual risks. The work aims to analyze emergencies in the selected local government and use methods of comparition, analysis, modeling, and qualitative risk analysis using their correlation to point out the most dangerous types of events and propose measures to reduce their impact.

Keywords: Analysis; Comparition; Emergencies; Zlin.

1. Introduction

Emergencies are events that threaten the health and lives of persons, property, animals, and the environment. Šin et al. 2017 divide emergencies into two categories:

- natural emergencies,
- anthropogenic emergencies.

These main categories are further divided into subcategories. Natural emergencies include:

- abiotic emergencies (origin in non-living nature),
- biotic emergencies (origin in living nature).

Anthropogenic emergencies include:

- man-made emergencies (originating in operational accidents and accidents connected with infrastructure),
- internal sociogenic emergencies (origin of national social, social, economic crisis),
- external sociogenic emergencies (origin of a military crisis),
- agrogenic emergencies (origin associated with agriculture and soil). [1]

The term emergency is also closely related to the term danger. Haddow, Bullock, and Coppola, 2020 define a hazard as "a resource that may or may not lead to an emergency or disaster. It is an event or physical condition that has the potential to cause death, injury, damage to property, infrastructure, and more." [2]. The issue of emergencies is dealt for example in the article "The Comparative Analysis of Safety in the Czech Republic and in Abroad" by Vichova, K. et al., where the dangers for the Czech Republic are compared. In the article "Preparedness in Crisis Management", the authors Brumova, L. et al deal with the general concept of security in the Czech Republic and collect conceptual and strategic materials. The most basic material includes the Threat Analysis for the Czech Republic from 2015. This document analyzes the most serious threats to the Czech Republic and the result is the identification of 22 types of dangers with unacceptable risks. [3], [4], [5]

The biggest issue that all these articles deal with is the devastating effects of these events. It is necessary to deepen knowledge not only among the authorities of crisis management and population protection but also among the general public. This work aims to point out the number of emergencies in the Zlin region when the municipality with the extended scope of Uherske Hradiste was chosen as an experimental model. To emphasize the importance, a qualitative risk analysis was created using their correlation, which points to the most serious risks.

2. Materials and methods

A total of four methods are used in this work. The first of them is the method of explanation, which is used to describe the theoretical basis of the work. Another method is the method of comparition, which is used to assess emergencies ts in the territory of the Zlin Region and the municipality with the extended jurisdiction of Uherske Hradiste. The third method is the modeling method, which is used to create map outputs in the QGIS software, thanks to which the dislocation of municipalities with extended jurisdiction in the Zlin Region is shown. Last but not least, the method of qualitative analysis of risks using their correlation is used for the evaluation and assessment of emergencies and the resulting graph of correlation.

The main goal of the work is the analysis of emergencies that were recorded in the Zlin Region in the Czech Republic and a closer analysis of the municipality with the extended scope of Uherske Hradiste, which falls into this region. The Zlin region is located in the eastern part of the Czech Republic on the border with the Slovak Republic and covers an area of 3,964 km2. The total number of inhabitants is reported to be around 587,000 and it is divided into a total of thirteen administrative districts with extended jurisdiction. [6]

Fig. 1. Dislocation of municipalities with extended jurisdiction in the Zlin region

Statistical yearbooks of the Fire Rescue Service of the Czech Republic were used for the analysis of emergencies in this region, with the help of which the resulting event values were calculated. These calculations showed that a total of 56,942 emergencies occurred in the Zlin region from the period 2010–2021. In graph number 1, you can see the number of emergencies for individual years. [7]

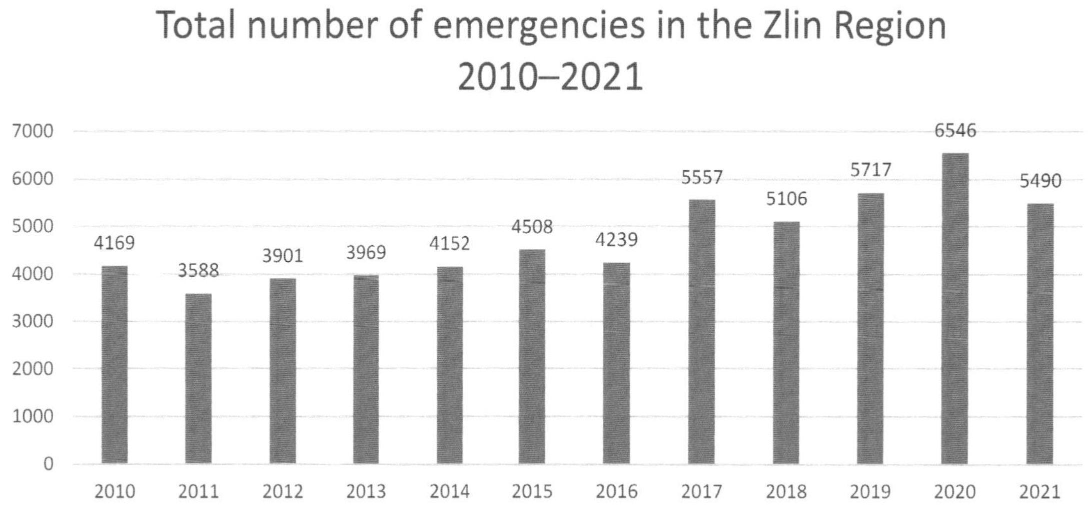

Graph 1. Total number of emergencies in the Zlin Region 2010–2021

For a more detailed analysis, individual types of emergencies were selected and their total numbers in the region were calculated. Between 2010 and 2021, a total of 56,942 emergencies were recorded in the Zlin Region. These events were divided into the following categories:

- Fires,
- Traffic accidents,
- Leakage of dangerous substances,
- Technical crashes,
- Radiation accident,
- Other emergencies,
- False alarms.

Graph number 1 shows the total number of emergencies in the Zlin Region for the period 2010–2021, and for comparition, the number of emergencies in the municipality with the extended scope of Uherske Hradiste, which was chosen for further analysis as a simplified model, is indicated. Individual emergencies are categorized according to the standards of the Fire and Rescue Service of the Czech Republic, just as they are listed in the statistical yearbooks. [7]

	Zlin region	The municipality with extended scope of UH
Emergencies	The number of emergencies	The number of emergencies
Fires	8 234	1 150
Traffic accidents	10 477	1 646
Leakage of dangerous substances	2 688	343
Technical accidents	33 446	6 613
Radiation accidents	2	0
Other emergencies	693	16
False alarms	4 134	706
Total amount	59 674	10 474

Table 1. Comparition of the number of emergencies between the Zlin region and the municipality with the extended jurisdiction of Uherske Hradiste

3. Results

To determine the most dangerous emergency events in the municipality with the extended jurisdiction of Uherske Hradiste, a qualitative risk analysis will be used using their correlation. The method is based on comparing individual events, where the values 0 and 1 are assigned according to the possibility of triggering another event or not. Based on the comparition of the number of emergencies, the emergencies were further included in the qualitative risk analysis using their correlation. In the first step, table number 2 was created, the so-called risk correlation table, where 7 branches of emergencies were assigned. [8]

	Risk	**1.**	**2.**	**3.**	**4.**	**5.**	**6.**	**7.**	**Sum**
1.	Fires	0	1	1	1	1	1	0	**5**
2.	Traffic accidents	1	0	1	1	0	1	0	**4**
3.	Leakage of dangerous substances	1	0	0	1	0	1	0	**3**
4.	Technical accidents	1	1	1	0	1	1	0	**5**
5.	Radiation accidents	0	0	0	0	0	1	0	**1**
6.	Other emergencies	1	1	1	1	0	0	0	**4**
7.	False alarms	0	0	0	0	0	0	0	**0**
	Total amount	**4**	**3**	**4**	**4**	**2**	**5**	**0**	

Table 2. Correlation of risks

After constructing the risk correlation table, the values 1 and 0 were assigned. If the risk Ri can cause the risk Rj, the value 1 is assigned. Otherwise, the value 0 was assigned.

Calculation of activity and passivity coefficients

For the assessment of emergencies in the municipality with the extended scope of Uherske Hradiste, it is necessary to calculate the coefficients of activity and passivity. Using these coefficients, the mathematical and graphical form of the method is then compiled.

- KARi – activity coefficient – is a percentage expression of the number of selected risks that are connected to the risk marked Ri. In case the risk Ri occurs,

so, these subsequent risks can be triggered.

- KPRi – passivity coefficient – is a percentage expression of the number of selected risks that are connected to the risk marked Ri and that can subsequently trigger the risk Ri.

To express the coefficients of activity and passivity, it is necessary to create several combinations. It is based on the formula that the number of risks is equal to x = 7, where the number of possible combinations is always x-1.

Calculation of the KARi activity coefficient for individual risks Ri:

$$K_{ARi} - \frac{\sum Ri}{x - 1} * 100\ [\%]$$

1. $K_{ARi} = \frac{\sum Ri}{x - 1} * 100\ [\%] = \frac{5}{7-1} * 100 = \frac{5}{6} * 100 = 83{,}33\ \%$
2. $K_{ARi} = \frac{\sum Ri}{x - 1} * 100\ [\%] = \frac{7}{7-1} * 100 = \frac{4}{6} * 100 = 66{,}66\ \%$
3. $K_{ARi} = \frac{\sum Ri}{x - 1} * 100\ [\%] = \frac{3}{7-1} * 100 = \frac{3}{6} * 100 = 50\ \%$
4. $K_{ARi} = \frac{\sum Ri}{x - 1} * 100\ [\%] = \frac{5}{7-1} * 100 = \frac{5}{6} * 100 = 83{,}33\ \%$
5. $K_{ARi} = \frac{\sum Ri}{x - 1} * 100\ [\%] = \frac{1}{7-1} * 100 = \frac{1}{6} * 100 = 16{,}66\ \%$
6. $K_{ARi} = \frac{\sum Ri}{x - 1} * 100\ [\%] = \frac{4}{7-1} * 100 = \frac{4}{6} * 100 = 66{,}66\ \%$
7. $K_{ARi} = \frac{\sum Ri}{x - 1} * 100\ [\%] = \frac{0}{7-1} * 100 = \frac{0}{6} * 100 = 0\ \%$

Calculation of passivity coefficients KPRi for individual risks Ri:

$$K_{PRi} - \frac{\sum Ri}{x - 1} * 100\ [\%]$$

1. $K_{PRi} = \frac{\sum Ri}{x - 1} * 100\ [\%] = \frac{4}{7-1} * 100 = \frac{4}{6} * 100 = 66{,}66\ \%$
2. $K_{PRi} = \frac{\sum Ri}{x - 1} * 100\ [\%] = \frac{3}{7-1} * 100 = \frac{3}{6} * 100 = 50\ \%$

3. $K_{PRi} = \frac{\sum Ri}{x-1} * 100\ [\%] = \frac{4}{7-1} * 100 = \frac{4}{6} * 100 = 66{,}66\ \%$

4. $K_{PRi} = \frac{\sum Ri}{x-1} * 100\ [\%] = \frac{4}{7-1} * 100 = \frac{4}{6} * 100 = 66{,}66\ \%$

5. $K_{PRi} = \frac{\sum Ri}{x-1} * 100\ [\%] = \frac{2}{7-1} * 100 = \frac{2}{6} * 100 = 33{,}33\ \%$

6. $K_{PRi} = \frac{\sum Ri}{x-1} * 100\ [\%] = \frac{5}{7-1} * 100 = \frac{5}{6} * 100 = 83{,}33\ \%$

7. $K_{PRi} = \frac{\sum Ri}{x-1} * 100\ [\%] = \frac{0}{7-1} * 100 = \frac{0}{6} * 100 = 0\ \%$

After calculating the coefficients of activity and passivity for individual risks, a table of coefficients of activity and passivity is drawn up, which makes it possible to more clearly create the axes of the resulting correlation graph.

Risk R_i	1	2	3	4	5	6	7
K_{ARi} [%]	83,33	66,66	50	83,33	16,66	66,66	0
K_{PRi} [%]	66,66	50	66,66	66,66	33,33	83,33	0

Table 3. Coefficients of activity and passivity

The resulting correlation graph

The task of creating a chart is to determine the significance of all risks and their correlation in a system. The graph is divided by two axes O_1 and O_2 into 4 categories, which are shown in graph No. 2.

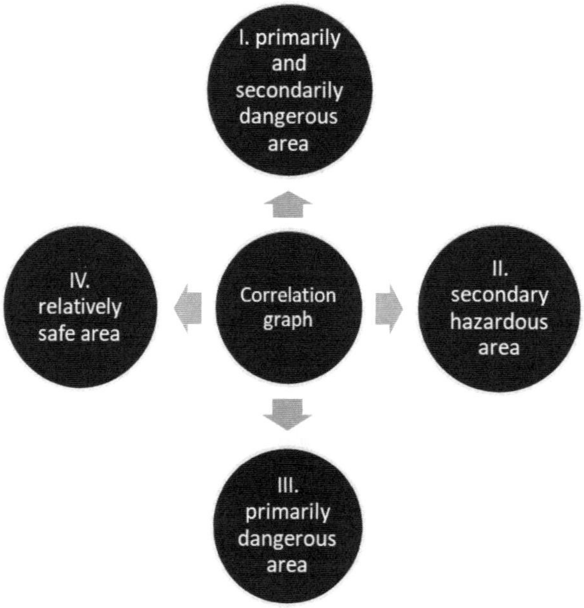

Graph 2. Correlation graph areas

Area I. in the resulting graph covers 80% of the total area where the assessed risks are located. For axis O_1:

$$K_{Amax} - K_{Amin} = 100\ \%$$

In the case of the construction of the O_1 axis, if the 80% condition is met, it will be parallel with the y-axis at a distance of:

$$O_1 = K_{Amax} - \frac{K_{Amax} - K_{Amin}}{100} * 80$$

$$O_1 = 83{,}33 - \frac{83{,}33 - 0}{100} * 80 = 83{,}33 - 66{,}66 \doteq 17$$

Result for O1 = 17%.

For the O_2 axis, if 80% of the condition is met, the parallel to the x-axis is at a distance of:

$$O_2 = K_{Pmax} - \frac{K_{Pmax} - K_{Pmin}}{100} * 80$$

$$O_2 = 83{,}33 - \frac{83{,}33 - 0}{100} * 80 = 100 - 66{,}66 \doteq 17$$

Result for O2 = 17%.

Evaluation of the KARS method

After calculating the O_1 and O_2 axes, the values for defining the quadrants of the correlation graph were obtained. This graph can be seen in figure 1.

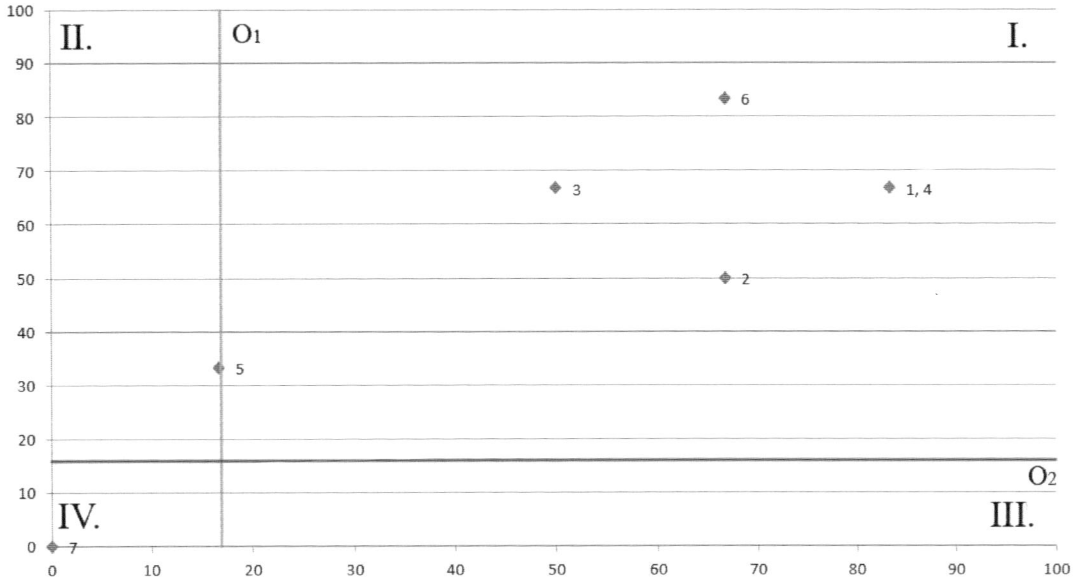

Fig. 2. Correlation of KARi and KPRi coefficients for Ri

It follows from the correlation graph that III. the quarter contains no risk. The most comprehensive is the area I. The others have one risk each. A more comprehensive list is described below.

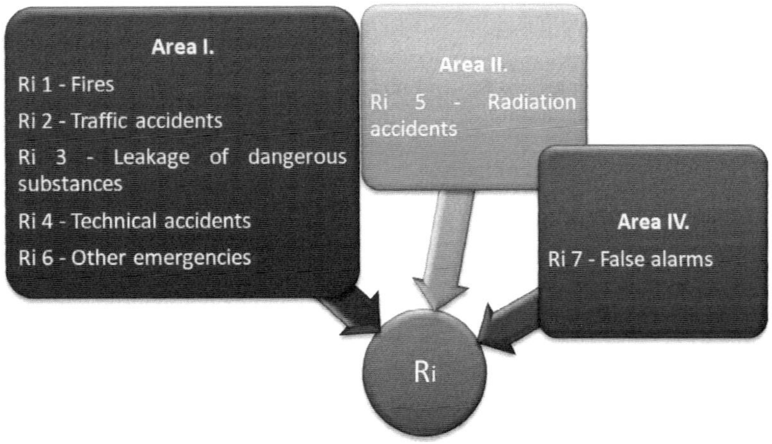

Graph 3. Result areas and risks of the correlation graph

Area I. Primary and secondary dangerous risks:

A total of five emergencies were included between Area I. primary and secondary dangerous risks, that is 71% of the total number of emergencies defined in the analysis.

Ri 1 – fires – are emergencies that cause unwanted burning, resulting in the death or injury of persons or animals, and damage to property or the environment. Fires are generally among the most destructive emergencies, and an average of 686 fires occur annually in the Zlin region and 96 times in the municipality with the extended scope of the Uherske Hradiste.

Ri 2 – traffic accidents – emergencies in which activities related to the elimination of the consequences of a collision of means of transport take place. This includes road traffic accidents, mass road traffic accidents, railway traffic accidents, air traffic accidents, and other traffic accidents (traffic accidents on country and forest roads).

Ri 3 - spill NL - emergencies that lead to the release of dangerous chemical substances, including petroleum products and other substances.

Ri 4 – technical accidents – emergencies in which dangers or dangerous conditions are eliminated. Technical breakdowns, technical assistance, technological assistance, and other assistance can be distinguished.

Ri 6 - other emergencies - represent interventions for events such as epidemics or contagions caused by a dangerous disease. The number of these events increased mainly during the period of the COVID-19 disease and the declaration of a state of emergency in the Czech Republic.

Emergencies that occurred in this area are the most dangerous for the municipality with the extended scope of the Uherske Hradiste, and it is necessary to reduce their impact to a minimum. As a solution, an increased number of exercises of the components of the integrated rescue system is offered, which can make rescue and liquidation work more efficient. So-called standard activities are processed for this purpose. These activities contain the procedures and coordination of all components of the integrated rescue system and are thus an indispensable part of coping with emergencies. Another measure is the increased awareness of the population about the nature of emergencies, information, and early warning using a unified warning and notification system.

Area II. Secondary dangerous risks:

Ri 5 – radiation accidents – emergencies in which radioactive substances or ionizing radiation are released. In the last 12 years, these events have occurred only 2 times, but in the analysis, the event came out among the most dangerous risks. Although these events do not occur as often and in number as other events that have been included in this area, they represent their place here due to the impact on the population and the possible emergence of a domino effect that would trigger other emergencies, see table no. 1.

Risks placed in this area are also a high risk for that area, but not as high a priority as the risks in the first area. In the case of radiation accidents, similar proposals are offered for the first area, i.e. a greater deepening of the knowledge of the population in the event of a radiation accident and the involvement of components of the integrated rescue system in more frequent exercises on the given issue.

Area IV. Relatively safe:

Ri 7 – false alarms – these are activities that are triggered due to reports of fire or other events that have not been confirmed. These false alarms cannot be estimated, so after an emergency call has been announced, the individual components of the EMS are obliged to arrive at the place of notification. Unfortunately, these false alarms have a growing tendency and people are not aware of the risks involved.

4. Discussion

Emergencies are an integral part of the lives of all people around the world. Thousands of incidents occur every day that require rescue and liquidation work. It is, therefore, necessary to pay increased attention to these events and to prepare not only the components of the integrated rescue system to deal with them but also the population itself. The focus of the work was oriented to the analysis of emergencies in the Zlin region and, for a closer specification, to the municipality with the extended scope of Uherske Hradiste.

Data from the statistical yearbooks of the Fire and Rescue Service of the Czech Republic were used, with the help of which the total number of emergency incidents was calculated and compared within the territorial self-governing units, i.e. the Zlin Region and the municipality with the extended jurisdiction of Uherske Hradiste. To evaluate the seriousness of emergencies, a qualitative analysis of risks using their correlation was primarily used.

5. Conclusion

The presented article aimed to analyze and compare emergencies in the Zlin region with a focus on a smaller model, namely the municipality with the extended scope of Uherske Hradiste. The work used the methods of explanation, comparition, modeling, and qualitative analysis of risks using their correlation. The first method of explanation is used to explain a process or phenomenon, and specifically, in this work, it was used to explain individual emergency events and their categorization that occur in the Zlin region and to explain the issue of the integrated rescue system.

The comparition method was used to compare emergencies both between the Zlin region and the municipality with the extended scope of Uherske Hradiste and also between themselves as such. The third method used was the modeling method. This method represents a simplified picture of reality and was used in this work to create a map output that shows the territorial distribution of municipalities with extended jurisdiction in the Zlin region. The fourth and last method is a qualitative analysis of risks using their correlation, which was used to compare individual emergencies and compile the resulting graph of correlation, which determined four basic areas and thus pointed to the most dangerous emergencies, and conversely also to the relatively safest ones.

The analysis showed that the most serious emergency events in the municipality with the extended scope of Uherske Hradiste are fires, traffic accidents, leaks of dangerous chemical substances, technical accidents, and other emergencies. It, therefore, reflects a fact that is also based on the calculation of individual data from the statistical yearbooks of the Fire Rescue Service of the Czech Republic. To be prepared for these emergencies, exercises are carried out by the components of the integrated rescue system, which aim to prepare for emergencies and carry out rescue and liquidation work. The limiting element of this work is the lack of published treatment of this area. Only a minimum of articles are devoted to the analysis of this area within the framework of emergencies. Future research will deal with deepening the area of exercises for the components of the integrated rescue system and streamlining rescue and liquidation work during emergencies in the Zlin Region.

6. Acknowledgments

This research was supported by the Internal Grant Agency of Tomas Bata University under project No. IGA/FAI/2022/003.

7. References

[1] ŠÍN, Robin, (2017). Medicine of disasters. Praha: Galén. ISBN 9788074922954.

[2] HADDOW, George D., Jane A. BULLOCK a Damon P. COPPOLA, (2020). Introduction to emergency management. 7. Oxford: Butterworth-Heinemann. ISBN 978-0-12-817139-4.

[3] Brumarová, Lenka & Brumar, Jakub & Tomanová, Kateřina. (2020). Preparedness in Crisis Management. 1915-1922. 10.3850/978-981-14-8593-0_4442-cd.

[4] Vichova, K.; Hromada, M.; Ficek, M. & Gracla, M. (2018). The Comparative Analysis of Safety in the Czech Republic and in Abroad, Proceedings of the 29th DAAAM International Symposium, pp.1181-1186, B. Katalinic (Ed.), Published by DAAAM International, ISBN 978-3-902734- 20-4, ISSN 1726-9679, Vienna, Austria DOI: 10.2507/29th.daaam.proceedings.170

[5] Analysis of Threats for the Czech Republic. In: . Prague: Ministry of the Interior, 2015, ročník 2015.

[6] Zlín Region, c2022. Zlindnes. Zlín: Zlindnes. Available from: https://www.zlindnes.cz/region/

[7] Statistical yearbook of the Fire and Rescue Service of the Czech Republic. (2022). Prague: Ministry of the Interior - General Directorate of the Fire and Rescue Service of the Czech Republic, 2022.

[8] PACINDA, Štefan, (2010). Network analysis and the KARS method. The Science for Population Protection. Available from: http://www.populationprotection.eu/prilohy/casopis/8/56.pdf.

DOI: 10.2507/33rd.daaam.proceedings.049

RISK ANALYSIS OF THE POST OFFICE

Nikola Cajkova & Martin Dzermansky

This Publication has to be referred as: Cajkova, N[ikola] & Dzermansky, M[artin] (2022). Risk Analysis of the Post Office, Proceedings of the 33rd DAAAM International Symposium, pp.0352-0357, B. Katalinic (Ed.), Published by DAAAM International, ISBN 978-3-902734-36-5, ISSN 1726-9679, Vienna, Austria
DOI: 10.2507/33rd.daaam.proceedings.049

Abstract

The issue of security accompanies humanity throughout civilized history. Since ancient times, ensuring security has been a priority. One of the main functions that every state must fulfill is security, not only internal but also external so that every citizen feels safe. Based on security, people choose the countries, states, cities, and buildings where they want to live. Security is one of the issues that is addressed in every area of life and is the second rung of Maslow's pyramid of needs. The present paper deals with the issue of protecting soft targets. Postal risk analysis has been chosen as the specific target. Explanation, modelling, and analysis methods have been used for the analysis. This article aims to analyze the environment of the reference object – the post office. Using the analysis of the object's surroundings, the object, the traffic situation, the location of the reference object, and the description of the location, risks are specified and specific safety measures are subsequently proposed.

Keywords: analysis; post office; reference object; risk analysis; soft target.

1. Introduction

This article analyzes the reference object of the post office in the Zlin region. It looks at the reference object from the point of view of the layout, the near and far surroundings of the post, the layout of the object, and the overall security of this soft target. A post office is a place that has our sensitive data, financial resources, and personal data and for several factors, it could be a very attractive target for various types of attacks.

Even though this is an attractive target, in the Czech Republic, for the most part, post offices have undergone almost no or minimal reconstruction, and money has not been invested even in higher security or security. In the case of this article, a specific post office in Malenovice was chosen, which is located a short distance from the regional town of Zlin. This post office is interesting in that it covers nine territorial areas. A post office is a place where several hundred people go every day. these people visit the post office regularly at similar times, which makes them a soft target. Vulnerability increases with the number of people in the facility. This article wants to point out the critical points of an object that is very well-known to everyone. The article aims to analyze possible risks and propose specific security measures. As described in the article by Vichova, K. et al. the world is constantly changing and new threats are coming, which also threaten the elements of critical infrastructure, which need to pay attention and increase their protection and attention to the issue of risk management. [1], [2, [3], [4], [5], [6]

Post Office

The post office building is located in one of the local parts of the city of Zlin, specifically in Malenovice. The building is located in the center of Malenovice on Masarykova street. It is a relatively large post office, as nine territorial areas fall under this post office - such as Karlovice, Tečovice, Chlum, and Lhotka.

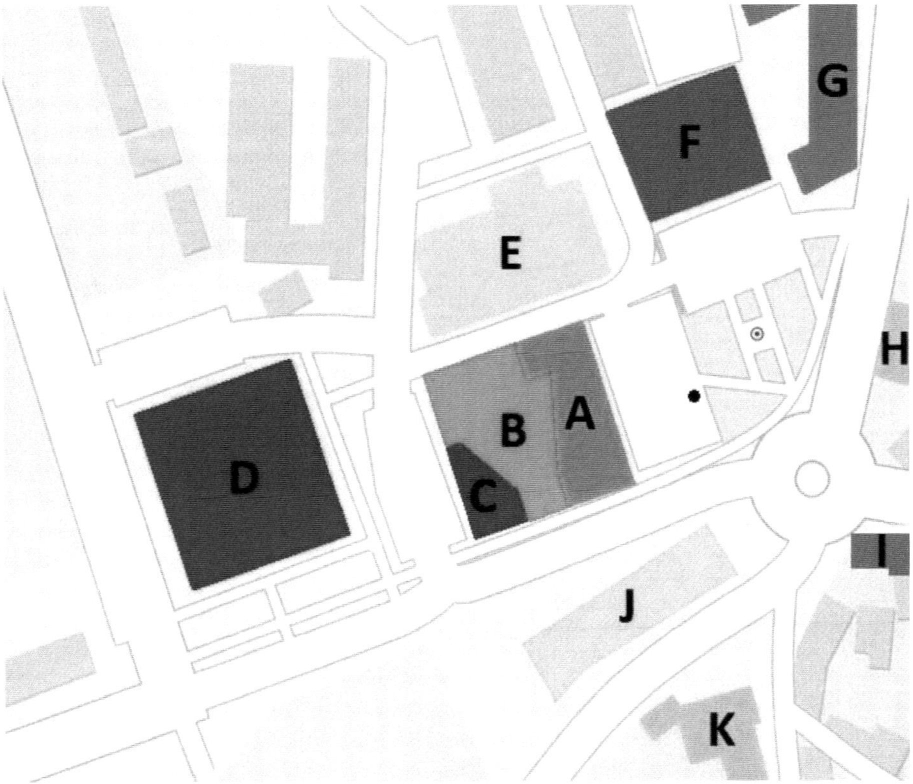

Fig. 1. The location of the building and its surroundings

Legend:

A – Post Office
B – Cinema
C – Police Office
D – Hospital
E – Sports Center
F – Grocery Store
G – Dorm
H – Goldsmithing
I – Restaurant
J – Apartments, Fast Food, Pizzeria, Wine Shop, Offices, Hairdresser
K – Bar
Unmarked -Appartments
Black point – Anti-nuclear Shelter

As already mentioned, the post office is located in the center of the village. In front of it is a recently built square with benches, grassy areas, some trees, and a water feature. Thanks to this, this space has become a popular place that many people visit every day. It is, therefore, constantly monitored by a "dome" camera located at the southeastern tip of the post office.

There is an anti-nuclear shelter underground, the entrance of which is right on this square and is shown by a small black dot on the map. The post office is part of a larger building, which also houses the office of the Zlin Municipal Police and the former Květen cinema, which is currently undergoing reconstruction and will become a cultural center. To the north, this building is adjacent to the recently renovated Sportcentrum Malenovice. The sports center contains many training rooms, gymnasiums, playing areas, wellness facilities, and two restaurants. To the northeast are a grocery store, nursing home, and the office of the local part of the Zlin Municipality in one building. The already mentioned square is the east of the post office, followed by a goldsmith's shop across the road.

Not far from it is also a recently built roundabout (picture 1), next to which the Restaurant Margherita is located. In the south, a large building includes offices, apartment units, a hair salon, fast food Mr. Grill, Pizzeria, a construction company, and a wine shop. Further south is the Hostinec Horní pub. To the west of the post office building is the local hospital, where there is a pharmacy and collection room in addition to doctors' offices. Unmarked buildings on the map are family houses or multi-story apartment buildings.

There is much traffic in this area every day, as approximately 250 m north of the roundabout is the main traffic route connecting Otrokovice and Zlin. There are parking spaces almost all around. These can be found at the Enapo store, the hospital, or in the street from the roundabout to building J. However, the post office building can only be accessed by vehicle with a permit on the road between it and the sports center. Everywhere else, bollards or other forms of barriers are erected. There is also a bike path right around the south side of the building. [1]

2. The Reference Object

The post office is one of three entities within the larger building. It also shares some walls with one of them, the former Květen cinema. These are the walls of rooms 5, 6, 7, 8, and 9, i.e., the upper walls in the picture below. The post office has 16 internal rooms, with the number 0 marking the outer corridor, which can also be used by wheelchair users or parents with prams.

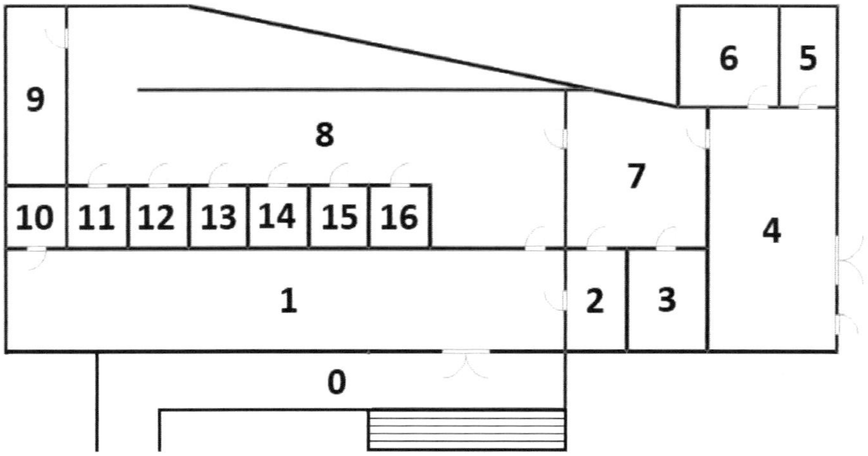

Fig. 2. Floor plan of the reference object

Legend:

0 – access for wheelchairs or prams
1 – entrance hall/waiting room
2 – office manager
3 – dressing room
4 – receiving and sending shipments
5 – WC
6 – bathroom, showers
7 – corridor
8 – corridor, warehouse
9 – partition of the discrete zone
10 – discrete zone
11 – partition No. 1
12 – partition No. 2
13 – partition No. 3
14 – partition No. 4
15 – partition No. 5
16 – partition No. 6

Visitors are usually only allowed access to two or three rooms. The building is slightly elevated from the square, so they have to use the stairs or the corridor marked with the number 0 to get to the main door. They enter the entrance hall/waiting room through this door, take a serial number from the device, and, based on the chosen activity, are then called to the corresponding counter. If they do not want someone to follow them at the counter, they can use the discreet zone with the number 10. [1]

From the entrance hall, specifically to rooms 2 and 8, these are locked, and only employees have access to the keys. It is in room 2 that the manager resides in her office so that she can at any time easily supervise the operation of the post office in both parts of the post office, the one accessible to visitors and the one inaccessible. Another door from the office leads to a corridor (7), from this corridor you can go to three more rooms (3, 4 and 8). Room 3 is a staff cloakroom. The space marked with the number 4 is used for receiving and sending shipments; it is also used for the entrance of employees into the building; there is a standard door just for employees, and next to it is a large door for easier handling of packages, etc. There is also a door to the toilet (5) and the bathroom and showers (6). The last door from corridor 7 leads to room 8, mainly used as a warehouse for all parcels stored at the post office. From the warehouse, the door leads to the partition of the discrete zone (9) and individual partitions 11 – 16. Counter No. 1 (11) serves as the Checkpoint public administration point of contact for verifying documents and signatures and deals with monetary services, ČSOB supplementary pension insurance, pensions, and building savings of the ČMSS. Counter No. 2 (12) deals with online betting, lottery tickets, phone coupons, and SIM cards, as well as postal savings bank services, Western Union money transfers, and also Checkpoints. [1], [2]

Counters No. 3 and 4 (13, 14) sell stamps and stamps, highway coupons, magazines, newspapers, and additional assortment. Counter No. 5 (15) deals with the submission of letters and parcels, the establishment of radio and television, the centralized collection of payments from the SIPO population, the debt collection of the DINO population, and the change of delivery of all incoming parcels (delivery). The last counter, i.e., No. 6 (16), issues letter items and packages.

3. Methods

There are many different areas to consider in risk analysis. Whether the location of the city, the location of the object and its surroundings, the premises of the object itself, the number of visitors, dangerous substances, the traffic situation, and the distance of the IRS. [7]

Location

Malenovice is located near the regional town of Zlin Region, Zlin, in the southeast of the Czech Republic. They are a local part of Zlin, located at an altitude of 208 meters above sea level and approximately 7,000 inhabitants live there. Crime is very rare in this village. [1]

Location of the reference object

The post office is located in the center of the village, where the village is divided into the so-called "old" and "new" Malenovice. Within the building, the post office is one of three entities, the other two being the former Květen cinema and the Zlin Municipal Police office. In addition, in the vicinity of the post office, there is a hospital, a sports center, a grocery store, a nursing home, and the office of the local part of the Zlin Municipality, then a goldsmith's shop, a restaurant, a pub, offices, a construction company, a wine shop, fast food, a hairdresser and a pizzeria. There is also an anti-nuclear shelter, the entrance of which is direct infront of the post office, and the premises of the shelter are partially under the post office itself.

Attendance

It is one of the most visited parts of the village, precisely because of the post office, hospital, pharmacy, grocery store, restaurants, pizzeria, bars, fast food, and also thanks to the recently reconstructed square and sports center. Dozens of people in every season of the year spend time on the square every day. This number increases to hundreds per day in the summer months, when people buy and consume ice creams in abundance in one of the restaurants of the sports center. In the warm months, more people from nursing homes also visit the square. Visits to the post office itself are relatively constant; the only significant increase is around Christmas.

The largest number of people at once appear in this locality during the annual lighting of the Christmas tree directly in the square infront of the post office, accompanied by a musical program. On this occasion, many stalls with alcoholic drinks and other snacks are displayed in the square. The lighting always takes place on the first Sunday in December, and attendance ranges from hundreds to thousands of people. It can be assumed that when the reconstruction of the cinema into a cultural center is completed, the number of visitors will increase even more. Especially on weekends, when this center will host various social events, celebrations, balls, entertainment, etc.

Structure of the reference object

The building has three entrances, but only one is accessible to visitors. In the event of danger, they would have only one possible escape route; otherwise, they would have to escape by breaking the windows, or the employees would let them into a section reserved just for them through the door to the manager's office or the warehouse. The post office is accessible to everyone; stairs and a corridor for wheelchairs or prams lead to the main entrance. However, this is not the case at the entrances for employees, which can only be accessed by stairs. A risk factor could be that the post office shares some of its interior walls with the renovated cinema. There are no hazardous substances in the building. The neighborhood is constantly monitored by a camera on the southeast corner of the post office.

Traffic situation

The post office is close to the main traffic route from Zlin to Otrokovice. A large number of vehicles move every day near the post office. Many people pass through here and head for Lhota u Malenovic, Salaš, Karlovice, or Bohuslavice u Zlin. The traffic situation in this area has recently improved with the construction of a roundabout as you can see in Fig. 1. There are many parking spaces around the building, and they are always almost all occupied. Some parking lots have a three-hour limit, which the Zlin Municipal Police monitor. For this reason, drivers must have a clock in their car with a set parking time. Otherwise, they risk receiving a fine. Cars can only get to the post office along the way, with exceptions; otherwise, there are various types of barriers in the way. However, if someone wanted to get to the post office instead of by car, they would get there, for example, via the bike path, the grassy area near the square, etc.

IRS

If necessary, IRS forces can be on site within a few tens of seconds. As has already been mentioned several times, the office of the Zlin Municipal Police is part of the building that also houses the post office. The presence of this office and frequent patrols in the area are also a form of deterrence against potential criminals.

The police of the Czech Republic would be on the scene within 5-10 minutes. Depending on the traffic situation, it could be determined whether he should come from Zlin or Otrokovice. The ambulance from Prštná could also be at the post office within ten minutes at the latest. As for the firefighters, the volunteer firefighters from Malenovice could be on the scene in a few minutes, and the professional firefighters from Otrokovice in 5-10 minutes.

Object category

Since the object of interest is the post office, the potential risks can be significant. Dozens of people (employees + visitors) are constantly moving in the building, and relatively large sums of money are also located there. For this reason, increased attention and emphasis on safety are definitely in place. The disadvantage is that there is no physical security guard in the building, but it is at least equipped with a camera system and emergency buttons. The post office opens at 8:00 a.m. and closes at 5:00 p.m. on weekdays. On Saturdays, it opens the same, but closes already at 10:00. It is closed on Sundays.

Risks

The main risk factors for this object are the increased number of people daily. It is even amplified in the summer months or in case of some events. Another factor is the layout of the building itself, where it is relatively difficult to escape in case of danger if the primary exit is blocked by, for example, fire or an attacker, etc. A more complex traffic situation can also be a risk, which could partially slow the arrival of the IRS units. A big risk can be that there are not enough barriers and obstacles infront of the post office so that the car cannot go off the road and get to the square. In the case of a large number of people, this could be catastrophic. The main factor can be marked high the amount of money deposited at each post office. This can be a "lure" for a would-be offender.

Security measures

Multiple entrances, especially for visitors, could facilitate escape in case of danger. It would help to build a second entrance door, for example, opposite partition No. 1 (11). Another measure could be the hiring of at least one physical security worker, who would be present in the entrance hall and ready to intervene if necessary. Surveillance would also be suitable after closing time (especially at night); combined with a camera system, it would already be high-quality security. The problem with the possible entry of a car or other means of transport could be solved by building pillars or a continuous railing or other elements in the entire vicinity of the square. If there were bollards, people would have no problem getting to the square, while a car would have no chance to pass through this barrier. Barriers could be built on access roads, which would, for example, drive into the ground after verification by remote control. In this way, entry would be ensured for those who have permission to do so, while at the same time, the entry would be denied to others. [8]

4. Conclusion

There is a need to proactively think and act in the security prevention industry these days. Nowadays, the spotlight is placed on objects with high turnover, but the goal of all of us should be a way to solve the issue of object safety not only in big cities and places with the highest concentration of people, but we should strive to set safety standards that should be observed across the board.

This article was focused on a small reference facility in a small town that showed us that the technical aspects of security and the overall security view of this facility were lacking. The securitization problem, threatening soft targets is a risk to everyday life. Analysis of current objects that are used daily and their corrective safety measures are a must and should be standardized across the board. This article deals with the analysis of the Post office object with specific examples of improving the security of the object. These security measures were individually chosen for a specific reference object. More objects are visited every day and have a low level of security. Experts are constantly working on security elements and improving the security of objects, however, it is important to standardize across the board, which would determine the necessary level of security. We would like to discuss the topic of normalization in a future article.

5. Acknowledgments

This article was supported by an internal grant project of Tomas Bata University under the number IGA/FAI/2022/003.

6. References

[1] Pošta: online [online]. Zlin, 2022.Available from: https://www.postaonline.cz/detail-pobocky/-/pobocky/detail/76302

[2] LUKÁŠ, Luděk. Teorie bezpečnosti: Teorie a praxe ochrany majetku a fyzické bezpečnosti. 2017. Zlin: VeRBuM, 2017. ISBN 978-80-87500-89-7.

[3] LUKÁŠ, Luděk. Teorie bezpečnosti: Teorie a praxe ochrany majetku a fyzické bezpečnosti. 2017. Zlin: VeRBuM, 2017. ISBN 978-80-87500-99-6.

[4] Vichova, K[aterina]; Hromada, M[artin]; Valasek, J[armil] & Paulus, F[rantisek] (2020). Comparison Analysis the Use of Modern Technologies by Fire Rescue Service, Proceedings of the 31st DAAAM International Symposium,

pp.0535-0541, B. Katalinic (Ed.), Published by DAAAM International, ISBN 978-3-902734- 29-7, ISSN 1726-9679, Vienna, Austria DOI: 10.2507/31st.daaam.proceedings.074

[5] Vulanovic, S[rdjan]; Zizakov, M[arina]; Vasic, S[tana]; Delic, M[ilan] & Sremcev, N[emanja] (2019). The Impact of Occupational Health and Safety (OH&S) Management Systems on Risk Management Practices , Proceedings of the 30th DAAAM International Symposium, pp.1188-1195, B. Katalinic (Ed.), Published by DAAAM International, ISBN 978-3-902734-22-8, ISSN 1726-9679, Vienna, Austria DOI: 10.2507/30th.daaam.proceedings.167

[6] Malatinsky, A[dam]; Drofova, I[rena]; Sousedikova, L[ucie] & Hromada, M[artin] (2021). Fire Safety Threat Risk Analysis for Soft Target, Proceedings of the 32nd DAAAM International Symposium, pp.0586-0592, B. Katalinic (Ed.), Published by DAAAM International, ISBN 978-3-902734-33-4, ISSN 1726-9679, Vienna, Austria DOI: 10.2507/32nd.daaam.proceedings.083

[7] KALVACH, Zdeněk. Základy ochrany měkkých cílů metodika. Praha: Ministerstvo vnitra České republiky, 2016.

[8] LUKÁŠ, Luděk. Bezpečnostní technologie, systémy a management. Zlin: Radim Bačuvčík - VeRBuM, 2015. ISBN 978-80-87500-19-4.

DOI: 10.2507/33rd.daaam.proceedings.050

RISK ANALYSIS OF A GAS STATION

Nikola Cajkova & Martin Dzermansky

This Publication has to be referred as: Cajkova, N[ikola] & Dzermansky, M[artin] (2022). Risk Analysis of a Gas Station, Proceedings of the 33rd DAAAM International Symposium, pp.0358-0364, B. Katalinic (Ed.), Published by DAAAM International, ISBN 978-3-902734-36-5, ISSN 1726-9679, Vienna, Austria
DOI: 10.2507/33rd.daaam.proceedings.050

Abstract

Gas stations have long been perceived as easy targets for petty theft, yet not much has changed regarding safety and security strategies. Each reference object is individual, which also applies to gas stations. However, vandals and attackers are still attracted to these places precisely because of what connects the given objects - how quickly the attacker can arrive at the given object and how quickly he can leave it. Whether by car, truck, or any other means. This fact makes gas stations an attractive and relatively quick target. This article aims to describe the issue of petrol station security and the risks arising from it. To achieve the objectives, the methods of exploration, analysis, and comparison are used. The paper includes an analysis describing incidents that have occurred over the last 3 years.

Keywords: Czech Republic; gas station; safety; security; soft targets.

1. Introduction

This article deals with the general safety of gas stations in the Czech Republic. Gas stations are subject to the specifics of weak security, loose people who do not need to be verified on arrival, and the number of stations varies depending on time, season, allocation, etc. Depending on the attractiveness of the target, these random persons may be physically attacked within the reference facility as the number of attackers may increase proportionally with the number of these random persons. From a security point of view, petrol stations can be described as soft targets. In an article by Malatinsky, A. et al. soft targets are defined as an "object, space, or action that is characterized by the presence of a large number of people and, at the same time, the absence or low level of security against attacks". [1] Another article that describes problematic of soft targets should be for example by authors Zema, T. [2] or Hofreiter, L. et. al. [3]

In addition, gas stations are not only a place where people can refuel their vehicles, but many gas stations offer a wide range of goods such as foodstuffs. Thus, people often use these services as a substitute for traditional shops, making gas stations more frequented areas. This trend was particularly observed during the COVID-19 pandemic. The number of visitors to gas stations increased due to the closure of establishments, and restaurants and reduced access to shops and shopping centers. The concept of emergency is also linked to safety and security. These events can be classified into two divisions, namely natural or man-made emergencies. The area of the article is influenced by anthropogenic activity, with a man being the source of all risks, mainly armed robberies, and threats to the lives of service station attendants. [4]

The aim and outcome of the present article are to protect gas stations and their security strategy. It includes the characteristics and location of the selected gas station. Urban gas stations may be located near residential areas that may be inhabited by inadaptable citizens. Service stations on main roads, busy transport links, intersections, etc. near industrial zones, and very busy traffic arteries with high turnover. The area is usually surrounded by gas stations on motorways and truck drivers have more traffic. This implies that the security strategy must be approached individually considering the near and far surroundings of the reference object.

2. Safety specifications of gas stations in the Czech Republic

Safety specifications also vary depending on the characteristics and location of the pumping station. However, general safety requirements should apply to all gas stations and their surroundings. These requirements include easy access for arriving vehicles but at the same time difficult access for those in vehicles with illegal intentions. In this case, perimeter security must maintain the necessary freedom for normal business activities and events around the gas station, but at the same time create a barrier. Examples include the application of concrete blocks or hedges or shrubs. [5]

The installed barrier shall be reinforced with a wire fence to delineate the area where the reference property is not adjacent to a public road. Outside of the Department's obvious pump station from the public right-of-way, consideration shall be given to clear directional signage for open access, unobstructed visibility, and appropriate lighting of the pump station. According to the Czech technical standard ČSN P CEN/TR 14383, the environment of a service station is divided into the parts mentioned in the diagram below.

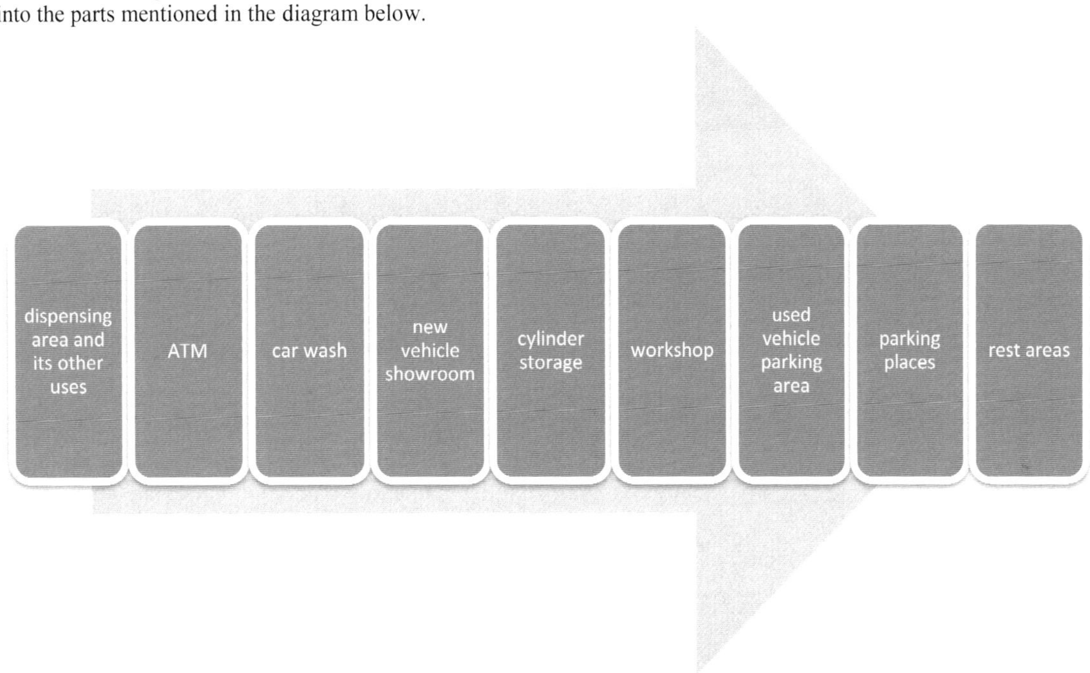

Chart 1. Division of the gas station environment

Dispensing area

The delivery area should be equipped with a device to prevent vehicles from backing out and must always be switched on. However, if necessary, the operator must be able to deactivate it (for example, if the vehicle needs to be released). Traffic barriers must be installed at the exit. The purpose of these barriers is to make access to the service station more difficult. The dispensing area must be equipped with CCTV, cabling, and optimal cameras. This camera system must be installed to allow surveillance of sensitive areas. Furthermore, their purpose and objective are to detect suspicious behavior and to make potential perpetrators fearful of conviction. However, in the event of an unusual event, these systems need to be unaffected by sunlight and be able to withstand vandalism. Fuel dispensers shall be fully automatic and effectively secured. [5]

ATM

In some European countries, ATMs are installed around gas stations, which, as well as the possibility of buying refreshments, attracts more people. However, their installation also has a negative impact, namely the possibility of an increase in crime. Thus, the presence of this device also increases the risk for gas station attendants. The requirements of EN1143-1 should therefore be met. In addition, the location of the ATM should be subjected to a study to include vandalism resistance and resistance to break-ins, particularly through motor vehicle intrusion. [5]

Car Wash

The space of the car wash only increases the risk in the case of coin-operated machines, where there is a risk of theft or vandalism. For this reason, a system working with tokens or special cards, which the customer receives after payment at the service station, is preferred for car washes. A camera system can be installed for possible detection near this area. Often the employees operating the car wash are part of the car wash. [5]

Storage of compressed gas cylinders

An outdoor rack protected by a metal cage should be used for the storage of compressed gas cylinders. The reason for this security is to combine anti-theft security with the necessary ventilation. In addition, this metal cage-protected outdoor rack is also locked with a high-quality locking system complying with EN 12320 or EN1303. The security itself can be increased by installing a full-width barrier or eye bolts and padlocks. Padlocks and eyelets should conform to the specified EN 12320 standard. [5]. In particular, the conditions it imposes are that the barrier should be strong and at least as strong as the padlock. Padlocks with a concealed shackle should be used. These locks provide the least possibility of being overridden. Fixing bolts and screws with protruding heads should not be able to be loosened from the vulnerable side. They should have coupling-type heads, or the heads should be threaded.

3. Analysis of Threats

The following chapter focuses on and discusses incidents at gas stations in the Czech Republic. Since December 2018, a total of 16 incidents at gas stations have been recorded in the Czech Republic. At this scale, only one incident resulted in the death of a gas station attendant. This was specifically an incident on 3 December 2018. The information on this incident describes the incident as occurring at 3.30 a.m. when two masked offenders entered the premises of the gas station to commit a robbery. During the robbery, a total of CZK 10 000 was stolen from the cash register as several packs of cigarettes. Unfortunately, the gas station attendant was murdered in the process.

Table 1 records all incidents recorded from 2018 to 2022. The table is broken down into several criteria. It lists the date, location of the incident, nature of the incident, the consequence of the act, and the weapon with which the act was committed. [6], [7], [8], [9], [10], [11], [12], [13], [14], [15], [16], [17]

Date	Location	Incident	Consequence	Weapon
06/01/2022	Písek	Armed robbery	Frightened staff	Gun
07/02/2022	Zlín	Armed robbery	Frightened staff	Knife
15/12/2021	Tišnov	Threats/Robbery	Frightened staff	Lighter
11/09/2021	Babice	Armed robbery	Frightened staff	Pepper spray
25/12/2020	Sokolov	Armed robbery	Frightened staff	Gun
15/12/2020	Praha 5	6x Armed robbery	Frightened staff	Gun
06/08/2020	Brno	Armed robbery	Frightened staff	Knife
07/05/2020	Ostrava-Poruba	Armed robbery	Frightened staff	Gun
26/03/2020	Rozhovice	Armed robbery	Frightened staff	Knife
Date	**Location**	**Incident**	**Consequence**	**Weapon**
05/12/2019	Vyškov	Armed robbery	Frightened staff	Gun
23/10/2019	Plasy	Armed robbery	Frightened staff	Gun
10/10/2019	Praha	Armed robbery	Frightened staff	Gun
02/08/2019	Lovosice	Armed robbery	Frightened staff	Gun
16/12/2018	Rozhovice u Heřmanova	Armed robbery	Frightened staff	Gun
12/12/2018	Rumburk	Armed robbery	Thud attendants	Metal Stick
03/12/2018	Nelahozeves	Armed robbery	Death of an attendant	Gun

Table 1. List of incidents in the Czech Republic

4. Technical Security Solutions

The specific security requirements of individual filling stations may vary depending on their location and wider surroundings. The proposal for specifying technical security is based on a table of attacks on gas stations in recent years. To specifically determine the appropriate technical security for a particular reference site, a risk analysis would need to be developed, with values reflecting the specifics of the gas station and its surroundings. The list of incidents shows that armed robbery resulting in intimidation and psychological harm to staff is the most significant. The information provided indicates a situation where an attacker may get close to the staff, bring a weapon into the premises of the service station, and use this weapon to threaten the service station staff. Figure 1 illustrates the use of event-tree analysis.

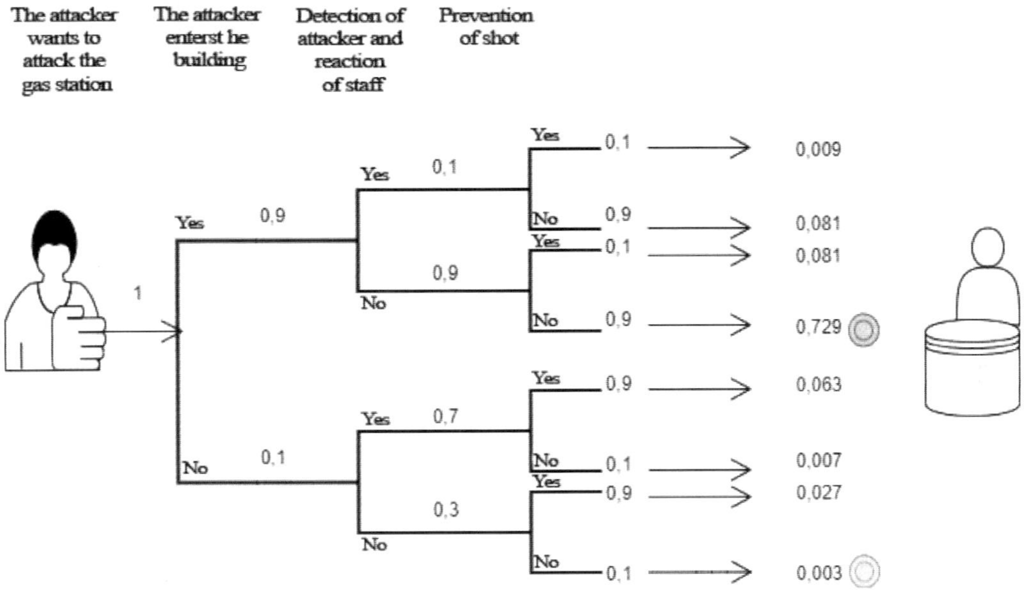

Fig. 1. Event-Tree Analysis

The diagram in Figure 1 shows the steps of the event tree analysis. The most likely scenario is the one shown in the diagram below.

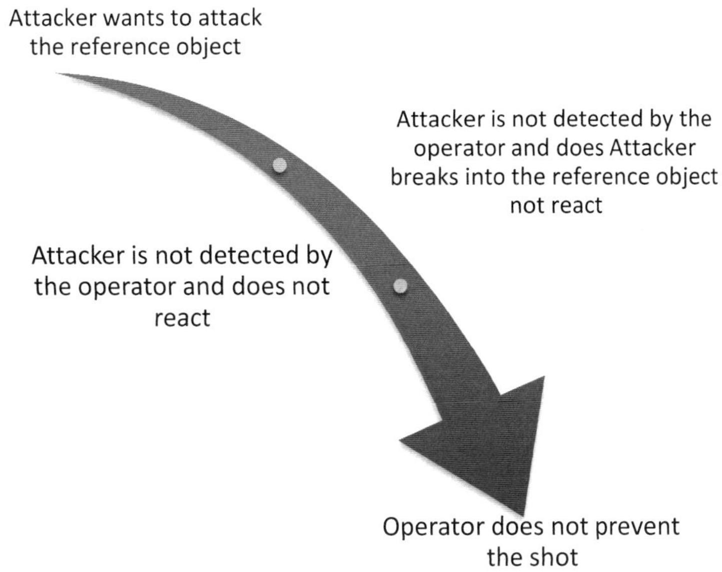

Chart 2. The most likely scenario

This scenario is not only the most likely under the current state of gas stations but can also result in death. It can also be seen from the diagram that the largest difference in values is in the first step of the "Attacker enters the facility" scenario. This step should be given special attention if it is nighttime or if the number of gas station personnel who could intervene in a critical situation is compromised.

Fig. 2. Dispensing window at the gas station in Portugal [7]

To ensure the highly probable "Attacker enters the building" step, which is particularly dangerous during night shifts, complete barring of all persons from entering the interior is suggested. The use of Sterling products' "Night-Pay Window" made of bulletproof glass with a special stainless-steel box built in for the exchange and transfer of people's belongings on both sides not only maximizes security but also makes the job of disinfecting viruses and bacteria (for example, during the COVID-19 pandemic) easier. For this reason, it appears to be one of the best options applicable to gas stations. [18]

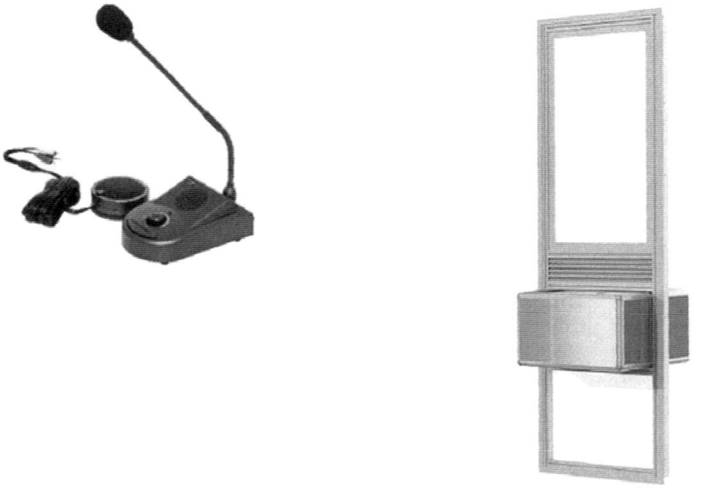

Fig. 3. Night-Pay Window with audio transmission [18]

After the introduction of bullet-proof glass with a special box for dispensing and receiving goods, there could be a problem with customer turnover at a given gas station. The slowdown in service could also cause a clustering of people waiting to pay for fuel.

As part of this issue, it is proposed to extend the QR code payment option using a smartphone to all gas stations. Currently, only the Benzina gas station chain has this option. This method of payment would significantly reduce the number of customers requiring any interaction with the service station attendant and would reduce the number of people waiting at the bulletproof glass dispensing window. [18], [19]

5. Conclusion

Gas stations have been a focus for potential vandalism in the past. In theory, it could be argued that if an attacker is prepared to attack completely unprepared, often tired gas station staff with long opening hours, they are an easy target. The attacker can quickly get to the gas station and escape on foot or in a motor vehicle after committing the crime. According to the table of attacks, 16 attacks have been committed against gas station operators in the last 3.5 years. These conditions are so impossible to monitor that relying on human factors and operator training to defend oneself in a crisis is not enough. Much more emphasis needs to be placed on prevention, technical security, and, if possible, maximum elimination of the threat, which will be helped by the security measures proposed above.

In the securitization of gas stations in the Czech Republic, the standard for technical safety elements should be increased. Since threats, security risks, and scenarios are similar in every EU country, their form should be standardized. Starting to see technical security solutions as a form of breaking the attack scenario is a start. In future work, we would like to focus on the possibilities of standardization and forms of security measures from the point of view of psychology. Mentally scaring the attacker is often enough to ward off an attack. This form could be considered as part of a practical benefit/cost comparison.

The methods of explanation, analysis, and comparison were used to reach the objectives of the article. Attacks on gas stations in the last three years in the Czech Republic were analyzed and both the methods of attacks and their impact on the operators were described. This article provides a perspective on the security of petrol stations and all the people around them. There is a need to address this issue and to protect smaller objects, which may be petrol stations. Security itself is an ever-growing and dynamic trend that is spilling over into all areas. Future research aims to further describe and specify the means of protecting petrol stations and protecting them from attack.

6. Acknowledgments

This article was supported by an internal grant project of Tomas Bata University under the number IGA/FAI/2022/003.

7. References

[1] Malatinsky, A[dam]; Drofova, I[rena]; Sousedikova, L[ucie] & Hromada, M[artin] (2021). Fire Safety Threat Risk Analysis for Soft Target, Proceedings of the 32nd DAAAM International Symposium, pp.0586-0592, B. Katalinic (Ed.), Published by DAAAM International, ISBN 978-3-902734-33-4, ISSN 1726-9679, Vienna, Austria DOI: 10.2507/32nd.daaam.proceedings.083

[2] Zeman, Tomáš. (2020). Soft Targets: Definition and Identification. Academic and Applied Research in Military and Public. 19. 109-119. 10.32565/aarms.2020.1.10.

[3] Hofreiter, Ladislav & Berezutskyi, Viacheslav & Figuli, Lucia & Zvaková, Zuzana. (2020). Soft Target Protection Theoretical Basis and Practical Measures: Theoretical Basis and Practical Measures. 10.1007/978-94-024-1755-5.

[4] Vichova, K[aterina]; Hromada, M[artin]; Ficek, M[artin] & Gracla, M[ichal] (2018). The Comparative Analysis of Safety in the Czech Republic and Abroad, Proceedings of the 29th DAAAM International Symposium, pp.1181-1186, B. Katalinic (Ed.), Published by DAAAM International, ISBN 978-3-902734- 20-4, ISSN 1726-9679, Vienna, Austria DOI: 10.2507/29th.daaam.proceedings.170

[5] ČSN P CEN/TR 14383-5. Preliminary Czech technical standard: Crime prevention. Urban planning and building design - Part 5: Gas stations. October 2011. Prague: Office for Technical Standardization, Metrology and State Testing, 2011.

[6] With a knife in his hand, he robbed a gas station in Podvesná. (2022). Zlín: Zlínský deník. Available from: https://zlinsky.denik.cz/zlociny-a-soudy/prepadeni-benzinka-zlin-podvesna-2022.html

[7] DEMONY, Catarina. Night pay window. In: Https://twitter.com/. Available from: https://pbs.twimg.com/media/ETJ9djqWkAAEK7s?format=jpg&name=small

[8] A robber held up a gas station. (2021).Prague. Blesk. Available from: https://www.blesk.cz/clanek/regiony-brno-brno-krimi/698570/lupic-prepadl-benzinku-a-vyhrozoval-volejte-policii-jinak-to-tady-podpalim.html

[9] Beginner robber robbed a gas station in a sweatshirt. (2021). Prague: Blesk. Available from: https://www.blesk.cz/clanek/regiony-brno-brno-krimi/651717/lupic-zacatecnik-prepadl-benzinku-v-mikine-naruby-a-obsluze-se-omlouval-20-tisic-v-tahu.html

[10] A masked man robbed a gas station: A masked man approached the cash register of a gas station in Ostrava-Poruba with a pistol in his hand. (2021). Prague: Blesk. Available from: https://www.blesk.cz/clanek/regiony-ostrava/643035/naval-prachy-maskovaneho-lumpa-ktery-prepadl-benzinku-nasla-policie-schovaneho-v-krovi.html

[11] Masked gunman robs gas station. (2020). Prague. Available from: https://www.blesk.cz/clanek/zpravy-krimi/638746/ozbrojeny-muz-s-rouskou-vyloupil-benzinku-utok-zachytila-kamera.html

[12] Shooter's 19-year sentence upheld by court. (2020). Prague: Blesk. Available from: https://www.blesk.cz/clanek/zpravy-krimi/632899/tecka-za-vrazdou-cerpadlarky-jany-strelci-soud-potvrdil-19lety-trest.html
[13] Armed criminals robbed six gas stations in Prague: They ordered a "taxi" to Radlicka, police are looking for the driver. (2020). Prague: Blesk. Available from: https://www.blesk.cz/clanek/regiony-praha-praha-krimi/631675/ozbrojeni-zlocinci-v-praze-prepadli-sest-benzinek-do-radlicke-si-objednali-taxi-ridice-hleda-policie.html
[14] Young man robbed gas station as he ran from police, crashed and ended up in hospital. (2019). Prague: Blesk. Available from: https://www.blesk.cz/clanek/regiony-plzen/623299/to-je-matla-mladik-prepadl-benzinku-kdyz-ujizdel-policii-boural-a-skoncil-v-nemocnici.html
[15] Another gas station robbed. (2020). Prague: Blesk. Available from: https://www.blesk.cz/clanek/regiony-praha-praha-krimi/622273/dalsi-prepadena-benzinka-v-praze-maskovany-zlodej-vytahl-na-obsluhu-pistoli.html
[16] The robber threatened the Lovosice pump operator with death. (2019). Prague: Blesk. Available from: https://www.blesk.cz/clanek/zpravy-krimi/597604/lupic-vyhrozoval-lovosicke-cerpadlarce-smrti-k-jeho-dopadeni-pomohla-verejnost.html
[17] Another gas station robbery: robber in the Chrudim region. (2021). Prague: Blesk. Available from: https://www.blesk.cz/clanek/zpravy-krimi/584001/dalsi-prepadeni-benzinky-lupic-na-chrudimsku-vzal-na-pumparku-nuz-policie-po-nem-patra.html
[18] Sterling: products. (2022). Austria: Sterling products. Available from: https://sterlingproducts.com.au/fuel-sites-industry/#nightpaywindows
[19] Gasoline: applications. (2022). Prague: Benzina. Available from: https://www.benzina-aplikace.cz/

DOI: 10.2507/33rd.daaam.proceedings.051

INNOVATION IN THE PROBLEMATICS OF TURNING WITH A LINEAR CUTTING EDGE

Ondrej Marsalek, Lukas Horak, Simon Syrovatka & Tomas Zatloukal

This Publication has to be referred as: Marsalek, O[ndrej]; Horak, L[ukas]; Syrovatka, S[imon] & Zatloukal, T[omas] (2022). Innovation in the Problematics of Turning with a Linear Cutting Edge, Proceedings of the 33rd DAAAM International Symposium, pp.0365-0370, B. Katalinic (Ed.), Published by DAAAM International, ISBN 978-3-902734-36-5, ISSN 1726-9679, Vienna, Austria
DOI: 10.2507/33rd.daaam.proceedings.051

Abstract

Thermal spray technology is a developing trend in the surface treatment of workpieces. This makes it possible to produce a very hard, abrasion-resistant, and corrosion-resistant surface while maintaining a tenacious core. There are several application methods of thermal spraying, which are not technologically demanding, but the final quality of the surface of the applied layer is not good. The surface is rough and dimensionally and geometrically inaccurate. So, it is necessary to machine it further. The current method of machining thermal sprays is grinding, which is quite slow and, in some cases, unnecessarily precise, which is increasing costs. Therefore, other machining methods are being developed, such as turning using a tool with a linear cutting edge. A significant advantage of this method is a significant reduction in machining times and the possibility of using the entire cutting edge of the insert. To accurately set the position of the cutting edge in the machine, a program was created in the Python programming language, which determines the number of possible cuts on the cutting edge depending on the depth of cut and the angle of the cutting edge.

Keywords: linear cutting edge; python programming; thermal sprays; depth of cut; inclination of the cutting edge

1. Introduction

Nowadays, when all technologies are moving forward, the requirements for machine parts are necessary to keep up with this. With this trend, the technology of applying thermal sprays is also increasing. This technology allows the application of a thin layer that will ensure increased resistance to mechanical wear, corrosion, and oxidation while leaving a tough core. The roughness Ra of the thermal spray after application to the surface is around 7 μm. Such a low-quality surface is not suitable for the functional purposes of the sprayed surface. In order to achieve smaller surface roughness and dimensional and geometric accuracy, subsequent processing of the thermal coating is necessary [1] [2] [4].

The principle of thermal spray technology is that a primarily molten powder is applied to the pre-prepared surface of the workpiece, which adheres to the surface and forms a layer of the desired properties. Thermal sprays are mainly used in the automotive and aerospace industries, where they are applied to internal combustion engine parts and act against wear at elevated temperatures. Those parts are, for example, various pistons, pins, cylinders, pistons, valves, crankshafts, and others. Furthermore, they are used significantly for turbine blades in the energy industry, as a thermal barrier and protection against corrosion and cavitation. In engineering, various types of thermal spraying are applied to components of pumping and hydraulic technology [3] [6] [7].

As far as machining is concerned, the vast majority of used thermal sprays are classified as hard-to-machine materials due to their composition. The composition contains many hard-to-machine elements such as nickel, tungsten, and vanadium. These elements increase the hardness and form very hard carbides in the microstructure together with carbon. That increases the wear of the cutting-edge during machining. Thermal sprays are currently machined using grinding technology for these reasons [3] [4] [5]. This method of processing achieves high machine times and unnecessary high-quality surfaces on certain types of products. This unnecessarily increases costs. For this reason, other machining methods are being developed, which will reduce the quality of the surface but will significantly reduce machining times. One method is, for example, turning with a tool using a linear cutting edge [1]. The main goal of this work is to devise and create a method for calculating the cross-section of the cut layer during machining with a linear cutting edge.

2. Linear cutting edge

When turning with a linear cutting edge is no tip present in the cut. Tools that do not have a cutting tip produce relatively smooth surfaces when cutting at speeds significantly higher than conventional tools [8]. The minor cutting-edge angle is zero and the inclination of the cutting angle is very positive or negative. The following figure (Fig. 1) shows a model to give an idea of the position of the insert depending on the workpiece. The insert, which is square positive, is yellow and the workpiece is grey. The image (Fig. 2a) shows a tool with a positive inclination of the cutting edge, and the image (Fig. 2b) shows a tool with a negative inclination of the cutting edge.

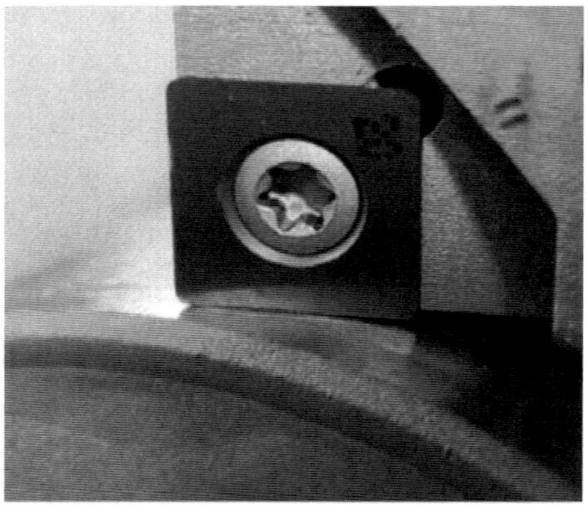

Fig. 1. Position of the cutting edge relative to the workpiece

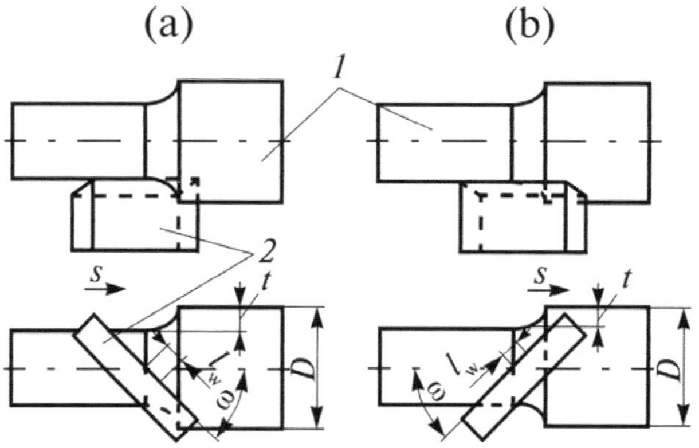

Fig. 2. Positive (a) and negative (b) inclination of the cutting-edge angle [8]

This principle of machining with tools with a linear cutting edge or otherwise with tools without cutting tips was dealt with earlier in 2016 by Filippov et al. He stated that theoretical analysis of turning by tipless tools with linear and curved cutting edges indicates that the basic parameters that determine the shaping by such tools are as follows: the cutting depth, the tool supply, and the tool inclination. He and his team conducted several experiments where they varied these parameters and measured the cutting forces. In the thesis, he presents a preliminary overview and states that this issue needs to be investigated even more thoroughly [8]. Due to this, our scientific team started to investigate it and decided to apply this method of machining with a linear cutting edge to the machining of hard-to-machine thermal sprays.

The advantage of this machining principle is that the tip in which heat accumulates is not present in the cut. It is generally known from machining theory that the depth of the cut layer should not be less than the corner radius because the surface is formed rather than machined. Because when machining with a linear cutting edge, only the cutting-edge radius is in the cut. This radius is significantly smaller than the radius of the tips, so it is possible to machine very small depths of cut. In this way, it is possible to make the coating layer thinner, thus saving the cost of coating. Only a small part of the total length of the cutting edge is used with a small depth of cut. This linear cutting-edge technology makes it possible to use the entire length of the cutting edge by moving the tool to the Y-axis of the machine. This is another great advantage of this method. For example, if one edge of a square insert is used 3 times, the total number of cuts on one insert is 12. The cross-section of the cut-off layer has the shape of an elliptical segment in turning with a linear cutting edge. This changes with a change in the depth of cut, the diameter of the workpiece, and especially with a change in the inclination of the cutting angle. As this angle decreases, the ellipse expands, and thus the line section of the segment increases. So, for example, to maintain the same number of cuts on the cutting edge, when reducing the inclination of the cutting angle setting, it is necessary to reduce the depth of the cut. During the change of the shape of the ellipse section, the centre of the line section moves, which needs to be known to set the tool on the Y-axis, so that the cuts on the cutting edge do not overlap. To speed up and facilitate the determination of the size of the cut-off layer in the shape of an ellipse section and the determination of the coordinates of the tool settings, it was advisable to invent a solution that would allow this. A solution that also evaluates the area of the cut layer. It is, for example, suitable for calculating cutting resistances.

3. Problem solving

The first solution was to create a model in a modelling program that evaluates the required parameters when the criteria are changed. With correctly set bonds, it was possible to achieve in the model to evaluate the cross-section of the cut layer when changing the diameter of the workpiece, the depth of cut, and the inclination of the cutting angle. However, the process of entering and evaluating parameters was not very user-friendly. In addition, this solution method requires a complex licensed CAD program to obtain the required values. The model for determining the cross-section of the cut layer is shown in the next figure (Fig. 3). In the picture, the colors correspond to those in the picture (Fig. 3). The cross-section of the cut layer has a purple color.

Fig. 3. Model for determining the cross-section of the cut layer

The second solution was to make a program in some software such as MATLAB or Python. Python was chosen because it is free and open-source software. Specifically, the Spyder program was chosen for programming. To create the program, it was convenient to draw what needed to be calculated. The images below show the task in graphical form. The figure (Fig. 4) shows a circular workpiece with a cut A-A that corresponds to the setting angle of the blade inclination λs. This angle with the workpiece radius is the variable criterion that will vary depending on the actual settings required. In section A-A, the depth of cut parameter and other dimensions needed for the equation are also displayed. They are the width of the ellipse a and its height b, which corresponds to the radius of the workpiece r, half of the line section Xn and lastly the cross-section area A. The cut depth parameter is the next and last input condition.

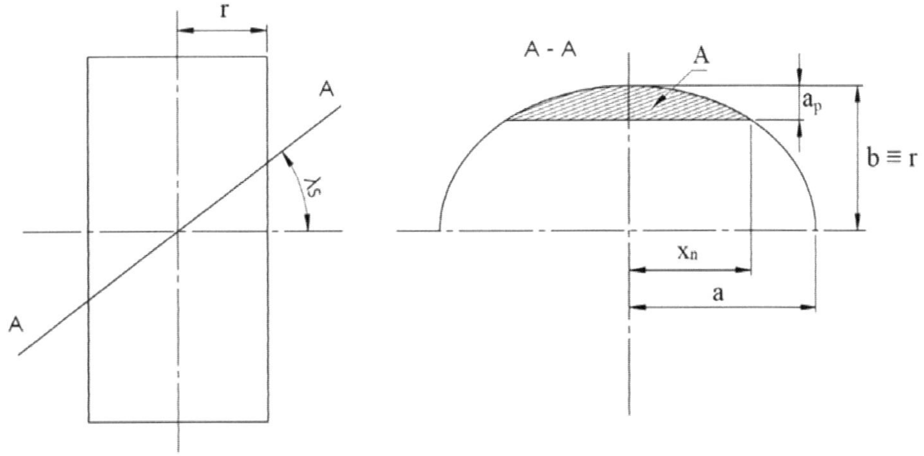

Fig. 4. Graphical definition of the task

The next picture (Fig. 5) is a division of the areas so that it is possible to define the required A area. This area is crosshatched in black in the picture. Another red area shows an ellipse from zero to xn. Last is the rectangular green area. To calculate the cross-section of cut layer A, it is necessary to integrate the red area and then subtract the green area from it. After further reflection, it was found that the calculation of area A could be done in another simplified way. This is such that when the ellipse is tilted ninety degrees to the vertical position, the cross-section of the finished surface can be calculated directly, without the need to subtract other areas. This variant is shown in the figure (Fig. 6). Using this expression, the calculation became more precise. Therefore, the following equation will be expressed just for this variant.

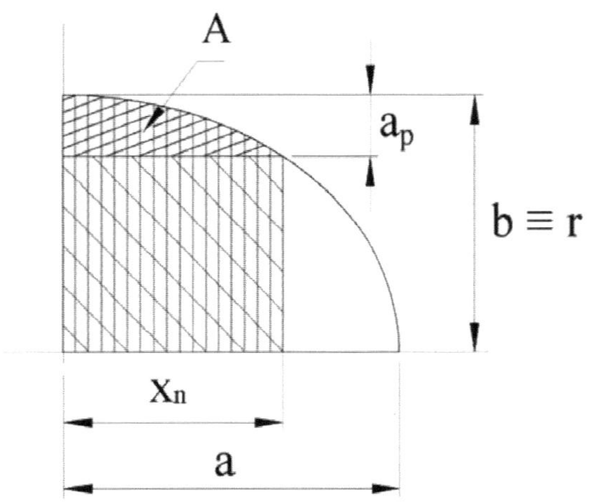

Fig. 5. The first variant of the calculation

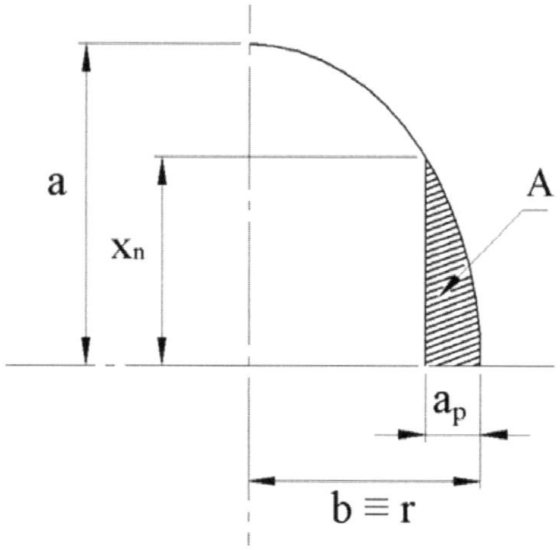

Fig. 6. The final variant of the calculation

The integral used to calculate the cross-section of the cut layer in the tilted variant is given in equation (1) below. In the integral x is the unknown, a is the height of the ellipse, and b is the width of the ellipse. Parameter X changes depending on the setting of the inclination of the cutting angle, parameter b on the change in the radius of the workpiece, and, of course, the parameter ap on the setting in the depth of cut. The whole integral is multiplied by two since it is calculated only for half the cross-section of the cut layer.

$$A(x) = 2\int_{b-a_p}^{b} a\sqrt{1-\frac{x^2}{b^2}}\,dx = a\left[x\sqrt{1-\left(\frac{x}{b}\right)^2} + b\cdot\sin^{-1}\left(\frac{x}{b}\right)\right]_{b-a_p}^{b} \quad (1)$$

By entering all the necessary equations into the Python program, it was achieved that when the required parameters were entered, the program evaluated the area of the cut layer. Both numerically and graphically. This can be seen in the graph in the picture (Fig. X). In the graph, the ellipse is shown in blue, and the straight line shows the edge in orange. The graph in the picture corresponds to the specified values of $r = 40$mm, $ap = 0.05$ mm and $\lambda_s = 60°$.

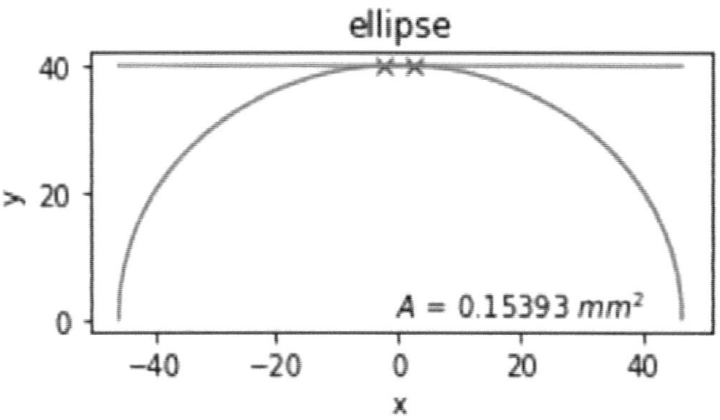

Fig. 7. Numerical and graphical evaluation of the cross-section of the cut layer

Furthermore, the program evaluates the depth of cut that must be set to fit n number of cuts on the cutting edge. For this, it is necessary to enter the number of cuts and the depth of the cut layer into the program. This can be seen in the graph in the figure (Fig. X), where the width of the blade L in mm is indicated on the X-axis and the depth of cut ap in mm on the Y-axis. The graph also shows the coordinates of the centres of the cross sections of the cut layers for setting the tool in the Y axis of the machine. The graph in the picture corresponds to the specified values of $r = 40$ mm, $\lambda_s = 60°$, $L = 12$ mm, and $n = 4$.

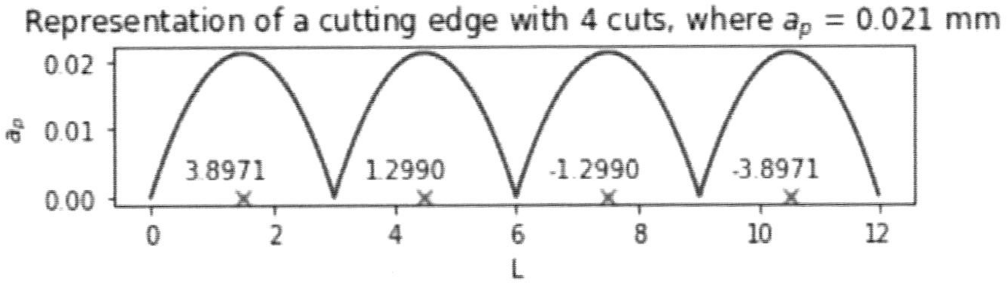

Fig. 8. Evaluation of ap for a certain number of cuts

There was also an inverse problem programmed into the program, which solves how many cuts n can be made on the cutting edge when a certain depth of cut is entered. The graph in the picture corresponds to the specified values of r = 40 mm, λs = 60°, L = 12 mm, and ap = 0.05 mm.

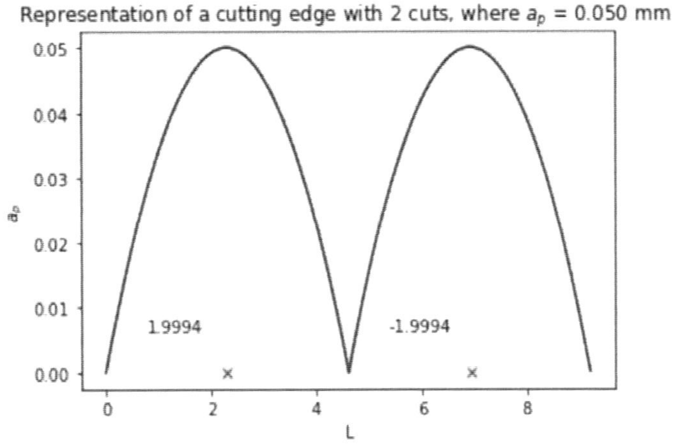

Fig. 9. Evaluation of the number of cuts for a certain a_p

4. Conclusion

This work aimed to create a program for calculating the area of the cut layer when machining with a linear cutting edge or otherwise with a tool without cutting tips, depending on the change in the depth of cut, the diameter of the workpiece, and the angle of the main cutting edge. In our case, for the machining of hard, difficult-to-machine hot sprays, but also for the machining of common materials. Because one of the main advantages of a linear cutting edge is the possibility of using the entire cutting edge, i.e. it is possible to move the tool in the Y axis so that the worn part of the tool gets outside the next cut on the same cutting edge. Therefore, one of the other requirements was to use the program to determine the coordinates of the centers of the cut layers to determine the coordinates in the Y-axis. This is so that the individual cuts on the blade do not overlap. Furthermore, the calculation of the depth of cut for a certain number of cuts on the cutting edge and vice versa. Thus, the inverse problem for determining the number of cuts for the required depth of cut. The Spyder software and the python programming language were chosen to create this program. The created program expresses the required values both numerically and graphically. All these values determined by the program can be used in practice.

A functional model was also created for this issue, but it turned out to be unsatisfactory for quickly determining the values. However, it can also be used to evaluate other parameters, such as the theoretical roughness of the machined surface at a certain displacement. The python program cannot solve this so far. This is another future research plan. Extend the Python program so that it evaluates the theoretical roughness of the surface using the cross-section of the cut layer and the displacement.

5. Acknowledgments

The article contribution has been prepared under project SGS-2022-007 - Research and Development for Innovation in Engineering Technology - Machining Technology IV.

6. References

[1] Zatloukal, T., Řehoř, J., Fulemová, J. & Povolný, M. (2020) Optimization of the machining process of NiCrBSi thermal spraying. In Annals of DAAAM and Proceedings of the International DAAAM Symposium. Vienna: DAAAM International Vienna, s. 632-637. ISBN: 978-3-902734-29-7, ISSN: 1726-9679

[2] Hoesel, D.; Wertheim, R.; Haenel, T. & Thurner, S. (2011). Milling of Cast Material and PTA-Hard-Facing Layers of the Cobalt-Base-Alloy Stellite 12, Annals of DAAAM for 2011 & Proceedings of the 22th International DAAAM Symposium, Published by DAAAM International, Vienna, 2011, ISBN 978-3-901509-83-4

[3] Cesanek, Z. (2015). Thermally Sprayed Coatings Used in Corrosive Environments, Ph.D. Dissertation, ZCU, FST, Pilsen, Czech Republic

[4] Houdkova, S., Vostrak, M. (2013). Comparison of NiCrBSi Coatings, HVOF Sprayed, Re-melted by Flame and by High-Power Laser. Available from: http://metal2013.tanger.cz/files/proceedings/12/reports/1407.pdf, Accessed: 2020-08-08

[5] Zaman, H. A.; Sharif, S.; Kim, D. W.; Idris, M. H.; Suhaimi, M. A. & Tumurkhuyag, Z. (2017). Machinability of Cobalt-based and Cobalt Chromium Molybdenum Alloys - A Review, Procedia Manufacturing, Volume 11, Pages 563-570, ISSN 2351-9789, Available from: https://doi.org/10.1016/j.promfg.2017.07.150.

[6] https://www.triboing.cz/zarove-nastriky.html, (2018). Thermal sprays, Tribo Engineering, Accessed: 2020-08-08

[7] https://www.plasmametal.cz/pouzite-technologie, (2019). Thermal spraying. Plasmametal, Accessed: 2020-08-08

[8] Filippov, A.V., Filippova, E.O. & Chazov, P.A. (2016) Cutting-force components in turning by tools with no cutting tip. Russ. Engin. Res. 36, 1040–1043. Available from: https://link.springer.com/content/pdf/10.3103/S1068798X16120066.pdf

DOI: 10.2507/33rd.daaam.proceedings.052

PROGRAMMING OF OBJECT CAPTURE SYNERGIES FOR MOBILE SERVICE ROBOTS

Plotnikov Aleksey & Pryanichnikov Valentin

This Publication has to be referred as: Plotnikov, A[leksey] & Pryanichnikov, V[alentin] (2022). Programming of Object Capture Synergies for Mobile Service Robots, Proceedings of the 33rd DAAAM International Symposium, pp.0371-0375, B. Katalinic (Ed.), Published by DAAAM International, ISBN 978-3-902734-36-5, ISSN 1726-9679, Vienna, Austria
DOI: 10.2507/33rd.daaam.proceedings.052

Abstract

When creating service mobile robots with manipulators used in cramped rooms (home/office/hospitals), there is a problem of capturing and transporting both small-sized objects, that make up a significant part of the robot's weight. When constructing the control of the lifting process of such objects, the task of dynamic capture arises while maintaining the stability of the robot. The article considers the solution of the problem of countering the overturning of the AMUR-307 robot with a two-armed manipulator circuit. The basis for the construction of such control is proposed to take the solution of the dynamics problem using the kinetostatics method. The article describes the conditions for tipping the robot. The application of the synergy of movements is considered. The synergy of movements consists of simultaneous acceleration of the transport platform and the gripping movement of the manipulator. Recommendations are given to increase the load capacity depending on the removal of the object outside the support surface of the robot.

Keywords: Robotics; Mobile Service robot; Movement synergy algorithms for heavy object manipulation; Kinetostatic dynamics analysis.

1. Introduction

Designers of modern mobile robots approach the problem of stability using several techniques. The most obvious way is to use a structure with the widest possible support area and limit the working area in such a way, that almost any movement of the robot could not lead to overturning. The disadvantage of this approach is the large dimensions of the robot, which negatively affects the mobility of the MR. An example of such a robot is the KUKA youBot robot (Fig 1a and [9]).

Another approach is to use a special design of the robot manipulator, which, in case of overturning, allows you to restore the vertical position of the robot. Otherwise, the problem is solved by using outriggers, when the robot is working. The advantage of such structures is an increase in the support area. A significant disadvantage of this approach is the inability to move the robot during the manipulator operations. An example of a robot with this approach is the Husqvarna DXR 310 disassembly robot (Fig 1b).

It is possible to use a multi-track design of variable geometry (Fig 1c) allows not only to increase the stability of the robot without losing the ability to move, but also to overcome various obstacles. However, these designs are relatively complex in technical execution.

a) b) c)

Fig. 1. Examples of the mobile robot design.

The topic of kinetostatic analysis of industrial robots is widely covered in the literature, for example, see the monograph [1] and the article [2]. However, in these works, the kinetostatic analysis of industrial robots of various configurations is consecrated. At the same time, the kinetostatic analysis of mobile robots is not considered. For mobile robots, dynamics analysis is often performed using third-party systems for solving the dynamics problem [3] or using the Lagrange method [4], [5]. The advantages of the kinetostatic robot model [6] are the absence of integration operations. This makes it possible to use this model as part of the on-board software of the robot both in the classical form [7] and in the form of separate modules [8].

2. Object capture strategy

The following object capture strategy is proposed (Fig.4):

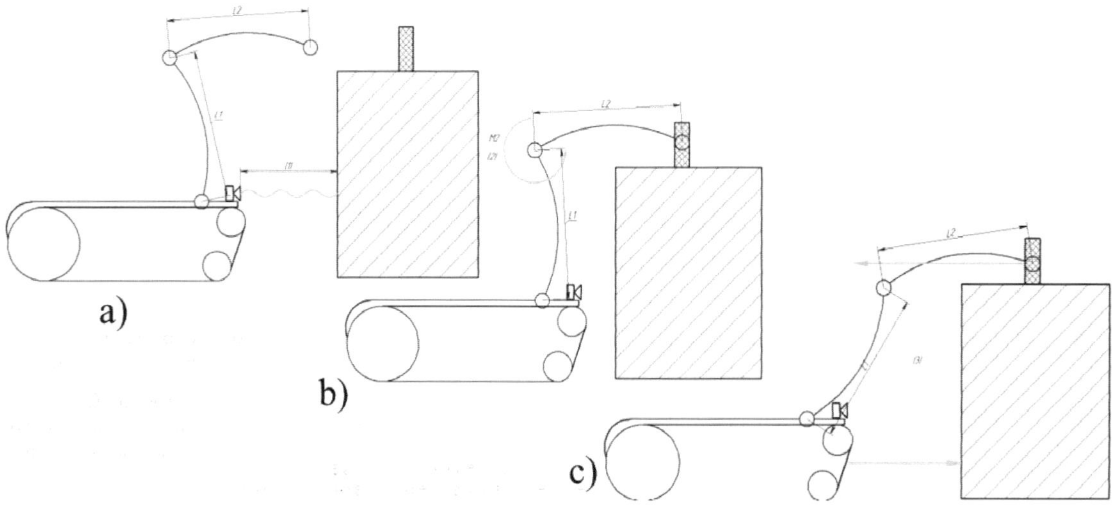

Fig. 2. Object capture strategy

Navigation system data is used to determine the safe area of movement during capture (Fig 2a). The mass of the manipulated object is determined, when it is captured, using a special test movement (Fig 2b). The object of manipulation is transferred inside the service area to the transport table, while the robot is moving forward in a coordinated manner to maintain balance (Fig 2c).

3. Robot rollover condition

Using the kinetostatic model of the robot's capture of an object, considered earlier, it is possible to determine the stability of the robot during this operation. The equations of moments, relative to the front and rear support points. Equation, relative to the front support point:

$$\vec{M}_{Nb} = -\left(\vec{M}_{MF} + \vec{r}_{FNf} \times \vec{F}_{10} + \vec{r}_{mNf} \times (m_0\vec{g}) - \vec{r}_{m0Nf} \times m_0\vec{a}_{m0}\right) \tag{1}$$

$\vec{M}_{Nb}$ the sum of the moments of the reaction forces of the supports relative to the front fulcrum;
$\vec{M}_{MF}$ – the sum of the moments of the manipulator forces acting relative to the front fulcrum;
$\vec{r}_{mNf} \times (m_0\vec{g})$ – the moment caused by the action of gravity on the transport platform;
$\vec{r}_{m0Nf} \times m_0\vec{a}_{m0}$ – the moment caused by the action of the inertia force on the transport platform;

If we assume, that the right coordinate system is used in the calculation, then it is obvious, that the stability condition of the robot is:

$$\vec{M}_{Nb} < 0 \tag{2}$$

If the reaction moments directed towards the support plane are necessary to maintain equilibrium, which is impossible, then the robot will lose stability and tip over. Similarly, it is necessary to write down the equation of moments relative to the rear support point:

$$\vec{M}_{Nf} = -\left(\vec{M}_{MB} + \vec{r}_{FNb} \times \vec{F}_{10} + \vec{r}_{mNb} \times (m_0\vec{g}) - \vec{r}_{m0Nb} \times m_0\vec{a}_{m0}\right) \tag{3}$$

The condition for the stable position of the robot will be the equation

$$\vec{M}_{Nf} > 0 \tag{4}$$

It is worth noting, that the cases $\vec{M}_{Nb} = 0$ and $\vec{M}_{Nf} = 0$ are boundary positions and theoretically stable. However, for a mobile robot, such a position is unsafe, so we can assume, that the robot is not stable.

4. Simulation results

The simulation of the object capture by a robot with objects in various masses was carried out. The image (Fig 3) shows the location of the robot's masses (gray points) and the direction of acceleration (green lines) at the initial (Fig 3a) and final positions of the manipulator (Fig 3b).

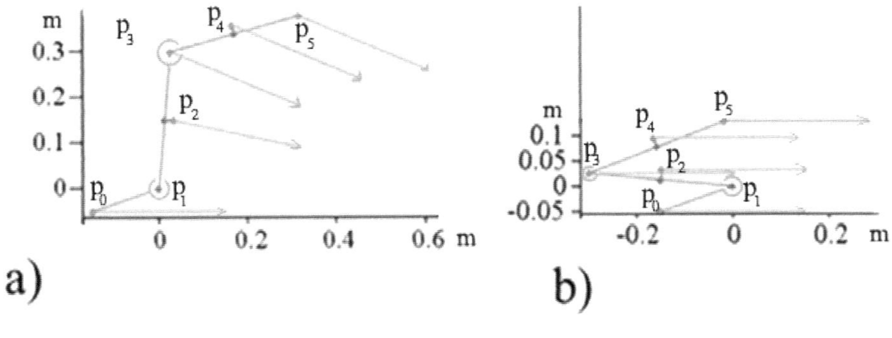

Fig. 3. The starting and ending positions of the manipulator (side view).

p_0 – position of the center of mass of the transport platform
p_1 – position of the first manipulator joint
p_2 – position of the center of mass of the first manipulator link
p_3 – position of the second manipulator joint
p_4 – position of the center of mass of the second manipulator link
p_5 – position of the center of mass of the manipulator

As a result, data on the maximum permissible weight of the object, that the robot is able to lift was obtained.

The object in forward position	Without acceleration	With platform acceleration 0.15 м/с^2	With platform acceleration 0.3 м/с^2
0.286 m	5.80 kg	6.35 kg	6.95 kg
0.312 m	5.15 kg	5.60 kg	6.10 kg
0.322 m	5.00 kg	5.35 kg	5.75 kg
0.403 m	4.00 kg	4.30 kg	4.60 kg
0.420 m	3.35 kg	3.55 kg	3.80 kg
0.530 m	2.45 kg	2.65 kg	2.80 kg

Table 1. Maximum mass of the captured objects

Based on the graphs of the reaction moment of the supports (4) relative to the front point (Fig 4), it can be concluded, that the acceleration of the robot platform must be maintained only at the initial stage of movement.

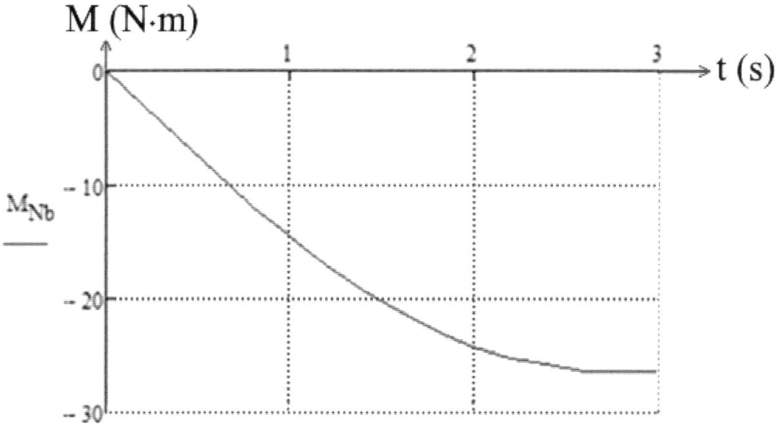

Fig. 4. The reaction moment of the supports, relative to the front point.

The simulation shows (Table 1), that the additional acceleration of the robot during object capturing allows you to get an increase in load capacity up to 14-20%. The amount of additional mass depends on the configuration of the manipulator during capturing. For example, the higher the output link is, the greater the gain will give additional acceleration of the platform.

5. References

[1] Mechanics of industrial robots: a textbook for higher education institutions: in 3 books. / Edited by K.V. Frolov and E.I. Vorobyov. Book 1: Kinematics and dynamics / E.I. Vorobyov, S.A. Popov, G.I.Sheveleva. – M.: Higher School, 1988. – 304p.

[2] [Механика промышленных роботов: учебное пособие для втузов: в 3 кн. / Под ред. К.В. Фролова и Е.И. Воробьева. Кн. 1: Кинематика и динамика / Е.И. Воробьев, С.А. Попов, Г.И.Шевелева. – М.: Высш. шк., 1988. – 304 с.]

[3] Anciporovich, P. P. Kinetostatic force analysis of a manipulator with 4 degrees of freedom / P. P. Anciporovich, V. K. Akulich, E. M. Dubovskaya // Theoretical and applied mechanics. Issue 28: International Scientific and Technical collection / edited by A.V. Chigarev ; BNTU. – Minsk, 2013. – pp. 224-227.

[4] [Анципорович, П. П. Кинетостатический силовой анализ манипулятора с 4 степенями свободы / П. П. Анципорович, В. К. Акулич, Е. М. Дубовская // Теоретическая и прикладная механика. Выпуск 28 : международный научно-технический сборник / под ред. А. В. Чигарева ; БНТУ. – Минск, 2013. – С. 224-227.]

[5] Damic, V[jekoslav]; Cohodar, M[aida] & Muratovic, M[unir] (2017). Dynamic Modelling of Mobile Robots Based on its 3D CAD Model, Proceedings of the 28th DAAAM International Symposium, pp.0144-0149, B. Katalinic (Ed.), Published by DAAAM International, ISBN 978-3-902734-11-2, ISSN 1726-9679, Vienna, AustriaDOI: 10.2507/28th.daaam.proceedings.019

[6] Du, Y. (2016). Kinematics and Dynamic Modeling and Simulation Analysis of Three-wheeled Mobile Robot, International Conference on Mechanics Design, Manufacturing and Automation (MDM 2016) ISBN: 978-1-60595-354-0

[7] A. A. Hatab and R. Dhaouadi, "Dynamic modelling of differential-drive mobile robots using Lagrange and Newton-Euler methodologies: A unified framework," Adv. Robot. Autom., vol. 02, no. 02, 2013.

[8] Plotnikov, A[leksey] & Pryanichnikov, V[alentin] (2021). Kinetostatic Analysis of the Service Mobile Robot with Manipulator, Proceedings of the 32nd DAAAM International Symposium, pp.0446- 0453, B. Katalinic (Ed.), Published by DAAAM International, ISBN 978-3-902734-33-4, ISSN 1726-9679, Vienna, Austria DOI: 10.2507/32nd.daaam.proceedings.065

[9] Pryanichnikov, V[alentin]; Kolesov, Y[uriy]; Tarasov, R[adomir] & Plotnikov, A[leksey] (2021). Integration Software for a Distributed Group of Mobile Service Robots, Proceedings of the 32nd DAAAM International Symposium, pp.0311-0315, B. Katalinic (Ed.), Published by DAAAM International, ISBN 978-3-902734-33-4, ISSN 1726-9679, Vienna, Austria DOI: 10.2507/32nd.daaam.proceedings.045

[10] Andreev, V[iktor]; Kim, V[alerii] & Eprikov, S[tanislav] (2020). ModRob: the Hardware-Software Framework for Modular Mobile Robots Prototyping, Proceedings of the 31st DAAAM International Symposium, pp.0391-0402, B. Katalinic (Ed.), Published by DAAAM International, ISBN 978-3-902734- 29-7, ISSN 1726-9679, Vienna, Austria DOI: 10.2507/31st.daaam.proceedings.054k

[11] https://roboting.ru/1189-mobilnyj-manipulyator-youbot-s-otkrytym-kodom.html, Mobile manipulator Kuka youBot, access 01.10.2022

DOI: 10.2507/33rd.daaam.proceedings.053

KINEMATICS OF RELATIVE DISPLACEMENTS OF CONTACT SURFACES OF THE PORT PLATE UNIT OF AN AXIAL PISTON HYDRAULIC MACHINE

Elchinsky Victor, Kuzmin Anton & Stazhkov Sergey

This Publication has to be referred as: Elchinsky, V[iktor]; Kuzmin, A[nton] & Stazhkov, S[ergey] (2022). Kinematics of Relative Displacements of Contact Surfaces of the Port Plate Unit of an Axial Piston Hydraulic Machine, Proceedings of the 33rd DAAAM International Symposium, pp.0376-0379, B. Katalinic (Ed.), Published by DAAAM International, ISBN 978-3-902734-36-5, ISSN 1726-9679, Vienna, Austria
DOI: 10.2507/33rd.daaam.proceedings.053

Abstract

Modern precision hydraulic drives are widely spread in mechatronics, robotics and heavy machinery. High energy consumption allows to minimize the whole drive mass-dimensional characteristics. However, such benefits require hydraulic machines to have high efficiency. This work presents the research of kinematics of the port plate unit, as friction forces and leakages in the cylinder block – port plate interface are significant for the axial piston swash plate hydraulic machine efficiency. The dependences obtained allow computing the hydrodynamics of this unit and also to conduct further research in order to improve the whole hydraulic drive performance.

Keywords: Kinematics; Hydraulic machines; Axial piston machine; Machinery; Hydraulic drive.

1. The port plate unit influence on the axial piston machine performance

Friction forces and volume losses in the units of hydraulic machines of a high-precision hydraulic drive have a significant impact on the characteristics of the entire system, in particular on the dead zone and, consequently, its control range [1]. One of the critical components of axial piston hydraulic machines is the port plate [2], [3], [4], [5]. Figure 1 shows a cylinder block unit (right) and the port plate unit (left). Under nominal operating conditions, the port plate is pressed against the end surface of the cylinder block.

However, when straining the hydraulic machine and at high dynamic loads, the gap between the mating elements can open and the surface of the port plate can make a complex spatial movement relative to the end surface of the cylinder block. The opening of the joint leads to an increase in volume losses, and the local contact of the conjugate surfaces leads to an increase in friction forces. To study the hydromechanical processes in the port plate interface, it is necessary to conduct a kinematic analysis of the relative motion of the conjugate surfaces [6].

Fig. 1. Port plate unit of an axial piston hydraulic machine

2. Kinematic analysis of the port plate unit

Let's use the principle of inversion of motion of conjugate elements. We will consider the cylinder block fixed in space, then the distribution disk can generally perform spherical motion relative to its support surface. Its centre is located in the centre of the spline joint of the block with the shaft of the hydraulic machine. There is also a reciprocating motion within the end gap between the support surfaces of the cylinder block and the port plate. Figure 2 shows an arbitrary position of the port plate relative to the support surface of the cylinder block.

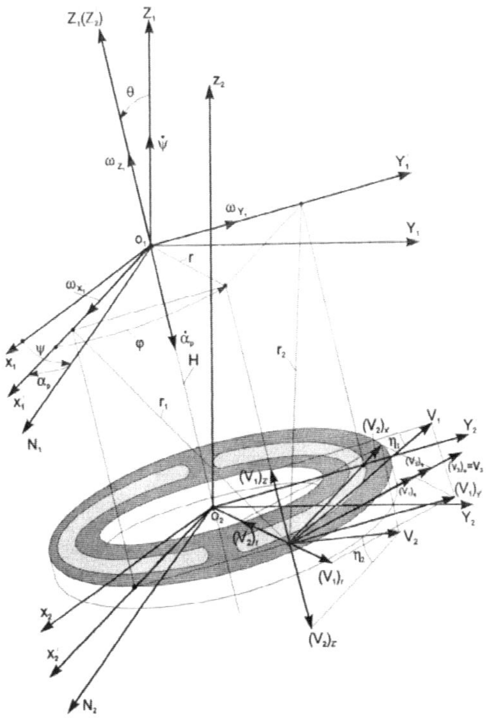

Fig. 2. Scheme of the velocities of a point on the port plate surface

$0_1X_1Y_1Z_1$ is a coordinate system rigidly connected to the cylinder block with the origin in the centre of its spline joint with the shaft of the hydraulic machine.

$0_2X_2Y_2Z_2$ is a coordinate system parallel to the coordinate system $0_1X_1Y_1Z_1$ with the origin in the centre of the plane of the port plate facing the cylinder block.

$0'_2X'_2Y'_2Z'_2$ is a coordinate system rigidly connected to the port plate and the origin coinciding with the origin of the $0_2X_2Y_2Z_2$ coordinate system.

$0'_1X'_1Y'_1Z'_1$ is a coordinate system parallel to the coordinate system $0'_2X'_2Y'_2Z'_2$ with the origin coinciding with the origin of the coordinate system $0_1X_1Y_1Z_1$.

θ- NUTATION ANGLE -the angle between the axis of the block and the port plate.

Ψ - PRECESSION ANGLE - the angle between the axis perpendicular to the plane of the port plate skew (node line N) and the X1 axis.

α_P - ROTATION ANGLE - the angle that characterizes the position of the port plate when it rotates relative to its own axis.

Figure 2 shows three components V_1 V_2 V_3 of the total velocity of an arbitrarily selected point A on the surface of the port plate with its spherical motion relative to the origin 0_1.

The projections of the components V_1 V_2 V_3 on the directions of the cylindrical coordinates r, z and ϕ can be written as follows

$$\begin{cases} (V_1)_r = V_1 cos\mu_1 sin\varphi \\ (V_2)_r = -V_2 cos\mu_2 cos\varphi \\ (V_3)_r = 0 \\ (V_1)_\varphi = V_1 cos\mu_1 cos\varphi \\ (V_2)_\varphi = V_2 cos\mu_2 sin\varphi \\ (V_3)_\varphi = V_3 \\ (V_1)_z = V_1 sin\mu_1 \\ (V_2)_z = -V_2 sin\mu_2 \\ (V_3)_z = 0 \\ \mu_1 = arctg(\frac{r}{H} sin\varphi) \\ \mu_2 = arctg(\frac{r}{H} cos\varphi) \end{cases} \tag{1}$$

H is the distance from the center of the spline connection of the cylinder block and the shaft of the hydraulic machine to the support surface of the port plate.

Grouping the velocity projections by the corresponding coordinates:

$$\begin{cases} (\boldsymbol{V_s})_r = (-\dot{\psi} \cdot sin\theta \cdot sin\alpha_P + \dot{\theta} cos\alpha_P)\sqrt{H^2 + r^2 sin^2\varphi} \cdot cos\left(arctg\left(\frac{r}{H} sin\varphi\right)\right) \cdot sin\varphi - \\ (-\dot{\psi} \cdot sin\theta \cdot cos\alpha_P + \dot{\theta} sin\alpha_P)\sqrt{H^2 + r^2 cos^2\varphi} \cdot cos\left(arctg\left(\frac{r}{H} cos\varphi\right)\right) \cdot cos\varphi \\ (\boldsymbol{V_s})_\varphi = (-\dot{\psi} \cdot sin\theta \cdot sin\alpha_P + \dot{\theta} cos\alpha_P)\sqrt{H^2 + r^2 sin^2\varphi} \cdot cos\left(arctg\left(\frac{r}{H} sin\varphi\right)\right) \cdot cos\varphi + \\ (-\dot{\psi} \cdot sin\theta \cdot cos\alpha_P + \dot{\theta} sin\alpha_P)\sqrt{H^2 + r^2 cos^2\varphi} \cdot cos\left(arctg\left(\frac{r}{H} cos\varphi\right)\right) \cdot sin\varphi + (\dot{\alpha}_P + \dot{\psi} cos\theta) r \\ (\boldsymbol{V_s})_z = (-\dot{\psi} \cdot sin\theta \cdot sin\alpha_P + \dot{\theta} cos\alpha_P)\sqrt{H^2 + r^2 sin^2\varphi} \cdot sin\left(arctg\left(\frac{r}{H} sin\varphi\right)\right) \cdot cos\alpha_P - (V_2)_\varphi = V_2 cos\mu_2 sin\varphi \\ (-\dot{\psi} \cdot sin\theta \cdot cos\alpha_P - \dot{\theta} sin\alpha_P)\sqrt{H^2 + r^2 cos^2\varphi} \cdot sin\left(arctg\left(\frac{r}{H} cos\varphi\right)\right) \cdot sin\alpha_P \end{cases} \tag{2}$$

The expressions obtained can be significantly simplified if we take into account that the angle of the tilt of the port plate is very small, that is, if we take $sin\theta = 0$, $cos\theta = 1$:

$$\begin{cases} (V_s)_r = \dot{\theta} H sin(\varphi - \alpha_P) \\ (V_s)_\varphi = \dot{\theta} H cos(\varphi - \alpha_P) - \dot{\alpha}_P r + \dot{\psi} r \\ (V_s)_z = \dot{\theta} r (cos^2\alpha_P sin\varphi - sin^2\alpha_P cos\varphi) \end{cases} \tag{3}$$

3. Conclusions and further work

The obtained dependences describing the mutual motion of the conjugate surfaces of the port plate and the cylinder block can serve as a basis for the subsequent mathematical description of hydrodynamic processes in the port plate unit of an axial piston hydraulic machine and determination of its energy characteristics under various operating conditions. Following steps should consider hydrodynamic processes, overall dynamics of the cylinder block and complete hydromechanical efficiency of the axial piston swash plate machine.

4. References

[1] Achten, Peter & Potma, Jeroen & Achten, Jasper. (2018). Low Speed Performance of Axial Piston Machines. V001T01A014. 10.1115/FPMC2018-8832

[2] Zecchi, M.; Ivantysynova, M. (2012) An Investigation of the Impact of Micro Surface Shaping on the Cylinder Block/Valve Plate Inter-face Performance through a Novel Thermo-Elasto-Hydrodynamic Model. In Proceedings of the 7th FPNI PhD Symposium, Reggio Emilia, Italy, 27–30 June 2012

[3] Xu, Ye & Zhang (2015). Effects of index angle on flow ripple of a tandem axial piston pump. Journal of Zhejiang University - Science A: Applied Physics & Engineering. 16. 404-417. 10.1631/jzus.A14e0309

[4] Bergada, J.M., Davies, D.L., Kumar, S. and Watton, J. (2012a), "The effect of oil pressure and temperature on barrel film thickness and barrel dynamics of an axial piston pump", Meccanica, Vol. 47 No. 3, pp. 639-654

[5] Stazhkov, S[ergey]; Kuzmin, A[nton]; Elchinsky, V[iktor]; & Korolev, V[ladimir] (2020). Investigation of the Effect of the Gap on the Friction Forces in the Piston Pair of the Axial Piston Hydraulic Machine, Proceedings of the 31st DAAAM International Symposium, pp.0384-0390, B. Katalinic (Ed.), Published by DAAAM International, ISBN 978-3-902734-29-7, ISSN 1726-9679, Vienna, Austria

[6] Jiang J. and Wang K., (2015) An integrated model of hydrodynamic lubricating for piston/cylinder interface, International Conference on Fluid Power and Mechatronics (FPM), Harbin, 2015, pp. 47-52. doi: 10.1109/FPM.2015.7337083

DOI: 10.2507/33rd.daaam.proceedings.054

THEORETICAL RESEARCH OF GANTRY TRACTOR TURNING

Volodymyr Bulgakov, Jüri Olt, Simone Pascuzzi, Valerii Adamchuk,
Jaroslav Gadzalo, Volodymyr Kuvachov, Hryhoriy Kaletnik,
Viktor Kaminskiy & Tormi Lillerand

This Publication has to be referred as: Bulgakov, V[olodymyr]; Olt, J[ueri]; Pascuzzi, S[imone]; Adamchuk, V[alerii]; Gadzalo, J[aroslav]; Kuvachov, V[olodymyr]; Kaletnik, H[ryhoriy]; Kaminskiy, V[iktor] & Lillerand, T[ormi] (2022). Theoretical Research of Gantry Tractor Turning, Proceedings of the 33rd DAAAM International Symposium, pp.0380-0389, B. Katalinic (Ed.), Published by DAAAM International, ISBN 978-3-902734-36-5, ISSN 1726-9679, Vienna, Austria

DOI: 10.2507/33rd.daaam.proceedings.054

Abstract

Turning the gantry tractor on the turning lane can be done in different ways. An analysis of publications on the use of gantry tractors in the track system showed that the most common turn schemes involve pneumatic control. The theory of static and dynamic rotation of a wheeled vehicle has been sufficiently studied by scientists around the world. The aim of the research is to increase the efficiency of the process of turning the gantry tractor in the coordinate transport system of its operation, by studying its static and dynamic manoeuvrability. Theoretical research was carried out by modelling the conditions of the gantry tractor operation on the PC using the provisions of theoretical mechanics and tractor theory. Experimental studies of the gantry tractor were carried out according to both generally accepted and developed methods and involved the use of modern and specially designed equipment. Processing of experimental data was performed on a PC using regression and correlation-spectral analyses. As a result of the research, new schemes of turning such tractor are justified. It is proven that the speed of its rotation by this method should not exceed 1 $m \cdot s^{-1}$, and the ratio of its wheelbase should be minimal.

Keywords: gantry tractor; permanent technological track; coordinate transport system; manoeuvrability; theoretical and experimental research.

1. Introduction

Gantry devices for routed agriculture [1], [2], [3], [4], [6], [7], [8] should be equipped with a wheeled engine, which should be considered the most acceptable. The use of a crawler or other known type of propulsion in the presence of an improved track or road is impractical. [5] The theory of motion and controllability of wheeled traction and transport machines has been sufficiently developed by us [2], [3], [4]. The operation of the wheeled machine takes place under the action of a large number of perturbing factors (forces and their moments), which change the position of the machine in space and divert its movement from a given trajectory. The steering of modern wheeled machines is adapted to manual control and is built on the kinematic or power principle of turning.

When kinematic rotation, the schemes of turning steered wheels (front, rear or both front and rear) relative to the car body, or changing the position of one part of the mobile machine relative to another in the horizontal plane (articulated frame). Power (on-board) rotation is realized by rotating the wheels of different sides of the machine at different speeds. The most widespread schemes of turn with steered wheels. At the same time, if the machine has all the steered wheels, they can turn simultaneously on the front and rear axles in different directions (full-turn mode or "path way") or to one side (crabbing mode) [13], [15]. Sometimes combined kinematic-force patterns are used or proposed for use, which improve the steerability of mobile machines [13], [15].

Traditional turning of wheeled machines is carried out on a certain area and includes several phases: entry into the turn, the actual turn, exit from the turn. Due to the presence of elasticities in the system "propulsion – support surface" the trajectory of the mobile machine when turning does not have a clear dependence on the rotation of the steering wheel, which complicates the construction of machines for turning [9], [10], [11].

Some scientists believe that effective in the automatic driving of vehicles within the coordinate transport system may be a way to change the direction of movement by turning the stopped car on the spot, for example, in a balanced state [12]. To implement this method of changing the direction of movement, the wheeled vehicle must be equipped with lifting and turning mechanisms, software device, position sensors relative to the intersection and control system. A machine turning pattern with all wheels steered has advantages over a weighted turning pattern: There is no need for a special lifting mechanism. But for automated machines with purely on-board turning and rigid mounting of all wheels, the preference must be given to a weighted headland. Therefore, the issue of modelling the static and dynamic manoeuvrability of the gantry tractor in the coordinate transport system of its operation is relevant and an enhanced scheme for turning such machinery is necessary. The purpose of research is to increase the efficiency of the process of turning the gantry tractor in the coordinate transport system of its operation, by studying its static and dynamic manoeuvrability.

2. Materials and methods

Theoretical research was carried out by modelling on a PC the conditions of operation of the gantry tractor using the provisions of theoretical mechanics and tractor theory. Experimental studies of the gantry tractor were carried out according to both generally accepted and developed methods and involved the use of modern and specially designed equipment. Processing of experimental data was performed on a PC using regression and correlation-spectral analyses.

Experimental studies were conducted in the greenhouse complex, which was equipped with a specially equipped laboratory for testing a gantry tractor. Turning of the gantry tractor was carried out by means of basic racks of power jacks (Fig. 1) placed on the bottom surface of its platforms for raising of the right or left boards accordingly. The time to perform the manoeuvre of turning (t_{tur}) of the gantry tractor in this way was determined by the sum of the duration of the individual actions:

$$t_{tur} = t_1 + t_2 + t_3 + t_4 + t_5 \tag{1}$$

where t_1 – the duration of the rise of the working bodies in the transport position, s;
t_2 – the duration of lifting one of the sides of the gantry tractor by extending the rod of the power jack, s;
t_3 – the duration of rotation of the gantry tractor around the support strut of the power jack, s;
t_4 – duration of lowering the board of the gantry tractor on the supporting surface of the traces of the constant technological track by the return stroke of the rod of the power jack, s;
t_5 – the duration of the transfer of working bodies from the transport position to the working one, s.

Fig. 1. Technical implementation of turning a gantry tractor

The angular velocity of rotation of the drive wheel was found in this way:

$$\omega_k = \frac{\pi \cdot n_k}{30} \tag{2}$$

n_k – the number of revolutions of the wheel of the gantry tractor (s^{-1}), which was defined as follows:

$$n_k = \frac{N_{ik} - 1}{2t_{ik}} \quad (3)$$

where N_{ik} – the total number of pulses received for analysis received from the speed sensor of the drive wheel of the gantry tractor during t_{ik}. Taking into account (3) the calculated working speed of the gantry tractor (V_{0p}, $m \cdot s^{-1}$):

$$V_{0p} = \omega_k \cdot R_k = \frac{L_k \cdot (N_{ik} - 1)}{120 \cdot t_{ik} \cdot (N - 1)} \quad (4)$$

Comparing the actual operating speed V_0 of the gantry tractor on the test section of the experimental laboratory with the theoretical V_{0p} estimated measurement error of this parameter:

$$\varsigma_V = \frac{V_0 - V_{0p}}{V_0} \cdot 100 \quad (5)$$

Experimentally, the slip coefficient δ of the engines of the gantry tractor was determined by expression:

$$\delta = 1 - \frac{n_{kx}}{n_{k0}} \cdot \frac{V_0}{V_x} \quad (6)$$

where n_{kx} and n_{k0} – the number of revolutions of the wheel of the gantry tractor on the same section of the road at idle and operating speeds, respectively;
V_x and V_0 – the speed of the gantry tractor at idle and operating strokes, respectively.

In the process of experimental research, the tangential thrust force P_k, developed by the wheel of the gantry tractor was analytically evaluated by the sum of two forces realized by it – traction and rolling resistance:

$$P_k = P_h + P_f \quad (7)$$

where P_h – traction force, which develops the wheel of the gantry tractor;
P_f – rolling resistance overcoming the wheel of the gantry tractor.

3. Results and Discussion

To conduct research, consider the scheme of rotation of the gantry tractor (Fig. 4), according to which its rotation is carried out by turning the chassis platform with steered wheels from one side around the centre, located in the centre of the wheelbase on the other side. According to this scheme of turning of the gantry tractor its wheels from one board can remain within the limits of the transport and technological track. Technically it can be implemented by our proposed second scheme of turning the gantry tractor (see Fig. 4) in at least two ways:
1) kinematic rotation of the platform by steered wheels on one side by means of a hinged rotary mechanism located in the centre of the wheelbase on the other side;
2) kinematic-power reversal of the gantry tractor by hanging one of its sides by means of lifting mechanisms.

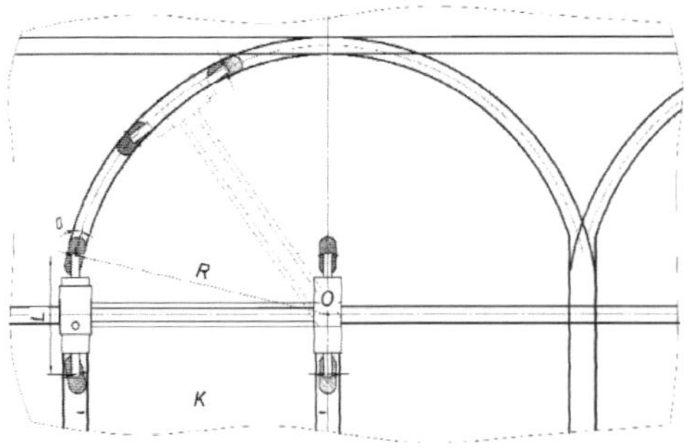

Fig. 2. The scheme of turning the gantry tractor around the centre, located in the centre of the wheelbase of one of its sides: O – centre of rotation; K – track width of the tractor; R – turning radius; L – wheelbase; α – the angle of rotation of the steered wheels

Complete information on the possibility of movement with specified kinematic parameters can be obtained by determining the power parameters (tangential thrust on the drive wheels, energy consumption of the gantry tractor to overcome the total torque, etc.), which characterize its static and dynamic manoeuvrability. To obtain the characteristics of static rotation, consider the circular motion of a flat model of a gantry tractor around the centre of rotation (point O) (Fig. 2) with a constant angular velocity $\omega = \text{const}$, $\dot{\omega} = 0$. The steered wheels of one of its sides (in this case – the left) are turned to the corners α_1 and β_1.

When compiling the equations of motion, it is based on the generally accepted notion of the correctness of the turn [8]. According to which – all the driving wheels of the tractor (in Fig. 2 – are the two wheels of the left side) move without sliding, and the intersection of their axes occurs at a point that is the centre of curvature of the path. The lateral input of the wheels is not taken into account, because the gantry tractor performs angular rotation around a fixed axis and its movement in the perpendicular direction is absent. It is considered only those elements of the gantry tractor which carry out plane parallel movement. Tangential forces of inertia and inertial moments of resistance to rotation of the gantry tractor and its technological part due to their small size are not taken into account in the study.

The moving coordinate system xSy is rigidly connected to the gantry tractor (see Fig. 2), the beginning of which will be placed in the centre of mass of its port side (point S). The x-axis is directed parallel to the longitudinal axis of the axle tractor, and its positive direction will be in the direction of movement.

In the turning mode, a number of appropriate forces act on the axle tractor (see Fig. 2). In order to carry out further actions, it is advisable to place the mass m_b divided into two parts (m_1 and m_2), falling on its left side exactly at these points. The impact of the resistance of the working elements of the technological part of the axle tractor is represented by the main vector F_{tr}, applied in the centre of resistance (point C), the direction of the action of which at the turn makes an angle γ with the longitudinal axis of the axle tractor. Reactions arising in the articulated pivot of an axle tractor can be represented by the forces F_{jx} and F_{jy}, applied at the points J, and the moment of resistance M_j.

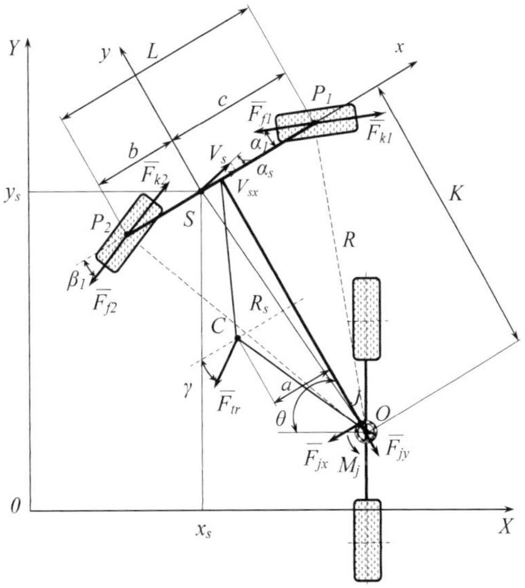

Fig. 3. Equivalent turning scheme of a (right-hand) gantry tractor with steered wheels on the left side

The equations of motion of an overhead tractor, describing the static rotation with respect to the moving coordinate system xSy, can be represented by the following system of known kinetostatic dependencies:

$$\left.\begin{aligned} m_b \cdot a_{Sx} &= \sum F_{xj}, \\ m_b \cdot a_{Sy} &= \sum F_{yj}, \\ M_R &= \sum M_{Si}, \end{aligned}\right\} \tag{8}$$

where a_{Sx}, a_{Sy} – the acceleration of the centre of mass (which is located at the point S) along the axes x and y the moving coordinate system;
$\sum F_{xj}$, $\sum F_{yj}$ – the sum of the projections of the j-th forces on the axes x and y the moving coordinate system;
M_R – total torque resistance;
$\sum M_{Si}$ – is the sum of the i-th moments with respect to the centre of mass (point S).

In expanded form, system (8) will have expression:

$$\left.\begin{aligned} m_b \cdot a_{Sx} &= \left(F_{k1} - F_{f1}\right) cos\,\alpha_1 + \left(F_{k2} - F_{f2}\right) cos\,\beta_1 - F_{tr}\,cos\,\gamma - F_{jx}\,cos\,\alpha_S + F_{jy}\,sin\,\alpha_S\,, \\ m_b \cdot a_{Sy} &= -\left(F_{k1} - F_{f1}\right) sin\,\alpha_1 + \left(F_{k2} - F_{f2}\right) sin\,\beta_1 - F_{tr}\,sin\,\gamma - F_{jy}\,cos\,\alpha_S - F_{jx}\,sin\,\alpha_S\,, \\ M_R &= c\left(F_{k1} - F_{f1}\right) sin\,\alpha_1 + b\left(F_{k2} - F_{f2}\right) sin\,\beta_1 - F_{tr}\,sin\,\gamma\,[0.5L + \mathrm{a}] + 0.5 \cdot K \cdot F_{tr}\,cos\,\gamma - M_j, \end{aligned}\right\} \tag{9}$$

where α_S, a, b and c – constructive parameters of the gantry tractor, the nature of which is clear from Fig. 3.

To study the dynamic steering of an overhead tractor, the computational scheme (Fig 5.) is used. The tractor motion will be considered in a stationary coordinate system XOY, connected with the plane of the turnstile. This allows to describe its motion by three independent generalized coordinates x_S, y_S, θ, where x_S, y_S – is the coordinates of the centre of mass (point S) when it moves in the stationary coordinate system XOY; θ – is the angle between the abscissa axes of the moving and stationary coordinate systems.

The equations of plane-parallel motion of the gantry tractor, describing its dynamic rotation, with respect to the stationary coordinate system XOY represent the known dependencies:

$$\left.\begin{aligned} m_b \cdot \ddot{x}_S &= \sum F_{Xj}\,, \\ m_b \cdot \ddot{y}_S &= \sum F_{Yj}\,, \\ J_{zS} \cdot \ddot{\theta} &= \sum \mathrm{M}_i\,, \end{aligned}\right\} \tag{10}$$

where $\ddot{x}_S$ – acceleration of the centre of mass (point S), directed along the axis X of the stationary coordinate system;
$\ddot{y}_S$ – acceleration of the centre of mass (point S), directed along the axis Y of the stationary coordinate system;
J_{zS} – moment of inertia relative to the vertical axis z, passing through the point S;
$\ddot{\theta}$ – angular acceleration of the gantry tractor.

To compile the dynamic equations of motion of the gantry tractor, the principle of D'Alembert is used:

$$\left.\begin{aligned} m_b \cdot \ddot{x}_S &= \left(F_{k1} - F_{f1}\right) \cdot cos(\theta + \alpha_1) + \left(F_{k2} - F_{f2}\right) \cdot cos(\theta - \beta_1) - F_{tr} \cdot cos(\theta - \gamma)\,, \\ m_b \cdot \ddot{y}_S &= -\left(F_{k1} - F_{f1}\right) \cdot sin(\theta + \alpha_1) + \left(F_{k2} - F_{f2}\right) \cdot sin(\theta - \beta_1) - F_{tr} \cdot sin(\theta - \gamma)\,, \\ J_{zS} \cdot \ddot{\theta} &= c \cdot \left(F_{k1} - F_{f1}\right) \cdot sin\,\alpha_1 + b \cdot \left(F_{k2} - F_{f2}\right) \cdot sin\,\beta_1 - F_{tr} \cdot sin\,\gamma \cdot [0.5L + \mathrm{a}] - M_R. \end{aligned}\right\} \tag{11}$$

Initial conditions (t = 0): $x_S = x_0$; $y_S = y_0$; $\theta = 0$; $\dot{x}_S = 0$; $\dot{y}_S = 0$; $\dot{\theta} = 0$.

From fig. 3, the correlation of kinematic radii R_S and R, translational V_{Sx} and circumferential velocity V_S of the centre of mass (point S) with the construction parameters of the overhead tractor is expressed:

$$R_S = \sqrt{(\mathrm{c} - 0.5 \cdot L)^2 + \mathrm{K}^2} = \sqrt{R^2 - c \cdot L + c^2} = \sqrt{K^2\left(1 + \frac{\mu^2}{4}\right) - c(\mu \cdot K + c)} \tag{12}$$

With the help of equations (11) and (12) influence of structural and kinematic parameters of the gantry tractor on the criteria of its dynamic rotation can be estimated, which will justify the optimal values of structural and force parameters when performing circular motion on a turn.

The driving forces F_{k1} and F_{k2} of the front and rear left wheels of the overhead tractor can be represented with sufficient accuracy for practice by the known dependence:

$$F_{k1} + F_{k2} = F_{f1} + F_{f2} + F_{tr} \tag{13}$$

The rolling resistance forces of the front and rear (left) wheels of the axle tractor F_{f1} and F_{f2} are obtained from the following well-known expression:

$$F_{f1} = f + m_1 \cdot g \tag{14}$$

$$F_{f2} = f + m_2 \cdot g \tag{15}$$

where f – rolling resistance coefficient;
g – free fall acceleration, m·s^{-2}.

Assuming that $F_{k1} = F_{k2}$ then, if $F_{f1} \neq F_{f2}$, the following dependencies are valid:

$$F_{k1} = F_{f1} + F_{tr} \cdot \left(\frac{F_{f1}}{F_{f1} + F_{f2}}\right) \tag{16}$$

$$F_{k1} = F_{f1} + F_{tr} \cdot \left(\frac{F_{f1}}{F_{f1} + F_{f2}}\right) \tag{17}$$

The mutual influence of the platform of the axle tractor chassis and the wheel bogie at the hinged mount (point O), in accordance with the generally accepted principle of replacing the rejected bonds by reactions, the forces F_{jx} and F_{jy}, as well as the moment of resistance M_j are represented. The following equations are valid for the above parameters:

$$F_j = \sqrt{F_{\mathrm{jx}}^2 + F_{jy}^2} \tag{18}$$

$$M_j = F_j \cdot f_j \cdot r_j \tag{19}$$

where F_j – the resulting reaction from the action of forces F_{jx} and F_{jy};
f_j – the coefficient of resistance of the rotating elements of the hinge;
r_j – pivot radius.

The projections of the circular velocity vector V_S of the left side centre of mass on the axes of the stationary coordinate system are:

$$V_{Sx} = V_c \cdot \cos \alpha_S \tag{20}$$

$$V_{Sy} = V_c \cdot \sin \alpha_S \tag{21}$$

where V_{Sx} – translational speed of the centre of mass of the board of the gantry tractor;
V_{Sy} – the speed of lateral displacement of the centre of mass of the board of the gantry tractor, $V_{Sy} = 0$;
α_S – angle between vectors of translational V_{Sx} and circular V_S velocity of centre of mass of the axle tractor.

In accordance with Fig. 3:

$$\tan \alpha_S = \frac{\mathrm{c} - 0.5 \cdot L}{\mathrm{K}} \tag{22}$$

The projections of the velocity components V_S on the axes of the stationary coordinate system will be:

$$V_{SX} = V_{Sx} \cdot \cos \theta \tag{23}$$

$$V_{SY} = V_{Sx} \cdot \sin \theta \tag{24}$$

By differentiating in time the velocities V_{Sx} and V_{Sy}, the acceleration of the centre of mass of the left side of the gantry tractor in the stationary coordinate system is obtained:

$$a_{SX} = \dot{V}_{Sx} \cdot \cos \theta - V_{Sx} \cdot \omega \cdot \sin \theta \tag{25}$$

$$a_{SY} = \dot{V}_{Sx} \cdot \sin \theta + V_{Sx} \cdot \omega \cdot \cos \theta \tag{26}$$

where $\dot{V}_{Sx}$ – linear acceleration during translational movement of the tractor;
$\omega = \dot{\theta} = V_{Sx} \cdot R_S^{-2}$ – angular speed of rotation of the gantry tractor;
R_S – turning radius of the gantry tractor.

To determine the accelerations of the left-side centre of mass in the moving coordinate system, projection of the acceleration vectors a_{SX} and a_{SY} on the moving x and y axes:

$$a_{Sx} = a_{SX} \cdot \cos \theta + a_{SY} \cdot \sin \theta \tag{27}$$

$$a_{Sy} = -a_{SX} \cdot \sin \theta + a_{SY} \cdot \cos \theta \tag{28}$$

Taking into account relations (25, 26) after the transformation it is obtained:

$$a_{Sx} = \dot{V}_{Sx} \tag{29}$$

$$a_{Sy} = V_{Sx} \cdot \omega \tag{30}$$

Consider the motion of the centre of the front wheel of the left side of the overhead tractor (point P_1) (Fig. 3). Denoting $V_{1x} = \dot{x}_{p1}$ and $V_{1y} = \dot{y}_{p1}$. The coordinates of point P_1 through the coordinates of the centre of mass of the left side of the overhead tractor are expressed:

$$x_{p1} = x_S + c \cdot cos\,\theta \tag{31}$$

$$y_{p1} = y_S + c \cdot sin\,\theta \tag{32}$$

Differentiating equation (31, 32) in time, the speed of the front wheel of the left side is obtained, expressed because of the speed of the centre of mass of the left side in projections on the stationary coordinate axes:

$$\dot{x}_{p1} = \dot{x}_S - c \cdot \dot{\theta} \cdot sin\,\theta \tag{33}$$

$$y_{p2} = \dot{y}_S + c \cdot \dot{\theta} \cdot cos\,\theta \tag{34}$$

Similarly, the coordinates and speed of the centre of the rear wheel of the left side of the overhead tractor (point P_2) (Fig. 3) are obtained through the coordinates and speed of the centre of the left side of the overhead tractor:

$$x_{p2} = x_c - b \cdot cos\,\theta \tag{35}$$

$$y_{p1} = y_c - b \cdot sin\,\theta \tag{36}$$

$$\dot{x}_{p2} = \dot{x}_c + b \cdot \dot{\theta} \cdot sin\,\theta \tag{37}$$

$$y_{p2} = \dot{y}_c - b \cdot \dot{\theta} \cdot cos\,\theta \tag{38}$$

To solve the static rotation problem, it is assumed that the translational velocity of the gantry tractor board is constant $V_{Sx} = const$. Then linear accelerations are equal to $a_{Sx} = 0$ and $a_{Sy} = (V_{Sx})^2 \cdot (R_S)^{-1}$. Thus, in the system of three equations (9) the unknowns are three parameters: reactions F_{jx} and F_{jy}, as well as the moment of resistance to rotation M_j, which allows to estimate the static rotationality of the axle tractor. To determine the moment of inertia J_z the well-known Huygens-Steiner theorem about moments of inertia about parallel axes is used:

$$J_{Zs} = J_S + m \cdot \left(\frac{K}{2}\right)^2 \tag{39}$$

where J_S – the moment of inertia of the gantry tractor relative to its centre of mass.

If the gantry tractor in the longitudinal-horizontal plane can be represented as a projection of a rectangle with sides K and L, the moment of inertia about the vertical axis passing through its geometric centre (assuming that this point is the centre of mass) can be calculated with sufficient accuracy by:

$$J_S = \frac{1}{12} m \cdot (K^2 + L^2) \tag{40}$$

Taking into account (40), the moment of inertia J_z can be determined from the dependence:

$$J_{ZS} = \frac{1}{12} m \cdot (K^2 + L^2) + m \cdot \left(\frac{K}{2}\right)^2 \tag{41}$$

Three equations of the initial system (9) contain the following unknown parameters: acceleration $\ddot{x}_S$, directed along the axis X; acceleration $\ddot{y}_c$ – along the axis Y; angular acceleration $\ddot{\theta}$, traction forces of driving wheels F_{k1} and F_{k2}. Thus, the system of output dynamic equations (9) is uncertain. Therefore, a particular case according to the following considerations is necessary. Turning the gantry tractor on the headland contains two characteristic phases: acceleration to a certain speed V_S and circular motion with constant forward and angular velocity ω (Fig. 4). It is clear that the smaller is the angle θ, the smaller is the distance to be covered by the gantry tractor on the turn with the desired acceleration.

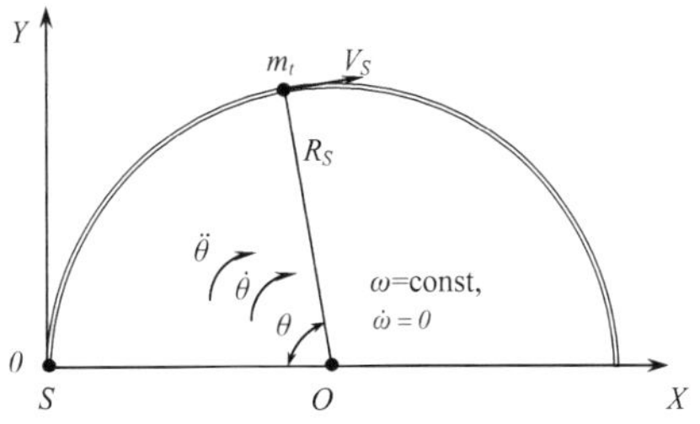

Fig. 4. Scheme explaining the mode phases of motion of the gantry tractor when turning in the fixed coordinate system $X0Y$

Under the initial conditions ($t = 0$) the origin of coordinates of the stationary system $X0Y$ (see Fig. 4) is connected with the centre of the mass side of the gantry tractor (point S). In this case, at the initial moment of time:
$x_s = 0;\ y_s = 0;\ \theta_0 = 0;\ \dot{x}_s = 0;\ \dot{y}_s = 0$.
At the end of the 1st phase of turning the gantry tractor ($t = t_1$):
$x_s = x_{s1};\ y_s = y_{s1};\ \dot{x}_s = \dot{x}_{s1(max)};\ \dot{y}_s = \dot{y}_{s1(max)};\ \theta = \theta_1$.

The acceleration $\ddot{x}_s$ and $\ddot{y}_s$ and in the system of equations (9) is represented by derivatives of velocity $\dot{V}_{SX}$ and $\dot{V}_{Sy}$, $\ddot{\theta}$ – by $\dot{\omega}$. Then, it is taken into account that the trajectory of the centre of mass of the left side of the axle tractor is given by the angle θ, then variable t is expressed through the variable θ:

$$\ddot{x}_S = \frac{dV_{SX}}{dt} = \frac{dV_{SX} \cdot d\theta}{d\theta \cdot dt} = \frac{dV_{SX}}{d\theta} \cdot \omega = \frac{dV_{SX}}{d\theta} \cdot \frac{V_S}{R_S} \tag{42}$$

$$\ddot{y}_S = \frac{dV_{Sy}}{dt} = \frac{dV_{Sy} \cdot d\theta}{d\theta \cdot dt} = \frac{dV_{Sy}}{d\theta} \cdot \omega = \frac{dV_{Sy}}{d\theta} \cdot \frac{V_S}{R_S} \tag{43}$$

$$\ddot{\beta} = \frac{d\omega}{dt} = \frac{d\omega \cdot d\theta}{d\theta \cdot dt} = \frac{d\omega}{d\beta} \cdot \omega = \frac{d\omega}{d\theta} \cdot \frac{V_S}{R_S} \tag{44}$$

After integrating the equations of system (9) it is obtained:

$$\begin{aligned} V_{SX}^{\ 2} = \frac{R_S}{2\mathrm{M}} \cdot \{ & (F_{k1} - F_{f1}) \cdot [sin(2\theta + \alpha_1) + 2\theta \cdot cos\,\alpha_1 + sin\,\alpha_1] - (F_{k2} - F_{f2})[sin(\beta_1 - 2\theta) \\ & - 2\theta \cdot cos\,\beta_1 - sin\,\beta_1] - F_{tr}[sin(2\theta + \gamma) + 2\theta \cdot cos\,\gamma + sin\,\gamma]\} \end{aligned} \tag{45}$$

$$V_{Sy}^{\ 2} = \frac{R_{п}}{2\mathrm{M}} \cdot \begin{cases} (F_{k1} - F_{f1}) \cdot [sin(2\theta - \alpha_1) + 2\theta \cdot cos\,\alpha_1 + sin\,\alpha_1] - (F_{k2} - F_{f2}) \cdot \\ [sin(\beta_1 - 2\beta) + 2\beta \cdot cos\,\beta_1 - sin\,\beta_1] - F_{tr}[sin(2\theta + \gamma) - 2\theta \cdot cos\,\gamma + sin\,\gamma]\} \end{cases} \tag{46}$$

$$\begin{aligned} (V_{SX}^{\ 2} + V_{Sy}^{\ 2}) = \frac{2R_S^{\ 2}}{J_z} \cdot \theta\{ & c \cdot (F_{k1} - F_{f1}) \cdot sin\,\alpha_1 + b \cdot (F_{k2} - F_{f2}) \cdot sin\,\alpha_2 - F_{кр} \cdot sin\,\gamma \\ & \cdot (0.5L + \mathrm{a}) - M_j \end{aligned} \tag{47}$$

Specific traction force p_t, which is developed by an axle tractor, necessary to make a turn, it is defined as the ratio of tangential traction force F_k to its operating weight G_a [7]:

$$\mathrm{P}_t = \frac{F_k}{G_{\mathrm{a}}} \tag{48}$$

The physical essence of the p_t parameter is that as it decreases, the power inputs consumed by the axle tractor during curving on the headland decrease. The solution of the system (45,46,47) allowed us to establish the regularities of the impact of the speed mode of the gantry tractor on the specific traction force p_t, needed to make a turn (Table 1).

Preliminary calculations have established that the studied gantry tractor develops the maximum specific traction force $p_t = 0.05$. According to the data of Table 1, it is clear that when the latter makes a turn at low speed (0.5 m·s^{-1}) the driving traction force realized by it allows to have the turning angle of the gantry tractor in the phase of its acceleration even less than 10 deg. With an increase in speed up to 1.0 m·s^{-1} the angle of rotation of the gantry tractor of the first phase of its acceleration is increased to 30 deg. Achieving the speed mode on the turn to 2.0 m·s^{-1} already becomes problematic. Because it is necessary to increase the acceleration phase of the specified axle tractor to the turning angle θ at least 140 deg. This means that almost all the way on the turn of the gantry tractor must move with acceleration, which is not desirable.

Specific traction force p_t, developed by an overhead tractor, is necessary to achieve the speed V_{Sx} on a turn			**Angle θ (deg) of rotation of the overhead tractor, accelerated to speed V_{Sx}**
$V_{Sx} = 0.5$ m·s^{-1}	$V_{Sx} = 1.0$ m·s^{-1}	$V_{Sx} = 2.0$ m·s^{-1}	
3.45E-02	1.35E-01	5.35E-01	10
1.79E-02	6.84E-02	2.68E-01	20
1.23E-02	4.62E-02	1.79E-01	30
9.55E-03	3.51E-02	1.35E-01	40
7.90E-03	2.85E-02	1.08E-01	50
6.75E-03	2.40E-02	9.06E-02	60
4.55E-03	1.52E-02	5.51E-02	100
3.60E-03	1.14E-02	3.99E-02	140
3.30E-03	1.02E-02	3.51E-02	160
3.10E-03	9.25E-03	3.14E-02	180

Table 1. Rotation characteristics of an overhead tractor according to the scheme in Fig. 4

Studies have shown that to obtain the optimal mode of rotation of the gantry tractor with increasing magnitude of their track, the value of the kinematic index of the mode of rotation also increases. In absolute value, the index of the turning mode varies from 5.8 m·rad^{-1} and reaches more than 30 m·rad^{-1} in the range of values of the track width of the gantry tractor 3-12 m.

The result of determining the specific traction force, which is necessary to perform a turn of an overhead tractor, according to the results of the study of dynamic turning, showed that when the latter makes a turn at low speed (0.5 m·s^{-1}) the driving traction force it realizes allows it to have an angle of turn in the phase of its acceleration even less than 10 deg, which is desirable. When the speed is increased to 1.0 m·s^{-1} the turning angle of the axle tractor's acceleration phase increases to 30 deg. Reaching the turning speed by 2.0 m·s^{-1} already becomes problematic. Because it is required to increase the acceleration phase of the specified axle tractor to a turning angle of not less than 140 deg. And this means that almost all the way on the turn the gantry tractor must move with acceleration, which is not desirable.

Turning an overhead tractor, the index of the turning mode significantly depends on the ratio of its wheelbase to the track width. When implementing the turning scheme by running all the steered wheels of the tractor around the centre of rotation, located in the centre of symmetry of its chassis, the index of the turning mode in the range of values of the track width of 3...12 m is in the range 2.3...20.0 m·rad^{-1}. When he realizes the rotation around the centre, located inside the interwheel space, this parameter varies from 5.8 m·rad^{-1} and reaches more than 30.0 m·rad^{-1}. Experimental timing for performing the power (onboard) turn manoeuvre (t_{tur}) of the axle tractor on the turning lane showed (Table 2) that the total duration of this process is about 1 min (0.0168 h).

Name of elementary time	**Duration, hours**
1. Lifting the working bodies in the transport position	0.0027
2. Lifting one of the sides of the tractor	0.0033
3. Rotate the tractor around the support jack of the power jack	0.0061
4. Lowering the tractor board	0.0025
5. Transfer of working bodies from the transport position to the working one	0.0022
Total time	0.0168

Table 2. Duration of the manoeuvre of power on-board rotation of the gantry tractor

This result is significant improvement in comparison with the duration of the turn manoeuvre of traditional large-capacity machine-tractor units. If it is possible to carry out this process in automatic mode, then the resulting duration of time to perform the turn manoeuvre with an overhead tractor can be reduced by at least half. The turning calculation and the new scheme can be used in automated movement and AI based applications, reducing the necessary time on non-effective movements.

4. Conclusion

The new scheme of turning the axle tractor on the headland by turning its chassis with steered wheels from one side around the centre located in the centre of the interwheel space from the other side, proposed by us, allows moving it simultaneously with turning to the next working position with better kinematic parameters. In practice, the best turning performance is achieved with an axle tractor construction when the ratio of its wheelbase to track width is less. The steering angle of its steering wheels is also the smallest under this condition.

The conducted researches show that in order to obtain the optimum mode of turning of an axle tractor with the increase of their path from 3 m to 12 m the kinematic index of the turning mode also increases from 5.8 $m \cdot rad^{-1}$ to 30 $m \cdot rad^{-1}$ and more. Practically, this requires an increase in the speed mode of their turning movement, which requires an increase in energy consumption for this process. The manoeuvre of the gantry tractor turn at low speed (0.5 $m \cdot s^{-1}$) allows you to have its turn angle in the acceleration phase even less than 10 deg. But when increasing the speed up to 1.0 $m \cdot s^{-1}$ the turning angle of the gantry tractor in the acceleration phase increases to 30 deg. Realization of the speed mode by an overhead tractor in a turn with speeds about 2.0 $m \cdot s^{-1}$ already becomes problematic. Because it is required to increase the phase of its acceleration to the turning angle not less than 140 deg. The new scheme could and will be used as an input to AI based and automated tractors.

5. References

[1] Bulgakov V., Adamchuk V., Kuvachov V. (2017). Investigation of technological properties of wide span tractors for controlled traffic farming, Proceedings 16 International Scientific Conference «Engineering For Rural Development» (24 – 26 May 2017, Jelgava, Latvia), Vol. 16. pp. 303–308.

[2] Bulgakov V., Adamchuk V., Kuvachov V. (2017). Theoretical justification of the turn of a wide span tractor (vehicle) for controlled traffic farming, INMATEH – Agricultural Engineering, Vol. 53, No. 3, pp. 159–167.

[3] Bulgakov V., Kyurchev V., Kuvachov V. (2017). The study of movement wide span tractor (vehicles) with kinematic (turn of wheels) method of its control, Mechanization in agriculture & conserving of the resources, Vol. 5, pp. 188 – 193.

[4] Bulgakov V., Kuvachov V., Nozdrovický L. (2018). The study of movement of the wide span tractor–based field machine unit with power method of its control, Acta Technologica Agriculturae, No. 4. pp. 163–168.

[5] Bulgakov, V.; Kuvachov, V.; Olt, J. (2019). Theoretical Study on Power Performance of Agricultural Gantry Systems, 30th International DAAAM Symposium "Intelligent Manufacturing & Automation", DAAAM International, pp. 167–175. DOI: 10.2507/30th.daaam.proceedings.022.

[6] Chamen W.C.T. (1992). Assessment of a wide span vehicle (gantry) and soil and cereal crop responses to its use in a zero traffic regime, Soil & Tillage Research, Vol. 24, No. 4, pp. 359–380.

[7] Chamen W.C.T. (2000). A new methodology for weed control and cereal crop production based on wide span vehicles and precision guidance: Biotrac, Processing 4th EWRS workshop on physical weed control (20–22 March 2000), Netherlands. pp. 51–55.

[8] Guskov V.V. (1984). Tractors: theory. Moscow: Mechanical Engineering, 374 p. (Russian)

[9] Onal I. (2012). Controlled traffic farming and wide span tractors, Agricultural Machinery Science, Vol. 8, No 4, pp. 353–364.

[10] Pedersen H.H. (2011). Harvest capacity model for a wide span onion bunker harvester, Automation and System Technology in Plant Production: CIGR section V & NJF section VII conference, pp. 27–36.

[11] Pedersen, H.H., Sørensen C.G., Oudshoorn, F.W. (2013). User requirements for a wide span tractor for controlled traffic farming, CIOSTA XXXV Conference «From effective to intelligent agriculture and forestr», International commission of agricultural and biological engineers (3–5 July 2013, Section V). Billund, Denmark, pp. 134–136.

[12] Pedersen, H.H., Oudshoorn, F.W., McPhee, J.E. (2016). Wide span – re–mechanising vegetable production, XXIX International horticultural congress on horticulture: sustaining lives, livelihoods and landscapes: international symposia on the physiology of perennial fruit crops and production systems and mechanisation, precision horticulture and robotics, Book Series: Acta Horticulturae, Vol. 1130, pp. 551–557.

[13] Uleksin V.O. (2008). Automation of driving of machine-tractor units, Proceedings of Kharkiv National Technical University, Vol. 2. No. 75, pp. 105–116. (Ukraine)

[14] Uleksin V.O. (2008). Gantry farming: a monograph, Dnepropetrovsk: Porogi, 224 p. (Ukraine)

[15] Uleksin V.O. (2011). Automation of vehicle control in gantry farming. Proceedings of Kharkiv National Technical University, Vol. 2, No. 10, pp. 101–110. (Ukraine)

DOI: 10.2507/33rd.daaam.proceedings.055

MOBILE ROBOT COBOT MANIPULATOR MOUNTING DIRECTION ADJUSTMENT DEPENDANT ON GRAVITY VECTOR

Indrek Virro, Tormi Lillerand & Jüri Olt

This Publication has to be referred as: Virro, I[ndrek]; Lillerand, T[ormi] & Olt, J[uri] (2022). Mobile Robot Cobot Manipulator Mounting Direction Adjustment Dependant on Gravity Vector, Proceedings of the 33rd DAAAM International Symposium, pp.0390-0396, B. Katalinic (Ed.), Published by DAAAM International, ISBN 978-3-902734-36-5, ISSN 1726-9679, Vienna, Austria
DOI: 10.2507/33rd.daaam.proceedings.055

Abstract

Mobile robots with manipulators are brought into more diverse situations, such as agriculture. With the introduction of more unstructured environments a more dynamic approach is needed for the manipulator setup as the previous static mounting direction may obstruct completion of the platform's planned task. The aim of this article is to propose a dynamic methodology for determining manipulator mounting angle before executing manipulator end effector motion. An example of the methodology was used on the precision fertilization robot prototype cobot manipulator to confirm the suitability of the proposed methodology.

Keywords: Mobile Manipulator; Robot Arm; Field Robot; Base Disturbances; IMU

1. Introduction

Robotics technology deployment is increasing, one of the emerging trends is field robotics. Where mobile robots [1], [2], [3], [4] are used to solve different tasks and their capabilities are also increased by using cobot manipulators [5], [6], [7], [8], [9]. The collaborative robots can share workspace safely with humans [10], because compared to traditional industrial manipulators they have heightened sensing capabilities, but this also increases the chance of triggering an emergency stop from overload. One reason for the overload can be that base disturbances [11] have occurred – like the mobile robot driving on uneven terrain. For example, uneven terrain on blueberry plantation is caused by the planting crater and plants position deviates from the plant row central axis up to 365 mm [12]. The aim of this article is to propose a dynamic methodology for determining manipulator mounting angle considering the actual gravity vector, because the mobile robot will be tilted while following the plant row, which means that the cobot manipulator mounting settings will be at an angle to the actual gravity vector and the setting must be adjusted to avoid triggering the emergency stop from overload. Field robots can increase productivity and decrease specific costs for the specific task and therefore reduce the realization price of harvested crops [13]. For the maximum effect the autonomous unit must execute planned tasks with high success rate and have a self-resetting capability in case of error states. For example, in horticultures need based farming one specific task is precision fertilizing and it can be carried out with solid fertilizer [14], but for the best success rate for spot fertilizing gravity direction must be considered to align the solid fertilizer discharge port axis to the gravity vector direction.

For the alignment and self-resetting capabilities of field robot cobot manipulator an easy to implement and robust solution is needed. This work will enable to enable to use a random cobot robot in field robotics and other applications where base disturbances correction is needed.

2. Materials and methods

As a platform a mobile field robot prototype [15] was used and xArm 6 cobot from UFACTORY company was mounted on it (Figure 1). The cobot tool fixture sensor was chosen an inertial measurement unit (IMU) 9DoF Razor IMU M0 from SparkFun company. Tool fixture sensor calibration can be done as a known process [16] proposed on an actual installed system. Intel NUC was used for development where Robot Operating System (ROS) distribution Noetic Ninjemys was installed. ROS environment has built in tools and packages to speed up robotics technology development and with good practice design guides try to have high reusability [17], [18], [19], [20]. This work leverages razor_imu_9dof, moveit and xarm_ros package to build a custom package with the necessary workflow. The imu stock firmware was changed to the ROS supported firmware and direction cosine matrix algorithm [21], [22] was used in data fusion.

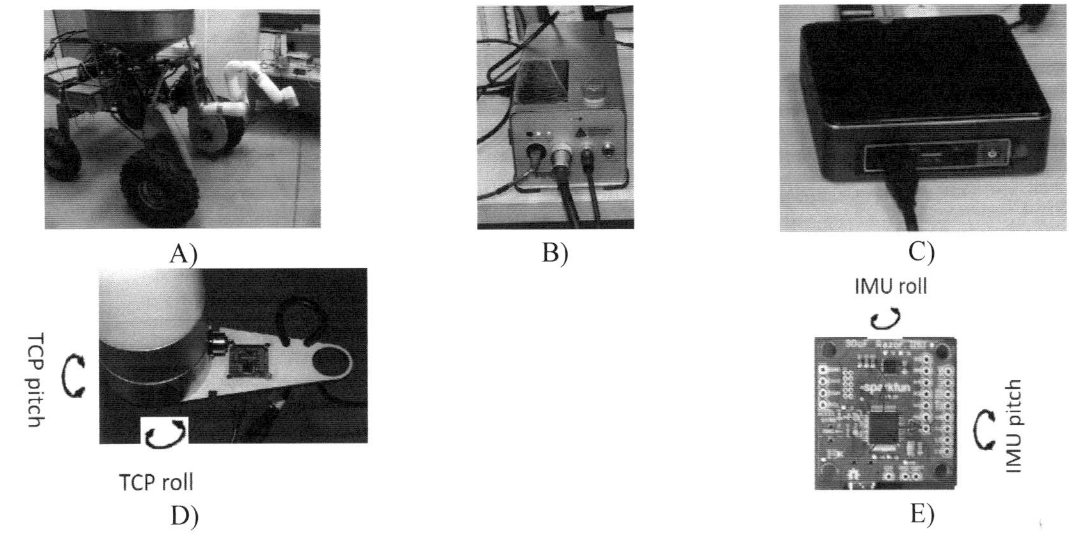

Fig. 1. Example setup. A) field robot prototype with xArm 6 cobot manipulator, B) cobot controller, C) Intel NUC with ROS Noetic, D) manipulator tool fixture with IMU 9DoF Razor IMU M0, TCP roll and pitch directions, E) IMU 9DoF Razor IMU M0 axis, roll and pitch directions.

Orientation of robot arm end effector tool centre point (TCP) and IMU are read when the mobile field robot is stationery. The read orientation is transformed to roll, pitch and yaw angles. In simple cases only base roll and pitch need to be considered [23]. The IMU orientation angles show the needed adjustment for the TCP orientation. Therefore, adjusted the new TCP orientation is calculated in roll and pitch by subtracting IMU orientation angles from TCP orientation angles. Then, new goal state for the TCP is put together from the TCP position and the adjusted new TCP orientation. Move to the new goal state is triggered to keep the end effector in the same x, y and z coordinates and minimize the roll and pitch in IMU read orientation. After archiving TCP alignment with the gravity vector, then the base real orientation can be found by relating the orientation between consecutive links [24] and the correct mounting direction angle can be sent to the cobot controller.

3. Results

With the completed method, the cobot manipulator end effector and base were aligned successfully with respect to the gravity vector under laboratory conditions. Different beginning states were chosen for the manipulator TCP orientation and the ending state was roll and pitch 0 degree angle with the updated base angle (Figure 2 - 5). This result can be transferred to halted field robot manipulator performing tasks in the real life, like automatic spot fertilizing with solid fertilizer as this task can be performed the mobile robot stop and go motion. The manipulators actual kinematic chain and link lengths must be considered finding the correct base mounting direction based on the TCP gravity vector, this means the method gives manipulator specific outcome and needs to be setup with the correct kinematic model.

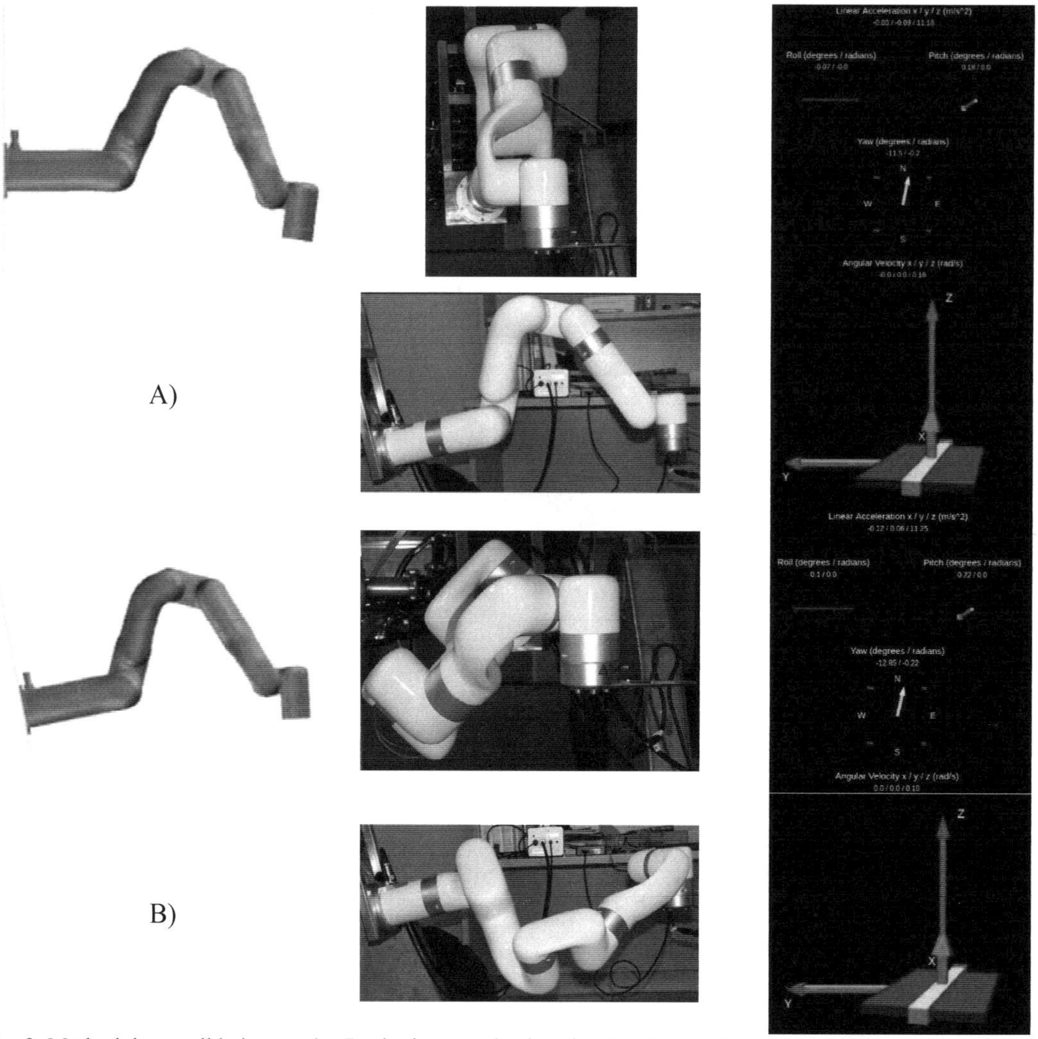

Fig. 2. Methodology validation results. Beginning state is given in A) roll angle 0 deg and pitch angle 0 deg and ending state is given in B. In frames from the left to right xArm base mounting orientation from xArm control GUI, front and side view of actual manipulator state, IMU data visualization razor_imu_9dof razor-pub-and-display.launch.

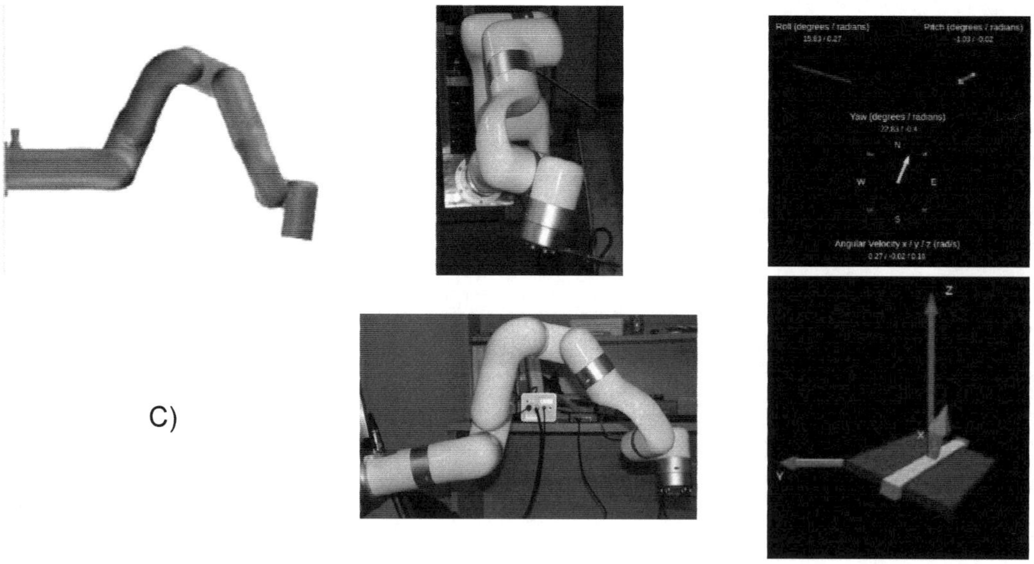

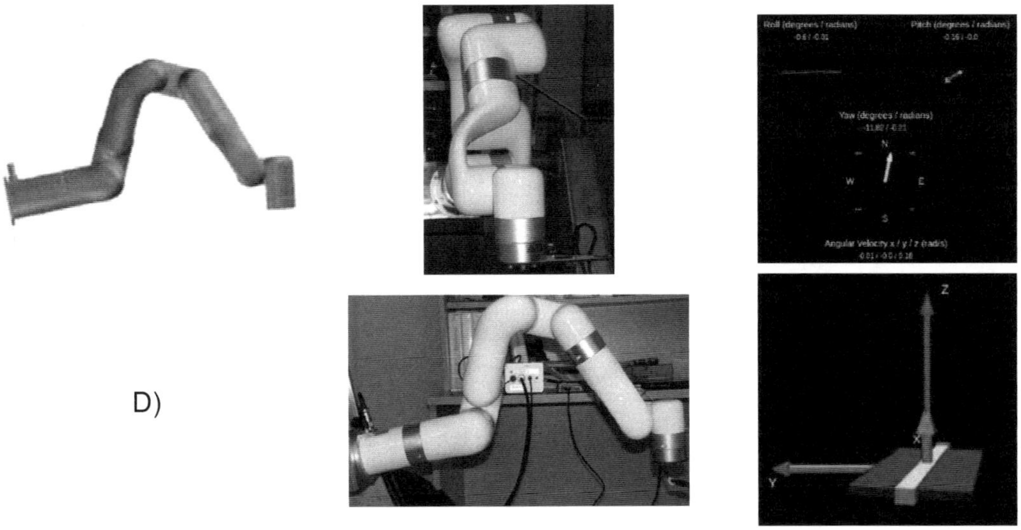

D)

Fig. 3 - Methodology validation results. Beginning state is given in C) roll angle 15 deg and pitch angle 0 deg, E and ending state is given in D). In frames from the left to right xArm base mounting orientation from xArm control GUI, front and side view of actual manipulator state, IMU data visualization razor_imu_9dof razor-pub-and-display.launch.

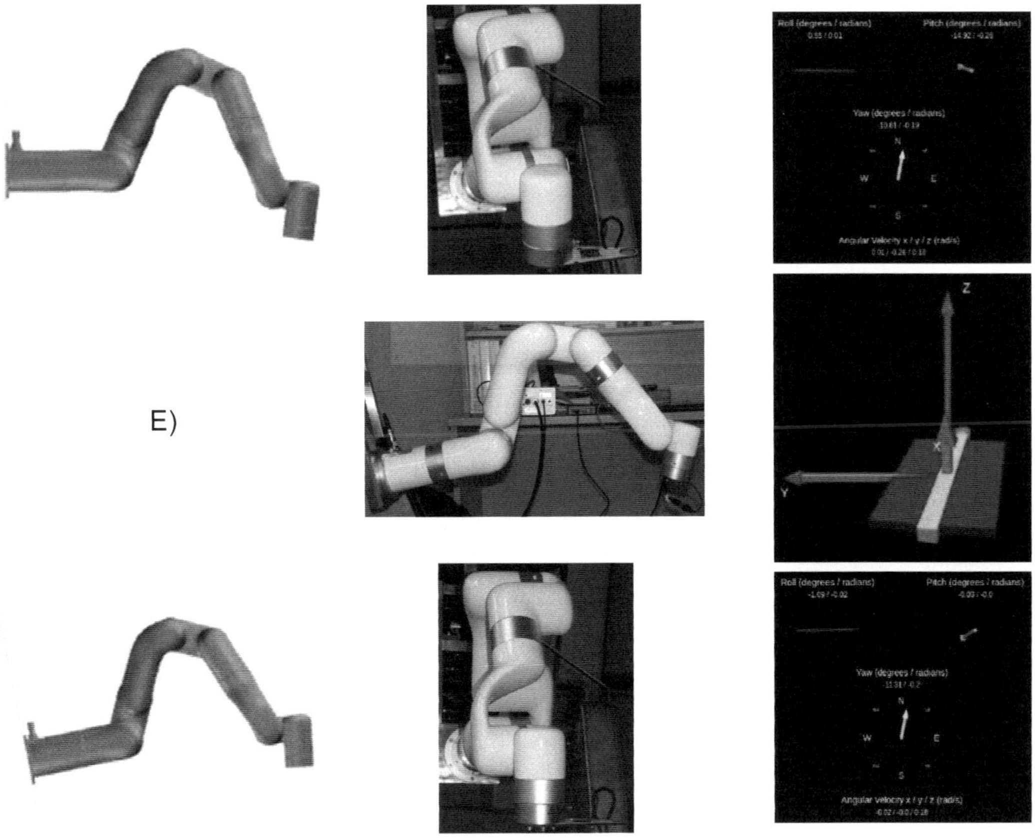

E)

F)

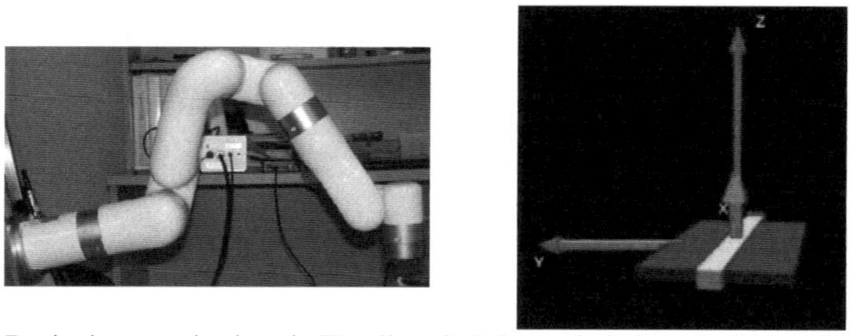

Fig. 4 - Methodology validation results. Beginning state is given in E) roll angle 0 deg and pitch angle 15 deg, G) roll angle 0 deg and pitch angle 0 deg and ending state is given in F). In frames from the left to right xArm base mounting orientation from xArm control GUI, front and side view of actual manipulator state, IMU data visualization razor_imu_9dof razor-pub-and-display.launch.

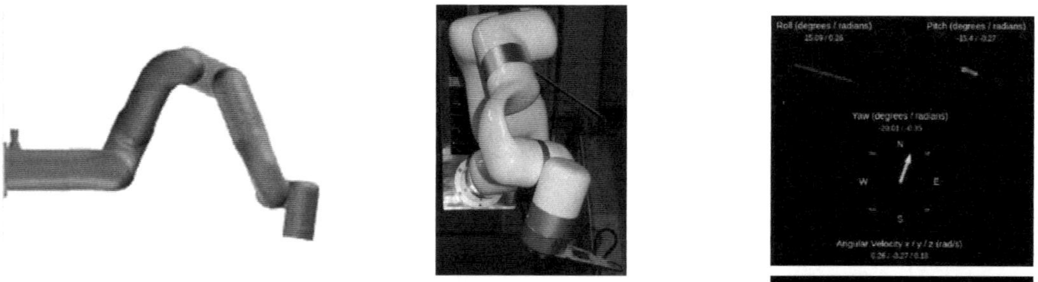

G)

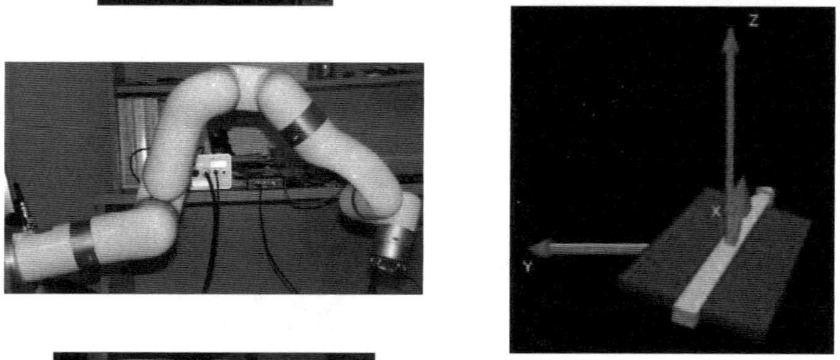

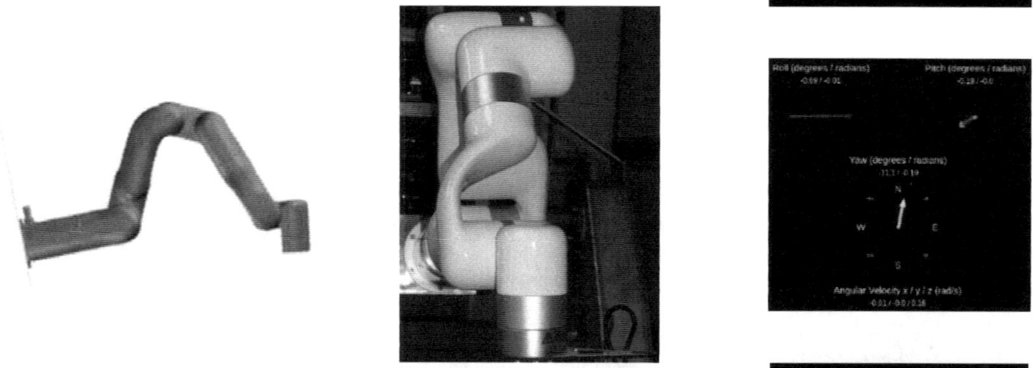

H)

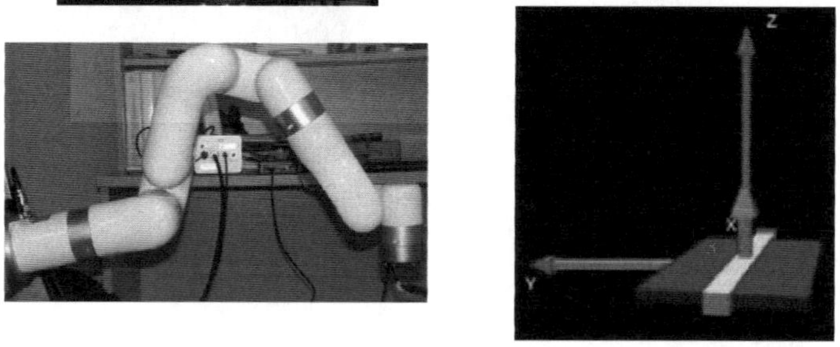

Fig. 5 - Methodology validation results. Beginning state is given in G) roll angle 15 deg and pitch angle 15 deg and ending state is given in H). In frames from the left to right xArm base mounting orientation from xArm control GUI, front and side view of actual manipulator state, IMU data visualization razor_imu_9dof razor-pub-and-display.launch.

Compared to fixed base methods [18], [25] that use IMU the proposed method allows for repositioning of the mounted cobot in the world and it is more universal as it can be used for any off-the-self cobot even those whose point-to-point gravity compensation is closed source. In the course of the work, it became evident that xArm ROS driver does not have an inbuilt way to change the base mounting orientation settings, therefore over Modbus Transmission Control Protocol / Internet Protocol (TCP/IP) message for set the gravity direction needs to be sent. The message contains Modbus TCP/IP header and parameters, where the base mounting angle adjustment is represented as gravity direction vector unit components. Also, other cobot manufactures ROS drivers seem to lack this feature, namely Universal Robots similarly over TCP/IP transmit a message, but this message runs a script that using function set_gravity(gravity_vector_x, gravity_vector_y, gravity_vector_z) adjust the base mounting angle with the gravity vector components and the expected magnitude of the gravity vector is 9.82 in value. In communicating with the cobot controller the changing of the base mounting orientation is not standardized and differs between manufacturers.

IMU data can be used to improve the forward kinematics estimates [25] and the end effector pose error can be reduced with the predictive base motion the manipulator [26]. The IMU based approach can have a drawback like [17] found in high-dynamic situations estimate became unstable and they needed to slow down the manipulator. As a possible solution for the problem is to try and find a high robustness data fusion algorithm, [27], [28] show promising direction for further investigation.

4. Conclusion

Mounting cobots on mobile robots will affect the systems reliability. Especially with the base disturbances that occur from the mobile robot driving on uneven terrain an overload event is likely. Therefore, a situation may arise where the cobot manipulator refuses to carry out planned movement, because the base is out of alignment with respect to the gravity vector. In this work a dynamic methodology for determining manipulator mounting angle was proposed. By using a TCP fixture mounted IMU data for adjusting the manipulator mounting direction dependant on gravity vector. In the course of the work, it was found out that the cobot ROS driver lacks the feature to affect the base mounting setting, but the manufacture has support for it in the controller. Using the proposed methodology solves overload issues derived from mobile robot base misalignment to the gravity vector. Further work needs to be done on field testing the proposed methodology, possibly finding out the shortcomings of used IMU data fusion algorithm and looking into other algorithms as a solution. In addition, the custom ROS package needs to be made into an easy-to-use tool by modifying the xArm default driver and implementing a generic base mounting angle adjustment calculator that takes advantage of existing manipulators kinematic chain that is utilized for the forward kinematics.

5. Acknowledgments

This publication was supported by project PM210001TIBT from the Estonian University of Life Sciences.

6. References

[1] Oliveira, L. F., Moreira, A. P., & Silva, M. F. (2021). Advances in agriculture robotics: A state-of-the-art review and challenges ahead. Robotics, 10(2), 52 DOI: http://dx.doi.org/10.3390/robotics10020052

[2] Hayashi, S., Shigematsu, K., Yamamoto, S., Kobayashi, K., Kohno, Y., Kamata, J., & Kurita, M. (2010). Evaluation of a strawberry-harvesting robot in a field test. Biosystems engineering, 105(2), 160-171 DOI: https://doi.org/10.1016/j.biosystemseng.2009.09.011

[3] Grimstad, L., Zakaria, R., Le, T. D., & From, P. J. (2018, October). A novel autonomous robot for greenhouse applications. In 2018 IEEE/RSJ International Conference on Intelligent Robots and Systems (IROS),1-9. IEEE DOI: https://doi.org/10.1109/IROS.2018.8594233

[4] Gonzalez-de-Santos, P., Fernández, R., Sepúlveda, D., Navas, E., Emmi, L., & Armada, M. (2020). Field robots for intelligent farms—Inhering features from industry. Agronomy, 10(11), 1638. DOI: https://doi.org/10.3390/agronomy10111638

[5] Kang, H., Zhou, H., & Chen, C. (2020). Visual perception and modeling for autonomous apple harvesting. IEEE Access, 8, 62151-62163 DOI: https://doi.org/10.1109/ACCESS.2020.2984556

[6] SepúLveda, D., Fernández, R., Navas, E., Armada, M., & González-De-Santos, P. (2020). Robotic aubergine harvesting using dual-arm manipulation. IEEE Access, 8, 121889-121904 DOI: https://doi.org/10.1109/ACCESS.2020.3006919.

[7] Birrell, S., Hughes, J., Cai, J. Y., & Iida, F. (2020). A field-tested robotic harvesting system for iceberg lettuce. Journal of Field Robotics, 37(2), 225-245. https://doi.org/10.1002/rob.21888.

[8] Botterill, T., Paulin, S., Green, R., Williams, S., Lin, J., Saxton, V., & Corbett-Davies, S. (2017). A robot system for pruning grape vines. Journal of Field Robotics, 34(6), 1100-1122 DOI: https://doi.org/10.1002/rob.21680

[9] Underwood, J. P., Calleija, M., Taylor, Z., Hung, C., Nieto, J., Fitch, R., & Sukkarieh, S. (2015). Real-time target detection and steerable spray for vegetable crops. In Proceedings of the International Conference on Robotics and Automation: Robotics in Agriculture Workshop, Seattle, WA, USA, 26-30

[10] Peshkin, M. A., Colgate, J. E., Wannasuphoprasit, W., Moore, C. A., Gillespie, R. B., & Akella, P. (2001). Cobot architecture. IEEE Transactions on Robotics and Automation, 17(4), 377-390 DOI: https://doi.org/10.1109/70.954751

[11] Dunnigan, M. W., & Wronka, C. M. (2011). Comparison of control techniques for a robotic manipulator with base disturbances. IET Control Theory & Applications, 5(8), 999–1012 DOI: https://doi.org/10.1049/iet-cta.2010.0331

[12] Arak, M., & Olt, J. (2020). Technological description for automating the cultivation of blueberries in blueberry plantations established on depleted peat milling fields. In Proceedings of the International Scientific Conference "Rural Development", 2019(1), 98-103 DOI: http://doi.org/10.15544/RD.2019.024

[13] Virro, I., Arak, M., Maksarov, V., & Olt, J. (2020). Precision fertilisation technologies for berry plantation. Agronomy Research, 18(S4), 2797-2810 DOI: https://doi.org/10.15159/AR.20.207.

[14] Lillerand, T., Virro, I., Maksarov, V. V., & Olt, J. (2021). Granulometric Parameters of Solid Blueberry Fertilizers and Their Suitability for Precision Fertilization. Agronomy, 11(8), 1576 DOI: https://doi.org/10.3390/agronomy11081576.

[15] Soots, K., Lillerand, T., Jogi, E., Virro, I., & Olt, J. (2021). Feasibility analysis of cultivated berry field layout for automated cultivation. In Engineering for Rural Development. 20, 1003–1008 DOI: https://doi.org/10.22616/ERDev.2021.20.TF222

[16] Axelsson, P., & Norrlöf, M. (2012). Method to estimate the position and orientation of a triaxial accelerometer mounted to an industrial manipulator. IFAC Proceedings Volumes, 45(22), 283-288 DOI: https://doi.org/10.3182/20120905-3-HR-2030.00066

[17] Quigley, M., Brewer, R., Soundararaj, S. P., Pradeep, V., Le, Q., & Ng, A. Y. (2010). Low-cost accelerometers for robotic manipulator perception. In 2010 IEEE/RSJ International Conference on Intelligent Robots and Systems, 6168-6174. IEEE DOI: https://doi.org/10.1109/IROS.2010.5649804

[18] Montalvo, W., Escobar-Naranjo, J., Garcia, C. A., & Garcia, M. V. (2020). Low-cost automation for gravity compensation of robotic arm. Applied Sciences, 10(11), 3823 DOI: https://doi.org/10.3390/app10113823

[19] Hernandez-Mendez, S., Maldonado-Mendez, C., Marin-Hernandez, A., Rios-Figueroa, H. V., Vazquez-Leal, H., & Palacios-Hernandez, E. R. (2017, November). Design and implementation of a robotic arm using ROS and MoveIt!. In 2017 IEEE International Autumn Meeting on Power, Electronics and Computing (ROPEC) ,1-6, IEEE DOI: https://doi.org/10.1109/ROPEC.2017.8261666

[20] Novotny, G., Emsenhuber, S., Klammer, P., Poschko, C., Voglsinger, F., & Kubinger, W., (2019). A Mobile Robot Platform for Search and Rescue Applications, Proceedings of the 30th DAAAM International Symposium, pp.0945-0954, B. Katalinic (Ed.), Published by DAAAM International, ISBN 978-3-902734-22-8, ISSN 1726-9679, Vienna, Austria DOI: 10.2507/30th.daaam.proceedings.131

[21] Wang, Y., & Rajamani, R. (2018). Direction cosine matrix estimation with an inertial measurement unit. Mechanical Systems and Signal Processing, 109, 268-284 DOI: https://doi.org/10.1016/j.ymssp.2018.02.038

[22] Hyyti, H., & Visala, A. (2015). A DCM Based Attitude Estimation Algorithm for Low-Cost MEMS IMUs. International Journal of Navigation & Observation DOI: https://doi.org/10.1155/2015/503814

[23] Wronka, C. M., & Dunnigan, M. W. (2011). Derivation and analysis of a dynamic model of a robotic manipulator on a moving base. Robotics and Autonomous Systems, 59(10), 758-769 DOI: https://doi.org/10.1016/j.robot.2011.05.010

[24] Roan, P., Deshpande, N., Wang, Y., & Pitzer, B. (2012). Manipulator state estimation with low cost accelerometers and gyroscopes. In 2012 IEEE/RSJ International Conference on Intelligent Robots and Systems, 4822-4827 DOI: https://doi.org/10.1109/IROS.2012.6385893

[25] Hedberg, E., Norén, J., Norrlöf, M., & Gunnarsson, S. (2017). Industrial robot tool position estimation using inertial measurements in a complementary filter and an EKF. IFAC-PapersOnLine, 50(1), 12748-12752 DOI: https://doi.org/10.1016/j.ifacol.2017.08.1828

[26] Woolfrey, J., Liu, D., & Carmichael, M. (2016, May). Kinematic control of an autonomous underwater vehicle-manipulator system (AUVMS) using autoregressive prediction of vehicle motion and model predictive control. In 2016 IEEE International Conference on Robotics and Automation (ICRA), 4591-4596 DOI: https://doi.org/10.1109/ICRA.2016.7487660

[27] Valenti, R. G., Dryanovski, I., & Xiao, J. (2015). Keeping a good attitude: A quaternion-based orientation filter for IMUs and MARGs. Sensors, 15(8), 19302-19330 DOI: https://doi.org/10.3390/s150819302

[28] Hol, J. D. (2011). Sensor fusion and calibration of inertial sensors, vision, ultra-wideband and GPS, Ph.D. Dissertation, Department of Electrical Engineering, Linköping University, Linköping, Sweden

DOI: 10.2507/33rd.daaam.proceedings.056

PREVENTIVE AND REPRESSIVE MEASURES FOR FIRE SAFETY

Adam Malatinsky & Martin Hromada

This Publication has to be referred as: Malatinsky, A[dam] & Hromada, M[artin] (2022). Preventive and Repressive Measures for Fire Safety, Proceedings of the 33rd DAAAM International Symposium, pp.0397-0403, B. Katalinic (Ed.), Published by DAAAM International, ISBN 978-3-902734-36-5, ISSN 1726-9679, Vienna, Austria
DOI: 10.2507/33rd.daaam.proceedings.056

Abstract

Fire safety, as one of the basic safety categories, has developed throughout history. The article focuses on determining preventive and repressive measures related to fire protection. The preventive and repressive model of ensuring safety is first described as a theoretical basis. Subsequently, an analysis of fire protection measures is defined, describing specific preventive and repressive fire protection measures. Based on these measures, individual phases of fire protection activities are described, followed by a description of proactive fire protection measures. The article's result is the proposal and application of fire prevention measures. These measures are proposed for a soft target - a shopping centre.

Keywords: Fire Protection; Measures; Fire Safety; Safety

1. Introduction

Fire protection is a primary type or category of security and a sense of security in society. It began to develop in ancient times and was transformed several times throughout history to its current form. Among the first major threats related to fire protection was the massive fire in Rome in CE 64. [1] The region's largest recorded fire in history was the Siberian Taiga fire in Russia in 2003, which spread over an area of up to 55 million acres (about 5,500 square kilometres). [2]. Fire safety is a highly resolved issue regarding security breaches. Primary fire threats include fire, explosion and leakage of hazardous substances. Of these fire threats, there are many other fire threats associated with them. The article deals with individual fire protection measures. In the first part, he describes the security model, which from the point of view of security, can be divided into preventive, repressive and combination. The next chapter describes individual preventive and punitive fire protection measures. In the next part, the individual phases of the fire protection activity process are defined and arranged chronologically. The following chapter describes specific proactive fire protection methods belonging to the second phase of the fire protection activity. The same results of the article are the design and application of proactive methods for fire protection. It is the creation of new or modification of current measures. As part of future research could be about creating and applying proactive fire protection measures intended for schools, hospitals, restaurants, stadiums and various cultural and social events.

2. Security model

The model expresses the main aspects of reality. It is a specific type of abstract understanding of the actual situation. It can be represented graphically, on a map, numerically (quantitatively), descriptively (qualitatively), programmatically or in other ways. The subjects and objects found in the given model are linked logically, hierarchically and structurally according to the given situation. The security model describes an actual situation oriented toward a possible security breach. This description and representation can also be done graphically, on a map, quantitatively, qualitatively, by a program or in other ways. It is distinguished between 3 basic security breach models: preventive, repressive, and combination.

The preventive security assurance model expresses knowledge, prevention methods and preparation for future possible security breaches. Performs individual activities before a security breach occurs. The repressive model of ensuring security expresses a set of knowledge, methods of repression and their implementation in already created states of security violation. Performs individual activities after the occurrence of a security breach. The combination of preventive and repressive models expresses a set of knowledge, methods of prevention and repression and their implementation from the period of preparation for the state of security assurance to the solution of the state of a security violation. [3]

3. Analysis of fire protection measures

This chapter analyses fire protection measures, describing preventive and repressive activities. Preventive and repressive activities are described together within fire protection units - state (professional) and simultaneously voluntary fire protection units, legal entities and natural persons - entrepreneurs and state authorities.

3.1 Prevention of fire protection

As part of the prevention of fire protection, the following activities are carried out:

Prevention of fires or explosions - repair and maintenance of equipment working with dangerous substances, creation of maps in case of fire, conducting preventive fire inspections and building evacuation drills in the event of a fire.

Fire protection technician - determining places with increased fire risk, carrying out fire protection training, activities regarding fire protection documentation and practising fire alarms.

Fire protection specialist - development of fire safety solutions in project documentation of buildings, assessment of conformity of products, placement of technological equipment and analysis of the risk of fire.

Preparation of documents - fire identification card, workplace fire regulations, fire alarm directives, fire evacuation plan, fire book, fire risk analysis, documents on the control of fire equipment according to a special regulation, documentation on the training of fire patrol employees, documentation on the training of fire patrols, documentation on the activities of the fire department, fire protection documentation for apartment buildings, municipality documentation on fire protection and the municipality's fire regulations.

Educational activities - educational training within primary schools, secondary schools and universities, information leaflets, radio, videos and other ways of presenting fire protection activities to the population, teaching the subject of fire protection in schools and vocational secondary or higher education focused on fire protection.

Training - training of fire patrols, conducting basic training for members of voluntary fire protection units, conducting professional training for members of the fire and rescue service (professional firefighters), training of a preventive officer, technician and fire protection specialist.

State fire supervision - carrying out comprehensive, thematic or follow-up inspection, assessment of building documentation, zoning proceedings, building proceedings, approval proceedings and ascertaining the causes of fires.

Fire patrol - creation and performance of fire patrol at the workplace, creation and performance of fire assistance patrol.

Control activities - conducting preventive fire inspections and a safety audit within the fire safety framework.

Employee training - employee training regarding fire protection and regular implementation of safety at work.

Fire evacuation plan - drawing up a fire evacuation plan for objects with text, graphics and a legend.

Exercises - regular performance of screening (not prepared in advance) and tactical (prepared in advance) exercises.

Prevention officer - prevention officer of the village within the voluntary fire protection unit, proper heating of households, proper placement of fire extinguishers in households, control of burning places during the ban.

Other activities - promotion of fire protection, recruitment of new fire protection members, implementation of various actions for children and adults aimed at fire protection. [4] [5]

3.2 Repression of fire protection

As part of the repression of fire protection, the following activities are carried out:

Firefighting - firefighting in a family house, apartment or cottage, firefighting in an industrial building, at the airport, firefighting in a vehicle, firefighting in the natural environment, and other types of firefighting.

Water transport - long-distance transport of water by shuttle or chain method.

Water removal - water removal in the natural environment, residential environment, water removal or industrial environment.

Liquidation activities - objects, animals, persons, dangerous substances, and other liquidation activities.

Activities within the framework of crisis management - the evacuation of people, animals and objects, hiding of people, animals and objects, emergency survival of people and animals and provision of life needs to people and animals.

Traffic accident: traffic accident in a natural environment, on a road, expressway or highway, traffic accident in another environment, traffic control on the road, removal of trees or objects from roads and public spaces.

Providing first aid and rescue - providing first aid to a person or an animal, rescuing people or animals from heights or depths, rescuing people or animals from inaccessible places.

Control activities - control of an emergency after intervention in a natural, residential or industrial environment.

Documents: report on the intervention of firefighters, annual report on the activities of the fire and rescue service, creation of statistics for the past periods.

Other activities: explosives, preparation for suicide, removal of stinging insects, removal of other animal species, the performance of other activities related to an emergency event, and performance of other activities related to a crisis event. [6] [7]

4. Phases of fire protection

When determining proactive fire protection methods, it is essential first to characterize the phase of proactive fire protection activity. The phases of fire protection activities can be divided into four parts: passive prevention of fire protection, active prevention of fire protection, active repression of fire protection and passive repression of fire protection.

4.1 Passive prevention of fire protection

The first phase of fire protection activities prepares for possible future security breaches. The most common categories of activities include the preparation of documents, educational activities, professional training, state fire supervision, training, prevention officer, fire protection technician, and fire protection specialists.

4.2 Active prevention of fire protection

The second phase of fire protection activities involves active and up-to-date preparation for future security breaches. It is also a proactive activity. The most common categories of activities include fire patrol, electric fire alarm, preventive-tactical exercises, and preparation for typical activities.

4.3 Active repression of fire protection

The third phase of fire protection activities is intended for active repression after a security breach. The most common categories include firefighting, hazardous substances, water transportation, water removal, liquidation, crisis management, traffic accidents, first aid and rescue, and other activities.

4.4 Passive repression of fire protection

The fourth phase of fire protection activities is characterized as passive repression after a security breach. It is also called an evaluation activity. The most common categories of activities include evaluation documents, control after the intervention, evaluation of the procedure, and creation of statistics. Individual phases follow each other. In the last fourth phase, the first phase of the process of fire protection activities is again transferred.

5. Proactive methods of fire protection

Proactive methods of fire protection belong to the group of active prevention. As part of the proactive activity, it must not be passive prevention but an active and timely reaction to the current situation of a possible security breach within a specific object. The main general proactive fire protection methods include fire patrol, fire alarm systems and other systems, preventive-tactical exercises, and preparation for typical and other activities.

5.1 Fire patrols

Fire patrols are divided into workplace fire patrol and fire assistance patrol.

5.1.1 Workplace fire patrol

The workplace fire patrol is established at workplaces with an increased fire risk. If it is only an occasional workplace (the employee stays here only for the occasional inspection, delivery of material, etc.), the establishment of a workplace fire patrol is not required.

Its primary tasks include monitoring compliance with regulations at the workplace and immediately reporting detected deficiencies to the workplace manager, and implementing measures in the event of a fire, especially for rescuing people, calling for help, fighting the fire, closing fire dampers and combustible material inlets, and turning off the electricity. The members of the fire patrol at the workplace ensure that after the end of working hours, the workplace is in impeccable condition from the point of view of fire safety, that the fire shutters, the supply of flammable substances and all appliances are turned off. Their task is also to check the workplace's equipment with firefighting equipment, fire extinguishers and connecting means, and they also check their completeness and access them. [8]

5.1.2 Fire assistance patrol

This type of patrol is established for activities with an increased risk of fire or events attended by many people. Among its primary tasks is the supervision of compliance with measures designed to prevent the occurrence of fire and the implementation of measures in the event of a fire, mainly for rescuing people, fighting a fire, closing fire valves and supplies of flammable substances and turning off the electric current. The members of the fire assistance patrol familiarise themselves with the nature of the activities, the nature of the secured event or the object related to the measures to ensure protection against fires. Before starting an activity or event, members of the patrol check whether the measures to ensure protection against fires have been met. The members also check compliance with the fire safety regulations of the building. Another of their activities is the inspection of the given premises before the event's start, during the course, and after the end for a specified period. In imminent danger, the patrol members immediately alert the head of the workplace or the event organiser. [9]

5.2 Fire alarm system

The fire alarm system is a security system designed for the fire protection of a given object. Nowadays, various institutions and other objects are necessary and needed in addition to workplaces, and it is also starting to be used in households.

The fire alarm system is an essential safety system, as it can perform several functions. When a fire detector detects a fire through the central electrical fire alarm system, it can perform signalling, whether acoustic or optical, automatically contacts the Fire and Rescue Service operation centre and can also perform object control (for example, start sprinklers, open fire doors, open windows, activate safety gate and so on. [10]

5.3 Fire protection control panel

The fire protection control panel is an element and part of the electrical fire alarm system. It is a device connected to the electrical fire alarm system through a remote transmission device for fire protection units. The advantage of this device is the essential operation of the electrical fire alarm control panel remotely. The fire protection control panel also indicates the operational state of the electric fire alarm system and enables the fire department to control the electric fire alarm system remotely and reliably. The fire protection control panel is usually placed at the entrances to buildings for easy accessibility. [11]

5.4 Review exercises

Review exercises are carried out by all fire protection units, whether professional, corporate or voluntary. Screening exercises are not pre-arranged. The individual components are not ready for the given type of intervention. In this exercise, the corps' ability to act, their interrelationship and cooperation during the given exercise are mainly tested. At the end of the exercise, the leader evaluates the progress of the given review exercise.

5.5 Tactical exercises

Tactical exercises are also carried out by all fire protection units, whether professional, corporate or volunteer firefighters. Tactical exercises are pre-planned exercises. The individual components of fire protection are informed about the place, time and date. Information is also provided on the type of intervention and what firefighting equipment will be used. At the end of the exercise, the leader will also evaluate the progress of the given tactical exercise.

5.6 Preparation for typical activities

It is a simulation of the course of a given event related to fire protection. As part of the process simulation, the components of the fire and rescue service, the medical rescue service and the police are most informed. All these emergency services perform their tasks within a simulated event. The most frequently simulated events include car, bus or train traffic accidents, plane crashes, the explosion of a dangerous substance, forest fire, flooding of an area, leakage of a dangerous substance into a river or lake, managing the evacuation of the population, simulating the process of solving an explosive device, preparing for a suicide attempt and others events.

5.7 Other activities

Other proactive fire protection methods activities may include new evaluation systems that actively respond to the occurrence of fire or the imminent danger of explosion caused by dangerous substances. Other activities may include software programs designed for fire protection.

6. Proposal and application of proactive methods of ensuring fire protection

Proposed proactive methods include fire patrol, private security service, software, tactical exercises, evacuation, typical activities and control activities.

6.1 Fire patrol

In this case, it would be the type of workplace fire patrol because this type of patrol is constantly at the workplace, unlike the fire assistance patrol, which carries out supervision within the framework of fire protection only in the case of shorter periods, such as social-cultural events or events and the like. In tables 1 and 2, there is a proposal for the number of fire patrols with a specific number of members and their working hours. The range of tables is divided into 6 degrees (0 to 5) according to the shop area. As the level increases, so do the number of fire patrols and the number of members for individual units. Table 1 shows these suggestions for the case when the shop is not in non-stop operation, and Table 2 shows these suggestions for shops in non-stop operation. A detailed breakdown of the degrees can be found in the following two tables.

Degree	Shop area [m^2]	Non-stop operation	Working hours of fire patrols	Number of fire patrols	Number of members in one fire patrol
0	until 10 000	no	x*	0*	0*
1	10 001 – 30 000	no	4 hours/day	1	1
2	30 001 – 50 000	no	8 hours/day	1	2
3	50 001 – 70 000	no	12 hours/day	1	3
4	70 001 – 100 000	no	8 hours/day	2**	2
5	over 100 000	no	8 hours/day	2**	3

Table 1. Fire patrol without non-stop operation

* At fire patrol degree 0, it is not necessary to create a fire patrol if there is without non-stop operation

** With two patrols, there is a regular change (morning and afternoon shifts) - that is, the working time of both patrols is a total of 16 hours per day

Degree	Shop area [m^2]	Non-stop operation	Working hours of fire patrols	Number of fire patrols	Number of members in one fire patrol
0	until 10 000	yes	2 hours/day	1	1
1	10 001 – 30 000	yes	4 hours/day	1	2
2	30 001 – 50 000	yes	8 hours/day	1	3
3	50 001 – 70 000	yes	12 hours/day	1	2
4	70 001 – 100 000	yes	12 hours/day	2*	3
5	over 100 000	yes	12 hours/day	2*	4

Table 2. Fire patrol for non-stop operation

* With two patrols, there is a regular exchange (day and night shifts) - that is, the working hours of both fire patrols are a total of 24 hours a day.

6.2 Private security service

The performance of a private security service is also significant for the proper operation of the shopping centre. The activities of the private security service are mainly focused on the physical security of the shopping centre. As part of fire protection, members of the private security service would ensure that there is no intentional fire and coordination of groups of people in case of evacuation.

6.3 Software

Another proactive method of fire protection would be the creation of software that would focus on specific criteria of individual operations in the shopping centre and, based on the established criteria, would evaluate the resulting degree based on which fire protection measures would be proposed. The evaluation process would refer to the following criteria: number of persons, shop area, type of shop, time mode of shop, working hours, holiday, number of floors, presence of fire alarm system and security alarm system, emergency exit, presence of fire patrol and statistics.

6.4 Tactical exercises

Tactical exercises are pre-planned exercises conducted by fire protection units. The main task of this type of exercise is excellent preparation, coordination and efficiency. In the case of a proactive method, it would be the performance of tactical exercises in the event of a fire or dangerous smoke in periods when there are more people in shopping centres (for example, Christmas and Easter) or in periods when there is a risk of fire occurring more frequently (for example, during the summer season).

6.5 Evacuation

The evacuation process is mainly carried out by the units of the integrated rescue system of the employer. In the case of a practice evacuation process within the shopping centre, the fire patrol would carry out the evacuation process in cooperation with the private security service. In a practice evacuation, persons would be moved to a gathering place outside the shopping centre. It would be ideal for carrying out practice evacuations at least two times a year.

6.6 Typical activities

This section covers different events that can occur in the shopping centre. It is about creating a simulation of the event with an immediate reaction. Among the most common types of events in the framework of fire protection in a shopping centre can be fire, generation of dangerous smoke, presence of gas or other dangerous substance, the explosion of gas or other dangerous substance, failure of electric fire alarm, malfunction of fire equipment, blocking off escape exit, process evacuations, the process of evacuation, in case the fire is located at the entrance. There is no other escape exit or other emergency activities.

6.7 Control activities

These are random inspections of compliance with fire regulations in the shopping centre. The check would be done randomly. The employer would not be informed about the inspection in advance. The control of compliance with fire regulations would be carried out by a member of the state fire supervision, a fire protection technician, or an authorized preventive officer (a member of the prevention department of the Fire and Rescue Service or voluntary fire protection).

7. Conclusion

Safety is one of the basic needs of human society. One of the basic categories of safety is fire safety. The article dealt with the analysis and proposal of new or modified methods. The first chapter described the security model. According to individual characteristics, the security model is generally divided into preventive, repressive and combined from the point of view of safety. Fire protection primarily deals with the repressive part, and the preventive part is not addressed to such an extent. Therefore, it is advisable to deal more with combined security models. The following part dealt with analysing individual preventive and repressive fire protection measures, which are also individually characterised. The next chapter describes the individual phases of the fire protection process.

These four phases of the process were timed and logically arranged. Typical activities and fire protection activities were defined for each phase. This chapter was followed by a section on proactive fire protection methods, which were dealt with in the second phase of the fire protection process. Proactive fire protection methods included fire patrols, fire alarm systems, fire protection control panels, review exercises, tactical exercises, and preparation for typical and other activities. The last chapter deals with the proposal and application of new or modified proactive methods intended for fire protection. The proposed methods included fire patrol modified by number and time requirement, private security service, software, tactical exercises, evacuation, typical activities and control activities. All these proposed methods are intended for a shopping centre object.

The disadvantage and limit of this proposal may be the time or cost required for the shopping centre. As part of future research, proactive methods could also be proposed for other objects such as schools, hospitals, restaurants, stadiums and various cultural and social events where many people are present.

8. Acknowledgments

This research was based on the support of the Internal Grant Agency of Tomas Bata University in Zlín, IGA/CebiaTech/2022/004 and The Department of Security Engineering, Faculty of Applied Informatics.

9. References

[1] Jul 18, 64 CE: Great Fire of Rome: National Geographic [online]. (2022). Available at: https://education.nationalgeographic.org/resource/great-fire-rome

[2] Igini, M. (2022) Top 12 Largest Wildfires in History: CLIMATE CHANGEPOLLUTION [online]. Available at: https://earth.org/largest-wildfires-in-history/

[3] Lukas, L. (2017). Theory of security, VeRBuM, ISBN 978-80-87500-89-7, Zlin, Czech Republic

[4] Vichova, K.; Hromada, M.; Valasek, J. & Paulus, F. (2020). Comparison Analysis the Use of Modern Technologies by Fire Rescue Service, Proceedings of the 31st DAAAM International Symposium, pp.0535-0541, B. Katalinic (Ed.), Published by DAAAM International, ISBN 978-3-902734-29-7, ISSN 1726-9679, Vienna, Austria DOI: 10.2507/31st.daaam.proceedings.074

[5] Malatinsky, A.; Drofova, I.; Sousedikova, L. & Hromada, M. (2021). Fire Safety Threat Risk Analysis for Soft Target, Proceedings of the 32nd DAAAM International Symposium, pp.0586-0592, B. Katalanic (Ed.), Published by DAAAM International, ISBN 978-3-902734-33-4, ISSN 1726-9679, Vienna, Austria. DOI: 10.2507/32nd.daaam.proceedings.083

[6] Ronchi, E. (2021). Developing and validating evacuation models for fire safety engineering. Fire Safety Journal 120. Available at: https://www.sciencedirect.com/science/article/pii/S0379711220300679?via%3Dihub

[7] Koutsomarkos, V.; Rush, D.; Jomaas, G. & Law, A. (2021). Tactics, objectives, and choices: Building a fire risk index. Fire Safety Journal 119. Available at: https://linkinghub.elsevier.com/retrieve/pii/S0379711220307591

[8] Furness, A. & Muckett, M. (2007). Introduction to Fire Safety Management, Routledge, ISBN 978-0-750-68068-4, New York, USA

[9] Kratochvil, V.; Navarova, S & Kratochvil, M. (2021) Fire Safety Devices in Buildings, ISBN 978-80-7385-238-2, Ostrava, Czech Republic

[10] Kucera, P.; Pokorny, J. & Pavlik, T. (2013) Fire engineering – Active Elements of Fire Protection, SPBI Spektrum, ISBN 978-80-7385-136-1, Ostrava, Czech Republic

[11] Lukas, L. (2013) Security technologies, systems and management III, VeRBuM, ISBN 978-80-87500-35-4, Zlin, Czech Republic

DOI: 10.2507/33rd.daaam.proceedings.057

EFFECT OF CUTTING LIQUIDS AND OILS IN MACHINING PROCESS ON ENVIRONMENTAL

Tetjana Tomaskova & Katerina Bicova

This Publication has to be referred as: Tomaskova, T[etjana] & Bicova, K[aterina] (2022). Effect of Cutting Liquids and Oils in Machining Process on Environmental, Proceedings of the 33rd DAAAM International Symposium, pp.0404-0410, B. Katalinic (Ed.), Published by DAAAM International, ISBN 978-3-902734-36-5, ISSN 1726-9679, Vienna, Austria

DOI: 10.2507/33rd.daaam.proceedings.057

Abstract

Sustainability is an urgent social development task that is increasingly coming into the limelight. The United Nation's Brundtland Commission Report of 1987 concluded that humanity could make development sustainable if it meets "the needs of the present without compromising the ability of future generations to meet their own needs." Industrial companies' implementation of the environmental management system on a voluntary basis and by implementing them undertake to carry out all their activities in an environmentally friendly manner. The aim of this contribution is to focus on the issue of the impact of the metalworking industry, specifically chip machining, on the environment. Based on the analysis of the proposed register of environmental aspects according to the requirements of ISO 14001, we focus on the aspects that threaten the environment the most and try to minimize their impact. In this case, it is specifically the cutting fluids used in machining. The effort is to prepare the basis for the preparation for the certification of the company according to ISO 14001 and mainly to propose corrective measures to minimize the impact on the environment.

Keywords: Environment; sustainable development; metalworking industry; cutting liquids; machining.

1. Introduction

Use Society changes over the years and with it the view of certain problems that previous generations did not solve as a priority. If we talk about the natural environment, man himself was an integral part of it at the beginning of human evolution. Among the reasons why it was so, in the first place is the effort to stay alive, which was later replaced by the fulfillment of human needs, which, however, resulted in the fact that man began to disturb nature with his artificial environment. Underneath all the above (i.e., natural environment, artificial environment, and socio-economic conditions) it is possible to imagine an interwoven system which is referred to as the environment or environment [1]. According to the ISO 14 001 standard, the environment is defined as the company's environment, including air, water, soil, natural resources, plants, animals, people, and the relationships between them. Because of this, the environment cannot be talked about as a business problem, but rather as a global system problem. The aim of this contribution is to focus on the issue of the impact of the metalworking industry, specifically chip machining, on the environment. Based on the analysis of the proposed register of Environmental Aspects (EA) according to the requirements of ISO 14001, we focus on the aspects that threaten the environment the most and try to minimize their impact.

In this case, it is specifically the cutting fluids used in machining. The effort is to prepare the basis for the preparation for the certification of the company according to ISO 14001 and mainly to propose corrective measures to minimize the impact on the environment.

1.1. Chip machining and the environment

In the introduction, it was mentioned that the environment is an issue that does not only concern the corporate sphere but is a global sphere. This is not the case in the machining environment either. The machining process itself also includes flows of energy and resources that are necessary for the given production. Other aspects influencing the impact of the machining process on the environment also include the factory itself or the company building with the heating, ventilation, and air conditioning system, and even the supply and customer chain [2].

It is necessary to evaluate all the environmental impact of chip machining and to know the exact production process of a certain product, because the environmental impacts can also differ from product to product, for example because other production methods or other operational means can be used [1].

1.2. Legislative requirements for environmental conformity assessment

EMS systems are currently the most suitable evidence for organizations to prove that they carry out their activities with respect to the environment and that the production or service itself is designed so that the environmental impact is as small as possible, ideally zero.

In this contribution, we focus on designing an environmental management system according to the ISO 14 001 standard, as it is one of the most widespread methods in the Czech Republic. This standard serves as a tool that enables the company to ensure the fulfillment of the declared ecological concept and at the same time to demonstrate this fulfillment. At the same time, however, this is not a one-time issue, but rather a spiral development. The very concept of this system is such that it can be applied to any organization, and it does not matter what activity the organization performs [1], [3], [4], [5].

1.3. Register of environmental aspects according to ISO 14001

In today's unstable and complex environment, there are many aspects that a manager must focus on, and it goes without saying that individual problems cannot be approached equally. How to approach these issues is shown directly in the ISO 14 001 standard. The standard therefore requires an organization to create and follow a procedure for identifying the environmental aspects of its activities, services, and products. The result is the aspects that the organization can directly control or have some influence on, but the most important thing is to determine those aspects that have or can have a significant impact on the environment.

According to the standard, all aspects of the company's activities affecting the environment were mapped. This was followed by their evaluation and determination of essential aspects. How to carry out this evaluation is again not directly determined by the standard and it is only up to the company itself which criteria and procedures it chooses. The criteria listed in the following table 1 were chosen for this work. However, the very evaluation of the significance of individual environmental aspects [1] is crucial for this contribution.

Assessment:

$$\text{Frequency} \cdot \text{Severity} \cdot \text{Crash} \cdot \text{Cost} = \text{Significance of EA} \tag{1}$$

Significance of EA:

Aspect type	**Point range**
Common aspect	1 - 50
Significant aspect	51 - 150
Alarming significant aspect	150 - 300

Table 1. Criteria for evaluating environmental aspects

There is no need to implement any corrective measures for common aspects. Significant aspects need to be further addressed and appropriate corrective measures should be implemented to reduce their significance.

Register of environmental aspects															
Machining hall															
Process	**Impact**					**Assessment**						**Significance of aspects**	**Control measures**	**Monitoring**
Environmental aspect	**Water**	**Soil**	**Air**	**Resources**	**Man**	**Frequency**	**Severity**	**Accident**	**Costs**	**Legislation**	**Evaluation**			
Process of the Production														
Generation of general waste		•			•	4	2	2	1	A	16	Common	Training	Continuous monitoring
Generation of hazardous waste	•	•	•		•	3	3	3	1	A	27	Common	Training	Continuous monitoring
Consumption of metallurgical material		•		•		4	2	1	1	N	8	Common	Training	Production records
Generation of waste from metallurgical material	•	•	•		•	4	3	3	2	A	72	Significant	Training	Production records
Use of cutting oil during machining	•	•	•	•		4	3	3	2	A	72	Significant	Training	Production records
Generation of waste oil	•	•	•		•	4	3	4	2	A	96	Significant	Training	Production records
Use of chemical substances			•	•	•	2	3	3	1	A	18	Common	Training	Continuous monitoring
Formation of inhomogeneous dust emission during grinding			•		•	4	3	1	1	A	12	Common	Training	Continuous monitoring
Noise generation	•	•	•		•	4	2	1	1	N	8	Common	Training	Continuous monitoring

Table 2. Register of environmental aspects for the machining hall

In the case of alarmingly significant aspects, immediate action is needed because they are so serious that ignoring them could mean irreversible consequences in the field of the environment. For the proposed register, the impact is related to the group of water, soil, air, resources, and people. In practice, the assessment takes place in a work team of people who work together to implement the EMS in the company.

2. Cutting fluids and their impact on the environment

Based on a detailed analysis of possible impacts, we decided to respond to the alarming possible impact of cutting fluids on the environment. Process fluids are an indispensable part of chip machining. In order to achieve the best results, the correct cutting fluid must be selected. There are two types of process fluids for metalworking. We divide them according to their water content into:

- water-miscible cutting fluids
- cutting oils

We can also use molten metals as cutting fluids, such as tin, bismuth, zinc, etc. These are used when machining hard-to-machine metals. Process fluids have two main effects, cooling and lubrication. Water-miscible cutting fluids have a cooling effect, while cutting oils have a lubricating effect. Therefore, it must always be considered for what purpose we need to use the given liquid [6], [7], [8].

2.1 Ecological aspects in the use of cutting fluids

Nowadays, environmental issues are increasingly being addressed. Due to global warming and the profiteering of non-ecological farming of the past years, more and more emphasis is being placed on the greening of industry. Current legislation has more and more stringent requirements, which is why even process fluids are increasingly controlled. The main controlled component is the disposal of process fluids and their handling. Each company must sign contracts during liquidation, which are subsequently controlled by the state, to prevent the discharge of process fluids into nature, as has already happened several times in the past. Transportation and storage play an equally important role here. Everything must be done with great care to avoid an accident that could contaminate nature and contaminate groundwater. Water is a very important component in the field of cutting fluids. Each business should consider the machining method and the cutting fluid used. To give you an idea, only 50 l of oil is enough to cover 1 km2 of water surface with a continuous layer with a thickness of approximately 0.05 μm. This causes a decrease in biodegradability, which means that the self-purification process of water takes place very slowly, and this can cause the extinction of aquatic organisms. Unfortunately, due to the use of process fluids, polydisperse aerosols are released, which contaminate the working atmosphere and thereby harm health [6].

2.2 The effect of cooling emulsion on the environment

In recent years, our planet has been affected by a major drought. Therefore, water consumption should be reduced. Thanks to the cancellation of machining using a cooling emulsion, 3 862,7 liters of water would be saved annually (on one CNC machine), which would provide water for 1 person for 3 and a half years.

Furthermore, the reduction of water consumption would improve the health of the employees, as there would be no skin contact with the cooling emulsion, which is often a significant source of allergic reactions and various eczemas. It is a hazard class 2, which means that it must not enter the water, as there would be a large contamination and pollution that could be fatal for organisms, especially when the water is contaminated, the destruction of aquatic organisms is a certainty. Since water makes up 95% of the emulsion, there is more evaporation, which is unfavourable for the human body, which can develop respiratory diseases when inhaling the vapours. Therefore, emphasis must be placed on high-quality vapor extraction and air purification [6].

2.3 The impact of cutting oil on the environment

Cutting oil greatly endangers the environment if disposed of improperly. It is enough for only a small amount to encounter water and there is already a big threat. Cutting oil is destructive to aquatic organisms, the water hazard level is 2. The same aspects apply to it as for cooling emulsion, only because of its 100%, it is treated with more respect than cooling emulsion. Its stability is higher, it does not undergo rapid degradation. May cause an allergic reaction if it encounters the skin. Therefore, it should be handled with gloves. The manipulator should wear long sleeves, long pants, and safety glasses. If the oil splashes into the eyes, it causes serious health problems. When using CNC machines, the risk of contact with cutting oil is significantly lower than with conventional machines. It must be dealt with according to legislation. Violations of decrees and regulations can often be dealt with by sanctions [6], [7], [8].

2.4 The effect of biodegradable oil on the environment

Biodegradable oil affects the environment only with some chemical additives. These serve to ensure that the given oil has the same properties as mineral cutting oil. No problem is expected here in case of contamination with human skin, the oil is considered suitable for work according to the legislation. As far as the environment is concerned, the consequences of contamination are not fatal, as it is an oil based on sunflower oil, which is fully biodegradable. However, it must not be poured into the sewer, etc. It can also be treated as a food product if it is filtered, and no chips of the processed material remain in it. However, it must be checked more often to prevent its degradation [6], [7], [8].

3. Proposal of corrective measures to reduce the impact on the environment

3.1 The problem with the release of dangerous substances in general

This environmental aspect refers to all hazardous substances (thinners, paints, oils), including natural gas. According to the register, it is known that such a leak can occur both in the warehouse and in the production hall.

Measures that a business can put in place include:

- At the workplace where these dangerous substances will be present, a list of these substances and their corresponding safety data sheets will be maintained, which include H-phrases (standard phrases about the danger of chemical substances and mixtures) and P-phrases (standardized instructions for safe handling of chemical substances and their mixtures).
- Hazardous substances will be stored in designated places that will not be accessible to unauthorized persons. Storage places must always meet the requirements for the given substance and all these places must be marked with warning symbols that are related to the stored substance (see Fig. 1).

Fig. 1. Warning symbols according to ES 1272/2008 [9]

- A responsible person will be designated in the company who will oversee the emergency kit intended for oil products.
- The company will develop controlled documentation dealing with leakage, pollution, and subsequent disposal of oil products.
- Hazardous substances will be stored using catch basins/collection tubs, which in case of leakage will prevent contamination of soil and waterways (see Fig. 2).

Fig. 2. Catch basins/Collection tubs [10]

- All equipment that distributes or burns natural gas will be regularly inspected and adequately vented to prevent the accumulation of gas that could result in an explosion.

3.2 Use of cutting oil during machining

Corrective measures for this aspect will be like those in the chapter on the release of hazardous substances.

3.3 Generation of waste oil

Waste oil is classified as hazardous waste. It is a waste that is produced from cutting oils during the machining process. Measures that a business can put in place include:

- To achieve maximum use of the oil, the metal shavings will be separated from the oil after use, using a centrifugal device (centrifuge) and filters (see Fig. 18). The result will be oil that will be reused. This process cannot be carried out indefinitely, because even with fine filtration, a small amount of microscopic metal particles will remain in the oil. This will cause a change in the original properties, which will result in a deterioration of the cutting oil's function, and it will be necessary to replace the oil after several uses.
- The company will start using suitable oil substitutes that will have less impact on the environment.

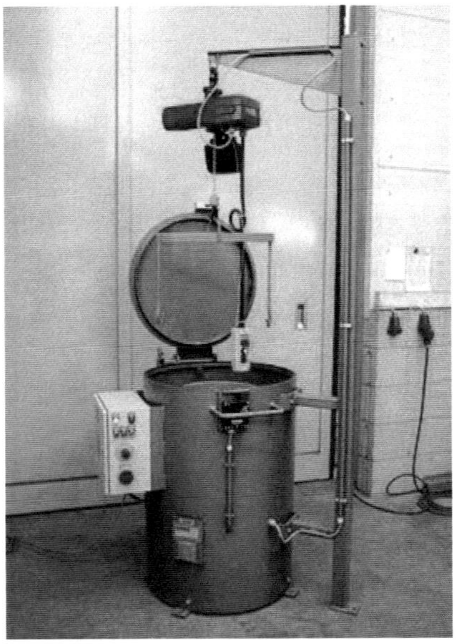

Fig. 3. Centrifuge with manual drum change [11]

3.4 Disposal

The disposal of cutting emulsions or cutting oils is carried out with the help of specialized companies. Larger machining companies can arrange their own disposal facilities, but this involves several strict and complex hygiene processes and controls that are not easy to pass. Because of this, the first option is chosen in most companies. To give an idea, about 26,000 tons are consumed annually (according to a survey by Fuchs oil). Most of them end up in demulsification stations, where the water is separated from the oil. The water goes to the WWTP (wastewater treatment plant) and the oil is mostly burned in incinerators.

4. Conclusion

There According to the created register, environmental goals were chosen for the identified significant aspects, for which possible corrective measures were also written down, which could help to reduce their significance. In this post, we focused mainly on cutting fluids and their impact on the environment. Cutting fluids must be handled in accordance with legislation and relevant decrees. They must be disposed of by specialized companies to prevent natural disasters. Workers should be sufficiently trained so that they do not knowingly contaminate cutting fluids and do not neglect their care. This should be monitored more, as the workers do not care much about quality. From an economic point of view, the cooling emulsion is the most advantageous, because water makes up more than 90% of the volume.

However, in terms of price, it is approached by biodegradable cutting oil, where disposal costs fall. Businesses should also consider machining with petroleum cutting oil, which can last for several years with proper care, again reducing costs. An important aspect to consider is dry machining or machining with a low amount of cutting fluid. However, the tools must be adapted to handle other operational requirements. Another aspect that needs to be analysed in detail is the generation of waste from metallurgical material. This aspect was chosen as important mainly because it is not directly known what metallurgical materials will be machined. It is considered that metallurgical materials can also contain dangerous substances such as lead, nickel or phosphorus. So far, a corrective measure has been proposed in the form of limiting the use of these materials.

However, this cannot always be ensured, because the processing of a certain material can be crucial for a given enterprise and it is impossible to limit its processing. For the possible prevention and reduction of the environmental impact of this aspect will be subjected to future investigation.

5. Acknowledgments

The article contribution has been prepared under project SGS-2022-007 - Research and Development for Innovation in Engineering Technology - Machining Technology IV.

6. References

[1] Tomaskova, T., & Bicova, K. (2021). IMPACT OF MACHINING ON THE ENVIRONMENT. Paper presented at the Annals of DAAAM and Proceedings of the International DAAAM Symposium, , 32(1) 623-630. Daaam 2021.
[2] Sihag, N. et al. (2019). The Influence of Manufacturing Plant Site Selection on Environmental Impact of Machining Processes, Available from: https://1url.cz/OzYmQ
[3] Cascio, J., Woodside, G. & Mitchell, P. (1996). ISO 14000 Guide, The New International Environmental Standars. McGraw-Hill, New York, pp. 3–217
[4] Kudlacek, I. (2002). Industry ecology (in Czech), Prague ISBN 80-01-02495-4, Prague
[5] Veber, J. et al. (2010). Management of quality, environment and work safety: legislation, systems, methods, practice (in Czech), Prague ISBN 978-80-7261-210-9, Prague
[6] Hůlová, G. (2019). Cutting fluids impact on the economics of the machining process and od the environment (in Czech), Bachelor these, ČVUT Prague
[7] Ojstersek, R.; Palcic, I. & Buchmeister, B. (2020). Simulation Modelling of Environmental and Economic Aspects in Manufacturing Systems, Proceedings of the 31st DAAAM International Symposium, pp.0436-0443, B. Katalinic (Ed.), Published by DAAAM International, ISBN 978-3-902734- 29-7, ISSN 1726-9679, Vienna, Austria DOI: 10.2507/31st.daaam.proceedings.061
[8] Blazevic, A.; Bibic, D.; Trobradovic, M.; Filipovic, I. & Gebert, K. (2020). Effective Reduction of NOx Emissions of Modern Medium Speed Diesel Engines, Proceedings of the 31st DAAAM International Symposium, pp.0812-0818, B. Katalinic (Ed.), Published by DAAAM International, ISBN 978-3-902734-29-7, ISSN 1726-9679, Vienna, Austria DOI: 10.2507/31st.daaam.proceedings.113
[9] Blažek, Jiří. Chemical substances and mixtures. Web vzdelavani-dh.cz. (in Czech)
[10] Skupina DENIOS. Catch basins. Web denios.cz. (in Czech)
[11] BELMET s.r.o. ME - Centrifuge with manual drum change. Web belmet.cz. (in Czech)

DOI: 10.2507/33rd.daaam.proceedings.058

MARKETING 4.0 – A SYSTEMATIC REVIEW

Martin Polívka & Lilia Dvořáková

This Publication has to be referred as: Polívka, M[artin] & Dvorakova, L[ilia] (2022). Marketing 4.0 – A Systematic Review, Proceedings of the 33rd DAAAM International Symposium, pp.0411-0420, B. Katalinic (Ed.), Published by DAAAM International, ISBN 978-3-902734-36-5, ISSN 1726-9679, Vienna, Austria
DOI: 10.2507/33rd.daaam.proceedings.058

Abstract

Industry 4.0 and the Fourth Industrial Revolution are buzzwords which frequently appear in the discussion about the economy of the future. From the extensions of their principles to the field of marketing the term Marketing 4.0 originated. According to its inventor, Marketing 4.0 should by characterized by the connection of real and virtual world. Nevertheless, unlike Industry 4.0, Marketing 4.0 appears in the current discussion much rarely. The systematic review of professional literature based on the Web of Science and Scopus databases was therefore conducted so as to assess the current state of Marketing 4.0 and to find out how it is understood by the relevant professionals. Relevant papers were analyzed both from the time and geographical perspective and from the perspective of their content. Based on this analysis we can conclude that the use of Marketing 4.0, even if not very frequent, has been slightly growing lately, mostly in the European countries. On the other hand, most of the papers dealing with Marketing 4.0 concentrate mainly on the e-commerce and virtual world, and therefore do not fully implement the definition of Marketing 4.0.

Keywords: Big data; Industry 4.0; Marketing 4.0; Online marketing; 5A model;

1. Introduction

Marketing has been constantly developing, which is best evidenced by the fact that the American Marketing Association, one of the main authorities in this field, has already updated its definition three times during its existence. From its original definition in 1937, which understood marketing primarily as a tool for optimizing the distribution of goods from producer to customer [1], the meaning of the term has shifted to the current one claiming that "*Marketing is the activity, set of institutions, and processes for creating, communicating, delivering, and exchanging offerings that have value for customers, clients, partners, and society at large.*" [2]. This contemporary definition no longer sees marketing as a narrowly focused field concentrating on optimizing a specific process, but as a much broader concept that brings value not only to the company and its customers, but also to society as a whole - Gundlach & Wilkie [3] speak of the so-called aggregate, inclusive concept of marketing in this context. Technological progress is undoubtedly one of the major factors influencing the marketing development. The development of information and communication technologies, which first began to be used in internal management, especially in manufacturing companies, in the 1970s, brought about fundamental changes in this respect. This phenomenon was called the "Third Industrial Revolution", which followed the First Industrial Revolution triggered by the discovery of the steam engine, and the Second Industrial Revolution associated with the development of electrical power engineering and assembly line production at the turn of the 19th and 20th centuries (for more information see, for example, [4]).

However, in the area of consumer goods, which is more relevant in terms of B2C marketing, information and communication technologies have come in more slowly; their massive diffusion in the form of personal computers, the Internet and cell phones has primarily been the matter of the turn of the 20th and 21st centuries. The development of information and communication technologies has both enabled the development of new products that could not have existed at the lower level of technological development (which, however, is true in principle for any technological shift), and has brought completely new possibilities, especially in the other two components of the marketing mix - promotion and place. The spread of the Internet and devices for its consumption is thus linked to the emergence and development of internet marketing as "*a qualitatively new form of marketing, which can be characterized as the management of the process to satisfy human needs with information, services or goods using the Internet*" [5], and hence the whole concept of e-business. The emergence of widely used social networks and platforms for sharing videos and text in the late 2000s then brought about viral marketing, influencer marketing, etc.

However, the integration of modern information and communication technologies into everyday life and business does not end with the use of smartphones, the Internet, and social networks. On the contrary, the next phase is currently underway, referred to as the "Fourth Industrial Revolution" or the emergence of "Industry 4.0". The basic ideas of the Industry 4.0 concept first became apparent at the Hannover Fair in 2011, and its pivot is in the close connection of the physical world with the virtual world through smart manufacturing machinery and equipment, as well as final products that will constantly monitor their condition using sensors and communicate with each other via fast mobile networks within the Internet of Things. This interconnection should result in the emergence of cyber-physical systems in which every physical element will have its virtual representative (cf., for example, [6]).

Naturally, modern technologies are not being deployed only in manufacturing, and so there are many studies on the impact of the Fourth Industrial Revolution on agriculture (in this context the "Agriculture 4.0" term is used), transport ("Logistics 4.0"), or healthcare ("Healthcare 4.0"). Therefore, it is not surprising that the term "Marketing 4.0" also appeared shortly after that. Its basic characteristic is the interconnection of the virtual and real worlds - e.g., Rathod et al. [7] define it as follows: "Marketing 4.0 is the confluence of offline and online marketing strategy." With this proclaimed confluence of online and offline elements, Marketing 4.0 has adopted the idea of Industry 4.0, but at the same time its name follows the three previous development phases of marketing, which Kotler had previously defined as "Marketing 1.0", "2.0" and "3.0". Even the authorship of Marketing 4.0 is therefore sometimes (e.g., [8]) mistakenly attributed to Kotler, who indeed published a pivotal book on this topic in 2017. However, in fact, the term of Marketing 4.0 and its basic definition had already been introduced five years earlier by Jara et al. [9], and later attempts to define Marketing 4.0, including the Kotler´s one, already built on their work.

Nevertheless, compared to the impact of the Fourth Industrial Revolution on other industries and processes, the issue of Marketing 4.0 has been given rather marginal attention in the professional literature so far. The Scopus citation database, which aggregates data on scientific papers from recognized journals and conferences, recorded a total of 2,886 papers with "Smart factory" in the title, keywords or abstract, 241 papers containing "Agriculture 4.0", 156 containing "Logistics 4.0" and 118 with "Healthcare 4.0", while "Marketing 4.0" was included in the same elements in only 31 papers. The fact that, similar to Industry 4.0 (see [10]), Marketing 4.0 is still, to a certain extent, a buzzword, which is used in different contexts without always being clear what exactly is meant by the term, is also problematic.

Therefore, this paper aims to present the results of the research that attempted to answer the following questions:

1. What is the state of the current professional debate on the issue of Marketing 4.0?
2. Which technologies and topics are currently seen as the core components of Marketing 4.0?

The systematic review of the available literature and the subsequent analysis of its results were chosen to be the research method. The paper is constructed as follows: Chapter 2 defines the exact methodology to conduct the review and the procedure to select the sources, Chapter 3 contains the basic classification of the examined papers, Chapter 4 briefly describes the content of the reviewed articles, and Chapter 5 subsequently analyzes and discusses the content.

2. The methodology of systematic review and the selection of sources

The systematic review procedure was based on the methodology introduced by Soni & Kodali [11] and Manoharan & Singal [12], consisting of six sequential steps:

1. Choosing the time frame of the study
2. Selecting source databases
3. Selecting sources - journals, books, and proceedings
4. Selecting specific works
5. Classifying papers
6. More detailed analysis of the classification.

As far as the time span of the study is concerned, it is possible to proceed from the consideration that the issue of the Fourth Industrial Revolution, and thus also Marketing 4.0, has only been studied in the last ten years, and therefore any paper dealing with this topic is still relevant.

Therefore, it was not necessary to define any starting date, which would exclude already outdated papers from the set of reviewed papers. To search for papers, the pair of major citation databases of scientific literature - the Web of Science and Scopus - were chosen. All types of sources registered in these databases were taken into account, i.e., articles in peer-reviewed journals as well as papers in proceedings and book chapters. Marketing 4.0 was used as a key term for the search and its occurrence was searched in the title of the paper, its keywords and abstract in both databases. The search was restricted to these elements in order to exclude papers that only mention Marketing 4.0 in some context but do not primarily address the topic from the resulting set.

As of February 28, 2022, 31 papers were found in the Scopus database for the given filtering criterion, and 22 papers were found in the Web of Science database. Of these, 15 papers were found in both databases, so a total of 38 unique papers dealing with the topic of Marketing 4.0 were found in both databases.

Of the 38 identified articles, the full text is not available for 8 of them and has not been made available even after the authors´ request to do so. The full text of one paper is available only in Chinese. Therefore, these papers were excluded from the selection, thereby reducing the set to 29 papers. However, two of these are not individual scientific papers, but only editorials of special issues of Frontiers in Psychology, and in one case it is a critical review. Thus, after excluding these entries, the selected set contained 26 scholarly papers. Furthermore, an article dealing with the education of seniors, which only indirectly addressed the issue of Marketing 4.0, and an article presenting the case study of a specific Portuguese company, which only stated that "in terms of Marketing 4.0, this company is still in the Stone Age" [13], were excluded from the set of papers. Therefore, in total, 24 papers were examined.

3. The basic classification of papers

This step involved analyzing the searched papers in terms of their time and geographical distribution. The time distribution and mainly its trend make it possible to assess how the interest in Marketing 4.0 has been developing in the professional community, especially whether this interest has been copying the trend of the growing interest in the Fourth Industrial Revolution as a whole. The geographical distribution of the papers then enables us to identify the countries that are the most advanced in the study of Marketing 4.0, and the countries in which this phenomenon is not yet understood as a topical issue. The distribution of articles related to Marketing 4.0 over time is shown in Fig. 1.:

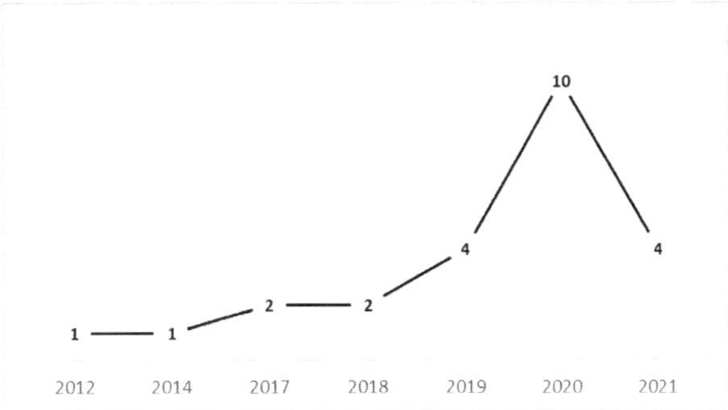

Fig. 1. The time distribution of the existing articles on Marketing 4.0.

As Fig. 1. shows, the first article addressing Marketing 4.0 was published as early as 2012, just one year after the ideas of Industry 4.0 themselves were formulated. However, Fig. 1. also shows that although the first reflection of the Fourth Industrial Revolution in marketing was very quick, the topic clearly did not really resonate much at that time and remained marginalized for the following few years.

Only since 2017, there was a gradual increase in the number of articles dominantly dealing with the issue of Marketing 4.0, which continued until 2020. In 2021, the number of such articles decreased, but it should be kept in mind that there is usually a time lag of several months or even years between writing an article and its publication and, above all, its registration in citation databases. Therefore, the search carried out in February 2022 is unlikely to include all articles from 2021, as some of them will only appear in the citation databases during 2022.

The geographical distribution of the identified articles is shown in Table 2, using the place of work of the main (corresponding) author of each paper as a classification criterion. Table 2 shows that the topic of Marketing 4.0 is mainly addressed by researchers working in Europe. The largest number of articles comes from Germany, which corresponds to the fact that the ideas of the Fourth Industrial Revolution were first formulated in this country, and also from Spain. The issue of Marketing 4.0 is also given attention in the Czech Republic, as evidenced by three articles of researchers working at Czech universities. On the other hand, somewhat surprisingly, none of the first authors of the identified articles is active in the US. This absence is all the more surprising because Philip Kotler, one of the world´s most reputable authors in the field of marketing working in the USA, published a book titled "*Marketing 4.0: Moving from Traditional to Digital*" back in 2017. However, apparently, other American authors have not adequately reflected on Marketing 4.0 yet.

Author´s place of work	Number of articles
Germany	6
Spain	4
Czech Republic	3
India	2
Indonesia	2
South Korea	2
Poland	2
China	1
Romania	1
Saudi Arabia	1

Table 1. The geographical distribution of existing articles on Marketing 4.0.

4. The content of the articles reviewed

As already mentioned in the introduction to this paper, Jara et al. were probably the first to come up with the term "Marketing 4.0" [9]. When defining it, they mainly emphasized the importance of the interaction between the customer and the producer or supplier. To the traditional three dimensions of brand differentiation, which include brand identity, brand image and brand integrity, they have also added the fourth dimension, namely brand interaction, meaning the enhanced interaction with the given product or brand using modern technologies. The authors give examples of using technologies such as NFC, RFID or IoT in retailing in general. For example, these technologies could be used to guide customers to the products of the desired brand using their smartphones already during the shopping process. Once the product has been found, customers could then use their mobile phone to scan the NFC tag or QR code on the product and thus be immediately directed to a website with the detailed description of the product.

Martín-Guart & Cavia [14] perceived Marketing 4.0 mainly in the context of the development of social networks and other new media. In their study, they focused on the impact of technological changes on the role of media agencies and on the media world in general, namely based on the questionnaire survey conducted among representatives of media and advertising agencies, as well as academic institutions. Based on their answers, they have concluded that the development of new technologies in both hardware (smartphones, tablets) and software (social networks, internet TV, etc.) is leading to the fragmentation of the media world and, therefore, marketing channels too. Hence, the product can no longer be promoted in the Marketing 4.0 environment using only a few "main" media (e.g., the most watched TV station or widely visited websites), but an approach using a broad spectrum of media with messages tailored to typical consumers of individual media must be chosen. The authors also highlight the potential of social networks, where existing customers can act as "ambassador" of products purchased by sharing photos or reviews of their use.

The idea that a firm can use the stories that its customers share on social media to promote its own products for free and in an authentic way has subsequently been developed by other authors, including Kotler [15]. According to him, the emergence of social networks has changed the customer journey from the original 4A model (Aware, Attitude, Act, and Act Again) to 5A - Aware, Appeal, Ask, Act, and Advocacy. In other words, after purchasing a product, the customer often further communicates this fact on social media, thereby putting the product in the subconscious of their contacts. In any discussion that may occur under such a social media post, in which some participants naturally inform about alternative products, the customer in question then usually gives reasons for buying that particular product and not one of the suggested alternatives. In doing so, he/she, de facto, unintentionally (and for free) promotes the product - he/she becomes its advocate. As it will become apparent from the summary of other papers reviewed, the 5A concept has been accepted by the professional community shortly after that.

Only a year later, Szymoniuk & Valtari [16] highlighted the importance of the 5A concept in their work, presenting their case study of the Marketing 4.0 use in a rather specific field - a community organic food store. They give the example of the Finnish REKO project, where the focus of communication between consumers and producers, as well as the organization of the store itself takes place on Facebook, always within local (in the geographical sense) groups. In this case, social networks enable the very existence of the business model and also serve as the main marketing channel used in the REKO system. In this case, satisfied customers, who are drawn into the role of "advocates" of the system, and thus help it to attract more customers by their activities on social networks, which is in line with the 5A concept, constitute the main promotional channel that allows the project to develop further. The authors even use the term "Prosumers" for these customers, which is a combination of Producers and consumers.

Hartono et al. [17], who have examined the use of social networks in the Marketing 4.0 environment, highlight the applicability of this concept in startups. They present the specific case study of Chilly Bin, a startup company, in which they have concluded, based on the questionnaire survey and analysis of publicly available sources, that marketing on social networks can be highly effective for a startup company because it allows them to raise awareness of their existence among a large number of their users quickly and at relatively low costs, precisely thanks to the functioning of 5A.

The issue of the relationship between social networks and customer journey in the 5A model was also addressed by Hwang & Kim [18], who focused their research on the specific environment of tourism, or, more precisely, restaurant services. Based on their questionnaire survey, they have concluded that social networks (and social interactions in general) are important when choosing a restaurant, especially in the Ask phase, i.e., when customers decide whether they will actually go to a particular restaurant that has attracted their attention in the previous phases. Therefore, communication and promotion of restaurants on social media should aim to capture their attention exactly at that phase so that customers are convinced by the information they have found they really want to visit the restaurant.

A situation where a group of tourism-focused businesses attempted to use Marketing 4.0 tools with the explicit aim of using the 5A concept was also examined in the case study of 14 small and medium-sized enterprises from a specific village in Bali by Karta et al. [19]. Those enterprises faced a substantial decline in sales volume in 2020 caused by the Covid-19 pandemic. They attempted to counter that decline by greater differentiation of their products and services, which they subsequently tried to promote through digital marketing. The intended aim was to first generate their potential customers´ interest in their products through online marketing and social media (the Aware and Appeal phases), get them to visit the village, and then use those satisfied customers as advocates to further promote their products and services (differentiated from their competitors). However, based on the data collected, the authors concluded that this strategy has not proven to be successful. They failed to elicit customers´ behaviour corresponding to the complete 5A cycle, and the decline in sales has not been stopped.

The issue of the customer journey, possible communication channels, and points of contact with customers in the context of Marketing 4.0 and the 5A model was also investigated by Székely et al. [20], but they focused on industrial marketing, where the combination of a questionnaire survey, interviews with salespeople and buyers, and long-term observations was used to examine the degree of use and effectiveness of individual communication channels. Their research has shown that, despite the increasing use of digital marketing in recent years, the greatest emphasis in industrial marketing is still placed on face-to-face contacts between salespeople and buyers, and that this communication channel is also the most preferred one by buyers. Therefore, they have concluded that the digitization of marketing (and thus the advent of Marketing 4.0) is delayed in industrial marketing compared to B2C, but they expect a further development in this area in the future.

Upamannyu et al. [21], addressing the interdependence of brand and company image, customer loyalty, and Word of Mouth (WOM), also implicitly (although not explicitly mentioned in the paper) draw on the 5A concept. Based on the analysis of questionnaire data with 272 respondents, they have documented the positive dependence of customer loyalty on brand and firm image, and also the positive dependence of spreading the WOM advertising on customer loyalty. Thus, the authors have concluded that the better the brand and company image, the more loyal the customer buying the product will be and the more intensively they will promote the product among their contacts at the same time. The authors emphasize the importance of WOM in the context of the existence of social networks and Marketing 4.0 in general.

Opresnik has been dealing with the issue of Marketing 4.0 for a long time and is represented in our review by 5 papers in total. In them, he views Marketing 4.0 from different perspectives, paying the most attention to the use of social networks. He is of the opinion that the emergence of social networks has fundamentally and irreversibly changed the paradigm of marketing communication from the "bowling" concept to the "pinball" concept [22]. While in the traditional "bowling" concept, marketing communication goes like a bowling ball directly from the company that represents the players to the bowling pins - the customers, the world of social networks is more like pinball, where the original marketing message, once sent out into space, is "bounced" between individual customers in different ways completely without the company´s control.

In his paper [23], he discussed the differences between creating a "regular" marketing plan and a digital marketing plan, which is focused on social networks. To create a digital marketing plan, he suggests a six-step process in which he places great emphasis on both auditing the existing use of social networks, selecting appropriate social networks for further marketing efforts, and on the individual use of each of those networks. According to Opresnik, in the course of digital marketing planning, companies should define a specific mission for each social network used, expressing what is to be achieved using that network.

In his article [24], he then discussed the possibilities of using modern technologies for marketing research purposes, mentioning, for example, the possibility to track the movement of customers around the store using the RFID technology or to evaluate the customer´s brain reactions to stimuli using neuroscience tools. However, the greatest attention in his work is given to the possibilities of online marketing research, where he sees its main advantage in the possibility of fast and inexpensive collection of large amounts of data, whether through web questionnaires or by analyzing data from social networks, metadata from the browsers of individual visitors to company websites, etc. This topic was followed up with another paper [25], where he addressed the issue of effective marketing on the Internet in the Web 3.0 environment, again emphasizing the use of data from cookies and other metadata from the browser of website visitors in order to build a personalized web presentation for the given customer.

In this paper [26], he then addressed opportunities that the development of IoT offers to marketing, especially in cooperation with social networks. He sees the potential of using IoT and social networks primarily in marketing research, where the combination of data obtained from sensors of individual products that are part of the IoT, and data mined from social networks should allow a much more detailed analysis of the customer journey of every customer, leading to the purchase of other products. He also sees the further potential of smart products in the fact that thanks to IoT it will be possible to gather detailed data on their use by customers and to use these data in the design of new products or their functions.

The issue of using IoT in Marketing 4.0 has also been addressed by Rathod et al. [7], who, just as Opresnik, see the potential in using data from smart devices to segment the market better and to target specific customers more precisely with specific products. However, they also point out that some IoT elements, e.g., in the form of wearable devices or home assistants, collect large amounts of data about their users, often of a very intimate nature, and therefore they warn against both their unethical use and the risk of data leakage.

Dash et al. [27] build on the aforementioned Marketing 4.0 model by Jara et al. [9], trying to identify the relationships between the individual elements of the model (Brand Identity, Brand Image, Brand Integrity, and Brand Interaction), their influence on customer satisfaction (Customer Satisfaction), and the customer´s intention to buy a certain product (Purchase Intention). These elements are used to construct a conceptual model in which the relationships between the elements are described by hypotheses, which are subsequently tested by the authors based on the analysis of the data obtained from the questionnaire survey. Having analyzed the data, the authors conclude that while Brand Identity and Brand Image have a significant positive impact on customer satisfaction, this impact is not found to be statistically significant for the other two components of the Marketing 4.0 model. The same conclusions have also been drawn from the analysis of the relationship between the individual components of Marketing 4.0 and the customer´s intention to buy the given product. Also, the hypothesis of the positive relationship between customer satisfaction and customer purchase intention has been accepted. Thus, according to Dash et al [27], marketers should primarily focus on building Brand Identity and Brand Image, as these are the elements that bring the greatest practical business effect.

Härting et al. [28] analyzed the possible use of modern technologies in online marketing. They have defined 6 modern technologies and approaches (artificial intelligence, data-driven marketing, dynamic product pricing, mobile marketing, the use of the latest Google services, and algorithm-driven ad publishing) and asked 102 experts in the field about their applicability in online marketing. Based on the statistical analysis of their responses, they have concluded that both the use of mobile marketing and data-driven decision making positively influence online marketing results. On the other hand, the use of the other technologies defined does not bring a significant positive effect in online marketing, according to the experts surveyed.

Online marketing was also addressed by Wang [29], who investigated the phenomenon of Live Broadcast Marketing (LBM). The author illustrates the phenomenal growth of this promotional method over the last few years, aided by the rapid growth in popularity of TikTok, a social network for sharing short videos. Nevertheless, the article also uses a number of examples to highlight the fact that the success of LBM promotional campaigns is not automatically guaranteed, even if marketers manage to enlist a well-known celebrity for the campaign. The author stresses that influencers must be chosen with regard to the product promoted so that the entire message is believable to content consumers.

Kolářová & Kolářová [30] investigated the use of Marketing 4.0 in the specific sector of microbreweries in the Czech Republic. Their questionnaire survey among microbreweries in the Capital City of Prague has confirmed the hypothesis that more than 80% of microbreweries use Marketing 4.0 tools to manage customer relationships, mainly through their activity on social networks, especially Facebook and Instagram. The majority of the microbreweries surveyed also try to utilize their websites for marketing outside of social networks, both using SEO to drive more visitor traffic and by analyzing data on visitor behavior on their website through Google Analytics. By contrast, the possibility of involving customers in the process of new product development (i.e., the possibility for customers to come up with new types of brewed beer) is, according to the survey, not used by any of the microbreweries surveyed, despite the fact that it is theoretically one of the related phenomena of Industry 4.0, which could be technically feasible in the environment of craft breweries brewing small batches of beer.

The potential benefits of customer involvement in the product design process that can be achieved in the Marketing 4.0 environment are also highlighted by Rubio et al. [31]. For these purposes, they primarily suggest using the analysis of big data from social networks, with the help of which firms can obtain accurate information about customers´ opinions on existing products or their requirements for new products in a non-intrusive and inexpensive way.

The issue of using Marketing 4.0 tools in the Czech economy was addressed by Petrů & Zich [32], who first focused on family businesses. Using a questionnaire survey among 109 mostly small family businesses, they tried to answer the question of whether these businesses are ready to use modern technologies in marketing. In addition to the questions on the use of modern technologies, the questionnaire also contained questions on marketing in general, where the authors reasoned that in order to be able to use modern technologies in marketing effectively, a business must first have a good grasp of "traditional" marketing. However, the questionnaire has shown that marketing in Czech family businesses is generally not at a very high level, with the use of modern technologies being one of the worst-rated indicators within this unsatisfactory level.

Petrů et al. [33] then followed up this work with a longer-term study, where they examined the development of marketing and the extent of using modern marketing tools by Czech businesses between 2016 and 2019 through semi-structured research interviews. Again, the authors primarily focused on the segment of small businesses, but in this case, they added other enterprises to family businesses for the purpose of comparison. The data obtained from the study confirm the conclusion of the previous article, i.e., the general level of marketing activities of Czech businesses was relatively low in the observed years. Moreover, activities consisting in the use of modern technologies were used rather less compared to other activities, especially by family businesses.

Furthermore, the trend to use modern technologies for marketing by family businesses was also decreasing over the years, however, it should be noted that the set of respondents varied significantly between the individual years, so the trend comparison is not entirely conclusive. In any case, the research by Petrů et al. [33] has shown that the use of Marketing 4.0 tools is not widespread yet among small Czech enterprises.

Blazquez-Resino et al. [34] discuss another phenomenon associated with the Fourth Industrial Revolution, namely the emergence of the so-called "Consumer 4.0". This "new kind of consumer" should be characterized by a desire for technological innovations, a desire not to be merely a passive recipient of a product, service, or information, but to actively participate in its development and inform other customers. This change from a passive recipient of information to an active propagator of information, especially through social networks, is also linked to another term, that is "netizen", or "citizen of the net". Blazquez-Resino et al. point out that this active collaboration and two-way communication between businesses and customers will require significant changes in the way businesses operate and work, and that it will therefore be necessary to first train all relevant employees for these new conditions.

The specifics of Consumers 4.0 and the two-way communication with them through the Marketing 4.0 tools were also addressed by Wereda & Wozniak [8] in their research focused on the specifics of Polish economy. The results of their survey among marketing managers of 100 companies show that companies understand the importance of active communication with Consumers 4.0 and try to develop it actively. At the same time, however, the research has shown that the companies surveyed still mostly use older forms of communication, such as phone calls and emails, for this communication. Conversely, more modern communication tools, mainly social networks, or chat on company websites, are still used by Polish companies to a lesser extent.

Kim & Kim [35] dealt with a somewhat different topic compared to the other papers reveiwed. While other authors have mainly addressed the world of social media and online marketing in the context of Marketing 4.0, Kim & Kim have focused on the impact of Marketing 4.0 on the physical layout of shopping malls. They argue that the layout of shopping malls must reflect the interests of the new kind of consumers (although they do not explicitly mention the Consumer 4.0 term) who want to share the shopping process with the others both physically and online through social networks. Thus, according to these authors, shopping malls in the Marketing 4.0 environment should not only be space for shopping, but for sharing experiences in a broader sense. They give the specific case study of the shopping mall in Seoul as an example of such changes.

5. Discussion

As the detailed analysis of the papers reviewed in the previous chapter clearly has shown, the vast majority of authors focus on online marketing, or more precisely e-commerce in general, when researching the issue of Marketing 4.0. Out of the 24 papers reviewed, only Kim & Kim [35], who dealt with the physical layout of the shopping mall (although they also emphasized the necessity of linking the physical world with the online one at the theoretical level), and Jarra et al. [9], who looked into the possibilities of using IoT in the process of physical shopping, had a different focus.

As for the specific topics and technologies related to Marketing 4.0 that have been discussed in the papers reviewed, their frequency of occurrence is shown in Table 2.

Topic	Frequency of occurrence
Social networks and media	14
5A Concept	5
IoT	4
Consumer 4.0	2
Big data	1
Artificial intelligence	1
Virtual reality	1

Table 2. The specific topics of Marketing 4.0 examined in the papers reviewed.

Table 2 shows that the largest number of authors dealing with Marketing 4.0 focuses mainly on the world of social media and networks and the possibility of using them to promote products or services. The second most frequently occurring topic is also related to social networks and the so-called C2C marketing in the form of the Kotler´s 5A concept describing the customer journey. In five of the papers reviewed, this model and the ability of social networks to serve as an effective medium for the Advocate step were explicitly highlighted, while in several other papers the same logic was mentioned without direct reference to the Kotler´s concept. The concept of Consumer 4.0, which was examined in two papers, is then related to the issue of social networks and the role of existing customers in product promotion to a certain extent. Quite surprisingly, the issue of augmented reality, the use of which has been gaining importance in the field of marketing lately (see e.g. the summarizing paper [36]), was not seriously discussed in any of the reviewed studies.

In other words, although the technology of augmented reality definitely is a part of Industry 4.0 and at the same time has its use in marketing, it is probably not understood as the part of Marketing 4.0 in its present state of art. Based on the conducted systematic review, it can therefore be concluded that although the theoretical definition of Marketing 4.0 emphasizes the necessity of interconnecting the virtual world with the physical one, the papers published so far addressing Marketing 4.0 have clearly focused primarily on the virtual world. Thus, the concept of Marketing 4.0 has not been approached in its entirety and its definition has not been completely fulfilled.

Of course, it is necessary to bear on mind the limitations of the chosen research approach. Firstly, the review focused only on the academic papers, which means, that it only describes the state of art of the Marketing 4.0 concept as it is understood by the academic professionals. The opinion of business professionals on this topic can be different, though. Finding out, how the business professionals understand the Marketing 4.0 concept, is therefore interesting topic for further study. The other limitation is connected to the choice of scientific databases, which were used as a source of reviewed articles. Although we believe that the two main citation databases of the scientific literature provide complex overview of relevant academic literature, some uncited papers, which deal with the topic of Marketing 4.0, may have therefore been deliberately omitted from the review. Again, the review based on the full-text search through journals and proceedings not indexed in these databases is potential direction of further research.

6. Conclusion

The idea of Marketing 4.0 emerged as early as 2012, not long after the emergence of the more general concept of Industry 4.0. These concepts are related to a large extent, their common element being the emphasis placed on the connection of the real and virtual worlds using modern technologies. However, the systematic review of the available professional literature, the results of which have been presented in this article, shows that so far, Marketing 4.0 has been given considerably less attention in the literature than Industry 4.0, and the growing interest in this issue can be observed only in recent years. Based on the content analysis of the existing papers dealing with this topic, it can also be concluded that the majority of researchers still associate Marketing 4.0 mainly with social networks and e-commerce in general, despite the fact that, according to its theoretical definitions, the main characteristic of Marketing 4.0 should be the interconnection and overlapping of the virtual world and the real one. Therefore, it is the interconnection of the real and virtual worlds for marketing purposes where there is undoubtedly a research gap that presents an opportunity for further research in this field.

7. Acknowledgement

This paper was created within the project SGS-2020-026 titled "Economic and financial transformation in the context of digital society".

8. References

[1] Thomas, J. (2020). Marketing Concept: Examining AMA Definitions, Evolution, Influences, Proceedings of the 17th AIMS International Conference on Management, January 2-4 2020, Kozhikode, ISBN 978-1-943295-14-2, Omprakash, G. & Shivprakash, A. (Eds.), pp. 957-964, AIMS International, Kozhikode

[2] https://www.ama.org/the-definition-of-marketing-what-is-marketing/, (2017). American Marketing Association, Accessed on: 2022-06-26

[3] Grundlach, G. T. & Wilkie, W. L. (2009). The American Marketing Association's New Definition of Marketing: Perspective and Commentary on the 2007 Revision. Journal of Public Policy & Marketing, Vol. 28, No. 2, pp. 259-264, ISSN 0743-9156

[4] Rojko, A. (2017). Industry 4.0 Concept – Background and Overview, International Journal of Interactive Mobile Technologies, Vol. 11, No. 5, pp. 77-90, ISSN 1865-7923

[5] Nondek, L. & Řenčová, L. (2000). Internet a jeho komerční využití, Grada Publishing a.s., ISBN 80-7169-933-0, Praha

[6] Vaidya, S.; Ambad, P. & Bhosle, S. (2018). Industry 4.0 – A Glimpse, Procedia Manufacturing, Vol. 20, pp. 233-238, ISSN 2351-9789

[7] Rathod, A. Y., Panday, S. & Doshi, N. (2020). IoT and Modern Marketing: Its Social Implications, Proceedings of ICACT 2020, 16-19 February 2020, Phoenix Park, ISSN 1738-9445, pp. 407-413, IEEE, DOI: 10.23919/ICACT48636.2020.9061210

[8] Wereda, W. & Wozniak, J. (2019). Building Relationships with Customer 4.0 in the Era of Marketing 4.0: The Case Study of Innovative Enterprises in Poland. Social Sciences, Vol. 8, No. 6, pp. 1-27, ISSN 2076-0760

[9] Jara, A. J., Parra, M. C. & Skarmeta, A.F. (2012). Marketing 4.0: A new value added to the Marketing through the Internet of Things, Proceedings of Sixth International Conference on Innovative Mobile and Internet Services in Ubiquitous Computing, 4-6 July 2012, Palermo, ISBN 978-0-7695-4684-1, Ilsun, Y. et al. (Eds.), pp. 852-857, IEEE, DOI: 10.1109/IMIS.2012.203

[10] Inkermann, D.; Schneider, D.; Martin, N.L.; Lembeck, H.; Zhang, J. & Thiede, S. (2019). A framework to classify Industry 4.0 technologies across production and product development, Procedia CIRP, Vol. 84, pp. 973-978, ISSN 2212-8271, DOI: 10.1016/j.procir.2019.04.218

[11] Soni, G. & Kodali, R. (2011). A critical analysis of supply chain management content in empirical research. Business Process Management Journal, Vol. 17, No. 2, pp. 238-266, ISSN 1463-7154, DOI: 10.1108/14637151111122338

[12] Manoharan, A. & Singal, M. (2017). A systematic literature review of research on diversity and diversity management in the hospitality literature. International Journal of Hospitality Management, Vol. 66, pp. 77-91, ISSN 0278-4319, DOI: 10.1016/j.ijhm.2017.07.002

[13] Morreira, T. D., Brandao, A. M. P. C. & Novais, R. A. (2018). Peeble in the shoe? Content marketing made in Portugal, Proceedings of 11th Annual Conference of the EuroMed Academy of Business, 12-14 September 2018, Valetta, ISBN 978-9963-711-67-3, Vrontis, D., Weber, Y. & Tsoukatos, E. (Eds.), pp. 1622-1624, EuroMed Press

[14] Martín-Guart, R. F. & Cavia, J. F. (2014). Advertising and media agencies in the face of the change in the media ecosystem. Cuadernos.info, Vol. 34, pp. 13-25, ISSN 0719-3661, DOI: 10.7764/cdi.34.572

[15] Kotler, P., Kartajaya, H. & Setiawan, I. (2017). Marketing 4.0: Moving from Traditional to Digital, John Wiley & Sons Inc., ISBN 978-1-119-34120-8, Hoboken

[16] Szymoniuk, B. & Valtari, H. (2018). The REKO System in Finland: A New Model of a Sustainable Marketing Channel. Problems of Sustainable Development, Vol. 13, No. 2, pp. 103-111, ISSN 1895-6912

[17] Hartono, B. D., Yusuf, B. P. & Arifiati, A. (2020). Implementation of Digital Marketing Strategies through Social Media Marketing, Supply Chain Management and Online Sales of Bill Chillly Product. International Journal of Supply Chain Management, Vol. 9, No. 4, pp. 348-354, ISSN 2050-7399

[18] Hwang, J. & Kim, J. (2020). Food tourists' connectivitiy through the 5A journey and advocacy: comparison between generations Y and X. Asia Pacific Journal of Tourism Research, Vol. 25, No. 1, pp. 27-38, ISSN 1094-1665, DOI: 10.1080/10941665.2019.1580756

[19] Karta, N. L. P. A., Wardani, R. K. & Suarthana, J. J. P. (2021). Differentiation and the Five As Concept in Digital Marketing of Small and Medium Enterprises During the Covid-19 Pandemic (a Study in Cepaka Village, Kediri Tabanan), Proceedings of International Conference on Enterpreneurship (ICOEN), virtual, pp. 566-578, Universitas Ciputra

[20] Székely, S., Csata, Z., Cioca, L. I. & Benedek, A. (2020). Industrial Marketing 4.0 – Upgrading the Industrial Customers' Path to The Digital Economy. Polish Journal of Management Studies, Vol. 22, No. 2, pp. 535-548, ISSN 2081-7452, DOI: 10.17512/pjms.2020.22.2.35

[21] Upamannyu, N. K., Rajput, S. & Bhakar, S. S. (2021). Antecedents of Customer Loyalty and Word of Mouth: A Study on Consumer Durables in Emerging Economies. FIIB Business Review, Vol. 10, pp. 1-20, ISSN 2319-7145, DOI: 10.1177/23197145211053398

[22] Opresnik, M. O. (2017). From Bowling to Pinball: Understanding How Social Media Changes the Generation of Value for Consumers and Companies, In: Social Computing and Social Media, Meiselwitz, G. (Ed.), pp. 190-199, Springer International Publishing AG, ISBN 978-3-319-58558-1

[23] Opresnik, M. O. (2018). Effective Social Media Marketing Planning – How to Develop a Digital Marketing Plan. In: Social Computing and Social Media, Meiselwitz, G. (Ed.), pp. 333-341, Springer International Publishing AG, ISBN 978-3-319-91520-3

[24] Opresnik, M. O. (2019). Digital Marketing Research – How to Effectively Utilize Online Research Methods. In: Social Computing and Social Media, Meiselwitz, G. (Ed.), pp. 444-451, Springer International Publishing AG, ISBN 978-3-030-21904-8

[25] Opresnik, M. O. (2020). Effective Online Advertising Strategy. In: HCI International 2020 – Late Breaking Papers, Stephanidis, C., Salvendy, G., Wei, J. et al. (Eds.), pp. 418-424, Springer International Publishing AG, ISBN 978-3-030-60151-5

[26] Opresnik, M. O. (2021). Effective Social Media Marketing Through Customer Journey Mapping. In: Social Computing and Social Media, Meiselwitz, G. (Ed.), pp. 461-469, Springer International Publishing AG, ISBN 978-3-030-77625-1

[27] Dash, G., Kiefer, K. & Paul, J. (2021). Marketing-to-Millenials: Marketing 4.0, customer satisfaction and purchase intention. Journal of Business Research, Vol. 122, pp. 608-620, ISSN 0148-2963, DOI: 10.1016/j.jbusres.2020.10.016

[28] Härting, R. Ch., Reichstein, Ch. & Müller, A. (2019). Potential Benefits of New Online Marketing Approaches. In: Business Information Systems, Abramowicz, W. & Corchuelo, R. (Eds.), pp. 109-120, Springer International Publishing AG, ISBN 978-3-030-20481-5

[29] Wang, Z. (2020). Analysis of The Problems Existing in Network Live Broadcast. E3S Web of Conferences, No. 218, pp. 1-4, ISSN 2267-1242, DOI: 10.1051/e3sconf/202021801025

[30] Kolářová, V. & Kolářová, E. (2020). An analysis of the use of marketing 4.0 principles for the managing customers relationships in microbreweries in the capital city of Prague. Potravinárstvo Slovak Journal of Food Science, Vol. 14, pp. 336-342, ISSN 1337-0960, DOI: 10.5219/1261

[31] Rubio, N., Villasenor, N. & Yagüe, M. J. (2020). Value Co-creation in Third-Party Managed Virtual Communities and Brand Equity. Frontiers in Psychology, ISSN 1664-1078, DOI: 10.3389/fpsyg.2020.00927

[32] Petrů, N. & Zich, F. (2017). Evaluation of Family Business Marketing Activities. In: MARKETING IDENTITY – Online rules – part II, Petranová, D., Čábyová, L. & Bezáková, Z. (Eds.), pp. 397-409, Faculty of Mass Media Communication, University of Ss. Cyril and Methodius in Trnava, ISBN 978-80-8105-918-6, Trnava

[33] Petrů, N., Kramoliš, J. & Stuchlík, P. (2020). Marketing tools in the era of digitization and their use in practice by family and other businesses. E+M Economics & Management, Vol. 23, No. 1, pp. 199-214, ISSN 2336-5064, DOI 10.15240/tul/001/2020-1-014

[34] Blazquez-Resino, J. J., Gutiérrez-Broncano, S., Ruiz-Palomino, P. & Jimenez-Esteves P. (2020). Dealing With Human Resources in the Age of Consumer 4.0: Aiming to Improve Service Delivery. Frontiers in Psychology, ISSN 1664-1078, DOI: 10.3389/fpsyg.2019.03058

[35] Kim, I. S. & Kim, Y. O. (2019). Emergence of a New Type of Consumption Space in the Era of Marketing 4.0: Change in Merchandising Strategy with Focus on Spatial Confguration & User Behavior, Proceedings of the 12th Space Syntax Symposium, 8-13 July 2019, Beying, pp. 154-170, Beying Jiao Tong University

[36] Mitrovic, K., Novakovic, N., Spajic, J. & Cosic, I. (2021). Augmented Reality in Marketing – State of Art, Proceedings of the 32nd DAAAM International Symposium, pp.0566-0575, B. Katalinic (Ed.), Published by DAAAM International, ISBN 978-3-902734-33-4, ISSN 1726-9679, Vienna, Austria DOI: 10.2507/32nd.daaam.proceedings.081

DOI: 10.2507/33rd.daaam.proceedings.059

RISK MANAGEMENT DURING A PROCESS OF DESIGNING A TECHNICAL PRODUCT

Katerina Kasova & Nikola Skrivanova

This Publication has to be referred as: Kasova, K[aterina] & Skrivanova, N[ikola] (2022). Risk Management During a Process of Designing a Technical Product, Proceedings of the 33rd DAAAM International Symposium, pp.0421-0428, B. Katalinic (Ed.), Published by DAAAM International, ISBN 978-3-902734-36-5, ISSN 1726-9679, Vienna, Austria DOI: 10.2507/33rd.daaam.proceedings.059

Abstract

This paper focuses on an application of risk management in a process of creating technical solutions on a specific case study. The case study will deal with manufacturing of a medical equipment made of unconventional materials, specifically a tibia protector. Basic theoretical starting points for managing risks will be described in the introduction of this article, these starting points will be used to determine the key risks which will be later described and evaluated on a base of their probability and impact. Defined risks will be then included into the designing phase in order to fully unlock the potential of risk management even for prevention purposes.

Keywords: Risk management; Tibia protector; Technical solution; Unconventional material.

1. Introduction

The number of authors studying risk management is increasing lately because companies have to reduce their error rate and make themselves as efficient as possible due to still raising competition on the market. [1;2] If organizations do not wish to suffer a loss in business, they can follow the methods of business risk management where various risks can be found. Such as manufacturing, technical, economic and investment risk and more. [3;4] The problem starts when the organizations are supposed to begin control risks in new products in early development, here it is up to the designer to take into account certain risks when creating a design. They also have to be familiar with the concepts of risks and acknowledge benefits from timely implementing it. Because there is an idea at the very beginning of designing a new product. An idea which the designer will transfer into the final product, the methods of risk management have to be in this case sufficiently easy to follow. [5]

We can find many different motives for such developments but if we focus solely on medical industry, these new concept designs should be united with only one goal in mind which is helping others, making their lives easier and being as useful as possible. In order to achieve this, it is necessary to also focus on risks which could endanger our efforts from the very beginning of the project. There are several questions that have to be answered before any advanced stage of a project. Which risks can we expect to occur, how grave the risks are and their impacts and consequences? What should be done with these risks? How the negative impacts and economical results should be approached? How the possibility of a risk occurring can be reduced and how to evaluate the experience obtained while dealing with a dangerously situation? [6] This paper follows up on managing risks during a process of designing a technical application.

2. Risk management

A risk is defined by ISO 9000:2015 norm as an effect of uncertainty where the effect is a deviation from what was expected, it can be either positive or negative and the uncertainty is a state caused by a lack of information. [7] From risk control point of view the risk will be viewed as a possible event which differs from expected state of development. Nevertheless, the risks should not be reduced to mere possibility because it includes both possibility itself and a quantitative range of the event. [8] If a risk occurs it does not immediately mean it will cause a negative effect with consequences. It is beneficial to have some corrective measures prepared in case the risk has to be mitigated.

Division of risks

- Expectable – easier to control and to work with, these risks are repeatable in many cases.
- Unexpectable – harder to control, impossible to predict and their repeatability is improbable. [9]

Risk management is no longer optional but it is rather established by norm requirements, legislation or by customers personal requests.

2.1 Risk management process

No matter the reason an organisation has for risk management, these logical steps have to be always followed.

- Risk analysis (finding out what are the threats and if there are any) – it is necessary to locate risks that can endanger the project, mark them and describe them in the most detailed manner possible. Only the risks that are a significant threat to the project should be chosen for monitoring because it is impossible to create a list of all possible threats.
- Evaluate risks (determine a level of risk and its acceptability) – qualitative and quantitative methods are used for risk analysis both of these methods are sorted according to the mean of expression of given values in the risks analysis. It is possible to use either one or combination of both methods.
- Secure risks (supervise the risks and set corrective measures in case some risks occur) – most often used applied methods of reducing and securing the risk are mainly preventive actions, risk retention, transferring the risks to other entrepreneurial subjects, insurance, avoiding risks (avoiding the activity during which the risks could occur), creating reserves. [10]

The risk management is not only beneficial to organizations and their owners for whom the knowledge of risk is important for its financial reasons but also project managers and end customers. (Customers get products they need for a fair price, in agreed time period with no extra expenses). [11]

Hard benefits	**Soft benefits**
More trustworthy plans with better information, time schedules, budgets	Gaining experience and improving communications
Objective comparison of alternative suggestions	Development of employees' ability to evaluate risks
Identification of responsibility for specific risks	Focusing on important issues
Realistic evaluation of possible risks	Responsible approach to customers

Table 1. Hard and soft benefits of risk management process

3. Case study

Case study was performed on the process of making a medical device made of unconventional materials in this case a custom-made shinbone protector designed for a leg of specific patient. The main purpose of this protector is spreading the pressure on the spot where the bone had been broken which is caused by the edge of skiing boot during manoeuvring.

3.1 Current status

The patient has recovered from fractures of his shin bone and calf bone of his left leg. Two surgical bolts were used for fixation of the shin bone during the surgery. The bone later outgrew the bolts making them impossible to take out, see Tab. 1.

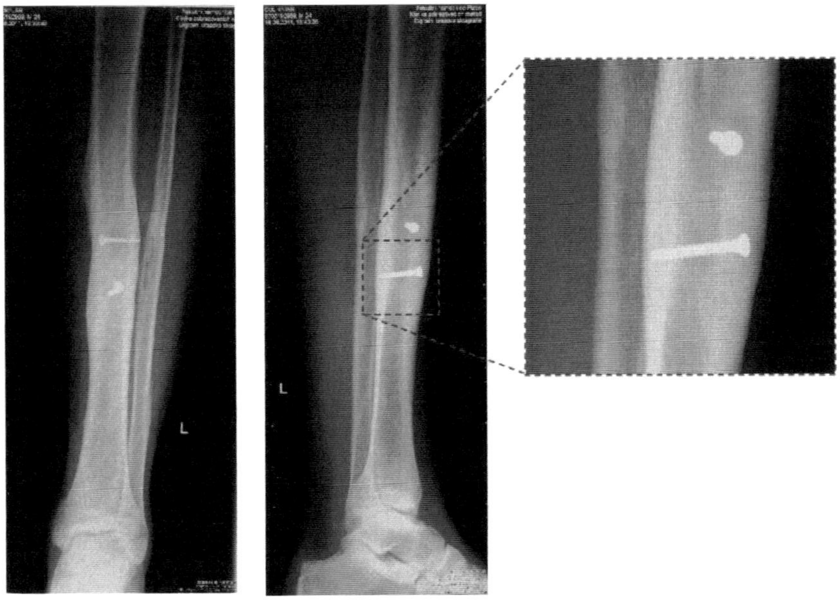

Fig. 1. Current state of the fractures of shin bone and calf bone. Side view on the left, front view on the right

The patient is restricted by pain only in certain activities, skiing being one of them. The pain is inflicted by increasing pressure on the contact point of the edge of the skiing boot and the leg itself when a skier performs a downhill stance (see Fig. 1.) and it basically prevents the patient from downhill ski. Numerous commercial shin bone protectors were tested for their ability of reducing pressure thus reducing the pain but none of them was found adequate. The most common causes of failure were following: a) the protector did not fit precisely. b) the protector was too thick and thus did not fit between the leg and the boot (in some cases inflicting pain to other areas of the leg). c) the protector lacked needed rigidity making it unable to effectively spread pressure in the stricken area.

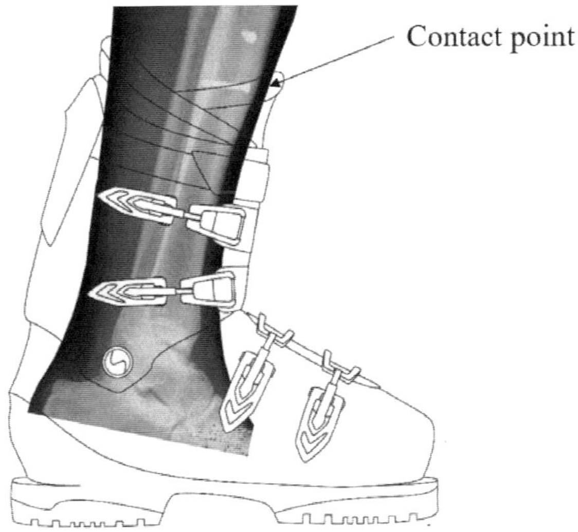

Fig. 2. A depiction of the pressure sensitive area of the leg shown in a skiing boot.

Based on the available pieces information these preliminary requirements for the design of the protector were set.

- Low thickness <=1,5 mm
- Sufficient bending stiffness
- Low weight <= 200 g
- Shape accuracy according to the patient's leg

3.2 Risk management

In order create a product as beneficial as possible a thorough risk analysis had been processed before the product designing began. Analysing the risks in a timely manner helped to understand the issue and to uncover further design requirements. Table 2. shows some of the key risks which can have negative effect on the quality of finished product.

Requirement	**Risk occurrece criterium**	**Risk**	**Probability**	**Impact**	**Value**
Safety/Complience with legislation	Possible defect formation, which would make the product unsafe for use with no warning sings or incompliant with legislation.	**R1**-Structure failure and loss of stability when the product is being used	2	4	8
	Possible defect formation, which would make the product unsafe for use despite obvious warnings or incompliant with legislation.	**R2**- the product is being used by a person for whom it was not designed	2	3	6
Accomplishing primary function	Loss of primary function (the product no longer serves its purpose, safety not compromised)	**R3**-The product is not rigid enough in required direction, but does not crack when used.	2	2	4
	Loss of effectiveness of the primary fiúnction (the product serves its purpose in a limited extent)	**R4**- Rigidity of the product allows it to be used for what it was designed but does not provide desired support in borderline situations.	3	2	6
Accomplishing secondary function	Loss of secondary function (the product serves its purpose, but comfort securing features are not working)	**R5**- The product servers its purpose, provides needed rigidity in required direction. Extremity injury will occur when use. (formation of scrapes, blisters)	1	4	4
	Loss of effectiveness of the secondary function (the product serves its purpose, but comfort securing features are limited)	**R6**-The product servers its purpose, provides needed rigidity in required direction, it was used in complience with given instructions, but leg proportions changed (e.g. change of muscle conditions). **R7**- The product was used in complience with given instructions, but when used for a longer period of time edges of the product damage the sock .	1	2	2
Amenity	Appearance, weight, the product serves its purpose, but it does not match all the amenity expectations of a common user	**R8**-The surface of the product is wrinkled because of used hand-handled process and used mold. **R9**- Total weight is over 200g.	3	1	3
	Appearance, weight, the product serves its purpose, but it does not match all the amenity expectations of a demanding user	**R10**-The manual laying of carbon fibers caused minimal distortion of the fabric. **R11**-Total weight is over150g.	4	1	4

Table 2. The key risks which can have negative effect on the quality of finished product

Based on evaluation of the impacts and probability each risk is assigned with numerical values, the resulting value is then set. This value is calculated as product of impact and probability. Final evaluation of the risk is then shown in the impact and probability matrix (see table below).

		Impact			
		1 - Insignificant	2 - Minor	3 - Medium	4 - High
Probability	1 (0-10%)		R6		R5
	2 (11-35%)	R10	R3, R9	R2, R4	R1
	3 (36-65%)	R7, R11			
	4 (66-100%)	R8			

Table 3. Risks matrix

3.3 The connection between risk and design process

- R1 – Due to high demands on rigidity, low weight and small thickness a carbon fibre reinforced polymer was used for manufacturing because it has the required properties. In order to avoid a structure failure and a loss of stability during use three variants with different rigidity and thickness were made. The products were then tested in operating conditions and based on results of the tests the most suitable variant was chosen.
- R2 – The formation of risk cannot be influenced during designing or manufacturing process, the risk was reduced with instructing the user about correct use of the equipment.
- R3 - The risk corresponds with R1. Each product variant was evaluated during testing.
- R4 - The risk corresponds with R1. Each product variant was evaluated during testing.
- R5 – This protective equipment was designed and manufactured in such way that when using it, there is always a functional ski sock 2-3mm thick between the leg and the protective equipment. The user was introduced to the method of use.
- R6 – The formation of risk cannot be influenced during designing or manufacturing. The risk was reduced with the user being sufficiently trained in using the product (e.g. if the user feels any discomfort or the product does not fit well it is necessary to produce a new piece).
- R7 – All sharp edges were sanded during manufacturing
- R8 – Due to a fact that the product is made by hand it was impossible to fully eliminated the wrinkle. Great emphasis was placed on laying the separation foil on the top and bottom of the product. The foil directly influences the quality of finished product.
- R9 – The weight is affected by the number of layers in relation to the R1 risk and matrix to fabric ratio. The ratio of the matrix cannot be greatly influenced in wet lamination without the infusion method. Production was carried out using a minimum amount of matrix during lamination.
- R10 – Because of the hand-made process and shape complexity it was not possible to fully eliminate the fabric distortion. Great emphasis was placed on laying each layer especially the top layer which then forms the view part of the product
- R11 – The weight is affected by the number of layers in relation to the R1 risk and matrix to fabric ratio. The ratio of the matrix cannot be greatly influenced in wet lamination without the infusion method. Production was carried out using a minimum amount of matrix during lamination.

3.4 Design

Material: Due to the required values of weight, thickness and stiffness, it was chosen as the basic construction material carbon fibre reinforced polymer. Composite orthotropic materials such as carbon fabric laminates are characterized by 9 material constants to determine the mechanical properties in all three directions. [12] Material CC200 carbon fabric twill with Toray T300 3K fibres with density of 1690 kg.m-3 was used for manufacturing tibia protection. The thickness of one layer is 0.26 mm. The material properties given by (Table 4) were taken from the article. [13] These values are for the production of laminate in an autoclave at a specific pressure and temperature.

Material properties		**Value**
Young's modulus	E_1	52.3 (GPa)
	E_2	48.7 (GPa)
	E_3	5 (GPa)
Shear modulus	G_{12}	2.32(GPa)
	G_{13}	2 (GPa)
	G_{23}	2 (GPa)
Poisson's ratio	μ_{12}	0.11 (-)
	μ_{13}	0.35 (-)
	μ_{23}	0.35 (-)

Table 4. Material properties of Twill CC200 – Toray T300 3K

Load: The tibia protection is loaded by bending stress. This bending moment is created by a torque moment with the axis of rotation in the ankle. This torque moment during the normal carving curve for 50th percentile of European male adult weight (85kg) is M_b=195 (Nm) according to the publication. [14] This torque produces a force F_c= 750 which is given by equation 1 and which is at a distance l=0.260 (m) from the ankle to contact point (see Fig. 3.)

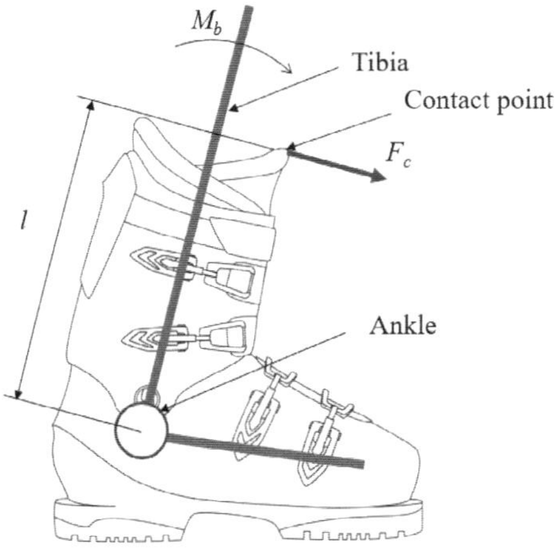

Fig. 3. Schematic action of forces at the contact point of the leg and ski boot

$$F_C = \frac{Mb}{l} = \frac{195}{0.26} = 750N \tag{1}$$

where F_c is force in the contact point (N), M_b is torque moment to ankle point (Nm), l is distance from ankle to contact point (m).

Shape: One of the requirements was the exact shape of the tibia bone protection in relation to the shape of the leg. The method of direct production on the patient's leg was chosen, because in this case it was a prototype intended for just one patient. This method of production ensures the shape accuracy of the produced parts.

Manufacturing: The product was manufactured using a method of wet lamination directly on patient's leg. To ensure the product will keep its ergonomic shape during the whole process a skiing sock was put on the leg. This also reduced a possibility of the risk R5 occurrence. An Unperforated Release Film was applied to the sock for preventing the resin from getting into contact with patient's skin while making separating the sock from the product significantly easier. Layers of CC200 carbon fabric twill with Toray T300 3K fibres were applied onto the release film. Direction of the threads is parallel to the axis of tibia which enriched the product with the highest possible level of bending stiffness in required direction. Polyester resin, the material chosen for manufacturing, is processable for 30 minutes. That provides enough time for the lamination process and at the same time it was not necessary to keep patient's leg fixed for hours. The release film was then applied to the saturated fabric layers again. The whole kit was fixed with a stretch foil in order to tighten the layers together around patient's leg (see Fig. 4. For cross section and Fig. 5 for real manufacturing final status).

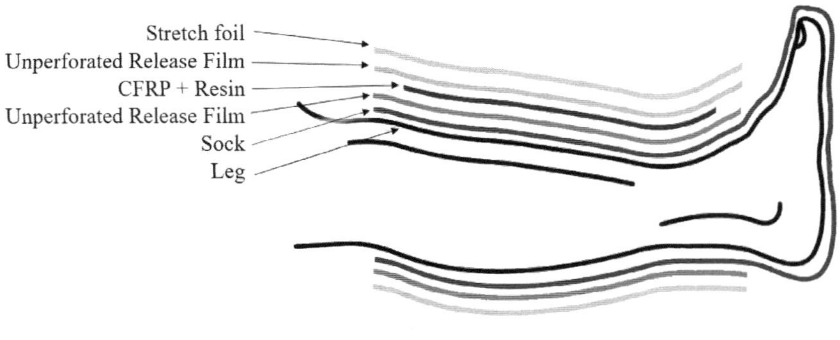

Fig. 4. A cross-section view of lamination layers

Fig. 5. Manufacturing of leg protection (final status with all layers)

The unfinished protector was removed from patient's leg after 30 minutes and then left in room temperature for 24 hours which allowed the resin to harden completely. Then the resin excesses were cut off and the surface was sanded smooth into a round shape as shown in the Fig. 6 in an effort of minimizing the risk of formation of scrapes, blisters and bruises, and keeping the sock from being damaged. This reduced a probability for risks R5 and R7.

Fig. 6. Left - Raw product with sharp edges, right – final product round shape

Three samples were produced using this technique, they differ from each other only in the number of layers of the fabric. The samples were tested in common and extreme conditions of downhill skiing, the results were recorded in the table 5.

The sample No. 1 with four layers of fabric was found unsatisfactory because its rigidity was not sufficient in both tested cases. The samples No. 2 & 3 with six and eight layers of fabric provided sufficient rigidity in both common and extreme skiing conditions and the shape was not causing any discomfort to the user. After evaluation of the results the sample No. 2 was chosen as the best fit because its weight was lower than the set goal of 150 g.

Sample No.	Layer count	Thickness (mm)	Weight (g)	Test results		
				Rigidity during common conditions	Rigidity during extreme conditions	Shape
1	4	1.5	105	Insufficient	Insufficient	Satisfactory
2	6	1.8	140	Sufficient	Sufficient	Satisfactory
3	8	2.5	182	Sufficient	Sufficient	Satisfactory

Table 5. The properties of each variant and test result

4. Conclusion

We can find a plethora of methods of risk evaluation on the market which are being used in various modifications. Unfortunately, utilising this tool usually comes later during the process instead of in the very beginning and that often brings lower effectiveness and some unnecessary efforts into the project. Theoretical starting points were set in the introduction of this paper which focused on an application of risk management in a process of creating technical solutions on a specific case study. Specifically, it described the said issue during a designing and manufacturing process of protective medical equipment for tibia.

At first, based on user's needs and other important factors, basic requirements for product structure were set. A table with chosen key risks was created. The risks marked as the most influential were risks R1, R2 & R4. During designing process, the risks R1 and R4 were reduced by choosing a suitable material, producing multiple variants and consequential testing. The risk R2 can be eliminated completely by sufficient introduction of the product to the user.

The risk R5, which had a great impact, influenced the technological procedure with using a functional ski sock during manufacturing. This procedure modification led to reducing the impact of risk. This case study showed that risk management in the design of medical equipment is applicable and very useful even in case of prevention. For future research will be focus on the use of risk management methods and tools for more complex medical protective equipment.

5. Acknowledgments

The article contribution has been prepared under project SGS-2022-007.

6. References

[1] Anderson, D., Sweeney, D. & Williams, T. (2001). Statistic for Business and Economics, South-Western, ISBN 978-03-2406-671-5, Mason, OH.
[2] Rados, B. & Rados, A. (2016). Risk Evaluation of Production and Implementation of the Project, Proceedings of the 27th DAAAM International Symposium, ISSN 1726-9679, ISBN 978-3-902734-08-2, B. Katalinic (Ed.), pp.0051-0058, Published by DAAAM International, Vienna, Austria DOI: 10.2507/27th.daaam.proceedings.008
[3] Svingerova, M. & Melichar, M. (2017). Evaluation of Process Risks in Industry 4.0 Environment, Proceedings of the 28th DAAAM International Symposium, ISSN 1726-9679, ISBN 978-3-902734-11-2, B. Katalinic (Ed.), pp.1021-1029, Published by DAAAM International, Vienna, Austria DOI: 10.2507/28th.daaam.proceedings.142
[4] Vulanovic, S., Zizakov, M., Vasic, S., Delic, M. & Sremcev, N. (2019). The Impact of Occupational Health and Safety (OH&S) Management Systems on Risk Management Practices, Proceedings of the 30th DAAAM International Symposium, ISSN 1726-9679, ISBN 978-3-902734-22-8, B. Katalinic (Ed.), pp.1188-1195, Published by DAAAM International, Vienna, Austria DOI: 10.2507/30th.daaam.proceedings.167
[5] Smejkal, V. & Rais, K. (2010). Řízení rizik ve firmách a jiných organizacích, Grada, ISBN 978-80-247-3051-6, Prague
[6] Culp, Ch. L., (2001). The risk management process: business strategy and tactics, John Wiley & Sons, ISBN 978-04-7140-554-2, New York
[7] ČSN EN ISO 9000. Quality management systems – Fundamentals and vocabulary. Prague: Czech Office for Standards, Metrology and Testing, 2016, p.35
[8] Balog, M. & Iakovets, A. (2020). Risk management: theory, background, practice, Leges Teoretik, ISBN 978-80-7502-463-3, Prague
[9] Kafol, C. (2016). Project Risk Management Methodology in Practice, Proceedings of the 26th DAAAM International Symposium, ISSN 1726-9679, ISBN 978-3-902734-07-5, Katalinic, B. (Ed.), pp.0445-0452, Published by DAAAM International, Vienna, Austria, DOI: 10.2507/26th.daaam.proceedings.059
[10] Vasilnakova, A. (2018). Risk Management in Accredited Testing Laboratories, Proceedings of the 29th DAAAM International Symposium, ISSN 1726-9679, ISBN 978-3-902734-20-4, Katalinic B. (Ed.), pp.1071-1075, Published by DAAAM International, Vienna, Austria DOI: 10.2507/29th.daaam.proceedings.153
[11] Fraser, J. R. S., Quail, R., & Simkins, B. (2021). Enterprise Risk Management: Today's Leading Research and Best Practices for Tomorrow's Executives, Wiley, ISBN 978-0-470-49908-5, USA
[12] Campbell, F. C. (2010). Structural Composite Materials, ASM International, ISBN:978-1-61503-037-8, USA
[13] Skovajsa, M., Kroft, R. & Sedlacek, F. (2021). Numerical and experimental modal analysis of a carbon fibre monocoque, MM Science Journal, June 2021, DOI 10.17973/MMSJ.2021_6_2021048
[14] Knye, M., Grill & T., Senner, V. (2016). Flexural Behavior of Ski Boots Under Realistic Loads – The Concept of an Improved Test Method, Procedia Engineering, Vol. 147, pp. 342-347, ISSN 1877-7058

DOI: 10.2507/33rd.daaam.proceedings.060

PRINCIPLES OF CRISIS COMMUNICATION AND INFORMATION VULNERABILITIES IN CRISIS MANAGEMENT

Miroslav Tomsu

This Publication has to be referred as: Tomsu, M[iroslav] (2022). Principles of Crisis Communication and Information Vulnerabilities in Crisis Management, Proceedings of the 33rd DAAAM International Symposium, pp.0429-0436, B. Katalinic (Ed.), Published by DAAAM International, ISBN 978-3-902734-36-5, ISSN 1726-9679, Vienna, Austria
DOI: 10.2507/33rd.daaam.proceedings.060

Abstract

Modern man is constantly under enormous time and financial pressure, increased uncertainty, and a significantly multiplied frequency of change. It must integrate new knowledge or information into the functional flow of personal or professional life. The ongoing crisis further escalates this. Therefore, an information-based society requires each member to pass on the correct information to the right people at the right time. The article discusses information behaviour in crisis management. Above all, he pays attention to the basic principles of crisis communication, describes the tools of crisis communication, and explains the basic concepts of information ecology, which is also crucial in crisis management. The final section deals with information vulnerabilities that can occur in the information environment of crisis management.

Keywords: Crisis communication; information vulnerability; information ecology; crisis management.

1. Introduction

Crisis and emergencies such as conflicts, natural disasters, accidents, terrorist attacks, or also problems of social, economic, political, and psychological problems are an inescapable part of human life. They impact not only the environment in which individuals' function but also the individual's physiological, affective, and social well-being. Crises drastically change the context of all individual and collective actions, including information activities. Crises and disasters disrupt our routines and condition our information behaviour, becoming more dynamic and sometimes chaotic while remaining focused on eliminating the problem to sustain and ultimately overcome the crisis. Information behaviour often becomes extremely varied and subject to many factors due to the sudden nature of a crisis or catastrophe. She tends to be anxious, determined by emotion, and time-sensitive [1]. A crisis is an experience and an event or situation that is usually unexpected, accidental, unacceptable, unprecedented, and often impervious. In addition, it has a significant impact on social beliefs and can reduce trust; it can trigger dangerous responses such as riots and wars that threaten the lives of many communities [2]. A crisis can also be a kind of stimulus that triggers specific information behaviour [3]. The crisis is the context for all possible activities, including solutions [4]. Disaster is often defined by its disruptive impact on entire communities, countries, and the natural environment.

This phenomenon is hazardous: it can cause a fatal disruption of social order and harmony and a sense of helplessness in individuals who lose their sense of security due to damage to property, loss of health, or economic and environmental impacts [2]. Crises and disasters undoubtedly affect a person's informational behaviour. In a crisis situation, habits often change as individuals adapt their active and passive processes of gathering information, seeking and retrieving information, and using or disseminating information. Managing these situations well by individuals, both on the part of the recipients of the information and the part of producers of the information, requires successful crisis communication and management of the transmission of information. Thus, a little-discovered information ecology field infiltrates the crisis management field.

This article aims to broaden the horizons and try to define the intersections of the fields of crisis communication and information ecology and point out the connections. The goal is also to find vulnerabilities in the information environment of crisis communication.

2. Crisis Communication

The academic dictionary of foreign words explains the notion of conflict as a conflict of conflicting, conflicting interests, etc. Conflict is an integral part of our lives, and conflict resolution is one of the most common communication activities in our professional and private lives. The same dictionary interprets the concept of crisis as a moment of maximum contradiction, a decisive break. Disaster is defined here as a disaster, an event with tragic consequences.

There may or may not be a causal link between conflict and crisis. For example, it can be said that an unresolved conflict can turn into a crisis because the balance previously maintained by the ongoing resolution of conflict situations has been lost. An uncontrolled crisis will gain momentum on its own and could turn into a disaster. However, natural disasters, in particular, can strike without warning. However, crises or disasters cannot be seen as single negative phenomena because they can be seen as opportunities and lead to improved measures. [5]

Crisis communication has a similar organizational objective to crisis management during a crisis. However, its primary focus is on gathering, coordinating, and disseminating information and views related to the crisis to stakeholders on time in order to protect and defend the organization in publicly questioning its reputation [6]. Communication is crucial in times of crisis. If crisis management forces did not have prepared crisis scenarios for how to communicate during these situations, they are likely to cause even more damage.

In the absence of adequate internal and external communication:

- The operational response will disintegrate.
- Stakeholders will not know what is happening and quickly be confused and react incorrectly.
- The ingredients will be perceived as incompetent.
- The time taken to resolve the issue fully will be extended.
- The effects on financial losses will be more severe. [7]

Crisis communication aims to provide timely, reliable, and persuasive information at the right time and place to achieve rapid and professional preparedness of all authorities and elements of crisis management (German, 1999).

3. Principles and methods of crisis communication in an integrated rescue system

The primary prerequisite and basis for effective emergency response are to ensure reliable communication between the different components of the integrated rescue system and the authorities involved in coordinating rescue and disposal efforts. For this reason, so-called crisis communication is used in preparation for emergencies and emergencies. Crisis communication, in this case, means the transmission of information between public authorities and components of an integrated rescue system using voice and data transmission via a telecommunications network, public, and selected parts of a non-public communications network.

The Ministry of the Interior provides technical means for crisis communication; it is also obliged to allow crisis communication between authorities and components of the telecommunications network according to the unique implementation of the Ministry of the Interior. Communication service providers are asked to cooperate with the Ministry of The Interior, which determines the continuity of the telephone line service for this number, in the preparation and resolution of means of communication on a single European crisis and emergency number. Anyone involved in crisis communications using appropriate technical equipment should be available within their jurisdiction when the emergency is declared or in their rescue and disposal, providing, where appropriate, statutory synergies.

Crisis communication in the integrated rescue system is organized according to needs into three levels of coordination (tactical, operational, and strategic) between ministries, ministries, other central administrative offices, competent authorities at the level of counties, or administrative offices at the level of districts. Extended municipalities, regional and municipal governments, as well as within each of these entities

For crisis communication, the following are used:

- Telecommunications networks specially designed by the Ministry of the Interior used to secure voice and data communications as well as to connect the integrated safety system's public radio communication network, operate and serve to operate the integrated safety system's public radio communication network;
- An integrated rescue system radiocommunication network set up by the Ministry of Interior for the regular operation of the components;
- A fixed public telecommunications network with priority connectivity;
- public mobile telecommunications network;
- A mobile network reserved and earmarked to ensure links between emergency management authorities and municipalities;
- A radio network to transmit messages to be used in the event of a failure of all technologies;
- Mobile telecommunications networks and equipment can only be permitted if the capacity of standard means of communication is damaged. [8]

3.1. Crisis communication in crashes and disasters

Time plays a vital role in the emergence of accidents and disasters. As a result, sudden and unexpected emergencies are arising with very rapid developments, including irreversible damage to company property and people's health or property damage. In this context, every organization must also be aware of its social responsibility towards its surroundings. Each organization positively and negatively impacts its surroundings through its activities. Crisis communication is seen here as a tool used by the organization in situations where it threatens or harms its surroundings with its activities, such as crashes, by releasing harmful substances. [9]

3.2. Communication in a process crisis

Crisis communication procedural has specific characteristics and differences from communication in accidents and disasters. This is a crucial crisis management tool in times of acute crisis, when a company may face financial problems that may affect not only employees but also contractors, state governments, banks, and other institutions. There is more time for communication in a procedural crisis than in the event of an accident and disaster, but this does not mean that crisis communication should be postponed. [10]

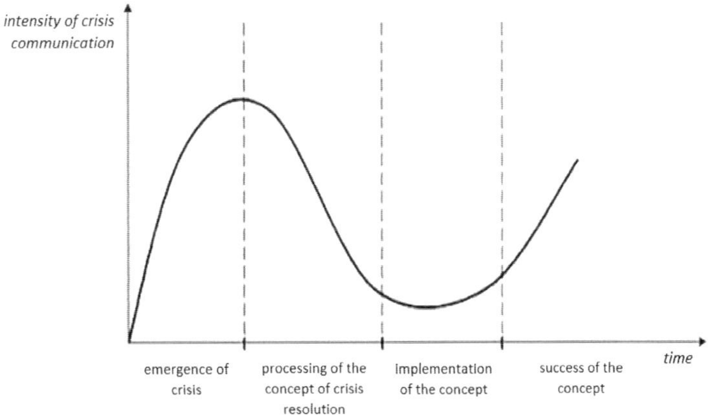

Fig. 1. The course of crisis communication [9]

3.3. Successful Crisis Communication

In rapid expression and subsequent reflection, the most challenging element of crisis communication is a ready response with a reasonable and accurate response. In emergencies, a late or disproportionate response may cause a negative response from the recipient of the information. This can cause embarrassment and lead to legal action by the company and any interested parties. Effective crisis communication is used to explain the impact of a crisis. The impact can also be mitigated by a timely response to, or even before, the crisis. It is, therefore, a positive approach to effective crisis communication. During a crisis, how communication is handled can positively and negatively affect the organization's future. An effective crisis communication program should reflect the organization's requirements (standards) and expectations. This is also reflected in the perception of all target groups. Crisis communication should also include managing the causes of the crisis, helping victims of the crisis, communicating with staff and stakeholders, and communicating with the public in a timely and accurate manner. [11]

Communication aims to pass on as much memorable information as possible to the audience. Knowledge of the other side is the basis of successful crisis communication. Here is a rundown of what is recommended for drafting the report:

- Provide messages that promote kindness, openness, and interaction,
- reports should not contain technical data and information,
- messages should build credibility,
- limit information to three main messages,
- reports must be structured, organized, and concise. [12]

3.4. Crisis Communication Planning

Since the crisis, in most cases, attracts the public's attention, it is necessary to create a crisis communication plan. We must be prepared for a crisis. However, crisis communication cannot be done based on current feelings. It must be managed and based on specific established rules and principles. Planning crisis communication is a long-term process based on estimating the likelihood of adverse events occurring and influencing people's behaviour. It includes activities to develop basic communication strategies and key messages for all target groups.

The general process of planning crisis communication should be as follows:

- Awareness of opportunities in terms of crisis communication objectives, media, social priorities, finances, international commitments, etc.
- Analysis and comparison of possible crises – analysis of the nature of risks, identification of actors in individual communication circles, analysis, and description of possible crises.
- Determination of essential elements of crisis communication – the quality of a relationship, information design, and dialogue design.
- Defining the total area of crisis communication – pre-crisis, crisis, and post-crisis communication.
- Selection and comparison of possible alternatives to achieve the goals.
- Development of a crisis communication plan. [12]

4. Information ecology

Information ecology protects the organization's information environment and its interconnected system of social, cultural, and political subsystems. Within the organization, the information ecology approach affects the information that is generated, stored, and made available. It focuses primarily on the human factor, not just technology. Constant developments and changes in the organization's environment are also considered.

The essence of information ecology is based on classical biological ecology. Therefore, the term can be understood similarly to classical ecology. Furthermore, given the previous definitions, we can see this, on the one hand, as an approach to the protection, study, and application of the principles of information ecology. Furthermore, information ecology can also be understood as an indication of the relationship between elements and components of the information environment and their interactions. [13]

The best approaches to information ecology and understanding can be used where we are most often found, where we play some active social role, hold some function, have some specific opinions, and can comment on what is happening there. It is an environment we know well, and we can recognize the appropriate technologies and techniques we can use to improve the functioning of this environment. It can be a different place for everyone: work, school, home, library, hospital, church, club, community centre, or civic association. Much depends on our local experiences, practices, goals, and values associated with the place, whether as individuals, organizations, or as a group. [14]

4.1. Information behaviour

Informative behaviour includes knowingly seeking information, a range of unintentional or passive behaviour, and behaviour unrelated to information retrieval, such as deliberately avoiding certain information. Information avoidance is a phenomenon that is also studied as part of information behaviour - it is generally thought that when searching for information, people tend to seek out information that is consistent with their knowledge, beliefs, and beliefs, their current beliefs and opinions, and try to avoid contradictory information. On the other hand, a high level of interest in a subject tends to increase the "exposure" of certain information - people with a high level of interest are motivated to gather more information on a topic of interest to them. [15]

We distinguish three levels of related terms. General information behaviour (when using information, communicating, and solving common problems), information behaviour in active and targeted information retrieval, and user exploratory interaction with a system or service. This is the third informative behaviour as an action. [16]

The following model shows this in a simplified way.

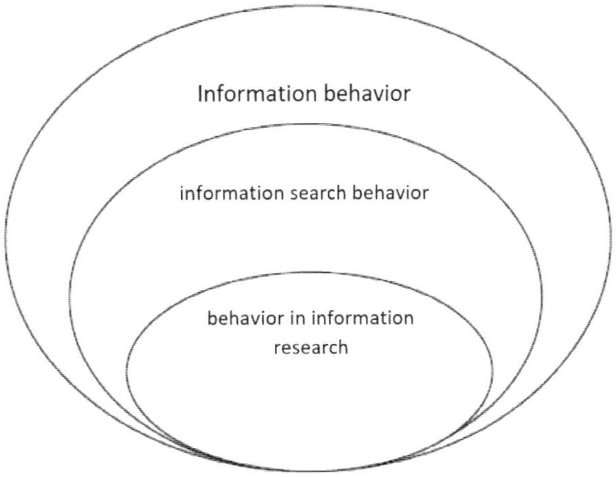

Fig. 2. Model of information behaviour [16]

4.2. Concepts of information behaviour

The field of information behaviour is often associated with many other concepts from the area in which information operates. These include, for example, the area of decision-making (evaluation and selection between different options; identification of problems that concern us) – it is characteristic of some professions, leadership positions, etc., and there is a large amount of research informing the behaviour of these groups. Another concept associated with information behaviour is the problem of information overload. This is a state where an individual cannot process such information. The most common situation where information overload can occur is searching for information on the Internet. Information overload can result from actively searching for information and passively receiving the information we already have without asking for it (spam, advertising, etc.). [15]. When exposed to large amounts of information, the human body reacts in different ways – for example, we choose which information to pay attention to and distinguish between what we need to know and what we "just" want to know or are interested in. By selective selection, we can determine seven different categories – of possible reactions of the body to supersaturation:

- omission – failure to process some inputs and stimuli;
- error – erred in processing information,
- sequential processing (queuing) – deliberate delay in the processing of some information to catch up over time,
- filtering – processing only information that is identified as high-priority information,
- approximation – lowering the standards of differentiation of individual information by reducing the accuracy of the evaluation of inputs and outputs,
- different channels – division of incoming information (inputs) into smaller units in order to divide responses (outputs),
- escaping – completely ignoring incoming information.

Information poverty can be defined as a state of an individual/group in which the following characteristics are fully or partially met:

- poor ability to process information (lack of comprehension or reading, speech/language impairment, hearing or vision impairment),
- the social situation in a minority culture, which (usually) leads to a misunderstanding of information known to the public, reliance on oral and traditional information, or media to transmit information in a more common form;
- fatalistic tendencies or helplessness, reducing the likelihood of positive information behaviour.

The term digital divide refers to the gap between those with regular and functional access to digital and information technology and those without. The approach involves one's technologies but, more generally, the resources and competencies needed to participate effectively in digital communication (e.g., the most severe problem is currently insufficient or underdeveloped infrastructure). [17]

5. Information vulnerabilities in crisis management

Safety plays a significant role in society. One of the crucial elements of an organization is its information environment. In an organization, security ensures that conditions are protected so that it can perform its intended function. A prerequisite for ensuring security is the identification of vulnerabilities in the information environment concerning the security environment and the organization itself.

5.1. Information overload

Decision-makers have limited cognitive ability to process information. Information overload occurs precisely when the amount of information exceeds this capacity. If there is a situation where whistle-blowers are overwhelmed, the quality of their decision-making may decrease. The problem of information overload arose during the Renaissance and industrial revolutions. However, with the information age and access to automated and efficient data collection that provides more information, the severity of this problem has increased.

5.2. Informational multitasking

Some people are very proud of their "multitasking" skills. Especially women, who are said to be doing well because the two hemispheres of their brains are more closely connected. If a person is constantly bombarded with information from different sources simultaneously, he cannot properly concentrate on them. It also affects short- and long-term memory and flexibility in solving new problems. Statistically, such people change jobs less frequently, even if they feel bored in their current job and it does not fulfil their career or financial dreams.

5.3. Inefficient way of working with information

The human brain works with information nonlinearly, using connections and the ability to see connections in a broader whole. Being able to accurately process and quickly find a large amount of information at any time is an increasingly important skill in the work of all managers. The most important thing is to be able to orient yourself, choose an excellent source of information, and meet the needs related to the performance of the task or the solution to the given task.

5.4. Little motivation to process information

Low motivation or demotivation is caused by a sense of threat, manifested mainly by concentration on one's problems, limited concentration on receiving new information, performance, inability to think creatively and find solutions, and searching for "escape" from the environment of receiving, processing, and remembering information.

5.5. Information fragmentation

Fragmentation of information that does not provide the full context of events and their interrelationships, which are generated by many different sources of information, leads to a fragmented perception of reality in many people. This may be the reason for their childlike view of reality. Especially at the moment, this damaging phenomenon is more noticeable and noticeable.

5.6. Chaotic organization of information

The problem is when a worker gets lost in the flood of documents and information and does not know where to find the necessary content or what version of the document is correct. The absence of information when it is just needed.

5.7. Verification of information

Much information appears in the information space. They attack us from all sides, and it is not very easy to navigate in them. Some information seems subliminally to us without us being aware of it. An essential skill of people is to be able to resist influences and to be able to filter out information. In other words, to distinguish between those who are severe and those whose intention is to manipulate.

5.8. Unavailability of information

The absence of information can cause stress and uncertainty and create room for speculation.

5.9. Misunderstanding of information

Misunderstanding can be understood as something often unreflective, something we are unaware of. We have limited means of language and communication to convey what we think and imagine. The transmission of information always entails communication noises. An important aspect is also the information literacy of the individual.

5.10. Communication barrier

In the process of communication (i.e., mutual sharing – communicating/receiving information), obstacles, i.e., communication barriers, are sometimes unavoidable. They interrupt successful communication. The obstacles to communication are as diverse and enduring as human communication itself. In some situations, the obstacles can be so severe that they cause armed conflict.

6. Quantification and evaluation of vulnerabilities

This chapter aims to quantify and evaluate vulnerabilities and their impact on the organization's operation. For the expert assessment, the assessment method will be chosen on a scale of 1-4; this scale will further assign vulnerability values, percentage expression of the problem, and verbal expression.

Degree	Abbreviation	Vulnerability	Expression in %	
			from	to
1	L	low	0 %	25 %
2	M	medium	26 %	50 %
3	H	high	51 %	75 %
4	C	critical	76 %	100 %

Table 1. Degrees of vulnerability and expression of vulnerability

Vulnerability description	Degree
Information overload	4
Informational multitasking	3
Inefficient way of working with information	1
Little motivation to process information	1
Information fragmentation	2
Chaotic organization of information	2
Verification of information	1
Unavailability of information	3
Misunderstanding of information	2
Communication barrier	2

Table 2. Description of vulnerabilities with expression of degree

Degree	Verbal expression
1	The vulnerability does not pose a particular security problem for the organization. The operation of the organization is not compromised. Special measures need not be introduced; they can be consciously accepted.
2	The vulnerability is a security problem for the organization, the operation of the organization is not significantly compromised, but the problem needs to be paid attention to and addressed.
3	A vulnerability is a security problem for the organization. The operation of the organization may be compromised. The problem needs to be addressed and given long-term attention.
4	Vulnerability represents an acute security problem for the operation of the organization. The operation of the organization is threatened. The problem must be solved and eliminated.

Table 3. Degrees of vulnerability and verbal expression

7. Conclusion

This article may be the first step towards a descriptive, conceptual analysis of the current state of information behaviour in crisis management. He gathered existing definitions in the fields of crisis communication and information ecology, the relationship between these disciplines, and created a brief overview of knowledge, a collection of terminology for future research. Informative behaviour in a crisis, which often occurs under the pressure of time, involves evaluating and responding immediately to existing circumstances. Many factors influence the collection and use of information, as well as other activities such as information avoidance and information manipulation; they also change the role of chance in information gathering. Information barriers that can occur in crises and information needs changes are still not given enough attention. Simple quantification and evaluation of vulnerabilities are the basis for further research into different information activities in crises of different types. This article dealt only with selected major crisis management and information ecology chapters and provided the basis for further research.

8. Acknowledgment

This article was supported by grant project IGA/FAI/2021/003 "Supporting the exercise of the Integrated Rescue System and other population protection authorities" - resolved in 2022.

9. References

[1] Allen, D., Karanasios, S. & Slavova, M. (2011). Working with activity theory: Context, technology, and information behavior, Journal of the American Society for Information Science and Technology, Vol. 62, No. 4, pp. 776-788., ISSN 1532-2882, DOI 10.1002/asi.21441

[2] Al-Rahmi, W. M., Alias, N., Othman, M. S., Marin, V. & Tur, G. (2018). A model of factors affecting learning performance through the use of social media in Malaysian higher education, Computers & Education, Vol. 121, pp. 59-72., ISSN 0360-1315, DOI 10.1016/j.compedu.2018.02.010

[3] Krakowska, M. (2020). Information behavior in crisis situations, Zagadnienia Informacji Naukowej - Studia Informacyjne, Vol. 58, No. 2A(116A), pp. 61-85., ISSN 2392-2648, DOI 10.36702/zin.716

[4] Goodman, M., Pang, A., Begam, N. & Chee Yang Chong, A. (2014). Negotiating crisis in the social media environment, Corporate Communications: An International Journal, Vol. 19, No. 1, pp. 96-118., ISSN 1356-3289, DOI 10.1108/CCIJ-09-2012-0064

[5] Holladay, S. J. (2012). The handbook of crisis communication, Wiley-Blackwell, ISBN 978-144-4361-902, Hoboken, New Jersey, United States

[6] Tomsu, M. & Pekaj, R. (2021). Information Support for Crisis Management in the Zlin Region with a Focus on Meteorological Information, Proceedings of the 32nd DAAAM International Symposium, pp.0701-0711, B. Katalinic (Ed.), Published by DAAAM International, ISBN 978-3-902734-33-4, ISSN 1726-9679, Vienna, Austria, DOI: 10.2507/32nd.daaam.proceedings.098

[7] Nemec, P. (1999). Public relations: communication in conflict and crisis situations, Management Press, ISBN 8085943662, Prague, Czech Republic

[8] Horak, R.; Krc, M.; Ondrus, R. & Danelova, L. (2004). Crisis Management Guide for Public Administration, Linde Prague, ISBN 80-7201-471-4, Prague, Czech Republic

[9] Zuzak, R. & Königova, M. (2009). Business crisis management 2, Grada, ISBN 978-80-247-3156-8, Prague, Czech Republic

[10] Zapletalova, S. (2012). Business crisis management for the 21st century, Ekopress, ISBN 978-80-86929-85-9, Prague, Czech Republic

[11] Walker, D. (2012). Mass Notification and Crisis Communications: Planning, Preparedness, and Systems, CRC Press, ISBN 978-80-86929-85-9, Boca Raton, Florida, United States

[12] Vymetal, S. (2009). Crisis communication and risk communication, Grada, ISBN 978-80-247-2510-9, Prague, Czech Republic

[13] Nardi, B. A. & O'Day V. L. (2000). Information Ecologies: Using Technology with Heart Paperback, MIT Press, ISBN 978-0262640428, Cambridge, Massachusetts, United States

[14] Rankov, P. (2006). Information society - perspectives, problems, paradoxes, Koloman Kertész Bagala, ISBN 8089129919, Bratislava, Slovakia

[15] Steinerova, J. (2005). Information behavior: perspectives from information science, Center for scientific and technical information, ISBN 80-85165-90-2, Bratislava, Slovakia

[16] Wilson, T. D. (1999). Models in information behaviour research, Journal of Documentation, Vol. 55, No. 3, pp. 249-270., ISSN 0022-0418, DOI 10.1108/EUM0000000007145

[17] Kuhlthau, C. C. (2004). Seeking meaning: A process approach to library and information services, Libraries Unlimited, ISBN 1-59158-094-3, London, United Kingdom

DOI: 10.2507/33rd.daaam.proceedings.061

RELIABLE INFORMATION SOURCES IN THE AGE OF PROPAGANDA

Miroslav Tomsu

This Publication has to be referred as: Tomsu, M[iroslav] (2022). Reliable Information Sources in the Age of Propaganda, Proceedings of the 33rd DAAAM International Symposium, pp.0437-0443, B. Katalinic (Ed.), Published by DAAAM International, ISBN 978-3-902734-36-5, ISSN 1726-9679, Vienna, Austria
DOI: 10.2507/33rd.daaam.proceedings.061

Abstract

One of the characteristics of the information society is the dramatic, difficult-to-manage increase in information. Nevertheless, there are negative aspects to the publication and information explosion. An increasing number of information and information sources do not guarantee the multiplication of knowledge to date. Not only do they often fail to benefit him, they sometimes harm him. One faces the problem of protecting oneself from misleading knowledge and the pernicious influence of ethically objectionable information. We often fail to navigate the amount of information and, in the absence of time, reliably distinguish information relevant from irrelevant, verified from unverified, and authentic from false. A particularly adolescent individual is not, without prior warning, often prepared to assume that publicly controlled media could provide false information. The article will try at least briefly to highlight tools, methods, or at least intuitive mechanisms that help to qualify the quality of information sources and make it more likely to achieve a higher level of reliability in their selection.

Keywords: Propaganda; information sources; information; information society.

1. Introduction

Spreading fake news is a significant problem because it poses epistemic and practical risks - influencing crucial decisions about one's life (e.g., health, finance, education, or governance), relying on inaccurate beliefs can lead to various harms (e.g., illness, bankruptcy, ignorance, or mismanagement). In response to this phenomenon, scientists have rightly highlighted the role of research in understanding fake news and developing literacy practices that contribute to educating information users. For example, Cooke (2017) argues that research into information behaviour on phenomena such as information retrieval, information selection, information avoidance, and the use of information can help to understand how to differentiate information properly. According to the European NGO Media Literacy for Citizenship, which is dedicated to increasing media literacy, when talking about fake news, we may come across one or more of ten types of misleading news that differ in their motivation and impact: propaganda, clickbait, sponsored content, satire and fraud, errors, partisans, conspiracy theory, pseudoscience, misinformation and false information (Steinberg, 2017). Three of these types of fake news (namely propaganda, guerrillas, and reports on conspiracy theories) form a unique grouping (which, for convenience, I call "propaganda"); they display an affective dimension and involve power relationships in a way that other types of fake news do not. Scientists across disciplines paid particular attention to propaganda fake news and the best practices available for epistemic self-defence.

For example, philosophers were interested in the recent emergence of a post-truth climate in our societies, where "faith may no longer be guided by reality" (Bermúdez, 2018) or "strong opinion is considered more valuable than facts" (Legg, 2018) and is often more or less explicitly discussed in the context of propaganda (e.g., Bermúdez, 2018; Fairfield, 2018; Gelfert, 2018; McIntyre, 2018; Stanley, 2015). These authors draw their intellectual resources and suggestions from traditional philosophical, epistemological frameworks (such as David Hume's empirical philosophy or Charles S. Peirce's semiotic theory) to develop and recommend epistemic strategies that individuals can use to cope with the spread of fake news (e.g., Fallis, 2004; Legg, 2018).

In this context, the article is seen as contributing to integrating these multidisciplinary perspectives on fake news and developing tools for individuals seeking to research them and defend their appeal. I am also examining the extent to which cases of fake news fit into these dimensions. Finally, the article presents several aspects that may serve as a basis for understanding the propaganda uses of fake news, building scientific literacy practices, and helping people acquire epistemic self-defence skills.

2. Propaganda theory

The term propaganda originally had a neutral meaning, defined as spreading an idea to influence public opinion. Over time, however, propaganda begins to be mainly viewed negatively and is associated, or confused, with various negative concepts. "Words often used as synonyms for propaganda are falsehood, distortion, deception, manipulation, brainwashing, and empty speech" [1]. I view the term negatively, mainly in liberal democracies, based on the use of propaganda by past regimes (e.g., fascism and Nazism) of the 20th century as a political tool. It is currently also viewed negatively because it is a hybrid warfare tool. Negatively, propaganda is not only perceived in the Czech language but also in English, Russian or German, for example. However, understanding of the term varies across cultures and languages. For example, propaganda is used as a synonym for advertising in Spanish and Portuguese, and an opposing point of view is absent [2].

The Catholic Church only mentioned the term in the 17th century at the Vatican. Many Protestants were speaking out against the church at the time, and their numbers continued to rise. In parallel, territories were discovered on the American and Asian continents, where the aim was to turn the population to faith by using the Sacred Congregation for the Propagation of Faith created by Pope Gregory XV. The word did not lose its neutrality of meaning until the 1890s when an English author dubbed the leaders of the French Revolution as propaganda disciples. Since then, propaganda has come to be considered, among other things, in a negative sense. This is demonstrated by an 1842 dictionary that presents propaganda as the activity of secret societies seeking to disseminate their opinions and ideas and is a dreaded activity for most governments. He is then mentioned in favorable terms a few years later by the poet Ralph Waldo Emerson, yet the word was not commonly used, as confirmed by the fact that its definition was not included in the Encyclopedia Britannica, which was published in 1911 [3].

Propaganda was not commonly used until the beginning of The First World War. The term has started to appear new in campaigns or printed form, for example, through posters and literature. The purpose was to use a false message to maintain the morale of his country's population and, in turn, to weaken the enemy. Based on this, propaganda has changed its meaning from trying to spread specific ideas to a means of gaining power. Even after the war ended, the propaganda continued to be viewed negatively and associated with acts of war [4]. During World War II, propaganda could be used on an even larger scale, thanks mainly to technological advances. Outside of posters and newspapers, the government began to use new communication channels, which were nothing more than movies and radio. The purpose of the communication was, as with the First World War, to strengthen the country's unity and weaken the enemy [3]. During the Cold War, propaganda became an instrument of ideological warfare. Both powers sought to define themselves against the enemy by making false claims, which propaganda helped them immensely. The aim was to cause fear in society [5]. In previous years, compared to now, propaganda was more recognizable. In the age of modern technology, it has become a mass tool used by the media and social network users. Propaganda thus becomes a professional device that can be very unobtrusive, even unrecognizable [3].

2.1. Definition of the term propaganda

Although the word propaganda has been around in its political sense since the 19th century, it is still relatively difficult to define, and its definitions differ among different authors. According to Paul Baines, propaganda is "in popular language an insidious ideological communicative force that attacks people's minds and is attributed to it more power than it has." [6] The definition is misleading, mainly due to the term "insidious." Sheryl Tuttle Ross provides a more general definition: "Any attempt to persuade anyone with any opinion." [7] While this definition is closer to the word's original meaning or persuasion, it is too general and could include the persuasion of children or other activities.

For comparison with the aforementioned general definitions, we look at an overview of definitions from other authors included, among others, in propaganda Vol. I. and also in dictionaries Oxford Dictionary and Merriam-Webster Dictionary. Oxford Dictionary provides a slightly different definition: "Information, predominantly of a distorted or misleading nature, used to promote political interests or opinions." [8] A certain preconception of the term resurfaces when propaganda is referred to as "misleading" or "distorted." However, propaganda is not necessarily misleading or deceptive.

In the Merriam-Webster Dictionary, the word propaganda is described as follows: "Ideas or statements which are often false or exaggerated and disseminated to help the dispute, political leadership, government, etc.". [9] Other meanings refer to the Congregation mentioned above for the Promotion of Faith, The Dissemination of Information or Opinions to help or Harm an Institution or Individual, as well as accusations disseminated in order to harm the other party [9]. It can be observed that the dictionary points to a more negative side of propaganda, above all in its political sense, above all to damage the other side.

Orbaugh offers a simplified definition of Stanley Cunningham in his publication Propaganda Performed. Cunningham's propaganda is predominantly used by the mass media, exploiting confusion and trying to develop the overall confusion of viewers, creating a system of opinion and presenting itself in the form of information and knowledge, discouraging rationality and encouraging the uptake of unverified information [10].

Looking at the above definitions, it appears that common elements include that this is a deliberate action to influence the mindset, persuade or manipulate the impressions of a specific group of people, thereby enriching the promoter or fulfilling his interest. Although this simplifies, it might be sufficient to understand the basic principle of propaganda.

2.2. Dividing propaganda

There are several divisions of propaganda. It can be divided into denigrating and celebratory. Ftorek divides propaganda into political, economic, didactic, wartime, and ideological. Political propaganda has a field of action in the political sphere. Ftorek describes it as an effort to "gain and maintain political power." [2]. Propaganda is divided by source and origin of information into white, grey, and black. This signals a progressively lower degree of transparency and identification [11].

- **White propaganda** contains the relatively correct information. The source of information is identified. It could be very close to the truth. It seeks to build trust in the final acceptors of information, as it can pay off in the future, and at some point, the built-up trust can bring benefits to the propagandist [12].
- **Black propaganda** uses misinformation and lies that seem plausible. It does not name the source or attribute the authorship to someone else. "Black propaganda is a big lie," including all types of creative deception [12]. It is used, for example, by an anti-government organization.
- **Grey propaganda** stands between white and black propaganda. The source may or may not be stated. The origin and veracity of the information are also unclear. Inaccurate content is supposed to be used for disinformation. Sitting propaganda is prevalent in many spheres of society. Jowett and O'Donnell cite a few examples of where grey propaganda is most prevalent. These are "companies that break the statistics on annual reports, advertisements that suggest a product can achieve results it cannot achieve, films made solely for product placement, and television evangelists who personally store the money they require for religious purposes." [13]

2.3. Propaganda techniques and tools

The techniques and tools of propaganda are intended to increase and enhance the effect of propaganda and its effect on society. For propaganda to be successful, the audience must accept it. Adoption is aided by special techniques to maximize the effect of propaganda.

The Propaganda Analysis Institute (IPA) has named and described essential propaganda tools. The IPA was established in 1937. It was created to educate the American public about the spread of propaganda. The IPA identified seven propaganda tools (name calling, glittering generalities, transfer, testimonial, plain folks, card stacking, bandwagon) (IPA, 1938).

[12] they work with the notion that propaganda, based on persuasion by the audience, promises to help people and meet their needs. A propagandist who responds to the needs of the people he intends to control and tries to meet their needs are always met with a favourable reception. Propagandists target beliefs, values, attitudes, behaviours, and social norms. Jowett and O'Donnell thought targets should evaluate propaganda. [12] described propaganda techniques that differ from the seven techniques described by the IPA Institute.

Donald A. Barclay cites another approach to manipulative propaganda techniques to increase the impact of manipulative messages. Calling the techniques that weaken our information guard [14]. The Information Guard describes it as a decision-making process through which we assess the messages that reach us. When the consumer reads the message, the information guard will evaluate it as either trustworthy or untrustworthy. It also describes techniques that weaken the ability to decode propaganda. Some of the techniques described by Barclay are similar to those described by the Propaganda Analysis Institute [12].

2.4. Misinformation

The emergence and dissemination of disinformation are also linked to the issue of propaganda. First, it is crucial to define what misinformation even means. Misinformation is "false, misleading, false information which aims to influence the judgment and opinion of an individual, multiple persons or the whole community. Significantly, it coincides with fake news as well as some hoaxes or conspiracies" [15]

Fake news is no different from disinformation. It is merely a new name for known disinformation. Contrary to misinformation, hoaxes are spread by chain using social networks or email, often alarms that aim to get messages to as many recipients as possible. Content can vary from amusing to a warning. The main thing is to get as much sharing as possible. Misinformation also differs from hoaxers in that it is created to influence people's opinions, and the purveyors are not just ordinary people but, for example, the governments of the countries in question. As I mentioned in the introduction, disinformation has become part of our daily lives, which we are constantly in contact with thanks to the media. The growth of misinformation was helped by the internet and, subsequently, by social networks. The purveyors of misinformation do not seek to inform the reader objectively and truthfully but mainly to engage [3]. Misinformation differs from misinformation in that it is not disseminated intentionally. It is, therefore, a piece of information that is wrong, false, or unbiased, but the purpose of the purveyor was not to deceive the recipient, as is precisely the case with misinformation. It gets among people by mistake, based on poorly verified data [16].

For the most part, the modern way of disinformation campaigning begins with reporting on a particular scandal spread through the not-so-well-known media. If misinformation is successful, it will become more widely distributed among readers, and more well-known media will take over [17]. An individual's attention is gained by misinformation mainly based on emotions that harassers enjoy. Messages that engage readers based on emotion have a better chance of success and subsequent remembrance. Given that misinformation may not be accepted or believed by the reader, it is highly likely to be remembered anyway. If the reader reencounters misinformation, it becomes more important to him, creating a danger of spreading it [3].

2.5. Summary

As is clear from the above, the meaning of the term propaganda is not entirely easily defined. We conclude that the definitions of the term are not uniform and differ among different authors. The basic principles of propaganda were summarized to give readers an idea of the concept. The primary propaganda model can be described as a deliberate action to influence the mindset, persuade, or manipulate the impressions of a particular group of people, thereby enriching the promoter or fulfilling his interest.

3. Trustworthy sources and trust in the media

The media is an important platform where journalists and professionals meet the general public. It seeks to inform and educate but also influence and manipulate them. Audience perceptions and attitudes determine the success or failure of media exposure. Trust and credibility are essential points. Credibility is central to trust [18], so it is one of the fundamental dimensions of trust. The media's credibility is a crucial prerequisite for the work of journalists and the legitimacy of the communications they present to the public in the form of media outlets. Concept of media credentials j e complex and multidimensional construct [19]. Media credibility research focuses on either the credibility of sources, such as a press officer, organization, or a specific journalist and the credibility of the medium through which the message is sent to the audience [20].

The resource's credibility is based on its integrity, referring to the resource's expertise and competence, ability to make correct claims and logical information based on knowledge, expertise, and professionalism, and the communicator's ability to speak truthfully and be reliable. Trust and confidence are linked and fundamental variables in the relationship between the media and the public [21].

Tsfati defines trust as a relationship over time between two parties: a trusting party8 and an innocent party, an innocent party. In a context, a confidant can be a medium (or a particular journalist), and a confidant is a student who consumes the medium. Trust in the media, according to Tsfati, should be essential for interacting with people and exposing themselves to the media [23]. The term "distrust of the media" then works with the feeling that "the mainstream media is not credible or reliable, that the news media is getting in the way of society rather than helping it" [22]. Tsfati also speaks of the concept of "media skepticism." It also explains to him the belief that journalists are not objective, that they do not always tell the whole story, and that they sacrifice accuracy and precision for personal gain [22]. In the context of trust, it also works with selective exposure, i.e., the assumption that audiences prefer media content that matches their beliefs, values, and opinions. According to Tsfati and Cappella (2003), people skeptical of the mainstream media will read less regular news. They expect these media skeptics to seek alternative news sources compared to their trusted counterparts [18].

3.1. Consumption of media content

Given the wide range of media, it is safe to assume that no one will read daily news on all media platforms. Each user consumes the media at different intensities and according to different habits. What matters is routinization, which is "deeply embedded in the structured practices of everyday life" [24]. Studies have found that users combine media differently [24]. The spectrum of media consumed and the time spent reading news can influence the ability to identify propaganda. Time is a contextual factor. It seems logical to predict that less time spent making decisions will lead to lower-quality decisions [24].

3.2. Media literacy

The term "media literacy" covers several types. We can even relate it to advertising, tv shows, etc. In addition to media literacy, literacy is also informational and digital. Media literacy can be understood as a set of knowledge and competencies to understand media content properly, be it print, television, radio, or internet media, and to create their messages. It is also about the ability to search for relevant media and information. Media education can be understood as the development of knowledge associated with media literacy, spreading awareness of media exposure, and verifying information. Definitions of media literacy vary across literature, and attitudes toward it tend to be ambivalent. William G. Christ and W. James Potter's essay Media Literacy, Media Education, and the Academy (1998) ask whether media literacy is a skill, an accumulation of knowledge, or a specific view of the world [25].

The assessment of the credibility of the reports is linked to the motivation and knowledge of the trustees. As Radic puts it, "the perception of trustworthiness is influenced by experience, knowledge and skills" [21]. It is, therefore, appropriate to examine the evaluation of propaganda texts, including in the context of knowledge of the Czech media landscape. Research by The National Association for Media Literacy Education's Maksla, Ashley, and Craft (2015) sets out a new approach to measuring media literacy. Developing the Hobbs and Frost measurement tool, they ask questions about respondents' knowledge of the media landscape in U S A. They also inquire about current affairs in the country. They build on research into media skepticism and trust [18].

4. Credibility, relevance, reliability of information and information resources

4.1. Link to source of knowledge

If a scientific article lacks references to sources, the reader may consider the text a consideration, a featurette, or an otherwise popularizing article, or worse, an attempt to feed into the results of the actual researcher's work. A reader who wants to get to the bottom of the matter should never rely on a single source of information but verify that a link (if one exists) is quoted accurately or that the author of the article has edited it to his advantage. A purposeful reader should strive to find a network of references to a particular prime source (such as Einstein's theory of relativity) from which the other theories derive and to arrive at its current implications from that source. Thanks to the hyperlink network, the Internet fully supports this option [26].

4.2. Resource Reliability Guarantee

The author's name and the institution he represents are essential indications of the document. The author's credibility can be verified by consulting an encyclopedia, a Dictionary like Who is Who, or by asking a professional authority. A scientific sector - Scientology - that uses exacting methods to identify factors that might automatically (for intellectually otherwise unmanageable amounts of information) distinguish quality information from poor quality information. Scintigraphy (a.k.a. infantry, naucommetry, etc.) represents a method of determining the quality of scientific work, which relies on the simple, intuitive assumption that the more the scientific community notices and cites a particular publication, the greater the value of the publication. The "value" of a document is the sum of the document values that refer to that document divided by the number of references that lead from the document. In order to obtain a high rating, a large number of references from low-level documents to a given document or even a smaller number from important documents is sufficient. The reliability of the information source can be inferred from the quotient2 found by the so-called impact factor (IF) of the journal in which the information is published. The IF value indicates how many times an article in a particular journal has been quoted in the first two years. A scientist's effort should be to publish in the highest quality journals with the maximum IF [26].

4.3. Architecture structure arrangement of ideas and arguments in the article

This can signal the appearance of apparent inconsistencies and contradictions in statements and allegations. The initial clue is the reader's intuition, which, through his experience and his acquired knowledge, can warn him of obviously unreliable information. The school supplies students with reliable and guaranteed knowledge fixed in encyclopedias and textbooks. However, in his private life, in contact with peers, adults, and the media, the student acquires experience that is often at odds with the lessons learned. However, even incoherent knowledge can become a base upon which a student can lean. Contradictions with the facts of the text and knowledge gained in natural science, natural history, physics, etc., are, despite their incompleteness, the first clues in distinguishing credible and implausible information [26].

4.4. Rule on economics of thought (Occam's razor)

Without this criterion, there would be no reliable and decisive way to choose between the different hypotheses. Occam's razor believes that if there are two or more explanations for a phenomenon, the simplest should be chosen. Often, ineffectual dissection and obfuscation are a symptom of theoretically low-benefit work. The rule recommends addressing what is essential, forcing authors to rely on substantiating facts and use substantiated arguments [26].

5. Rules for verifying information

In the past, responsibility for the veracity of the information and its verification was placed on the publisher of a particular medium, which had to be assessed by the ethical codes of reporting and audience. In principle, what is written is true. This is given (in the original context, it applied to contracts). Most of the population was primarily the recipients of the information, not its creators. Producing the information also required finance (capital), without which expanding the content en masse was impossible.

With the advent of the information society and the opening-up of the world of the internet and social networks, all barriers have fallen away - producing content can be produced by anyone at any time, without the need for funding and without the requirement to verify the information. Only some information is taken from verified sources - press agencies and offices. The reader is currently also the creator/producer of content (blogs, Youtube channels, alternative sites), which makes him more demanding - he must be able to judge which information is accurate and which is false.

5.1. Common sense

Very often, concerning misinformation, we are faced with the view that, quite simply, common sense will help us assess the veracity of information. However, this cannot rely upon 100% of the information disseminated online.

5.2. Authorship

We verify whether the author is an expert in the given field, e.g., according to its performance as a university facility or according to the library authors database, as well as whether the documents have undergone peer review assessment by other experts.

5.3. Publisher's reputation

We assess whether the publisher is listed, whether it is an educational organization or a professional company, which is one of the guarantees of trustworthiness.

5.4. Breadth and depth of coverage of the issue

The information should not only be objective but should also include all contexts and contexts. The information provided should be of a high standard.

5.5. Topicality

We are looking at whether the creation date is given. We note whether links can be found to the sources, whether there are also expert sources in the links, and whether the bequeathed information is still valid.

5.6. Objectivity of information

The description tries to be as close as possible to its subject and thus limit the influence of the recognizing person as much as possible. It is desirable precisely because it does not depend on persons and is therefore valid, applicable, and acceptable to everyone. Minimal switching and trying to influence the reader.

5.7. Verifiability of sources

We check that the originator of the information indicates the sources from which it is drawn and that direct citations are given when the information is received.

6. Conclusion

Propaganda in various forms has been part of war conflicts for centuries. Today, in the age of the internet and the multitude of unverified information, the issue of propaganda is becoming ever more topical. One of the features of the internet makes it very easy for a single person to send the same message to hundreds, thousands, and even millions of people, making the internet almost instantly a compelling and potentially dangerous means of spreading propaganda. Nevertheless, new media theorists often neglect the propaganda work carried out through mass media or in connection with PR and advertising. With the development of the information society and the growing popularity of the internet, we must admit that our lives depend on the daily flow of information. Probably this fact is not usually widely perceived by people, but essential information can be of priceless value. Information obtained in a web environment can save a life, prevent material damage, and ease a lifestyle. For companies, making the right decision means money or other benefits. They say that a message becomes valuable if it quickly comes at the right time and its content matches reality.

Uncovering unserious approaches to the truth of a fact does not always have as conclusive a means as those mentioned in this article. The mimicry used by some authors manages to hide actual knowledge so successfully that even a severe scientist can be deceived. I want to go from the unequivocal and easily conclusive documents we have presented to cases in which scientific truths are drowned out by the commercial interests of individuals (or entire corporations) who, in their interest, are even willing to turn the results of science against themselves. Defending against these approaches requires more patience and radical approaches on the part of science advocates.

7. Acknowledgments

This article was supported by grant project IGA/FAI/2021/003 "Supporting the exercise of the Integrated Rescue System and other population protection authorities" - resolved in 2022.

8. References

[1] Jowett, G. & O'Donnell, V. J. (2018). Propaganda & Persuasion, SAGE Publications, ISBN 9781506371320, Los Angeles, California, United States

[2] Ftorek, J. (2017). Manipulation and propaganda, Grada, ISBN 978-80-271-0605-9, Prague, Czech Republic

[3] Gregor, M. & Vejvodova, P. (2017). The best book about fake news, disinformation and manipulation, CPress, ISBN 978-80-264-1805-4, Prague, Czech Republic

[4] Fellows, E. W. (1959). Propaganda: History of a Word, American Speech, Vol. 34, No. 3, pp. 182-189

[5] Taylor, P. M. (2003). Munitions of the mind: A history of propaganda, Manchester University Press, ISBN 9780719067679, Manchester, England

[6] Baines, P. & O'Shaughnessy, N. J. (2013). Propaganda, SAGE Publications, Thousand Oaks, California, United States

[7] Ross, S. T. (2002). Understanding Propaganda: The Epistemic Merit Model and Its Application to Art, Journal of Aesthetic Education, Vol. 36, No. 1, ISSN 00218510, DOI 10.2307/3333623

[8] Propaganda – Definition of Propaganda in English from the Oxford Dictionary. Oxford

[9] Propaganda. Merriam-Webster

[10] Orbaugh, S. (2015). Propaganda Performed: Kamishibai in Japan's Fifteen-Year War, Z Japanese Visual Culture, Vol. 13, p. 7., ISBN 978-90-04-24944-8

[11] Kunczik, M. & Schweitzer, E. J. (2008). The use of public opinion research in propaganda, The SAGE handbook of public opinion research, pp. 496-503

[12] Jowett, G. S. & O'Donnell, V. J. (2019). Propaganda & Persuasion, SAGE Publications, ISBN 9781506371344, Los Angeles, California, United States

[13] Jowett, G. S. & O'Donnell, V. J. (2015). Propaganda & Persuasion, SAGE Publications, ISBN 9781412977821, Los Angeles, California, United States

[14] Barclay, D. A. (2018). Fake News, Propaganda, and Plain Old Lies: How to Find Trustworthy Information in the Digital Age, Rowman & Littlefield Publishers, ISBN 9781538108895, Lanham, Maryland, United States

[15] Nutil, P. (2018). Media, lies and too fast a brain, Grada, ISBN 978-80-271-0716-2, Prague, Czech Republic

[16] Fetzer, J. H. (2004). Disinformation: The Use of False Information, Minds and Machines, Vol. 14, pp. 231-240

[17] Reifova, I. (2004). Dictionary of media communication, Portal, ISBN 978-80-7178-926-7, Prague, Czech Republic

[18] Tsfati, Y. & Cappella, J. N. (2003). Do People Watch what they Do Not Trust?: Exploring the Association between News Media Skepticism and Exposure. Communication Research, Vol. 30, No. 5, pp. 504-529., ISSN 0093-6502, DOI 10.1177/0093650203253371

[19] Berlo, D. K., Lemert, J. B. & Mertz, R. J. (1970). Dimensions for evaluating the acceptability of message sources. Public Opinion Qu-arterly, Vol. 33, No. 4, pp. 563-576., ISSN 0033362X, DOI 10.1086/267745

[20] Golan, G. J. (2010). New Perspectives on Media Credibility Research. American Behavioral Scientist, Vol. 54, No. 4, pp. 3-7., ISSN 0002-7642, DOI 10.1177/0002764210376307

[21] Heath, R. (2013). Encyclopedia of Public Relations, SAGE Publications, ISBN 9781452240794, Los Angeles, California, United States, DOI 10.4135/9781452276236

[22] Tsfati, Y. (2010). Online News Exposure and Trust in the Mainstream Media: Exploring Possible Associations. American Behavioral Scientist, Vol. 54, No. 1, pp. 22-42., ISSN 0002-7642, DOI 10.1177/0002764210376309

[23] Taneja, H., Webster, J. G., Malthouse, E. C. & Ksiazek, T. B. (2012). Media consumption across platforms: Identifying user-defined reper toires. New Media & Society, Vol. 14, No. 6, pp. 951-968., ISSN 1461-4448, DOI 10.1177/1461444811436146

[24] Fitzgerald, M. A. (2005). Skills for Evaluating web-based information. Internet Credibility Symposium, University of Washington, Seattle

[25] Almazova, N.; Barinova, D. & Ipatov, O. (2018). Forming of Information Culture With Tools of Electronic Didactic Materials, Proceedings of the 29th DAAAM International Symposium, pp.0587-0593, B. Katalinic (Ed.), Published by DAAAM International, ISBN 978-3-902734-20-4, ISSN 1726-9679, Vienna, Austria, DOI: 10.2507/29th.daaam.proceedings.085

[26] Vymetal, J. (2000) Current information environment. Universe, Vol. 79, pp. 335-336., ISSN 1214-4029

DOI: 10.2507/33rd.daaam.proceedings.062

THE CURRENT TECHNICAL CONCEPTION OF VIRTUAL AND AUGMENTED REALITY

Valášek Pavel, Milánek Adam & Drofová Irena

This Publication has to be referred as: Valasek, P[avel]; Milanek, A[dam] & Drofova, I[rena] (2022). The Current Technical Conception of Virtual and Augmented Reality, Proceedings of the 33rd DAAAM International Symposium, pp.0444-0451, B. Katalinic (Ed.), Published by DAAAM International, ISBN 978-3-902734-36-5, ISSN 1726-9679, Vienna, Austria
DOI: 10.2507/33rd.daaam.proceedings.062

Abstract

This article discusses the overall conception of virtual and augmented reality considering modern technological approaches to the practical application of these concepts. The main aim was to assess available technology in respective fields of simulation. It analyses a consumer approach to the general ideas of these phenomena. A basic technical overview is included. The current state is reviewed using commercial sources of devices focused on delivering stimuli necessary for sensory experiences connected to simulated environments. A short discussion of future development concludes the presented study.

Keywords: virtual reality; augmented reality.

1. Introduction

As the main aim of the associated longer study, the suitability of augmented and virtual reality as a training tool for police forces is being researched. The idea of a simulated or semi-simulated environment appears to be suitable for training tasks of dangerous nature or in a hazardous environment. An example of learning and training activities in virtual reality already in use can be found in NASA's EVA (extravehicular activity). In its complexity, it represents tasks outside the relative safety of a cosmic vehicle, notably exterior repairs. Mistakes and out-of-ordinary events during these activities - possible on-site training included - may increase risks the astronaut and the whole crew are already facing. From a logistical point of view, it is unwise to perform the exercise of this character on the actual objects. In this case, we try to replicate an utterly different environment and its attributes, i.e., gravity or optical conditions. [1]

Virtual reality can also be used to simulate known and accessible environments. In these environments, we focus on the simulation of possibly dangerous situations. As a typical representative of this fact, training of combatant and non-combatant military units is suitable. Exercises in the simulated environment called "Virtual Battlespace" by Bohemia Interactive Simulations aim at tactic and procedural processes in the context of combat situations. [2]. Following the stated logic, we can deduce the suitability of these technologies for exercises and training in selected aspects of police work. Before suggesting an applicable simulation training model, it is appropriate to assess the current condition of available simulation means and define attributes of widely used terms virtual reality and augmented reality.

2. Virtual reality

The literal definition of the name "virtual reality" may seem like an oxymoron, for it combines words with the meaning of "pretended, unreal", and "real". This conceptual mismatch may be explained using the practical application of virtual reality, where its "virtual" or pretended part, meaning abstract environment and objects, overwrite perception of the real world. It would be more precise to call this concept a "substitute reality ". [3]

To "overwrite" perception of the real world, we must establish what senses are used to perceive reality and what to overwrite to achieve the desired effect. In modern practical applications, the interaction between the user and substitute reality can be divided into sensory and motor levels. [3]. On a sensory level, we can discuss what senses are used to perceive the real world (sight, hearing, touch, etc.) Motor level deals with ways of environmental interaction of the user - position and orientation of the user as well as his direct interaction with objects of a simulated environment. [3]

Based on this analysis, a conceptual description of the defined cyberspace of the virtual reality user can be formulated. The user is in a virtual environment, which he perceives using a motor interface. A motor interface captures his acts (movements and gestures) and transfers them into a computational unit (computer), where they are interpreted as purposeful changes in an environment. The same unit evaluates changes in an environment and mediates them to a user using a sensory interface. [3]. Environments with such capabilities may correspond with the physical properties of the real world and thus be considered a type of real-process simulation. By overwriting some senses, a simulated environment and its processes are perceived. A user can feel his presence in a simulated environment. His perception is "taken" out of the real world, and we may claim that the user is in a space defined as a virtual reality.

2.1. Overwritten senses

When overwriting a sensory perception of the real world to achieve a feeling of presence in a simulated world, it is necessary to limit the perception of the real world. The current practical use of virtual reality doesn't require overwriting all senses. By practical use, we mean commercially available resources of virtual reality. The required hardware is worn directly by a user. It integrates sensory and motor interfaces of the user's head and usually the motor interface of the user's hands. In figure no. 1, we can see such a device under the consumer branding "Oculus Quest 2". [4], [5], [6], [7]

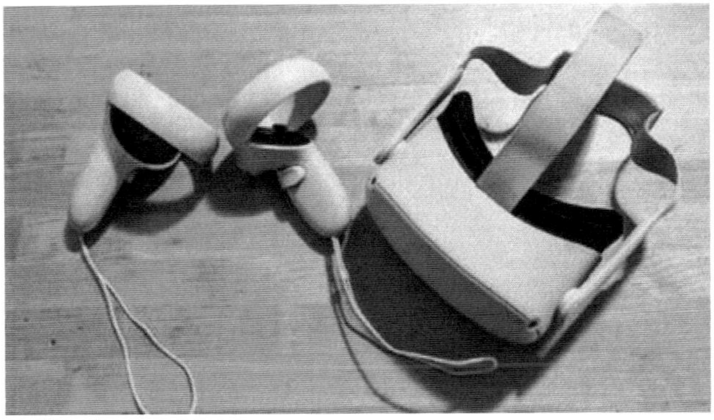

Fig. 1. Ocullus Quest 2 [9]

The primary aim is visual and hearing projection of a simulated environment. [4], [5], [6], [7]

2.2. Hearing

Hearing overruling in the environment of discussed purposes doesn't require customized hardware. Already high demands on audio projection, including dampening sensory perception of the real world, allow its utilization for virtual reality. One of the attributes concerning an undamaged auditory system is the ability to distinguish the direction, in some cases even distance, of the sound source relative to the observer. Simulation of an environment with a natural feeling requires focusing on this fact. Based on current knowledge, we can claim that humans use the following acoustic phenomena [8]:

- **Time difference**: If the sound source is not directly in front of or behind the observer, each auditory system pair organs receives acoustic waves with different feedback or interauricular time difference. [8]
- **Intensity difference**: Suppose the sound intensity lowers with a longer distance to the auditory organ. In that case, the position of the sound source out of the direct view axis of the observer can be perceived as a difference in incoming sound intensities. [8]
- **Acoustical shade**: If there is an obstacle between a sound source and the observer, the acoustic attributes of the said obstacle are usually exhibited as dampening or reflection of acoustic waves. [8]

A human perceives these observed acoustic phenomena in a close connection with visual perception. In some cases, previous experiences with an expected sound source may influence perception experience. For example, the assumed intensity of the sound source can affect its perceived distance. While using appropriate recording technology and techniques, necessary acoustic characteristics of sound can be recorded and reproduced in a way simulating surround sound. It can be achieved using multiple loudspeakers, as in the 10.2 configurations, meaning ten loudspeakers and two subwoofers (Fig. 2.). [8]

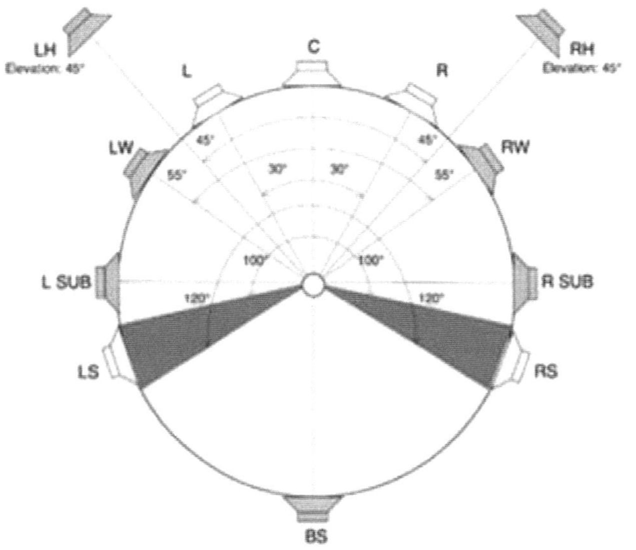

Fig. 2. Spatial sound projection in 10.2 configurations [8]

In this scheme, individual loudspeakers are marked with a corresponding direction letter L (left) and R (right), excluding the center speaker (C) and back surround speaker (BS). All other loudspeakers are marked with a letter determining their acoustic character. H means high speaker, ideally above the observer, W labels wide acoustic spectrum speakers, SUB subwoofers, and S for surround speakers.

A comparable effect may be achieved using headphones. In this case use of technology is limited by the size of the reproductory device. Fig. 3. shows a chart of an earphone shell equipped with multiple drivers. Using this method, we can achieve the desired acoustic range to simulate the sound experience in its original environment. [8]

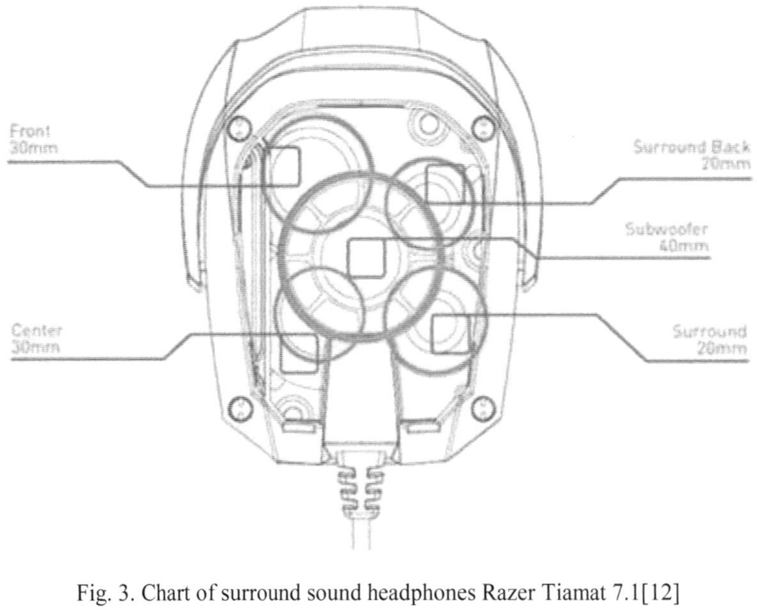

Fig. 3. Chart of surround sound headphones Razer Tiamat 7.1[12]

2.3. Sight

As for visual perceptions, higher levels of sensory overwrite can be achieved more efficiently while real-world visual perception is blocked altogether. For this purpose, a harness-like device is used. Its main goal is to hold displaying device in the desired position for the user and block his visual stimuli of the real world. This set is called a "head-mounted display" (HDM). Apart from display or displays and optics, it also includes various hardware, depending on before mentioned technology conceptions. [7]

Sight may be considered the primary human sense for spatial orientation. Approximating the distance between observer and object and a similar practical distance estimation between two things allows humans basic spatial orientation. Said tasks are processed using the following principles: [8]

- **Static distance estimation**
  - These are common elements for a static and dynamic scene. It includes basic physic and geometry optic phenomena suggestive of an object's distance. The brain perceives objects' relative position, the picture's size, linear perspective, shades, brightness, and changes in texture perception. [8]
- **Movement parallaxes**
  - Perception of a speed difference between objects at various distances. [8]
- **Physiologic indexes of distance**
  - An eye of a person, looking at objects at various distances, is moved by respective muscles to achieve an optimal picture. Concerning distance, these movements are:
    - ***Accommodation***
      - Bending of the eye lens using muscles. These muscles are registered by the brain and utilized to estimate the distance of the observed object. [8]
    - ***Convergence***
      - Convergence of eyesight. The whole eyeball is moved, and this process is again registered by the brain and interpreted as an index of the distance between the object and the observer. [8]
- **Stereoscopic sight**

Physiologic eye placement allows for a slightly different picture perception by each eye. It varies in the angle between the eye and the observed scene. A three-dimensional image is perceived by a subconscious combination of these different pictures. [8]. Knowledge of these principles can be used in creating an artificial digital scene where its purpose is for a user to perceive this scene as a fully three-dimensional space. Generally, the difference between the perception of each eye of the observer needs to be determined, and accordingly, various images should be presented to them.

This task can be achieved using multiple methods can be used: [8]

- **Passive**
  - ***Anaglyphic method***
    - The principle of this method uses a combination of two pictures with vastly different colors and shifted perspectives. While watching such pictures with accordingly tinted lenses, a three-dimensional image is imagined. This method requires a strict color palette, which narrows its applications. [8]
  - **Polarization method**
    - A projected image includes scenes for both left and right eye. While observing such pictures through dedicated polarization filters, each picture is perceived by the corresponding eye. Contrary to the anaglyphic method, color use is not significantly limited. [8]
- **Active**
  - ***Active shutter method***
    - An observer is subjected to alternating scenes for the left and right eye. He wears goggles with liquid crystal shutters (LCD shutters) that are synchronized with a projection device allowing only images for each eye to be seen by a respective eye, left or right. [8]
  - ***Direct calculation***
    - Modern HMDs contain two separate displays (in some cases, one display may be used for both eyes, such as Google Cardboard). This way, the device can project individual pictures directly to each eye. Principles of spatial awareness and orientation are used for modeling a realistic scene.

2.4. Construction of HMD

The construction of HDMs varies between manufacturers, depending on used conceptions and the hardware price range. The following scheme (Fig. 4.) shows the schematic principle of HDM construction.

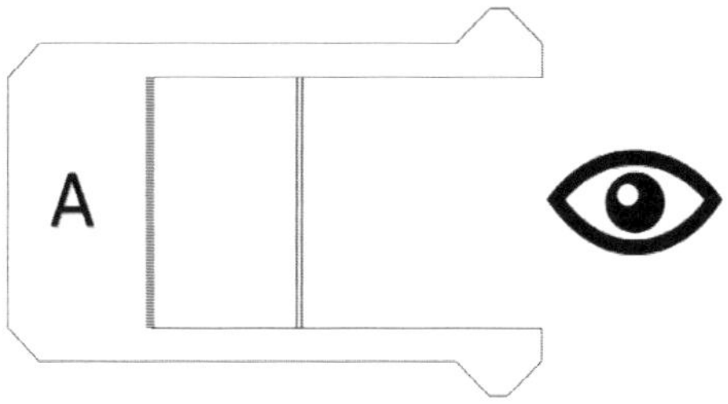

Fig. 4. Construction of virtual reality HMD

Depending on the individual type of virtual reality device, HMD contains most of its hardware in the area labelled A in Fig. 4. Based on used hardware, especially considering "standalone" versions of HMDs (these include naturally heavy technology, such as batteries) an ergonomics of the device and interaction with a simulated environment can be affected negatively. Due to its placement on the head of the user, this limitation can be more profound during more extended use. The outer shell of the HMD usually contains the device's interface and sometimes tracking cameras. [7]

HMD also contains display or multiple displays (in Fig. 4. marked with green colour) and lenses (in Fig.4. marked with blue colour.) Distance between lenses and eyes depends on the preferences of individual users as well as their respective sight irregularities. The distance between the lens and the display is not permanently fixed. It can vary based on many factors. Predispositions of the users can be one of them. Therefore, some HMDs allow for an adjustment of this distance. The distance can also be different with the used lens and display type. This distance is in the order of centimeters. [9]. The display unit and user's eye isolation achieve more realistic immersion and limitations of real-world perception.

2.5. Examples of current HMDs

	VRgineers XTAL 3 Mixed Reality	**Pico Neo 3 Link**	**Varjo Aero**	**Skyworth W1 Pro**	**Huawei VR Glass 6DoF**
		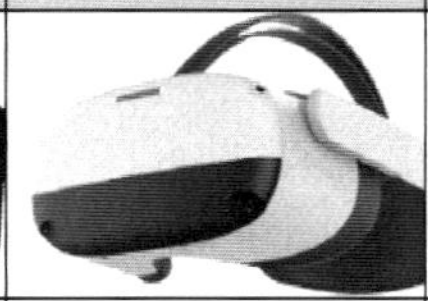			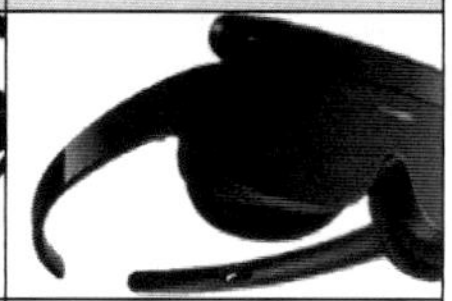
Computing Hardware	PC	Standalone	PC	PC	Smartphone, PC
Release	31. 5. 2022	24. 5. 2022	20. 1. 2022	31. 12. 2022	17. 11. 2022
Optical correction	-	Fresnel lenses	Aspheric lenses	Flat lenses	-
Real world view	Yes, HD color	Yes	No	No	-
Display type	2x LCD	1x LCD	2x Mini LED	2x LCD	2x LCD
Subpixel layout					
Resolution	3840x2160 / eye	1832x1920 / eye	2880x2720 / eye	1600x1600 / eye	1600x1600 / eye
Refresh rate	120 Hz	90 Hz	90 Hz	72 Hz	90 Hz
Visible field of view	180° horizontally 90° vertically	98° horizontally 90° vertically	102° horizontally 73° vertically	94° diagonally	90° diagonally

Preferential rendering	-	No	Dynamic in direction of view	No	No
Motion tracking	6 degrees of freedom, through HMD	6 degrees of freedom, through integrated cameras	6 degrees of freedom, through external stations	6 degrees of freedom, no position tracking	6 degrees of freedom, through integrated cameras
Eye tracking	Yes	No	Yes	No	No
Hand tracking	No	Yes	No	No	No
Controllers	-	2x 6 degrees of freedom	-	1x 3 degrees of freedom	2x 6 degrees of freedom

Table 1. Selected commercially available virtual reality headsets

3. Augmented reality

If we can place (at least visually) virtually simulated objects, we can talk about augmented reality (AR). Unlike virtual reality, perception of the real world is not entirely blocked. Instead, it is expanded by adding simulated objects from a virtual environment. Depending on the desired application of augmented reality, these expansions may extend information about the real world. We can say that the virtual environment of augmented reality copies and expands the natural world. [10]. Without the need to limit real-world perception, two main options can be deduced:

- **Handheld devise with a camera and display**
  - Such a device should be able to process an image in real-time. An essential part of processing is the addition of simulated objects. In consumer applications, we see smartphones and tablets primarily.
- **HMD**
  - The main difference between virtual and augmented reality headsets is the inherited need to perceive the real world and simulated environment. The most utilized concept is, therefore, a see-through display. The basic scheme of such HMD is in Fig. 5.

3.1. Construction of HMDs

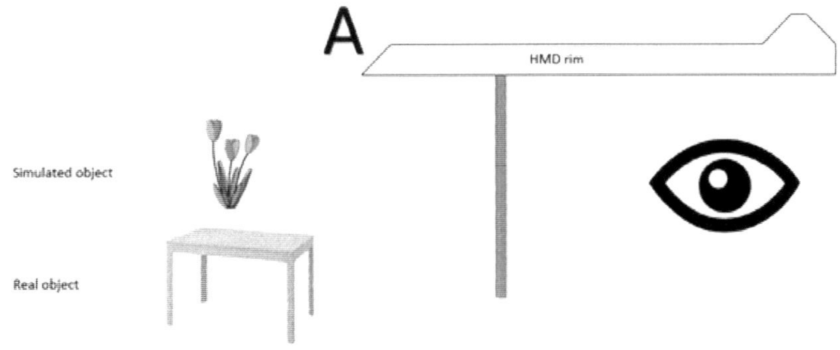

Fig. 5. Construction of augmented reality HMD

HMD-type augmented reality devices strive to achieve the best user comfort concerning the perception of the real world. The idea is for the user to be limited in the least way by the HMD hardware. For this reason, most HMDs have the look and feel of glasses with a thick rim around the user's head. This rim holds inside necessary hardware. In its front (labelled A in Fig. 5.), there are sensors - usually cameras - assessing their surroundings and providing the required data for the correct image of the simulated object. There is a see-through display device (green rectangle with a blue contour in Fig. 5.) between the real world (object) and an eye of a user. It shows an appropriately distorted image of a simulated object to the user. Distortion aims to liken size, proportion, and perspective to natural conditions. [7]

3.2. Examples of current HMDs

	VRgineers XTAL 3 Mixed Reality	**Pico Neo 3 Link**	**Varjo Aero**	**Skyworth W1 Pro**	**Huawei VR Glass 6DoF**
					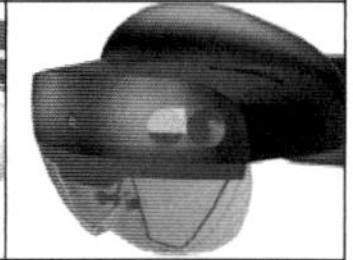
Computing hardware	Smartphone	Smartphone, tablet, PC	Standalone	Standalone	Standalone
Release	4. 3. 2022	4. 3. 2022	3. 3. 2022	24. 1. 2022	7. 11. 2019
Display type	2x Micro OLED	2x Micro OLED	Single Micro-LED		2x LBS
Subpixel layout					
Resolution	1920x1080 / eye	1920x1080 / eye	640x480 / eye	1920x1080 / eye	1440x936 / eye
Refresh rate	90 Hz	75 Hz			60 Hz
	VRgineers XTAL 3 Mixed Reality	**Pico Neo 3 Link**	**Varjo Aero**	**Skyworth W1 Pro**	**Huawei VR Glass 6DoF**
Visible field of view	46° diagonally	43° diagonally	28° diagonally	90° diagonally	43° horizontally 29° vertically 52° diagonally
Motion tracking	3 degrees of freedom, no position tracking	3 degrees of freedom, no position tracking	3 degrees of freedom, no position tracking	-	6 degrees of freedom, through integrated cameras
Eye tracking	No	No	No	No	Yes
Hand tracking	No	No	No	No	No
Controllers	No	No	No	1x 3 degrees of freedom	No

Table 2. Selected commercially available augmented reality headsets

4. Conclusion

In this study, we have tried to assess the currently available technology and terminology of simulated environments, namely virtual and augmented reality. Following formulated ideas and views are formed according to the presented findings. Currently, options for virtual and augmented reality simulations are increasing. As for now, it is possible to create visual and auditory simulations complemented with motoric elements. As can be observed throughout currently available HMDs for augmented and virtual reality, as presented in tables 1. and 2., the development of consumer technology is directed at improving visual perception. [13], [14]. Considering the overall topic of the overarching study, the immersion achieved so far is unsuitable for training physical aspects of police work. Lack of motoric feedback from environment interaction, considering mainly the sense of touch, renders such simulated environments more appropriate for managerial and logistic aspects of police work. Our future research will consider virtual and augmented reality options in this branch of police departments and organization elements. Options for other senses simulations are being researched to increase immersion in a simulated environment. Other senses possibly simulated in discussed virtual environments are touch and smell. [15], [16]

Consumer touch simulation devices focus on force feedback created by pneumatic pressure. To the user, this gives a feeling of being touched or hit. More precise technology, using actuators and allowing for the feel of handling a presented object, is still being developed. [17]. Olfactory simulation relies on aroma cartridges being excited upon pre-programmed events. Simulated smells are primarily used in a therapeutic context. [16], [18]

Both technologies are required for deeper simulation considering police forces and emergency services training. Unfortunately, from available open sources, it seems that the commercial part of the technology segment is not momentarily focusing on their further development. Simulation of olfactory senses appears to have a stable place on the market. The current conception of the touch simulation does not appear to be widely spread, but unlike in the case of smell, more research towards new concepts is more active. [15], [16], [17], [18]

5. Acknowledgments

This article was supported by an internal grant project of Tomas Bata University under the number IGA/CebiaTech/2022/004.

6. References

[1] Loftin, R. B. 1994 "Virtual environments for aerospace training," Proceedings of WESCON '94, 1994, pp. 384-387, DOI: 10.1109/WESCON.1994.403567.

[2] VBS4 [online]. [cit. 2022-03-24]. Available at: https://vbs4.com/

[3] FUCHS, Philippe, Guillaume MOREAU a Pascal GUITTON, ed. Virtual reality: concepts and technologies. 2e édition. Boca Raton, FL: CRC Press, 2011. ISBN 04-156-8471-4.

[4] Oculus for developers [online]. Meta, 2022 [cit. 2022-07-10]. Available at: https://developer.oculus.com/resources/oculus-device-specs/

[5] VIVE developers [online]. HTC, 2022 [cit. 2022-07-10]. Available at: https://developer.vive.com/resources/vive-wave/documentation/

[6] Google VR [online]. Google, 2022 [cit. 2022-07-10]. Available at: https://arvr.google.com/intl/cs_cz/cardboard/manufacturers/

[7] VR Compare [online]. Rory Brown, 2022 [cit. 2022-07-10]. Available at: https://vr-compare.com/

[8] JUN, Lukáš. Virtual reality in the design of production machines, systems, and robotics. Brno, 2017. Bachelor thesis. The Brno University of Technology. Thesis supervisor Doc. Ing. Radek Knoflíček, Dr.

[9] Google Cardboard Technical Specifications [online]. Google, 2015 [cit. 2022-07-11]. Available at: https://gstatic.com/cardboard_assets/cardboard_manufacturers_kit.zip

[10] HÁJEK Petr, DARMOVZAL Tomáš, POSLT Jan: Current options of augmented reality, PA158 Computer graphics - workshop, 2003

[11] Meta Quest 2 Review [online]. PCMag, 2022 [cit. 2022-07-11]. Available at: https://www.pcmag.com/reviews/oculus-quest-2

[12] Razer Tiamat: First Legit 7.1 Surround Gaming Headset. Gizmodo, 2018. Available at: https://www.gizmodo.com.au/2011/08/razer-tiamat-the-first-legit-7-1-surround-sound-gaming-headset/

[13] Drofova, I[rena]; Adamek, M[ilan]; Sousedikova, L[ucie]; Malatinsky, A[dam] & Valasek, P[avel] (2021). Comparison of the Lighting Condition of the Interior to Create a 3D Background in Virtual Reality, Proceedings of the 32nd DAAAM International Symposium, pp.0377-0383, B. Katalinic (Ed.), Published by DAAAM International, ISBN 978-3-902734-33-4, ISSN 1726-9679, Vienna, Austria DOI: 10.2507/32nd.daaam.proceedings.055

[14] Dzermansky, M[artin]; Snopek, L[ukas]; Vichova, K[aterina]; Ficek, M[artin] & Rak, J[akub] (2021). Use of Augmented Reality Technology in Population Protection and Crisis Management, Proceedings of the 32nd DAAAM International Symposium, pp.0408-0414, B. Katalinic (Ed.), Published by DAAAM International, ISBN 978-3-902734-33-4, ISSN 1726-9679, Vienna, Austria DOI: 10.2507/32nd.daaam.proceedings.060

[15] BHaptics. bHaptics, 2022. Available at:: https://www.bhaptics.com/

[16] 22. Gilbert, Avery (2008). Hollywood Psychophysics" What the Nose Knows. Crown Publishers. p. 159. ISBN 978-1-4000-8234-6

[17] Paul. Tactile body suit could let virtual reality users feel the action they are seeing. In: Futuristech [online]. 2016 [cit. 2017-05-20]. Available at: https://futuristech.info/posts/tactile-body-suit-could-let-virtual-reality-users-feel-theaction-they-are-see

[18] "US Patent Application for NANO EMULSION PROCESS FOR SCENTED LIQUIDS Patent Application (Application #20210121835 issued April 29, 2021) - Justia Patents Search". patents.justia.com. Retrieved 2022-08-17.

DOI: 10.2507/33rd.daaam.proceedings.063

ENERGY CONSUMPTION IN MILLING ALUMINIUM WITH DIFFERENT TOOLPATHS

Carlos Mario Flores Lazo, Edwing Isaac Rosales Molina,
Juan Manuel Vázquez-Martínez & Irene Del Sol

This Publication has to be referred as: Carlos Mario, F[lores] L[azo]; Rosales Molina, E[dwing] I[saac]; Vazquez Martinez, J[uan] M[anuel] & del Sol Illana, I[rene] (2022). Energy Consumption in Milling Aluminum with Different Toolpaths, Proceedings of the 33rd DAAAM International Symposium, pp.0452-0458, B. Katalinic (Ed.), Published by DAAAM International, ISBN 978-3-902734-36-5, ISSN 1726-9679, Vienna, Austria
DOI: 10.2507/33rd.daaam.proceedings.063

Abstract

This study analyses the impact of the toolpath selection on energy consumption, machining time, and surface quality when milling 6061-aluminium alloy. Ten cavities were milled using a 9.525 mm flat milling cutter, with a three-axis CNC. Five toolpaths were defined: Spiral, Zigzag, NT Spiral, NT Zigzag and Vortex. Each cavity's energy consumption and time were recorded while performing the milling process. Once the block was milled, the surface roughness was measured using the Ra parameter. Roughness results were in the range of 1.0 to 2.5 μm. It was found that the cutting parameters had a greater influence on the increase in surface roughness. Toolpath does not seem to have a remarkable influence on roughness, using the same cutting conditions. The results showed that the Spiral trajectory took up to 69% less time than the other trajectories, it is among the trajectories that consumed the least energy (up to 13%) and the roughness was up to 33% lower. Spiral seems to be the better alternative according to the variables analysed in this paper.

Keywords: milling; toolpath; energy consumption; surface quality; machining; sustainable.

1. Introduction

Nowadays, the environmental impact of every human activity is a constant concern. It is well known that production processes can cause significant environmental impacts. Therefore, different approaches are used to reduce or eliminate them. The Sustainability of machining processes may be achieved by improving their efficiency from an energy perspective. This approach is included in the Sustainable Development Goals, where it is proposed that "By 2030, double the global rate of improvement in energy efficiency". Where it should be considered that the industry, one of the greater energy consumers, has recently exceeded 50% of worldwide consumption [1]. The environmental impact of conventional machining covers energy consumption, auxiliary materials, and transportation of materials and waste, among others, that could pollute the air, ground, and water directly or indirectly [2]. There are different ways to influence machining sustainability and reduce negative effects, for example: eliminating, substituting, or reducing metalworking fluids; tool life extension; and energy consumption optimization [3], [4]. For milling processes, several researchers have evaluated the energy consumption for different materials.

It is possible to find studies that generate models to calculate consumption [3], [4], [5] or focused on establishing relationships between the cutting conditions [6], [7] or tool wear influence [8], [7] and the energy consumed. In particular, the toolpath selected for cutting is a key factor to determine energy consumption. There were found in the literature two definitions of a toolpath. One of them refers purely to geometry while the other identifies a type of geometry among the strategies available within a CAM software. The advantage of this approach is that it allows directly obtaining a recommendation to implement in the daily work in a workshop.

Several studies have presented advances in this topic. Aramcharoen et al. [9] designed an analytical model of the energy consumption based on tool trajectory, which identifies opportunities for reducing energy demand. Pavanaskar et al. [10] proposed an application that estimates the energy consumption based on the NC code; nonetheless, this proposal has limitations. Ma et al. [4] proposed a model that considers different options to estimate the energy consumed depending on the toolpath (i.e., curve or straight line). This study established the relationship between power consumption and tool movement. However, it does not help to select a particular toolpath in the CAM software. Zhao et al. [11] elaborated a model in which the complexity of the part is quantified. They also concluded that the complexity of the piece from the geometrical point of view is related to energy consumption.

Rangarajan et al. [12] presented a particular analysis of the trajectory. These researchers studied the best orientation of a given toolpath and a workpiece for the Y-axis of the machine based on the cycle time. Although, this case is not related to energy consumption. Similarly, Del Sol et al. [13] compared different trajectories and its effects on the macro and micro geometrical characteristics of the parts. The orientation of the trajectory is also addressed by Campatelli et al. [14], who explained why the energy consumption of the axes varies. The authors reported savings between 3% and 4%, in global terms, and up to 23% if only the energy consumed by the axes is considered. On the one hand, Edem et al. [15] followed this approach and found out the best alignment corresponds to the alignment of the piece with the axis that has less weight. On the other hand, they also analysed the type of trajectory, concluding that a reduction in the retractions lowers the energy values. In addition, they pointed out that helical paths reduce energy consumption and improve better surface finish.

Kong et al. [16] developed a software to evaluate energy consumption considering the milling path. This research found that the pattern followed to machine a part influenced the energy consumption. Nevertheless, no specific toolpath is selected. The study presented by Pervaiz et al. [17], analysed four types of trajectories. It pointed out the spiral strategy as the one with the lowest consumption due to the empty movements of the tool. Minquiz et al. [18] proposed a toolpath evaluation for a face milling. In this case, the so-called dynamic strategy is more favourable from the energetic point of view than the zigzag alternative. Altıntaş et al. [3] studied the impact of the trajectory on two features, for an open pocket and a closed pocket, in both cases, the energy consumption is lower with the zigzag trajectory. These authors also proposed a model to predict the real consumption, but the cutting component fits worst to the proposed energy model. Rahman Hemdi et al. [19] presented a design of experiments for several types of trajectories, finding that the feed rate is the more influential factor in energy consumption. However, it is not concluded which is the best trajectory based on this parameter.

Therefore, the combined influence of the toolpath and the cutting parameters on the energy consumption has not been completely stablished. For this reason, this paper analyses the influence of five milling toolpaths on the energy consumption, time cycle and surface quality of an aluminium part under conventional machining parameters.

2. Experimental Procedure

Toolpaths were compared in terms of energy consumption, time cycle, and surface quality. For this purpose, twenty pockets (27mm wide, 47mm high and 8mm deep) were milled. An HSS flat mill, two flutes and 9.525 mm diameter (Fig. 1, a). The tests were performed on a CNC Milltronics MM3017, with a three-axis, spindle power of 13 kW, a maximum spindle of 8000 rev/min, and a maximum feed rate of 7.6 m/min (Fig. 1, b &. Fig. 1, c) illustrates the milling process, which was performed under lubrication. The machined material was an aluminium alloy 6061 T6511 plate (Fig. 1, d.).

Five toolpaths (Fig. 2.) were selected and each one was machined under four sets of cutting parameters (Table 1), modifying the feed per tooth, the cutting speed, and the depth of cut, keeping constant the average chip load with the recommended value of 0.203 mm/tooth [20]. The NC code was obtained using Autodesk Feature CAM Ultimate 2022. The impact of the trajectories on power consumption was measured acquiring the signal of the servo controllers of the spindle motor. A Fluke 435-II power analyser was installed (Fig. 3.) to acquire the electrical voltage, current measurements, and the energy consumption. Roughness was measured using a roughness tester Mitutoyo SURFTEST SJ-210 following the standard ISO 4287:1997. Roughness Average (Ra) was selected as surface quality parameter.

Cutting parameters	Feed per tooth f_z (mm/tooth)	Cutting speed v_c (m/min)	Depth of cut a_p (mm)
1	0.103	139.4	1.5
2	0.141	158.3	2.0
3	0.173	174.2	2.5
4	0.194	196.1	3.0

Table 1. Cutting parameters

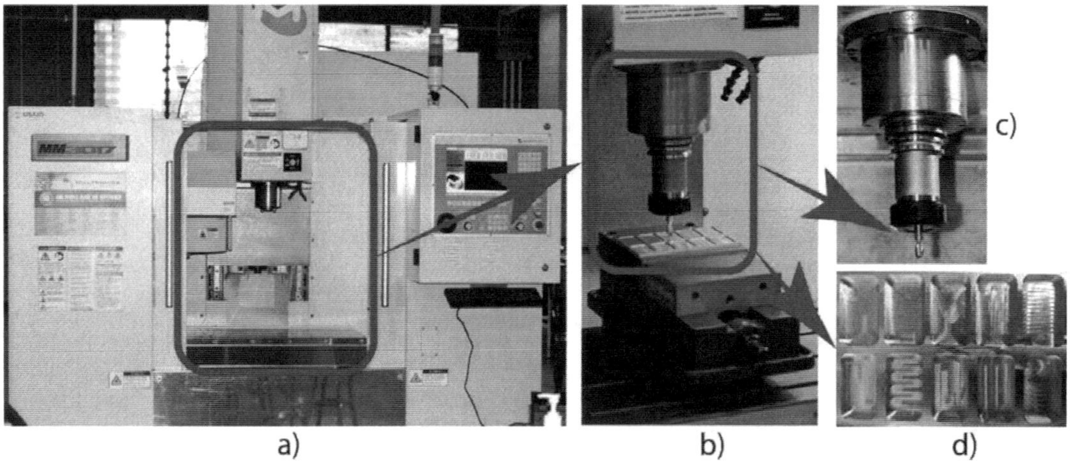

Fig. 1. Assembly of the workpiece and tool used for the tests: a) CNC Milltronics MM3017, b) process disposition, c) flat mill HSS and d) workpiece

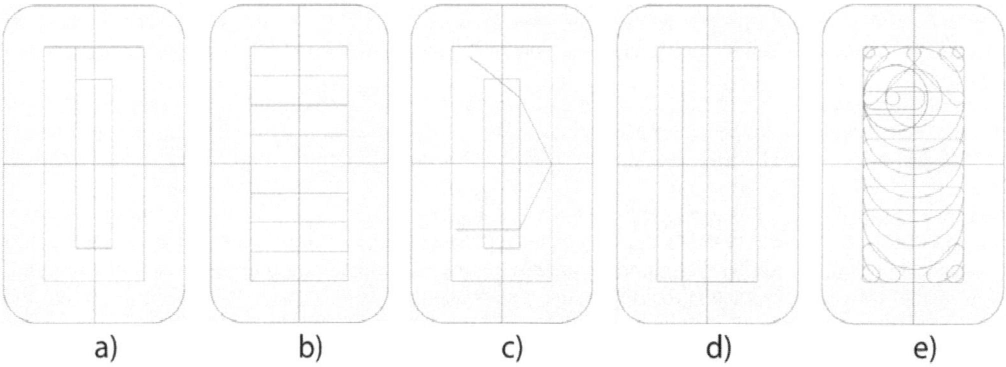

Fig. 2. Schematic illustration of toolpaths a) Spiral, b) Zigzag, c) NT Spiral, d) NT Zigzag and e) Vortex

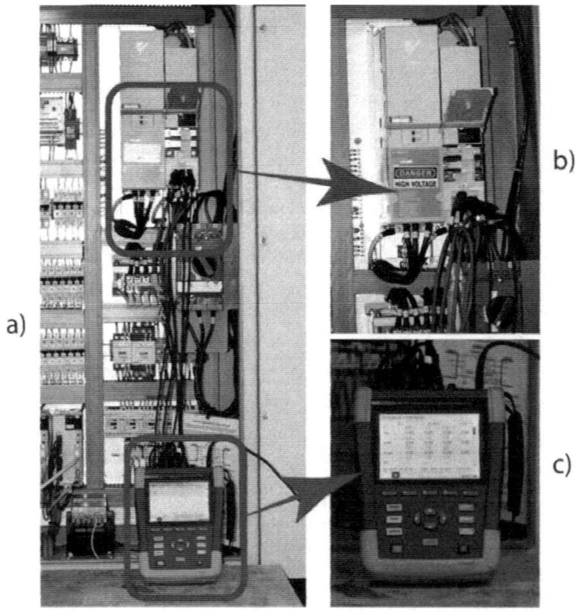

Fig. 3. Measurements of electrical variables on CNC: a) CNC Milltronics MM3017 control cabinet, b) Yaskawa CACR-JU065ADAB Drives-AC Servo, c) Fluke 435-II

3. Results and Discussion

3.1. Roughness

Fig. 4. shows Ra values for every test performed. It was observed that increasing cutting parameters (cutting speed, feed, and depth of cut) produced higher roughness values.

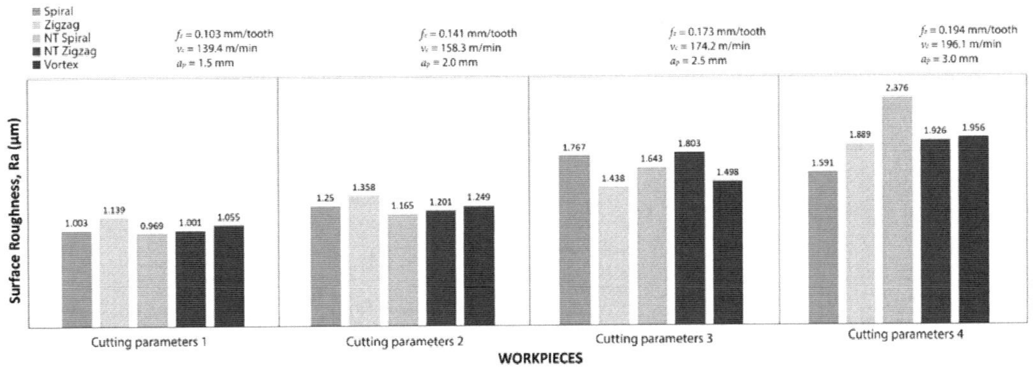

Fig. 4. Bar charts as a panel of surface roughness measurements Ra (µm) in all toolpaths.

The best quality surface was obtained for the lowest cutting parameters (f_z= 0.103 mm/tooth, v_c=139.4 m/min, a_p=1.5 mm). It is remarkable, under these conditions, that toolpath does not seem to have a significant effect on roughness. The trend of the results is in agreement with those presented by Minquiz et al. [18], who analysed this problem for steel milling. However, these authors used higher speed and lower feed and depth of cut. Therefore, Ra values were 0.28 µm lower, and the researchers did not select any particular toolpath based on this parameter. It must be noted that the difference between the lowest and the highest values obtained by Minquiz et al. was 0.08 µm when comparing toolpaths. As expected, the highest Ra was reached using the highest cutting parameters (f_z= 0.19 mm/tooth, v_c=196. m/min, a_p=3.0 mm). For higher cutting parameters, it seems that the chosen trajectory affects the surface quality. The worst results were obtained with NT spiral toolpath (2.376 µm) and the best with Spiral (1.591 µm). Ra was increased by 49% for the most aggressive parameters.

3.2. Energy consumption

Fig. 5. shows the results regarding energy consumption. Generally, the energy consumption decreased for an increase on the cutting parameters. Comparing the toolpaths, the lowest energy consumption is obtained for the Spiral, while the NT Zigzag milling strategy required 27% more energy compared to the Spiral. However, there is a difference for high values of f_z, v_c and a_p. In this case, Spiral and Zigzag trajectories registered the lowest consumption and the difference compared to NT Zigzag, which goes up to 36%. Looking at Cutting parameters 1 and Cutting parameters 4, it was found that the lowest energy consumption occurred in Spiral and Zigzag toolpath. The highest cutting parameters in these trajectories produced 22% lower energy consumption in comparison to the higher ones. These results agree with those of Pervaiz et al. [17], where Spiral toolpath got the lowest energy consumption (180 kWh) for a cutting speed of 100 m/min and feed rate of 0.05 mm/tooth.

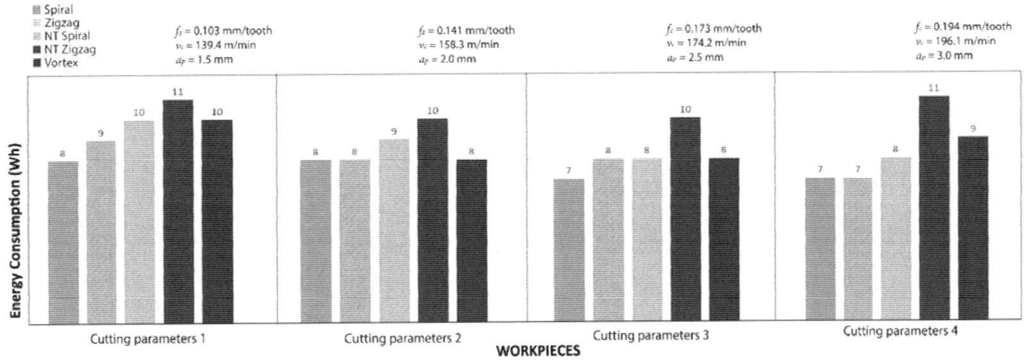

Fig. 5. Bar charts as a panel of energy consumption (Wh) measurements in all toolpaths

3.3. Time cycle

Fig. 6. presents the time cycle for every toolpath and set of cutting conditions. As it was expected, an increasing on the cutting parameters decrease the machining time. For instance, Spiral strategy time was reduced up to 69% for (f_z= 0.194 mm/tooth, v_c=196.1 m/min, a_p=3.0 mm). compared to the lower cutting parameters (f_z=0.103 mm/tooth, v_c=139.4 m/min, a_p=1.5 mm). Nevertheless, their selection of the cutting parameters should consider the surface quality requirements of the final part.

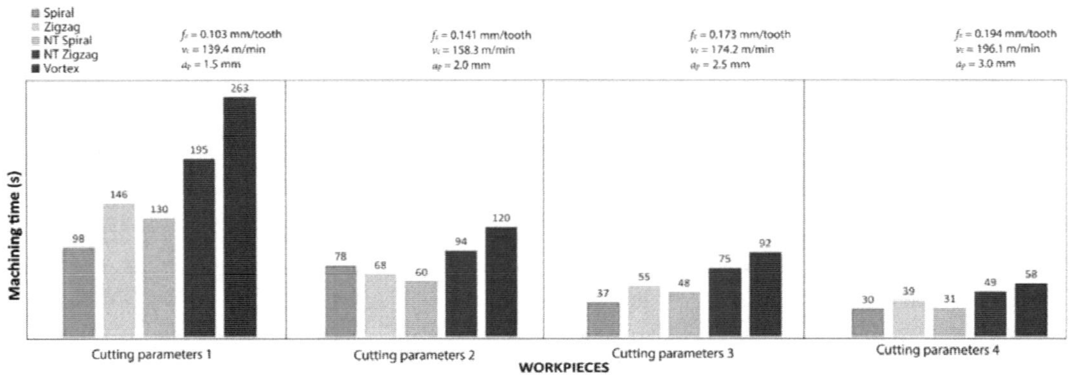

Fig. 6. Bar graphs as a panel of machining time (s) measurements in all toolpaths

Regarding the strategy selection, Vortex strategy presented the maximum time cycle values for every cutting condition, which took up to 63% longer than Spiral. These results are comparable to those by Rahman Hemdi et al. [19]; who also found that spirals had the lowest milling times.

3.4. Global comparison

A resume of the values obtained for the three evaluation parameters (roughness, energy consumed and machining time) is presented in Fig. 7.

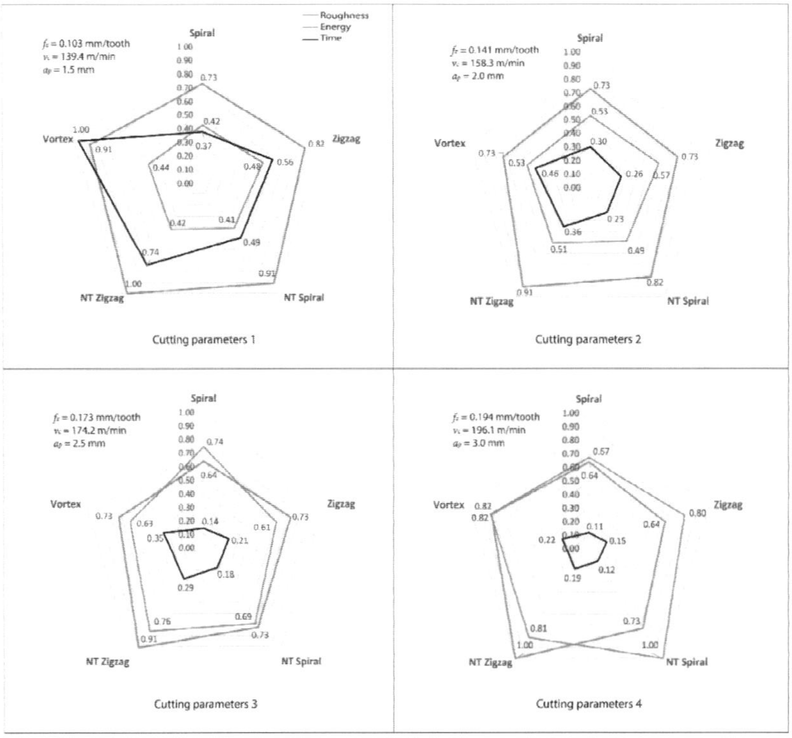

Fig. 7. Plots of normalized results of surface roughness, energy consumed, and milling time variables.

In general roughness values are similar for all the trajectories and the main difference is produced by the cutting parameters. NT Zigzag generally increase the energy consumption while the Vortex strategy increases the machining time. Spiral and Zigzag toolpath could be selected at the higher cutting conditions to get the lowest energy consumption and cycle time when roughness values are not considered critical. NT Spiral was found among the average values for every evaluation parameter.

NT zigzag and Vortex would not be recommended when lower cutting parameters are used because the evaluated performance variables seem to worsen. Finally, the Spiral trajectory is the one that registered the lowest values of roughness, time, and energy. It considerably improves the machining time for low-cutting parameters if compared to Vortex strategy.

4. Conclusions

This paper analyses the impact of the trajectories on the energy consumption, the machining time, and the surface quality to select the better trajectory, considering the effect under conventional machining parameters for the milling of 6061 aluminium alloy. The analysis of the experimental results leads to the following conclusions:

1. Trajectories are not the most influential factor for roughness values. Therefore, surface quality may be determined through the cutting parameter selection.
2. A combination of the cutting parameters and the toolpath strategy allows to obtain acceptable values in both variables, energy, and roughness.
3. Machining time is highly influenced by the feed rate and the cutting speed but also by the toolpath strategy. Spiral trajectory optimized the machining time.
4. Spiral trajectory and low cutting parameters (f_z= 0.103 mm/tooth, v_c=139.4 m/min, a_p=1.5 mm) were selected as the better machining conditions after a cross-comparison to optimize the three variables, energy consumption, surface quality and machining time.

Future research may explore other variables, such as lubricant or more in-depth cutting parameters optimization, which may be considered to improve the environmental impact of the machining process and ensure the sustainability of the process.

5. References

[1] Girouard, N. & Tam, C. (2011). Environment: Transforming the energy sector to deliver greener growth – OECD, Available from: https://www.oecd.org/newsroom/environmenttransformingtheenergysectortodelivergreenergrowth.htm. Accessed: 2022-04-26.

[2] Tomaskova, T. & Bicova, K. (2021). Impact of machining on the environment, Annals of DAAAM and Proceedings of the International DAAAM Symposium, Vol. 32, No. 1, pp. 623–630, DOI: 10.2507/32nd.daaam.proceedings.088.

[3] Altıntaş, R.; Kahya, M. & Ünver, H. (2016). Modelling and optimization of energy consumption for feature based milling, International Journal of Advanced Manufacturing Technology, Vol. 86, No. 9–12, pp. 3345–3363, DOI: 10.1007/s00170-016-8441-7.

[4] Ma, F.; Zhang, H.; Cao, H. & Hon, K. (2017). An energy consumption optimization strategy for CNC milling, International Journal of Advanced Manufacturing Technology, Vol. 90, No. 5–8, pp. 1715–1726, DOI: 10.1007/s00170-016-9497-0.

[5] Pawar, S.; Bera, T.; & Sangwan, K. (2022). Energy consumption modelling in milling of variable curved geometry, International Journal of Advanced Manufacturing Technology, Vol. 120, No. 3–4, pp. 1967–1987, DOI: 10.1007/s00170-022-08854-5.

[6] Liu, Z.; Sealy, M.; Li, W.; Zhang, D.; Fang, X.; Guo, Y. & Liu, Q. (2018). Energy consumption characteristics in finish hard milling, Journal of Manufacturing Processes, Vol. 35, pp. 500–507, DOI: 10.1016/j.jmapro.2018.08.036.

[7] Del Sol, I.; Rivero, A. & Gámez, A. (2019). Effects of machining parameters on the quality in machining of aluminum alloys thin plates, Metals (Basel), Vol. 9, No. 9, DOI: 10.3390/met9090927.

[8] Zhang, X.; Yu, T.; Dai, Y.; Qu, S. & Zhao, J. (2020). Energy consumption considering tool wear and optimization of cutting parameters in micro milling process, International Journal of Mechanical Sciences, Vol. 178, DOI: 10.1016/j.ijmecsci.2020.105628.

[9] Aramcharoen, A.; & Mativenga, P. (2014). Critical factors in energy demand modelling for CNC milling and impact of toolpath strategy, Journal of Cleaner Production, Vol. 78, pp. 63–74, DOI: 10.1016/j.jclepro.2014.04.065.

[10] Pavanaskar, S. & McMains, S. (2015). Machine Specific Energy Consumption Analysis for CNC milling toolpaths, Available from: http://www.asme.org/about-asme/terms-of-use.

[11] Zhao, J.; Li, L.; Wang, Y. & Sutherland, J. (2019). Impact of surface machining complexity on energy consumption and efficiency in CNC milling, International Journal of Advanced Manufacturing Technology, Vol. 102, No. 9–12, pp. 2891–2905, DOI: 10.1007/s00170-019-03334-9.

[12] Rangarajan, A. & Dornfeld, D. (2017). Efficient Tool Paths and Part Orientation for Face Milling, 2004.

[13] Del Sol, I.; Rivero, A.; Salguero, J.; Fernández-Vidal, S. & Marcos, M. (2017). Tool-path effect on the geometric deviations in the machining of UNS A92024 aeronautic skins, Procedia Manufacturing, Vol. 13, pp. 639–646, DOI: 10.1016/j.promfg.2017.09.134.

[14] Campatelli, G.; Scippa, A.; Lorenzini, L. & Sato, R. (2015). Optimal workpiece orientation to reduce the energy consumption of a milling process, International Journal of Precision Engineering and Manufacturing - Green Technology, Vol. 2, No. 1, pp. 5–13, DOI: 10.1007/s40684-015-0001-3.

[15] Edem, I. & Mativenga, P. (2017). Energy Demand Reduction in Milling Based on Component and Toolpath Orientations, Procedia Manufacturing, Vol. 7, pp. 253–261, DOI: 10.1016/j.promfg.2016.12.060.

[16] Kong, D.; Choi, S.; Yasui, Y.; Pavanaskar, S.; Dornfeld, D. & Wright, P. (2011). Software-based tool path evaluation for environmental sustainability, Journal of Manufacturing Systems, Vol. 30, No. 4, pp. 241–247, DOI: 10.1016/j.jmsy.2011.08.005.

[17] Pervaiz, S.; Deiab, I.; Rashid, A. & Nicolescu, M. (2012). An Experimental Analysis of Energy Consumption in Milling Strategies.

[18] Minquiz, G.; Borja, V.; López-Parra, M.; Reivich, A.; Huerta, L.; Lázaro, R.; Sánchez, A.; Leal, H.; Solana, M. & Méndez, J. (2020). Machining Parameters and Toolpath Productivity Optimization Using a Factorial Design and Fit Regression Model in Face Milling and Drilling Operations, Mathematical Problems in Engineering, Vol. 2020, DOI: 10.1155/2020/8718597.

[19] Hemdi, A.; Ali, U.; Noor, R.; Yahaya, M.; Mahadzir, M.; Zubair, A.; Othman, M.; Yola, M.; Saman, M. & Sharif, S. (2020). Cutting Path-Associated Energy Consumption of Milling Machining Process, IOP Conference Series: Materials Science and Engineering, Vol. 1003, No. 1, DOI: 10.1088/1757-899X/1003/1/012071.

[20] Oberg, E.; Jones, F.; Horton, H. & Ryffel, H. (2016). Machinery´s Handbook, Industrial Press.

DOI: 10.2507/33rd.daaam.proceedings.064

IDENTIFYING THE POTENTIAL OF VIRTUAL COMMISSIONING AS A PRACTICE FOR UNDERGRADUATE PROGRAMS

Robert Fellner[1], Sebastian Felix Rauh[2], Horst Orsolits[3] & José Garcia[4]

[1]Robert Fellner is with the Department of Industrial Engineering, University of Applied Sciences Technikum Wien, 1200 Vienna, Austria robert.fellner@technikum-wien.at

[2]Sebastian Felix Rauh is with the Department of Industrial Engineering, University of Applied Sciences Technikum Wien, 1200 Vienna, Austria sebastian.rauh@technikum-wien.at

[3]Horst Orsolits is with the Department of Industrial Engineering, University of Applied Sciences Technikum Wien, 1200 Vienna, Austria horst.orsolits@technikum-wien.at

[1]José Garcia is with the Department of Industrial Engineering, University of Applied Sciences Technikum Wien, 1200 Vienna, Austria jose.garcia@technikum-wien.at

This Publication has to be referred as: Fellner, R[obert]; Felix Rauh, S[ebastian]; Orsolits, H[orst] & Garcia, J[ose] (2022). Identifying the Potential of Virtual Commissioning as a Practice for Undergraduate Programs, Proceedings of the 33rd DAAAM International Symposium, pp.0459-0465, B. Katalinic (Ed.), Published by DAAAM International, ISBN 978-3-902734-36-5, ISSN 1726-9679, Vienna, Austria
DOI: 10.2507/33rd.daaam.proceedings.064

Abstract

Virtual commissioning provides opportunities for educational use cases, although it was not designed for them. Originally it was introduced to improve the process of designing, constructing, and implementing a production process or machine. Current works do not offer analysis of virtual commissioning practices intended for undergraduate study settings. In this paper, eight theses in the bachelor's program at the UAS Technikum Wien Mechatronics/Robotics have been retrospectively analysed. The analysis results were used to identify virtual commissioning practices in different undergraduate program settings. Further results suggest that a prerequisite for undergraduate students to accept the given tasks and keep their motivation high, is to provide software and environments to enable them to get started after only a short period of time. Thus, software and virtual environments must be adapted to the prior knowledge of the students. The results lead to the assumption that teaching single identified practices can lead to a more purposeful learning of designing and programming a machine.

Keywords: Virtual commissioning; engineering education; digital twin; simulation.

1. Introduction

In the course of the digital transformation and constant optimisation, the established processes of the manufacturing industry are also facing upheaval. A significant change in processes is emerging in the planning and implementation process, the so-called commissioning of plants, among other things. By digitally simulating the processes at an early stage, they can be implemented faster, more error-free and ultimately more cost-effectively.

The two main reasons for this are that the necessary work steps of the various disciplines for the realisation (or modification) of a plant can be carried out in parallel instead of one after the other. On the other hand, by simulating the processes, errors can be detected more quickly and corrected more cost-effectively. The number of prototypes usually required can thus also be drastically reduced, which has an effect on the costs of development. Due to these digitalised working tools, the term virtual commissioning is increasingly used in this context. By integrating these methods into teaching, it would be possible to practise low-cost, but nevertheless complex mechatronic and mechanical engineering development activities at the UAS Technikum Wien.

Virtual Commissioning (VC) describes various practices to simulate mechatronic systems while developing or changing that system to reduce real commissioning time. Depending on the task, the main simulation scenarios are described as Model-in-the-Loop (MiL), Software-in-the-Loop (SiL) and Hardware-in-the-Loop (HiL) [1]. While VC has been subject to research for several years [1], [2] and scholars have exemplified how to introduce VC practices in education [3], [4], little work has been conducted to investigate how to systematically introduce the subject to higher education. This paper aims to contribute to this issue by analysing achievements in bachelor theses and describing various practices used in this educational context.

2. Related work

Engineering education is an ongoing research topic [3], [4], [5]. As Eichinger and Richter describe in their paper [6], they face the challenges of digitalization to offer mechatronic courses on three levels: basics, development methodology and development tools and software tools. Focusing on VC, Brazina et.al. describe [7] VC as part of an educational process using different tools from ABB and Siemens as well as physical machines such as a robot and CNC machines. Describing a similar approach, Fernández et. al. describe [8] using a robotic cell to be used in educational settings to learn VC processes.

Many other platforms such as described in [9], [10], [11], [12], [13] strive to bring digital transformation aspects into educational settings. All these works implemented various settings where they use VC to convey knowledge for single learning objectives. To date no work known to the authors has been published, which identifies how single practices during VC can be used as tools for undergraduate study programs.

3. Methodological approach

Before implementing VC simulations in educational settings, VC practices should be broken down to specific tasks and analysed in perspective to effort and difficulty for students. Thus, in eight bachelor theses conducted within three years, students were tasked to implement and use various aspects of VC practices. This work reflects on the approaches students chose, the results and the learnings made by supervising theses.

The results of those projects as well as the challenges documented in the theses are the basis to retrospectively gaining insight using qualitative and descriptive data analysis [14]. Convenience sampling method [15] is applied in order to identify potential VC practices for undergraduate programs for industrial engineering settings. Used data were anonymized and can be found referenced as Student A, Student B, etc., in random order within the same year of implementation. The chosen sample only include male students, all between the ages of 23 till 26 years. The reason for the lack of female students in that sample is that within the analysed period, no female student chose one of the referenced bachelor projects in the study program (Mechatronics/Robotics) in which these were offered, most students is male.

4. Accompanied projects to set up VC practices

All virtual commissioning related projects are carried out in a lab infrastructure called digital miniature factory (DMF), see Fig 1. The digital miniature factory located in the university of applied sciences Technikum Wien is a smart manufacturing system consisting of different miniaturized production stations as well as mobile robots fitting all together on a table size 2m x 3m. It consists of different stations developed for certain production tasks like sorting, quality control or storing finished products. The dedicated product is a 3D printed carabiner which are produced colour-customised and then automatically packaged. All student projects in the context of the DMF were developed with industry related hardware such as PLCs from different vendors as well as image processing, encoders and several other sensors connected via IO-Link.

The communication and networking of each individual station is based on the Industrial Internet of Things platform PTC Thingworx and the communication standard OPC UA. As all relevant information per station are gathered in the cloud platform (sensors, status, available services) intelligent decision making based on the order backlog, current data and next step strategies can be investigated. As the whole system is developed together with industry partners and students all development data (CAD, E-Wiring, software) is available to form the basis for research activities in the field of virtual commissioning.

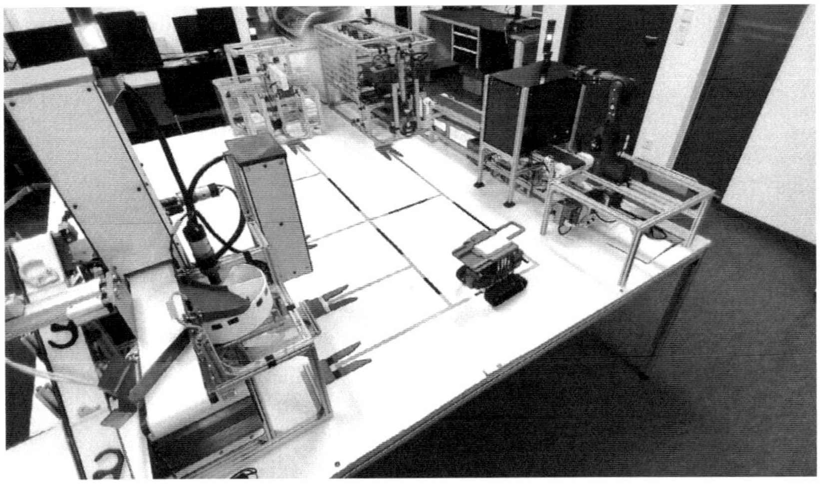

Fig. 1. Digital Miniature Factory Layout

Since 2020, eight students from bachelor's program for Mechatronics/Robotics (BMR) within the faculty of Industrial Engineering at University of Applied Sciences Technikum Wien have been supervised in VC-related projects to develop various aspects using VC simulation practices, see

Table 1. As the Goals in this table suggest, implementing VC simulations requires various tasks to be practiced. Since the practices of VC require knowledge in mechanical, electrical as well as electronic engineering of a machine, students from study program BMR were selected since they fulfil the required criteria. Within a scope of 250-300 hours (equivalent to 10 ECTS) in 6th semester of their study program, BMR students are tasked to implement bachelor projects and document findings as part of a bachelor thesis. For each of these bachelor projects from this sample, targets to implement VC practices have been expressed to see how far candidates would proceed and what goals they would come up with.

Student	Year	Target given	Simulation provided	Goals achieved	Documented challenges	Simulation type
Student A	2020	implement VC	-	Kinematics, circuit scheme, SiL logic	To implement the kinematic process in the form of a state machine.	SiL
Student B	2020	implement VC	-	Kinematics, MiL logic, statistics	To implement the kinematic process.	MiL
Student C	2021	implement VC	Kinematics & SiL logic for different station	Kinematics, SiL logic, statistics	To simulate expected kinematic behaviour	SiL
Student D	2022	compare VC tools	Kinematics & MiL logic	Comparing tools	To argue scoring scheme to compare tools	MiL
Student E	2022	implement continuous simulation	Kinematics & MiL logic	advancing given MiL model, statistics	To understand and use given model, implement kinematic processes	MiL
Student F	2022	implement digital shadow	Kinematics & MiL logic, access for data exchange	advancing given MiL model, digital shadow	To understand and use given model, implement kinematic processes, set up connectivity to external data	MiL
Student G	2022	Implement and test alternative soft PLC	Kinematics & MiL logic	implemented SiL logic, virtually tested alternative PLC, statistics	To understand, use and advance given model, setup connectivity	SiL
Student H	2022	advance and test station	Kinematics & SiL logic	Implement changes of CAD model, statistics	To understand, use and change given model	SiL

Table 1. List of BMR theses selected for analysis in the paper.

In 2020, Student A and Student B were independently asked to implement any VC simulation for a given CAD model of a miniature storage station. Further, they were asked to implement any VC practice using different software tools. They chose iPhysics from Machineering and Visual Components. Both students developed their simulation from scratch and during supervisory meeting stated that they invested most of the time to implement and being in exchange with their software support.

Student A was creating a SiL simulation by implementing kinematics for the given CAD model of a miniature storage station as well as the corresponding circuit scheme and program logic, see Fig. 2. This logic control was programmed with structured text implementation on a virtual Programmable Logic Control (PLC) using CodeSys. Connecting the virtual PLC and the kinematics of the virtual storage station enabled this SiL simulation.

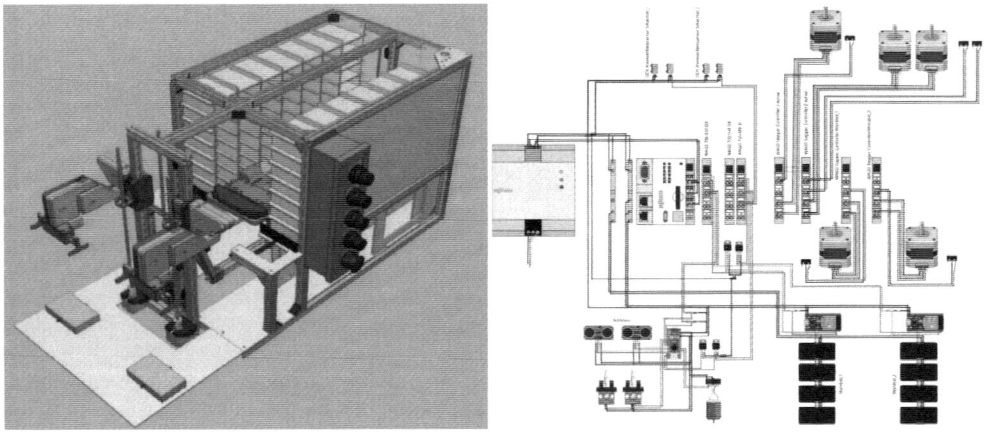

Fig. 2. implemented VC practices and circuit scheme, Student A

Student B was aiming for a MiL simulation and also implemented kinematics for the given CAD model of the same miniature storage station. He implemented an abstract programming logic enabling this MiL simulation. Further, student B implemented a statistic tool for demonstrating potential follow-up projects.

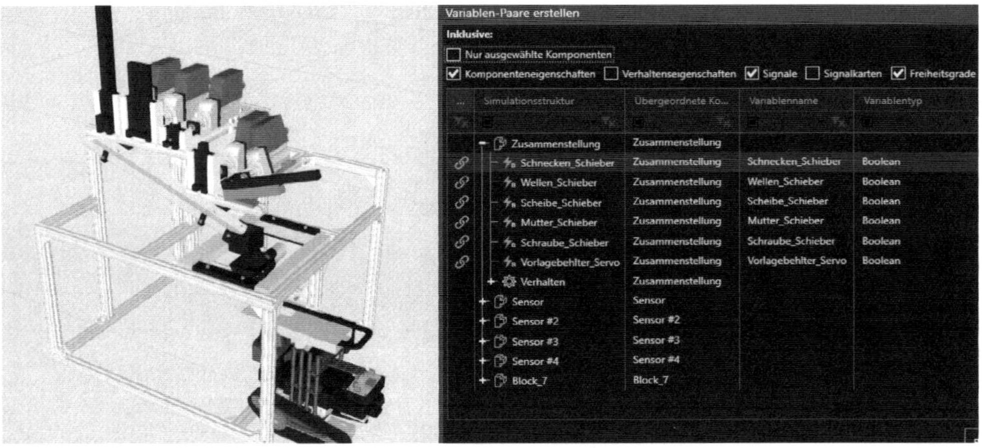

Fig. 3. Implemented VC practice and connection to soft PLC, Student C

Student C was asked to implement any VC simulation for a given CAD model of a miniature separation station, see Fig. 3. He was capable to implement a SiL simulation for the separation station using Visual Components as tool to implement kinematics. To implement a program logic he used ladder logic, to simulate a plc CodeSys software was used. Additionally, he analysed implemented kinematic behaviour of the separation station. In 2022, five students chose to implement a VC practice. Based on the learning in the previous two years, for these five students the targets given were more specific, therefore being able to use prior implementations: Student D was asked to compare different tools that can implement VC practices. Besides implementing a SiL simulation using NX MCD, he came up with self-defined metrics to compare chosen tools NX Mechatronic Concept Designer (NX MCD) and Visual Components. Therefore, he argued four parameters to compare on: integration of the tool, connectivity capabilities, support and functionalities.

He argued that Visual Components would score higher for integration, connectivity and support parameters. Defined criteria to argue functionalities of the tools were rated equally high. Student E and Student F were first asked to simulate all processes within the digital miniature factory, see Fig. 4. Therefore, the advanced given MiL models for the storage and separation station using Visual Components. After implementing missing CAD models, kinematics and abstract programming logic, Student E focused on evaluating each stations' workloads and overall equipment effectiveness (OEE) for the DMF. Student F advanced the common built-up simulation to be used as a digital twin of the real DMF. He therefore changed the previous program logic and connected the simulation to the real DMF's stations via the industrial network communication standard OPC UA.

Student G was asked to advance the model for the sorting station to change and analyse an alternative virtual PLC. He therefore received a MiL simulation setup for the station in Visual Components. He advanced the model to be tested in SiL simulation and analysed the current PLC logic on a virtual simulation of the current PLC. After that, the PLC logic was adapted and tested for an alternative soft PLC.

Student H was asked to adapt the CAD model of the storage station, revise and test the PLC code accordingly. He therefore received a SiL simulation of the storage station. After implementation of the changed model and program logic of the simulated storage station, the implemented change could be analysed, the improvements in comparison to the previous setup could be argued.

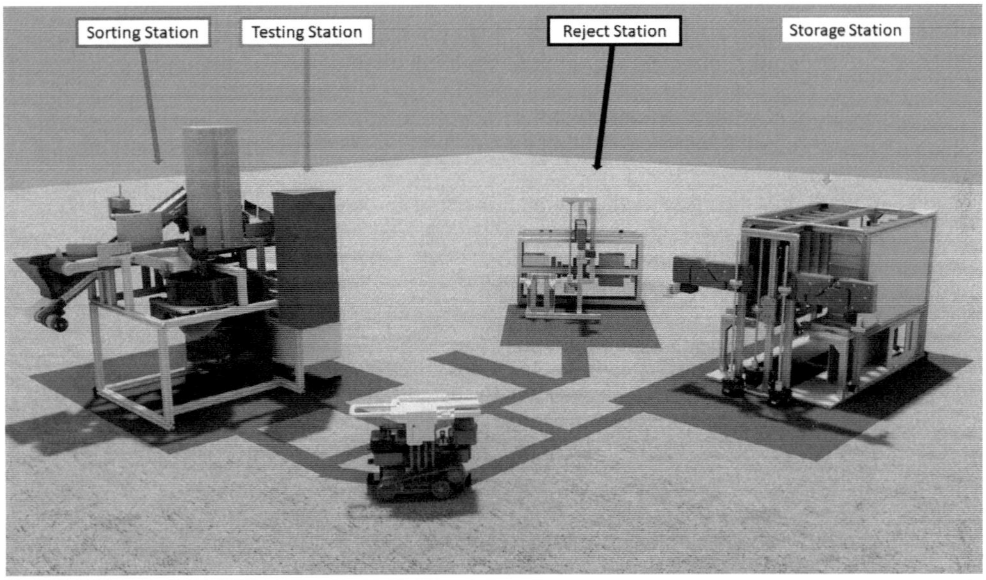

Fig. 4. Simulation of digital miniature factory (DMF)

5. Results and Discussion

As stated in section 3, convenience sampling method was chosen to collect the results. Therefore, the described peer group does not inevitably represent all undergraduate engineering students, but an excerpt based on the availability of students willing to do a bachelor thesis project in the field of VC. Since that students could influence which thesis project they were assigned to, a higher motivation can be expected compared to when all students would be randomly assigned to one bachelor project. It can be assumed that based on the higher motivation, these students perform better than the average student. In case of his analysis, this circumstance helps to unveil more advantages and challenges of using VC in the context of bachelor thesis. The insights collected therefore may also help identifying practices that can be useful not only for students in undergraduate mechatronic study programs.

The main objective of this work is to identify, whether and what practices during VC could be used for different undergraduate programs. Following practices could be identified:

- Implementing kinematics: The implementers need to cinematize mechatronic systems in the VC tool. To do so, planned movement spaces of the designed mechatronic system need to be correctly interpreted. Using VC tools allows to verify implemented movements since they can provide direct feedback on input.
- Implementing abstract program logics (MiL): The implementers need to program abstract program logics. To do so, present actuators need to be activated depended on signal inputs. Using VC tools allows to simulate various setups of actuators and sensors and their corresponding software logic to explore approaches to connect actuators and sensors.
- Implementing PLC program: The implementers need to program PLC program logics. To do so, simulation variables for simulated sensors and actuators need to be implemented to describe the same information exchange that would be

expected from real hardware. Using VC tools allows to learn on advantages and challenges of different system designs to identify the approach, which addresses the goal the most.
- Implementing connections between different tools: The implementers need to setup a connection, usually between heterogeneous tools. Using different tools allows overcome natural limitations of single tools to e.g., implement model behaviour to the cinematized model. Using multiple VC tools allows to extend the simulation capabilities for single tools and to learn on different communication standards.
- Using and arguing through statistical evaluation of a mechatronic system: The implementers need to find meaningful parameters and interpret them accordingly to derive findings from existing mechatronic system setups simulations. VC tools provide means to collect data and to evaluate it or to export it to other software.
- Comparing hardware virtually: The implementers need to decide what hardware setup would fulfil the set requirements to verify each hardware setup. Using VC tools allow evaluating different hardware in simulation to make a decision on which hardware might be more suitable to achieve project goals, informed by own experience.
- Implementing a digital shadow: The implementers need to set up a simulation interpreting incoming variables correctly to verify and run implemented simulations. With VC tools, implementers can run the digital shadow within an environment which can collect, replay, and validate data directly.

The following statements collected during supervisory meetings with students describe further aspects that have a positive impact on the results:
- Students who received a prepared simulation to base their work on, seemed to achieve more results within the same time.
- The motivation and focus of students appeared to be higher when specific targets were suggested.
- The most popular tool seemed to be one to get going the fastest not the one being most versatile.
- No matter if VC was set as a target or not, VC simulations were practiced reaching various goals.
- Only one out of eight projects (conducted by Student H) could have been realised without using VC practices by shifting to the use of real hardware setups. The need to adapt or change hardware due to targets or failures, could not only have cost more time but also resources in terms of material and space.

Single practices identified in the context of above stated theses could be adopted considering student's skills and learning targets. Using a single practice at a time may lead to a more purposeful learning of designing and programming a machine. Further, the results of this analysis indicate that a providing software and clear objectives can lead to higher motivation among undergraduate students. This increase in motivation can be connected to the fact that familiarizing with the topic of VC and the tools to implement VC is facilitated in terms of time and comprehensibility.

6. Conclusion and Future Work

From discussing documented bachelor thesis results in the study program of BMR, this paper concludes that various aspects throughout the process of designing and programming a machine seems to be beneficial not only for students but also for saving construction material. To potentially implemented in teaching, software and virtual environments must be adequately adapted to the prior knowledge and abilities of the students. To deliver better descriptions of the effort and difficulty to implement VC practices to be used by undergraduate students as well as for lecturers, further investigation should be carried out. We would like to follow on these findings by implementing and evaluating selected practices in future courses at our institution.

7. Acknowledgments

This work was supported by the city of Vienna (MA 23 – Economic Affairs, Labour and Statistics) through the research project INVIS (MA23 project 29-11). The authors further thank all participants for their contributions.

8. References

[1] Wünsch, G. (2008). Methoden für die virtuelle Inbetriebnahme automatisierter Produktionssysteme, Herbert Utz Verlag, München

[2] Drath, R.; Weber, P. & Mauser, N. (2008) An evolutionary approach for the industrial introduction of virtual commissioning, in 2008 IEEE International Conference on Emerging Technologies and Factory Automation, Hamburg, Germany, pp. 5–8. DOI: 10.1109/ETFA.2008.4638359

[3] Ronkova V. & Dimitrov, Y. (2019). Education in Engineering – Challenges and Contributions, in DAAAM Proceedings, 1st ed., vol. 1, B. Katalinic, Ed. DAAAM International Vienna, pp. 0717–0723. DOI: 10.2507/30th.daaam.proceedings.098

[4] Álvarez-Marín, A.; & Velázquez-Iturbide, J. Á. (2021). Augmented Reality and Engineering Education: A Systematic Review, IEEE Trans. Learn. Technol., vol. 14, no. 6, pp. 817–831, DOI: 10.1109/TLT.2022.3144356

[5] Miranda, J.; et al. (2021). The core components of education 4.0 in higher education: Three case studies in engineering education, Comput. Electr. Eng., vol. 93, p. 107278, DOI: 10.1016/j.compeleceng.2021.107278

[6] Eichinger, P.; Höfig, B. & Richter, C. (2017). Education 4.0 for mechatronics – agile and smart. p. 7. DOI: 10.1109/REM.2017.8075250

[7] Brazina, J.; Vetiska, J.; Stanek, V.; Bradac, F. & Holub, M. (2020). Virtual commissioning as part of the educational process, in 2020 19th International Conference on Mechatronics - Mechatronika (ME), Prague, Czech Republic, pp. 1–7. DOI: 10.1109/ME49197.2020.9286613

[8] Fernandez, I. A.; Eguia, M. A. & Echeverria, L. E. (2019). Virtual commissioning of a robotic cell: an educational case study, in 2019 24th IEEE International Conference on Emerging Technologies and Factory Automation (ETFA), Zaragoza, Spain, pp. 820–825. DOI: 10.1109/ETFA.2019.8869373

[9] Macías García, M. E.; Cortés Pérez, A. A. & Izaguirre Alegría, A. R. (2020). Cyber-Physical Labs to enhance engineering training and education, Int. J. Interact. Des. Manuf. IJIDeM, vol. 14, no. 4, pp. 1253–1269, DOI: 10.1007/s12008-020-00704-6

[10] Mahmood, K. et al. (2021). Advancement in production engineering education through Virtual Learning Factory Toolkit concept, Proc. Est. Acad. Sci., vol. 70, no. 4, p. 374, 2021, DOI: 10.3176/proc.2021.4.02

[11] Prieto, M. D.; Sobrino, A. F.; Soto, L. R.; Romero, D.; Biosca, P. F. & Martinez, L. R. (2019). Active Learning based Laboratory towards Engineering Education 4.0, in 2019 24th IEEE International Conference on Emerging Technologies and Factory Automation (ETFA), Zaragoza, Spain, pp. 776–783. DOI: 10.1109/ETFA.2019.8869509

[12] Rojas A. M. & Barbieri, G. (2019). A Low-Cost and Scaled Automation System for Education in Industrial Automation, in 2019 24th IEEE International Conference on Emerging Technologies and Factory Automation (ETFA), Zaragoza, Spain, pp. 439–444. DOI: 10.1109/ETFA.2019.8869535

[13] Oliveira, P. M. & Hedengren, J. D. (2019). An APMonitor Temperature Lab PID Control Experiment for Undergraduate Students, in 2019 24th IEEE International Conference on Emerging Technologies and Factory Automation (ETFA), Zaragoza, Spain, pp. 790–797. DOI: 10.1109/ETFA.2019.8869247

[14] Nassaji, H. (2015). Qualitative and descriptive research: Data type versus data analysis, Lang. Teach. Res., vol. 19, no. 2, pp. 129–132, DOI: 10.1177/1362168815572747

[15] Sedgwick, P. (2013). Convenience sampling, BMJ Online, vol. 347, p. f6304

DOI: 10.2507/33rd.daaam.proceedings.065

DEVELOPMENT OF PNEUMATIC TESTER FOR LIFE TESTING OF INDUSTRIAL ELECTRONIC ROTARY LATCHES

Selahattin Mert Aydin, Emre Keklik, Engin Gunes, Anil Akdogan & Ali Serdar Vanli

This Publication has to be referred as: Aydin, S[elahattin] M[ert]; Keklik, E[mre]; Genes, E[ngin]; Akdogan, A[nil] & Vanli, A[li] S[erdar] (2022). Development of Pneumatic Tester for Life Testing of Industrial Electronic Rotary Latches, Proceedings of the 33rd DAAAM International Symposium, pp.0466-0470, B. Katalinic (Ed.), Published by DAAAM International, ISBN 978-3-902734-36-5, ISSN 1726-9679, Vienna, Austria
DOI: 10.2507/33rd.daaam.proceedings.065

Abstract

Since electronic rotary latches operate under difficult conditions, there is always the possibility of errors. In this study, a pneumatic test system was designed to detect these errors and reduce them to the minimum number. Electronic rotary latches may fail under conditions such as; it may not be triggered electronically, it may not be physically opened in case of load on its latch, etc. These situations should be simulated and controlled, and necessary actions should be taken. In this study, a pneumatic life test device for electronic rotary latch was studied. This device performs the mechanical locking and unlocking operations of the lock, thanks to a specially designed mechanism with a pneumatic system. Electrical locking is achieved with a signal given repeatedly within the set time. On the LCD screen, the number of repetitions can be monitored and in case of a possible error, the system stops itself and gives a warning. As a result of this study, the life test of the latch locks is carried out and the errors are observed. Based on this study, similar studies can be carried out for other locks in the future, and the study can be developed and expanded.

Keywords: Life Testing; Automation; Electronic; Rotary Latches; Pneumatic.

1. Introduction

Today, industrial locks are used for locking panels and cabinets [1] in different sectors such as automotive, white goods, data center, railway, HVAC (air conditioning plant) [2], cargo delivery points [3]. Due to sectoral needs, electronic locks, the use of which has increased in the industrial field with the development of technology, can vary in different application areas according to the needs. Along with the electronic tiling systems used in the industry; due to the necessity of application areas and environmental conditions, the life span of the locks used in the field and their resistance to external effects have gained importance [4]. In this study, it is aimed to discover possible problems beforehand by simulating the working environment of electronic locks that need to work continuously. After this study, the damaged parts of electronic rotary latches, which are formed as a result of long-term continuous work, will be determined and improvements will be made on these parts.

2. Pneumatic Tester Design and Commissioning

There was a need for a test device against the repeated use of locks. The design here has been made according to the working mechanics of the electronic rotary latches we currently use. The design with the same criteria will not be suitable for different locks. For other locks, a design with a new working mechanic suitable for that lock is required. In this study, this test device was used with a pneumatic system [5]. Pneumatic system will make continuous linear motion to lock the electronic rotary latch. One of the disadvantages of the pneumatic system can be noise, which is mostly caused by the release of used compressed air. This noise problem can be solved with noise silencers attached to the pneumatic system nozzles [6]. In this way, the noise from the pneumatic system is prevented and can continue for a long time.

2.1 Mechanical Design

In our research, the concept model was designed with the 2021 version of SolidWorks, one of the 3D design programs. (Fig.1)

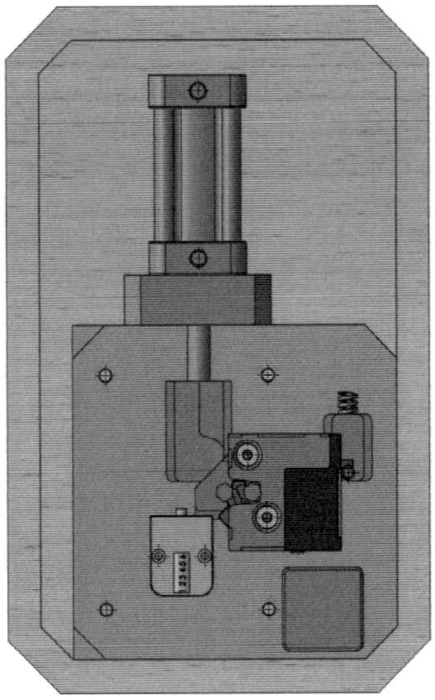

Fig. 1. Project Design Prepared in Solidworks

The concept design of the life test device consists of the following parts;

- Test platform
- Steel Bodies and Fasteners
- Piston and drive elements
- Life Test Counter
- Industrial electronic lock to be tested for life

The designed steel blocks are fixed to the test platform, and they are bearing the industrial electronic lock, piston, counter and drive elements, which will be tested for life. The industrial electronic latch lock to be tested is mounted inside the steel block in the open position. Pneumatic piston mounted on the test platform provides a pushing force with 0.3 bar air pressure and locks the electronic latch lock [7]. The process of opening the electronic lock; carried out with the electronic control card designed within the scope of the research study. The opening signal sent from the control card triggers the electronic lock and the electronic valve to open it. This on-off process is repeated continuously, and a life test is carried out. A counter has been added to the concept design to determine the number of on-off operations while the test process is in progress.

2.2 Electronical Design

The electronic board and schematic diagram of the test device are designed. Arduino Nano[8] model was used as the main processor. 1 LCD screen is used to show the necessary information and 1 buzzer is used to give the necessary warnings. The working voltage of the electronic rotary latch to be tested is 12 Volts, the operating voltage of the valve and sensor used in the system is 24 volts and the I/O's of Arduino Nano are 5 volts. 3 relays are used to control the electronic latch and valve and to control the necessary information from the sensor in a way that the processor can understand.

The pins that control the electronic lock signal, the buzzer and the valve are conditioned as output, the pins used for the sensor and the switch showing the open or closed position of the latch in the electronic lock are conditioned as Input. For the electronic lock to be easily connected to the test device, a 6-pin PCB type female socket suitable for the socket of the electronic lock was selected. Screw sockets are used for easy assembly for power input and other connections. The printed circuit board (PCB) of the described electronic system is shown in Fig.2, and its schematic diagram is shown in Fig.3.

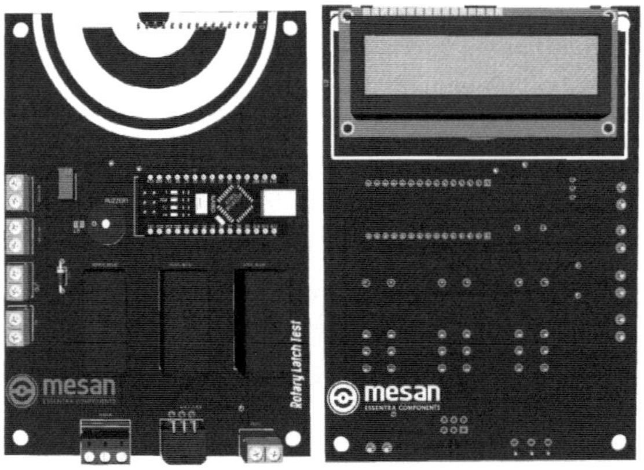

Fig. 2. PCB of tester

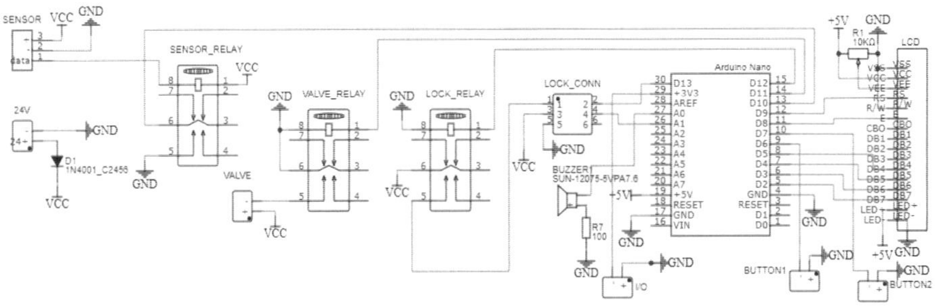

Fig.3 Schematic of tester

2.3 Software

Arduino IDE[8] was used as the programmer for the software of the system, since the microcontroller Arduino Nano was used. Electronically triggering the lock, Valve control, processing the sensor status, writing to the LCD screen, etc. Arduino Nano, which is used to perform operations, can control this process thanks to the software written for all these situations. A part of the code that prints the number of tests on the LCD screen as a result of a repetition of the system is given below.

```
#define sensor 10
#define buzzer 14

unsigned int count = 0;
#include <LiquidCrystal.h>
const int rs = 9, en = 8, d4 = 5, d5 = 4, d6 = 3, d7 = 2;
LiquidCrystal lcd(rs, en, d4, d5, d6, d7);

void setup() {
 Serial.begin (9600);
 lcd.begin(16, 2);
 pinMode(sensor, INPUT_PULLUP);

 }
void loop() {
```

```
  if (digitalRead(sensor) == HIGH) {
    lcd.clear();
    count = count + 1;
    lcd.setCursor(0, 0);
    lcd.print("Test Qty = ");
    lcd.print(count);
  }
}
```

2.4 General Structure and Operation of the System

For the system to work, the lock must be fixed to the system in the open position. When it is wanted to be operated in the closed position, the LCD screen is told to turn it on, and it cannot be operated until it is turned on. The pneumatic piston moves the part made to move the latch and locks the latch. After locking, the pneumatic piston retracts itself. The lock is triggered electronically, and the latch opens. Latch returns the piece to its original position. Thanks to the metal sensor, it seems that this part is in the right position. When all this scenario is completed, one test is considered completed and the number of tests becomes 1. Each time this scenario completes without errors, the number of tests is incremented by 1. The locking scenario of the system is shown in Fig.4.

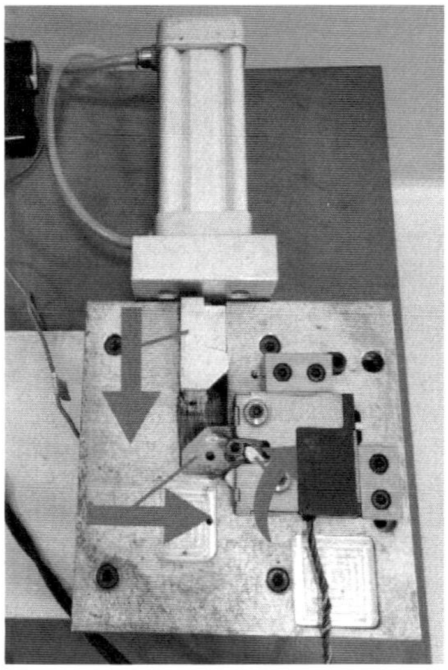

Fig. 4. Locking scenario of tester

In case of a possible error such as the spring in the lock is broken, the latch is worn out and cannot be locked anymore, it cannot be triggered electronically, "ERROR" is written on the LCD screen and the test number at the time of the error is written under the "ERROR" text on the screen. The test stops progressing so that the detection at the time of error can be made correctly. Meanwhile, the buzzer on the electronic card gives an audible warning and warns the person concerned. Error moment is shown in Fig.5. In this way, when the operator removes the electronic rotary latch from the assembly, it can be seen in which number of tests it failed.

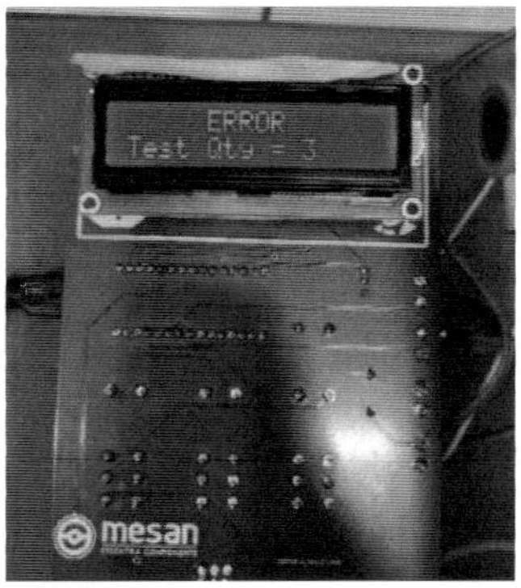

Fig.5. Error moment

3. Conclusion

Prior to the study, there was no system for testing electronic rotary latches. As a result of a few manual tests, the locks could encounter problems in the field of use. This tester system was made in order to detect and reduce these problems in advance. As a result of our work, we can test our electronic rotary latches and take the necessary actions according to the results. We become able to see which problems we encounter in which number of repetitions. This enables us to detect and solve problems that may arise in customers' use of electronic rotary latch.

When the results are compared with the old manual test mechanics without the tester, it is observed that the problems can be detected beforehand. In the manual testing system, it was very difficult to detect possible errors as very few repetitions were possible. With this tester, the rate of finding errors increased as a result of more repetitions. The problem addressed by the article is solved by the results represented. For this tester, it is aimed to remotely monitor the system and the status of electronic rotary latch by adding a wi-fi module or sim card module to the electronic card in the future. This work can be used for different locks using a similar mechanism system. For other existing locks or new locks, similar developments can be made in future.

4. Acknowledgements

This study was conducted to contribute to the development of Mesan Locks Inc.'s laboratory.

5. References

[1] Lin, L., Du, S., Ouyang, B., & Yan, M. (2021, December). Intelligent Cabinet Lock Control System Based on Cloud Technology. In Journal of Physics: Conference Series (Vol. 2136, No. 1, p. 012051). IOP Publishing.

[2] Es-sakali, Niima, (2022) et al. "Review of predictive maintenance algorithms applied to HVAC systems." Energy Reports 8: 1003-1012.

[3] Orozonova, A., Gapurbaeva, S., Kydykov, A., Prokopenko, O., Prause, G., & Lytvynenko, S. (2022). Application of smart logistics technologies in the organization of multimodal cargo delivery. Transportation Research Procedia, 63, 1192-1198.

[4] Keklik, E., Güneş, E., Aydin, S. M., Yaşar, M., & Akdoğan, A. (2022). Modular Smart Electronic Swinghandle With Remote Control Based Internet Of Things (Iot). Acta Technica Napocensis-Series: Applied Mathematics, Mechanics, And Engineering, 65(1s).

[5] Abela, K., Refalo, P., & Francalanza, E. (2022). Analysis of pneumatic parameters to identify leakages and faults on the demand side of a compressed air system. Cleaner Engineering and Technology, 6, 100355.

[6] Votrubec, R[adek] & Vavrousek, M[iroslav] (2018). Behaviour Identification and Linearization of Pneumatic Rotational Engine, Proceedings of the 29th DAAAM International Symposium, pp.0287-0293, B. Katalinic (Ed.), Published by DAAAM International, ISBN 978-3-902734-20-4, ISSN 1726-9679, Vienna, Austria DOI: 10.2507/29th.daaam.proceedings.041

[7] Ben-Dov, D., & Salcudean, S. E. (1995). A force-controlled pneumatic actuator. IEEE Transactions on Robotics and Automation, 11(6), 906-911.

[8] https://docs.arduino.cc/software/ide-v2, (2022). Arduino IDE, Accessed on: 2022-08-13

DOI: 10.2507/33rd.daaam.proceedings.066

SELECTION OF AN APPROPRIATE PRIOR DISTRIBUTION IN RISK ASSESSMENT

Dubravka Božić & Biserka Runje

This Publication has to be referred as: Bozic, D[ubravka] & Runje, B[iserka] (2022). Selection of an Appropriate Prior Distribution in Risk Assessment, Proceedings of the 33rd DAAAM International Symposium, pp.0471-0479, B. Katalinic (Ed.), Published by DAAAM International, ISBN 978-3-902734-36-5, ISSN 1726-9679, Vienna, Austria
DOI: 10.2507/33rd.daaam.proceedings.066

Abstract

The Bayesian approach, which combines prior information about the quantity to be measured, available before the measurement, and additional information obtained from the measurement, is used in risk assessment in metrology. According to the binary decision rule in risk assessment, there are four outputs: the number of accepted and rejected measurements and the number of falsely accepted and falsely rejected measurements. A falsely rejected measurement represents the producer's risk, while a falsely accepted measurement represents the consumer's risk. These four cases in risk assessment lead us to confusion matrix. In this paper, we evaluate the most suitable prior distribution for modelling the risk for roundness deviation of the inner ring of the bearing. This quantity is always positive; therefore, the choice of prior is limited to those distributions that take only the positive value of the argument. The assessment of the most appropriate distribution was performed by measures derived from confusion matrix and ROC - AUC analysis.

Keywords: consumer's risk; producer's risk; binary decision rule; confusion matrix; ROC curve

1. Introduction

There is no perfect measurement. Different sources of variability affect measurement results and measurement uncertainty. The measured values for the item of interest may or may not be within the interval of allowed values provided by the specifications for the specific product. The verification process of accepting or rejecting inappropriate measurement is the so-called conformity assessment rule. In this decision-making process, wrong decisions may occur that must be considered. There is a producer's risk of rejecting of conforms measurement and consumer's risk of accepting of non-conforms measurement [1]. The reference document [2], adopted by the Joint Committee for Guides in Metrology prescribes the procedure for calculating the producer's risk and the consumer's risk. When calculating the risk, a Bayesian approach is used. This approach combines information obtained by measurement and assumptions about the distribution of parameters that describe the measured data. Measurement data are usually assumed to belong to a normal distribution and are modelled by using a likelihood function. The risk assessment depends on the a priori distribution of the parameters. Different numbers of accepted, rejected, falsely accepted (consumer's risk) and falsely rejected products (producer's risk) are obtained for different priors. Based on these four outputs, it is possible to form a confusion matrix, a well-known term from machine learning. The aim of this paper is to use metrics associated to confusion matrix to estimate the most suitable prior distribution for risk assessment. The prior distribution selection was carried out for a real study case of risk modelling for roundness deviation of the inner ring of the bearing.

2. Background of risk calculation

The Bayesian approach to risk assessment in metrology combines two sources of information. One type of information about the measured variable Y comes from the a priori experience of the measurer himself. The item of interest, measured quantity Y, can take the values denoted by η, and is treated as a random variable with a probability distribution function (PDF), denoted by $g_0(\eta)$. Two parameters are associated to the measured value Y: best estimate $\bar{y} = \mu$, and standard uncertainty $u_0 = s$. This information can be evaluated before measurement performing. Another source of information are the measurement results given by the random variable Y_m. According to [3], the standard measurement uncertainty u_m, and the value of the measured quantity η_m, are associated to the measurement results. Measurement data are modelled via likelihood function $h(\eta_m|\eta)$, by normal PDF. This function, for given η_m, depends on η, and it can be calculated from:

$$h(\eta_m|\eta)=\frac{1}{u_m\sqrt{2\pi}}\exp\left[-\frac{1}{2}\left(\frac{\eta_m-\eta}{u_m}\right)^2\right] \quad (1)$$

As a result of Bayes' rule, the posterior distribution is written as

$$g(\eta|\eta_m) = Cg_0(\eta)h(\eta_m|\eta), \quad (2)$$

where C is the normalization constant chosen such that $\int_{-\infty}^{\infty} g(\eta|\,\eta_m)d\eta = 1$.

There are two types of risk: specific and global risk. The specific producer's risk is related with conformance probability p_c, the probability that the item of interest is within the interval of permissible value, i.e., within the tolerance interval $[T_L, T_U]$. Symbols T_L and T_U are for the lower and upper limit of the tolerance interval, respectively. Conformance probability is given by

$$p_c = \int_{T_L}^{T_U} g(\eta|\eta_m)d\eta \quad (3)$$

Probability $\overline{p_c}$ that an item of interest is non-confirmed can be calculated as

$$\overline{p_c} = 1 - p_c \quad (4)$$

Another important interval is the acceptance interval $[A_L, A_U]$, that is, the interval of permissible values for the measured quantity. Symbols A_L and A_U stands for the lower and upper limit of the acceptance interval, respectively. Depending on the problem under consideration, the tolerance interval and the acceptance interval can be in several different relationships [4]. The acceptance interval may be entirely within the tolerance interval, and separated from it, from both sides, by a guard band of width w. On this way, the consumer's risk is minimized. If the tolerance interval is within the acceptance interval, the producer's risk is minimized. By placing a guard band between the acceptance interval and the tolerance interval, the probability of making a wrong decision is reduced. From the natural requirements set for the measured quantity, it is possible to define one side tolerance interval when the measured values are limited from below with lower tolerance limit T_L, or from above with upper tolerance limit T_U (Figure 1).

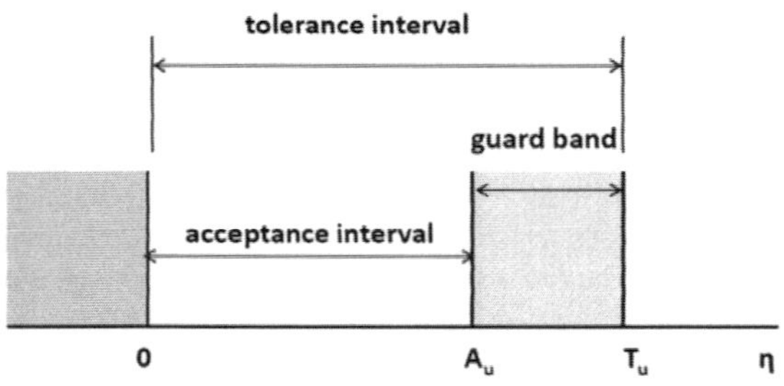

Fig. 1. One-sided tolerance interval bounded with upper limit [4]

The situation shown in Figure 1 corresponds to the case of risk assessment for roundness deviation of the inner ring of the bearing. For the given upper limit of the tolerance interval T_U, the upper limit of the acceptance interval A_U is determined. The width of guard band is equal to

$$w = T_U - A_U = 2ru_m, \quad (5)$$

where the multiplier r is in the range from -1 to 1.

3. Global consumer's and producer's risk

If the true value Y of the item of interest is outside the tolerance interval, and the measured value Y_m is within the acceptance interval, the global consumer's risk is given by

$$R_C = \int_{-\infty}^{T_L} \int_{A_L}^{A_U} g_0(\eta) h(\eta_m|\eta) d\eta_m d\eta + \int_{T_U}^{\infty} \int_{A_L}^{A_U} g_0(\eta) h(\eta_m|\eta) d\eta_m d\eta \tag{6}$$

If the true value Y of item of the interest is within the tolerance interval, and the measured value Y_m is outside the acceptance interval, producer's risk is given by

$$R_P = \int_{-\infty}^{A_L} \int_{T_L}^{T_U} g_0(\eta) h(\eta_m|\eta) d\eta_m d\eta + \int_{A_U}^{\infty} \int_{T_L}^{T_U} g_0(\eta) h(\eta_m|\eta) d\eta_m d\eta \tag{7}$$

Consumer's risk is associated with false accepted or false positive (FP) measurements while producer's risk is associated with false rejected or false negative (FN) measurements. For labelling are used the notation FP and FN, the standard notation in machine learning. The result of the consumer's and producer's risk calculation is most often expressed in percentages. The number of false positive measurements or products could be found by multiplying the R_C value by 100, 1000 or 10000, depending on how many products have to be checked. In this paper, the calculation was performed for 10000 products. In this case, the number of false positive products is equal to $FP = 10000 \cdot R_C$. The number of falsely rejected products is counted as $FN = 10000 \cdot R_p$. The total number of conformed products is equal to $10000 \cdot p_c$. From here, the number of accepted, or true positive (TP) products is equal to $TP = 10000 \cdot p_c - FN$, while the number of rejected, i.e., true negative products (TN) is equal to $TN = 10000 \cdot \overline{p_c} - FP$. According to this binary decision rule, four values that form the confusion matrix are obtained (Figure 2).

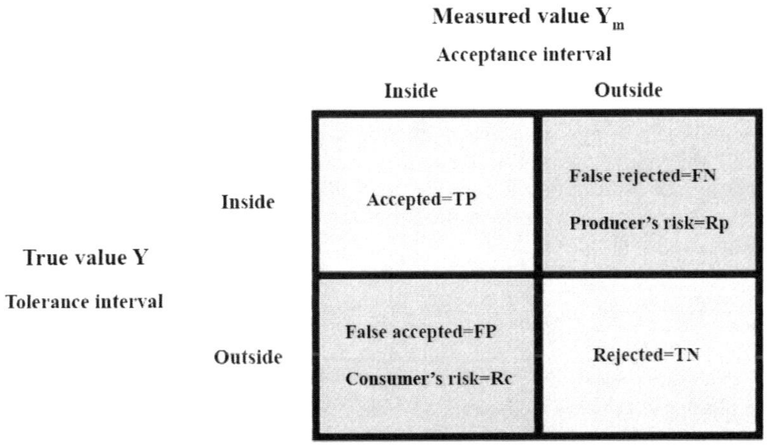

Fig. 2. Confusion matrix for risks calculation

4. Basic assumptions

The roundness deviation of the inner ring of the bearing, with a diameter of $D = 80$ mm, was measured. According to the specifications for these rings, the tolerance for roundness deviation is in the range from $0 - 25$ µm. The determined deviation from roundness amounts to $\bar{y} = 16$ µm, with a standard deviation equal to $u_0 = 5$ µm. Measurement of the deviation from roundness was carried out by an automated system for contact dimensional measurement of the bearing ring. The system enables 100 % control, storage and analysis of data, and separation of bad bearing rings from good ones. By monitoring the mean values and ranges with a control chart, it was determined that the process is stable (under control). The standard uncertainty of measurement results for roundness deviation is $u_m = 1$ µm.

For the given data, nine prior distributions were chosen: four non-parametric distributions, two one-parameter distributions and three two-parameter distributions. The parameters of the selected distributions were determined by using the arithmetic mean of the roundness deviation $\bar{y}$ and the standard deviation u_0. These two quantities are ignored with non-parametric distributions. With one-parameter distributions, the standard deviation is ignored. The estimation of parameters, for one-parameter and two-parameter distributions is determined according to the formulas that can be found in [5]. Parameters calculated for the selected distributions are shown in Table 1

Prior	Parameters	Arguments	p_c	IR_T	AUC
Uniform, U[0, 26.3213]	–	$\eta \geq 0$	0.9498	0.0529	0.9847
Uniform, U[0, 50]	–	$\eta \geq 0$	0.50	1	0.9988
Cauchy	–	$\eta \geq 0$	0.9873	0.0129	0.9999
Gibrat	–	$\eta > 0$	0.9994	0.0006	1
Rayleigh, R(α)	$\alpha = 12.8$	$\eta \geq 0$	0.8520	0.0174	0.9972
Maxwell, M(α)	$\alpha = 10.025$	$\eta \geq 0$	0.8986	0.1128	0.9963
Gamma, Γ(α, λ)	$\alpha = 10.24\,, \lambda = 0.64$	$\eta > 0$	0.9498	0.0529	0.9964
Beta, B(α, β)	$\alpha = 10.24, \beta = 639978.52$	$\eta > 0$	0.9498	0.0529	0.9964
Truncated normal, TN(μ, σ)	$\mu = 16.0119\,, \sigma = 4.9809$	$\eta \geq 0$	0.9643	0.0370	0.9949

Table 1. Prior information

Given that the risk of roundness deviation is assessing, chosen are distributions that take positive values of the argument, and that allows a small amount of measured values equal to zero so that $\eta \geq 0$. Also are selected prior distributions where the probability that the measured values are equal to zero is negligible, and $\eta > 0$.

Next step in risk assessment is the equidistant subdivision of the [−1, 1] interval, which is the domain of the multiplicative factor r. The subdivision with the threshold value equal to 0.1 gives the 21 nodes. For each node, the upper limit of the acceptance interval is found out. According to the (5), the step of changing the value for the upper limit of the acceptance interval is equal to 0.2 μm. The method allows estimation of the limit of the acceptance interval, which is lower than the upper limit of the tolerance interval, as well as the determination of the limit of the acceptance interval, which is higher than the upper limit of the tolerance interval. In this case, the range for the acceptance limit goes from 23 μm, in the case when $r = 1$, to 27 μm, for $r = -1$. Considering the given tolerance interval for risk calculation of roundness deviation, allowed values are in the range from 23 μm to 25 μm. All data required for risk calculation according to the (6) and (7) were determined on this way. For the selected prior distribution, and for each of the 21st subdivision nodes, one confusion matrix is generated.

5. Comparison with machine learning

The creating a confusion matrix in risk assessment does not need a large amount of data divided into a training set and a test set, as in machine learning. Only four data are using for the creating confusion matrix: best estimate $\bar{y}$, and standard uncertainty u_0 for the item of interest Y (which can be set up before measurement according to our prior beliefs), the standard measurement uncertainty u_m, associated to measurement data, and upper tolerance limit T_U. In comparison with machine learning, can be said that in risk assessment, these four data are formed a training set. Based on these four data and given threshold of the upper limit of the acceptance interval, by using (6) and (7), confusion matrixes are generated. In this paper, each confusion matrix contains 10000 data. In risk assessment, confusion matrix gives a comparison between true and measured values. Equation (6) and (7) for the consumer's and the producer's risk estimating are classifiers used to orders data into classes. And these are also the rules that enable the prediction of the number of TP, TN, FP and FN products for a given size of the output data set. A confusion matrix of type 2 × 2 has two classes. In risk assessment, we can talk about the classes "inside" and "outside" of a specific interval. The difference compared to machine learning is that when assessing risk, we distinguish between the intervals to which true and measured values belong. We are talking about measurements that are "inside tolerance interval" and "outside tolerance interval" for true values and "inside acceptance interval" and "outside acceptance interval" for measured values. Several different algorithms are used for binary classification of data in machine learning [6]. Some of the most famous are: Naive Bayes algorithm, logistic regression, k-nearest neighbours, support vector machine, decision tree and neural network. Equation (6) and (7) for ordering into classes, are most similar in origin to the Gaussian Naive Bayes classifier. Both, machine learning and risk assessment combine prior distribution, and the likelihood function for the normal distribution.

Depending on the prior distribution, confusion matrix in risk assessment can contain balanced or imbalanced data. In machine learning, imbalanced data represent a problem. Almost all standard machine learning algorithms give a good and reasonable result only for the balanced data [7]. In metrology, imbalanced data are not only allowed, but they are also desirable. With a well-performed measurement, there is always a disproportion between the number of accepted and the number of rejected products. In the production process, it is required that the number of accepted products that meet the specifications greatly exceeds the number of products that do not meet the required standards. We will define several different ways in which we can verify that data are imbalanced. In metrology, we are interested in conformed measurement or products. In this paper, we take the "inside tolerance interval" class for the majority class, and the "outside tolerance interval" class for the minority class. In that case, the size of the majority class is the same as the number of confirmed products, and the size of the minority class is equivalent to the number of non-conformed products. We define imbalanced ratio, or skew, as the ratio of the number of products in the minority class to the number of products belonging to the majority class [8]. The ratio refers to data that is inside or outside the tolerance interval and is denoted with IR_T. This ratio provides data on the balance of data in risk assessment. If the IR_T is equal to 1, data are balanced, and if the IR_T is close to zero then data are imbalanced. The results for IR_T, for selected priors, are presented in Table 1.

The IR_T determined in this way is unique for all confusion matrix generated for different thresholds of the upper limit of the acceptance interval. Data balance can also be checked by determining the ratio of the number of products that are "outside the acceptance interval" to the number of products or measurements that are "within the acceptance interval". This ratio is denoted with IR_A. The ratio is not unique and changes with the change in the value of the upper limit of the acceptance interval (Figure 3a). The third way of checking data balance is possible by forming the TN/TP ratio. This ratio is also not unique and its value changes with a change in the value of the upper limit of the acceptance interval (Figure 3b). For imbalanced data, the ratios IR_A and TN/TP take on small values close to zero, and for balanced data they can be greater than 1. In metrology, as a result of the risk assessment, we want to have as many true positive products and as few true negative products as possible. Which means that we are always interested in imbalanced data. On the other hand, a perfect separation of products into those that are TP and those that are TN in metrology is not allowed, in the sense that all measurements are classified into these two groups. This means that it is not allowed that the values of FP and FN are both at the same time equal to zero. It is always assumed that there is a certain measurement uncertainty and a certain number of unacceptable measurements, whether they are of the FP or FN type.

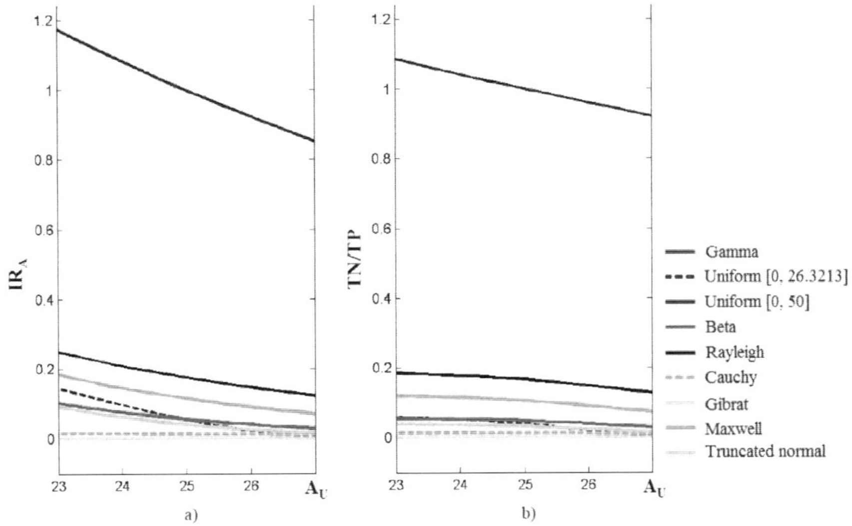

Fig. 3. Imbalanced ratio, a) IR_A, b) TN/TP

6. Results and analysis

The values of TP, TN, FP, and FN in the generated confusion matrix are different for different values of the upper limit of the acceptance interval. These matrices serve for the selection of the appropriate prior distribution. The assessment of the appropriate prior distribution was performed using metrics associated with confusion matrix. Three characteristic metrics: accuracy, precision and recall are calculated as follows:

$$\text{accuracy} = (\text{TP} + \text{TN})/(\text{TP} + \text{FN} + \text{FP} + \text{TN}) \tag{8}$$

$$\text{precision} = \text{TP}/(\text{TP} + \text{FP}) \tag{9}$$

$$\text{recall} = \text{TP}/(\text{TP} + \text{FN}) \tag{10}$$

Accuracy is metrics that tell how many rings are correctly classified into TP and TN categories in relation to the total amount of data. This metric is extremely sensitive to imbalanced data [9]. In the case of imbalanced data desirable in metrology, the number of data classified in the TN category is negligibly small, and it is valid that TP≫TN. According to (8), for such a value relationship between TP and TN, the accuracy actually provides a prediction of the number of rings classified in the TP category in relation to the total amount of data, Figure 4a. This is true for all chosen priors except for the uniform distribution defined on the interval [0, 50]. According to Table 1, Figure 3a and Figure 3b, the uniform distribution U[0, 50] is the only one of the selected distributions that has balanced data. The IR_T value for this distribution is the highest and is equivalent to 1. Compared to all other distributions, this distribution takes the highest values for the ratios IR_A and TN/TP over the entire range for the upper limit of the acceptance interval. According to Figure 4a, the accuracy for U[0, 50] is generally higher than the accuracy of the one-parameter distributions: Rayleigh and Maxwell distributions, but it is lower than the accuracy of the two-parameter distributions: truncated normal, gamma and beta.

Although the accuracy metric has a high value for U[0, 50], according to the requirements set in metrology, we cannot say that this prior is good for risk assessment in the case of roundness deviation of the inner bearing ring. In each of the figures presented in this paper, the graphs for the gamma and beta distributions completely coincide on the interval [24, 26]. On the interval [23, 24], the gamma distribution has higher values for R_c, that is, a larger number of FP rings compared to the beta distribution, and therefore a smaller number of TP rings. On the interval of [26, 27] gamma compared to beta distribution has smaller values for R_P, that is, the number of FR rings. Given that this difference is negligible, of the order of 10^{-4}, we can consider that the gamma and beta distributions have the same behaviour over the entire range for A_U.

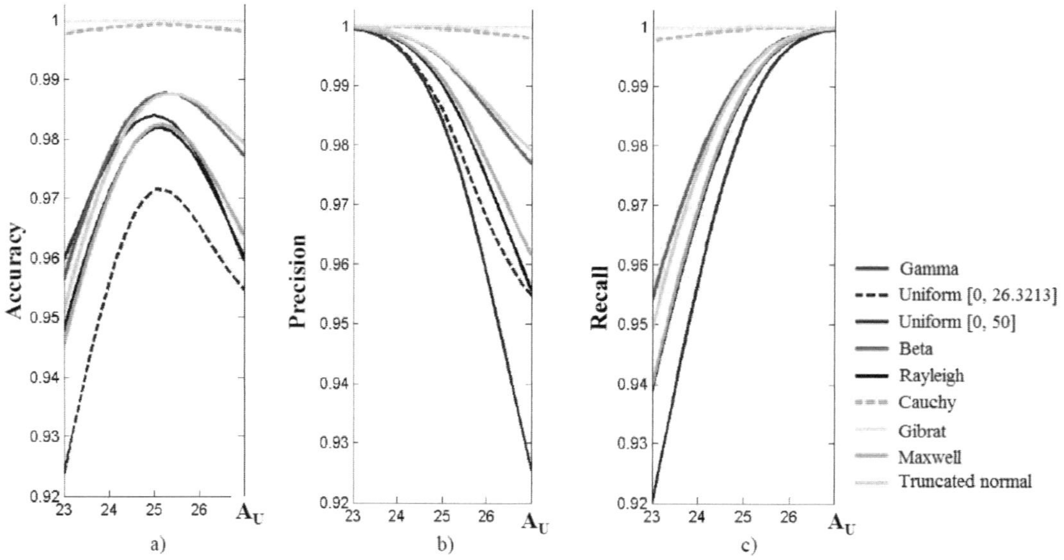

Fig. 4. a) Accuracy, b) Precision, c) Recall

For all distributions, accuracy increases for $A_U \in [23, 25]$ and decreases for $A_U \in [25, 27]$. The Rayleigh's distribution on the interval [23, 24.5] has a slightly higher accuracy than the Maxwell's distribution. Of the two-parameter distributions, beta and gamma distributions have higher accuracy than the truncated normal distribution. The Cauchy's and Gibart's distribution, among the all chosen non-parametric distributions, achieve the highest values for all three metrics, Figure 4. The smallest values for accuracy over the entire range for upper acceptance limit are achieved by the uniform distribution U[0, 26.3213]. This distribution is artificially generated so that the conformance probability, on the interval [0, 25], for this distribution, is equal to 0.9498 and has the same value as the conformance probability for the gamma and beta distributions.

The precision measure, according to (9), gives a prediction of the number of rings classified in the TP category in relation to the total number of rings from the "inside acceptance interval" class. As the upper limit of the acceptance interval increases and the width of the guard band decreases, the number of rings classified in the TP category increases, but the number FP in the denominator of expression (9) also increases, therefore the precision decreases for all distributions on the interval [23, 27]. Given that data for U[0, 50] are balanced, the precision for this distribution is the lowest, Figure 4b. Rayleigh's distribution achieves slightly lower values compared to Maxwell's. Truncated normal distribution achieves slightly higher values for precision, compared to beta and gamma distribution.

The recall measure, according to (10), gives a prediction of the number of rings classified in the TP category in relation to the total number of conformed rings. As the upper limit of the acceptance interval increases, the number of TP increases, but a bearing ring number classified in the FN category decreases, therefore the recall for all priori distribution increases on the interval [23, 27]. According to the recall measure, both uniform distribution U[0, 50] and U[0, 26.3213] show almost the same behaviour. Their graphs overlap, unlike the graphs for accuracy and precision. The difference between the precision metrics values, taken by these two distributions, is negligible, and amounts to 10^{-4}.

The true negative rate (TNR) metric provides a prediction of the number of rings classified as TN in relation to the total number of non-conformed rings and can be calculated as:

$$\mathrm{TNR} = \mathrm{TN}/(\mathrm{FP} + \mathrm{TN}) \qquad (11)$$

Given that the number of rings classified in the TN category decreases as the upper limit of the acceptance interval increases, and the number of rings from the FP category increases, the TNR decreases on the interval [23, 27], for all distributions, Figure 5a.

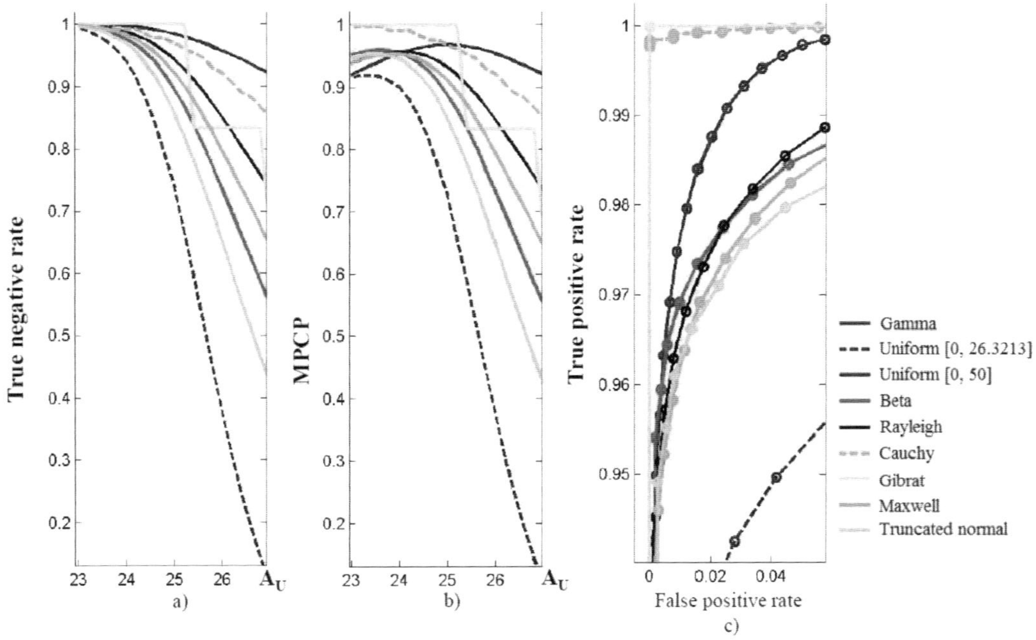

Fig. 5. a) True negative rate, b) Maximum probability of correct prediction, c) ROC curves

On the interval [23, 25], the Gibrat's distribution has the highest values for TNR, compared to all other distributions. The TNR for Gibrat's distribution decreases "stepwise" on the interval [25, 27], which, according to the assumptions, is not an interval of allowed values. Among all selected distributions, according to all selected metrics, the Gibrat's distribution is the distribution that best separates rings into categories TP and TN, with a few rings in categories FP and FN. Unlike other distributions, the number of FP and FN rings, for the Gibrat distribution, is either 0 or 1 or 2. That is why the graph of this distribution is stepwise, Figures 5a, 5b, 6b, 6c. A similar stepwise shape, but to a lesser extent, have the Cauchy's distribution, Figures 5a and 5b.

In order to determine the maximum probability of correct prediction (MPCP) the false negative rate (FNR) and the false positive rate (FPR) are defined as following:

$$\mathrm{FNR} = \mathrm{FN}/(\mathrm{TP} + \mathrm{FN}) \tag{12}$$

$$\mathrm{FPR} = \mathrm{FP}/(\mathrm{FP} + \mathrm{TN}) \tag{13}$$

The false negative rate gives the number of rings classified in the FN category in relation to the total number of conformed rings. The false positive rate gives the number of rings classified in the FP category in relation to the total number of non-conformed rings. Maximum probability of correct prediction is a measure associated with the confusion matrix that tells us what the maximum probability is that we did not choose the FP and FN ring. It is defined as:

$$\mathrm{MPCP} = (1 - \mathrm{FPR})(1 - \mathrm{TNR}) \tag{14}$$

This measure allows us to determine the maximum upper limit of the acceptance interval for each individual distribution, Figure 5b. The results are as follows: According to the MPCP, the upper limit of the acceptance interval for U[0, 26.3213] is $A_U = 23.4$, the associated risks are $R_C = 0.0864$ %, and $R_P = 6.17$ %. For the uniform distribution, U[0, 50] holds $A_U = 25$, $R_C = 0.8$ %, $R_P = 0.8$ %. It is about balanced data, so this result is expected. For the Cauchy distribution holds $A_U = 23.6$, $R_C = 0.0037$ %, $R_P = 0.16$ %. Gibrat distribution: $A_U = 25$, $R_C = 0.0033$ %, $R_P = 0.0040$ %. In this case, MPCP is equal to 1. Rayleigh distribution: $A_U = 24$, $R_C = 0.18$ %, $R_P = 2.71$ %, Maxwell's distribution: $A_U = 24$, $R_C = 0.17$ %, $R_P = 2.77$ %. The Rayleigh and Maxwell distributions are both one-parameter distributions that reach the maximum value for MPCP at $A_U = 24$, but for that value, the consumer risk for Maxwell distribution is lower compared to Rayleigh distribution. For the gamma distribution holds $A_U = 23.6$, $R_C = 0.0522$ %, $R_P = 2.93$ %, Beta distribution: $A_U = 23.6$, $R_C = 0.0523$ %, $R_P = 2.94$ %, and truncated normal distribution: $A_U = 23.4$, $R_C = 0.0331$ %, $R_P = 3.77$ %.

One of the methods by which we can evaluate the behaviour of the prior is Receiver Operating Characteristic (ROC) analysis, Figure 5b. Area under ROC Curve (AUC) is a measure which tells how TPR changes with increasing of FPR. This measure also isn't sensitive to imbalanced data.

If the ROC curve is closer to the upper left corner, the AUC value is higher. If the AUC is equal to 0.5, it represents chance, while the value 1 corresponds to perfect classification. Values below 0.5 are not considered [10]. For the selected priors, the AUC values are shown in Table 1. The AUC number for the Gibrat distribution is equal to 1. It is followed by the Cauchy distribution with an AUC number equal to 0.9999. The uniform distribution U[0.50] with balanced data has a surprisingly large value for the AUC number, equivalent to 0.9988. As said before, this distribution is not suitable for risk assessment. The behaviour of priors is also analyzed for three other known metrics associated with confusion matrix: F1 score, kappa statistics and Matthew's correlation coefficient (MCC), Figure 6.

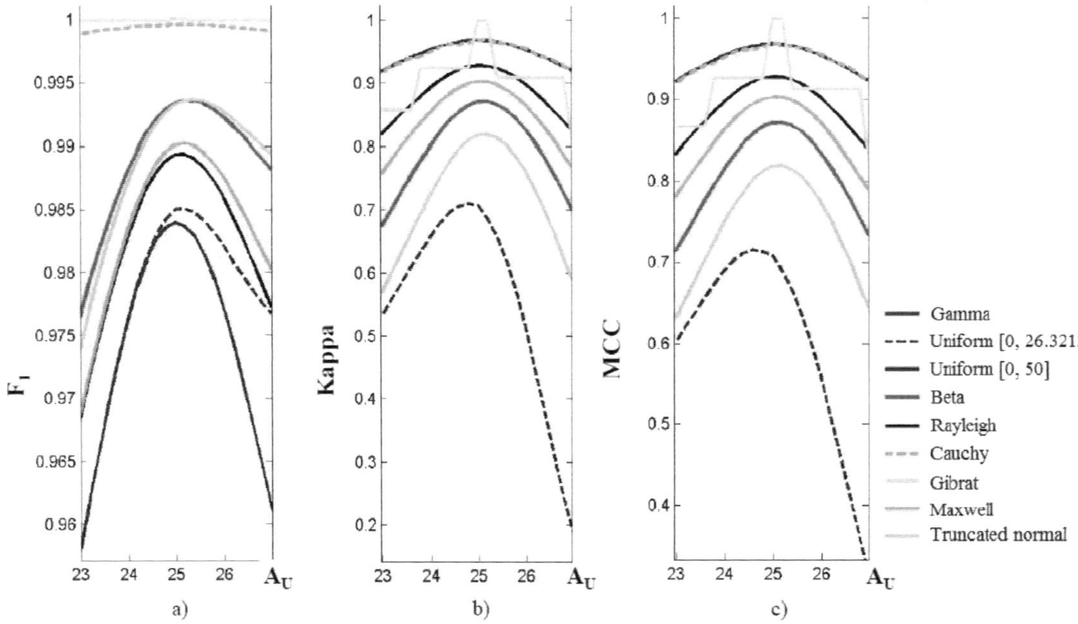

Fig. 6. a) F1 score, b) Kappa, c) Matthew's correlation coefficient

These metrics are calculated as follows:

$$\mathrm{F1} = 2 * (\mathrm{precision} * \mathrm{recall})/(\mathrm{precision} + \mathrm{recall}) \tag{15}$$

$$\mathrm{kappa} = 2 * (\mathrm{TP} * \mathrm{TN} - \mathrm{FN} * \mathrm{FP})/((\mathrm{TP} + \mathrm{FP}) * (\mathrm{FP} + \mathrm{TN}) + (TP + FN) * (\mathrm{FN} + \mathrm{TN})) \tag{16}$$

$$\mathrm{MCC} = ((\mathrm{TP} * \mathrm{TN}) - (\mathrm{FP} * \mathrm{FN}))/\sqrt{(TP + FP) * (TP + FN) * (TN + FP) * (TN + FN)} \tag{17}$$

The F1 score is the harmonic mean of the precision and recall metrics. It is a metric that is suitable for unbalanced data such as those encountered in metrology, in the case when the number of rings of the bearing classified in the TP category is much higher compared to the number of rings classified in the TN category. The values for the F1 score are in the range from 0 to 1. If the F1 score is equal to one, we are talking about the perfect separation of each ring into the appropriate class. If the F1 value is equal to 0, the model poorly classifies the measurements (rings) into the appropriate classes. According to the F1 score metric, Gibrat's and Cauchy's distributions best classify the rings into the corresponding classes, Figure 6a. They are followed by two-parameter distributions, and then followed one-parameter distributions. In the end are, as were expected, uniform distributions as poor classifiers.

The kappa statistic is a measure of agreement between predicted and observed data. Both measures: Kappa statistic and MCC take values in the range from -1 to 1. All values of those measures greater than 0.8 are considered to represent strong agreement [11]. The priors with values lower than 0.8 are U[0, 26. 3213], and partially truncated normal distribution, Figures 6a, 6b. The kappa statistics and MCC gives the same arrangement of priors for the ring of bearing classification, only the values of the kappa statistic are slightly smaller compared to the MCC measure. The balanced data associated with U[0, 50] distribution have higher values of the MCC and kappa statistic. The uniform distribution U[0, 50] and the Cauchy's distribution assume almost the same values. The maximum absolute difference between the values taken by these two distributions is of the order 10^{-3} in the favour of the U[0, 50] distribution.

7. Conclusion

Among all selected priors and according to all selected metrics associated with confusion matrix, the Gibrat distribution is imposed as the best choice for prior distribution in risk assessment. This prior is also the best choice among all those priors where it is implied that the argument of the distribution is $\eta > 0$. If a small measurement value equal to zero is allowed, that is, $\eta \geq 0$ is valid, then the Cauchy distribution is the best choice for the prior distribution. The Gibrat's and Cauchy's distribution are also the best choice among all the tested non-parametric priors. In metrology, we choose a non-parametric uniform prior if we do not want to influence the results of the risk assessment with our personal belief or if we believe that each of the measured values is equally likely, and we do not want to favour any one value over another. According to the performed analysis, the uniform distribution U[0, 50] is not a recommended distribution for the risk assessment because it generates balanced data in the confusion matrix. The uniform distribution U[0, 26. 3213] is also not appropriate prior for risk estimation. On all tested metrics, this prior is almost always low-graded. The AUC number for this prior is the lowest one compared to all other priors. One-parameter distributions for risk assessment should only be chosen when data for the standard uncertainty u_0 are not available, and there is no other choice. For all analyzed metrics, the results for the Rayleigh and Maxwell distributions are in between of the results for the chosen two-parametric and one-parametric distributions. If all the necessary data are available, as the best estimate $\bar{y}$ and standard uncertainty u_0, and if it is necessary to include them in the risk assessment, then two-parameter distributions are the best choice. The gamma or beta distribution are appropriate for risk assessment if the argument values are strictly positive values, and the truncated normal distribution is appropriate if the proportion of measured values equal to zero is significant.

It remains to be seen how changing the basic parameters used in risk assessment will affect the appearance of the confusion matrix and the selected metrics. And will changing the default parameters affect the choice of prior. Also, one of the future stages in the research will be the application of machine learning techniques to the data generated in this way. Models trained on imbalanced data will easily recognize data belonging to the majority class, but will not recognize data belonging to the minority class. Therefore, it will be necessary to apply machine learning algorithms that take into account the unbalancing of the data or to apply data balancing techniques such as data class weighting and various over and under-sampling techniques.

8. References

[1] Runje, B.; Horvatić Novak, A.; Razumić, A.; Piljek, P.; Štrbac, B. & Orošnjak, M. (2019). Evaluation of Consumer and Producer Risk in Conformity Assessment Decisions, Proceedings of the 30th DAAAM International Symposium, Zadar, Croatia, ISSN 1726-9679, ISBN 978-3-902734-22-8, Katalinic, B. (Ed.), pp. 0054-0058, Published by DAAAM International, Vienna, DOI: 10.2507/30th.daaam.proceedings.007

[2] https://www.bipm.org/en/committees/jc/jcgm/publications, (2012). JCGM 106:2012 Evaluation of measurement data – The role of measurement uncertainty in conformity assessment, Accessed on: 2022-09-19

[3] https://www.bipm.org/en/committees/jc/jcgm/publications, (2018). JCGM 100:2008 Evaluation of measurement data – Guide to the expression of uncertainty in measurement, Accessed on: 2022-09-19

[4] https://www.eurolab.org/pubs-techreports, (2017). EUROLAB Technical Report No.1/2017 - Decision rules applied to conformity assessment, Accessed on: 2022-09-19

[5] Đorić, D.; Mališić, J.; Jevremović, V. & Nikolić-Đorić, E. (2007). Atlas raspodela, Građevinski fakultet Univerziteta u Beogradu, ISBN: 978-86-7518-077-7, Beograd

[6] Mandić, M. & Kraljević, G. (2020). Two-Layer Architecture of Telco Churn Auto-ML, Proceedings of the 31st DAAAM International Symposium, Mostar, BiH, ISSN 1726-9679, ISBN 978-3-902734-29-7, Katalinic, B. (Ed.), pp. 0788-0792, Published by DAAAM International, Vienna, Austria, DOI: 10.2507/31st.daaam.proceedings.109

[7] He, H. & Garcia, E. A. (2009). Learning from Imbalanced Data, IEEE Transactions on Knowledge and Data Engineering, Vol 21, No 9, pp. 1263-1284, DOI: 10.1109/TKDE.2008.239

[8] Jeni, L. A.; Cohn, J. F. & De La Torre, F. (2013). Facing Imbalanced Data-Recommendations for the Use of Performance Metrics, International Conference on Affective Computing and Intelligent Interaction (ACII), Geneva, Switzerland, pp. 245-251, DOI:10.1109/ACII.2013.47

[9] Buda, M., Maki, A. & Mazurowski, M. A. (2018). A systematic study of the class imbalance problem in convolutional neural networks, Neural Networks, Vol. 106, pp. 249-259., ISSN 0893-6080, DOI: 10.1016/j.neunet.2018.07.011

[10] Fawcett, T. (2006). An introduction to ROC analysis, Pattern Recognition Letters, Vol. 27, No. 8, pp. 861-874., ISSN 0167-8655, DOI: 10.1016/j.patrec.2005.10.010

[11] McHugh, ML. (2012). Interrater reliability: the kappa statistic, Biochemia Medica, Vol. 22, No. 3, pp. 276-272., DOI: 10.11613/BM.2012.031

DOI: 10.2507/33rd.daaam.proceedings.067

PEACH FIRMNESS PREDICTION USING OPTIMIZED REGRESSION TREES MODELS

Tomislav Ivanovski, Xiaoshuan Zhang, Tomislav Jemrić,
Marko Gulić & Maja Matetić

This Publication has to be referred as: Ivanovski, T[omislav]; Zhang, X[iaoshuan]; Jemric, T[omislav]; Gulic, M[arko] & Matetic, M[aja] (2022). Peach Firmness Prediction Using Optimized Regression Trees Models, Proceedings of the 33rd DAAAM International Symposium, pp.0480-0489, B. Katalinic (Ed.), Published by DAAAM International, ISBN 978-3-902734-36-5, ISSN 1726-9679, Vienna, Austria
DOI: 10.2507/33rd.daaam.proceedings.067

Abstract

The paper focuses on creating accurate model for peach firmness prediction. For this purpose, multiple machine learning models and their optimized variations are developed and compared. Because of its simplicity and robustness, multiple linear regression is used as a base-line model for predicting peach firmness. It assumes a linear relationship between numerical predictors and the outcome. Regression trees is the second developed model. It is a flexible data-driven model that can be used for predicting numerical outcome. The experiment aims to investigate the possibility of improving regression tree model using various metaheuristic optimization techniques implemented in *metaheuristicOpt* and *GA* R packages. As a proof of concept, prediction accuracy between multiple linear regression, regression trees and optimized regression trees models is compared. The results shows that it is possible to improve the peach firmness prediction accuracy of regression trees model using metaheuristic algorithms.

Keywords: Peach firmness prediction; regression trees; global optimization; machine learning; metaheuristics

1. Introduction

Reduction of food waste is a crucial part of global food security and leads to more sustainable food production. Determining the correct fruit harvest time is critical for food waste reduction along the supply chain, and it ensures that the fruit meets the ripeness specifications [1]. Therefore, an early assessment of fruit ripeness using machine learning models trained on data obtained from non-destructive measurement techniques can lead to precise prediction of harvest date and extension of fruit shelf life. Colorimetry, visible imaging, electrical impedance spectroscopy, and spectroscopy imaging [2] are some of the non-destructive techniques that produce high-quality data. Hardware innovations, quality datasets and algorithmic advances are three main factors driving advances in machine learning [3]. Research on sensing technologies in the agriculture sector is being conducted [4] which will result in more quality data for analysis. With increased availability of quality sensor data, variations of machine learning models can be created for data analysis and future predictions [4]. The work presented in this paper aims to extend the research conducted in [29], where simple linear regression, multiple linear regression, and backpropagation neural network models were developed for peach firmness prediction.

In this experiment, a larger dataset is used for the development of the machine learning models, and the new regression tree model is created for firmness prediction. In addition, a special focus is given to the application of nature inspired metaheuristic algorithms as global optimizers for the machine learning models. Evolutionary algorithms (EA) are global optimization techniques used to solve optimization problems. All variants of EAs have the common underlying logic, where an initial random set of candidate solutions is created, and fitness function is applied to each candidate. Based on the fitness values, the best candidate solutions are selected as parents for the offspring. This paper is organized in seven chapters. The review of the papers with the similar topic of research is given in Ch. 2. Features that make up the data set are described in Ch. 3, as well as the machine learning and metaheuristic algorithms used in the experiment. In Ch. 4 the results of the experiment are described, and the most accurate model is presented. The conclusion is given in Ch. 5.

2. Related works

In the previous research [29] simple linear regression, multiple linear regression, and backpropagation neural network models were developed for peach firmness prediction. The dataset used consisted of four measured features: density, firmness, ratio of soluble solids content and titratable acidity, and magnitude of impedance. Experiment results show that multiple linear regression model is the most accurate, which is expected since the dataset is small and the features in it have large variance. This research extends the [29] by including two additional features: angle of impedance and colour. In addition, regression trees model was created, and various nature inspired metaheuristic algorithms were applied in order to additionally optimize the model. Another peach firmness prediction research by G. Zhang et al. is described in [7]. The correlation analysis in [7] shows that the highest correlation coefficients are between the dielectric properties and firmness of the peach. Using the measured peach dielectric properties as inputs to multi-layer neural network model, peach firmness is accurately predicted. The research also shows that the proportion of peach red surface area has small relationship with the quality scores calculated by the grading model developed in [7]. Similar to [29], in [8] D. Jamaludin et al. study the dielectric characteristic of banana fruit at different ripening stages using impedance measurement. Using the measured impedance, it is possible to differentiate the unripe, ripe, and overripe banana [8]. The experiment shows that the impedance correlates with the soluble solid content. This discovery is used to develop simple linear regression model to predict soluble solid content value based on measured impedance, which indicates that banana ripeness can be predicted by its dielectric properties. The study described in [9] uses image data of melon skin colour to distinguish ripe melons from unripe ones. The developed linear discriminant analysis model can be used to accurately predict melon harvesting time using the colour image data. Such model will prove useful as a preharvest tool for making decisions about whether the melon is ready to harvest. Like the study described in [9], in [10] areal images of apples are used for development of an apple ripeness classifier model. The described classifier model is based on the artificial neural network model optimized using the genetic algorithm. The model classifies the apples to one of the following stages of ripeness: unripe, half-ripe, ripe, and overripe.

Nature inspired metaheuristic algorithms are applicable in wide range of applications. Optimization of liner and exponential regression models to forecast global CO2 emissions using bat algorithm is presented in [11], while in [12] the bat algorithm is used to optimize fuzzy PD speed controller for brushless DC motor. Another application of BA is described in [13], where BA is used to optimize photo-voltic systems to operate at the maximum power point. Optimization of photo-voltic model is also described in [14], where optimization is realized using combination of DE algorithm and reinforcement learning. The study in [16] describes the use of DE algorithm for determining the appropriate number of wells and the maximum oil processing capacity in an oil and gas industry. In [15] an improved adaptive differential evolution (IADE) algorithm which adopts its parameters during execution is used to optimize the flight path trajectory of unmanned arial vehicles. The application of DE for CoCoMo and CoCoMo II model optimization is described in [17], while a modified DE algorithm that is used to optimize recommendation systems is described in [18]. In [19] the PSO algorithm is used to improve the accuracy and convergence speed of radial basis function neural network (RBFNN) for estimating the battery pack SOC. Similar in [20] the improved-hybrid particle swarm optimization (IH-PSO) algorithm is used to optimize the batch size in batch processing, while in [21] a modified PSO algorithm is used to optimize collaborative beam folding process in wireless sensor networks. Another application of metaheuristic algorithms in wireless sensor networks is described in [22], where the energy consumption was successfully optimized by using GA. The demonstration of GA use in the optimization of nuclear reactor core load, including fuel reloads is described in [23], while the study in [24] uses a GA as an optimization technique for creating a household level hybrid renewable energy system. In [25] the neuro-fuzzy model in combination with the GA are used to optimize the train routes on the railway network, where in [26] an optimization method based on GA is used to optimize ship voyage and therefore minimizing fuel consumption and greenhouse gas emissions. The study in [27] gives insight on the use of GA for optimizing structural design of deck concrete arch bridges, while the study in [40] utilizes GA for obtaining optimal design for I-beams.

3. Dataset description and methods

3.1. Data set description

The dataset on which the experiment is conducted consists of various peach feature measurements. Peach impedance is one of the measured features and it is measured with 500 mV and 10 kHz frequency alternate voltage.

Insight into the structural characteristics of biological tissues is obtained using electrical impedance measurements [28]. The dielectric properties are related to changes in the membrane structure [29], and changes in the membrane structure reflect changes in fruit ripening [30][31][32]. Since impedance is a complex number and can be represented in the polar form using magnitude and phase angle, the two main predictors in the dataset are *Zs* and *Angle* respectively. By combining juice extracted from a peach with alkaline solution, the titratable acidity *TA* of the peach is obtained. The soluble solids content *SSC* is obtained using a refractometer and juice extracted from a peach, while peach firmness is measured using penetrometer. Because the firmness is a good maturity indicator [33], it is used as outcome in the machine learning models described in the paper. The colour of the peach is represented by Delta E value. Initial dataset consisted of twenty-one measured features. To increase the efficiency of machine learning algorithms the dimension of the data set had to be reduced [34] using the expert's domain knowledge. An example of feature combination is quotient of SSC and TA which is used in the final dataset. The final dataset used in the experiment consists of 200 observations and 6 features. Fig. 1 shows the distribution of the measured features used in the experiment. It is possible to see that the colour feature follows symmetric normal distribution, while the ratio of soluble solids content and titratable acidity and density are slightly skewed to the right. Firmness, phase angle of impedance, and the impedance magnitude follow bimodal distribution. By examining the distribution of firmness, it is possible to get an insight into the characteristics of peaches that make up the data set. The firmness distribution is skewed towards lower values, and peaches with lower firmness tend to be riper [33].

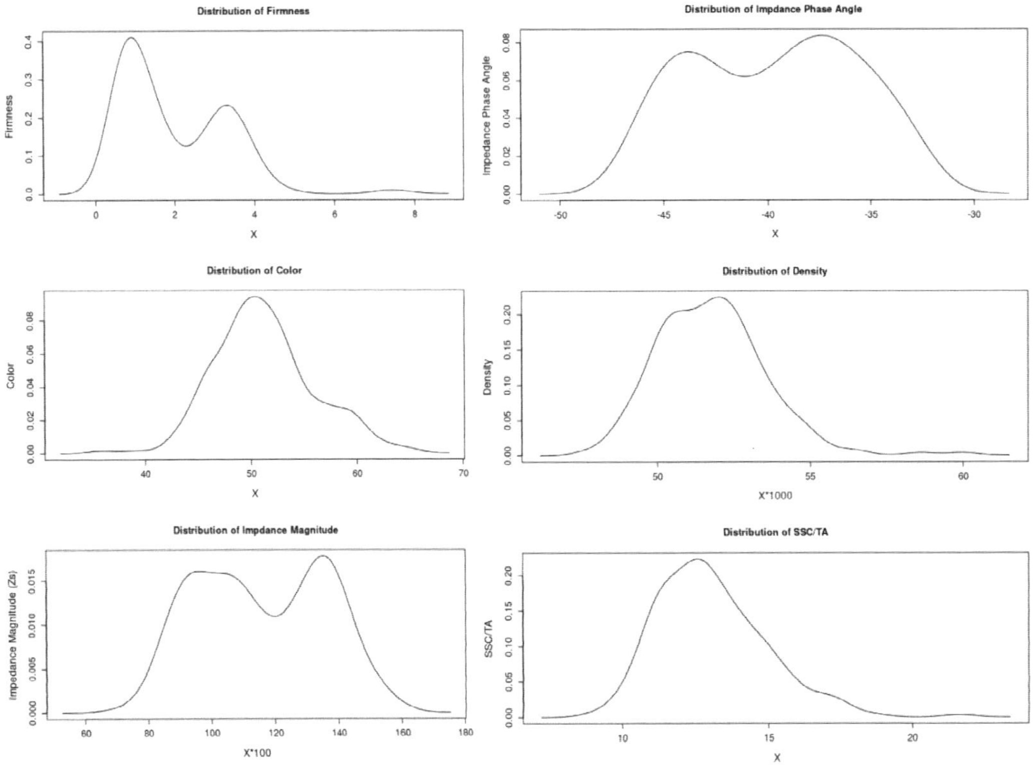

Fig. 1. Distribution of features in dataset

3.2. Classification and regression trees

Classification and regression trees (CART) are a flexible data driven method that can be used for classification or numerical prediction [34]. Segmentation of the predictor space into multiple simple regions using recursive partitioning and pruning processes is the main idea behind CART algorithm [35]. The resulting subgroups should be more homogeneous in terms of the outcome feature, thereby creating useful prediction rules that are easily interpreted by humans. The most prominent advantage of CART is its simplicity and transparency. CART model consists of terminal and decision nodes. The decision nodes give the splitting value on the specific predictor while the terminal nodes contain predicted value. For the regression program, the values of terminal nodes are the mean of the outcome feature of training observations in the adequate region to which the observations belong. Fig. 2 shows an example of regression tree model created using *Hitters* dataset included in *ISLR* R library. Terminal nodes are represented by round rectangles, while sharp rectangles represent decision nodes.

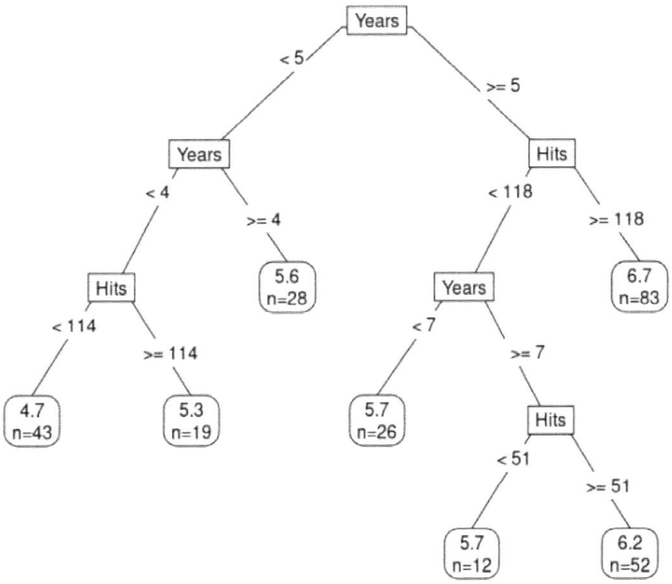

Fig. 2. Regression tree model example

Regression tree models are not very robust, meaning a small change in the data can cause a large change in the final estimated tree. However, by aggregating many decision trees, using methods like bagging, random forests, and boosting, the predictive performance of trees can be substantially improved [35].

3.3. Multiple linear regression

Multiple linear regression (MLR) method assumes a linear relationship between multiple predictors and outcome feature. It is a straightforward approach for predicting a numerical outcome based on multiple predictors [35]. Multiple linear regression, just like simple linear regression uses least square approach for adjusting the coefficients [35]. A generic expression for multiple linear regression model that approximates relationship between the predictors and the outcome is an extension of the simple linear regression model [35] and is shown in Eq. (1):

$$Y = \beta_0 + \beta_1 X_1 + \beta_2 X_2 + \beta_3 X_3 \dots + \beta_n X_n + e \quad (1)$$

Where:
- $\beta_{0\dots n}$: Multiple linear regression coefficients
- $X_{1\dots n}$: Feature values
- e: Mean-zero random error term

3.4. Genetic algorithm

Genetic algorithm is a metaheuristic algorithm that belongs to a family of evolutionary algorithms and can be applied in solving global optimization problems. Inherent characteristic of all metaheuristics algorithms is that they make few or no assumptions about the problem being optimized, and therefore can easily be applied to a wide range of optimization problems. GA is one of multiple variants of evolutionary algorithms that is inspired by Charles Darwin's theory of evolution. Given a population of individuals within some environment that has limited resources, competition for those resources causes natural selection, which in turn causes a rise in the fitness of the population [5]. By defining a quality function to be maximized and creating a population of initial candidate solutions, it is possible to apply the quality function to candidate solutions as an abstract fitness measure - the higher the better [5]. Based on calculated fitness values some of the better candidates are chosen as parents to the next generation. The offspring is created by applying recombination and mutation operations on parent candidates. Recombination operator is applied to two or more selected parents, producing one or more new candidates. The mutation operator is applied to a single candidate and results in one new candidate. Both parents and offspring have their fitness evaluated and then compete based on their fitness (and possibly age) for a place in the next generation. This process can be iterated until a candidate with sufficient quality is found or a previously set computational limit is reached [5]. Fig. 3 shows rise in fitness value of the best candidate solution (green) and the population (blue) during the run time of genetic algorithm used in decision tree model optimization, which is described in Ch. 4.

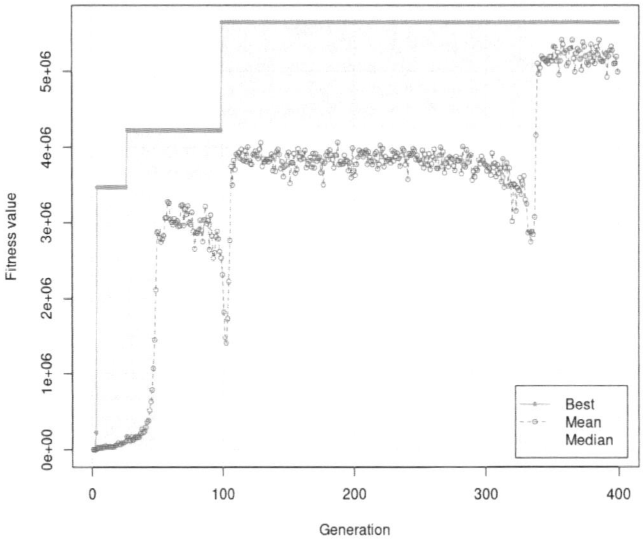

Fig. 3. Change in fitness value over time

3.5. *Bat algorithm*

Like the genetic algorithm, the bat algorithm (BA) is a metaheuristic algorithm for global optimization. The bat algorithm and particle swarm optimization algorithm can be classified as swarm intelligence (SI) algorithms. Like EA, SI algorithms are global optimization techniques that use a swarm of multiple, interacting agents to generate search moves in the search space. A wide spectrum of SI based algorithms have emerged in the last decades, but the lack of mathematical framework and in-depth understanding of how such algorithms may converge are still some of the important issues [6]. The bat algorithm was inspired by the echolocation behaviour of microbats, with varying pulse rates of emission and loudness. BA essentially uses a frequency tuning technique to increase the diversity of the solutions in the population, while the balance between exploration and exploitation can be controlled by tuning algorithm-dependent parameters [36]. Each bat i is associated with a velocity v_i^t and a location x_i^t, at iteration t, in a d-dimensional search space. The best solution in a population in any given iteration is marked with x^*. Following are equations for x_i^t and velocities v_i^t. The B [0,1] is a random vector drawn from a uniform distribution.

$$f_i = f_{min} + (f_{max} - f_{min})\beta \quad (2)$$

$$v_i^t = v_i^{t-1} + (x_i^{t-1} - x^*)f_i \quad (3)$$

$$x_i^t = x_i^{t-1} + v_i^t \quad (4)$$

The BA can be considered as a frequency-tuning algorithm to provide a balanced combination of exploration and exploitation of the search space. The loudness and pulse emission rates essentially provide a mechanism for automatic control and auto-zooming into the region with promising solutions. In order to provide an effective mechanism to control the exploration and exploitation and switch to exploitation stage, when necessary, the loudness A_i and the rate of pulse emissions r_i must change during algorithm iterations. In following equations α and γ are constants.

$$A_i^{t+1} = \alpha A_i^t \quad (5)$$

$$r_i^{t+1} = r_i^0[1 - \exp(-\gamma t)] \quad (6)$$

3.6. *Differential evolution*

Differential evolution (DE) algorithm is part of the family of evolutionary algorithms and is used for solving real parameter global optimization problems. DE is robust and simple algorithm and like other EAs. It can produce new offspring solutions through three mechanisms: mutation, crossover, and selection [37]. It utilizes directional information from the population, where each individual in the current generation is allowed to breed with multiple other randomly selected individuals from the population.

The algorithm starts by initializing all agents in the search space with random positions, after which mutation is applied to create a vector v_i^t [38]. The DE standard mutation operator needs three randomly selected different individuals from the current population for each individual to create a mutated vector. The goal of standard mutation operator is to recognize good variation directions and to increase the number of generations having fitness improvement [38].

$$v_i^t = x_j^t + F(x_k^t + x_l^t) \tag{7}$$

Where F is a differential weight real constant whose values is between 0 and 1. Offspring individual u_i^t is created by mating of the mutated individual v_i^t with x_i^t. The genes *m* of u_i^t are determined by the crossover probability $C_r \in [0, 1]$ and are inherited from both x_i^t and v_i^t [38].

$$u_{i,m}^t = \begin{cases} v_{i,m}^t, \ if\ rand(m) \leq C_r\ or\ m = rn(i) \\ x_{i,m}^t, if\ rand(m) > C_r\ and\ m \neq rn(i) \end{cases} \tag{8}$$

Where m=1,..., N corresponds to the mth gene of an individual vector. The expression rand(m) ∈ [0, 1] is the mth evaluation of a uniform random number generator and rn(i) ∈ {1,...,N} is a randomly chosen index which ensures that u_i^t gets at least one element from v_i^t [38].

The selection process in which fittest individuals are selected as parents for the next generation is conducted between each individual x_i^t and its offspring u_i^t. The winner is selected based on objective function values and promoted to the next generation. The old generation is replaced by the new one and the search process continues until the stopping condition is fulfilled.

3.7. Particle swarm optimization

Particle swarm optimization (PSO) is a metaheuristics algorithm inspired by the social cooperation of organisms observed in nature [39]. It solves an optimization problem by having a population of candidate solutions (particles) that work under social behaviour in swarms. The most prominent characteristic of PSO algorithm is the socio-cognitive learning process that is based on a particle's own experience and the experience of the most successful particle in the swarm. For an optimization problem of *n* variables, a swarm of N_P particles is defined, where each particle is assigned an initial random position in the n-dimensional space. Each particle has its own trajectory, namely position x_i and velocity v_i. In every iteration, each particle is updated by following the two best values: the best solution each particle has achieved so far x_i^*, and the best value obtained so far by any particle in the population x_g [38]. At iteration $t + 1$, the swarm can be updated by following equations.

$$v_i(t+1) = v_i(t) + cr_1[x_i^*(t) - x_i(t)] + cr_2[x_g(t) - x_i(t)] \tag{9}$$

$$x_i(t+1) = x_i(t) + v_i(t+1), i = 1, \dots, N_P \tag{10}$$

Where the acceleration constant $c > 0$, and r_1 and r_2 are uniform random numbers within [0, 1]. The intuitively simple representation and low number of adjustable parameters of PSO algorithm make it a popular choice for solving optimization problems [39]. Prominent characteristic of PSO algorithm is that it can locate the optimum region faster than EAs, but once in this region it progresses slowly due to the fixed velocity step size [38].

4. Results

Eighty percent of data set observations make the training and validation data, while the remaining twenty percent make the test data. The training data consists of 160 observations that are used to create the machine learning models, while the test data consists of 40 observations used to estimate their accuracy. Multiple linear regression (MLR) is used as a baseline machine learning model for predicting peach firmness. The root-mean-square error (RMSE) is used to estimate the accuracy of the model. The RMSE value for multiple linear regression model on test data is 1.717526. The GA, BA, DE, and PSO global optimization algorithms were applied to the regression model, but no significant change in the β coefficients and final test RMSE was obtained. Next, the CART model is trained. The default pruned tree is automatically built by selecting the complexity parameter that yields smallest error on validation data based on cross validation procedure using *rpart* R library. Using the *rpart* R library the cross validation and pruning of the tree is done automatically. From the Fig. 4 it is possible to see that the default pruned tree consists of seven decision nodes and eight terminal nodes. Note that recursive partitioning and pruning algorithms responsible for creating the default pruned tree did not include the colour feature in the decision nodes of the final model. By including it, additional complexity would be added to the model, and it would be prone to overfitting [35]. Below each decision node is the yes-no conditional based on which the split is made. Inside each decision and terminal node is the mean value of firmness that is calculated using observations that satisfy conditions specified by the decision nodes. The RMSE of default pruned tree on test data is 1.747593, which is slightly higher then base-line linear regression model.

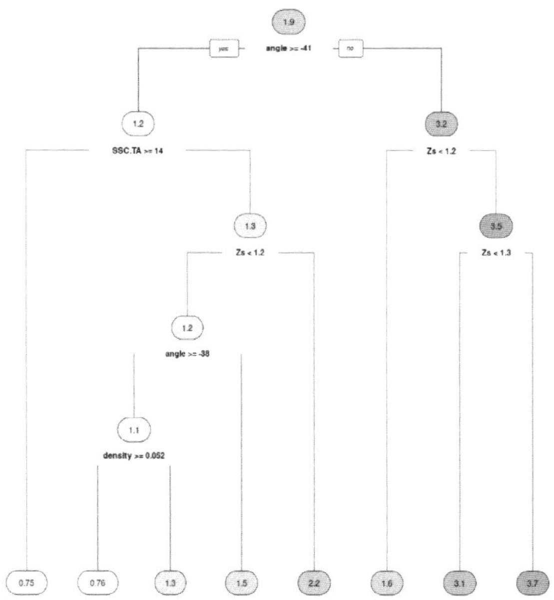

Fig. 4. Default pruned tree created using *rpart* R library

Using various metaheuristic algorithms, it is possible to optimize the splitting value on the predictors in the default pruned tree, thus obtaining modified CART model. By modifying the splitting value, the mean value of firmness in terminal nodes will change, hence predictions on test data will differ. For GA the upper bounds of the search space are defined as 50 percent above the splitting value of decision node in the default pruned tree, while the lower bounds are defined as 50 percent below the splitting value. The goal is the maximization of the objective function, thus the individual with the highest fitness value will be selected as the best one. The population size is set to 200 and the number of best individuals to survive at each generation is set to 40 percent. The *GA* R library is used to implement the genetic algorithm optimization of the splitting values. The model modified using GA is shown on Fig. 5, and its estimated accuracy is 1.589032, which is an improvement from the default pruned tree.

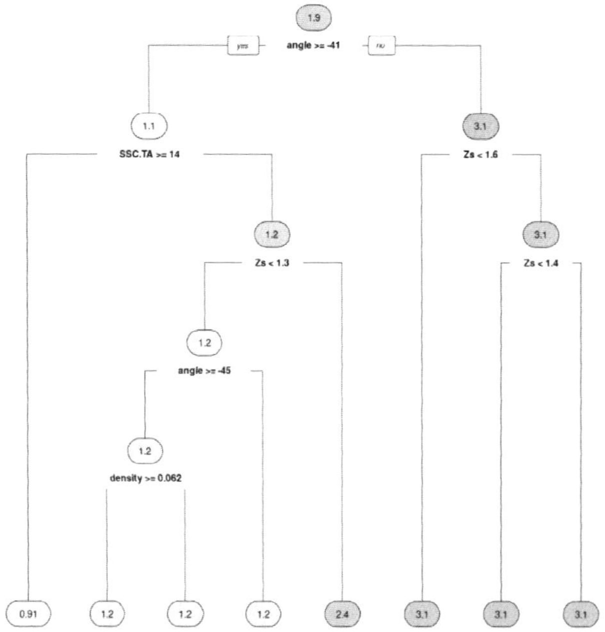

Fig. 5. Default pruned tree modified using GA

For BA, DE and PSO optimization techniques the upper bounds of the search space are defined as 10 percent above the splitting value of decision node in the default pruned tree, while the lower bounds are defined as 10 percent below the splitting value. For the BA, the population size is set to 40, while the maximum and minimum frequencies are set to the value of 0.1. The factor responsible for increasing the pulse rate is set to its default value of 1, and the factor used to decrease loudness is set to 0.1. The *metaheuristicOpt* R library is used to implement the BA, DE and PSO optimizations of the splitting values. The estimated accuracy for the model modified using BA is 1.741407, which is approximately the same as the default pruned tree.

The DE evolutionary algorithm defines that each individual in the generation is allowed to mate with other randomly selected individuals. The scaling factor for mutation operator is set to 0.8, while the factor determining the crossover probability is set to 0.5. The population size is set to 20. The estimated accuracy for the model modified using DE is 1.635962, which is an improvement from the default pruned tree.

When implementing PSO as an optimization algorithm for splitting values of default pruned tree model, the inertia weight is set to 0.729. Both the individual and group acceleration constants are set to 1.49445. As with DE, the population size is set to 20. The estimated accuracy for the model modified using PSO is 1.735599, which is slightly better than the default pruned tree. The Table 1 summarizes all the created models.

Model	RMSE
MLR	1.717526
Default CART	1.747593
GA Optimized CART	1.589032
BA Optimized CART	1.741407
DE Optimized CART	1.635962
PSO Optimized CART	1.735599

Table 1. Test RMSE summary

5. Conclusion

The aim of this research is to successfully predict the correct peach harvest time by predicting its firmness using machine learning models and measured peach features. Multiple linear regression model, default CART model, and CART model optimized using GA, BA, DE, and PSO metaheuristic algorithms are used for predicting peach firmness. Prediction results are compared, and the most accurate model was found. The results showed that CART model optimized using GA gave the most accurate predictions on the new data. This empirical experiment shows that by using various metaheuristic optimization techniques implemented in *metaheuristicOpt* and *GA R* libraries it is possible to improve the accuracy of the default CART model. It is important to note that the experiment is limited by the used *R* libraries and the optimization approach previously described. Hence different results could be obtained if other software packages or optimization approaches are used.

The future research will focus on applying additional nature inspired metaheuristic algorithms as optimizers for CART and other machine learning models. In addition, an adaptive network-based fuzzy inference system (ANFIS) model will be created, and its performance will be compared to the performance of models described in this paper.

6. Acknowledgement

This research was funded by the University of Rijeka, grant number uniri-drustv-18-122.

7. References

[1] Bo Li, J. Lecourt, G. Bishop. (2018). Advances in Non-Destructive Early Assessment of Fruit Ripeness towards Defining Optimal Time of Harvest and Yield Prediction - A Review. Plants 2018, 7, 3

[2] Satyam Srivastava, Shashikant Sadistap. (2017). Non-destructive sensing methods for quality assessment of on-tree fruits: a review. Springer Science+Business Media, LLC 2017

[3] F. Chollet., J.J. Allaire. (2017). Deep Learning with R - Version 1. Manning Publications (2017)

[4] Sonka, Steve. (2016). Big data: fueling the next evolution of agricultural innovation. Journal of Innovation Management 4.1: 114.

[5] A.E. Eiben, J.E. Smith. (2015). Introduction to Evolutionary Computing Second Edition. Springer (2015)

[6] Xin-She Yang. (2020). Nature-inspired optimization algorithms: Challenges and open problems, Journal of Computational Science 46 (2020)

[7] G. Zhang, Q. Fu, Z. Fu, X. Li, M. Matetić, M. B. Bakaric, T. Jemrić (2020). A Comprehensive Peach Fruit Quality Evaluation Method for Grading and Consumption. Appl. Sci. 2020, 10, 1348;

[8] D. Jamaludin, S. Abd Aziz, N. U. A. Ibrahim. (2014). Dielectric Based Sensing System for Banana Ripeness Assessment. International Journal of Environmental Science and Development, Vol. 5, No. 3, June 2014

[9] U. Ahmad, D. P. Bermani, Mardison. (2018). Color Distribution Analysis for Ripeness Prediction of Golden Apollo Melon. Telkomnika, Vol.16, No.4, August 2018, pp. 1659 1666

[10] S. Sabzi, Y. Abbaspour-Gilandeh , G. García-Mateos, A. Ruiz-Canales, J. M. Molina-Martínez, J. I. Arribas. (2019) An Automatic Non- Destructive Method for the Classification of the Ripeness Stage of Red Delicious Apples in Orchards Using Aerial Video. Agronomy 2019, 9, 84

[11] M. Bahmani, A. GhasemiNejad, F. N. Robati, N. A. Zarin. (2020). A novel approach to forecast global CO 2 emission using Bat and Cuckoo optimization algorithms, MethodsX 7 (2020) 100986

[12] K. Premkumar, B.V. Manikandan. (2016). Bat algorithm optimized fuzzy PD based speed controller for brushless direct current motor. Engineering Science and Technology, an International Journal 19 (2016) 818–840

[13] Ali M. Eltamaly, M. S. Al-Saud, Ahmed G. Abokhalil. (2020). A Novel Bat Algorithm Strategy for Maximum Power Point Tracker of Photovoltaic Energy Systems Under Dynamic Partial Shading. IEEE Access, Digital Object Identifier 10.1109/ACCESS.2020.2964759

[14] Z. Hu, W. Gong, S. Li. (2021). Reinforcement learning-based differential evolution for parameters extraction of photo-voltaic models. Elsevier, Energy Reports 7 (2021) 916–928

[15] Jian-li Su, H. Wang. (2021). An improved adaptive differential evolution algorithm for single unmanned aerial vehicle multitasking. KeAi, Defence Technology 17 (2021) 1967 - 1975

[16] Bilal, M. Pant, M. Stanko, L. Sales. (2021). Differential evolution for early-phase offshore oilfield design considering uncertainties in initial oil-in-place and well productivity. Elsevier, Upstream Oil and Gas Technology 7 (2021) 100055

[17] P. Singala, A C. Kumarib, P. Sharmaa. (2020). Estimation of software development effort: A differential evolution approach. Elsevier, Procedia Computer Science 167 (2020) 2643–2652

[18] U. Boryczka, M. Bałchanowski. (2021). Speed up differential evolution for ranking of items in recommendation systems. Elsevier, Procedia Computer Science 192 (2021) 2229–2238

[19] G. Zhang , B. Xia, J. Wang , Bo Ye , Y. Chen , Z. Yu , Y. Li. (2022). Intelligent state of charge estimation of battery pack based on particle swarm optimization algorithm improved radical basis function neural network. Elsevier, Journal of Energy Storage 50 (2022) 104211

[20] J. M. Powera, C. Ou-Yanga, Y. Juanb. (2021). Optimize batch size combination using improved hybrid particle swarm optimization. Elsevier, Sixth Information Systems International Conference (ISICO 2021)

[21] R. M. Mainaa, P. K. Lang'ata, P. K. Kihatoa. (2021). Collaborative beamforming in wireless sensor networks using a novel particle swarm optimization algorithm variant. Elsevier, Heliyon 7 (2021) e08247

[22] W. Yan, S. Xin-xin, S. Yan-ming. (2011). Study on the application of Genetic Algorithms in the optimization of wireless network. Elsevier, Procedia Engineering 16 (2011) 348-355

[23] A.V. Sobolev, A. S. Gazetdinov, D.S. Samokhin. (2017). Genetic algorithms for nuclear reactor fuel load and reload optimization problems. Elsevier, Nuclear Energy and Technology 3 (2017) 231–235

[24] M. J. Mayer, A. Szilágyi, G. Gróf. (2020). Environmental and economic multiobjective optimization of a household level hybrid renewable energy system by genetic algorithm. Elsevier, Applied Energy 269 (2020) 115058

[25] P. Dolgopolov, D. Konstantinov, L. Rybalchenko, R. Muhitovs. (2019). Optimization of train routes based on neuro-fuzzy modeling and genetic algorithms. Elsevier, Procedia Computer Science 149 (2019) 11–18

[26] H. Wang , X. Lang, W. Mao. (2021). Voyage optimization combining genetic algorithm and dynamic programming for fuel/emissions reduction. Elsevier, Transportation Research Part D 90 (2021) 102670

[27] M. Z. Abd Elrehim, M. A. Eid, M. G. Sayed. (2019). Structural optimization of concrete arch bridges using Genetic Algorithms. Elsevier, Ain Shams Engineering Journal 10 (2019) 507–516

[28] I. Jocsak, G. Vegvari, E. Vozary. (2019) Electrical impedance measurement on plants: a review with some insights to other fields. Theor. Exp. Plant Physiol. (2019) 31:359–375

[29] T. Ivanovski, Z. Guoxiang, T. Jemrić, M. Gulić, M. Matetić. (2020). Fruit firmness prediction using multiple linear regression. Proceedings of the 43rd International Convention MIPRO, (2020) 1570-1575

[30] Pietro Ibba, Aniello Falco, Biresaw Demelash Abera, Giuseppe Cantarella, Luisa Petti, Paolo Lugli. (2020). Bio-impedance and circuit parameters: An analysis for tracking fruit ripening. Postharvest Biology and Technology 159 (2020) 110978

[31] Marco Grossi, Bruno Riccò. (2017) Electrical impedance spectroscopy (EIS) for biological analysis and food characterization: a review. J. Sens. Sens. Syst., 6, 303–325, 2017

[32] G. M. Weaver, H.O. Jackson. (1966). Electric impedance, in objective index of maturity in peach. Canadian journal of plant science Vol. 46

[33] F. Roger Harker, John H. Maindonald. (1994). Ripening of Nectarine Fruit Changes in the Cell Wall, Vacuole, and Membranes Detected Using Electrical Impedance Measurements. Plant Physiol. (1994) 106: 165-171

[34] Galit Shmueli, Peter C. Bruce, Inbal Yahov, Nitin R. Patel, Kenneth C. Lichtendahl Jr. (2018). Data mining for business analytics. John Wiley & Sons, Inc. (2018)

[35] Gareth James, Daniela Witten, Trevor Hastie, Robert Tibshirani. (2017). An Introduction to Statistical Learning with Applications in R. Springer Science+Business Media New York 2013 (Corrected at the printing 2017)

[36] Xin-She Yang. (2013). Bat Algorithm: Literature Review and Applications. International Journal of Bio-Inspired Computation, August 2013. DOI: 10.1504/IJBIC.2013.055093
[37] M. F. Ahmad, N. A. Mat Isa, W. H. Lim, K. M. Ang. (2022). Differential evolution: A recent review based on state-of-the-art works. Elsevier, Alexandria Engineering Journal (2022) 61, 3831–3872
[38] Ke-Lin Du, M.N.S. Swamy. (2016). Search and Optimization by Metaheuristics - Techniques and Algorithms Inspired by Nature. Birkhäuser (2016)
[39] S. Sengupta, S. Basak, R. A. Peters. (2019). Particle Swarm Optimization: A Survey of Historical and Recent Developments with Hybridization Perspectives. Machine learning and knowledge extraction 2019, 1, 157–191; doi:10.3390
[40] Raspudic, V. (2020). Optimal Design of Laterally Unrestrained I-Beams Using Genetic Algorithm, Proceedings of the 31st DAAAM International Symposium, pp.0683-0691, B. Katalinic (Ed.), Published by DAAAM International, ISBN 978-3-902734-29-7, ISSN 1726-9679, Vienna, Austria DOI: 10.2507/31st.daaam.proceedings.095

DOI: 10.2507/33rd.daaam.proceedings.068

ONLINE MULTI – SENSOR PROCESS MONITORING DURING ROBOTIC SANDING OF THIN-WALLED STRUCTURES

Dora Bagaric*, Tomislav Staroveski, Miho Klaic & Damir Ciglar

This Publication has to be referred as: Bagaric, D[ora]; Staroveski, T[omislav]; Klaic, M[iho] & Ciglar, D[amir] (2022). Online Multi – Sensor Process Monitoring during Robotic Sanding of Thin-Walled Structures, Proceedings of the 33rd DAAAM International Symposium, pp.0490-0495, B. Katalinic (Ed.), Published by DAAAM International, ISBN 978-3-902734-36-5, ISSN 1726-9679, Vienna, Austria
DOI: 10.2507/33rd.daaam.proceedings.068

Abstract

The goal of the proposed research is the development of indirect sanding paper wear level and surface quality estimation models for sanding of thin-walled structures. The models should be applicable in robotic sanding systems in real time. Since the dynamics of tool wear appears highly nonlinear even in the conventional machining processes, it can be assumed that an array of impacting factors will occur due to lower rigidity of the robot, elastic deformation of the thin-walled workpieces, and due to a damping effect of the tool interface. Additionally, it can also be expected that the factors will have a negative influence on the development of the models. Therefore, to realize the research, a number of experiments will be conducted in order to determine the effect of machining parameters on the sanding paper wear dynamics. Subsequently, determination of the process parameters sensitive to change in tool state and/or workpiece surface quality will be conducted. Lastly, design of the sanding paper wear and workpiece surface quality estimation models will be carried out. By successfully realizing the aforementioned models, preconditions for achieving higher autonomy in automated robotic sanding systems will be established.

Keywords: sanding; tool condition; surface quality; multi-sensor monitoring; artificial intelligence

1. Introduction

Application of robots in finishing processes is one of significant trends in the development of state-of-the-art machining systems, and sanding processes are among the most performed finishing processes. Such systems are, in most cases, equipped with an axial force control drive, on which a main drive with a sanding tool with an elastic interface is mounted. Sanding tools are subjected to intensive wear resulting from mechanical, thermal, and chemical loads during sanding. Negative impact of tool wear on the workpiece surface quality is noticeable even in the early stages of tool wear. Consequently, sanding paper is frequently replaced, potentially causing unnecessarily frequent process interruptions and hence, unnecessary costs.

The moment in which the tool needs to be replaced is, in most cases, determined based on human experience. Nevertheless, today, when the highest possible level of machining process autonomy is sought for, such type of estimation is unacceptable for a number of reasons. If a certain amount of tool wear is exceeded, workpiece surface quality will deteriorate.

On the other hand, if the maximum threshold value of tool wear is not achieved, sanding paper is going to be prematurely discarded and the manufacturing process frequently interrupted. For that reason, a reliable and efficient method of indirect process monitoring, applicable during the process in real time, is of great importance for achieving high productivity, as well as high product quality [1], [2].

Although research done so far in the field of indirect process monitoring showed that robust tool wear models are attainable, the studies were often conducted on machining processes with geometrically defined cutting tools, and with the interfaces of tool, machine tool, and workpiece being rigid [3]. Even in those conditions, achieving a universal and generally applicable model for tool wear or surface quality monitoring was not possible due to the highly nonlinear and stochastic nature of tool wear. For that reason, such monitoring systems are almost always developed according to the specific characteristics of a machining process [1].

There are two main approaches in machining process monitoring – monitoring based on direct and indirect methods [4]. In direct methods, the information on tool wear level or state of the surface in acquired by directly measuring the tool or the machined surface features. Even though the accuracy of the measured value is the advantage, the main disadvantage of the method is the inability to be used during machining [5]. In other words, frequent interruptions to the machining process are caused in order to carry out the measurements.

Indirect methods can be applied in real time, during the machining process but the correlation between process parameters and tool or surface state must be established beforehand. Any machining process characteristic measurable or determinable during machining can be considered as a process parameter. These methods generally facilitate multiple different types of sensors. Amongst most commonly used are force [6], [1], [7], current [8], vibration [9], [1], acoustic emission [10], [1], and sound [11], [7], [1] sensors. After the signal acquisition, a number of features sensitive to change in tool state and/or workpiece surface quality, are extracted and used in tool wear and surface quality estimation models. Unlike direct, indirect methods do not require a disruption to the machining process and can in some cases significantly shorten overall machining time, which makes them suitable for industrial environments.

When it comes to tool condition monitoring in robotic finishing processes, research has been mostly done on belt grinding and polishing. Ren et al. [7] developed a method for machine learning based workpiece heat input monitoring in robotic belt grinding of Inconel 718 using force and sound sensors. They concluded that the heat input is influenced by grinding parameters and the state of the grinding tool. Between the state of the grinding tool and the force and sound signals, a clear correlation was discovered. The result of work done in [12] is a robust method of grinding belt condition monitoring in a robotic grinding system of superalloys, accomplished by utilizing sound signals and a machine learning method. An evaluation of the grinding belt condition was successfully performed by a neural network based multi-sensor system developed in [1], in which vibration, force (F_x and F_z) and acoustic emission sensors were used. In a study done by Chen et al. [13] a grinding belt condition monitoring system was proposed. The system uses a model based on random forest (RF) classifier and multiple linear regression (MLR) which use acoustic emission signal features as input data. Acoustic emission signals and tool condition revealed considerable correlation. Pandiyan et al. [6] have suggested a method of identifying the compliant belt grinder condition by means of force, acoustic emission, and vibration signals whose features were employed to train the support vector machine algorithm. As a result, a high-accuracy tool condition prediction model was attained. Meier et al. [14] developed a model for polishing stone breakage detection and determination of the optimal time for tool replacement in robotic polishing based on acoustic emission and force sensors. In [15] acoustic emission, force and current sensors were used in a robot-assisted polishing process to estimate the correct time for tool replacement (end-point detection), in order to maintain the desired surface roughness. A robust surface roughness classification model based on an artificial neural network was obtained.

In [16] a parameter optimization for manual sanding of black pine was conducted based on the relationship between process parameters (feed rate, grit size, spindle speed), and the achieved surface roughness. Motivated by the exposure of workers to harsh working environments (orbital sanding tool simultaneously executes rotary and eccentric motion to attain better surface quality and consequently, produces dust, noise, and vibrations), Nagata et al. [17] developed a robotic system for sanding furniture with free-formed surfaces. Robotic sanding systems were also utilized in naval industries [18] as a solution to problems with large, stationary parts. In research by Putz et al. [19] a step closer to an autonomous robotic system has been made by presenting a successful method of automated tool alignment, necessary for automated sanding paper replacement.

The industry has shown considerable interest in robotic machining despite the challenges they pose [20], [21]. Robotic machining systems have considerably lower rigidity compared to machine tools. However, due to their ability to perform complex tool paths in a relatively large working area they are facilitated in finishing processes that require lesser forces and accuracies during machining. Such systems are also commonly utilized in sanding of large, thin-walled structures.

The sanding process is performed with a geometrically undefined tool with an elastic interface that, on one hand, allows for better adherence to the workpiece surface, but on the other, causes a dampening effect. In addition, workpieces such as thin-walled structures are prone to elastic deformation. Therefore, despite the many advantages robotic cell application offers, it is expected that the aforementioned disturbances (e.g. vibration signals that could result from change in tool and/or surface quality state, would be dampened by the elastic interface) may make the existing process monitoring models difficult to apply. Additionally, even though machining process monitoring has been widely studied, not many have dealt with robotic finishing, and hardly any research has been done on robotic sanding. Hence, research on the possibility of attaining sanding paper wear level and workpiece surface quality estimation models specifically designed for the robotic sanding process of thin-walled structures is proposed. The goals of the research would be:

1. To determine the effect of cutting parameters (spindle speed, feed velocity, and axial force) on the dynamics of the sanding paper wear.
2. To determine the effect of cutting parameters and level of sanding paper wear on workpiece surface quality.
3. To determine the process parameters (sensor signal features) sensitive to changes in the sanding paper wear level.
4. To determine the process parameters (sensor signal features) sensitive to changes in workpiece surface quality.
5. To develop robust sanding paper wear level and workpiece surface quality estimation models applicable in real time during the sanding process.

The experimental setup and the research method will be described in the following section. Lastly, in section 3., a conclusion will be given.

2. Experimental system

The research will be carried out in three phases. In the first phase, the experimental setup (Fig. 1) will be developed. The setup will consist of two industrial robots of which the first one (ABB 6660) is equipped with a main drive for sanding.

Fig. 1. Robotic sanding cell (left: ABB 6660, middle: IRBP A with the clamping system, right: ABB 4600)

The main drive for sanding is realized through a precise force control unit with a mounted orbital sander (Fig. 2). The orbital sander includes an elastic interface pad for the sanding paper. Different sorts and granulations of sanding paper will be used.

Fig. 2. Mirka® AIROS 650CV orbital sander for industrial robots [22]

Vibration (Fig. 3, left), acoustic emission (Fig. 3, right), and sound sensors (Fig. 4), as well as a thermal imaging camera (Fig. 5) will be mounted in proximity to the main drive.

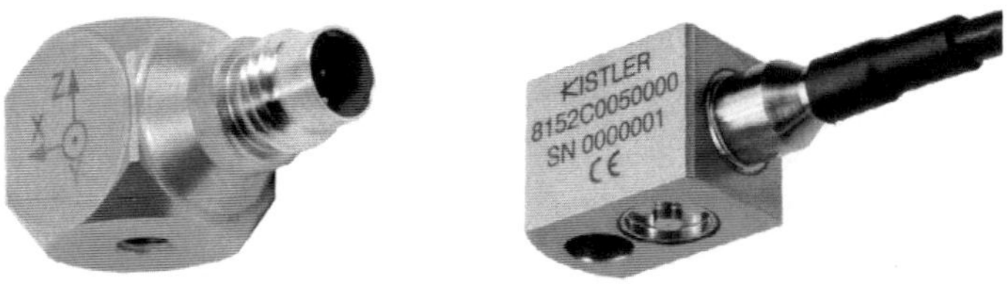

Fig. 3. KISTLER triaxial accelerometer [23] (left), and acoustic emission sensor [24] (right)

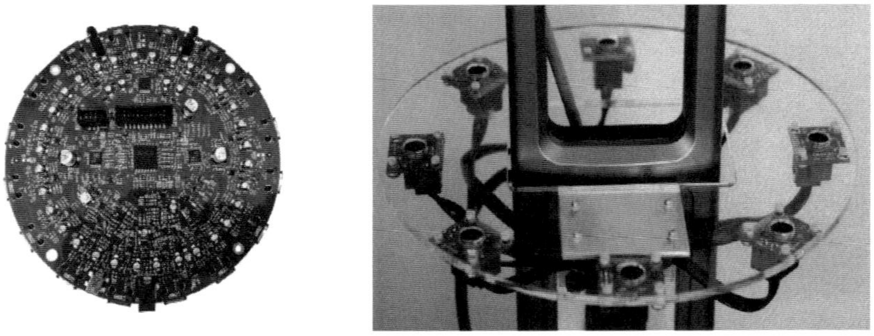

Fig. 4. 16SoundsUSB sound card [25] (left); example of an 8SoundsUSB system microphone array [26] (right)

Fig. 5. Teledyne thermal imagining camera Flir A700-EST, 24° Lens [27]

In between the two robots, a rotary-swivel table (IRBP A) with a suitable clamping system for the workpiece will be placed. This research will be conducted on relatively large and thin-walled composite structures. On the second robot (ABB 4600), an optical 3D measurement instrument for automated form and roughness measurements (Alicona IF-SensorR25) (Fig. 6) will be mounted.

Fig. 6. Alicona IF-SensorR25 [28]

The experiment will cover a large number of sanding cycles including different:

a. combinations of cutting parameters (feed velocity, spindle speed, axial force),
b. types of sanding papers (at least three different granulations) – that will be worn out to different degrees (at least three levels of wear),
c. types of workpieces with dissimilar geometrical features (at least two types of workpieces – homogenous and composite).

Tool wear level and workpiece surface quality will be determined with the Alicona optical sensor, before and after each cycle of sanding. Vibration, acoustic emission, sound signals, and a thermal image of the workpiece surface will be recorded during the sanding cycles. In the second phase (after completing all tests) the acquired information will be processed and analysed. Features will be extracted from recorded signals and images and examined to determine their sensitivity to changes in tool wear and/or workpiece surface quality. In the third phase, a portion of the extracted features will be used for the tool wear estimation model formation, and the rest will be used to obtain the workpiece surface quality estimation model. In the proposed research the estimation models will be based on artificial intelligence algorithms since the high complexity of the tool wear process makes statistical models unsuitable for this application [29].

3. Conclusion

Machining process monitoring models are not universal and generally applicable, and their successful implementation depends on specific process characteristics. Although robotic machining process monitoring has been studied, research on robotic sanding is quite scarce. Therefore, the formation of sanding paper wear level and workpiece surface quality estimation models would be conducted for the set process – robotic sanding. Robotic sanding has its advantages, but problems such as: lower rigidity of the machine performing the process, commonly used large, thin-walled workpieces prone to elastic deformation, undefined geometry of the sanding paper, and the dampening effect of the sanding paper elastic interface are expected to contribute to the difficulty of applying the existing models obtained so far and are also expected to be a challenge in obtaining the new models.

In the proposed research the possibility of developing a multi-sensor based sanding paper wear level and workpiece surface quality monitoring system will be studied. The aim is to determine the effect of cutting parameters on sanding paper wear and the effect of cutting parameters alongside the level of sanding paper wear on workpiece surface quality. Lastly, the objective would also be to determine the process parameters sensitive to change in sanding paper and/or workpiece surface quality and to develop robust sanding paper wear level and workpiece surface quality estimation models applicable during sanding process in real time. The suitability of signal features extracted from vibration, acoustic emission, and sound signals, along with the thermal image would be investigated and potentially used for the formation of estimation models based on artificial intelligence.

Successful realization of the proposed models would create preconditions for development of a robotic sanding monitoring system in real time. By obtaining the models, the level of robotic sanding system autonomy would be significantly increased and a foundation for adaptive control of cutting parameters (feed velocity, cutting speed, and axial force) would be established.

4. Acknowledgements

This research is supported by the European Regional Development Fund (ERDF) through the grant KK.01.1.1.07.0016 (ARCOPS).

5. References

[1] Qi. J.; Chen, B. & Zhang, D. (2020). Multi-information fusion-based belt condition monitoring in grinding process using the improved-Mahalanobis distance and convolutional neural networks, Journal of Manufacturing Processes, vol. 59, pp. 302-315, ISSN 1526-6125, DOI: 10.1016/j.jmapro.2020.09.061.

[2] Kuntoğlu, M.; Aslan, A.; Pimenov, D. Y.; Usca, Ü.A.; Salur, E.; Gupta, M.K.; Mikolajczyk, T.; Giasin, K.; Kapłonek, W. & Sharma, S. (2021). A Review of Indirect Tool Condition Monitoring Systems and Decision-Making Methods in Turning: Critical Analysis and Trends, Sensors 2021, Vol. 21, No. 1: 108, DOI: 10.3390/s21010108

[3] Klaic, M.; Staroveski, T. & Udiljak, T. (2014). Tool Wear Classification using Decision Trees in Stone Drilling Applications: a Preliminary Study, Proceedings of the 24th International DAAAM Symposium "Intelligent Manufacturing & Automation", ISSN 1877-7058, Katalinic, B. (Ed.), pp. 1326-1335, Published by Elsevier Ltd., Oxford, UK, DOI: 10.1016/j.proeng.2014.03.25.

[4] Zhou, Y. & Xue, W. (2018). Review of tool condition monitoring methods in milling processes, The International Journal of Advanced Manufacturing Technology, vol. 96, pp. 2509-2523, DOI: 10.1007/s00170-018-1768-5

[5] Jourani, A.; Hagege, B.; Bouvier, S.; Bigerelle, M. & Zahouani, H. (2013). Influence of abrasive grain geometry on friction coefficient and wear rate in belt finishing, Tribology International, vol. 59, pp. 30-37, ISSN 0301-679X, DOI: 10.1016/j.triboint.2012.07.001.

[6] Pandiyan, V.; Caesarendra, W.; Tjahjowidodo, T. & Tan, H. H. (2018). In-process tool condition monitoring in compliant abrasive belt grinding process using support vector machine and genetic algorithm, Journal of Manufacturing Processes, vol. 31, pp. 199-213, ISSN 1526-6125, DOI: 10.1016/ j.jmapro.2017.11.014.

[7] Ren, X.; Chai, Z.; Xu, J.; Zhang, X.; He, Y.; Chen, H. & Chen, X. (2020). A new method to achieve dynamic heat input monitoring in robotic belt grinding of Inconel 718, Journal of Manufacturing Processes, vol. 57, pp. 575-588, ISSN 1526-6125, DOI: 10.1016/ j.jmapro.2020.07.018

[8] Tian, Y.; Liu, F.; Wang, Y. & Wu, H. (2017). Development of portable power monitoring system and grinding analytical tool, Journal of Manufacturing Processes, vol. 27, pp. 188-197, ISSN 1526-6125, DOI: 10.1016/j.jmapro.2017.05.002.

[9] Wu, D.; Jennings, C.; Terpenny, J.; Gao, R. X. & Kumara, S. (2017). A comparative study on machine learning algorithms for smart manufacturing: tool wear prediction using random forests, Journal of Manufacturing Science and Engineering, vol. 139, no. 7, DOI: 10.1115/1.4036350

[10] Liao, T. W. (2010). Feature extraction and selection from acoustic emission signals with an application in grinding wheel condition monitoring, Engineering Applications of Artificial Intelligence, vol. 23, no. 1, pp. 74-84, 2010., ISSN 0952-1976, DOI: 10.1016/j.engappai.2009.09.004.

[11] Ghosh, N.; Ravi, Y. B.; Patra, A.; Mukhopadhyay, S.; Paul, S.; Mohanty, A. R. & Chattopadhyay, A. B. (2007). Estimation of tool wear during CNC milling using neural network-based sensor fusion, Mechanical Systems and Signal Processing, vol. 21, no. 1, pp. 466-479, ISSN 0888-3270, DOI: 10.1016/j.ymssp.2005.10.010.
[12] Zhang, X.; Chen, H.; Xu, J.; Song, X.; Wang, J. & Chen, X. (2018). A novel sound-based belt condition monitoring method for robotic grinding using optimally pruned extreme learning machine, Journal of Materials Processing Technology, vol. 260, pp. 9-19, ISSN 0924-0136, DOI: 10.1016/ j.jmatprotec.2018.05.013.
[13] Chen, J.; Chen, H.; Xu, J.; Wang, J.; Zhang, X. & Chen, X. (2018). Acoustic signal-based tool condition monitoring in belt grinding of nickel-based superalloys using RF classifier and MLR algorithm, The International Journal of Advanced Manufacturing Technology, Vol. 98, pp. 859-872, DOI: 10.1007/ s00170-018-2270-9
[14] Meier, N.; Papadoudis, J. & Georgiadis, A. (2020). Prediction of the tool change point in a polishing process using a modular software framework, Proceedings of the 13th CIRP Conference on Intelligent Computation in Manufacturing Engineering, ISSN 2212-8271, ISBN 978-1-713814-59-7, Teti, R. & D'Addona, D.M. (Ed.), pp. 341-345, Published by Elsevier B.V., Amsterdam, Netherlands, DOI: 10.1016/j.procir.2020.05.059.
[15] Segreto, T. & Teti, R. (2019). Machine learning for in-process end-point detection in robot-assisted polishing using multiple sensor monitoring, The International Journal of Advanced Manufacturing Technology, vol. 103, pp. 4173-4187, DOI: 10.1007/s00170-019-03851-7.
[16] Hazir, E.; Koc, K. H. & Hiziroglu, S. (2017). Optimization of sanding parameters using response surface methodology, Maderas. Ciencia y tecnologia, Vol. 19, pp. 407-416, ISSN 0717-3644, DOI: 10.4067/S0718-221X2017005000034.
[17] Nagata, F.; Kusumoto, Y.; Fujimoto, Y. & Watanabe, K. (2007). Robotic sanding system for new designed furniture with free-formed surface, Robotics and Computer-Integrated Manufacturing, vol. 23, pp. 371-379, ISSN 0736-5845, DOI: 10.1016/ j.rcim.2006.04.004.
[18] Subrin, K.; Garnier, S.; Bressak, T. & Furet, B. (2019). Digital chain development for sanding application with a kinematically redundant robotic system, Proceedings of the 17th CIRP Conference on Modelling of Machining Operations, ISSN 2212-8271, ISBN 978-1-7138-0094-1, Ozturk, E.; Mcleay, T. & Msaoubi, R. (Ed.), pp. 515-520, Published by Elsevier B.V., Amsterdam, Netherlands, DOI: 10.1016/J.PROCIR.2019.04.063.
[19] Putz, V.; Stangl, M; Kohlberger, C. & Naderer, R. (2019). Computer Vision Approach for the Automated Tool Alignment of an Orbital Sanding Robot, Proceedings of the 8th IFAC Symposium on Mechatronic Systems, ISSN 2405-8963, ISBN 978-1-7138-0790-2, Jadachowski, L. (Ed.), pp. 19-24, Published by Elsevier Ltd., Oxford, UK, DOI: 10.1016/J.IFACOL.2019.11.643.
[20] Nam, H. H.; Verlinden, O. & Rivière-Lorphèvre, E. (2017). Robotic Machining Simulation using a Simplified Multibody Model, Proceedings of the 28th International DAAAM Symposium "Intelligent Manufacturing & Automation", ISSN 1726-9679, ISBN 978-3-902734-11-2, Katalinic, B. (Ed.), pp. 0885-0894, Published by DAAAM International, Vienna, Austria, EU, DOI: 10.2507/28th.daaam.proceedings.123.
[21] Pérez, R.; Gutiérrez, S. C. & Zotovic, R. (2018). A Study on Robot Arm Machining: Advance and Future Challenges, Proceedings of the 29th International DAAAM Symposium "Intelligent Manufacturing & Automation", ISSN 1726-9679, ISBN 978-3-902734-20-4, Katalinic, B. (Ed.), pp. 0931-0940, Published by DAAAM International, Vienna, Austria, EU, DOI: 10.2507/29th.daaam.proceedings.134.
[22] https://www.mirka.com/en/p/MIA6502012. Mirka®. Accessed on: 2022-9.
[23] https://www.kistler.com/en/product/type-8763b/. KISTLER. Accessed on: 2022-9.
[24] https://www.kistler.com/en/product/type-8152c/. KISTLER. Accessed on: 2022-9.
[25] https://github.com/introlab/16SoundsUSB/blob/master/README.md. github. Accessed on: 2022-9.
[26] https://introlab.3it.usherbrooke.ca/sam/ca/description-du-projet/hardware/8soundsusb/. github. Accessed on: 2022-9.
[27] https://www.flir.eu/products/flir-a700-est/. TELEDYNE FLIR. Accessed on: 2022-9.
[28] https://www.alicona.com/en/products/if-sensorr25/. Bruker alicona. Accessed on: 2022-9.
[29] Lv, L.; Deng, Z.; Liu, T.; Li, Z. & Liu, W. (2020). Intelligent technology in grinding process driven by data: A review, Journal of Manufacturing Processes, vol. 58, pp. 1039-1051, ISSN 1526-6125, DOI: 10.1016/ j.jmapro.2020.09.018.

DOI: 10.2507/33rd.daaam.proceedings.069

COMPARATIVE ANALYSIS OF POLYMERIC COMPOSITE REINFORCED WITH GLASS FIBRE, MAT AND VISCOSE RAYON

Johnson Ugochukwu Ezenwankwo, Josef Žak & Iva Petríková

This Publication has to be referred as: Ezenwankwo, J[ohnson]; Zak, J[ozef] & Petrikova, I[va] (2022). Comparative Analysis of Polymeric Composite Reinforcedwith Glass Fibre and Viscose Rayon, Proceedings of the 33rd DAAAM International Symposium, pp.0496-0502, B. Katalinic (Ed.), Published by DAAAM International, ISBN 978-3-902734-36-5, ISSN 1726-9679, Vienna, Austria
DOI: 10.2507/33rd.daaam.proceedings.069

Abstract

Natural fibre-reinforced polymer composites (NFRPC) have taken centre stage in materials and engineering. They present many advantages over their synthetic counterparts, including environmental concerns, ease of disposal and reuse, and a high strength-to-weight ratio. Aerospace engineering and other critical engineering sectors are actively turning to NFRPCs for various reasons. Reinforcement in polymeric composites plays a vital role in forming the framework support system upon which the matrix is based; this, in turn, affects the vital mechanical properties of the composite formed and its impact-bearing characteristics. This paper evaluates selected mechanical properties of glass fibre and rayon yarn. The laminates were made so that the thickness was approximately equal. Five tensile tests each were performed following ASTM 3039/D3039 - 08 standard, and the average of each set is plotted and compared accordingly. The material is weighed, and the fibre and matrix volume fractions are calculated. The standard failure location, area and mode are described with the ASTM standard. A simplified micromechanical analysis is employed to demonstrate the homogenization of the laminate's representative volume element (RVE) to obtain the relevant engineering constants.

Keywords: Matrix; rayon; yarn; laminates

1. Introduction

From time immemorial, science and technology have greatly attributed processes' success and functionality to materials. Material selection plays a big role in systems' operation and useful life. For instance, a simple automobile engine can have materials made of aluminium, copper, iron, plastic, rubber and a mixture of two or more in the form of alloys. The same goes for composites reinforced with various natural/synthetic fibres singly or as a mixture (hybrid). There is now a global clarion call to switch from harmful synthetic fibres to bio-degradable. There are numerous evidential advances in composite materials as their application begins to foray into other areas of engineering. Recently, a study was carried out investigating fabric's usability and suitability and the electromagnetic compatibility of a chosen composite element. The study further delved into determining the fibre content to be effective for the particle shielding test [1]. Another study used fused deposition modelling process in additive manufacturing samples infused with carbon fibre.

The study investigated the mechanical properties of two polyesters reinforced with carbon fibre by making comparisons of type and print angle of orientation according to specified standards [2]. Fibre-reinforced polymer composites present a huge advantage in engineering practice and natural fibres because of their many desirable properties compared to their synthetic counterparts. In the past two decades, the ease of making quality composites into various shapes has also increased its desirability and market chase. Low density and cost-effectiveness are reasons natural fibres are popular, especially in automobile industries [3]. Other advantages possessed by natural fibres include relative non-abrasiveness, high specific strength and modulus, ease of processing, and high flexibility. Generally, the manufacturing of polymer matrix composites (PMC) and natural fibre-reinforced polymer composites (NFRPC) can be fundamentally classified into three broad groups: short-fibre suspension methods, squeeze flow methods and porous media methods [4]. Most of the known manufacturing procedures for PMCs and NFRPCs fall into at least one of these categories. These desirable properties make up for the drawbacks presented by natural fibres reinforced polymer composites which include relatively low impact strength and hydrophilic tendencies (due to the hydroxyl ion, cellulose and hemicellulose components), which encourages high water absorption [5]. Material scientists and engineers have developed many design-based approaches and models formulated to address these challenges [6][7][8][9][10][11]. One special technique is fibre hybridization, where fibres of varying geometric properties are used singly or woven together to enhance the mechanical properties of composites. Kumar *et al.* [12] performed a comparative analysis of kenaf/glass fibre hybrid composite. The study used woven kenaf/glass fibre and epoxy resin as the matrix while investigating such properties as tensile strength, Young's modulus, and the stress-strain curve of samples with different percentage contents of fly ash. The choice of fabrication method is vital during the manufacturing of composites. In their book, Safarabadi *et al.* [13] presented the effect of fibre waviness on the compressive strength of a laminate. Yahaya *et al.* [14] studied the effect of layering sequence and chemical treatment on the mechanical properties of woven kenaf-aramid hybrid laminated composites. The results showed marked improvement in the tensile properties of hybrid composites in 3 layers compared to 4. Also, hybrid composites containing kevlar at the outer layers displayed overall better mechanical properties than the others. In their research, Tagarielli *et al.* [15] addressed the issue of an efficient and reliable method to measure the through-thickness of a material owing to the sheer difficulty in manufacturing thick laminates with relatively high thickness. The study also proclaimed that the onset of failure, which occurs in through-thickness cases in composite laminates, largely depends on direct and shear inter-laminar stresses. Paul S. *et al.* [16] loaded banana composites in tension and monitored the effect of chemical alteration by treatment of the composites. The study showed the role of chemical treatment and how it improves the thermo-physical properties of the material tested. Vijayan *et al.* [17] experimented with the effect of hybridization by infusing aloe vera fibre with roselle and glass in various formations. The study also demonstrated the flexural (bending) test on the 3-point method up to a breaking criterion. The flexural strength and modulus were tabulated for the different sample configurations. Meenakshi *et al.* [18] investigated, via experiments and FEM, the mechanical response of glass, aloe vera, and kenaf in various configurations produced by the vacuum-assisted resin transfer moulding method. The study aims to study the mechanical behaviour of four different natural fibres when used as a reinforcement material in NFRPC/PMC. Its novelty lies in the material composite produced. Viscose rayon, in particular, is unique beacuse viscose rayon is not commonly used, as shown in many similar works on both natural and synthetic fibre. Hence, the problem addressed, or the specific novelty of the paper, lies in the fabrication of laminates and the comparative investigation of the mechanical properties of the different laminates produced according to stacking sequence. A simplified micromechanical analysis of the composite is performed in *Ansys r22 Material Model* Module using the hexagonal geometry type with a fibre diameter of 2μm. A comparison is drawn between the properties of E-Glass (Fibre Glass) with the resin used as they relate to the engineering constants, and relevant plots are obtained.

2. Materials and Method

The method chosen is dependent on the type of material involved in the manufacturing of the composite. The selection of suitable resin also plays an important role. Some of the manufacturing processes include the simple method called hand layup, compression moulding, injection moulding, pultrusion etc. The mechanical properties of composites largely depend on reinforcement, matrix, and the binding dynamics between them. Material preparation techniques and process is described in detail in section 2.1 below.

2.1 Composite Laminate Preparation

Preparing the sample involved obtaining the reinforcement fabric, cutting it into the specified dimension of 250mm x 250mm, mixing the epoxy and hardener mixture at a specified ratio, and following the hand layup procedure for making the laminate before thermal compression and curing. This section describes the hand layup method for preparing the laminate at the KMP-TUL lab. The glass fibre and mat were obtained from Havel Composites Company, and the viscose rayon (2440 dtex) yarn in simple weave architecture was prepared at the KMP lab. The epoxy-hardener mixing ratio was 2:1. The gauge length of the textile composite sample was 78.8 mm, as marked from the Instron 9000 series testing machine gauge kit. Prevailing environmental conditions were 24°C, 37.5% RH and the mean value of the tests were considered. As described before, the three different reinforcement configurations prepared are given as follows: $[0_{2gf}]s$, $[0,45_{gf}]s$, $[0_{gf}, Mat]s$, $[Mat_2]s$ and $[0_{2ry}]s$.

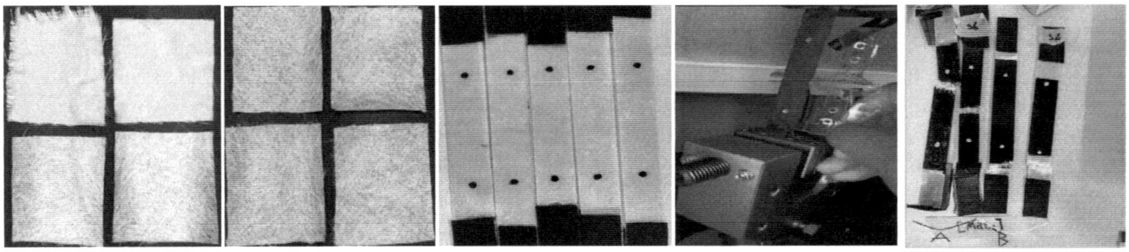

Fig. 1. (a) [0_{gf}, Mat]s and (b) [Mat_2]s configurations (c) Mat Composite Test Samples (d) Strain Gauge Marker (e) Proper Tab Configuration and Effects

The fibre materials are cut 250mmx250mm. The nylon material serves as the enclosure for the layup system and is laid open-leaf. The bleeder material is placed on one side of the leaf, and the separator material, which helps in easy detachment of the laminate after curing, is placed over the bleeder material. A reasonable quantity of matrix is applied, and the first fibre is laid on it; the process is repeated until all fibres are laid. Then the second side of the leaf is used to close the arrangement. The hot compression chamber consists of a top and lower platen with separate temperate controls. The platens are pre-heated to 75°C for both platens. The material is finally rammed and pressed manually with a hand roller to crush local bubbles to reduce the occurrence of voids. The setup is placed inside the compression chamber, and the pressure is set to 20kN, as shown in Figures 7 and 8. A glass enclave forms the sliding door that keeps the temperature inside in equilibrium through the curing process. The process continued for 1.5 hours, after which the sample was allowed to cool inside the chamber by adjusting the setpoint temperature to 24°C (room temperature) for about 12 hours.

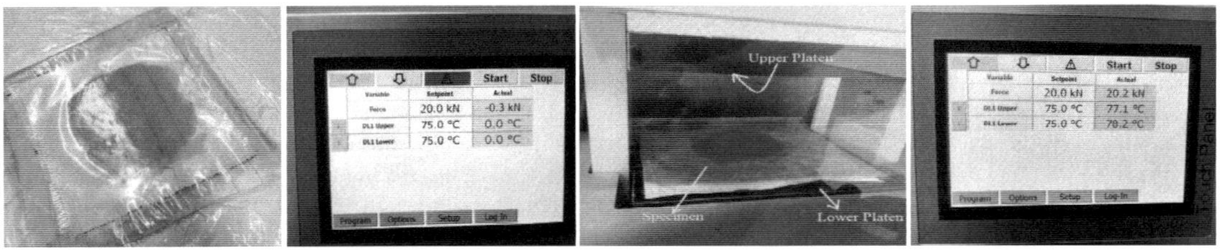

Fig. 2. (a) Fibre Material Hand Layup (b) Compression Controls Setpoint (c) Material Inserted for Compression (d) Compression Controls Actual Values

Fibre	Stacking Sequence	Orientation	Sample Thickness (mm)
Glass	G-G-G-G	[0]	1.02
Glass	G-G-G-G	$[0,45]_S$	1.02
Mat	M-M-M-M	-	1.01
Viscose-Rayon	R-R-R-R	[0]	1.45
Galss-Mat	G-M-M-G	[0_{gl}, -]	1.01

Table 1. Stacking Sequence, Orientation and Sample Thickness of Composite Laminate

3. Governing Equations

In the book by Autar Kaw K. [19], the concept of composite mechanics analysis, theory, experimental procedures and failure mechanics is expounded. The surface area of the composite is given as follows:

$$S_f = a \cdot b \cdot n \tag{1}$$

where a and b are the geometric dimensions of the specimen and n is the number of plies (layers) mass of fabric,

$$m_f = S_f \, . \, \gamma \tag{2}$$

In relation to laminate density,

$$\rho = \rho_f V_f + \rho_m V_m \text{ and } V_m + V_f = 1 \tag{3}$$

Where ρ is the density of the composite.

$$\text{Hence, } V_f = \frac{\rho - \rho_m}{\rho_f - \rho_m} \tag{4}$$

The longitudinal ultimate tensile strength is given as

$$(\sigma_1^T)_{ult} = (\sigma_f)_{ult} V_f + (\varepsilon_f)_{ult} E_m (1 - V_f)\,; \tag{5}$$

The transverse ultimate tensile strength is given as

$$(\sigma_2^T)_{ult} = E_2(\varepsilon_2^T)_{ult} \tag{6}$$

The longitudinal ultimate compressive strength is calculated using two formulas, of which the lesser value is adopted. The relations are given below

$$(\sigma_1^C)_{ult} = \frac{E_1(\varepsilon_2^T)_{ult}}{\nu_{12}} \text{ or} \tag{7}$$

Using the shear stress failure mode;

$$(\sigma_1^C)_{ult} = 2[(\tau_f)_{ult} V_f + (\tau_m)_{ult} V_m] \tag{8}$$

The transverse compressive strength is given as

$$(\sigma_2^C)_{ult} = E_2(\varepsilon_2^C)_{ult} \tag{9}$$

The ultimate shear stress is given as

$$(\tau_{12}) = G_{12}(\gamma_{12})_{m-ult} \tag{10}$$

4. Results and Discussion

The result and discussion session present the test outcome from the Instrom 9000 series machine. The samples are loaded with the appropriate tab material applied according to ASTM 3039/D3039-08 standard [20]. The crosshead speed is determined based on displacement as 1mm/min. The various laminates produced were tested for five samples each and up to failure of the specimen. The tab configurations before loading are shown in Fig. 1e above. The tabs play a crucial role as specified in the ASTM standard. The tabs help to avoid localized stress concentration around the edge/ zone where the crosshead grips the material, thereby limiting the occurrence of failure around this zone, which in practice, would be wrong. An in-depth study into the effect of tabs on specimens can be found in the work of Belingardi et al. and Tabrizi et al. [21, 22], where the authors investigated the effect of various tab configurations during tensile tests.

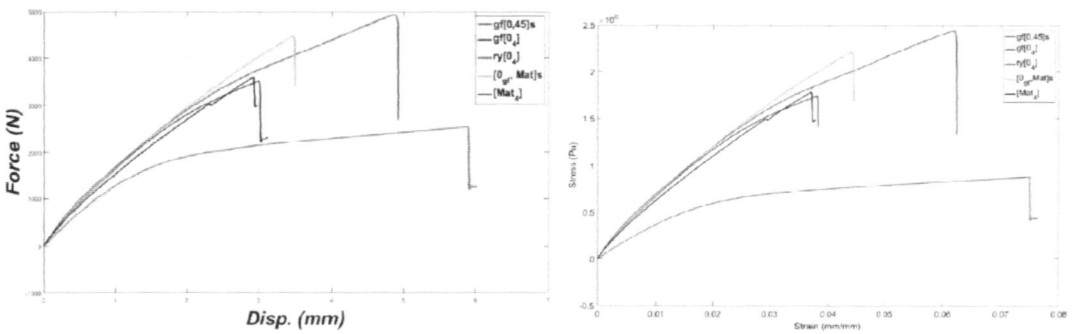

Fig. 3. (a) Average Force-Displacement Plot of Different Samples (b) Stress-Strain Average Plot of Different Samples

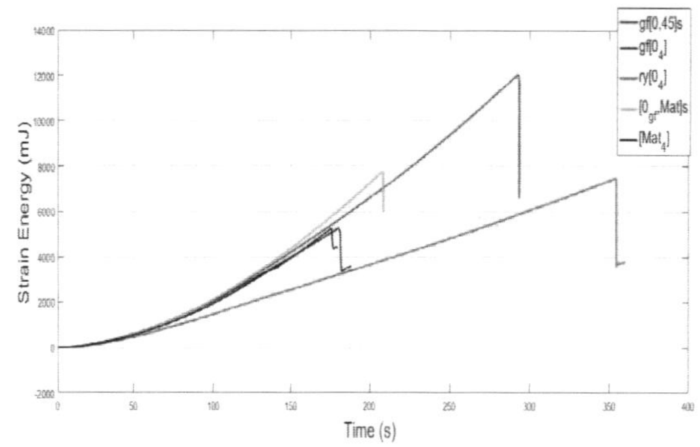

Fig. 3. (c) Strain Energy Vs Time

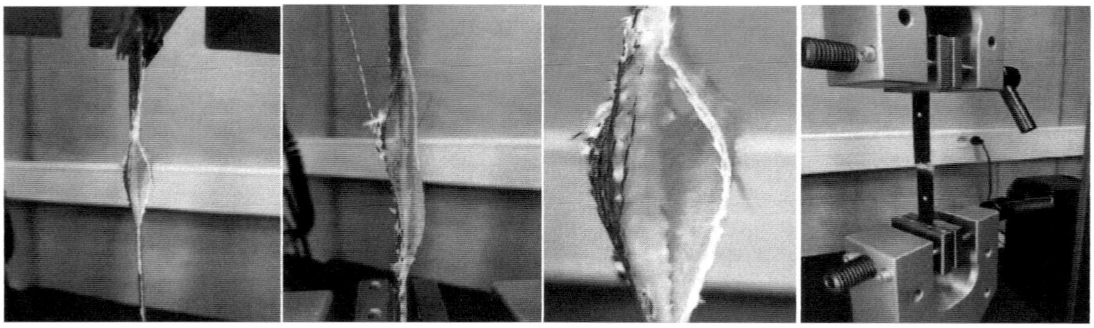

(a) (b) (c) (d)

Fig. 4. (a,b,c) Mid-zone edge delamination [Mat_4] (d) Mid-zone long splitting.

Figure 3 shows the various plots of displacement and force, stress-strain and strain energy as it relates to time, whereas Figure 4 captures the various failure profile discussed in ASTM 3039/D3039M - 08 standard.

4.1. FEM Micromechanical Homogenization of a Laminate

Homogenization is a critical step towards constructing and designing composite structures [23]. The fundamental idea is to establish the macroscopic behaviour of composites knowing that they are heterogeneous at the microscopic level to generate engineering constants E_{11} (axial direction), E_{22} (transverse direction), G_{12} (in-plane shear modulus), G_{23} (out of plane shear modulus), and v_{12} (in-plane Poisson's ration). Voigt, 1889 and Reuss, 1929 devised the earliest method, called the rule of mixture (ROM), for deducing the values of the engineering constants of a laminate. The more recent Halpin_Tsai model succeeded but is still limited in predicting the value of G_{23}, as is ROM. However, since the inception of finite element method (FEM)/finite element analysis (FEA) has been extensively used by researchers and scientists to determine the elastic properties of composite materials [24][25][26].

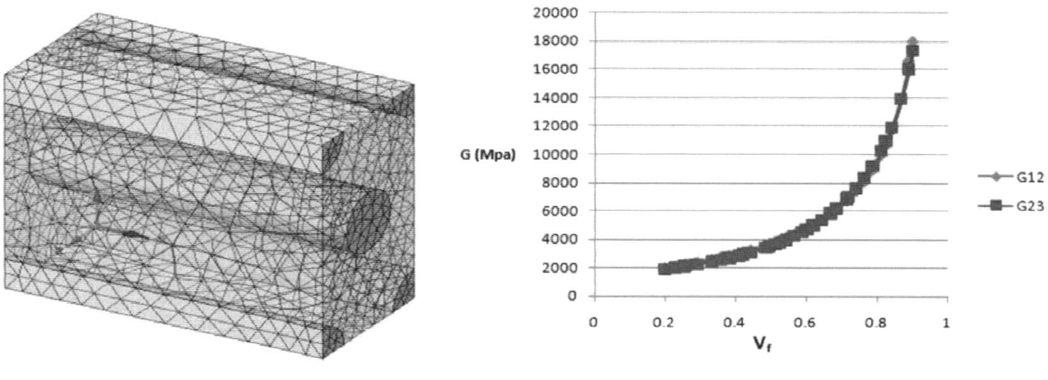

Fig. 5. a) 3D RVE of a laminate (hexagonal type) b) Plots of G(long. and transv.) and Vf

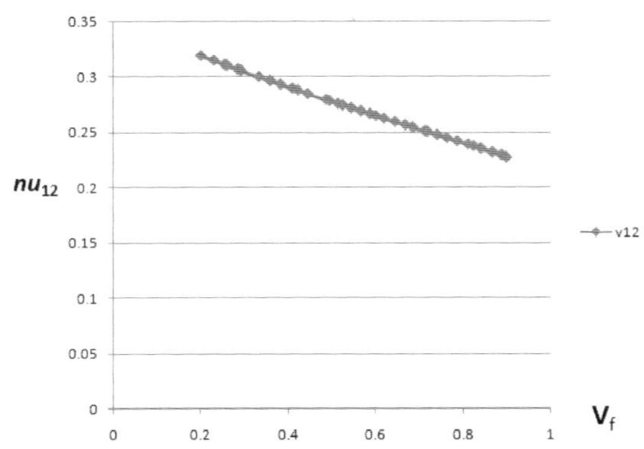

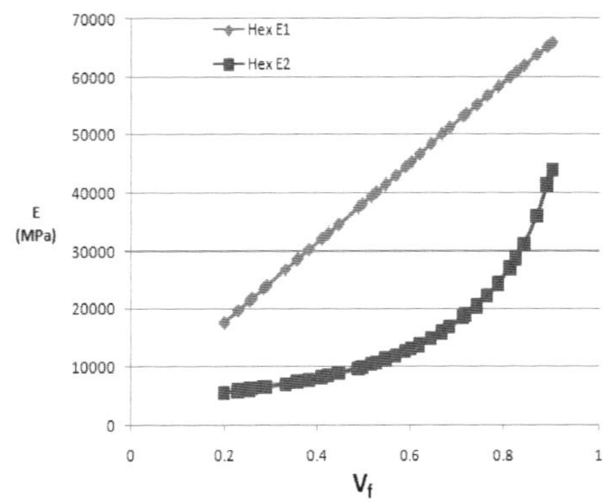

Fig. 6. a) Plots of v_{12} and V_f b) Plots of *E (long. and transv.)* and V_f

Although further work already in progress will compare more results by looking at two other model types different from Figure 6a. It will compare the results from other resin types, including polyamide and polyester resins. This method presents a suitable means of calculating the material properties by homogenization.

5. Conclusion

As shown in Figure 4, the failure profile is described considering three criteria: *failure type,* which could be angled, edge delamination, grip/tab, lateral, multi-mode or explosive. The second criterion is *failure area,* denoting that failure could occur inside the grip/tab, at the grip/tab, within the gauge zone, multiple areas, etc. The last criterion is failure location, which specifies the failure domain on the composite and includes bottom, top, left, right, middle, and various locations combined. Further work will be in the simulation of the failure of laminates using Comsol-multiphysics macromechanical analysis of laminates and the development of failure envelopes in terms of failure indices and safety factors. The layering configuration affects the mechanical behaviour and response of the laminate. The laminate with rayon yarn showed comparably higher strain than the other laminates making it a good fit for applications requiring strain applications. A cloud-based tool called CDMHub is also proposed to be used for modelling the laminates. Other areas of further research will be the hybridization of fabrics used for reinforcement, as mentioned in Section 1. Two different types of yarn are woven to form one fabric; this has been shown by research to improve the mechanical characteristics of fabrics taken from the individual yarns which make up the hybridized fabric. Finite element tools have proven to be productive enough to determine the properties of unidirectional composite materials in homogenization. Other areas of further research and continuation are on macromechanical analysis of laminates in Comsol and Ansys ACP Pre.

6. Acknowledgements

This publication was written at the Technical University of Liberec as part of the project "Research of advanced materials, and application of machine learning in the area of control and modelling of mechanical systems" nr. SGS-2022-5003 with the support of the Specific University Research Grant, as provided by the Ministry of Education, Youth and Sports of the Czech Republic in 2022.

7. References

[1] Skocik, Petr & Korbel, Jiri (2017) Assessment of shielding effectiveness and percolation threshold of composite materials, Proceedings of the 28th DAAAM International Symposium, pp.0054- 0062, B. Katalinic (Ed.), Published by DAAAM International, ISBN 978-3-902734-11-2, ISSN 1726-9679, Vienna, Austria DOI: 10.2507/28th.daaam.proceedings.007

[2] Hodzic, D. & Pandzic, A. (2019). Influence of carbon fibres on mechanical properties of materials in FDM technology, Proceedings of the 30th DAAAM International Symposium, pp.0334-0342, B. Katalinic (Ed.), Published by DAAAM International, ISBN 978-3-902734-22-8, ISSN 1726-9679, Vienna, Austria. DOI: 10.2507/30th.daaam.proceedings.044

[3] Ramesh, M.; Sri Ananda, A. T.; Aswin, U. S.; Eashwar, H. & Deepa, C. (2014) Processing and mechanical property evaluation of banana fibre reinforced polymer composites; 12th Global Congress on Manufacturing and Management; Procedia Engineering, pp. 97:563 – 572.

[4] Astrom, B. T. (2001) Introduction to manufacturing of polymer-matrix composites. ASM, International, Materials Park, OH, pp. 421 – 422.

[5] Gupta, M. K. & Srivastava, K. R. (2015) Mechanical properties of hybrid fibres-reinforced polymer composite: a review. Jour. of Pol. & Plas. Techn. and Eng., vol. 55, No.6, pp. 626 – 642. DOI:10.1080/03602559.2015.1098694.

[6] Todor, P. M.; Bulei, C. & Kiss, I. (2018) Composite materials manufacturing using textile inserts with natural origins fibres. IOP Conference Series: Mat. Sci. & Eng., 393 012088. DOI:10.1088/1757-899X/399/1/012088.

[7] Zaman, H. U.; Khan, A.; Khan, A. R.; Huq, T.; Khan, A. M.; Shah, Md.; Rahman, Al-Mamun Md & Poddar P. (2010) Preparation and characterization of jute fabrics reinforced urethane based thermoset composites: effect of UV radiation. Jour. of Fiber and Polymers, vol. 11, No.2, pp. 258--265. DOI 10.1007/s12221-010-0258-7.

[8] Idriss, A. K. M.; Khan, M. A. & Ali, A. M. (1997) Study on jute material with urethane acrylate by UV curing. Radiation Physics Chemistry. vol. 49, No. 3, pp. 383 – 388.

[9] Zhu, J.; Zhu, H.; Njiguna, J. & Abhyankar, H. (2013) Recent development of flax fibres and their reinforced composites based on different polymeric matrices. Materials Conference, vol. 6, pp. 5171 -5198.

[10] Bahrami, M.; Abenojar, J. & Martinez, M. A. (2020) Recent progress in hybrid bio-composites: mechanical properties, water absorption, and flame retardancy. www.mdpi.com/journal/materials. Material, vol. 13, pp. 5154, DOI:10.3390/ma13225145.

[11] Nguyen, H., Zatar, W. & Mutsuyoshi, H. (2017) Hybrid polymer composites for structural applications. Woodhead Publishing, Elsevier. DOI:10.1016/B978-0-08-100785-3.00002-4; Corpus ID: 136205525.

[12] Kumar, P. A. & Mohamed, N. M. (2017) A comparative analysis on tensile strength of dry and moisture absorbed hybrid kenaf/glass polymer composites. sagepub.co.uk/journals 0(00) 1 – 24, DOI:10.1177/1528083717720203.

[13] Safarabadi, M. & Shokrieh, M. M. (2014) Understanding residual stresses in polymer matrix composites. Science Direct. DOI:10.1533/9780857098597.2.197.

[14] Yahaya, R.; Sapuan, S. M.; Jawaid, M.; Leman, Z & Zainudin, E. S. (2015) Effect of layering sequence and chemical treatment on the mechanical properties of woven kenaf-aramid hybrid laminated composites. Jour. of Materials and Design vol. 67: pp. 173 – 179. http://dx.doi.org/10.1016/j.matdes.2014.11.024.

[15] Tagarielli, V. L.; Minisgallo, G.; McMillan, A. J. & Petrinic, N. (2010) The response of a multi-directional composite laminate to through-thickness loading, Jour. of Comp. Sci. and Tech. vol. 70, pp. 1950 – 1957. doi: 10.1016/j.compscitech.2010.07.013.

[16] Paul, S. A.; Boudenne, A.; Ibos, L., Candau, Y., Joseph, K. & Thomas, S. (2008) Effect of fibre loading and chemical treatments on thermophysical properties of banana/polypropylene commingled composite materials; Composites Part A: Applied Science and Manufacturing; vol. 39(9), pp. 1582-1588.

[17] Vijayan, R. & Krishnamoorthy, A. (2018) Experimental analysis of hybrid (roselle, aloe vera and glass) natural fibre-reinforced composite material. International Journal of Mechanical and Production Engineering Research and Development, vol. 8, Issue 4, pp. 303-314.

[18] Meenakshi, M. C. & Krishnamoorthy, A. (2019) Experimental and finite element analysis on the mechanical performance of hybrid fibre-reinforced epoxy composites. International Journal of Mechanical and Production Engineering Research and Development, vol. 9, Issue 4, pp. 1195 - 1202.

[19] Autar K. K. (2006) Mechanics of composite Materials; ISBN:10:0-8493-1343-0, 13: 978-0-8493-1343-1; Taylor and Francis; TA418.9.C6K39-2005.

[20] ASTM 3039/D3039-08. Standard test method for tensile properties of polymer matrix composite materials. ASTM Int'l: PA 19428 – 2959.

[21] Belingardi, G.; Paolino, D. S. & Koricho, E. G. (2011) Investigation of the influence of tab types on tensile strength of E-glass/epoxy fibre reinforced composite materials. Comp A: vol. 10, pp. 3279-3284. https://doi.org/10.1016/j.proeng.2011.04.541.

[22] Tabrizi, E. I. & Yildiz, M. (2019) Determining tab materials for tensile test of CFRP laminates with a combined usage of digital image correlation and acoustic emission techniques. Comp A: vol. 127:105623. https://doi.org/10.1016/j.cpompositesa.2019.105623.

[23] Yan, C. (2019) On homogenization and de-homogenization of composite materials. PhD Thesis, Drexel University, PA, USA.

[24] Alexander, A. & Tzeng, J. (1997) Three-dimensional effective properties of composite materials for finite element applications. Journal of Composite Materials. Vol. 31(5), pp. 446—485.

[25] Michel, J.; Moulinec, H & Suquet, P. (2009) Effective properties of composite materials with periodic microstructure: a computational approach. Computer Methods in Applied Mechanics and Engineering, vol. 172(1—4), pp.109—143.

[26] Sun, H.; Di, S.; Zhang, N.; & Wu, C. (2001) Micromechanics of composites materials using multivariable finite element method and homogenization theory. International Journal of Solids and Structures, vol. 38(17), pp.3007 – 3020.

DOI: 10.2507/33rd.daaam.proceedings.070

MACHINE LEARNING FOR DATA ANALYSIS IN FOOTBALL: A SURVEY OF METHODS AND PROBLEMS

Saša Tokić, Ante Panjkota & Maja Matetić

This Publication has to be referred as: Tokic, S[asa]; Panjkota, A[nte] & Matecic, M[aja] (2022). Machine Learning on Spatiotemporal Data in Football: A Survey of Methods and Problems, Proceedings of the 33rd DAAAM International Symposium, pp.0503-0510, B. Katalinic (Ed.), Published by DAAAM International, ISBN 978-3-902734-36-5, ISSN 1726-9679, Vienna, Austria
DOI: 10.2507/33rd.daaam.proceedings.070

Abstract

Machine learning is growing exponentially, and its applications are gaining more traction in the sports analysis community in recent years. The application of machine learning methods on spatiotemporal data in sports like football is getting attention from football clubs, academics, and amateur analysts and is the focus of this survey. This survey analyses and identifies current trends in research papers and literature to determine current and future applications in football analytics using spatiotemporal data.

Keywords: spatiotemporal data; sports analytics, event data; deep neural networks, machine learning

1. Introduction

Sports analytics in sports clubs has been evolving ever since "Moneyball" [1] showed how applied statistics could be used to select less know but valuable baseball players. While many sports, including football, have had rich data for decades until recently, spatiotemporal data was unavailable, and the advancement of technical capabilities enabled the evolution of available data. While so-called match sheet data certainly has its' value and is mainly understood by experts and laymen alike, spatiotemporal data opens new possibilities for clubs and scouts. Unfortunately, football clubs focus primarily on video analysis, which is highly time-consuming and introduces bias. Spatiotemporal event data helps in answering many practical questions of much interest in football clubs like:

- Calculate the probability of scoring a goal from a given situation?
- How to evaluate a particular pass?
- What are common tactics opponents use?
- Which players have similar playing styles?

This paper analyses and identifies current trends in research papers and literature to determine current and future applications in sports analytics using machine learning and spatiotemporal data. The goal of this research was to research the current state of data types, data availability, and research methods and applications of machine learning in sports analytics, focusing on applied research in football. Before the literature overview was conducted, we proposed the following research questions:

- Research question 1: How to categorize current research in machine learning in football analytics
- Research question 2: What are common methods of machine learning in football analytics
- Research question 3: What are possible further research directions

To answer our research questions, we've conducted a comprehensive literature overview using the Systematic Literature Review approach [2] to gather recent literature. The main goal was to classify recent research and analyse then categorize it by applying the content analysis method. The result of the analysis showed that the literature could be classified broadly as applying machine learning to spatiotemporal data in football to:

- Evaluate individual players
- Evaluate passes/actions
- Evaluate and classify teams
- Game result prediction

This paper is organized as follows: first, we introduce the domain of sports analytics and systemize the types of data and their availability; second, we provide an overview of recent research in the field of machine learning and spatiotemporal data in football; finally, we give a conclusion of the current state and suggest further research directions possible.

2. Data types and availability

Data suitable for football (soccer in the USA) analytics can be divided into three categories:

- Match sheet data
- Event data
- Tracking data

All types of data are obtained from specialized companies or websites. For example, service providers like FBref [3] usually provide match sheet data without fees, while event and tracking data are only available under a proprietary license. Some existing open datasets of event data are provided by providers like Wyscout [4] and Statsbomb [5]. Tracking data providers include Opta [6], Signality [7], SecondSpectrum [8], Metrica [9], and others.

Match sheet data provides a high-level statistic of the game, player, or club. In general, this data is freely accessible, and some websites even encourage the public to "scrape" the data from them. Many sources like FBref [3] provide additional statistical data, such as adding Expected Goals and Expected Assists. Still, this kind of data doesn't help answer complex questions club stakeholders might have. Event data is a sparser type of data, like tracking data; it is generated from video by human annotators and consists of current player and ball positions. Data points occur after a particular event like a pass, goal, foul or other.

Although this data also is usually not freely available, some available datasets are appropriate for research. The most significant difference between event and tracking data is that tracking data tracks the position of all players on the pitch. In contrast, event data only records the event, disregarding other players' positions. Pappalardo describes the biggest available dataset to date in [10]. Data covers the seven most significant European leagues, World Cup 2018. and European cup 2016. Event data consists of events like pass, foul, and others with subtypes like cross-pass.

Tracking (spatiotemporal) data represents a type of data that consists of both time and space with unique properties regarding the modeling of spatio-temporal relations [11]. For example, in sports, there are usually 10-30 data points per second (10-30Hz) representing players' current positions on the pitch plus the ball. Tracking data can be obtained in multiple ways – static cameras, commercial video broadcasts, and GPS devices. However, due to its high commercial value, it is challenging for researchers to obtain this data type. Therefore, this data type is usually reserved for analysts in clubs with access to their league data. Khaustov and Mozgovoy [12] propose "a rule-based algorithm for identifying several basic types of events in soccer, including ball possession, successful and unsuccessful passes, and shots on goal." Direct potential benefits of applying their algorithm to tracking data could enhance additional event information usually unavailable as part of tracking data., p.

3. Research

The research questions in the introduction are self-evident, bearing in mind that the research is broadly interested in the state-of-art machine learning applications in spatiotemporal data analysis in football. Systematic Literature Review guidelines [2] are used to answer these questions. In that sense, a literature search was conducted extensively in the IEEE Xplore Digital Library (IEEE), Scopus database, Web of Science database (WoS), ScienceDirect, and Google Scholar. Furthermore, due to the specificity of the research topic, search terms include a specific combination of targeted keywords, e.g., "Machine learning" AND "spatiotemporal" AND "football" OR "soccer" and many others. In the next step, we created a table where we classified all the research papers by data type used (event or tracking), machine learning method, practical application, and broad categories that could group relevant papers. Some research papers were included in more than one category.

The research categories that were most apparent from the literature overview of 45 most recent or most important research papers with a corresponding number of papers are as follows:

- Evaluate individual players (13)
- Evaluate passes/actions (12)
- Evaluate and classify teams (11)
- Game result prediction (8)

4. Results

After the literature overview it became evident that the application of machine learning models to football analysis varies from so-called classical methods like a k-nearest neighbours to complex neural networks. One of the most compelling papers which provide football-applicable CNN architecture [13] is "Soccermap" [14] shown in Figure 1

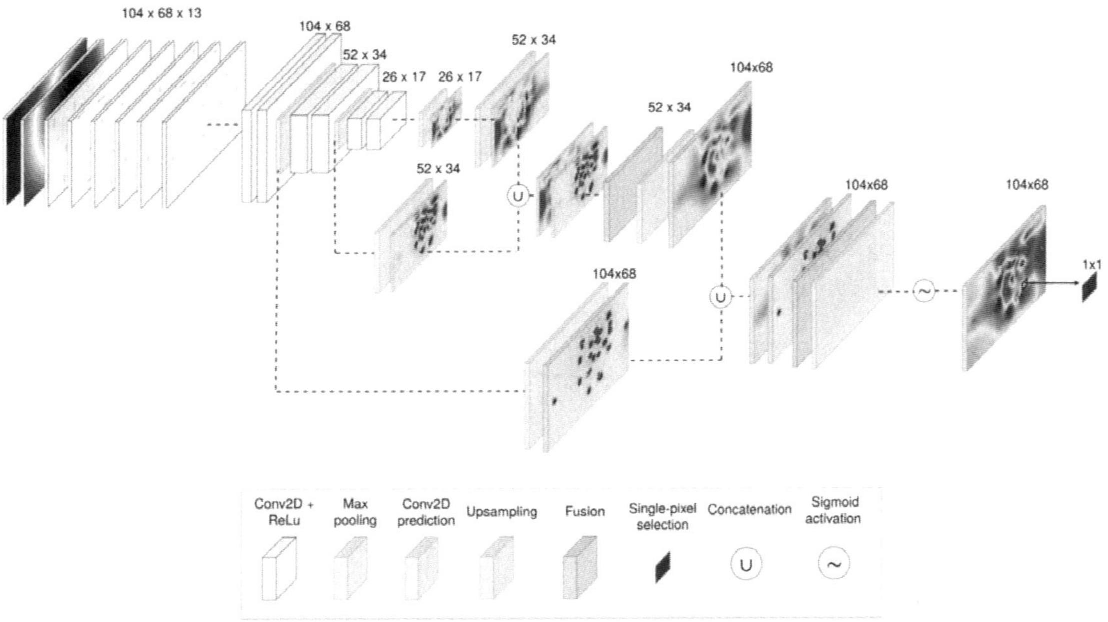

Fig. 1. Soccermap CNN, [14]

This architecture demonstrates that it is possible to apply modern machine learning approaches to tracking data with practical applications in football. By changing only, the output function, this network can calculate the probability surfaces of potential passes, estimate the pass-selection likelihood, and predict the expected value of the pass. In Table 1, we've organized the most recent and influential research in machine learning in football, mainly focusing on spatiotemporal data. Correspondingly, we provide an overview of data types, methods, practical applications, and identified tasks.

Reference	Data	Method	Application	Task
Cintia et al. [15]	Event	k-means, autoencoder	capturing and analysing the playing style of players, teams, and coaches in an automatic way	Evaluate and classify teams
Fernández et al. [16]	Tracking	Markov decision process	ability to evaluate the impact of observed and potential actions, both visually and analytically, expected possession value	Evaluate passes/actions
Hyeonah et al. [17]	Event	Convolutional Autoencoder	characterizing player's styles	Evaluate passes/actions
Kim et al [18]	Tracking	CNN, embeddings	Detecting players and their styles	Evaluate individual players
Lee and Jung [19]	Match sheet	DNN	predicting soccer tactics	Evaluate and classify teams
Malamatinos et al. [20]	Match sheet	k-Nearest Neighbour (k-NN), LogitBoost (LB), Support Vector Machine (SVM), Random Forest (RF), and CatBoost (CB)		Game result prediction

Kim et al. [21]	Tracking	graph based CPD	distinguishes tactically intended formation and role changes from temporary changes in soccer matches	Evaluate and classify teams
Clijmans et al. [22]	Event	Markov model	analysing playing style	Evaluate and classify teams
Goes et al. [23]	Tracking	KMeans	Classify attacks as successful	Evaluate passes/actions
Fernàndez et al. [14]	Tracking	CNN	probability surfaces of potential passes, the expected value of a pass	Evaluate passes/actions
Goes et al. [24]	Tracking	KMeans	defenders create space for attackers are strongly dependent on those attacks' success	Evaluate passes/actions
Raabe et al. [25]	Event	Graph neural networks	Analysing tactical patterns	Evaluate and classify teams
Verstraete et al. [26]		CPD (canonical polyadic decomposition) [27]		Evaluate individual players
Nunez and Dagnino [28]		pitch control, expected possession value and expected goals in a weighted function	Google Research Football competition [29]	-
Liu et al. [30]	Event	Deep Reinforcement Learning, LSTM	developed a new metric called GIM, to evaluate teams	Evaluate and classify teams
Decroos et al.[31]	Event	non-negative matrix factorization [32]	identifying players with a similar style	Evaluate individual players
Beal et al. [33]	Event	walks within graphs, mixed integer programming	forming teams with	Evaluate individual players
Pappalardo et al. [34]	Event	Linear Support Vector Classifier	role-aware player performance evaluation	Evaluate individual players
Groll et al. [35]	Event	random forests and Poisson ranking		Game result prediction
Goes et al. [36]	Tracking	Principal component analysis [37]	evaluating pass value	Evaluate passes/actions
Dick et al. [38]	Tracking	deep reinforcement learning	valuations of multiple players positioning	Evaluate individual players
Yam [39]	Tracking	deep reinforcement learning	valuations of multiple players positioning	Evaluate individual players
Decroos et al. [40]	Event	Generalized Additive Model [41]	improvement to VAEP model [42]	Evaluate passes/actions
Bransen et al. [43]	Event	distance-weighted k-nearest neighbours search	a new metric that aims to measure players' contribution in creating goal-scoring chances	Evaluate individual players
Zambom-Ferraresi et al. [44]	Tracking	Bayesian model averaging [45]	discover determinants of sports performance	Evaluate passes/actions
Steiner et al. [46]	Tracking	binary logistic regressions	estimate the effects of contextual features on passing decisions	Evaluate passes/actions
Pappalardo et al. [47]	Tracking	OLS regression and logit classification	finds team ranking in the future season by using data from previous seasons	Evaluate and classify teams
McHale et al. [48]	Tracking	probability of a successful pass and network centrality measures	help trainers and scouts identify vital players in either opposition teams when recruiting new talents	Evaluate individual players
Giancola et al. [49]	-	CNN, ResNet-152	detecting events in football broadcast videos	-
Decroos 20/12/2022 23:23:00et al. [49]	Event	clustering	discover tactics of football teams	Evaluate and classify teams
Decroos et al. [42]	Event	Clustering, exponential-decay-based	find top performing players in a league or a particular match	Evaluate individual players
Steiner 20/12/2022 23:23:00et al. [46]	Event	regression model	whom the player is most likely to pass the ball	Evaluate passes/actions
Horton et al. [50]	Tracking	computational geometry features fed into different classifiers (MLR, SVM, RUSBoost	classify the quality of passes in football	Evaluate passes/actions
Brooks et al. [51]	Event	L2-regularized Support Vector Machine (SVM) model	rank players based on the value of their passes	Evaluate individual players / Evaluate passes/actions
Brooks et al. [52]	Event	KNN	create a unique team identification	Evaluate and classify teams

McHale et al. [53]	Event	mixed-effects model	model for identifying the ability of football players to score goals	Evaluate individual players
Bialkowski et al. [54]	Tracking	minimum entropy data partitioning and expectation-maximization (EM) algorithm	analysing both individual players and teams	Evaluate and classify teams / Evaluate individual players
Bialkowski et al. [55]	Tracking	LDA/ k-NN	identifying teams from spatiotemporal tracking data	Evaluate and classify teams
Tewari et al. [56]		XGboost, SVM, Logistic regression	-	Game result prediction
Baboota et al. [57]		Gaussian naive Bayes, SVM, Random forest, and gradient boosting		Game result prediction
Razali et al. [58]		Bayesian networks		Game result prediction
Danisik et al. [59]		LSTM regression model		Game result prediction
Cho et al. [60]		gradient boosting		Game result prediction
Ulmer et al. [61]		Naive Bayes, Hidden Markov Model, Multinomial Naive Bayes, RBF SVM, Random forest, Linear SVM, One vs. All SGD		Game result prediction

Table 1. Literature overview, Source: authors contribution

5. Conclusion

This review aimed to evaluate the current state of machine learning in sports analytics, particularly interested in applications on data analytics in football. While the field might seem like a niche category, the number of research papers and non-academic content was pretty significant, so we've decided to focus on the most recent studies most representative of the current state-of-the-affairs. Besides many possible research directions from the literature, this review detects four main research tasks. These tasks are:

- players evaluation
- passes and actions evaluation
- team evaluation
- game result prediction

It is no surprise that most tasks have practical applications that are of interest to football club stakeholders. Such claims include scouting, which stems from player evaluation, overall team formation improvement by combining player evaluation and team evaluation, and team success expectations by combining all four tasks are just a few of many. The coaching staff has many benefits, including tactical preparations of matches from passes/actions evaluation, players evaluation, and opposing team evaluation. In addition, game result predictions are also of interest to sports betting as in academic research.

On the other hand, conducted results show that methods in this area span from so-called classical machine learning to deep learning, which has been the dominant method in recent times. Also, data types are closely related to the method used; in that sense, CNN models have shown the most promising research direction. We intentionally omit the area of simulation environments methods from this survey because it is the emerging approach to addressing problems. Among them, we would like to point out possible research directions in simulated multi-agent environments like Google Research Football as a method for coping with defined tasks.

6. References

[1] Lewis, M. (2004). Moneyball: the art of winning an unfair game, Norton, 978-0-393-05765-2, New York, NY

[2] Kitchenham, B. & Charters, S. (2007). Guidelines for performing Systematic Literature Reviews in Software Engineering, Keele, Staffs, UK, School of Computer Science and Mathematics, Keele University

[3] https://fbref.com/en/ (2020). FBref - Football Statistics and History, FBref.com, Accessed on: 2020-10-9

[4] https://wyscout.com/ (2020). Wyscout - Football Professional Videos and Data Platform, Wyscout, Accessed on: 2020-10-10

[5] http://statsbomb.com/ (2020). StatsBomb | Football Like Never Before, StatsBomb, Accessed on: 2020-10-10

[6] https://www.optasports.com (2020). Opta, Opta Sports, Accessed on: 2020-10-11

[7] https://www.signality.com/ (2020). Signality, Accessed on: 2020-11-10

[8] https://www.secondspectrum.com/index.html (2020). Second Spectrum, Accessed on: 2020-11-10

[9] https://github.com/metrica-sports/sample-data (2020). Metrica Sports Sample Data, Accessed on: 2020-10-10

[10] Pappalardo, L.; Cintia, P.; Rossi, A.; Massucco, E.; Ferragina, P.; Pedreschi, D. & Giannotti, F. (2019). A public data set of spatio-temporal match events in soccer competitions, Scientific Data, Vol. 6, No. 1, pp. 236, 2052-4463, https://doi.org/10.1038/s41597-019-0247-7

[11] Salibekyan, S. & Panfilov, P. (2016). Database Architecture for Specifying and Modeling Spatio-Temporal Relations, In B. Katalinic (Ed.), DAAAM Proceedings, (1st ed., Vol. 1, pp. 0589–0598) DAAAM International Vienna, 978-3-902734-07-5 https://doi.org/10.2507/26th.daaam.proceedings.080

[12] Khaustov, V. & Mozgovoy, M. (2020). Recognizing Events in Spatiotemporal Soccer Data, Applied Sciences, Vol. 10, No. 22, pp. 8046, 2076-3417, https://doi.org/10.3390/app10228046

[13] LeCun, Y.; Boser, B.; Denker, J. S.; Henderson, D.; Howard, R. E.; Hubbard, W. & Jackel, L. D. (1989). Backpropagation Applied to Handwritten Zip Code Recognition, Neural Computation, Vol. 1, No. 4, pp. 541–551, 0899-7667, 1530-888X, https://doi.org/10.1162/neco.1989.1.4.541

[14] Fernández, J. & Bornn, L. (2020). SoccerMap: A Deep Learning Architecture for Visually-Interpretable Analysis in Soccer, Presented at the The European Conference on Machine Learning and Principles and Practice of Knowledge Discovery in Databases, ECML-PKDD, Virtual

[15] Cintia, P. & Pappalardo, L. (2021). Coach2vec: autoencoding the playing style of soccer coaches, arXiv https://doi.org/10.48550/arXiv.2106.15444

[16] Fernández, J.; Bornn, L. & Cervone, D. (2021). A framework for the fine-grained evaluation of the instantaneous expected value of soccer possessions, Machine Learning, Vol. 110, No. 6, pp. 1389–1427, 1573-0565, https://doi.org/10.1007/s10994-021-05989-6

[17] Cho, H.; Ryu, H. & Song, M. (2022). Pass2vec: Analyzing soccer players' passing style using deep learning, International Journal of Sports Science & Coaching, Vol. 17, No. 2, pp. 355–365, 1747-9541, https://doi.org/10.1177/17479541211033078

[18] Kim, H.; Kim, J.; Chung, D.; Lee, J.; Yoon, J. & Ko, S.-K. (2022). 6MapNet: Representing Soccer Players from Tracking Data by a Triplet Network, In U. Brefeld, J. Davis, J. Van Haaren, & A. Zimmermann (Eds.), Machine Learning and Data Mining for Sports Analytics, (pp. 3–14) Springer International Publishing, 978-3-031-02044-5, Cham https://doi.org/10.1007/978-3-031-02044-5_1

[19] Lee, G. J. & Jung, J. J. (2022). DNN-based multi-output model for predicting soccer team tactics, PeerJ Computer Science, Vol. 8, No. 1, pp. e853, 2376-5992, https://doi.org/10.7717/peerj-cs.853

[20] Malamatinos, M.-C.; Vrochidou, E. & Papakostas, G. A. (2022). On Predicting Soccer Outcomes in the Greek League Using Machine Learning, Computers, Vol. 11, No. 9, pp. 133, 2073-431X, https://doi.org/10.3390/computers11090133

[21] Kim, H.; Kim, B.; Chung, D.; Yoon, J. & Ko, S.-K. (2022). SoccerCPD: Formation and Role Change-Point Detection in Soccer Matches Using Spatiotemporal Tracking Data, In Proceedings of the 28th ACM SIGKDD Conference on Knowledge Discovery and Data Mining, (pp. 3146–3156) Association for Computing Machinery, 978-1-4503-9385-0, New York, NY, USA https://doi.org/10.1145/3534678.3539150

[22] Clijmans, J.; Roy, M. V. & Davis, J. (2022). Looking Beyond the Past: Analyzing the Intrinsic Playing Style of Soccer Teams, In ECML PKDD 2022,Presented at the European Conference on Machine Learning and Principles and Practice of Knowledge Discovery in Databases

[23] Goes, F. R.; Brink, M. S.; Elferink-Gemser, M. T.; Kempe, M. & Lemmink, K. A. P. M. (2021). The tactics of successful attacks in professional association football: large-scale spatiotemporal analysis of dynamic subgroups using position tracking data, Journal of Sports Sciences, Vol. 39, No. 5, pp. 523–532, 0264-0414, https://doi.org/10.1080/02640414.2020.1834689

[24] Goes, F. R.; Brink, M. S.; Elferink-Gemser, M. T.; Kempe, M. & Lemmink, K. A. P. M. (2020). The tactics of successful attacks in professional association football: large-scale spatiotemporal analysis of dynamic subgroups using position tracking data, Journal of Sports Sciences, Vol. 0, No. 0, pp. 1–10, 0264-0414, https://doi.org/10.1080/02640414.2020.1834689

[25] Raabe, D.; Nabben, R. & Memmert, D. (2022). Graph representations for the analysis of multi-agent spatiotemporal sports data, Applied Intelligence, 1573-7497, https://doi.org/10.1007/s10489-022-03631-z

[26] Verstraete, K.; Decroos, T.; Coussement, B.; Vannieuwenhoven, N. & Davis, J. (2020). Analyzing Soccer Players' Skill Ratings Over Time Using Tensor-Based Methods, In P. Cellier & K. Driessens (Eds.), Machine Learning and Knowledge Discovery in Databases, (Vol. 1168, pp. 225–234) Springer International Publishing, 978-3-030-43886-9, Cham https://doi.org/10.1007/978-3-030-43887-6_17

[27] Sorber, L.; Van Barel, M. & De Lathauwer, L. (2013). Optimization-Based Algorithms for Tensor Decompositions: Canonical Polyadic Decomposition, Decomposition in Rank-$(L_r,L_r,1)$ Terms, and a New Generalization, SIAM Journal on Optimization, Vol. 23, No. 2, pp. 695–720, 1052-6234, 1095-7189, https://doi.org/10.1137/120868323

[28] Nunez, J. C. & Dagnino, B. (2020). Exploring the application of soccer mathematical models to game generation on a simulated environment, , pp. 10,

[29] https://kaggle.com/c/google-football (2020). Google Research Football with Manchester City F.C., Accessed on: 2020-10-4

[30] Liu, G.; Luo, Y.; Schulte, O. & Kharrat, T. (2020). Deep soccer analytics: learning an action-value function for evaluating soccer players, Data Mining and Knowledge Discovery, Vol. 34, No. 5, pp. 1531–1559, 1384-5810, 1573-756X, https://doi.org/10.1007/s10618-020-00705-9

[31] Decroos, T. & Davis, J. (2020). Player Vectors: Characterizing Soccer Players' Playing Style from Match Event Streams, In U. Brefeld, E. Fromont, A. Hotho, A. Knobbe, M. Maathuis, & C. Robardet (Eds.), Machine Learning and Knowledge Discovery in Databases, (Vol. 11908, pp. 569–584) Springer International Publishing, 978-3-030-46132-4, Cham https://doi.org/10.1007/978-3-030-46133-1_34

[32] Lee, D. D. & Seung, H. S. (2001). Algorithms for Non-negative Matrix Factorization, (p. 7)Presented at the NIPS

[33] Beal, R.; Changder, N.; Norman, T. & Ramchurn, S. (2020). Learning the Value of Teamwork to Form Efficient Teams, Proceedings of the AAAI Conference on Artificial Intelligence, Vol. 34, No. 05, pp. 7063–7070, 2374-3468, 2159-5399, https://doi.org/10.1609/aaai.v34i05.6192

[34] Pappalardo, L. & Cintia, P. (2018). Quantifying the relation between performance and success in soccer, Advances in Complex Systems, Vol. 21, No. 03n04, pp. 1750014, 0219-5259, 1793-6802, https://doi.org/10.1142/S021952591750014X

[35] Groll, A.; Ley, C.; Schauberger, G. & Van Eetvelde, H. (2019). A hybrid random forest to predict soccer matches in international tournaments, Journal of Quantitative Analysis in Sports, Vol. 15, No. 4, pp. 271–287, 1559-0410, 2194-6388, https://doi.org/10.1515/jqas-2018-0060

[36] Goes, F. R.; Kempe, M.; Meerhoff, L. A. & Lemmink, K. A. P. M. (2019). Not Every Pass Can Be an Assist: A Data-Driven Model to Measure Pass Effectiveness in Professional Soccer Matches, Big Data, Vol. 7, No. 1, pp. 57–70, 2167-6461, 2167-647X, https://doi.org/10.1089/big.2018.0067

[37] Wold, S.; Esbensen, K. & Geladi, P. (n.d.). Principal Component Analysis, , pp. 16,

[38] Dick, U. & Brefeld, U. (2019). Learning to Rate Player Positioning in Soccer, Big Data, Vol. 7, No. 1, pp. 71–82, 2167-6461, 2167-647X, https://doi.org/10.1089/big.2018.0054

[39] https://statsbomb.com/2019/02/attacking-contributions-markov-models-for-football/ (2019). Attacking Contributions: Markov Models for Football, StatsBomb, Accessed on: 2020-10-4

[40] Decroos, T. (2020). Soccer Analytics Meets Artificial Intelligence: Learning Value and Style from Soccer Event Stream Data, Ph.D. Dissertation, KU Leuven – Faculty of Engineering Science, Heverlee,

[41] Hastie, T. & Tibshirani, R. (1986). Generalized Additive Models, Statistical Science, Vol. 1, No. 3, pp. 297–310, 0883-4237, https://doi.org/10.1214/ss/1177013604

[42] Decroos, T.; Bransen, L.; Van Haaren, J. & Davis, J. (2020). VAEP: An Objective Approach to Valuing On-the-Ball Actions in Soccer (Extended Abstract), In Proceedings of the Twenty-Ninth International Joint Conference on Artificial Intelligence, (pp. 4696–4700)Presented at the Twenty-Ninth International Joint Conference on Artificial Intelligence and Seventeenth Pacific Rim International Conference on Artificial Intelligence {IJCAI-PRICAI-20}, International Joint Conferences on Artificial Intelligence Organization, 978-0-9992411-6-5, Yokohama, Japan https://doi.org/10.24963/ijcai.2020/648

[43] Bransen, L.; Van Haaren, J. & van de Velden, M. (2019). Measuring soccer players' contributions to chance creation by valuing their passes, Journal of Quantitative Analysis in Sports, Vol. 15, No. 2, pp. 97–116, 1559-0410, 2194-6388, https://doi.org/10.1515/jqas-2018-0020

[44] Zambom-Ferraresi, F.; Rios, V. & Lera-Lopez, F. (2018). Determinants of sport performance in European football: What can we learn from the data?, Decision Support Systems, Vol. 114, , pp. 18–28, 01679236, https://doi.org/10.1016/j.dss.2018.08.006

[45] Volinsky, C. T.; Raftery, A. E.; Madigan, D. & Hoeting, J. A. (1999). Bayesian model averaging: a tutorial, Statistical Science, Vol. 14, No. 4, pp. 382–417, 0883-4237, https://doi.org/10.1214/ss/1009212519

[46] Steiner, S.; Rauh, S.; Rumo, M.; Sonderegger, K. & Seiler, R. (2018). Using position data to estimate effects of contextual features on passing decisions in football, , pp. 9,

[47] Pappalardo, L.; Cintia, P.; Ferragina, P.; Massucco, E.; Pedreschi, D. & Giannotti, F. (2019). PlayeRank: Data-driven Performance Evaluation and Player Ranking in Soccer via a Machine Learning Approach, ACM Transactions on Intelligent Systems and Technology, Vol. 10, No. 5, pp. 1–27, 21576904, https://doi.org/10.1145/3343172

[48] McHale, I. G. & Relton, S. D. (2018). Identifying key players in soccer teams using network analysis and pass difficulty, European Journal of Operational Research, Vol. 268, No. 1, pp. 339–347, 03772217, https://doi.org/10.1016/j.ejor.2018.01.018

[49] Decroos, T. (2018). STARSS: A Spatio-Temporal Action Rating System for Soccer, , pp. 10,

[50] Horton, M.; Gudmundsson, J.; Chawla, S. & Estephan, J. (2017). Classification of Passes in Football Matches using Spatiotemporal Data, ACM Transactions on Spatial Algorithms and Systems, Vol. 3, No. 2, pp. 1–30, 23740353, https://doi.org/10.1145/3105576

[51] Brooks, J.; Kerr, M. & Guttag, J. (2016). Using machine learning to draw inferences from pass location data in soccer: Drawing Inferences from Pass Location Data Using Machine Learning, Statistical Analysis and Data Mining: The ASA Data Science Journal, Vol. 9, No. 5, pp. 338–349, 19321864, https://doi.org/10.1002/sam.11318

[52] Brooks, J.; Kerr, M. & Guttag, J. (2016). Developing a Data-Driven Player Ranking in Soccer Using Predictive Model Weights, In Proceedings of the 22nd ACM SIGKDD International Conference on Knowledge Discovery and Data Mining - KDD '16, (pp. 49–55)Presented at the the 22nd ACM SIGKDD International Conference, ACM Press, 978-1-4503-4232-2, San Francisco, California, USA https://doi.org/10.1145/2939672.2939695

[53] McHale, I. G. & Szczepanski, L. (2014). A mixed effects model for identifying goal scoring ability of footballers, Journal of the Royal Statistical Society: Series A (Statistics in Society), Vol. 177, No. 2, pp. 397–417, 09641998, https://doi.org/10.1111/rssa.12015

[54] Bialkowski, A.; Lucey, P.; Carr, P.; Yue, Y.; Sridharan, S. & Matthews, I. (2014). Identifying Team Style in Soccer Using Formations Learned from Spatiotemporal Tracking Data, In 2014 IEEE International Conference on Data Mining Workshop, (pp. 9–14)Presented at the 2014 IEEE International Conference on Data Mining Workshop (ICDMW), IEEE, 978-1-4799-4274-9, Shenzhen, China https://doi.org/10.1109/ICDMW.2014.167

[55] Bialkowski, A.; Lucey, P.; Carr, P.; Yue, Y.; Sridharan, S. & Matthews, I. (2014). Large-Scale Analysis of Soccer Matches Using Spatiotemporal Tracking Data, In 2014 IEEE International Conference on Data Mining, (pp. 725–730)Presented at the 2014 IEEE International Conference on Data Mining (ICDM), IEEE, 978-1-4799-4302-9, Shenzhen, China https://doi.org/10.1109/ICDM.2014.133

[56] Tewari, A.; Parwani, T.; Phanse, A.; Sharma, A. & Shetty, A. (2019). Soccer Analytics using Machine Learning, International Journal of Computer Applications, Vol. 181, No. 50, pp. 54–56, 09758887, https://doi.org/10.5120/ijca2019918773

[57] Baboota, R. & Kaur, H. (2019). Predictive analysis and modelling football results using machine learning approach for English Premier League, International Journal of Forecasting, Vol. 35, No. 2, pp. 741–755, 01692070, https://doi.org/10.1016/j.ijforecast.2018.01.003

[58] Razali, N.; Mustapha, A.; Yatim, F. A. & Aziz, R. A. (2017). Predicting Football Matches Results using Bayesian Networks for English Premier League (EPL), https://doi.org/10.1088/1757-899X/226/1/012099

[59] Danisik, N.; Lacko, P. & Farkas, M. (2018). Football Match Prediction Using Players Attributes, In 2018 World Symposium on Digital Intelligence for Systems and Machines (DISA), (pp. 201–206)Presented at the 2018 World Symposium on Digital Intelligence for Systems and Machines (DISA), IEEE, 978-1-5386-5102-5, Kosice https://doi.org/10.1109/DISA.2018.8490613

[60] Cho, Y.; Yoon, J. & Lee, S. (2018). Using social network analysis and gradient boosting to develop a soccer win-lose prediction model, Eng. Appl. Artif. Intell., https://doi.org/10.1016/j.engappai.2018.04.010

[61] Ulmer, B. & Fernandez, M. (2014). Predicting Soccer Match Results in the English Premier League, (p. 5)Stanford University Retrieved from https://www.semanticscholar.org/paper/Predicting-Soccer-Match-Results-in-the-English-Ulmer-Fernandez/d4dfbb3d1454416799b72004d3d88bae6bbd660a

DOI: 10.2507/33rd.daaam.proceedings.071

COMPARISON OF OPTIMIZED AND CONVENTIONAL COOLING WHEN TURNING STAINLESS STEEL

Adam Lukas & Jindrich Sykora

This Publication has to be referred as: Lukas, A[dam] & Sykora, J[indrich] (2022). Comparison of Optimized and Conventional Cooling When Turning Stainless Steel, Proceedings of the 33rd DAAAM International Symposium, pp.0511-0518, B. Katalinic (Ed.), Published by DAAAM International, ISBN 978-3-902734-36-5, ISSN 1726-9679, Vienna, Austria
DOI: 10.2507/33rd.daaam.proceedings.071

Abstract

Stainless steels are materials whose consumption increases constantly. However, their machining can be problematic. A cutting tool is under higher mechanical and thermal stress than in the case of carbon steels, thereby rapid tool wear can occur. To avoid this, cutting speeds are often kept low, increasing tool life despite lower productivity. The application of cutting fluid has also a beneficial effect, nevertheless, it can be increased when coolant is supplied directly to the cutting edge. Optimized tools with internal channels allow directing fluid precisely there. The main aim of this paper is to compare tool wear when precision and conventional cooling systems are used. The surface roughness of turned specimens made of AISI 316 stainless steel was also recorded. Based on the findings, it can be noticed that the precision cooling system has a significant influence on tool wear and the cutting speed can be set higher. In other words, the productivity of machining is improved. What is more, tool wear is more predictable, and the cutting process is reliable.

Keywords: internal cooling; tool wear; surface roughness; turning; stainless steel

1. Introduction

Steel alloying is a way to improve its properties. As a result, material requirements such as strength, high-temperature properties, ductility, toughness, corrosion resistivity and others can be achieved. Stainless steels are highly alloyed iron-based materials where mainly chromium is added to provide corrosion resistivity. It makes them suitable in applications where long life, maintenance free and human health are crucial. Additional elements, particularly nickel, molybdenum and manganese, mainly promote structure changes and improve corrosion resistivity in specific environments and strength [1]. According to chemical composition, three main structures of stainless steels can be achieved – austenitic, ferritic, and martensitic, where the austenitic structure is the most used [2] [3]. During the last 15 years, stainless steel production increased about twice, creating requirements for efficient processing [3].

When machining, complex technological property of material named 'machinability' is plays significant role. It describes how is difficult to cut (machine) material and is used to determine cutting conditions (cutting speed v_c, feed rate f, depth of cut a_p). Nevertheless, it was observed that if steel corrosion resistivity increases, its machinability decreases [4]. In comparison with non-alloyed steels, there are several differences which cause decrease in machinability, thereby rapid tool wear and consequently lower productivity because of frequent insert changes and lower cutting conditions.

Firstly, thermal conductivity is about three times lower, causing high tool temperature and a rapid degradation process. In addition, thermal expansion is higher, so it is harder to maintain tight tolerance. Secondly, work-hardening phenomena occurs. It means that after deformation (e.g. machining), there is a thin surface layer, which is strengthened and hardened hence tool is more stressed in the next cut. Finally, high ductility decreases chip breakability and promotes formation of build-up edge (BUE) which can damage the machined surface after its break off [5]. Mentioned problems cause rapid tool wear, thus low cutting speeds are often used to maintain high tool life. The influence of cutting speeds on tool wear and surface roughness was explored by Korkut et al. when turning The American Iron and Steel Institute (AISI) 304 austenitic stainless steel. The experiment was carried out without cutting fluid and they optimized the cutting speed at 180 m/min where the lowest tool wear and surface roughness Ra was obtained [6].

Every machining process happens in some cutting environment usually in the form of neat oil, water-based fluid (emulsion), gas, or mist [7]. Cutting fluids are commonly used when machining steels with carbide tools. It is because machining is a process where most of the energy is transformed into heat [8]. Application of these fluids cools down the cutting zone, lowering tool wear. A lubrication effect is also required to decrease friction on the tool-chip and tool-workpiece interface. As a result, cutting forces, heat generation and tool wear are lower. Cutting fluids also help to accomplish better surface. Last but not least, fluids control BUE formation, wash away chips and protect a machine from rust [5].

Different cooling systems can be used to ensure the above-mentioned effects. Flood cooling is conventional. Cutting fluid flows through a nozzle outside the tool holder in large amounts, where a pipeline and flow rate are adjustable [9]. Fluid immerses a tool and workpiece in the cutting zone, absorbing generated heat. Conversely, modern tools are equipped with internal channels where a cutting fluid flows. An exit of these channels is designed in a clamp or insert holder body near the cutting zone, directing fluid precisely to both – the rake and flank face of an insert [10]. Furthermore, the usage of high pressure penetrates the interface between the rake face and the chip more effectively, which results in lower friction, tool temperatures, and better chip breaking [10]. In order to accomplish this, the machine must be equipped with high-pressure pump.

Naves et al. evaluated tool wear when turning AISI 316. High pressure (10, 15 and 20 MPa), dry turning and overhead cooling approaches were compared. The result is that high pressure cooling significantly reduces adhesion wear, by contrast, dry turning attained the highest flank wear [11]. Bleicher et al. investigated tool wear when turning P750 austenitic stainless steel with low cutting speeds. External cooling, internal cooling through the insert, and both external and internal cooling were used. It was concluded that combined external and internal cooling decreased tool wear the most [12]. Kostadin enquired about the difference between emulsion and chilled air cooling when machining martensitic stainless steel. The result was that there is no surface roughness difference between emulsion and chilled-air cooling. In addition, "the feed rate f has the greatest influence in reducing the surface roughness parameter Ra, followed by depth of cut a_p, cutting speed v_c and cooling method" [13]. Janda and Fulemova studied surface roughness when milling ferritic-martensitic stainless steel with external cooling. Six different tool materials were used (4 cemented carbides, and 2 cermets). Carbide tools showed lower tool wear, and surprisingly, surface roughness was better at lower cutting speeds [14].

This paper compares tool flank wear of tools when conventional flood and precision cooling strategies are used during turning experiments. In addition, the surface roughness parameter Ra was monitored. Turned material was AISI 316 austenitic stainless steel.

2. Materials and methods

Fig. 1. a) Tool with conventional cooling b) Tool with precision cooling

The used specimens, round bars with an initial 200 mm diameter and 300 mm length of AISI 316 (for chemical composition see

Table 1) austenitic stainless steel were turned. The final machined length was 282 mm due to clamping with jaws.

[%]	Fe	Cr	Ni	Mo	Mn	Si	N	C	P	S
Min.	62	16	10	2	0	0	0	0	0	0
Max.	72	18	14	3	2	0,75	0,1	0,08	0,045	0,03

Table 1. Chemical composition of AISI 316 [15]

The experiments were conducted with the following cutting conditions:

- cutting speeds vc = 160; 250; 300 m/min
- feed rate f = 0.25 mm/rev
- depth of cut ap = 1.5 mm

The turning process was carried out using DMG MORI CTX BETA 1250 TC turn-milling machine with a maximum of the main spindle power of 32 kW . The 6% cutting emulsion was used at a maximum pressure of 40 MPa. The cemented carbide cutting inserts CNMG 120408E-NMR:T7325 with MT-CVD coating and negative rake were used. According to the manufacturer, the inserts are suitable for the versatile turning of stainless steel and interrupted cuts [16]. The surface roughness was measured with a SURTRONIC DUO tester. The flank wear of used inserts was analysed after every cut using a KEYENCE VHX-6000 digital microscope with 100× magnification. The ISO 3685 standard divides flank wear into four possibilities (Fig.2.) - VB_A (at tool corner), VB_{Bmax} (maximum flank wear)., VB_B (average flank wear), VB_C (notch wear) [17]. VB_B wear was measured.

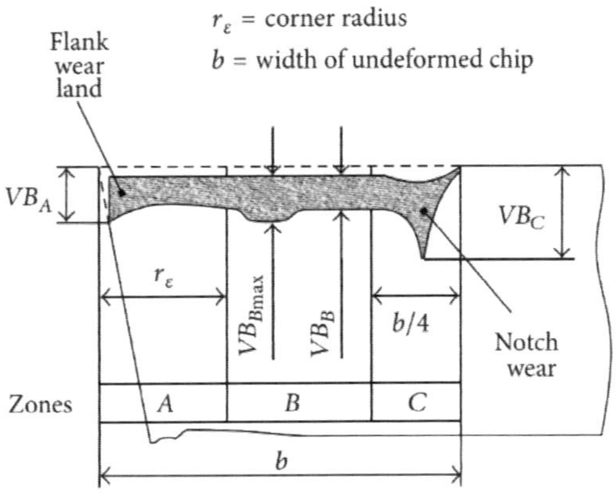

Fig. 2. Flank wear division [17] (adapted)

The development of flank wear progresses typically in three phases (Fig. 3). An initial wear increases rapidly due to the running-in of a new insert. Subsequently, a steady state occurs with a uniform rate. Finally, the failure region is characterised by accelerated wear [18].

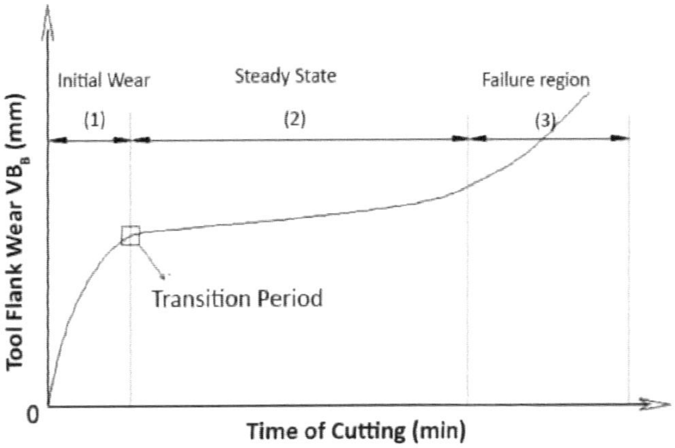

Fig. 3. Typical flank wear curve [18]

The experiment was stopped if one of these criteria came:
- Insert flank wear VBB is greater than VBcrit.= 0.4mm
- Insert failure
- Too low tool flank wear acceleration

3. Experimental results and discussion

3.1. Test 1: vc=160 m/min

The cutting speed v_c=160 m/min was defined according to the maximal recommended value by the insert manufacturer. Although criteria flank wear $VB_{BCRIT.}$ was not reached, the test was discontinued after 20 minutes of cut, due to low flank wear rate. It will not reach the criteria wear in a brief time if the test continues (Fig. 4.). The figures show that there was only a slight difference between flood and precision tool flank wear. Despite this, there was observed a significant difference in notch wear. Flood cooled tool showed a significant notch, whereas there was almost no notch at the precision cooled tool.

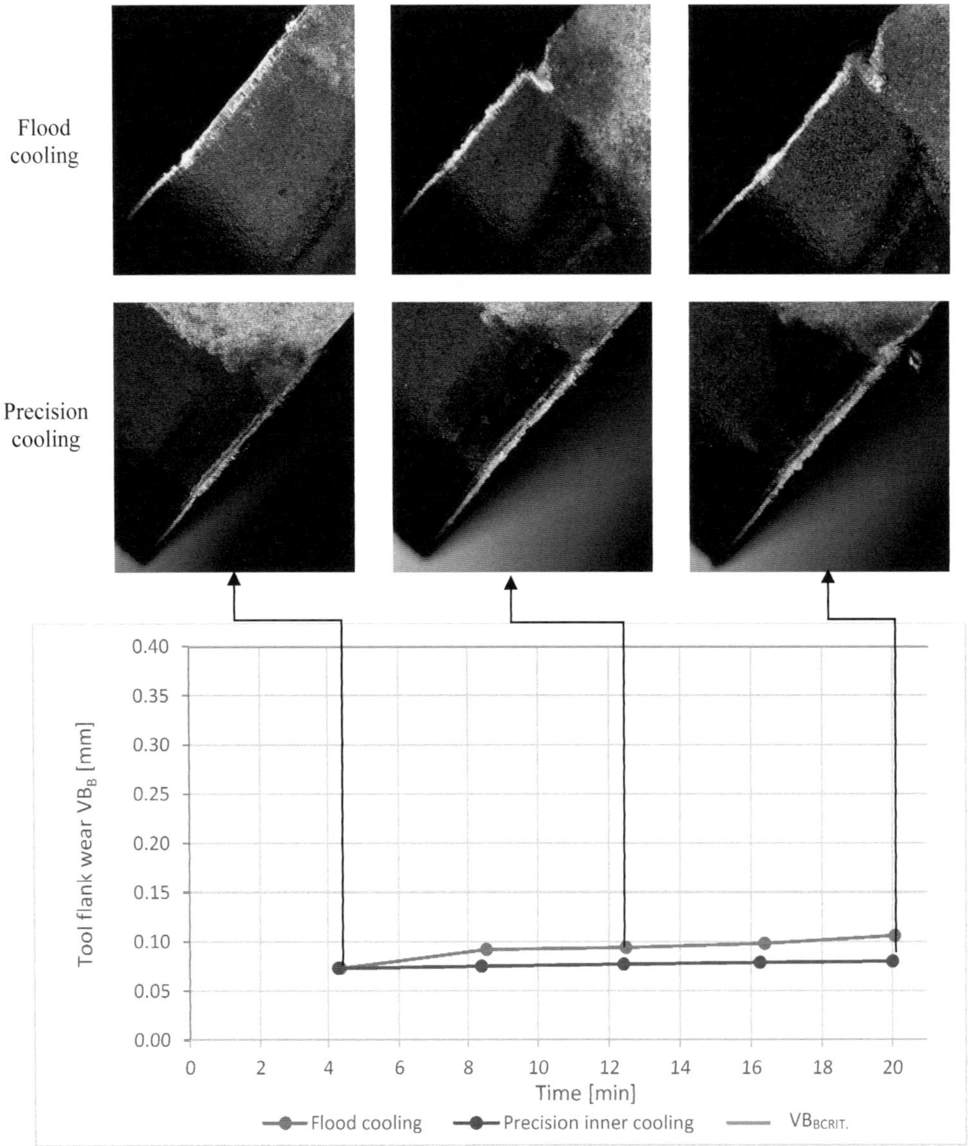

Fig. 4. Development of flank wear, v_c=160 m/min

3.2. Test 2: vc=300 m/min

As a result of slow flank wear increments during the previous test, the new test and insert with cutting speed of v_c = 300 m/min was set to accelerate it. These tests had to be ended due to tools failures, where flood cooled tool failed during the 3rd cut and precision cooled tool during the 7th cut. Excessive tool wear in combination with high mechanical and thermal stress led to tool tip break off.

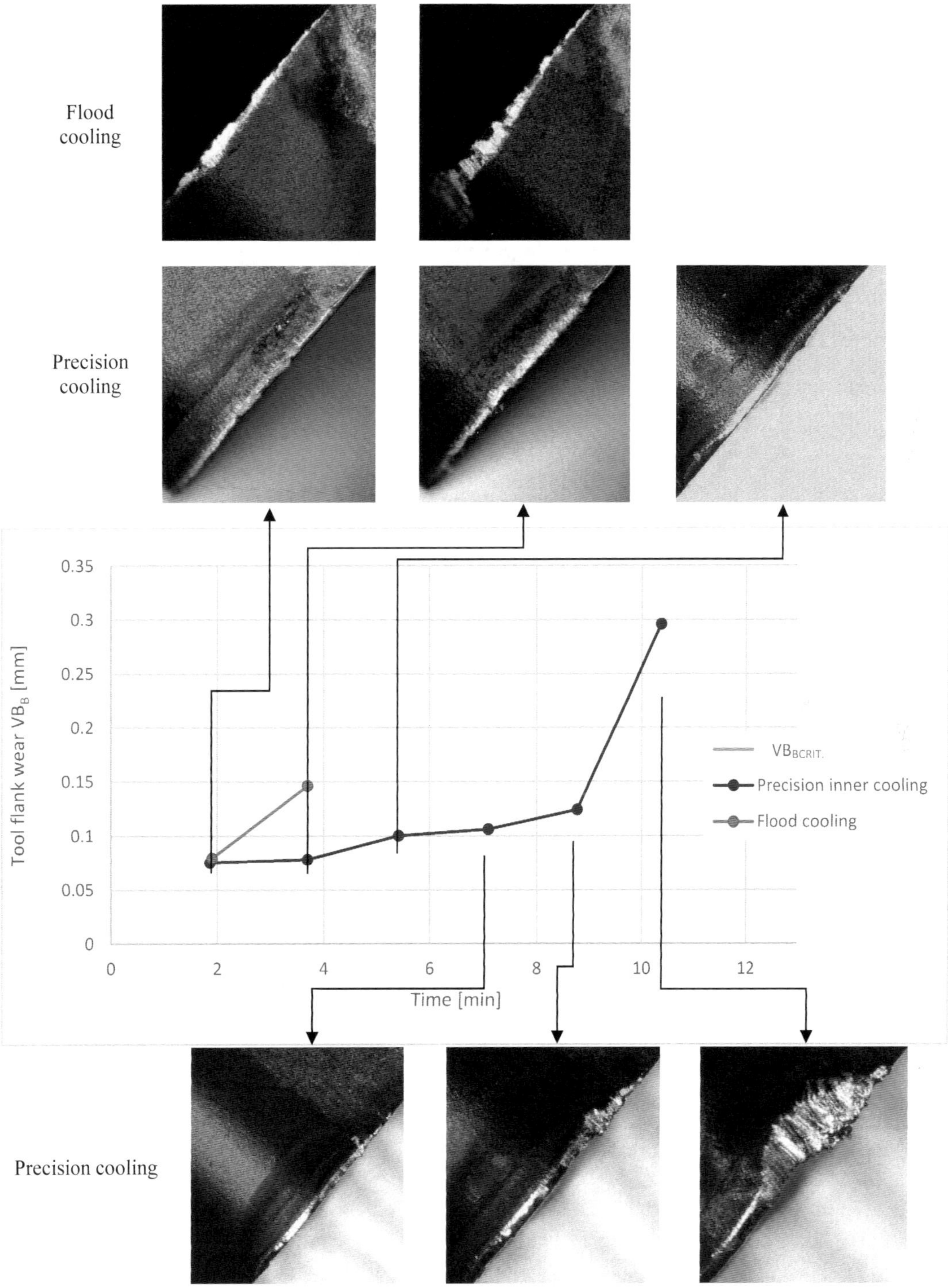

Fig. 5. Development of flank wear, v_c=300 m/min

3.3. Test 3: vc=250 m/min

During this test, the criteria flank wear $VB_{BCRIT.}$ was reached. The following graph (Fig.6.) shows the significant exponential rise of flood cooled tool after the twelfth minute, whereas the precision cooled tool started its exponential wear after 25 minutes.

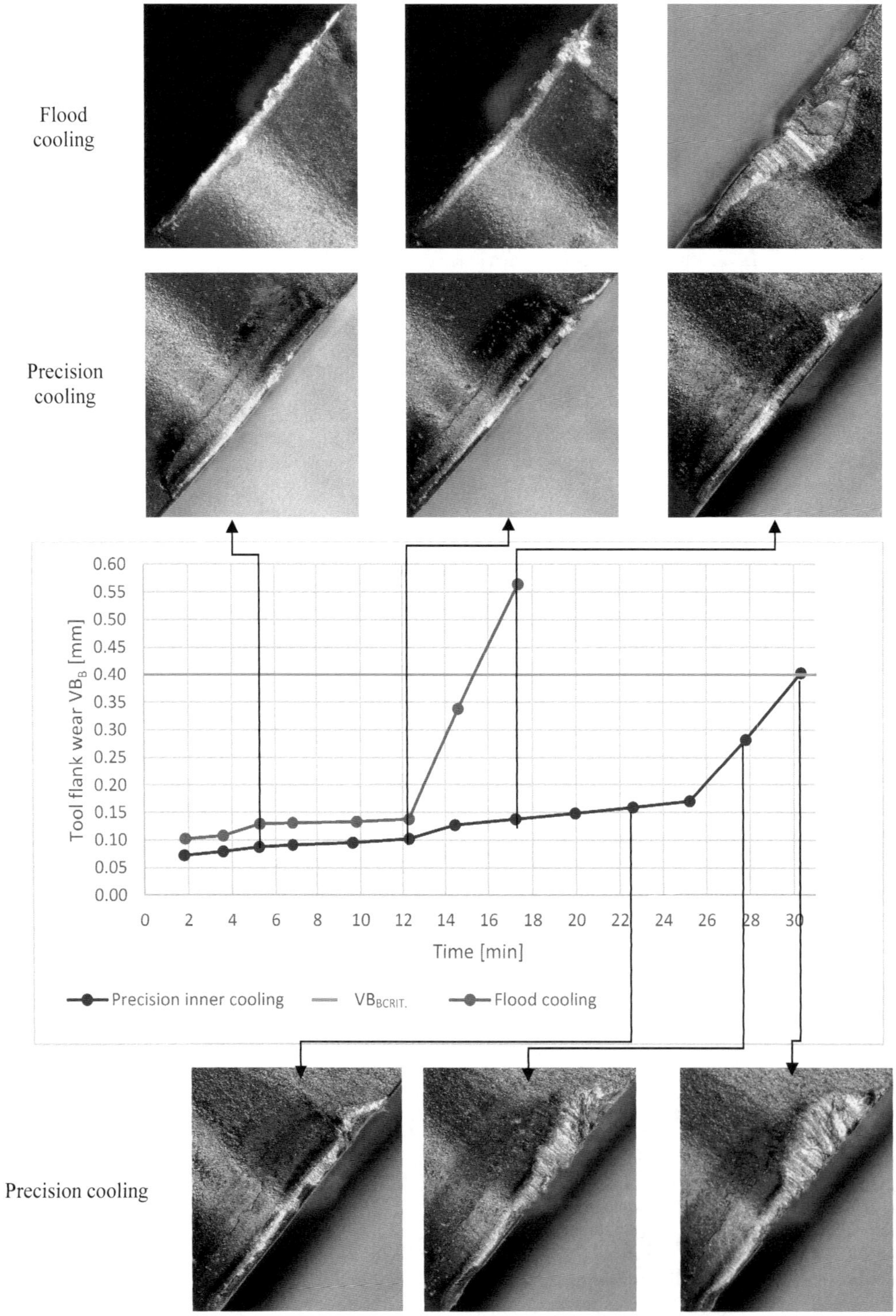

Fig. 6. Development of flank wear, v_c=250 m/min

Using linear interpolation, flood cooled tool reached criteria flank wear after 15 minutes, whilst precision cooled after the 30 minutes. In other words, tool life with the precision cooling system proved to be about 2× longer. This result was achieved with a 1.56× higher cutting speed than the maximum cutting speed recommended by the insert manufacturer. In the failure region, the flank wear of the flood cooled tool shoots up, whereas the precision cooled tool increases more gradually.

3.4 Surface roughness

It was concluded that at a cutting speed 160 m/min surface roughness ranged from 1.88 and 2.13 μm where the precision cooled tool showed a slightly better surface. However, the experiment with a cutting speed 250 m/min showed the opposite result. Flood cooled tool showed better surface even as flank wear increased, furthermore, it was better than precision cooling. This can be explained by higher temperature in the cutting zone which would promote plastic deformation of the surface. Although the exact mechanism would have to further investigated. In the case of v_c = 300 m/min, surface roughness went down continuously.

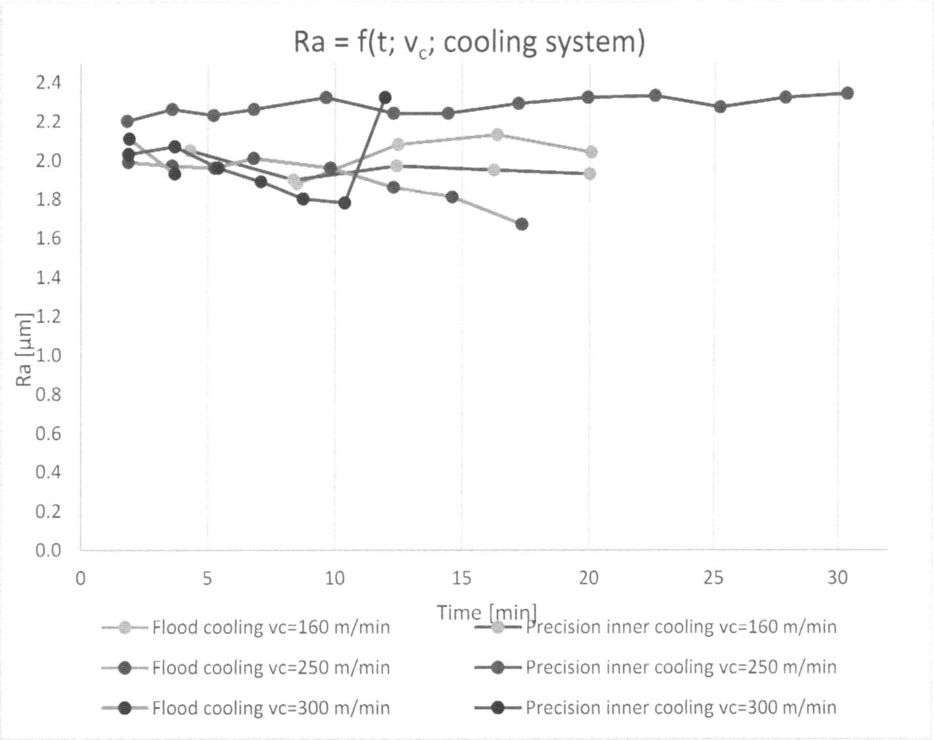

Fig. 7. The effect of cutting speed and cooling on the surface roughness

4. Conclusion

Stainless steels are various materials which show worse machinability than carbon steel and, therefore, cause rapid tool wear. Innovations in the cooling system allow to utilise cooling emulsion more effectively. Therefore, cutting speed or tool life can be increased, improving machining economics.

The purpose of this investigation was to compare tool life when conventional and precision inner cooling approaches were used. Stainless steel was chosen due to its low thermal conductivity which thermally and mechanically loads tools. Precision cooled tool always showed lower tool wear, however with higher cutting speeds the difference was more significant, namely tool life was about double when a 1.56 higher cutting speed than usual was used. On the other hand, during lower cutting speed there was small difference after the first 20 minutes in the cut. Although, it can be argued that the tool wear progression might look similar as with the cutting speed of 250 m/min. The benefits of precision cooling system would then be significant only when higher tool wear is reached. However, this research will be followed by investigation of tool wear behaviour at lower cutting speeds to verify this hypothesis. Moreover, precision cooled tool wear rises slower than conventional one, thus machining process is more predictable and reliable. As far as surface roughness, there was no clear difference between conventional and precision cooling. That is why these experiments should be repeated with more specimens.

5. Acknowledgement

This paper was created under the project SGS-2022-007: Research and Development for Innovation in the Field of Manufacturing Technology - Machining Technology III.

6. References

[1] http://www.steel-stainless.org/media/1546/outokumpu-stainless-steel-handbook.pdf, (2013). Outokumpu Oyj. Handbook of stainless steel, Accessed on: 2021-10-07

[2] Číhal, V. (1999). Korozivzdorné oceli a slitiny, Česká matice technická (Academia), ISBN 80-200-0671-0, Prague

[3] https://www.worldstainless.org/statistics/stainless-steel-meltshop-production/stainless-steel-meltshop-production-2014-2020/, (2020). ISSF Website - Stainless steel meltshop production 2014 - 2020. Home of stainless steels, Accessed on: 2021-11-10

[4] https://www.youtube.com/watch?v=PsXWMEUXmNs, (2021). Seco Tools & Patrick de Vos, Getting technical series – machining of stainless steel, Accessed on: 2022-03-26

[5] Helmi, Y. (2015). Machining of stainless steels and super alloys: traditional and nontraditional techniques, John Wiley and Sons, ISBN 978-1-118-91936-1, Hoboken

[6] Korkut, I.; Kasap, M.; Ciftci, I. & Seker, U. (2004). Determination of optimum cutting parameters during machining of AISI 304 austenitic stainless steel. Materials & Design, Vol. 25, No. 4, pp. 303–305 , ISSN 02613069, doi:10.1016/j.matdes.2003.10.011

[7] Stephenson, D. & Agapiou, J. (2016). Metal cutting theory and practice, CRC Press/Taylor & Francis Group, ISBN 978-1-4665-8753-3, Boca Raton

[8] Přikryl, Z. & Musílková, R. (1971). Teorie obrábění, Nakladatelství technické literatury, Prague

[9] https://www.sandvik.coromant.com/en-gb/knowledge/general-turning/pages/how-to-apply-coolant-and-cutting-fluid-in-turning.aspx, (2022), Sandvik Coromant, How to apply coolant and cutting fluid in turning, Accessed on: 2022-09-14

[10] https://www.walter-tools.com/cs-cz/tools/standard_products/turning/overview/iso_turning/turn-precision-cooling/pages/default.aspx, (2022). Walter Tools, Walter Turn, precizní chlazení, Accessed on: 2022-05-15

[11] Naves, V.T.G.; Da Silva, M.B & Da Silva, F.J. (2013). Evaluation of the effect of application of cutting fluid at high pressure on tool wear during turning operation of AISI 316 austenitic stainless steel. Wear, Vol. 302, No.1-2, p.p. 1201–1208, ISSN 00431648, doi:10.1016/j.wear.2013.03.016

[12] Bleicher, F.; Finkeldei, D. & Siller, A. (2016). Machining of Difficult-To-Cut Materials. Proceedings of the 27th International DAAAM Symposium 2016. DAAAM International Vienna, p.p. 0473–0479. ISBN 978-3-902734-08-2., doi:10.2507/27th.daaam.proceedings.071

[13] Kostadin, T., (2018). Influence of Cutting Fluids on the Corrosion Resistance of X20Cr13 Martensitic Stainless Steel, Vol. 13, No. 12, International Journal of Electrochemical Science,p.p. 11986–11999, ISSN 14523981, doi:10.20964/2018.12.81

[14] Fulemova, J. & Janda, Z. (2012). Selection of cutting material for milling of ferritic-martensitic stainless steel tool wear vs. roughness of machined surface. Proceedings of the 23rd International DAAAM Symposium 2012. DAAAM International Vienna, p.p. 0361–0364. ISBN 978-3-901509-91-9, https://www.daaam.info/Downloads/Pdfs/proceedings/proceedings_2012/083.pdf

[15] https://www.makeitfrom.com/material-properties/AISI-316-S31600-Stainless-Steel, (2022). MakeItFrom.com, AISI 316 (S31600) stainless steel : Accessed on: 2022-05-15

[16] https://www.replenex.com/Product/391590, (2022). Replenex, Pramet 7156198 Turning Insert, ANSI Code: CNMG 432-NMR:T7325, CNMG Insert, Material Grade: M, P, S, 432 Insert, Diamond Shape, 12 Seat, Negative Rake, For Use On Stainless Steel, Steel and Titanium/Super Alloys, Carbide, Manufacturer's Grade: T7325, Accessed on: 2022-09-15

[17] Jozić, S.; Lela, B. & Bajić, D. (2014). A New Mathematical Model for Flank Wear Prediction Using Functional Data Analysis Methodology, Advances in Materials Science and Engineering, p.p. 1–8, ISSN 1687-8434, 1687-8442., doi:10.1155/2014/138168

[18] Memariapour, M.; Niknam, S.A.; Turenne, S. & Balazinski, M. (2021) Study of the Effects of Initial Cutting Conditions and Transition Period on Ultimate Tool Life when Machining Inconel 718, Materials, Vol. 14, No. 3 ISSN 1996-1944, doi:10.3390/ma14030592

DOI: 10.2507/33rd.daaam.proceedings.072

ADDITIVE MANUFACTURING APPLICATION FOR THE INCLUSION OF THE VISION IMPAIRED POPULATION IN PUBLIC TRANSPORT

Mustafa Bevrnja, Zejd Imamović & Muhamed Bisić

This Publication has to be referred as: Bevrnja, M[ustafa]; Imamovic, Z[ejd] & Bisic, M[uhamed] (2022). Additive Manufacturing Application for the Inclusion of the Vision Impaired Population in Public Transport, Proceedings of the 33rd DAAAM International Symposium, pp.0519-0525, B. Katalinic (Ed.), Published by DAAAM International, ISBN 978-3-902734-36-5, ISSN 1726-9679, Vienna, Austria
DOI: 10.2507/33rd.daaam.proceedings.072

Abstract

This paper shows how additive manufacturing applications can be used to help people with vision impairments. There are many 3D printed tactile maps, books or individually customized hearing devices that can help for their inclusion. As a potential solution, a 3D printed eco box was designed and created. Small box with a separable and replaceable 3D printed QR code on the front side, tactile map on top and space for the integrated speaker inside of the box. 3D printing process allows each box and QR code to be unique, quickly designed and produced. They can be placed on tram, bus, train, or metro stations or at the airport. Scanning QR code with specialized app on the phones, gives detailed voice instructions about current location, timetable, guide marks and other important information. Product can be also used by general community, for the same purposes. This paper provides detailed instructions for product design and 3D printing.

Keywords: 3D printing; tactile; QR code; assistive products

1. Introduction

Global estimates number of people with visual impairment in 2020 is 1.1 billion people, where approx. 45 million people are completely blind and approx. 300 million have moderate to severe vision impairment. These individuals face broad challenges that affect their quality of life, which make difficulties in performing many activities in their life. Fig. 1 clearly shows increase in the number of people with vision disability through the past decades, which will unfortunately continue to increase in the future [1].

Public transport play one of the most important roles in the life of the visually-impaired and blind person in their productivity, community involvement, seeking education, medical care, work, independence and total integrity. Despite all of the advantages, visually-impaired persons experience difficulties in using public transport [2]. Some of the biggest issues they challenge are: identifications of correct transport vehicle, access to the timetables, routes other important information, non-adaptive signs, lack of sound information and so on.

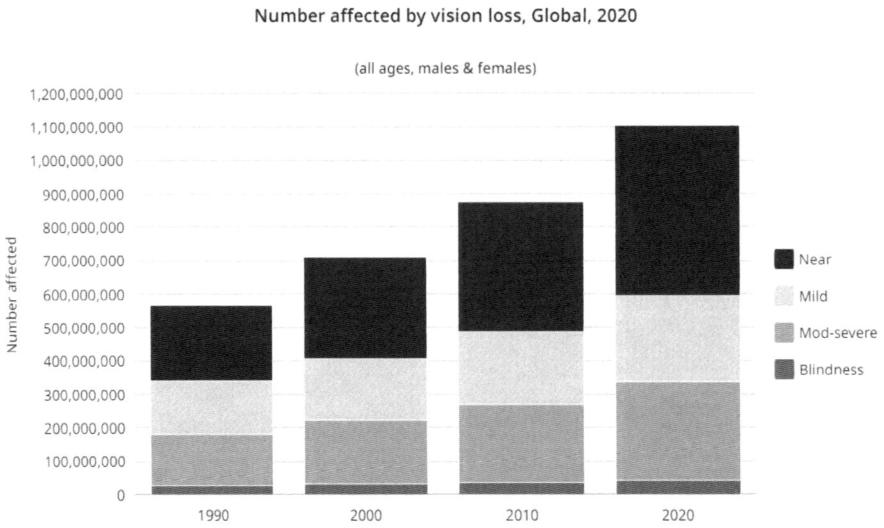

Fig. 1. Number of people affected by vision loss in 2020

Visual loss, unfortunately, greatly reduces information access to the visual impaired people in today's world of knowledge and information. About 90% of the all information processed by brain is visual, so efforts should aim for the development of public transport navigation for the people with visual disability. Advances in technology especially in information technology (IT) and mobile technology synchronized with cheap, simple and quickly developed products, using additive manufacturing (AM), can greatly improve inclusion of visually impaired population in public transport systems. Technology has the potential to enhance individual's ability to participate fully in societal activities and to live independently [3].

Although, some may not expect for visually impaired person to use smartphones, tablets and internet due to their disability, researches show they are catching up to that of non-disabled consumers as shown in Fig. 2. These graphs indicate, that indeed assistive technologies should focus on the mobile and internet technologies combined with traditional solutions for inclusion of visually impaired people which relies on the haptic and audio senses [4].

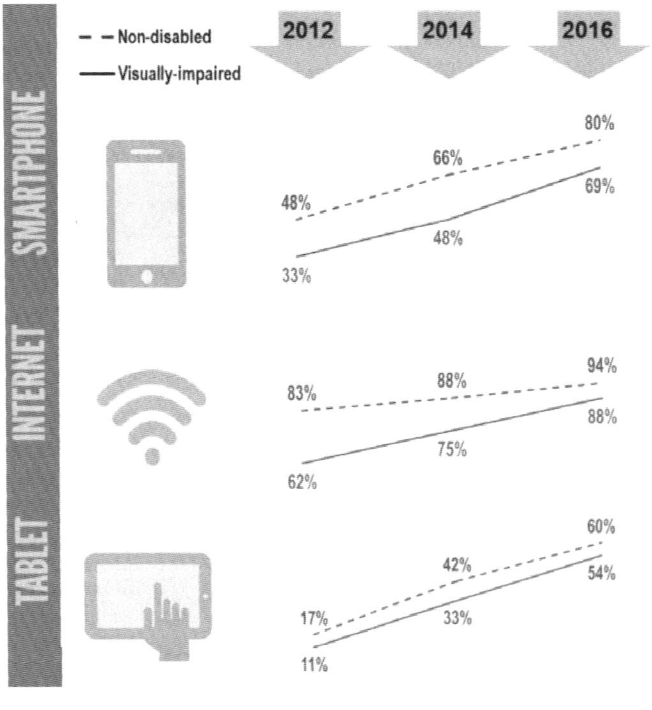

Fig. 2. Access to devices and services in the period from 2012 to 2016

Mobile phones, in particular smartphones, offer a massive cluster of features and applications which helps people with visual impairments. Innovations in these areas has explored usage of sensory modalities other than vision – for example speech recognition, screen reader, haptic feedback, text-to-speech – to reduce dependence on demanding visual which are helping people with visual impairments navigating through their mobile phones [5].

There are some systems which are integrated into public transport systems. For example *RAMPE* system is based on Wi-Fi enabled smart devices carried by users, fixed base-stations installed at bus stop and central server system that is connected with both bus stop and buses. This system allows user to choose stop where he wants to connect in order to get relevant information about timetables, routes, list of stops along the bus line and other. Position of the bus would be tracked using GPS (Global Positioning System) [6]. Similarly, mobile navigation and orientation systems for visually impaired users where introduced in a *Metrobus* environment. The systems is based on usage of smartphones, GPS and compass device which communicate via Bluetooth, which provides audible interface to help visual impaired user to listen relevant information about location and orientation of the user within *Metrobus* environment [7].

On the other hand, GPS can be inaccurate and can have slow response time. Some researches propose other technology for connections between visual impaired users and public transport system, like Radio Frequency (RF) communication to improve the success rate of visually impaired to board correct buses and getting to the right destination. One of the main purposes of using RF is to take advantage of its short distance communication, since the communication between a bus and a bus stop will be possible only when the bus approaches the stop. During the journey of the person with visual disability, a GPS-based application could inform the person about his/her location, but when approaching the stop the GPS could fail, and that is where the short range communication of the RF technology plays a key role [2].

From the above, knowing that the visual impaired people use smartphones in a great scale via assistive applications, and knowing that there are possibility of connecting visual impaired users with desired public transport vehicle, central database system via mentioned technologies, our task was to make product using AM which can help visually impaired person for their inclusion.

Our solution is printable 3D QR code which is integrated in the printable box, which would be fixed at the pole at the bus/metro/tram or metro station, or even at the airport and would be easily accessed by visually impaired person following the tactile paving. This product consist of changeable frontal plate and box with u clamp. Frontal plate has 3D QR code and short information about that station written in braille. Exterior of the box, consists of holes for speaker, USB or mobile charger, while inner space is meant for electronics, loudspeaker and charger. Once visual impaired person approach and examines the box, by a mobile phone can scan QR code and can get connected to the central system or get all available information about timetable, routes and approaching public transport. User could also communicate with public transport central system via application and signals could be sent to the driver of desired vehicle, which can trigger additional features. In this paper detailed instructions about design and manufacture of product using additive technologies are presented, as well as examples of installation in the public transport.

2. Product design

For the proper design of the product, a project plan is set, which includes problems, possible solutions, design, 3D printing and the finished product show in Fig. 3. Primary goals of designing such a device is easy and cheap manufacturing, assembling and installation on site, and to be adaptable to different locations. In addition, when designing, attention was paid to the fact that the housing of the device, alone, can provide some information to visually impaired person using braille.

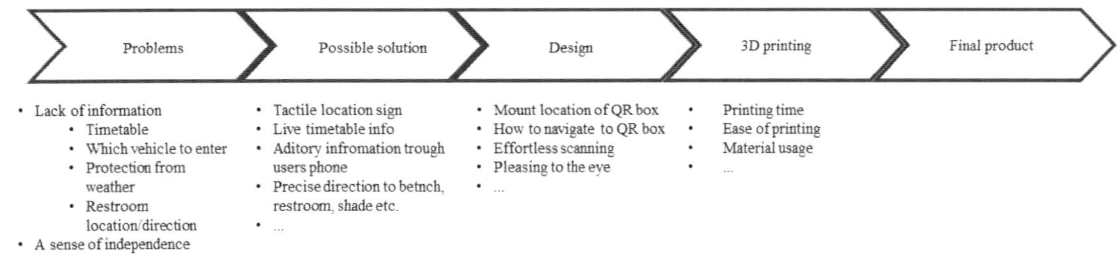

Fig. 3. Design flow

After analysing the problems and possible solutions, two different solutions were designed. The first solution can be mounted on flat surfaces. Second design are meant for mounting on cylindrical surfaces such as pillars using additional u clamp, which is also printed. Different solutions are shown in Fig. 4.

Fig. 4. Two different designs – first for flat surface (left), second for pole mounting (right)

The dimensions of the box are designed so that both solutions have space for the implementation of a micro-computer, a Wi-Fi module, a loudspeaker and a USB port for charging smartphones. After defining the dimensions of the box and the QR code, the case is designed to meet two additional conditions. Protection against vandalism and ease of 3D printing. Protection against vandalism was achieved by sliding the QR code plate into the housing prior to installation on a pole or wall so that it would be secured. In order to achieve the simplicity of 3D printing, overhangs were avoided so that it is printed without the need for support material, which would mean saving on printing time and material. Also, 3D printed support structures are a common cause of failed prints.

In Fig. 5 exploded-view of an assembly is shown. Steps for correct assemble of the product which will be mounted on the pole are:

1. Insert elecronic components inside box,
2. slide in QR plate from bottom,
3. slide in lower lid from behind,
4. place it on pole,
5. screw in u-clamps.

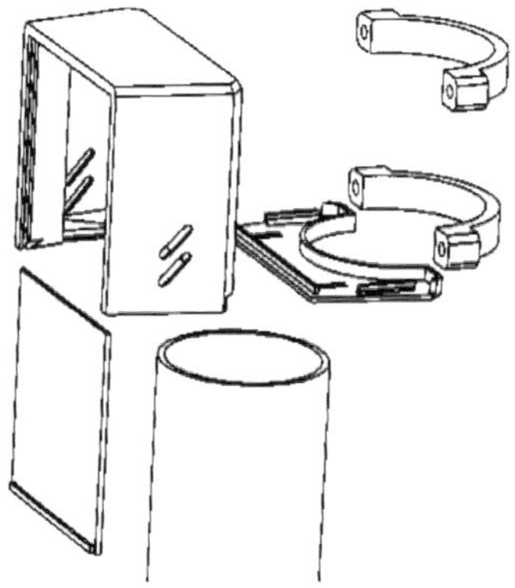

Fig. 5. Exploded-view of an product assembly

3. 3D Printing and materials of the product

Additive manufacturing is technology that is often used for producing of different products with complex geometries. One of the most popular AM technology today is fused deposition modelling (FDM) [8]. It is one of the types from 3D printing technique group, where a molten thermoplastic material (filament) is laid out layer-by-layer, creating a full-scale part in shorter cycle time and lower costs compared to the traditional manufacturing processes [9]. After creating a 3D CAD model of the product, next step is converting it to 3MF or another similar format, and using "Slicer" software, which generates G-code with all printing parameters. G-code is then used by 3D printer in order to create a final piece, layer by layer [10].

There are many 3D printers and slicing software available on the market, but for purpose of this research, Ultimaker S5 Pro was used for 3D printing and Ultimaker Cura software for generating the G-code. The main idea throughout the whole design process was to create a model that could be easily 3D printed by any FDM 3D printer, with basic knowledge about 3D printing in general. Both box and QR code plate don't have any overhangs and can be 3D printed without any supports and its' volumes are perfect size for most of 3D printers used worldwide. An important thing for this research is understanding of filaments for FDM printing. Material has to be easy to print, resistant to standard environmental conditions and, probably most important, be eco-friendly.

Combining everything listed with the economic side, there are two potential solutions – PLA and PETG, any brand. So, after creating the final model and choosing right material, next step is choosing right parameters for 3D print. The most important ones are represented in Table 1.

	PLA	**PETG**
Layer height [mm]	0.2	0.2
Printing Temperature [°C]	205	245
Print Speed [mm/s]	70	60
Fan Speed [%]	100	20

Table 1. Important printing parameters

Based on standard parameters, printing time is shorter when using PLA, but it is important to mention that PETG is a better option when it comes to printing boxes for outdoor use (public bus stops, tram stations etc.). Next step, after finalising the parameters, is to automatically generate a G-code using the „Slice" button. File is now ready to be sent to a 3D printer, using an SD-card, USB stick or simillar method to communicate with the 3D printer. Printer itself must be correctly calibrated and ready to use with enough material for 3D printing process.

The best option for 3D printing of the QR code plate is printing it in one piece, not connecting the black and white parts aftewards. To achieve that, there are two options; using a dual extruder 3D printer (like Ultimaker S5) where both colors can be used at the same time, or, the most common way because most of 3D printers are single extruder – using a script for filament change at predefined height. In Cura, scripts can be found under „Extensions" tab where user selects „Modify G-Code" option. For this particular case, plugin is used as presented in Fig. 6.

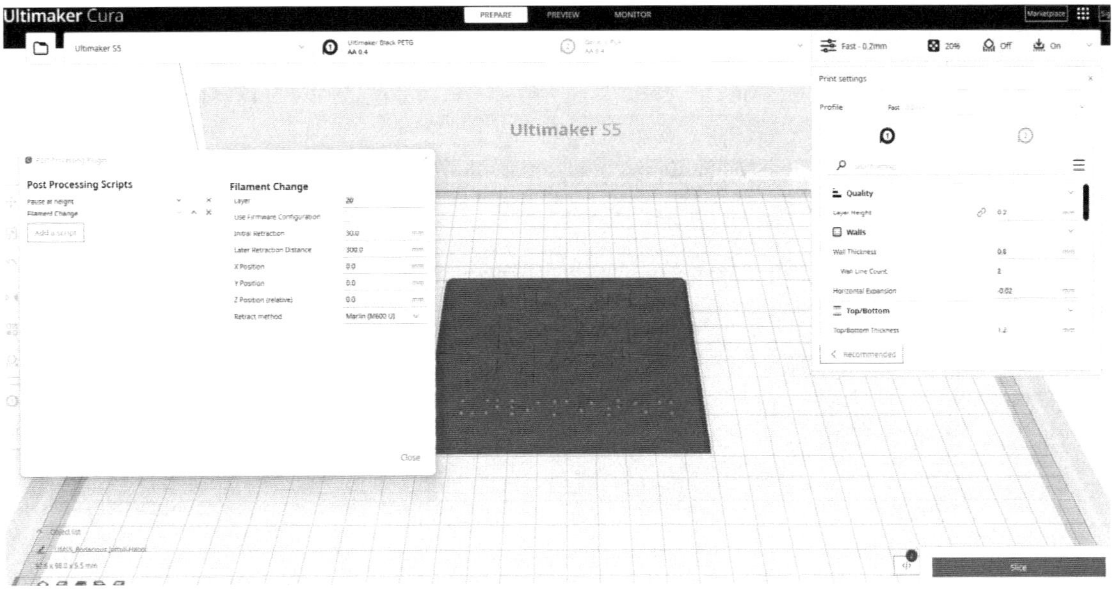

Fig. 6. Post Processing Scripts

It is important to understand that two different type of filaments can not be combined when performing filament change, because of their different melting temperatures. If a slicer software doesn't have an option to perform filament change by using a script, it can be done manually while printing by pausing the print at certain height, changing the filament and resuming the process. After the printing is finished and build plate is cooled down, product is ready to use, no post-processing required. In Fig. 7**Error! Reference source not found.**, modeling and final preparation of the frontal plate with 3D QR code is shown, as well as creating it with AM technology.

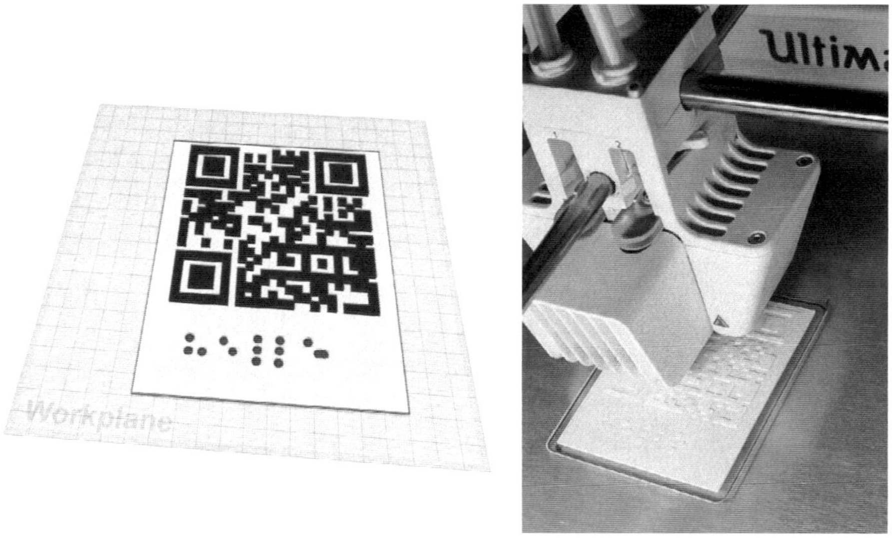

Fig. 7. Modeling and printing frontal plate with QR code

4. Examples of product applications

An essential feature of the design is that it enables easy scanning of the QR code even for those who are completely blind. Therefore, an appropriate QR code size was selected that could be easily read from the length of an average cubit – distance from elbow to fingertips. It would be scanned by holding the phone on the inside of the elbow facing the fingers of the other hand resting on the tactile letters on the case. Demonstration of scanning and example of application at metro station or airport is shown in

Fig. *8***Error! Reference source not found.** left, as well as example of applications at the bus station in Sarajevo Fig. *8* right.

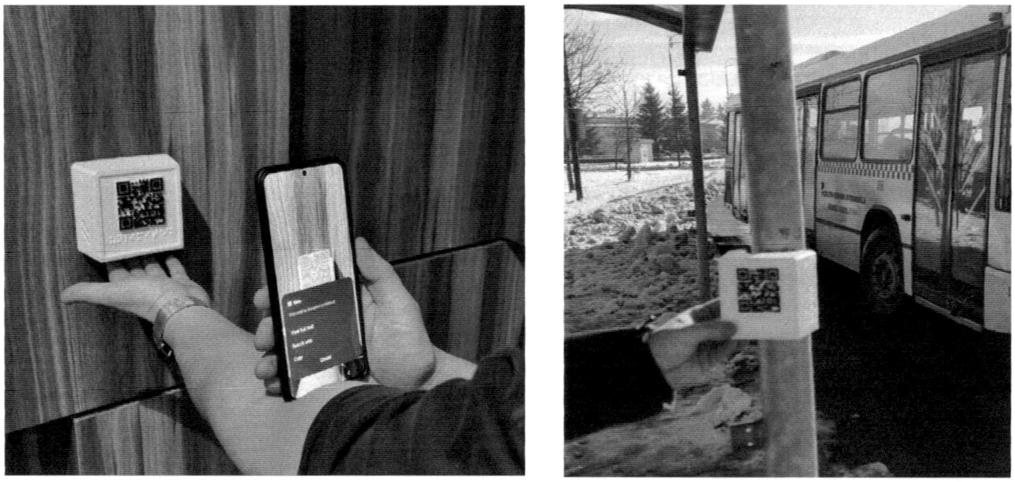

Fig. 8. Demonstration of scanning QR code using average cubit distance (left); Example of applications (right)

5. Conclusion

This research successfully showed how additive manufacturing can be used in order to create a friendly environment for vision impaired population. Product described in this research represents an innovation in fields of customizable product design and pointed out advantages of AM through practical application. It provides a great opportunity in massive usage of AM for creating final solutions for inclusion of visually impaired people in public transportation. Producing of presented 3D printed boxes and QR-code plates is a process that could be done by institutions on every level of government system by following our research. A very important thing is that the product is fully customizable for every place, eco-friendly and has short 3D printing time. This type of research is a solid base for exploiting additive manufacturing benefits in upcoming years and upgrading existing products that help humanity in many different ways.

It will be interesting to follow constant exponential growth of AM science and connecting it with everyday life for whole population. AM based products can be combined with works of experts from different professions to create powerful devices with unlimited possibilities, such as creating a complete interface for 3D printed QR-codes and turn them into entrances to virtual guiding world for all users. Main future goals are to provide instructions and 3D models to as many associations for visually impaired people as possible and help them with implementation of product in cooperation with governments. In this way, the network of users will grow, creating a large database of feedback. Additionally, reviews will help us in recognition of the downsides, allowing us to create a more powerful version of design and final product. Furthermore, presented product need a software applications that will be custom made for helping visually impaired people to find their way around in public transport better, so this could be next step in this research.

6. Acknowledgments

Authors are grateful to Faculty of Mechanical Engineering – University of Sarajevo Grants Commission for providing necessary financial assistance to carry out this research work.

7. References

[1] Bourne, R.;Steinmetz, J.; Flaxman, S.; et al. (2020), Trends in prevalence of blindness and distance and near vision impairment over 30 years: an analysis for the Global Burden of Disease study. Lancet Glob Health.

[2] Sáez, Y.; Parera, J.; Canto, F. & García, A. (2019). Assisting visually impaired people in the public transport system through RF-communication and embedded systems," Sensors.

[3] Steel, E. & Luc De W. (2011). Advances in European assistive technology service delivery and recommendations for further improvement," Technology and Disability.

[4] https://www.ofcom.org.uk/__data/assets/pdf_file/0034/98485/visually-impaired-consumers-access.pdf (2017), Accessed on 2022-10-19

[5] Hakobyan, L.; Lumsden, J.; O'Sullivan, D. & Bartlett, H. (2013). Mobile assistive technologies for the visually impaired, Survey of Ophthalmology.

[6] Baudoin, G.; Venard, O. & Uzan, G. (2005). The RAMPE project: interactive, auditive information system for the mobility of blind people in public transports.

[7] Mata, F.; Jaramillo, A. & Claramunt, C. (2011). A mobile navigation and orientation system for blind users in a Metrobus environment, Proceedings of Web and Wireless Geographical Information Systems - 10th International Symposium.

[8] Pandzic, A.; Hodzic, D.; Hajro, I. & Tasic, P. (2020). Strength Properties of PLA Material Obtained by Different Models of FDM 3D Printer, Proceedings of the 31st DAAAM International Symposium, pp.0315-0322, B.Katalinic (Ed.), Published by DAAAM International, ISBN 978-3-902734-29-7, ISSN 1726-9679, Vienna, Austria, DOI: 10.2507/31st.daaam.proceedings.044

[9] Ali, H.; Alani, T. & Othman, F. (2018). Influence of process parameters on mechanical properties and printing time of FDM PLA printed parts using design of experiment, Journal of Engineering Research and Application, vol. 8, no. 7, pp. 65-69, 2018.

[10] Pandzic, A. (2021). Influence of Layer Height, Build Orientation and Post Curing on Tensile Mechanical Properties of SLA 3D Printed Materials, Proceedings of the 32nd DAAAM International Symposium, pp.0075-0083, B. Katalinic (Ed.), Published by DAAAM International, ISBN 978-3-902734-33-4, ISSN 1726-9679,Vienna, Austria, DOI: 10.2507/32nd.daaam.proceedings.030

DOI: 10.2507/33rd.daaam.proceedings.073

ENSURING HEALTH AND SAFETY AT UNIVERSITIES

Petra Dostálová & Gabriela Králíčková

This Publication has to be referred as: Dostalova, P[etra] & Kralickova, G[abriela] (2022). Ensuring Health and Safety at Universities, Proceedings of the 33rd DAAAM International Symposium, pp.0526-0530, B. Katalinic (Ed.), Published by DAAAM International, ISBN 978-3-902734-36-5, ISSN 1726-9679, Vienna, Austria
DOI: 10.2507/33rd.daaam.proceedings.073

Abstract

This thesis deals with the issue of rules for the protection of occupational safety and health protection in colleges and universities in a few simple points. Identification and categorization of risky workplaces and, finally, proposals for improvement overall security status. The aim of this work is to educate the public about the issue of occupational safety at universities, which is often grossly underestimated.

Keywords: ensuring health; safety; university; students safety; employees safety

1. Introduction

Every company can be exposed to the risk of accident or injury to employees, and therefore the management of each of them sets its own safety rules and measures for its employees. There are a lot of students at the university, and many employees work in various activities, both in the main field of education, complementary or in multiple activities when participating in projects, further within the activities within the framework of economic agreements and research activities. The same goes for students whose education is involved in these activities differently. The purpose of this article is to present the possible risks that employees and university students may face. At the same time, the methods and obligations the employer must fulfill and provide to avoid risks will be listed.

2. Ensuring work safety

Ensuring work safety at universities is considered one of the most important tasks of an employer. In addition to several hundred to thousands of students, the university campus is home to the employees themselves, i.e. professors, technical workers, officials, etc., but also to various guests, be it parents of students, visitors with various goals. and interests (whether it is partners from other schools or possible supplier companies who came here on business, officials, people interested in possible studies, etc.) and possibly also employees or trusted persons from other companies who, for example, share a building with university, or they do their work here (we can mention, for example, bricklayers, plumbers and many others). There are really many options and the employer should consider them all. Occupational safety and health protection at work are protected by separate laws and decrees in the Czech Republic. Their compliance is strictly required and controlled by state administration authorities. Every company is obliged to train its employees and familiarize them with the possibility of certain specific risks. Employees should thus be able to react correctly in a critical situation and thus protect their health and colleagues' health.

2.1 Students

The content of studying for college and university students is not only sitting safely in desks and studying, but also many other activities. Above all, in technically oriented fields, they encounter activities that have nothing to do with the conditions of their future employment. Working on computers, but also mechanical devices and machines, research in laboratories, working with electronics, but also chemicals, hot material, machine parts and a lot of other not very safe looking things. The phenomenon is not exceptional when students have to work either on old devices, because newer ones are not available for various reasons or excuses, or on the contrary, on new but not very reliable or inappropriately located or equipped machines. [3]. And even if students sometimes go outside the school premises as part of their professional practice, many of them end up in the workplace, where they face the same danger as the surrounding employees. But unlike these students, they are at least clearly defined in all kinds of contracts and regulations, and they also usually have a lot more expertise and experience than someone who hasn't graduated yet. Thus, ensuring the safety of pupils is a very difficult task that requires due attention.

Students do not have to have any professional or medical qualifications during their studies. Professional competence can be achieved precisely by studying and gaining first experience, which is exactly what studying is for. Medical fitness is also not addressed much, and the university does not require students to undergo any medical examinations or tests. Restrictions resulting from the Labor Code apply to the work of minors (that is, practically no one at universities) and pregnant women who may appear during their studies. Pregnant students usually have exceptions for more complex work, for example in workshops or laboratories, they are also taken into account during studies, etc. [4]

Students will receive introductory training at the time of starting their first year of university. The content of the training is usually general information about the functioning of the school and study, about the basics of safety and about the rules that must be followed. The training is carried out by an authorized representative, usually an employee of the study department. If a student gets into some specialized classrooms during his studies, e.g. with a focus on computers, he should be instructed by his professor about the special cases in which this classroom and the work in it differ from ordinary studies, about special instructions on how to avoid risks and how to respond to it, if necessary, about the risks that students face and about the correct procedure for working on the assigned machines.

Generally, different behaviour is expected from university students than children in primary and possibly secondary and vocational schools. University students are often adults who are fully aware of possible risks and naturally try to avoid them. Nevertheless, universities should also take out suitable insurance to cover injuries financially. Air temperature affects the productivity and health of students and teacher. Long-term components of comfort include air temperature, mean radiant temperature, humidity, and room air velocity, along with the personal factors of clothing insulation and activity level. But the understanding of what makes a space comfortable is still evolving, and these components, we're finding, are only part of the thermal comfort puzzle. It is important that students and teachers have the best possible conditions. [11]

2.2 Employees

Employees Fortunately, the issue of security for the faculty and other employees of the university is relatively well addressed. As for regular employees, they are subject to clear rules and specific laws. When they enter the job, they undergo initial training, which is also repeated at regular intervals. They are professionally prepared for the performance of their jobs and know what dangers they and those around them face during work.

Each employee must demonstrate appropriate professional medical fitness, according to the requirements of the place where he applies. Medical fitness proves when starting work and then with regular medical examinations, arranged by the school. A physician examining a prospective or current employee must assess whether he would be exposed to special risks arising from them while performing his work state of health and whether he is even fit to do this job. If it doesn't find any risks or other obstacles, will confirm their findings by signing the "Request for medical assessment the status of the agreed work", which he hands over to the employee, who then immediately hands it over to his employer employers.

They must then keep lists of all risk workplaces (divided into category) and employees working on them, including medical documentation tours so as not to lose track. In addition to medical qualification, there is also professional qualification. Here they have to this must be demonstrated again by the employee upon starting employment and thereafter at each transition to a new job risky workplace. Professional competence is measured both by the age of the employee, so above all according to the education achieved in a certain field, taking into account his
work experience. Expertise is very important because different machines and according to the Labor Code, only professionally qualified persons can operate the equipment. Each employee receives initial training during the start of employment, further professional training directly at your workplace from a senior employee and periodic safety training, which is repeated at regular intervals according to his job. An employee can receive further training whenever the content of his job changes or switches to another workplace. Furthermore, repeated training is carried out periodically once every two years, which has the goal is to remind employees of the principles of health and safety in the workplace. [8] All the mentioned steps, right down to the correct categorization of work and the related frequency of medical checks, as well as the decision about the need to provide personal protective equipment, are the activities of a professionally qualified person who determines the rules of work safety at specific universities. At the same time, this person trains employees and draws attention to possible risky situations and their possible solutions.

2.3 Visitors

However, not only students and staff move around the school grounds, but also many people who seemingly have little to do with the school itself. Each school, for example, organizes several events, whether it is open house days, benefits, graduation and graduation ceremonies, or even presenting the work of its students, there are many options and many spectators from among the students' families, those interested in studying, come to see each of them, curious passers-by, etc. In addition, each school also has a lot of cooperating entities and suppliers or contracted external workers who also have to visit the school campus in order to perform their jobs. [9]

Most colleges and universities are so-called public, in their public spaces so even people who have nothing to do with the school can move freely. These visitors must follow the visitor regulations, which are available at the gatehouses and contain quite logical points such as no smoking, no feeding animals (except guide dogs), but also a ban on open fire or launching drones. Every visitor would with it he should have been informed of the rules and must follow them. [1] They have no right to enter the workplace of school employees or in any way disrupt the activities of the university.

It is the openness of university premises that can be a security risk. Even less of a chance in occupational safety and health protection. However, universities are responsible for the safety of visitors on their premises, and even here, the university management should decide on arranging suitable liability insurance.

3. Dangerous university workplaces

Concierge - control room must carry out activities associated with the monotony of monitoring security and monitoring systems, rounds after assigned area. They are also exposed to great responsibility in monitoring the EPS system, security systems, control of entrance ramps to the premises, camera systems, systems detection of dangerous substances. When running errands, they are influenced by weather conditions. And in the winter months, they must be prepared for snow removal in the entrance area object. At the same time, the gate operator provides the fire alarm and is the first group employees involved in solving not only fire alarms but solving them in general emergency situations. [6]

However, there are certain risks in all workplaces. They are primarily associated with humans a factor that cannot be largely influenced and must be taken into account. Among the most serious mistakes are insufficient knowledge or, conversely, underestimation rules and regulations, which can lead to major damages. Employee his actions can endanger himself and those around him, including students. This risk even senior employees or the management of the company (university) are not protected. Thanks to their recklessness, they can, for example, carry out activities that bring an unreasonable risk or are restricted or outright prohibited by law. [2]

Universities include laboratories focused on the fields offered. These laboratories often work, for example, with high-power lasers. A laser of sufficient power can cause serious and irreversible injuries careless users. In addition, they work with devices of great value, which also contributes to their stress. To ensure health protection, it is necessary to have an appropriate number of mobile-first aid kits and other tools for providing workplace first aid.

4. Proposal to improve security at universities

A person professionally qualified in risk prevention is only an advisory function without managerial ones powers, but equipped with the duties specified in the law. Its provisions in an organizational structure close to the statutory representative, would prevent a conflict of interests and ensuring safety and health protection at work would be part of the school's top management, universities. This would avoid the lengthy transition between risk identification and elimination when the security engineer must convince management of the urgency of taking appropriate action. [7]

Therefore, the security technician not only does not enjoy the appropriate respect that should be the profession was supposed to enjoy from the point of view of the contribution to society (which the security engineer his does not make work much easier), but the question of safety is not among the more important questions school management (ie, for example, the distribution of finances).

Subconscious awareness of imminent risks in the workplace is also present among senior employees sometimes very small or even minimal. He delights in the nature of each person optimism (or perhaps pride or a sense of security) is to downplay the warning and adjust one's work according to one's own, not paying proper attention to training topics, etc. Nothing can be done about it, however, the subconscious awareness of impending risks needs to be increased, for example by already proposed measures ensure the management of safety and health protection from the highest places and the policy set in this way implement security. [7]

The system of safety and health protection at work at universities should be constantly improved. Many times, however, providing new or updating existing safety regulations is not enough for development universities (construction of new workplaces, introduction of new work activities, etc.) or there is usually not enough funding or time to develop them. The management of universities therefore should have given due attention to this problem. They often spent resources moreover, it also returns in the form of a safe working environment, a low incidence of accidents, and openness more options or more satisfied employees. In addition, with each individual, over time, there is an inevitable resignation and acceptance of risks. Therefore, it is crucial to constantly remind this area of safety as part of regular training.

5. Basic points of safety and health protection when working at universities

Universities should fulfill the following minimum and basic obligations in the field of occupational safety: [5]

- provide information on health and safety to employees, teachers and pupils and educate them regularly,
- manage health and safety risks - search for, assess and evaluate risks related to local activities and the environment,
- when determining measures, consider the danger to pupils during teaching,
- create a school education program that familiarizes pupils with dangers and risks,
- record and document work accidents,
- conduct safety and protection training when working with professional accessories for universities,
- carry out regular health and safety inspections of all employees,
- ensure regular revisions and inspections of all machines and equipment,
- to process and update occupational health and safety documentation,
- observe special hygiene principles and rules for schools,
- introduce students to the risks at work and establish measures,
- introduce students to personal protective work equipment and strictly observe their use,
- introduce students to the principles of fire safety,
- to remind the ban on the consumption of alcohol and addictive substances.

The risks that occur most at universities. The most common risk factors of working conditions include in particular: [5]

- mental and physical stress,
- Chemicals,
- dust, noise and vibration,
- electrocution,
- poisoning when working with chemicals,
- visual strain,
- insufficient work ergonomics,
- non-ionizing device,
- electromagnetic fields,
- work with biological agents,
- heat or cold stress,
- work with pressure cylinders and gases,
- work at heights.

As employers have a legal obligation to carry out risk assessments that are suitable and adequate in accordance with the Occupational Health and Safety Act, so do schools and colleges. [10]

6. Conclusion

Ensuring security at universities is a rarely addressed topic. The problem is usually in legal and internal regulations, which often remain misunderstood, ignored or belittled, leading to many downstream problems, especially when introducing new workplaces and activities into operation. The risk of stopping these activities from operation or the occurrence of other extraordinary events in the future is disproportionate big. A significant problem for university visitors. They are rarely trained and fully prepared for the pitfalls they may encounter during their stay at school, above all during planned school events, where as part of the tour they can also get to specialized workplaces where they lack thorough safety training and awareness about the dangers that occur here.

The issue of occupational safety at universities is because in their university campuses there are both students, employees and visitors, very comprehensive. This article focused on the specification of these areas and the proposal of possible measures that could lead to improvement current state. The article generally presented the environment of Czech universities related to the primary, mandatory provision of occupational safety and health protection. The territory was divided according to the type of people who appear and work most often at universities. Using their examples, individual points specified that universities must ensure and comply with. Furthermore, personal risks faced by persons on the premises and working for universities were written down. Research in this area will continue by analysing the actual state of documentation and individual measures at the chosen university.

7. Acknowledgments

This study was prepared with the help of funds provided by Tomas Bata University in Zlín - Faculty of Applied Informatics within the project IGA/FAI/2022/005.

8. References

[1] OSH info [online]. (2015). Prague: Research Institute of Labor Safety, [cit. 2022-09-10]. Available from: http://www.bozpinfo.cz/kategorizace-praci-1

[2] Health and safety in school facilities. (2013). Ostrava: University of Ostrava, ISBN 978-80-7464-374-3.

[3] ČSN 01 8003 (018003). (2017). Principles for safe work in chemical laboratories. 2nd edition.

[4] Bílek E. (2006). A practical example with commentary on how to evaluate workplace risks. OSH info [online]. Prague: Research Institute of Labor Safety [cit. 2022-09-21]. Available from: http://www.bozpinfo.cz/prakticky-priklad-s-komentarem-jak-vyhotnotit-rizika-na-pracovisti

[5] BOZP.cz. (2019). [online]. Prague: Occupational safety for universities. Health and safety requirements, obligations and risks [cit. 2022-09-24]. Available from: https://www.bozp.cz/aktuality/bezpecnost-prace-pro-vysoke-skoly/

[6] Rudolf, Ladislav. (2013). Health and safety in school facilities. Ostrava: University of Ostrava, ISBN 978-80-7464-374-3.

[7] Romaněenko, Jan and Pavel Skácelík. (2018). OSH and PO in examples of school practice. 2nd edition. Karviná: Paris, ISBN 978-80-8717-340-4.

[8] BOZP.cz [online]. (2015). Prague: OSH training for school employees, [cit. 2022-09-27]. Available from: https://www.skolenibozp.cz/aktuality/skoleni-bozp-pro-zamestnance-skoly/

[9] University of Tomas Bata in Zlin -student guide, (2021). [online]. Avaible from: https://pruvodcestudenta.utb.cz/wp-content/uploads/2021/10/pruvodce_utb_2021_2022.pdf

[10] Safetywallet.co.za. (2021). [online]. Managing health and safety schools and colleges. [cit. 2022-10-20]. Avaible from: https://www.safetywallet.co.za/Blog/managing-health-and-safety-schools-and-colleges

[11] Králíková R., Wessely E., Impact of the environmental factors and its effects on productivity and health. (2018). DAAAM International, ISBN 978-3-902734-20-4, ISSN 1726-9679, Vienna, Austria. DOI: 10.2507/29th.daaam.proceedings.016

DOI: 10.2507/33rd.daaam.proceedings.074

AN APPROACH TO AUTOMATING THE MOVEMENT MODE CHANGE OF A REHABILITATION EXOSKELETON, DEPENDING ON EXTERNAL CONDITIONS

A. Johara, D. Klimov*, Pismennaya E., & Y. Poduraev

This Publication has to be referred as: Johara, A[hmad]; Klimov, D[aniil]; Pismennaya, E[lena] & Poduraev, Y[uri] (2022). An Approach to Automating the Movement Mode Change of a Rehabilitation Exoskeleton, Depending on External Conditions, Proceedings of the 33rd DAAAM International Symposium, pp.0531-0535, B. Katalinic (Ed.), Published by DAAAM International, ISBN 978-3-902734-36-5, ISSN 1726-9679, Vienna, Austria
DOI: 10.2507/33rd.daaam.proceedings.074

Abstract

The main application of exoskeleton robots in the modern world is focused on the services of medicine, industry, and military applications. A particularly important application is the rehabilitation and improvement of the quality of life for patients with central nervous system damage and spinal cord injuries. In this paper, we propose an approach to automating the movement mode change of a rehabilitation exoskeleton, depending on external conditions. The exoskeleton's movement are modelled in a simulated environment based on the robot operating system (ROS) using external sensors.

Keywords: Exoskeleton; Simulation; Robot Operating System (ROS); Rehabilitation; Automation

1. Introduction

Gait disturbance is one of the main factors that determine the quality of life and the development of possible complications in patients with acute cerebrovascular accidents, patients with the consequences of a craniocerebral or spinal injury, multiple sclerosis, and degenerative diseases of the central nervous system (parkinsonism) with total hip arthroplasty and cerebral palsy. Gait disturbance is caused by disturbances in spatial movements in each joint, both in the sagittal and the frontal and horizontal planes, which leads to the formation of pathological synergy. For example, cerebral palsy is a common cause of childhood neurological disability. Although cerebral palsy is characterized by disorders - not only of motor, but also of perceptual, speech, and often cognitive functions- but it is the pathology of voluntary motor skills which is the main cause of social maladaptation, limitation of life activity, and integration of patients into society [1]. Modern methods for correcting motor disorders in cerebral palsy are aimed at functional neuroplastic reorganization of the structures of the motor analyser, when the nervous system is restructured and adapted, which is based on the formation of new changes in existing neural connections [2], [3]. The researchers note that neuroplasticity can be stimulated by creating a variety of environments, as well as by creating innovative equipment, the action of which is aimed at performing the required movements and increasing the patient's attention to ongoing rehabilitation activities aimed at learning or restoring walking skills.

Till now, the most advanced direction of motor rehabilitation is rehabilitation using robotic devices or complexes with visual feedback from sensors of the lower and upper extremities. Such devices allow you to train the functions of the upper and lower extremities, involving almost all the muscles of the body, and, as a result, activate the cognitive functions of controlling your own body when moving in space.

The advantage of robotic therapy is a higher quality of training compared to classical therapeutic gymnastics due to: their longer duration, the accuracy of repeatedly repeated cyclic movements that implement a real walking pattern with the ability to reproduce the reaction forces of the support under the patient's feet during walking, which excite and send to the cerebral cortex afferent signals, as well as the creation of a training program designed for the individual characteristics of the patient. It should be especially noted that the existing and newly created tools for assessing the success and effectiveness of ongoing rehabilitation measures, when a comparative analysis of the clinical and biomechanical parameters of walking in dynamics in patients at various stages of rehabilitation or diagnosis, makes it possible to create an evidence base for the use of robotic equipment and substantiate the protocol for its use. Such a toolkit makes it possible to operate not only with qualitative but also with objective quantitative data obtained by instrumental methods without elements of subjectivity. Also, the benefit of showing the patient the dynamics of his achievements cannot be ruled out. The parameters obtained with the help of such instrumental methods, which characterize the dynamics of the rehabilitation process, will allow the creation of an evidence base for the effectiveness and safety of the used equipment and methods (protocols) for its use in nosology. Medical exoskeletons for recovery, ~~or~~ walking assistance, support, or other purposes are becoming more common. An increasing number of studies are demonstrating efficacy and safety, as well as showing relevant results in the field of rehabilitation in nosology.

Exoskeletons typically perform pre-programmed movements triggered by switches or sensors. Users (hereinafter referred to as Pilots) can exercise only limited control over their leg trajectory and walking speed to avoid obstacles. In this work, we will use ROS robot operating system [4], [5], [6], [7] and the Gazebo simulator to build a model of an exoskeleton for the rehabilitation of the lower extremities, which moves and interacts with objects similar to a "smart home", depending on the data of an external information device. The purpose of the proposed system is to replace the standard manual movement control by the pilot, who usually uses the buttons on the handles of the exoskeleton for this, with an automated exoskeleton movement mode for typical movements.

2. Relevance of automating the movement mode change

The relevance and practical usefulness of automating the movement mode change of a rehabilitation exoskeleton of the considered type are determined by three main factors:

- Improving the efficiency and reliability of the use of the exoskeleton of the lower extremities
- Improvement of the pilot's psychological state in the process of movement during rehabilitation
- Economic expediency.

In the case of conventional manual control, the pilot must maintain balance and is forced to stop movement to change the movement mode. To do this, he has to use a set of buttons on the handles of the exoskeleton or a special watch that is connected using a Bluetooth. This option is inconvenient, especially when the pilot must change modes of motion to avoid obstacles. Automated motion mode change makes these movements faster and more reliable.

As a result, it improves the psychological state of the pilot when using the exoskeleton. Some people feel embarrassed when they stop and cannot move, but with the help of an automated system, they feel more confident. A pilot can rent an exoskeleton for a fixed period instead of buying the device or only using it in the hospital, he can rent it and use it at home. And by automating the change of the mode of movement of the exoskeleton inside the house, the pilot will increase the active time of its use, which will make the rehabilitation process more efficient and, as a result, save money and time.

3. Building a model of the rehabilitation exoskeleton

First, we will design an exoskeleton model by defining the necessary joints and links (the main components of the robot model), as well as a "smart home" model containing obstacles [8]. Then we install an external information device into the model - a special type of LiDAR sensor (HOKUYO) [9]. To control the movement of the hinges, a control program (a special "node" [10]) was written in the Python programming language [11] to represent 5 typical modes of movement of the rehabilitation exoskeleton of the lower extremities [12]:

- Straight movement
- Transition to a sitting position
- Transition to a standing position
- Climbing stairs
- Descending stairs

Based on the LiDAR sensor signal, the exoskeleton can determine the type of the next obstacle and determine the necessary movement mode to bypass this obstacle.

The simulation contains a model of an exoskeleton in the form of a mobile robot that interacts with the virtual world (Figure 1). To design a model, an XML file (.xacro) must be defined that represents the shape and components of the model, the obstacle house, and the LiDAR sensor. A model of any robot featured in ROS and Gazebo, showing its links, the joints that connect those links, and the transmission that describes the relationship between the actuator and the link. The exoskeleton model (Figure. 2) is built on 7 links, shown in Table 1 [13],[14], [15]:

№	Exoskeleton's	Label of link	Label of joint	Connection between	Type of joint
1	Waist	waist_link	RU_joint	1 and 2	rotary
2	Right thigh	RU_link	LU_joint	1 and 3	rotary
3	Left thigh	LU_link	RD_joint	2 and 4	rotary
4	Right leg	RD_link	LD_joint	3 and 5	rotary
5	Left leg	LD_link	RF_joint	4 and 6	fixed
6	Right foot	RF_link	LF_joint	5 and 7	fixed
7	Left foot	LF_link	UP_joint	-	prismatic

Table 1. The exoskeleton model joints, links and connections

In addition, there are virtual links and hinges required for a plugin (libgazebo_ros_control) [16] provided by Gazebo that gives the model the ability to move along the X-axis and Y-axis in linear and angular directions. We can control the model movement in these directions using the computer keyboard.

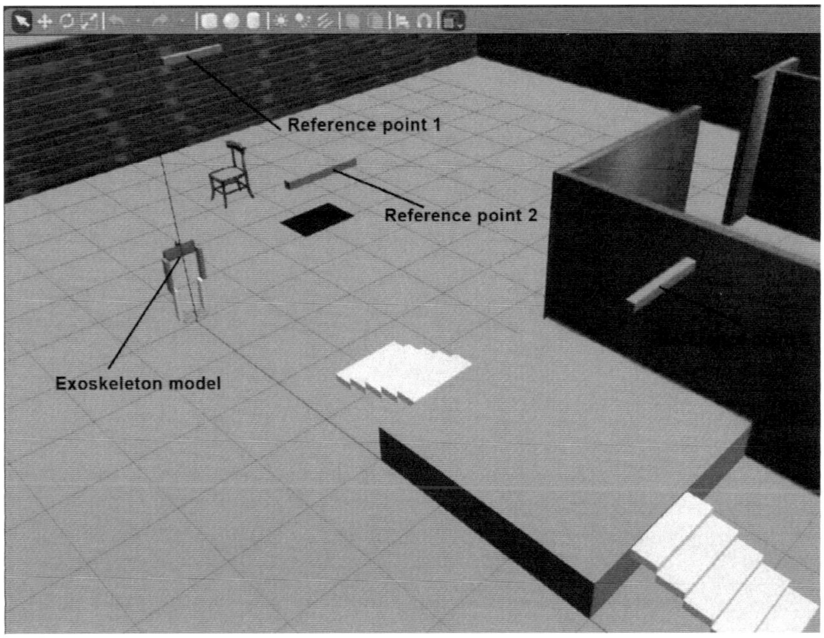

Fig. 1. The exoskeleton model interacts with the constructed world

Since the pilot will move in a specific room [17], the reference points have also been added over all the obstacles that the pilot needs to overcome and move. For example, we added the reference point 1 above the chair where the pilot will sit in the exoskeleton, the reference point 2 above the starting point of stair climbing, and the reference point 3 above the starting point of stair descent. As shown in Figure 1, the distance between the reference point and the sensor differs from one obstacle to others.

Based on the sensor signal that determines the distance between the exoskeleton and an obstacle, it is possible to determine the type of obstacle that the pilot will encounter and the appropriate exoskeleton movement system. The place or room in which the pilot with the exoskeleton will walk is equipped with these obstacles with appropriate reference points, which are installed in the ceiling of the room at a given height.

As for the used sensor, is a LiDAR (HOKUYO) sensor, which is used to detect the obstacles and determine the distance between the sensor and the obstacle using a laser beam. The sensor is directed in the direction of the Z-axis (the reference points used to calculate the distance between it and the sensor are located directly above the exoskeleton). Typically, the detection range of LiDAR (HOKUYO) is 180 degrees, but in the model, the required detection process occurs only horizontally and directly above the model.

Therefore, we only need one laser beam from the sensor to read the distance between the sensor and the reference point (all sensor parameters can be edited through the LiDAR file plugin (HOKUYO). This type of sensor was chosen because of its ease of use and handling in the working environment of ROS and Gazebo, on the other hand, it withstands shock and vibration, which is important since the designed model is a moving object [14].

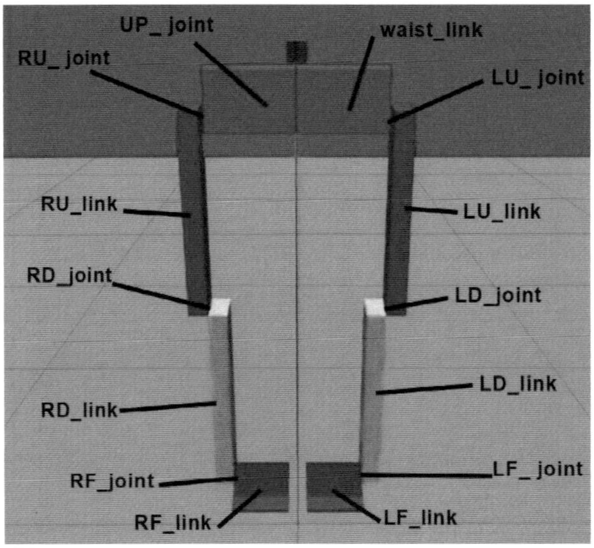

Fig. 2. Joints and links of the model of the rehabilitation exoskeleton

4. Approach to the control of the exoskeleton model

After building a model using an XML file, we can define a Python (.py) file [18] and write an algorithm for managing the model and interacting with surrounding objects. To control any joint, we must apply the necessary force to the transmission. There are 4 transmissions [13], one transmission for each movable joint:

- tran_RU to control the RU_joint;
- tran_LU to control the LU_joint;
- tran_RD to control the RD_joint;
- tran_LD to control the LD_joint.

As we said earlier, the modelled exoskeleton has 5 movement modes and based on the sensor signal, LiDAR determines the type of the next obstacle and the required mode to overcome this obstacle. Depending on the signal from the LiDAR, we can send a certain amount of force to each joint at every moment to simulate the required movement to overcome a certain obstacle.

For example, after starting the ROS environment and calling the package that contains all the files needed to run the designed model, the movement of the model can be controlled. When the model passes under the reference point 2, the LiDAR sensor detects this reference point and calculates the distance between the model and the reference point 2. Accordingly, the type of obstacle that the pilot will encounter is determined (in this case, it is "climbing stairs"). In addition to the corresponding mode of movement of the exoskeleton, the amplitude of this movement is calculated, which must be transmitted to each joint in accordance with the size of the obstacle to be overcome.

5. Conclusion

This study presented an approach to automate the rehabilitation exoskeleton motion mode change using an external LiDAR sensor. The model was defined by an XML script and implemented in the Gazebo environment. The required trajectories for each joint are determined depending on the sensor signal and a certain obstacle that the model must bypass. In addition, the Gazebo simulation is useful for visualizing the movement of a prototype exoskeleton. In conclusion, our new model, modelled in Gazebo, is suitable for building a gait training device. The main goal of engineering in general and robotics in particular is to help humans perform certain tasks with less time and effort, and by automate the movement mode change of the rehabilitation exoskeleton for lower extremities, the pilot's mental and psychological state of the will improve due to the confidence that he will gain by using the simulated model. Moreover, the time of the rehabilitation sessions will be more efficient, since the period of time needed by the pilot to change the movement mode of the exoskeleton will be zero. In the future we can improve the sensing system of the model to overcome other obstacles

6. Acknowledgments

Scientific Research by Educational Organizations in 2020–2022 Project under Grant NoFSFS-2020-0031.

7. References

[1] Titarenko, N.Yu.; Titarenko, K.E.; Levchenkova, V.D.; Semenova, K.A.; Voronov, A.V. & Dotsenko, V.I. (2014). Quantitative evaluation of motor functions disorders in cerebral palsy patients by means of videoanalysis of movements with the use a two-dimensional biomechanical model, Rossiyskiy pediatricheskiy zhurnal, №5,

[2] Semenova, K.A. (1999). Medical rehabilitation of patients with residual stage of cerebral palsy. Moscow: Antidor; (in Russian)

[3] Baranov, A.A.; Klochkova, O.A.; Kurenkov, A.L.; Namazova-Baranova, L.S.; Nikitin, S.S.; Artemenko, A.R. & Mamedyarov, A.M. (2012). The role of brain plasticity in the functional adaptation of body at cerebral infantile paralysis with the affection of hands. Pediatric pharmacology; 9(6):24-32. (In Russ.) https://doi.org/10.15690/pf.v9i6.515

[4] Robot Operating System — The Complete Reference (Volume 4), CISTER Research Center, 2019. — CISTER-TR-190702.

[5] Kalevatykh. I. A. & Leskov, A. G. (2012). Using a robot operating system (ROS) to create a semi-natural modeling complex for manipulation robots, N.E. Bauman, No. 11, 5 p.

[6] Sirazetdinov, R.T.; Devaev, V.M.; Kamalov, A.R. & Katsevman, E.M. (2015). Software complex for modeling and virtualization of the anthropomorphic robot AR-601 based on ROS and gazebo systems, Seventh All-Russian scientific and practical conference on simulation modeling and its application in science and industry «Simulation Theory and Practice» (IMMOD-2015), No. 7, pp. 328–331.

[7] Lavrenov, R.O. & Magid, E.A. (2019). programming in the robotic operating system (ROS), Kazan, 38 p.

[8] http://wiki.ros.org/xacro wiki.ros, Accessed 2022-01-20

[9] http://wiki.ros.org/hokuyo_node wiki.ros, Accessed 2022-01-20

[10] Balakina, E.S. (2019) Integration of a robotic OS (ROS) with a TRIK cybernetic controller, St. Petersburg State University, 23 P.

[11] Lentin, D. (2019). Learning Robotics with PYTHON, transl. from Eng. A. V. Koryagina. M.: DMK Press, 250 p.

[12] Kotov, S.V.; Isakova, E.V.; Pismennaya, E.V.; Petrushanskaya. K.A.; Lizhdvoy, V.Yu. & Sekirin, A.B. (2020). The method of training patients after acute cerebrovascular accident using the ExoAtlet medical simulator. GBUZ MO MONIKI im. M.F. Vladimirsky (Minutes of the meeting of the Academic Council), No 3, 91 p.

[13] Mohammad, S.A.; Rizauddin, R.; MohdAizat, A.T.; Mohd, F.I. & Khashayar, D.N. (2020) Simulation and Control of a Six Degree of Freedom Lower Limb Exoskeleton, Jurnal Kejuruteraan, No 32(2), pp. 197–204.

[14] GOST R 60.0.0.4 — 2018 / ISO 8373:2012. Robots and robotic devices terms and definitions, National standard of the Russian Federation; 2018, 42 p.

[15] Ivanov, A. A. (2012). Fundamentals of robotics: textbook. Allowance for university students, A.A. Ivanov, M.: Forum, 222 p.

[16] https://russianblogs.com/article/1041288878/ Official site Russian Blogs, ROS and GAZEBO real-time hardware simulation (2) — attribute gazeburd, accessed 2022-07-08

[17] Yang, W.; Tan, Y.; Hu, S.; Gaku, R. & Takakuwa, S[oemon] (2021). Modeling and Analysis of Healthcare Facilities Using RFID Data, Chapter 07 in DAAAM International Scientific Book 2021, pp.085-094, B. Katalinic (Ed.), Published by DAAAM International, ISBN 9783-902734-31-0, ISSN 1726-9687, Vienna, Austria DOI: 10.2507/daaam.scibook.2021.07

[18] http://wiki.ros.org/ROS/Tutorials/, Accessed 2022-08-27

DOI: 10.2507/33rd.daaam.proceedings.075

ANALYSIS OF RAIL ROUGHNESS ON TRAMWAY TRACK

Krešimir Burnać & Ivo Haladin

This Publication must be referred as: Burnac, K[resimir] & Haladin, I[vo] (2022). Analysis of Rail Roughness on Tramway Track, Proceedings of the 33rd DAAAM International Symposium, pp.0536-0540, B. Katalinic (Ed.), Published by DAAAM International, ISBN 978-3-902734-36-5, ISSN 1726-9679, Vienna, Austria
DOI: 10.2507/33rd.daaam.proceedings.075

Abstract

Rail corrugation is an irregularity that often occurs on a railhead running surface and can cause an increase in the noise and vibration levels as well as faster track deterioration. The possibilities of direct and indirect rail corrugation measurements are described in this paper as well as direct rail corrugation measurement that was carried out on a tramway track network in the city of Zagreb. Railhead corrugation was measured on several locations on the tramway track and the values were analysed with respect to the exploitation period of the tramway sections and the type of horizontal alignment. Analysis has shown that rail corrugation occurs more often in the curves than in the straight track and that the wavelength and amplitude of rail corrugation are higher on tramway track that has longer exploitation period. The conclusions of this paper are as expected, but as a next step towards automating the rail corrugation data analysis, the indirect measurements model is tested with the purpose of application on the whole tram track network using tramway vehicle monitoring.

Keywords: tramway vehicle; track; rail corrugation; measurement; rail irregularities

1. Introduction

Rail running surface subjected to dynamic traffic loading can develop unevenness or corrugation of the running surface. It can cause elevated vibration levels in vehicles, railway track and its's surroundings as well as speed up the degradation of the track components and railway vehicles [1]. Such phenomenon can also be described as rail surface acoustic roughness if it contributes to elevated noise levels of rail vehicles in operation. Detection of these irregularities is very hard using only visual inspection of the track, especially in the early phase of corrugation development, and additionally requires sophisticated measurement methods and rail track occupation. Furthermore, grinding as a method for corrugation removal does not prevent the rail corrugation to reappear after it has been exploited for some time [2]. For that reason, monitor of rail corrugation development is proposed in controlled environment and to define the parameters that have an impact on its increase [3]. Rail corrugation can be measured directly (using manual measurement devices, directly on the track) or indirectly (using a railway vehicle equipped with an inertial measurement system) [4]. In [5], the authors describe the development of a measurement trolley for continuous measurement of the railhead surface roughness (direct measurement) with which data connected to the appearance of corrugation mechanisms was collected.

As a result of the analysis, with the help of the rail corrugation measurements that were conducted using a corrugation measurement trolley (CAT) in a combination with a portable device for corrugation monitoring (that was installed on a railway vehicle), an efficient grinding management method was developed to optimise and reduce the costs of maintenance.

The use of railway vehicles with inertial measurement system (accelerometers) opens several possibilities for irregularities and damage detection on the rail track, such as rail breakage, poorly executed welds, isolated joints, and rail corrugation and acoustic roughness [6]. A mathematical model that was built as part of the study [1], predicts an increase of the corrugation and optimization of the rail grinding schedule, thereby increasing the efficiency of the method and as a result, represents a significant reduction of the rail track maintenance costs. The study confirmed that functions of the behaviour curves (growth) of corrugation over time can be modelled using the rail roughness measurements over a longer measurement period. Vibrations and dynamic forces generated by wheel – rail contact contains the most important information used to assess the rail track condition and predict the behaviour of damage and irregularities on the track, to ensure a high degree of efficiency for such methods.

Considering the research of wheel – rail contact, two different indirect measurements of vibrations on the track can be used: mounting the sensors on a railway vehicle bogie to monitor the rail or/and installing the sensors on a rail for wagon monitoring. As a part of the MaVico project [7], a system prototype for rail corrugation monitoring was developed to prevent the consequences that corrugation can have on the rail track and wagon equipment. Algorithms were made based on acceleration measurements on the bogie used to determine the state of rail corrugations, which showed satisfactory results of rail roughness measurements using the indirect method, compared to the classic, direct measurement method (using corrugation measurement trolley). The described system for indirect measurement of rail roughness can measure more than 100 kilometres of railway track in one day, while measurements on the same railway section using the direct method would last for several weeks [8].

This paper aims to present the data of railhead surface corrugation measurements using the direct method with the purpose of measured corrugation data analysis automatization in future research using the indirect method of rail roughness measurements and the possibility of applying the method on the whole tramway track network using monitoring on tramway vehicles.

2. Rail corrugation measurements on tramway tracks in the city of Zagreb

The first part of the research contains direct (manual) railhead corrugation measurements on the Zagreb tram track network. Railhead corrugation was measured according to HRN EN 15610 "Railway applications – Acoustics – Rail and wheel roughness measurements related to noise generation" standard [9], using the device for railhead corrugation measurement (Fig. 1d) which enables measurement of waveforms with wavelength between 1 and 40 centimetres. After the measurement, analysis of the results was made according to the mentioned standard because it gives a more detailed insight in the railhead corrugation wavelengths than the standard HRN EN 13121-3 "Railway applications - Track - Acceptance of works - Part 3: Acceptance of rail grinding, milling and planning work in track " [10]. Measurement of railhead corrugation was carried out on Zagreb tramway network at locations: Horvaćanski zavoj (location L1, Fig. 1a), Horvaćanska street – Hrgovići street (location L2, Fig. 1b.), and at the intersection Horvaćanska street – Hrvatskog sokola street (location L3, Fig. 1c.), on both tramway tracks with a total of 54 measuring points (
Table 1).

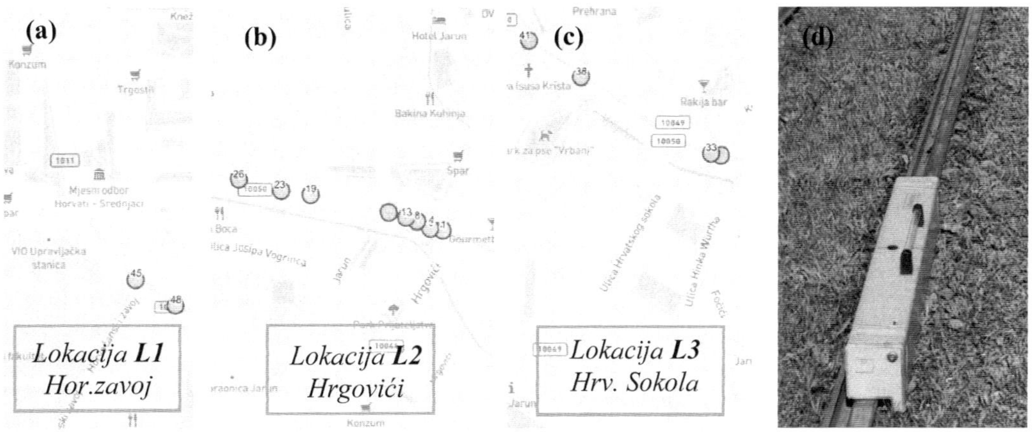

Fig. 1. Locations of railhead corrugation measurement (a, b, c) and manual device for rail roughness measurement (d)

The locations were chosen based on the visual inspection of the track, during which corrugation was noticed. Following the visual inspection, increased levels of vibration and noise were detected near the tram track.

Additional criteria selection of the locations was various exploitation time of the rails on the observed locations, various fastening systems, and geometrical elements of the track (straight line and curve) to cover as many track and rail parameters as possible that could be analysed. At each of the locations (L1, L2, and L3), measuring points were located both in a straight line and in a curve which should additionally contribute to the analysis related to the geometrical elements of the track. After the measurements, result analysis was made with the aim of a qualitative evaluation of the railhead corrugation and a comparison regarding different parameters of the track.

Location	Year of reconstruction (exploitation)	Fastening system	Measuring point	Radius	Rail
L1	2017 (4)	PPE	**MP 1**	straight line	-
			MP 4	348 m	outer
			MP 7	348 m	inner
L2	2004 (17)	DEPP	**MP 2**	straight line	-
			MP 5	1025 m	outer
			MP 6	straight line	-
			MP 8	1025 m	inner
L3	2000 (21)	DEPP	**MP 3**	straight line	-

Table 1. Overview of measurement locations, measuring points, and their characteristics

3. Results analysis

Measured values of the railhead corrugation were analysed, and their amplitudes (dB re 1µm) and wavelengths (cm) were processed using Fourier transformations and presented in third-octave frequency bands with wavelengths between 1.25 and 40.0 centimetres, following the standard [9]. The occurrence and properties of the rail corrugation were analysed depending on the parameters that have an influence on the corrugation. Those parameters include the horizontal position of the track (straight line or curve), the size of the radius of the curve, and the position of the rail to the central axis (outer or inner rail). In the graphs below, along with the analysed railhead corrugation, a limit curve according to the HRN ISO 3095 "Acoustics – Railway applications – Measurement of noise emitted by rail bound vehicles" [11] standard is presented. According to the standard, values of railhead corrugation that are below the defined limit will not have an impact on the overall noise level of a passing rail vehicle, while the values above the limit will have an impact on noise levels. Rail corrugation wavelengths (in centimetres) are shown on the horizontal axis of the graph, while the amplitude values (in decibels, with the referent value of 1µm) are shown on the vertical axis. The basic comparison of the railhead corrugation is made for three different sections (L1, L2, and L3), analysing the exploitation period of the track.

3.1 Railhead corrugation and exploitation period

The time-dependent form and development of railhead corrugation are being inspected with the analysis. It is important to emphasize that all the locations are situated on the same tram track section, so the traffic volume is proportional to the exploitation period. Railhead corrugation of the track in a straight line on locations L1 (MP1), L2 (MP2), and L3 (MM3) are displayed in Fig. 2. On the section which has the longest exploitation time (L3, MP3), a higher number of expressed corrugation patterns at wavelengths λ = 2cm, λ = 4 cm and λ = 8 cm appeared.

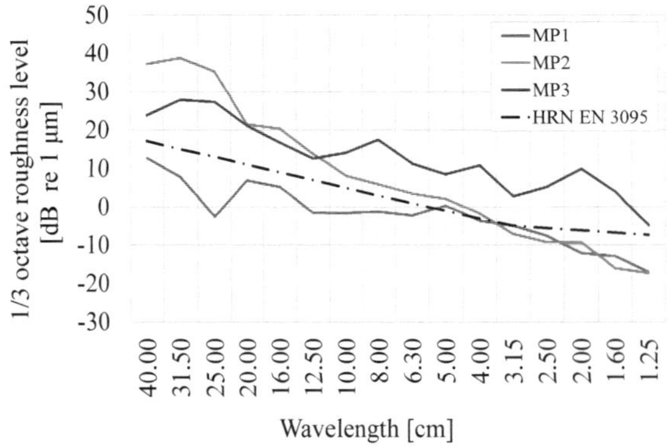

Fig. 2. Railhead corrugation comparison considering the exploitation period

On location L2 (MP2), dominant corrugation wavelengths were from 25 to 40 centimetres (with the highest amplitudes a = 38.7 dB for λ = 31.5 cm. Considering the short exploitation period on location 1 (L1, MP1), there was no railhead folds with dominant wavelengths spotted.

3.2 Railhead corrugation and horizontal track alignment

Outer rail railhead corrugation was analysed for different curve radii R = 348m (L1, MP4) and R = 1025m (L2, MP5) (Fig. 3.) in comparison to railhead corrugation for track in a straight line at the same measurement locations (L1, MP1 and L2, MP6). With this kind of analysis, it is noticeable that a significant corrugation difference is present with the smaller radius curves as opposed to the corrugation on tracks in a straight line. On the outer rail track with the smaller radius (MM4, R = 348m), a few dominant corrugation wavelengths were detected (λ = 25 cm, λ =16 cm, λ = 4 cm), while on tracks with bigger radius (MM5, R = 1025m) only corrugation wavelengths around 25 cm with amplitudes a = 33.2 dB occur on the tracks. Looking at the rail track in the straight line, almost none of the corrugation wavelengths were dominant in location L1 (MM1), but at location 2 (MM6) there was a visible corrugation development, like the one in the curve with a big radius (MM5), with an additionally emphasized wavelength of 5 centimetres.

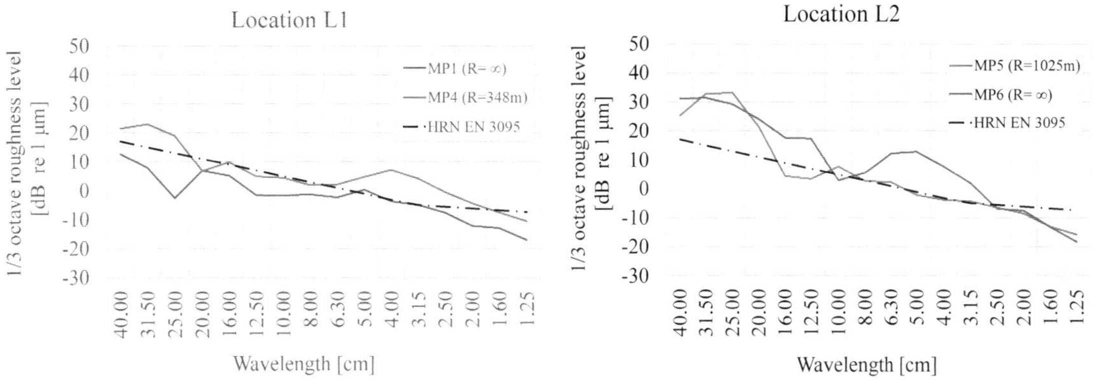

Fig. 3. Railhead corrugation comparison of outer rail in a curve and straight line on location L1 (a) and L2 (b)

It is also interesting to observe the corrugation difference between the inner and outer rail on measuring points in a curve because of the different effects of vehicle load coming from the impact of centrifugal force that is present during the circular motion. In Fig. 4., the railhead corrugation of the inner and outer rail for tracks with different curves radii (R = 348m, R = 1025m) is shown. Independent of radius size, bigger corrugation wavelengths were dominant in the 31.5 (a = 23 dB) and 25 centimetres (a = 33.2 dB) frequency bands on the outer rail (MP4, MP5). On the inner rail (MP7, MP8), shorter corrugation wavelengths from 5 to 10 centimetres with amplitudes a = 18.7 dB (for R = 1025m) and a = 12.7 dB (for R = 348m) were dominant. Bigger amplitudes occurred on the track in curves with bigger radii (which was not expected). The explanation for this could be the fact that the tram track at the location L2 (MP5, MP 8) has a longer exploitation period than the one at location L1 (MP4, MP 7) and the exploitation period parameter has more effect on the size of the amplitude than the radius size.

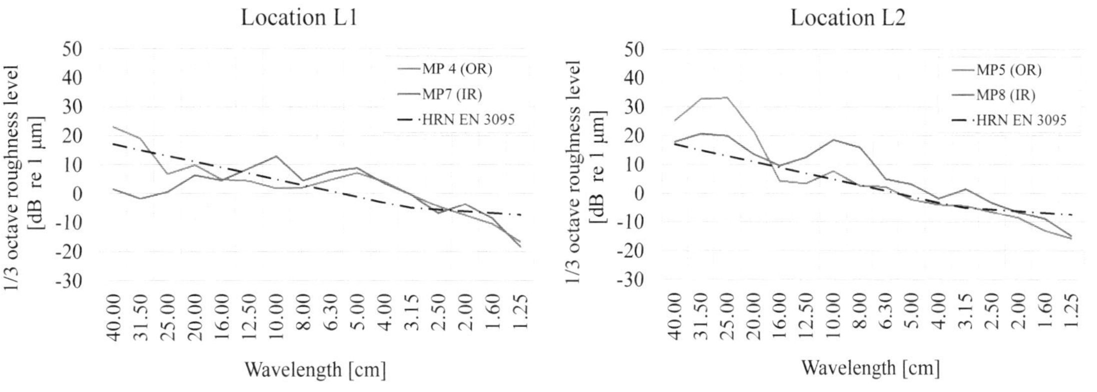

Fig. 4. Railhead corrugation of inner (IR) and outer (OR) rail in curves with different radii on locations L1 (a) and L2 (b)

3.3 Future research

The second part of the research implies the conversion of wavelengths (measured with direct railhead corrugation measurement method) into the frequency domain to make the values comparable to the vibration signal values recorded onboard an instrumented tramway vehicle (indirect measurement method). Finally, direct, and indirect measurement methods comparison would represent the basis for further research and making of the rail corrugation development assessment model (considering the amplitude and wavelength of the fold). That kind of model would serve for the development of optimal track maintenance (grinding) schedule and would be based on the indirect measurements using accelerometers installed on in-service tramway vehicle, which would represent not only a big step ahead and simplification of data acquisition (no need for manual (direct) measurement) but also additional financial relief for track maintenance and decrease of noise and vibration levels.

4. Conclusion

Observing the dependence of corrugation on the exploitation period, it is concluded that the number of dominant wavelengths will increase on the rail tracks with the increased exploitation period. In the case of curved tracks, it is noticeable that the radius size is inversely proportional to the corrugation development and that the corrugation development on the tracks with bigger radii is similar to the one recorded on the track in a straight line. Higher values of outer rail corrugation wavelengths are more dominant than those on the inner rail, and dominant wavelengths are not dependable on radii value. The results of direct measurements, which were carried out on 3 locations in the city of Zagreb according to the EN 15610 standard, will serve as the basis for corrugation detection using indirect measurement method and for further research on the correlation of indirect and direct rail corrugation measurement methods.

5. References

[1] Tanaka, H. & Miwa, M. (2020). Modeling the development of rail corrugation to schedule a more economical rail grinding, Proceedings of the Institution of Mechanical Engineers, Part F: Journal of Rail and Rapid Transit, Vol. 234, No. 4, pp. 370–80., ISSN: 0954-4097, https://doi.org/10.1177/0954409719894833

[2] Fan, W.G.; Hou, G.Y.; Wang, W.X.; Zhang, X.L. & Wang, J.D. (2019). Desing and dynamic analysis of a new rail grinding device using closed abrasive belt, International Journal of Simulation Modelling, Vol. 18., No. 3, pp. 531 – 541, ISSN 1726-4529, https://doi.org/10.2507/IJSIMM18(3)CO14

[3] Ng, A.K.; bin Alias, Z.; Chassin, J.F. & Yebra, J.H. (2016). Managing rail corrugation through modelling, simulation, and instrumentation technologies, Proceedings of IEEE International Conference on Intelligent Rail Transportation (ICIRT), 23. – 25.08.2016., Birmingham, UK, ISBN:978-1-5090-1555-9, Institute of Electrical and Electronics Engineers Inc. (Ed.), pp. 307–14., IEEE, Birmingham, https://doi.org/10.1109/ICIRT.2016.7588748

[4] Tanaka, H.; Shimizu, A. & Sano K. (2019). Development and verification of monitoring tools for realizing effective maintenance of rail corrugation, Structure and Infrastructure Engineering, Vol.15, No. 10, pp. 1308 – 1318.

[5] Tanaka, H. & Shimizu, A. (2016). Practical Application of Portable Trolley for the Continuous Measurement of Rail Surface Roughness for Rail Corrugation Maintenance, Railway technical research institute, Quarterly Report of RTRI, Vol. 57., No. 2, https://doi.org/10.2219/rtriqr.57.2_118

[6] Molodova, M., Li, Z. & Dollevoet, R. (2011). Axle box acceleration: Measurement and simulation for detection of short track defects, Wear, Vol. 271, pp. 349–56., https://doi.org/10.1016/j.wear.2010.10.003.

[7] Vibratec Group (2022).MaVico Project, Available from: https://www.vibratecgroup.com/en/blog/news/connected-mainenance-using-vibration-measurements/, Accessed: 05.07.2022.

[8] Augez, R; Cloix, A & Clerc, C. (2019). Autonomous Embedded Vibroacoustic Measurements: an efficient tool for railway monitoring, Surveillance, Vishno and AVE conferences, INSA-Lyon, Université de Lyon, Lyon, France

[9] Croatian standard HRN EN 15610 (2019). Railway applications – Acoustics – Rail and wheel roughness measurements related to noise generation, Croatian Standards Institute

[10] Croatian standard HRN EN 13231-3 (2016). Railway applications - Track - Acceptance of works - Part 3: Acceptance of rail grinding, milling and planning work in track, Croatian Standards Institute

[11] Croatian standard HRN ISO 3095 (2013). Acoustics – Railway applications – Measurement of noise emitted by rail bound vehicles, Croatian Standards Institute

DOI: 10.2507/33rd.daaam.proceedings.076

DEMONSTRATING THE A3 TECHNIQUE USING A LEAN PRODUCTION MODEL

Bojan Jerinic, Nemanja Sremcev, Nebojsa Ralevic,
Nemanja Tasic, Jovan Mandic & Ilija Cosic

This Publication has to be referred as: Jerinic, B[ojan]; Sremcev, N[emanja]; Ralevic, N[ebojsa]; Tasic, N[emanja]; Mandic, J[ovan] & Cosic, I[lija] (2022). Demonstrating the A3 Technique Using a Lean Production Model, Proceedings of the 33rd DAAAM International Symposium, pp.0541-0546, B. Katalinic (Ed.), Published by DAAAM International, ISBN 978-3-902734-36-5, ISSN 1726-9679, Vienna, Austria
DOI: 10.2507/33rd.daaam.proceedings.076

Abstract

Structured problem solving has become one of the key important skills in every industry as problems are a daily occurrence, mastering a structured problem-solving technique can greatly increase one's chances of employment but unfortunately structured problem-solving techniques are underrepresented in education. This paper aims to provide a way of demonstrating the A3 technique with a model of a Lean production system thus providing a simple way of teaching the A3 method in educational circumstances while linking together industrial problems.

Keywords: A3; Lean; Problem Solving; Teaching; Jidoka

1. Introduction

Structured problem solving represent an integral part of everyday work in industry, but is not often taught or represented in education therefore there is a gap in young individuals about the proper usage of problem-solving tools such as the A3, 8D, Why-Why techniques etc. The reason behind this might lie in the illusion that to demonstrate the proper usage of a structured problem technique you need a real-life situation or an industrial related problem [1]. But this is not the case as for example Toyota uses movies such as Apollo 13 to teach structured problem solving, also simulations and games can be used for teaching and demonstrating different techniques [2]. With problem solving being not a need bur a requirement of industry there is need of adjusting to the needs of the industry be educational institutions [3]. This paper aims to use a physical Lean model of a production system do demonstrate the usage of the A3 technique, showing that problem solving can be taught with using modelling and a simulation of production processes by applying the A3 technique to a problem achieving production targets and number of Jidoka stoppages on an assembly line.

2. A3 technique

A3 represents a structured problems solving technique which follows a series of steps such as (1) establish the business context and importance of a specific problem or issue; (2) describe the current conditions of the problem; (3) identify the desired outcome; (4) analyse the situation to establish causality; (5) propose countermeasures; (6) prescribe an action plan for getting it done; and (7) map out the follow-up process [4]. When trying to understand a process fully the A3 method could be helpful perhaps even when mining a process [5], the method has been used for several different applications such as quality improvement [6] and product development [7].

Although the number of steps which an A3 report contains varies [8] the structure remains the same from problem definition to root cause analysis, defining countermeasures/actions and validating the results, other than for problem solving the technique can be used for projects as a structured approach to give an overview of the goal, status, and timeline of the project as well as a summarization of a project. The technique represents a very powerful tool when applied correctly for productive dialogue and helping people learn from one another, as it involves a cross functional team [4].

When the technique is applied correctly it can lead to several benefits such as process improvement [9] [10], and environmental development [11], but further applications of A3 are possible, perhaps it is possible to apply it event protocol development [12]. In this paper the technique will be applied to a model of assembly line which is one more way of the application of the technique [13].

Demonstrating the A3 techniques using models of production systems allows the demonstrator the freedom to not only show problem solving in a way that best suits the needs of the students but also allows the actual problem to be customized to the needs of the audience and allows the audience to interact with the model and problem. One of the key advantages of using models, more specifically physical models is that they allow reusability, meaning that the same model could be used to demonstrate the same problem multiple times or even to demonstrate a different problem each time, depending on the size and parts of the model.

With the usage of models problem solving could be taught in a structured way providing a clear understanding of what structured problem solving is to the students and thus preparing them for the challenges they will face in industry. The negative side of using models is the problems that will demonstrated will often be simplistic and therefore might lead to the understanding the structured problem solving only works on simple problems which are rapidly solved and don't require detailed analysis, also physical models are that they are inactive meaning they are not systems with continuously moving parts and while they can simulate real life problems they are not as complicated as real-life problems.

The limitations of using models to demonstrate problem solving are that building the model requires a lot of time and effort for the model to be as similar to a real system which would require the model to be prepared beforehand the demonstration of the problem solving. Another limitation of using models for problem solving is that it heavily relies on the creativity and experience of the demonstrator as the demonstrator needs to create the situation or problem where the problem solving can be applied, these issues could be solved using augmented reality and building a models which is interactive [14].

3. Demonstrating the A3 technique

The A3 technique which will be used for the demonstration represents a modified version which consists of four parts:

A. Problem definition – in this part the problem is clearly defined, all historical data is represented if available as well as countermeasures performed. In this part of the technique there the key questions which should be answered are:
   a. What is the problem?
   b. Where is it?
   c. What has been done to solve it?
   d. What are the effects of the problem?
B. Root cause analysis – when the problem is clearly defined possible causes which lead to the consequences of the problem must be listed and evaluated to define possible countermeasures.
C. Countermeasures – for the causes which are deemed most likely to cause the problem countermeasures must be defined and evaluated, the measures which provide the most effect with the least amount of effort should be prioritized.
D. Validation – the countermeasures which have been applied must be validated be comparing the starting state to the current one, if the problem is solved and there have been no reoccurrence of the problem it will be deemed that the problem is solved.

The model which is used has five stations on the final assembly line which perform the following processes (1) setting the bodywork, (2) positioning the steering wheel, (3) positioning the front bumper, (4) setting the wheels right side, (5) setting the wheels left side (Fig. 1).

In the next part of the paper the A3 technique is applied to a problem on the assembly line. The A3 technique has been split into four tables, with each table representing one of the steps of the technique, problem definition is shown in Table 1 where the problem and its trends are described, in Table 2 the root cause analysis is demonstrated using the Ishikawa technique and the Why-Why method, Table 3 shows the proposed countermeasure developed based on the root cause analysis, while Table 4 validates that the given countermeasure solved the defined problem and that the root cause was eliminated therefore an improvement of performance is achieved which is shown in the results.

Fig. 1. Assembly line layout

A. Problem definition

For the past four days the production target has not been achieved, also there has been at least one Jidoka stop in each day, also the average cycle time has gone up due to the frequent stops (Fig. 2). Also, all the Jidoka stops have been on station number 2 (Fig.3).

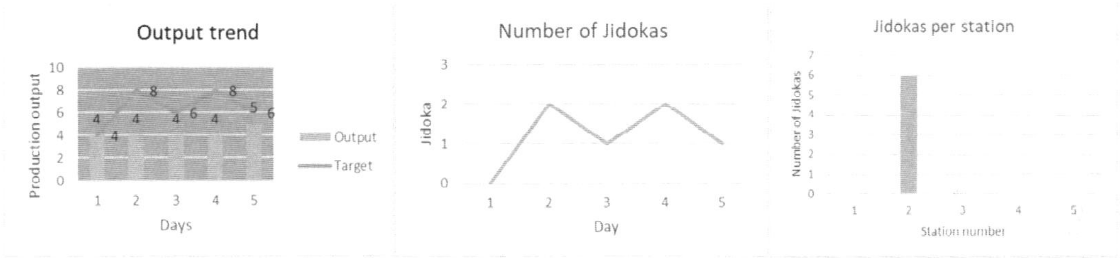

Fig. 2. Problem trends

Fig. 3. Jidoka stops

Containment action: Each time the Andon cord has been pulled the production supervisor has assisted the assembly worker the error has been corrected and the process was continued (Fig. 4).

Fig. 4. Containment action

Table 1. Problem definition

B. Root Cause analysis

Since the issue is on station 2 possible root causes to the problem are analysed using the Ishikawa diagram (Fig. 5) and Why-Why method (Fig. 6). This part demonstrates one of the possible ways of performing root cause analysis, where possible causes regarding the machine, man, material and method are tested.

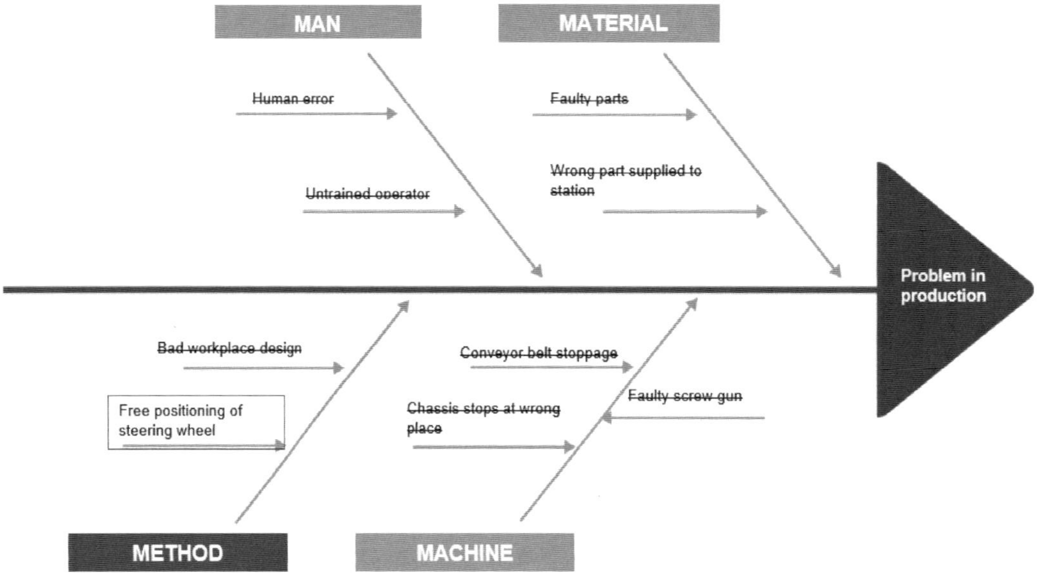

Fig. 5. Ishikawa diagram

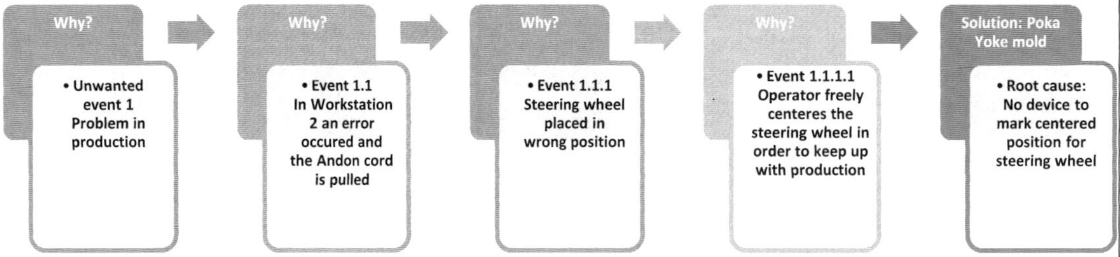

Fig. 6. Why-Why analysis

The result of the Ishikawa diagram and Why-Why analysis is that the root cause to the issue was that the worker has no marks or tool to help him centre the steering wheel in the right place.

Table 2. Root cause analysis

C. Countermeasures

As the proposed countermeasure a Poka Yoke device which reveals the position where the steering wheel should be placed is built and placed on the station.

Fig. 7. Poka Yoke device

Table 3. Countermeasures

D. Validation

Since the implementation of the Poka Yoke device the production targets have been reached each day for the past five days and there have been no Jidoka stops on the assembly line (Fig. 8).

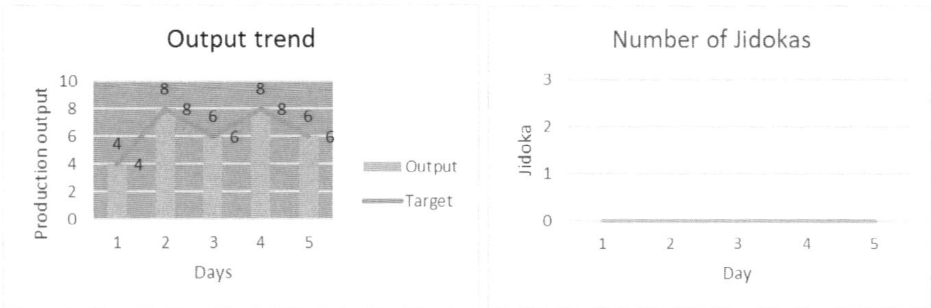

Fig. 8. Validation

Table 4. Validation

The result of the A3 technique which was applied to the problem is the elimination of the error which was causing multiple stops on workstation 2 and in turn reducing the production output and putting the customer at risk, the result of the A3 method is a Poka Yoke device which shows the worker the exact position where to assemble the steering wheel, the solution proved effective as shown in Table 4 and alleviated the deviation at the start of the method.

With the demonstration all the basic elements of structured problem solving were demonstrated from problem description to the validation of the proposed solution showing that by using a model of a Lean production system to set up a problem the A3 technique can be applied to it and demonstrated in a simple but effective way, showing that teaching can be done through games [2] [15], as well as simulations which have shown to have a positive trend with students especially when it comes to Lean methods [16] and coincides with the evolving role which teachers will have in the future [17].

4. Conclusion

Through this paper it has been shown that the A3 method can be demonstrated using simple problems and techniques without the need of problems directly from the industry, and that using models of Lean production systems can be used to not only show the A3 technique but also promote creativity especially when the technique is taught. The way the technique demonstrated in the paper is simplistic and therefore surpasses some of the problems of teaching structured problem solving such as missing real problems from the industry. While the demonstration is simple it can be effective, but in the future using methods such as augmented reality for problem solving can be effective and can show that problem solving can be applied even out of the physical world.

5. References

[1] Oversluizen, G. & Slomp J. (2021). "Use of A3-method by engineering students in industry projects," Production Planning and Control, Vol. 32, No. 6, pp. 489-496., DOI: 10.1080/09537287.2020.1742372.

[2] Mathew R.; Malik S. I. & Tawafak R. M (2009). "Teaching problem solving skills using an educational game in a computer programming course," Informatics in Education, Vol. 18, No. 2, pp. 359–373., DOI: 10.15388/infedu.2019.17.

[3] E. Shevtshenko et al., "Dissemination of engineering education at schools and its adjustment to the needs of enterprises," Proceedings of the 28th DAAAM International Symposium, 08-11th November 2017, Zadar, Croatia, Volume 28, No.1, ISSN 1726-9679, ISBN 978-3-902734-11-2, Katalinic, B. (Ed.), pp.0044-0053, DAAAM International Viena, Austria, DOI:10.2507/28th.daaam.proceedings.006

[4] Shook J. (2009). "Toyota's secret: The A3 report," MIT Sloan Manag Rev, Vol. 50, No. 4, pp. 30-33.

[5] Stefanovic D.; Dakic D.; Stevanov B.; Lolic T. & Marjanovic U. (2021). "Process mining in the manufacturing context: Review and recommendations," International Journal of Industrial Engineering : Theory Applications and Practice, Vol. 28, No. 4, DOI: https://doi.org/10.23055/ijietap.2021.28.4.7287

[6] Shahroudi P. & Aarabi A. (2021). "Quality improvement through lean A3 method for foot traffic in operating room," Perioper Care Oper Room Manag, Vol. 23, PP. 1-8., DOI: 10.1016/j.pcorm.2021.100155..

[7] Lorenzi C. I. & Ferreira J. C. E. (2018). "Failure mapping using FMEA and A3 in engineering to order product development: A case study in the industrial automation sector," International Journal of Quality and Reliability Management, Vol. 35, No. 7, pp. 1399-1422., DOI: 10.1108/IJQRM-10-2016-0179.

[8] Bassuk J. A. & Washington I. M. (2013). "The a3 problem solving report: a 10-step scientific method to execute performance improvements in an academic research vivarium.," PLoS One, Vol. 8, No. 10, pp. 1-8., DOI: 10.1371/journal.pone.0076833.

[9] Visich J. K.; Wicks A. M. & Zalila F. (2010). "Practitioner Perceptions of the A3 Method for Process Improvement in Health Care," Decision Sciences Journal of Innovative Education, Vol. 8, No. 1, pp. 191-213., DOI: 10.1111/j.1540-4609.2009.00251.x.

[10] Chakravorty S. S. (2009). "Process Improvement: Using Toyota's A3 Reports," Quality Management Journal, Vol. 16, No. 4, pp. 7-26., DOI: 10.1080/10686967.2009.11918247.

[11] Lenort R.; Staš D.; Holman D., & Wicher P. (2017). "A3 Method as a Powerful Tool for Searching and Implementing Green Innovations in an Industrial Company Transport," Procedia Engineering, 2017, Vol. 192., pp. 533 – 538., DOI: 10.1016/j.proeng.2017.06.092.

[12] Sarang S.; Stojanović G. M.; Stankovski S.; Trpovski Ž. & Drieberg M. (2020). "Energy-Efficient Asynchronous QoS MAC Protocol for Wireless Sensor Networks," Wirel Commun Mob Comput, Vol. 2020, pp. 1-10., DOI: 10.1155/2020/8860371.

[13] Pereira J.; Silva F. J. G.; Bastos J. A.; Ferreira L. P. & Matias J. C. O. (2019). "Application of the A3 methodology for the improvement of an assembly line," 29th International Conference on Flexible Automation and Intelligent Manufacturing (FAIM 2019), June 24-28, 2019, Limerick, Ireland, Beyond Industry 4.0: Industrial Advances, Engineering Education and Intelligent Manufacturing, ISSN: 2351-9789, Ryan A. (Ed.) Vol. 38., pp. 745-754, Procedia Manufacturing, DOI: 10.1016/j.promfg.2020.01.101.

[14] Tegeltija S. S.; Lazarević M. M.; Stankovski S. v.; Ćosić I. P.; Todorović V. v. & Ostojić G. M. (2016). "Heating circulation pump disassembly process improved with augmented reality," Thermal Science, Vol. 20, pp. 611-622, DOI: 10.2298/TSCI150927043T.

[15] Perutka K. & Vymazal M., "APPLICATION OF GAME-BASED LEARNING IN MATLAB USING OBJECT-ORIENTED PROGRAMMING," Proceedings of the 32nd DAAAM International Symposium, 28-29th October 2021, Hosted from Vienna University of Technology, Vienna, Austria, EU, Vol. 32, No. 1. ISBN 978-3-902734-33-4, ISSN 1726-9679, Katalinic, B. (Ed.), pp. 0022-0032, DAAAM International Viena, Austria DOI: 10.2507/32nd.daaam.proceedings.004.

[16] Badurdeen F.; Marksberry P.; Hall A.; & Gregory B. (2010). "Teaching lean manufacturing with simulations and games: A survey and future directions," Simul Gaming, Vol. 41, No. 4, pp. 465-486 DOI: 10.1177/1046878109334331.

[17] Milicevic J.; Sremcev N.; Cosic I.; Lazarevic M. & Tasic N.; "THE NEW ROLE OF TEACHERS AFTER THE COVID19 PANDEMIC AND THE APPLICATION OF THE LEAN CONCEPT IN EDUCATION," ," Proceedings of the 32nd DAAAM International Symposium, 28-29th October 2021, Hosted from Vienna University of Technology, Vienna, Austria, EU, Vol. 32, No. 1. ISBN 978-3-902734-33-4, ISSN 1726-9679, Katalinic, B. (Ed.), pp. 0402-0407, DAAAM International Viena, Austria, DOI: 10.2507/32nd.daaam.proceedings.059.

DOI: 10.2507/33rd.daaam.proceedings.077

ADOPTION OF AI IN EU MANUFACTURING. GAPS AND CHALLENGES

Siaterlis George, Nikolakis Nikolaos, Alexopoulos Kosmas & Makris Sotiris

This Publication has to be referred as: Siaterlis, G[eorge]; Nikolakis, N[ikolaos]; Alexopoulos, K[osmas] & Makris, S[otiris] (2022). Adoption of AI in EU Manufacturing. Gaps and Challenges, Proceedings of the 33rd DAAAM International Symposium, pp.0547-0550, B. Katalinic (Ed.), Published by DAAAM International, ISBN 978-3-902734-36-5, ISSN 1726-9679, Vienna, Austria
DOI: 10.2507/33rd.daaam.proceedings.077

Abstract

In the context of Industry4.0, several digital solutions have been included in the production line. Especially, ICT technologies have as a result the collection of huge amounts of data. The aforementioned, enable the use of several artificial intelligence methodologies such as predictive maintenance, scheduling, and recommendation systems. However, some challenges, as well as gaps, are observed. This study aims to document the gaps as well as the challenges that have been observed due to the adoption of Artificial Intelligence (AI) in the production lines.

Keywords: artificial intelligence; manufacturing; gaps; challenges; Industry4.0

1. Introduction

The fourth industrial revolution and the underlying digital transformation, known as Industry 4.0, are progressing exponentially. In particular, manufacturing is entering a period of substantial innovation and change driven by the increased integration of sensors and the Industrial-Internet-of-things (IIoT), increased data availability, and advances in robotics and automation. This leads to pervasive digitalization of the factory and challenges manufacturing enterprises to reconsider, re-examine, and re-evaluate their present operations and future strategic directions in the new era known as Smart Manufacturing and Industry 4.0, hence, several services have been included in the production lines, most of them are using and processing huge amounts of data and apply several AI methodologies in order to provide meaningful results or support to the manufacturers [1]. In particular, the new trends focusing on condition-based monitoring highlight the maintenance as well as the production planning sector [2]. In particular, smart technologies such as deep learning techniques, ontologies, and machine learning are used in order to predict an unexpected failure as well as to assess and schedule the production status [3],[4],[5],[6],[7],[8],[9].

Moreover, the trend focusing also on operator support and especially personalized support by using virtual assistance or AR and VR technologies [11],[12]. This study presents the gaps as well as the challenges regarding the adoption of AI in manufacturing. Moreover, it also presented an analysis that has been implemented in some European research projects. The analysis aims to classify and categorize the European research projects in categories based on the gaps that are focused on.

2. Gaps of adoption of AI in manufacturing

The existing gaps hinder the seamless adoption of AI in manufacturing. The gaps have been identified according to the literature [13],[14],[15],[16],[17]and are described in detail below:

2.1 Lack of talent with appropriate skillsets for AI work

The adoption of AI in manufacturing sites has as a result several complex applications such as predictive analytics, decision making, and condition monitoring to be included in the production line to reduce the cost and improve the quality of the products. The skills gap is thus denoted as the absence of specialized personnel for maintaining, updating, and mostly understanding these applications.

Lack of AI-related standards for manufacturing
Currently, there is no standardized approach for developing/testing/deploying AI tools in manufacturing sites to ensure the secure co-existence of such tools with production operations as well as the coexistence with human personnel. Moreover, the current approach of handling them in a similar approach to other heterogeneous systems without considering their unique characteristics along with the uncertainty they introduce in deterministic production sites poses significant threats to their operation.

Lack of mature IoT technologies
Several AI-driven applications are based on cloud technologies regarding the storing and manipulation of data. However, cloud technologies are at a low level regarding the technology maturity in conjunction with the ERP, SCADA, etc. Furthermore, there is a concern regarding security since the operating systems of the factories operate under a local network and may not be designed to face cyber threats.

Lack of data quality and efficiency
Data-driven applications can have a significant impact on manufacturing sites. However, a limitation that is identified in several of them is the lack of data quality. The data quality is related to the amount and the content of the collected data. In particular, high amounts of data contain information (noise from the sensors, errors, etc.) that is not useful for AI systems. AI systems are needed clean data with adequate variation to adequately create/train an AI/ML model. Many companies still do not have the appropriate knowledge on how to obtain the quality and quantities of data required to feed such models. Moreover, the high amounts of data affect the efficiency of the AI/MK methodologies that are used.

3. Gaps address and ethics challenge

3.1 Ethics Challenge

Ethics in AI becomes a global topic of interest for both policymakers and academic researchers The challenges related to the ethics that are observed in the EU regarding the adoption of AI in manufacturing are quite important, based on [10] the most critical challenges that the developers, as well as the software engineers, should take into consideration during the design of an autonomous software are the following. First transparency of operations. Transparency answers how and why a specific decision is made by the system and further triggers the other constructs, including interpretability and explainability. Next is privacy, AI/autonomous systems must assure user and data privacy throughout the system lifecycle. Accountability is the third most frequently reported principle, which specifically focuses on liability issues. Subsequently, fairness. In particular, discrimination between individuals or groups made by the decision-making systems leads to ethical fairness problems.

3.2 Appropriate skillsets for workers to adopt AI

Several, studies focused on the address of the skills gap that is observed during the adoption of AI in the production lines. In particular, Nikolakis [12] presents a solution that provides personalization support to the users. In particular, a two-level collaborative filtering framework is described as able to be adapted to the human operator's needs and motivating him through gamification features to implement. This approach aims to increase the skill set of the employers and bridge the gap among the different experience levels. Moreover, in the [18] survey the authors created an AR application in order to support operators in the assembly process of an engine. The result after all the aforementioned technologies in the production lines was that all the technologies that are supporting operators are useful and decrease the errors as well as increase the productivity

3.3 Lack of data quality

The lack of data quality is a significant gap that is observed in AI applications and methodologies. Thus, several studies aim to address this gap.

In particular, in the survey [19] the authors are using edge intelligence in order to collect the appropriate. In particular, the data are processed on the edge by transforming them from the time to the range of the frequency. The authors in the aforementioned study observed that by filtering the data in the edge the result of the predictive analytics was improved. Another approach s the transformation of big data to Smart data as described in [20], [21]. It includes many techniques mainly data cleaning, integration, discretization, and dimensionality reduction techniques like the principal component analysis. Especially the authors present some statistical methods of data pre-processing such as R functions median values in order to reduce the dimensionality of the data and to define a sufficient number of observations in a way that does not affect the data sets.

3.4 Research projects & Gaps

The authors have analyzed and classified in specific categories twenty (20) research projects. In particular, the aim of the analysis was the classification of the projects according to the gap(s) that aim to bridge. The research projects are coming from European projects. The classification consists of 5 (five) different categories. The five categories are the following, Skills Gap, AI Standards for manufacturing, Data Quality gap, Mature IoT technology & social legal, and ethical. The authors observed that most of the projects aim to address more than one gap.

Project	Topic	Skills gap	AI standards for manufacturing	Data quality gap	Mature IoT technology	Social, Legal, and Ethical
SERENA	FOF-09-2017	X	X	X	X	
SERRANO	ICT-40-2020		X		X	
ColRobot	ICT-24-2015		X			X
SPEXOR	ICT-24-2015	X	X			X
CROWDBOT	ICT-25-2016-2021		X			X
PICKPLACE	ICT-25-2016-2028		X			X
ROBINS	ICT-25-2016-2030		X	X		
HARMONY	ICT-46-2020		X	X		
ODIN	ICT-46-2020		X		X	X
FAROS	ICT-47-2020	X	X			X
FIT4FoF	DT-FOF-01-2018	X	X	X		
BETTER	EO-2-2017		X	X	X	
MAS4AI	ICT-38-2020		X		X	
TEAMING.AI	ICT-38-2020		X			X
INBOTS	ICT-38-2017		X			X
Robotis4EU	ICT-46-2020		X			X
MERGING	DT-FOF-12-2019		X		X	X
DIVERSITY	FoF-2014-2015		X	X		
Micro-ROS	ICT-25-2016-2026		X		X	
LOGISTAR	MG-5-2-2017		X		X	

Table 1. Projects categorization

5. Conclusion

Most of the aforementioned projects as well as surveys focus on the bridge of the gaps of AI in manufacturing. Is also important to be mentioned that all the projects and studies contributing to the improvement of standards regarding AI as well as all the analyzed projects and studies are focusing on bridging more than one gap. Furthermore, the adoption of AI in manufacturing remains promising, however, further investigation and extra experiments are needed in order to enhance the sustainability as well as the efficiency of AI developments. Is needed the researchers to provide mature and sustainable solutions based on standards.

In conclusion, the aim of this study is to identify and provide the status of the gaps in the adoption of AI in manufacturing. The authors provided some alternatives on how the community aims to bridge the gaps that have been observed. Subsequently, the authors provided a classification of twenty research projects based on the gaps that are focused on. As a next step, the authors consider updating and including additional projects and surveys in the analysis as well as classifying more projects. The authors aim to implement the same analysis for Japan and finally to implement a comparison between Europe and Japan.

6. Acknowledgments

This research has been partially supported by the European project "EU-Japan.AI: Advancing Collaboration and Exchange of Knowledge Between the EU and Japan for AI-Driven Innovation in Manufacturing" funded by the European Commission under grant agreement No 957339.

7. References

[1]. M. Ghobakhloo, "Industry 4.0, digitization, and opportunities for sustainability," J. Clean. Prod., vol. 252, p. 119869, Apr. 2020, doi: 10.1016/J.JCLEPRO.2019.119869.

[2]. N. Nikolakis et al., "Services to Facilitate Predictive Maintenance in Industry4.0," pp. 75–95, 2021, doi: 10.1007/978-981-16-2940-2_4.

[3]. K. Sipsas, K. Alexopoulos, V. Xanthakis, and G. Chryssolouris, "Collaborative Maintenance in flow-line Manufacturing Environments: An Industry 4.0 Approach," Procedia CIRP, vol. 55, pp. 236–241, Jan. 2016, doi: 10.1016/J.PROCIR.2016.09.013.

[4]. J. F. Arinez, Q. Chang, R. X. Gao, C. Xu, and J. Zhang, "Artificial Intelligence in Advanced Manufacturing: Current Status and Future Outlook," J. Manuf. Sci. Eng. Trans. ASME, vol. 142, no. 11, Nov. 2020, doi: 10.1115/1.4047855.

[5]. Q. Cao, F. Giustozzi, C. Zanni-Merk, F. de Bertrand de Beuvron, and C. Reich, "Smart Condition Monitoring for Industry 4.0 Manufacturing Processes: An Ontology-Based Approach," https://doi.org/10.1080/01969722.2019.1565118, vol. 50, no. 2, pp. 82–96, Feb. 2019, doi: 10.1080/01969722.2019.1565118.

[6]. I. de Pater, A. Reijns, and M. Mitici, "Alarm-based predictive maintenance scheduling for aircraft engines with imperfect Remaining Useful Life prognostics," Reliab. Eng. Syst. Saf., vol. 221, p. 108341, May 2022, doi: 10.1016/J.RESS.2022.108341.

[7]. T. Zonta, C. A. da Costa, F. A. Zeiser, G. de Oliveira Ramos, R. Kunst, and R. da Rosa Righi, "A predictive maintenance model for optimizing production schedule using deep neural networks," J. Manuf. Syst., vol. 62, pp. 450–462, Jan. 2022, doi: 10.1016/J.JMSY.2021.12.013.

[8]. D. Bowden et al., "A Cloud-to-edge Architecture for Predictive Analytics," undefined, 2019.

[9]. T. Hafeez, L. Xu, and G. McArdle, "Edge intelligence for data handling and predictive maintenance in IIoT," IEEE Access, vol. 9, pp. 49355–49371, 2021, doi: 10.1109/ACCESS.2021.3069137.

[10]. A. A. Khan et al., "Ethics of AI: A Systematic Literature Review of Principles and Challenges," ACM Int. Conf. Proceeding Ser., pp. 383–392, Sep. 2021, doi: 10.48550/arxiv.2109.07906.

[11]. K. Lotsaris et al., "Augmented Reality (AR) based framework for supporting human workers in flexible manufacturing," Procedia CIRP, vol. 96, pp. 301–306, Jan. 2021, doi: 10.1016/J.PROCIR.2021.01.091.

[12]. N. Nikolakis, G. Siaterlis, and K. Alexopoulos, "A machine learning approach for improved shop-floor operator support using a two-level collaborative filtering and gamification features," in Procedia CIRP, Jan. 2020, vol. 93, pp. 455–460, doi: 10.1016/j.procir.2020.05.160.https://doi.org/10.1016/j.ifacol.2021.08.100

[13]. D. Zhang et al., "The ai index 2021 annual report," arXiv preprint arXiv:2103.06312, 2021.

[14]. F. – European Union Agency for Fundamental Rights, "Data quality and artificial intelligence-mitigating bias and error to protect fundamental rights Helping to Make Fundamental Rights A Reality for Everyone in the European Union Fra Focus Contents," doi: 10.2811/615718.

[15]. "Bringing AI to the device: Edge AI chips come into their own." https://www2.deloitte.com/cn/en/pages/technology-media-and-telecommunications/articles/tmt-predictions-2020-ai-chips.html (accessed Aug. 25, 2021).

[16]. "Deloitte Survey on AI Adoption in Manufacturing | Deloitte China | Consumer & Industrial Products." https://www2.deloitte.com/cn/en/pages/consumer-industrial-products/articles/ai-manufacturing-application-survey.html (accessed Aug. 25, 2021).

[17]. "Adoption of AI advances, but foundational barriers remain | McKinsey." https://www.mckinsey.com/featured-insights/artificial-intelligence/ai-adoption-advances-but-foundational-barriers-remain (accessed Aug. 26, 2021).

[18]. O. Danielsson, A. Syberfeldt, M. Holm, and L. Wang, "Operators perspective on augmented reality as a support tool in engine assembly," in Procedia CIRP, 2018, vol. 72, pp. 45–50, doi: 10.1016/j.procir.2018.03.153.

[19]. G. Siaterlis et al., "An IIoT approach for edge intelligence in production environments using machine learning and knowledge graphs," Procedia CIRP, vol. 106, pp. 282–287, 2022, doi: 10.1016/J.PROCIR.2022.02.192.

[20]. A. Souifi, Z. C. Boulanger, M. Zolghadri, M. Barkallah, and M. Haddar, "From Big Data to Smart Data: Application to performance management," IFAC-PapersOnLine, vol. 54, no. 1, pp. 857–862, Jan. 2021, doi: 10.1016/J.IFACOL.2021.08.100.

[21]. D. García-Gil, J. Luengo, S. García, and F. Herrera, "Enabling Smart Data: Noise filtering in Big Data classification," Inf. Sci. (Ny)., vol. 479, pp. 135–152, Apr. 2019, doi: 10.1016/J.INS.2018.12.002.